PULSE, DIGITAL,
AND SWITCHING WAVEFORMS

McGRAW-HILL ELECTRICAL AND ELECTRONIC ENGINEERING SERIES

Frederick Emmons Terman, *Consulting Editor*
W. W. Harman and J. G. Truxal, *Associate Consulting Editors*

Ahrendt and Savant · Servomechanism Practice
Angelo · Electronic Circuits
Aseltine · Transform Method in Linear System Analysis
Atwater · Introduction to Microwave Theory
Bailey and Gault · Alternating-current Machinery
Beranek · Acoustics
Bracewell · The Fourier Transform and Its Applications
Brenner and Javid · Analysis of Electric Circuits
Brown · Analysis of Linear Time-invariant Systems
Bruns and Saunders · Analysis of Feedback Control Systems
Cage · Theory and Application of Industrial Electronics
Cauer · Synthesis of Linear Communication Networks
Chen · The Analysis of Linear Systems
Chen · Linear Network Design and Synthesis
Chirlian · Analysis and Design of Electronic Circuits
Chirlian and Zemanian · Electronics
Clement and Johnson · Electrical Engineering Science
Cote and Oakes · Linear Vacuum-tube and Transistor Circuits
Cuccia · Harmonics, Sidebands, and Transients in Communication Engineering
Cunningham · Introduction to Nonlinear Analysis
D'Azzo and Houpis · Feedback Control System Analysis and Synthesis
Eastman · Fundamentals of Vacuum Tubes
Evans · Control-system Dynamics
Feinstein · Foundations of Information Theory
Fitzgerald and Higginbotham · Basic Electrical Engineering
Fitzgerald and Kingsley · Electric Machinery
Frank · Electrical Measurement Analysis
Friedland, Wing, and Ash · Principles of Linear Networks
Geppert · Basic Electron Tubes
Ghausi · Principles and Design of Linear Active Circuits
Ghose · Microwave Circuit Theory and Analysis
Greiner · Semiconductor Devices and Applications
Hammond · Electrical Engineering
Hancock · An Introduction to the Principles of Communication Theory
Happell and Hesselberth · Engineering Electronics
Harman · Fundamentals of Electronic Motion
Harman · Principles of the Statistical Theory of Communication
Harman and Lytle · Electrical and Mechanical Networks
Harrington · Introduction to Electromagnetic Engineering
Harrington · Time-harmonic Electromagnetic Fields
Hayashi · Nonlinear Oscillations in Physical Systems
Hayt · Engineering Electromagnetics
Hayt and Kemmerly · Engineering Circuit Analysis
Hill · Electronics in Engineering
Javid and Brenner · Analysis, Transmission, and Filtering of Signals
Javid and Brown · Field Analysis and Electromagnetics
Johnson · Transmission Lines and Networks
Koenig and Blackwell · Electromechanical System Theory
Kraus · Antennas
Kraus · Electromagnetics

PULSE, DIGITAL, AND SWITCHING WAVEFORMS

Devices and circuits for their generation and processing

Jacob Millman, Ph.D.

Professor of Electrical Engineering
Columbia University

Herbert Taub, Ph.D.

Professor of Electrical Engineering
The City College of the
City University of New York

McGRAW-HILL BOOK COMPANY
New York   St. Louis   San Francisco
Toronto   London   Sydney

PULSE, DIGITAL, AND SWITCHING WAVEFORMS

Library of Congress Catalog Card Number 64-66293

ISBN 07-042386-5
1112131415 HDBP 76543

To our wives
SALLY
and
ESTHER

PREFACE

This book describes active and passive devices and circuit configurations used for the generation and processing of pulse, digital, and switching waveforms. These nonsinusoidal signals find extensive application in such fields as computers, control systems, counting and timing systems, data-processing systems, digital instrumentation, pulse communications, radar, telemetry, television, and in many areas of experimental research.

Methods are presented for the generation of very narrow (nanosecond or microsecond) pulses and of wider (millisecond or second) gates or square waves. Techniques are also given for the generation of a variety of other waveforms. These include a step, an exponential, a pulse code, a staircase, a precisely linear ramp, etc. Having been generated, a waveform must be processed in some way in order to perform a useful function. For example, it may be necessary to transmit the signal from one location to another, to amplify it, to select a portion of it in voltage, to choose a section of it in time, to combine it with other signals in order to perform a logic operation, to use it to synchronize a system, and so forth. All these processes are studied in detail in this text.

The book begins with a review of those topics in electronic circuit theory which will be most useful throughout the text. The first section defines a uniform system of notation which is equally applicable to transistors and tubes and which differs in a minimal way from present standards. The reader, after becoming acquainted with this notation, may wish temporarily to omit the rest of the first chapter. He may, instead, prefer to review each selected topic individually when a reference is made to it at a later point in the text. The subjects covered in Chapter 1 include network theorems, the small-signal equivalent circuits of tubes and transistors (including the correlation between transistor low-frequency h parameters and the high-frequency hybrid-II circuit elements), some very elementary feedback amplifier considerations, and the graphical methods of analysis. The book then

ix

continues with a study of how pulse-type signals are transmitted, shaped, or amplified by linear circuits. Included are resistive, capacitive, and inductive networks (Chapter 2), pulse transformers and delay lines (Chapter 3), and amplifiers (Chapters 4 and 5). A particularly detailed study of transistor wideband amplifiers (including compensation techniques) is made. As background material for the nonlinear wave-shaping circuits which are to follow, an extensive summary of the steady-state switching characteristics of devices is given (Chapter 6). Included are the semiconductor diode, the avalanche diode, the vacuum diode, a lengthy study of the transistor at cutoff and in saturation, the avalanche transistor, and the vacuum tube. Analyses of wave-shaping and switching functions which can be performed with nonlinear elements are introduced in the next two chapters: clipping and nonregenerative comparator circuits (Chapter 7) and clamping and switching circuits (Chapter 8). The study of digital operations begins in Chapter 9 with logic circuits, including Boolean algebra. Bistable multivibrators are treated in Chapter 10. The generation of gating signals and square waves by monostable and astable multivibrators is considered in Chapter 11. Negative-resistance devices are treated in Chapter 12. These include the tunnel diode, the unijunction transistor, the four-layer diode, the silicon-controlled switch (and its variants), and the avalanche transistor. Switching circuits constructed from these negative-resistance devices are discussed in Chapter 13. The next two chapters treat voltage and current time-base generators (including the phantastron circuit, the Miller integrator, and the bootstrap circuit). Chapter 16 discusses the blocking oscillator and includes the multiar configuration. Chapter 17 considers gates for sampling or transmission of signals and introduces the field-effect transistor as an important device for these applications. The next two chapters deal with counting, timing, synchronization, and frequency division. The final chapter (20) treats the transient switching characteristics of diodes and transistors, including the snap-off diode and the hot-carrier diode. The emphasis throughout this chapter is on the charge-control method of analysis.

In summary, this book presents a thorough study of the following basic circuits or techniques: transmission networks, differentiating circuits (including the transmission-line differentiator), clippers (limiters), comparators (discriminators), clampers (d-c restorers), the transistor or tube as a switch, logic circuits (AND, OR, NOT, NAND, diode matrices, etc.), bistable multis (flip-flops), monostable multis (one-shots), astable multis (square-wave generators), negative-resistance devices and circuits, time-base generators, counting, synchronization, and pulse amplification (including transient response and the effects of driving a transistor into saturation). The signals considered range from the very slow (millisecond or longer) to the very fast (nanosecond).

Semiconductor and tube circuits are presented side by side throughout the text, but with the principal emphasis on transistors. The basic philosophy adopted is to analyze a circuit on a physical basis in order to provide a clear understanding and intuitive feeling for its behavior. Only after the physical

analysis is established is mathematics used to express quantitative relationships. It is assumed that the student has a background in mathematics that
includes the study of linear differential equations with constant coefficients.
In order to avoid distractions from the principal concern of the analysis of
electronic circuits, algebraic and other mathematical manipulation has been
kept to a minimum. Solutions to differential equations which describe the
circuits under study are given without analysis, but the response indicated by
these equations has been plotted and studied in detail.

The piecewise linear and continuous model is introduced wherever such
an approximation is useful, particularly in a generalized discussion. However,
for the most part, real (commercially available) device characteristics are
employed. In this way, the reader may become familiar with the order of
magnitude of practical device parameters, the variability of these parameters
within a given type and with a change in temperature, the effect of the inevitable shunt capacitance in circuits, the consequence of minority-carrier storage
in semiconductor devices, the precautions which must be taken when dealing
with nanosecond pulses, the effect of input and output impedances and loading
on circuit operation, etc. These considerations are of utmost importance to
the student or practicing engineer since the circuits to be designed must function properly and reliably in the physical world rather than under hypothetical
or ideal circumstances.

There are a large number of examples worked out in the text in detail in
order to illustrate how theory may be applied to obtain quantitative results
and to emphasize the order of magnitude of the effects under consideration.
In addition, the 700 homework problems give the student experience in the
analysis and the design of the circuits discussed in the text and of other configurations to perform similar functions. In almost all numerical problems
realistic parameter values and specifications have been chosen. Considerable
care has been exercised in the development of these problems, which the authors
consider an integral and important part of the text.

There are many ways of implementing a pulse or digital system designed
to perform a particular function. It is hoped that through a study of this text
and through the experience gained from solving a goodly number of problems,
the reader will develop facility with these circuits and sharpen his creativity
and ingenuity so that he can arrive at a fairly optimum implementation of the
system under consideration.

To cover all the material in the book requires three semesters, at least
one of which should be part of an undergraduate electronics sequence. The
instructor has a wide range of topics to choose from, and he need not follow the
exact sequence in the book. For example, the two chapters (4 and 5) on wideband amplifiers may be considered too specialized for an undergraduate
program (or it may be desired that these topics be studied in a communications
course), and they may be omitted without particularly disturbing the sequence.

This book was planned originally as a second edition of the authors'
"Pulse and Digital Circuits" (McGraw-Hill Book Company, New York,

1956).   However, so much new material has been added and so extensive and thorough have been the revisions that a new title for the present text seems much more reasonable.   About half the topics in this book did not appear in the earlier work, and of the material that was presented there, almost every section has been completely rewritten.   This very major overhaul has been made necessary by the rapid developments which have taken place recently in this field, and particularly by the shift in emphasis from vacuum tubes to transistors and other semiconductor devices.

It may be of some interest to note that consideration was given to the advisability of splitting this work into two volumes, each of more moderate size.   A questionnaire that sought recommendations concerning the division of the material was addressed to a large number of our academic colleagues. The responses were so divergent that there seemed no alternative but to include all of the topics in one volume.

Considerable thought and effort were given to the pedagogy of presentation, to the explanation of circuit behavior, to the use of a consistent system of notation, and to the care with which detailed waveforms and other diagrams have been drawn in order to facilitate the use of this book in self-study.   It is hoped that the practicing engineer will find this book of service in updating himself in this field.

The authors are grateful to the many companies who supplied information in the form of device characteristics, application notes, and instrument instruction manuals.   The General Electric Company, Fairchild Semiconductor, Hewlett-Packard Company, Philco Corporation, Radio Corporation of America, Raytheon Company, Tektronix, Inc., Texas Instruments, Inc., and Transitron Electronic Corporation were particularly helpful.

We are pleased to acknowledge our indebtedness to our colleagues at Columbia University, at The City College, and in industry for many fruitful discussions.   In particular, the following persons read various portions of the manuscript and offered a great deal of constructive criticism: G. J. Clemens, R. C. Gebhardt, J. Hahn, V. I. Johannes, A. B. Marcovitz, P. T. Mauzey, I. M. Meth, A. C. Ruocchio, and L. Packer.   Mr. Mauzey merits our special gratitude because of the many valuable suggestions he offered and because of the diligence with which he assisted in the chore of proofreading.

We express our particular appreciation to Miss S. Silverstein, administrative assistant of the Electrical Engineering Department at The City College, for her most skillful service in the preparation of the manuscript.   We also thank W. I. H. Chen, A. B. Glaser, J. T. Millman, and J. N. Taub for their assistance.

Jacob Millman
Herbert Taub

CONTENTS

1/ REVIEW OF SELECTED TOPICS IN ELECTRONIC CIRCUIT THEORY

This book is concerned principally with the generation and processing of nonsinusoidal waveforms. A voltage or current waveshape having been generated, it may require processing in some manner. For example, it may be necessary to transmit it from one location to another, to amplify it, to shape it by clipping the top or bottom, to shift its d-c level, to select a portion (in time) of the waveform, to use it as a gate in connection with some other waveform, to perform with it some logic operation, and so forth. For the most part, the devices and circuits capable of accomplishing the functions referred to above must operate in a highly nonlinear manner, referred to as the *switching mode* of operation.

No previous acquaintance is assumed, on the part of the reader, with the switching behavior of devices or with the processing of the nonsinusoidal waveforms thereby generated. However, it is assumed that the reader has completed an introductory course in linear circuit analysis and a course in the small-signal theory of electronic devices—diodes, vacuum tubes, and transistors. For the sake of convenient reference, we summarize in this chapter certain network theorems, device models, concepts, and techniques from electronic circuit theory to which we shall have occasion to make reference. We assume the reader's familiarity with these topics and therefore present them without proof or elaboration.

1-1   NOTATION

We shall often consider, side by side, a tube and an analogous transistor circuit that perform identical functions. Since tube and

1

TABLE 1-1   Notation

	Grid (plate) voltage with respect to cathode	Base (collector) voltage with respect to emitter
Instantaneous total value.....................	v_G (v_P)	v_B (v_C)
Quiescent value............................	V_G (V_P)	V_B (V_C)
Instantaneous value of varying component........	v_g (v_p)	v_b (v_c)
Effective value of varying component (phasor, if a sinusoid)................................	V_g (V_p)	V_b (V_c)
Supply voltage (magnitude)....................	V_{GG} (V_{PP})	V_{BB} (V_{CC})

transistor will appear in the same discussion, it is very important that we use a system of notation which is applicable to either device and which deviates in a minimal way from contemporary practice. These requirements are met by adhering to the IEEE standards[1]† for semiconductor symbols and adopting[2] these standards as well for electron tubes. Only three modifications are required in the tube standards. First, the symbol e (E) is dropped, and v (V) is used for voltage. Second, the plate subscript b is no longer used, but is replaced by P. Third, the grid subscript c is replaced by G. Note that b and c are now reserved for *base* and *collector*, respectively. This notation is summarized in the following six statements and in Table 1-1.

1. Instantaneous values of *quantities* which vary with time are represented by lowercase letters (i for current, v for voltage, and p for power).

2. Maximum, average (d-c), and effective, or root-mean-square (rms), values are represented by the uppercase letter of the proper symbol (I, V, or P).

3. Average (d-c) values and instantaneous total values are indicated by the uppercase subscript of the proper electrode symbol (B for base, C for collector, E for emitter, G for grid, P for plate, and K for cathode).

4. Varying components from some quiescent value are indicated by the lowercase subscript of the proper electrode symbol.

5. A single subscript is used if the reference electrode is clearly understood. If there is any possibility of ambiguity, the conventional double-subscript notation should be used. For example, v_{ce} = instantaneous value of the varying component of voltage drop between collector and emitter and is positive if the collector is positive with respect to the emitter at a given instant of time. If the emitter is grounded and all voltages are understood to be measured with respect to ground, then the symbol v_{ce} may be shortened to v_c. The ground symbol is N (for *neutral*). For example, v_{PN} = instantaneous value of total voltage from plate to ground.

6. The *magnitude* of the supply voltage is indicated by repeating the electrode subscript.

† Superscript numerals are keyed to the References at the end of the chapter.

1-2   NETWORK THEOREMS

The following theorems are used frequently in the analysis of the circuits discussed in this book.

Kirchhoff's Current Law (KCL)   The sum of all currents toward a node must be zero at all times.

Kirchhoff's Voltage Law (KVL)   The sum of all voltage drops around a loop must be zero at all times.

These two theorems are valid even if the network contains nonlinear devices. The following laws are applicable only for linear circuits, but are valid even if dependent sources are present. A *controlled* or *dependent generator* is one whose voltage or current is a function of the voltage or current elsewhere in the circuit. A word of caution is in order when considering impedance in a network containing controlled sources. To find the impedance Z seen between two points, an external voltage generator V is considered to be applied between these points and the current I drawn from the source is determined. Thereafter, $Z = V/I$ provided that in the above procedure each independent (externally applied) source is replaced by its internal impedance—an ideal voltage source by a short circuit, and an ideal current source by an open circuit. *All dependent sources, however, must be retained in the network.*

Superposition Theorem   The response of a linear network containing several independent sources is found by considering each generator separately and then adding the individual responses. When evaluating the response due to one source each of the other *independent* generators is replaced by its internal impedance.

Thévenin's Theorem   Any linear network may, with respect to a pair of terminals, be replaced by a voltage generator (equal to the open-circuit voltage between the terminals) in series with the impedance seen at this port.

Norton's Theorem   Any linear network may, with respect to a pair of terminals, be replaced by a current generator (equal to the short-circuit current) in parallel with the impedance seen at this port.

From Thévenin's and Norton's theorems it follows that a voltage source V in series with an impedance Z is equivalent to a current source I in parallel with Z, provided that $I = V/Z$. These equivaler t circuits are indicated in Fig. 1-1.

Open-circuit Voltage—Short-circuit Current Theorems   As corollaries to Thévenin's and Norton's theorems we have the following relationships. If V represents the open-circuit voltage, I the short circuit current, and $Z(Y)$ the

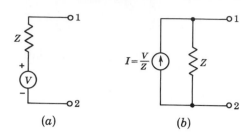

(a)   (b)

Fig. 1-1 As viewed from terminals 1 and 2, the Thévenin's circuit in (a) is equivalent to the Norton's circuit in (b). Note: Throughout the text a circle with a ± sign represents an ideal voltage source, whereas a circle with an arrow in it signifies an ideal current source.

impedance (admittance) between two terminals in a network, then

$$V = IZ = \frac{I}{Y} \qquad I = \frac{V}{Z} = VY \qquad Z = \frac{V}{I} \qquad (1\text{-}1)$$

In spite of their disarming simplicity, these equations (reminiscent of Ohm's law) should not be overlooked because they are most useful in analysis. For example, the first equation, which states "open-circuit voltage equals short-circuit current divided by admittance," is often the simplest way to find the voltage between two points in a network, as the following problem illustrates.

EXAMPLE Find the voltage V between nodes 1 and 2 of Fig. 1-2.

Solution The current flowing in a short circuit placed between 1 and 2 is, using superposition,

$$I = \tfrac{25}{10} - \tfrac{10}{5} = 0.50 \text{ mA}$$

If the sources are replaced by their internal resistances (assumed zero), then the three resistors are in parallel between 1 and 2. Hence

$$Y = \tfrac{1}{10} + \tfrac{1}{5} + \tfrac{1}{20} = 0.35 \text{ mA/V} = 0.35 \text{ millimho}$$

and

$$V = \frac{I}{Y} = \frac{0.50}{0.35} = 1.43 \text{ V}$$

The third relationship in Eq. (1-1), which states that "the impedance between two nodes equals the open-circuit voltage divided by the short-circuit current," is frequently the simplest way to calculate the output impedance of a circuit.

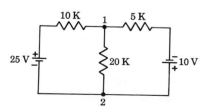

Fig. 1-2 An illustrative problem.

1-3   LOW–FREQUENCY SMALL–SIGNAL TRANSISTOR MODEL[3]

The incremental terminal behavior of a transistor is best described in terms of the h parameters for the following reasons. These hybrid parameters are *real numbers* at low frequencies, are particularly easy to measure, can also be obtained from the transistor static characteristic curves, and are convenient to use in circuit analysis and design. Furthermore, in present practice a set of h parameters is customarily specified for each transistor by the manufacturer, and it is a simple matter to convert from the h parameters for one configuration, say the common-base (CB) circuit, to another arrangement, say the common-emitter (CE) configuration.

The hybrid small-signal model, valid in the active region of the transistor, for any configuration at low frequencies is indicated in Fig. 1-3. The input (output or load) voltage is V_i (V_L), the input (output) current is I_i (I_L), and the impedance loading the output is Z_L. Note that this model contains two dependent sources, one ($h_r V_L$) controlled by the output voltage and the other ($h_f I_i$) controlled by the input current.

The quantities of most interest when using the transistor as an amplifier are the current gain A_I, the input impedance Z_i, and the voltage gain A_V. From Fig. (1-3) we can obtain the following formulas for these quantities:

$$A_I \equiv \frac{I_L}{I_i} = \frac{-h_f}{1 + h_o Z_L} \tag{1-2}$$

$$Z_i \equiv \frac{V_i}{I_i} = h_i + h_r A_I Z_L \tag{1-3}$$

$$A_V \equiv \frac{V_L}{V_i} = \frac{A_I Z_L}{Z_i} \tag{1-4}$$

The beautiful simplicity of the above amplifier equations is evident. Numerical calculations for any configuration may be carried out quite rapidly.[3] Note that the expression for A_V does not contain the h parameters explicitly

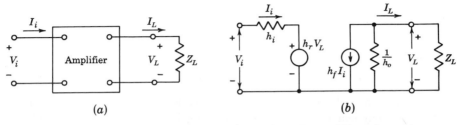

(a)   (b)

Fig. 1-3   (a) A transistor amplifier in either the CE, CB, or CC configuration; (b) the hybrid-parameter model for small-signal variations from the quiescent operating point.

TABLE 1-2   Typical h-parameter values
(at $I_E = 1.3$ mA)

Parameter	CE	CC	CB
h_i	1,100	1,100	21.6 Ω
h_r	2.5×10^{-4}	~1	2.9×10^{-4}
h_f	50	−51	−0.98
$1/h_o$	40 K†	40 K	2 M‡

† K = kilohms.
‡ M = megohms.

and hence is valid regardless of what equivalent circuit is used for the transistor. In particular, Eq. (1-4) applies even at high frequencies, where the h parameters are functions of the frequency or where we may prefer to use another model for the transistor (Sec. 1-4).

If a specified configuration is under consideration it is identified by a second subscript on the h parameters. Thus, for the CE configuration, h_i, h_r, h_f, and h_o are replaced by h_{ie}, h_{re}, h_{fe}, and h_{oe}, respectively. The negative of the current transfer ratio with the output short-circuited (often referred to simply as the *short-circuit current gain*) for the CE configuration is often called the *beta* of the transistor, or $h_{fe} = \beta$. For the CB configuration this quantity is called the *alpha* of the transistor, or $h_{fb} = -\alpha$.

Tables of conversion formulas of h parameters for the three configurations are available.[3,4] For example, the CC parameters are given in terms of the CE hybrid values by the following nearly exact relationships:

$$h_{ic} = h_{ie} \qquad h_{fc} = -(h_{fe} + 1) \qquad h_{rc} = 1 \qquad h_{oc} = h_{oe} \tag{1-5}$$

An *emitter follower* is a CC circuit with the load in the emitter leg so that Z_L is replaced by Z_e. From the above equations it is found that for the emitter follower

$$A_I = \frac{h_{fe} + 1}{1 + h_{oe}Z_e} \tag{1-6}$$

$$Z_i = h_{ie} + A_I Z_e \tag{1-7}$$

$$1 - A_V = \frac{h_{ie}}{Z_i} \tag{1-8}$$

The voltage gain of an emitter follower may be very close to unity, and Eq. (1-8) is a nearly exact expression for the deviation from unity. We shall make use of these equations in Sec. 14-15, where the Darlington cascade is discussed. Representative values of the hybrid parameters for a low- or medium-power junction transistor are indicated in Table 1-2.

We should like to point out for future reference that the CE input impedance under a reasonable load does not differ greatly from the short-circuit input resistance h_{ie}. Thus, for a 5-K load and for parameter values in Table 1-2 we find $Z_i = 1,045\ \Omega$. This value is only 5 percent smaller than $h_{ie} = 1,100\ \Omega$.

1-4   THE HYBRID–PI (Π), HIGH–FREQUENCY, SMALL–SIGNAL, COMMON–EMITTER MODEL[3,5]

An equivalent circuit which gives excellent agreement with experiment in the range from d-c to frequencies where the effectiveness of the transistor begins to be limited is indicated in Fig. 1-4. The h-parameter circuit is simpler at low frequencies and hence the hybrid-Π model is usually reserved for high-frequency calculations. We shall exploit it in Chaps. 4 and 5, where wide-band amplifiers are considered. All parameters (resistances and capacitances) in the model are assumed to be independent of frequency. They may vary with the quiescent operating point but under given bias conditions are reasonably constant for small signal swings. The internal node B' is not physically accessible. The ohmic base resistance, the so-called *base-spreading resistance* $r_{bb'}$, is represented as a lumped parameter between the external base terminal and B'. The transistor transconductance g_m is linearly related to the emitter current I_E and varies inversely with the absolute temperature T as follows:

$$g_m = \frac{h_{fe}}{1 + h_{fe}} \frac{|I_E|}{\eta V_T} \quad \text{mhos} \tag{1-9}$$

where $\eta = 1$ for germanium and approximately 2 for silicon and where $V_T = T/11{,}600$ (Sec. 6-1). Hence, for a germanium transistor at room temperature (with $h_{fe} \gg 1$), *with I_E in milliamperes,*

$$g_m \approx \frac{|I_E|}{26} \quad \text{mhos} \tag{1-10}$$

If the CE h parameters at low frequency—h_{fe}, h_{ie}, h_{re}, and h_{oe}—are determined

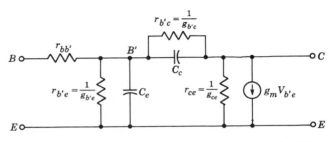

Fig. 1-4   The hybrid-Π model for a transistor in the CE configuration.

at a given emitter current I_E, then the resistances in the hybrid-Π circuit are calculable from the following four equations in the order given:

$$r_{b'e} = \frac{h_{fe}}{g_m} \quad \text{or} \quad g_{b'e} = \frac{g_m}{h_{fe}} \tag{1-11}$$

$$r_{bb'} = h_{ie} - r_{b'e} \tag{1-12}$$

$$r_{b'c} = \frac{r_{b'e}}{h_{re}} \quad \text{or} \quad g_{b'c} = \frac{h_{re}}{r_{b'e}} \tag{1-13}$$

$$g_{ce} = h_{oe} - (1 + h_{fe})g_{b'c} = \frac{1}{r_{ce}} \tag{1-14}$$

For the typical h parameters in Table 1-2 we find at $I_E = 1.3$ mA and room temperature that for a germanium transistor

$$g_m = 50 \text{ mA/V} \quad r_{b'e} = 1 \text{ K} \quad r_{bb'} = 100 \ \Omega$$
$$r_{b'c} = 4 \text{ M} \quad r_{ce} = 82 \text{ K}$$

The collector junction capacitance $C_c = C_{b'c}$ is the measured CB output capacitance with the input open ($I_E = 0$) and is usually specified by manufacturers as C_{ob}. The emitter-junction capacitance $C_e = C_{b'e}$ is determined from a measurement of the frequency f_T at which the CE short-circuit current gain drops to unity. We verify in Sec. 4-6 that

$$C_e \approx \frac{g_m}{2\pi f_T} \tag{1-15}$$

Reasonable values for these capacitances are

$$C_c = 3 \text{ pF}† \quad C_e = 100 \text{ pF}$$

1-5   SMALL–SIGNAL TUBE MODELS

For small variations from the quiescent operating point a vacuum tube (Fig. 1-5a) may be replaced by either the Thévenin's model of Fig. 1-5b or the Norton's equivalent of Fig. 1-5c. Both of the dependent sources, the Thévenin's voltage generator and the Norton's current generator, are controlled by the voltage V_{gk} from grid to cathode. In these circuits r_p = plate resistance, μ = amplification factor, and g_m = transconductance. These three parameters are not independent because $\mu = g_m r_p$. The model of Fig. 1-5 is equally valid for a triode or a pentode provided that the screen and suppressor are held at fixed voltages. A network containing tubes may be analyzed by replacing each device with its equivalent circuit and by disregarding all those features, such as supply and bias voltages, which have an influence only on the quiescent state.

† The abbreviation pF = picofarad = $\mu\mu$F = micromicrofarad.

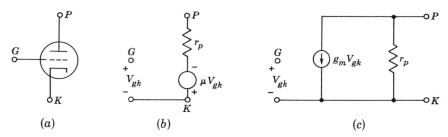

Fig. 1-5   The vacuum tube in (a) may be replaced by either the Thévenin's model, as in (b), or the Norton's equivalent circuit, as in (c).

1-6   VOLTAGE AND CURRENT AMPLIFICATIONS

In Chaps. 4 and 5 we shall deal with the gain, frequency response, and time-domain response of vacuum-tube and transistor amplifiers and amplifier stages.   To avoid the need to digress in that discussion, we shall, at this point, define a number of voltage and current amplifications, introduce symbols for them, and derive certain useful relationships among them.

In Fig. 1-6a a signal I is applied to an amplifier stage (transistor or tube) from a generator of source impedance Z_s, and, as a result, a signal I_L is delivered to the load Z_L.   We have chosen here to represent the input generator by its Norton's equivalent.   The current entering the input terminals of the amplifier is I_i.   We define the four current gains

$$A_I \equiv \frac{I_L}{I_i} \qquad\qquad A_{Is} \equiv \frac{I_L}{I} \qquad\qquad\qquad (1\text{-}16a)$$

$$A_i \equiv \frac{I_L}{I_i}\,(Z_L = 0) \qquad A_{is} \equiv \frac{I_L}{I}\,(Z_L = 0) \qquad\qquad (1\text{-}16b)$$

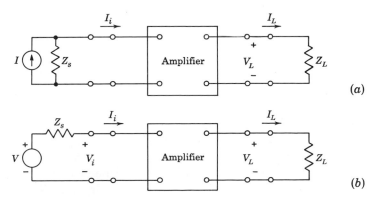

Fig. 1-6   An amplifier driven by (a) a current source and (b) a voltage source.

The gain A_{Is} takes account of the source impedance Z_s although A_I does not. The gain A_I is the ratio of load current to the current that would be furnished to the amplifier input by an ideal current source for which $Z_s = \infty$. For such an ideal source $A_I = A_{Is}$. The gains A_i and A_{is} are the corresponding gains when the amplifier output terminals are shorted.

Alternatively, we might have chosen to have represented an input generator by its Thévenin's equivalent (a source V in series with its impedance Z_s), as in Fig. 1-6b. The input and output (load) voltages of the amplifier are V_i and V_L, respectively. In this case it is appropriate to define the voltage gains

$$A_V \equiv \frac{V_L}{V_i} \qquad\qquad A_{Vs} \equiv \frac{V_L}{V} \qquad\qquad (1\text{-}17a)$$

$$A_v \equiv \frac{V_L}{V_i}\,(Z_L = \infty) \qquad A_{vs} \equiv \frac{V_L}{V}\,(Z_L = \infty) \qquad (1\text{-}17b)$$

As in the case of the current amplifications, the gain A_{Vs} has the "practical" aspect that it takes account of the generator impedance, whereas A_V does not. The gain A_{Vs} is equal to A_V when the input signal is applied by an ideal voltage generator of zero impedance. The gains under open-circuit conditions (when the load is disconnected from the output terminals) are A_v and A_{vs}.

In general, the relationships among these various gains depend on the amplifier. If the input impedance is Z_i, then it follows from the defining equations (1-16) and (1-17) and from Fig. 1-6 that

$$A_{Is} = A_I \frac{I_i}{I} = A_I \frac{Z_s}{Z_s + Z_i} \qquad\qquad (1\text{-}18)$$

and

$$A_{Vs} = A_V \frac{V_i}{V} = A_V \frac{Z_i}{Z_s + Z_i} \qquad\qquad (1\text{-}19)$$

We observe that these equations are consistent with the statements made above that $A_{Is} = A_I$ if $Z_s = \infty$ and $A_{Vs} = A_V$ if $Z_s = 0$. Combining Eqs. (1-4) and (1-19),

$$A_{Vs} = \frac{A_I Z_L}{Z_s + Z_i} \qquad\qquad (1\text{-}20)$$

Equations (1-18) and (1-20) indicate how to calculate current and voltage gains taking source impedance into account if the current gain A_I from an ideal source is known. In order to evaluate these amplifications we must be able to evaluate the input impedance Z_i of the amplifier under consideration.

There is one relationship which is independent of the amplifier (since it does not require a knowledge of Z_i) and which we shall find useful. Since

$$A_{Is} = \frac{I_L}{I} \qquad A_{Vs} = \frac{V_L}{V} \qquad V_L = I_L Z_L \qquad V = I Z_s$$

we have that

$$A_{Vs} = A_{Is} \frac{Z_L}{Z_s} \tag{1-21}$$

This same result is obtained by dividing Eq. (1-20) by Eq. (1-18). It is important to keep in mind that Eq. (1-21) applies only provided that A_{Vs} and A_{Is} *correspond to the same source and load impedances.* But the result is independent of whether the source impedance appears as an element in series with the input terminals and voltage generator, as in Fig. 1-6b, or in shunt with the input terminals and current generator, as in Fig. 1-6a.

Suppose now that the ratio Z_L/Z_s is independent of frequency (or, more generally, of the Laplace transform variable). Then from Eq. (1-21) we have the result that the *frequency response and time-domain response of the voltage gain A_{Vs} and of the current gain A_{Is} are precisely the same in form.* Such would be the case if, for example, the source and load impedances were both resistive. However, even under this circumstance note that the transient response of A_V ($Z_s = 0$) may not be the same as that of A_I ($Z_s = \infty$).

Note that none of the equations in this section depends upon the particular model chosen for the amplifier. They are equally applicable to a tube or to a transistor amplifier of any number of stages. Of course, in evaluating the symbols, such as Z_i, in the equations we must refer to an equivalent circuit valid for the active device over the frequency range under consideration.

We shall often have occasion to refer to an amplification in the midband region; this quantity will be indicated explicitly by an additional subscript o. Two examples follow:

A_{Iso} = *midband current gain under load taking the source impedance into account*

A_{vo} = *midband open-circuit voltage amplification fed from an ideal voltage source* ($R_s = 0$, $R_L = \infty$)

A few final important observations with respect to the frequency response of an amplifier: If the entire circuit can be reduced at low frequencies to one containing a single time constant τ_1, then the frequency response is given by

$$A_1 = \frac{A_o}{1 - j/\omega\tau_1} = \frac{A_o}{1 - j(f_1/f)} \tag{1-22}$$

where A_1 represents amplification (either current or voltage) as a function of frequency, A_o is the midband gain, and $f_1 = 1/2\pi\tau_1$ is the lower 3-dB frequency.

Similarly, for a single-time-constant (τ_2) circuit at high frequencies the response A_2 is given by

$$A_2 = \frac{A_o}{1 + j\omega\tau_2} = \frac{A_o}{1 + j(f/f_2)} \tag{1-23}$$

where $f_2 = 1/2\pi\tau_2$ is the upper 3-dB frequency.

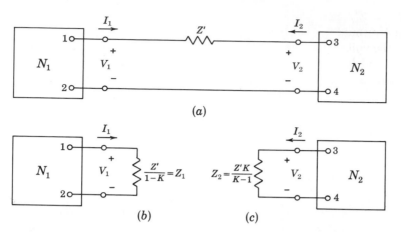

Fig. 1-7 Pertaining to Miller's theorem. By definition, $K \equiv V_2/V_1$.

The following procedures allow a very simple evaluation of A_o and τ (τ_1 or τ_2). If all independent voltage sources are short-circuited and if all independent current sources are open-circuited, the resulting network contains only resistors and capacitors. Since we have assumed a single-time-constant circuit we should now be able to evaluate τ by inspection. To calculate A_o we must solve a *resistive network* for the ratio of output-to-input voltage (or current). This midband region contains no reactive elements because in the low-frequency circuit this region is defined by $f \to \infty$ and in the high-frequency case by $f \to 0$.

1-7 MILLER'S THEOREM

Consider an arbitrary circuit configuration which can be split into two networks N_1 and N_2 interconnected with an impedance Z', as indicated in Fig. 1-7a. More specifically, a set of terminals 1-2 is selected in N_1 and a second set 3-4 is designated in N_2 so that 1 and 3 are connected through Z' and 2 and 4 through a short circuit. We *postulate* that we know the ratio V_2/V_1, where V_1 is the voltage between terminals 1-2 and V_2 is the voltage between terminals 3-4. Designate the ratio V_2/V_1 by K, which in the sinusoidal steady state will be a complex number and more generally will be a function of the Laplace transform variable s. We shall now show that the current I_1 drawn from N_1 can be obtained by disconnecting terminal 1 from Z' and simply bridging an impedance $Z'/(1 - K)$ across 1-2, as indicated in Fig. 1-7b.

The current I_1 is given by

$$I_1 = \frac{V_1 - V_2}{Z'} = \frac{V_1(1 - K)}{Z'} = \frac{V_1}{Z'/(1 - K)} = \frac{V_1}{Z_1} \tag{1-24}$$

Therefore if $Z_1 \equiv Z'/(1 - K)$ were shunted across terminals 1-2 the current I_1 drawn from N_1 would be the same as that from the original circuit. Accordingly, as far as the voltages and currents in N_1 are concerned, the two configurations (Fig. 1-7a and b) are indistinguishable from one another.

In a similar way it may be established that the correct current I_2 drawn from N_2 may be calculated by replacing the connections to terminals 3-4 by an impedance Z_2, given by

$$Z_2 \equiv \frac{Z'}{1 - 1/K} = \frac{Z'K}{K - 1} \tag{1-25}$$

In other words, as far as N_2 is concerned Fig. 1-7a and c are equivalent. It must be emphasized that this theorem will be useful in making calculations only if it is possible to find the value of K by some independent means.

EXAMPLE Consider a triode-tube amplifier stage, taking interelectrode capacitances into account. (a) What is the effective input impedance? (b) What is the effective capacitive loading at the output?

Solution a. As indicated in Fig. 1-8a, network N_1 may be taken as the source V_1 in parallel with C_{gk}, and network N_2 as the tube including R_p and C_{pk} and $Z' =$

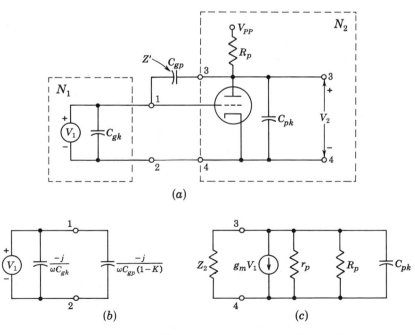

(a)

(b)   (c)

Fig. 1-8 (a) A triode stage; (b) the equivalent input circuit; (c) the equivalent output circuit.

$-j/\omega C_{gp}$. Using the above theorem the loading on N_1 is as indicated in Fig. 1-8b, where $K = V_2/V_1$ represents the voltage gain of the stage. The calculation of K is made by replacing the tube by its small-signal model and solving the resulting network. In the midband region we find, approximately,

$$K = -A_o \qquad \text{where } A_o = \frac{g_m R_p r_p}{R_p + r_p}$$

is the magnitude of the voltage gain. In this region the load on the source V_1 (or on the stage preceding the one under consideration, if a multistage amplifier is under consideration) is a capacitance

$$C_i = C_{gk} + C_{gp}(1 + A_o) \tag{1-26}$$

Since A_o may be a large positive number, C_i may be very much larger than C_{gk}. This exaggeration of the grid-plate capacitance is called the *Miller effect*. If, typically, $C_{gp} = 3.5$ pF $= C_{gk}$ and $K = -10$, then the input capacitance is 42 pF. The Miller effect operates in pentodes as well, but the capacitance C_{gp} in a pentode is smaller than in a triode by a factor in the range 100 to 1,000.

b. In Fig. 1-8c we have indicated N_2 with the tube replaced by its Norton's small-signal model and with Z_2 given by Eq. (1-25), namely,

$$Z_2 = \frac{-j}{\omega C_{gp}} \frac{K}{K-1} \approx \frac{-j}{\omega C_{gp}} \tag{1-27}$$

if the voltage gain K is much greater than unity. Under these circumstances Z_2 simply represents a capacitance equal to C_{gp}. Hence, the total capacitance loading this stage is $C_{gp} + C_{pk}$ (plus, of course, the input capacitance of the next stage if this amplifier is followed by another stage). Note that in this case we do not need to know the exact expression for K, but we merely must be certain that $|K| \gg 1$.

The Miller effect[6] was originally enunciated in connection with the input capacitance of vacuum tubes. However, the transformation presented above, by which the Miller effect is deduced, is generally useful in the analysis of other circuits, and we shall refer to the transformation as *Miller's theorem*.

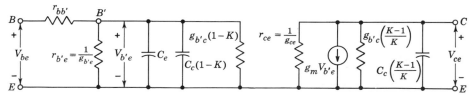

Fig. 1-9 The circuit which results from an application of the Miller theorem to the hybrid-Π circuit of Fig. 1-4. The impedance Z' of Fig. 1-7 is now equal to the parallel combination of $g_{b'c}$ and C_c.

An application of Miller's theorem to the hybrid-II model of Fig. 1-4 yields the circuit of Fig. 1-9. Here $K \equiv V_{ce}/V_{b'e}$. Observe that the use of Miller's theorem has resulted in two isolated networks, one at the input and one at the output. We discuss further simplifications of this circuit in Sec. 4-7, where we also inquire into the determination of the value of K in cases of interest.

1-8   THE OPERATIONAL AMPLIFIER

An operational amplifier is constructed in the manner indicated in Fig. 1-10a. Here an amplifier, with input terminals 1 and 2 and output terminals 3 and 4, whose gain is negative, real, and large has been augmented by the addition of two impedances Z and Z'. The impedance Z_i represents the input impedance of the amplifier. A simplified representation of the amplifier is shown in Fig. 1-10b. The amplifier within the box in Fig. 1-10a is called the *base amplifier* and may consist of one or more vacuum-tube or transistor stages in cascade.

We define $A_V \equiv V_o/V_i$ to be the voltage amplification with Z' in place. Comparing Fig. 1-10 with Fig. 1-7 we see that Miller's theorem is directly applicable with $K \equiv A_V$. Hence, an equivalent circuit of the operational amplifier, which gives the same input current I from the source V_s, the same amplifier input voltage V_i, and consequently the same output voltage V_o as in Fig. 1-10, is indicated in Fig. 1-11. From this figure we see that if

$$\left| \frac{Z'}{1 - A_V} \right| \ll |Z_i| \tag{1-28}$$

then $I' \approx I$. Under these circumstances the output voltage is

$$V_o = A_V V_i = A_V I \frac{Z'}{1 - A_V} \tag{1-29}$$

Even for a transistor, for which the input impedance is much smaller than

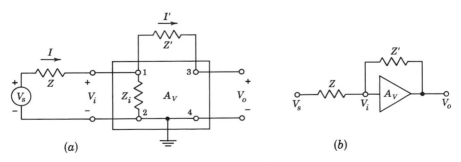

Fig. 1-10 Two representations of an operational amplifier.

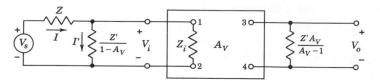

Fig. 1-11 An equivalent circuit of the operational amplifier.

for a vacuum-tube amplifier, the inequality (1-28) will be satisfied provided that A_V is made sufficiently large. As $|A_V| \to \infty$ the impedance across terminals 1 and 2 approaches zero (a short circuit), and $I \approx V_s/Z$. Also, as $|A_V| \to \infty$, we see from Eq. (1-29) that the output is

$$V_o \approx -IZ' = -\frac{Z'}{Z} V_s$$

and the overall voltage gain is

$$A_f \equiv \frac{V_o}{V_s} = -\frac{Z'}{Z} \tag{1-30}$$

The operation of the circuit may now be described in the following terms. At the input to the amplifier proper there exists a virtual short circuit or virtual ground. The term "virtual" is used to imply that, while the feedback from output to input through Z' serves to keep the voltage V_i at zero, no current actually flows through this short. The situation is depicted in Fig. 1-12, where the virtual ground is represented by the heavy double-headed arrow. The current furnished by the generator V_s continues past this virtual short through the impedance Z', so that $V_o = -IZ'$.

An operational amplifier may be used to perform many mathematical operations. This feature accounts for the name which has been assigned to this type of amplifier configuration. Among the basic configurations are the following:

Sign Changer or Inverter If $Z = Z'$, then $A_f = -1$, and the sign of the input signal has been changed. Hence such a circuit acts as a phase inverter. If two such amplifiers are connected in cascade, the output from the second stage equals the signal input without change of sign. Hence, the outputs from the two stages are equal in magnitude but opposite in phase, and such a system is an excellent paraphase amplifier.

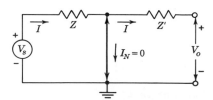

Fig. 1-12 Virtual ground in the operational amplifier.

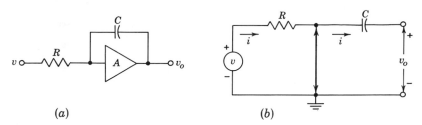

Fig. 1-13 (a) Operational integrator; (b) equivalent circuit.

Scale Changer If the ratio $Z'/Z = k$, a real constant, then $A_f = -k$, and the scale has been multiplied by a factor $-k$. Usually, in such a case of multiplication by a constant, -1 or $-k$, Z and Z' are selected as resistors.

Phase Shifter Assume Z and Z' are equal in magnitude but differ in angle. Then the operational amplifier shifts the phase of a sinusoidal input voltage while at the same time preserving its amplitude. Any phase shift from 0 to 360° (or $\pm 180°$) may be obtained.

Integrator If $Z = R$ and a capacitor C is used for Z', as in Fig. 1-13, we can show that the circuit performs the mathematical operation of integration. The input need not be sinusoidal and hence will be represented by the lower-case symbol $v = v(t)$. (The subscript s will now be omitted, for simplicity.) Correspondingly, the current as a function of time is designated by $i = i(t)$. In Fig. 1-13b (analogous to Fig. 1-12) the double-headed arrow represents a virtual ground. Hence $i = v/R$ and

$$v_o = -\frac{1}{C} \int i \, dt = -\frac{1}{RC} \int v \, dt \tag{1-31}$$

The amplifier, therefore, provides an output voltage proportional to the integral of the input voltage.

If the input voltage is a constant $v = V$, then the output will be a ramp $v_o = -Vt/RC$. We shall discuss this *Miller integrator* in detail in Chap. 14, where the errors introduced because A_V cannot be infinite are analyzed.

Differentiator If Z is a capacitor C and if $Z' = R$, then we see from the equivalent circuit of Fig. 1-14 that $i = C \, dv/dt$ and

$$v_o = -Ri = -RC \frac{dv}{dt} \tag{1-32}$$

Hence, the output is proportional to the time derivative of the input.

We note that when the gain A_V is large enough, the overall gain of the operational amplifier is $A_f = -Z'/Z$. Therefore, provided only that the amplifier has adequate gain, the overall amplification depends only on the

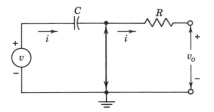

Fig. 1-14 Equivalent circuit of the operational differentiator.

impedances Z' and Z and not on other elements nor on the characteristics of the active devices used in the base amplifier. If Z' and Z are selected to be stable components the overall gain will be similarly stable. The operational amplifier suppresses the effect of the variability of gain with operating point, with age, with replacement, with temperature, etc. Similarly, the operational amplifier suppresses the effect of active element nonlinearity, for nonlinearity may be viewed as a variation of device parameters with operating point.

These features of stability as well as other useful features which are characteristic of the operational amplifier result from the fact that this amplifier configuration incorporates *feedback*. The input signal to the amplifier V_i in Fig. 1-10a is a linear combination of the external signal V_s and the output signal V_o. Since the signal which is fed back to the input is proportional to the output voltage, the amplifier is described as incorporating *voltage feedback*. The feedback is negative (or degenerative) in the sense that the gain with feedback $|A_f|$ is less than the gain without feedback $|A_V|$. In a physical circuit, the base amplifier consists of a cascade of common-emitter or common-cathode stages. The input is applied at the base or grid of the first stage and the output taken from the last collector or plate. Therefore, this type of feedback is often called *collector-to-base* or *plate-to-grid feedback*.

1-9 A CURRENT FEEDBACK AMPLIFIER

A second feedback-amplifier configuration of a type we shall encounter in Chap. 15 is shown in Fig. 1-15. Here the feedback voltage is taken across

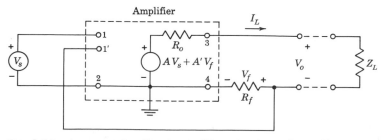

Fig. 1-15 A current-feedback-amplifier circuit configuration.

the resistor R_f, and this voltage $V_f = I_L R_f$ is now proportional to the output *current* I_L rather than to the output voltage, as was the case with the operational amplifier. As a matter of convenience, separate inputs 1 and 1' are provided for the external signal and feedback signals, respectively. For example, the two inputs may be the bases of the two transistors of a difference amplifier. The superposition of the two signals which appears at the collector continues thereafter through a common channel to the output. A Thévenin's replacement of the base amplifier has been made with respect to terminals 3 and 4. The output impedance without feedback is R_o, and provision has been made to account for the possibility that the gains A and A' for the external signal and the feedback signal, respectively, may not be the same.

From Fig. 1-15 we have

$$V_o = A V_s + A' V_f - (R_o + R_f) I_L = Z_L I_L \tag{1-33}$$

Eliminating V_f through $V_f = I_L R_f$ and solving for I_L we find

$$I_L = \frac{-A V_s}{A' R_f - (R_o + R_f + Z_L)} \tag{1-34}$$

If the gain $|A'|$ is made large enough so that $|A'| R_f \gg |R_o + R_f + Z_L|$, then

$$I_L \approx - \frac{A V_s}{A' R_f} \tag{1-35}$$

Suppose further that we arrange that the feedback and external signals proceed, for the most part, through a common channel. Then $A \approx A'$ and

$$I_L \approx \frac{-V_s}{R_f} \tag{1-36}$$

Equation (1-36) indicates that the load current will be proportional to the input signal and will be stable if R_f is a stable resistor. As indicated by Eq. (1-36), the load current does not depend on the parameters of the amplifier or circuit elements, active or passive, other than R_f. The result is independent of any nonlinearity of amplifier or the load Z_L. These advantageous features result from the *negative current feedback* present in the circuit when the gain A is large, real, and negative. Thus in the present case the current feedback assures stability and lack of distortion in the output current just as, in the operational amplifier, the voltage feedback secures these same benefits for the output voltage.

1-10   GRAPHICAL CALCULATION FOR A TUBE CIRCUIT

A computation which must often be made in pulse circuits is that of calculating the quiescent voltages and current in a triode tube circuit. Many practical one-tube circuits have a resistor R_k in series with the cathode in addition to the

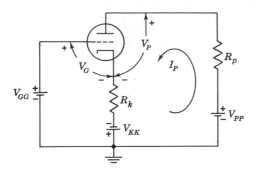

Fig. 1-16 A tube circuit with both plate and cathode resistors.

resistor R_p in series with the plate. The resistor R_k is returned either to ground or to a negative supply $-V_{KK}$, as indicated in Fig. 1-16.

We consider now how to use the characteristic curves of the vacuum tube to analyze the circuit for a fixed (d-c) grid bias V_{GG}. Kirchhoff's law (KVL) applied to the plate circuit yields

$$-V_{PP} - V_{KK} + V_P + I_P(R_p + R_k) = 0 \tag{1-37}$$

Similarly, KVL around the grid circuit gives

$$-V_{GG} + V_G - V_{KK} + I_P R_k = 0 \tag{1-38}$$

Equation (1-37) represents a load line corresponding to an effective supply voltage $V_{PP} + V_{KK}$ and a total resistance $R_p + R_k$. This line is plotted on the plate characteristics in Fig. 1-17. From Eq. (1-38) the current is given by

$$I_P = \frac{V_{GG} + V_{KK} - V_G}{R_k} \tag{1-39}$$

For each value of V_G for which there is a plotted plate characteristic the current I_P is calculated from Eq. (1-39). The corresponding values of I_P and V_G are plotted on the characteristics, as indicated by the dots in Fig. 1-17.

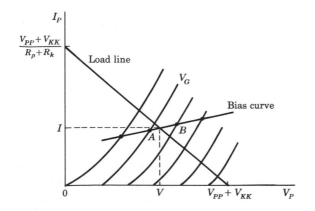

Fig. 1-17 The intersection of the load line and the bias curve gives the operating point.

The locus of these points is called the *bias curve*. The intersection of the bias curve and the load line gives the plate voltage V and current I corresponding to the given input voltage V_{GG}.

The procedure outlined above is extremely simple to carry out. It is really not necessary to use all values of V_G; it suffices to find two adjacent grid curves which give currents above and below the load line, as indicated by points A and B in Fig. 1-17. The intersection of a straight line drawn through A and B with the load line gives the desired operating point. In particular it should be noted that if $V_{GG} + V_{KK}$ is large compared with the range of values of V_G (called the *grid base*), then I_P will be almost constant and hence the curve connecting the dots in Fig. 1-17 will be approximately a horizontal straight line.

The cathode follower is a special case of the circuit of Fig. 1-16 with $R_p = 0$. Often the cathode resistor is connected to ground, so that $V_{KK} = 0$. Sometimes neither V_{KK} nor V_{GG} is used, and *self-bias* is obtained from the quiescent drop across R_k. The construction is the same as that indicated in Fig. 1-17, with a load line corresponding to V_{PP} and R_k (or to $R_k + R_p$ if $R_p \neq 0$) and a bias curve given by $I_P = -V_G/R_k$.

Methods for calculating the quiescent operating condition in a transistor circuit both in the active region and in saturation are given in the text where needed and therefore are not discussed here.

1-11   INPUT AND OUTPUT IMPEDANCE OF A TUBE STAGE

Figure 1-18 shows a tube stage with plate and cathode resistors. Our concern here is with small-signal operation; therefore all bias and supply voltages have been omitted. The plate and cathode resistors have been placed to suggest that these may be viewed, when we please to do so, as external loads on the

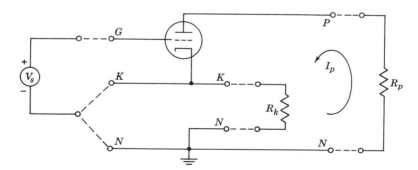

Fig. 1-18 A vacuum-tube stage with a plate and cathode resistor. The signal source V_s is connected between grid and cathode or between grid and ground.

amplifier and not part of the amplifier. Provision has been made so that the external signal source V_s may be connected between grid and cathode or between grid and ground.

Let us now consider the Thévenin's equivalent circuit with respect to the output terminals P-N when the signal source V_s is connected between grid and cathode. We consider that R_k is in place but R_p is not connected. Replacing the tube by its small-signal equivalent of Fig. 1-5b we find that the Thévenin's equivalent source V and output impedance Z (Fig. 1-1a) are

$$V = -\mu V_s \qquad Z = r_p + R_k \tag{1-40}$$

Next we determine the Thévenin's equivalent when the source V_s is connected between grid and ground. To make this calculation we connect R_p so that a current I_p will flow. Using again the small-signal equivalent circuit of the tube we find

$$I_p = \frac{\mu V_s}{r_p + (\mu + 1)R_k + R_p} \tag{1-41}$$

The output voltage $V_o \equiv V_{pn}$ is

$$V_o = -I_p R_p = \frac{(-\mu V_s)R_p}{[r_p + (\mu + 1)R_k] + R_p} \tag{1-42}$$

If a load Z_L were connected across the terminals 1 and 2 of Fig. 1-1a then the output in that case $V_o \equiv V_{12}$ would be

$$V_o = \frac{V Z_L}{Z + Z_L} \tag{1-43}$$

Comparing Eqs. (1-42) and (1-43) we note that if we consider R_p an external load like Z_L, then in the Thévenin's equivalent with respect to terminals P-N,

$$V = -\mu V_s \qquad Z = r_p + (\mu + 1)R_k \tag{1-44}$$

Comparing Eq. (1-44) with (1-40) we see that the change in connection of the input source has not modified the Thévenin's source voltage but has resulted in the addition of a term μR_k to the output impedance.

When the output is taken between K and N and when the signal source is connected between G and K, with R_p in place, but R_k not connected, then the Thévenin's circuit consists of a source V and output impedance Z given by

$$V = \mu V_s \qquad Z = r_p + R_p \tag{1-45}$$

With the source V_s connected between grid and ground we proceed as before and obtain Eq. (1-41) for I_p. Hence $V_o \equiv V_{kn}$ is given by

$$V_o = I_p R_k = \frac{\mu V_s R_k}{r_p + R_p + (\mu + 1)R_k} \tag{1-46}$$

or

$$V_o = \frac{\mu V_s}{\mu + 1} \frac{R_k}{\dfrac{r_p + R_p}{\mu + 1} + R_k} \tag{1-47}$$

Comparing Eq. (1-47) with (1-43), with $Z_L = R_k$, we find that the Thévenin's source voltage and output impedance are given by

$$V = \frac{\mu V_s}{\mu + 1} \qquad Z = \frac{r_p + R_p}{\mu + 1} \tag{1-48}$$

Thus the change in connection, in this case, results in both V and Z being divided by $\mu + 1$.

If, in Fig. 1-18, the grid draws no current, then the input impedance is infinite. In a physical tube, however, even with a negative bias on the grid some small grid current will flow because of residual gas in the envelope. Hence an equivalent resistance R_g appears between grid and cathode owing to this gas current or because the grid is driven positive or simply because a grid-leak resistor is connected between grid and cathode externally to the tube. If the source V_s is connected between the grid and cathode, then the source sees an input impedance R_g. However, if the source V_s is connected between grid and ground, then, in accordance with Miller's theorem of Sec. 1-7, the input resistance is

$$R_i = \frac{R_g}{1 - A_V} \tag{1-49}$$

where A_V is the voltage gain from the external signal to the cathode, that is, $A_V = V_{kn}/V_s$. Since A_V represents the gain of a cathode follower (except for the additional resistance R_p in the plate circuit), then A_V is positive and usually not much less than unity. Hence, Eq. (1-49) indicates that the input impedance may be very much larger when V_s is connected between grid and ground than when connected between grid and cathode.

1-12   INPUT AND OUTPUT IMPEDANCE OF A TRANSISTOR STAGE

Figure 1-19a shows a transistor stage with an emitter and a collector resistor. Again, bias and supply voltages have been omitted. For a base current I_b, the collector and emitter currents are written in terms of the current gain A_I, which is defined as $A_I \equiv -I_c/I_b$. If the emitter resistor were not present, the input impedance, seen between base and ground, would be $R_i \approx h_{ie}$. This result follows from Fig. 1-3b and from the consideration that h_{re} is so small (see Table 1-2) that ordinarily the voltage $h_{re}V_L$ is negligible in comparison with the voltage across h_{ie}. This approximation is justified more quantitatively below. Similarly, the output impedance between collector and ground (considering R_L an external load) is $1/h_{oe}$ if $R_e = 0$.

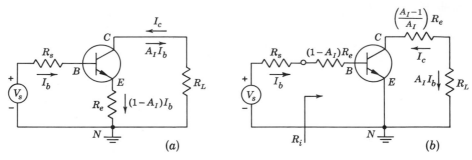

Fig. 1-19 (a) A CE transistor circuit with an emitter resistor R_e. (b) An equivalent circuit which gives the same currents as in (a).

The input and output impedances in the presence of the emitter resistor may be determined very simply by recognizing the equivalence of the circuits in Fig. 1-19a and b. This equivalence may be established by writing the KVL equations for the base mesh and for the collector mesh with the transistor junction voltages in Fig. 1-19a set equal to the transistor junction voltages in Fig. 1-19b. The set of mesh equations for Fig. 1-19a and b are identical. Hence, for a given source and load the input and output currents will be identical.

We observe that, since A_I is large and $(A_I - 1)/A_I \approx 1$, the collector circuit resistance is increased by R_e Ω. If $R_e + R_L < 4$ K, then the current gain differs from its short-circuit value by less than 10 percent. (We shall justify this approximation below.) From inspection of Fig. 1-19b, we have, with $A_I = -h_{fe}$,

$$R_i = h_{ie} + (1 + h_{fe})R_e \tag{1-50}$$

Typically, $h_{ie} \approx 1$ K, $h_{fe} \approx 50$, and $R_e \approx 1$ K. In this case, the addition of the emitter resistance increases the input resistance from 1 to 52 K, an impressive change.

The output impedance seen at the collector is also greatly increased by the presence of the emitter resistance. For values of R_e which are large compared with $R_s + h_{ie}$, the emitter may be assumed to be an open circuit and the base to be grounded. Under these conditions, the output resistance from collector to ground approaches the resistance from collector to base with an open emitter, or $R_o = 1/h_{ob} \approx 2$ M. This value greatly exceeds the output resistance for $R_e = 0$, namely, $1/h_{oe} \approx 40$ K. Clearly, R_o depends upon R_e and R_s. This dependence is given in Prob. 1-32.

Consider now the case where the output is taken between emitter and ground. We assume again that $A_I \approx -h_{fe}$ and that the impedance between base and emitter is h_{ie}. Then from the application of KVL around the base mesh of Fig. 1-19a we find

$$I_b = \frac{V_s}{R_s + h_{ie} + (1 + h_{fe})R_e} \tag{1-51}$$

and

$$V_{en} = (1 + h_{fe})I_b R_e \tag{1-52}$$

Substituting from Eq. (1-51) into Eq. (1-52) and dividing numerator and denominator by $1 + h_{fe}$, we obtain

$$V_{en} = \frac{V_s R_e}{\dfrac{R_s + h_{ie}}{1 + h_{fe}} + R_e} \tag{1-53}$$

Comparing this expression with Eq. (1-43), we see that if R_e is considered to be an external load, the Thévenin's source voltage V and output impedance Z with respect to the emitter and ground terminals are

$$V = V_s \qquad Z = \frac{R_s + h_{ie}}{1 + h_{fe}} \tag{1-54}$$

Thus, looking into the emitter we see that the impedance in the base mesh of Fig. 1-19a is divided by the factor $1 + h_{fe}$. The reason for this reduction is that of the current drawn from the emitter terminal only a fraction $1/(1 + h_{fe})$ is supplied as base current.

We shall now justify the approximation used above that $A_I \approx -h_{fe}$. The effective load impedance Z_L is seen from Fig. 1-19b to be

$$Z_L = R_L + \frac{A_I - 1}{A_I} R_e \tag{1-55}$$

The exact expression for the current gain may be found by using Eq. (1-2), with Z_L given by Eq. (1-55). Solving for A_I we find

$$A_I = \frac{-h_{fe} + h_{oe} R_e}{1 + h_{oe}(R_L + R_e)} \tag{1-56}$$

Using the parameter values in Table 1-2 we can verify that, if $R_L + R_e < 4$ K, A_I differs from $-h_{fe}$ by less than 10 percent. In a similar manner we can justify the assumption that if $R_L + R_e < 4$ K, the impedance between base and emitter differs from h_{ie} by less than 5 percent.

REFERENCES

1. IEEE Standard Letter Symbols for Semiconductor Devices, *IEEE Trans. Electron Devices*, vol. ED-11, no. 8, pp. 392–397, August, 1964.

2. Reich, H. J.: Standard Symbols for Electron Devices, *Proc. IEEE*, vol. 51, no. 2, pp. 362–363, February, 1963.

3. Millman, J., and C. C. Halkias: "Electronic Devices and Circuits," McGraw-Hill Book Company, New York, 1967.

4. "Transistor Manual," 7th ed., p. 53, General Electric Company, Syracuse, N.Y., 1964.

5. Giacoletto, L. J.: Study of p-n-p Alloy Junction Transistors from dc through Medium Frequencies, *RCA Rev.*, vol. 15, no. 4, pp. 506–562, December, 1954.

6. Miller, J. M.: Dependence of the Input Impedance of a Three-electrode Vacuum Tube upon the Load in the Plate Circuit, *Natl. Bur. Std. Sci. Papers*, vol. 15, no. 351, pp. 367–385, 1919.

2 / LINEAR WAVE SHAPING: RC, RL, AND RLC CIRCUITS

If a sinusoidal signal is applied to a transmission network composed of linear elements, then, in the steady state, the output signal will have a waveshape which is a precise reproduction of the input waveshape. The influence of the circuit on the signal may then be completely specified by the ratio of output to input amplitude and by the phase angle between output and input. With respect to this feature of preserving waveshape in all linear networks, the sinusoidal signal is unique. No other periodic waveshape preserves its form precisely, and, in the general case, the input and output signal may bear very little resemblance to one another. The process whereby the form of a nonsinusoidal signal is altered by transmission through a linear network is called "linear wave shaping."

In pulse circuitry there are a number of nonsinusoidal waveforms which appear very regularly. The most important of these are the *step, pulse, square wave, ramp,* and *exponential* waveforms. The responses to these signals of certain simple *RC, RL,* and *RLC* circuits are described in this chapter.

2-1   THE HIGH–PASS RC CIRCUIT

The capacitive coupling network of Fig. 2-1 is a rudimentary high-pass filter. Since the reactance of a capacitor decreases with increasing frequency, the higher-frequency components in the input signal appear at the output with less attenuation than do the lower-frequency components. At very high frequencies the capacitor acts almost as a short circuit and virtually all the input appears at the output. This behavior accounts for the designation "high-pass filter."

At zero frequency the capacitor has infinite reactance and hence

27

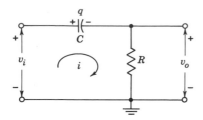

Fig. 2-1   The high-pass RC circuit.   [If the input is sinusoidal, the lowercase letters should be replaced by capitals to represent sinor (phasor) quantities.   For example, v_o is replaced by V_o.]

behaves as an open circuit.   Any constant (d-c) input voltage is "blocked" and cannot reach the output.   Therefore C is called a "blocking capacitor." The basic configuration of Fig. 2-1 is the most common coupling circuit used to obtain d-c isolation between input and output.

Sinusoidal Input   For a sinusoidal input V_{in}, the output signal V_o increases in amplitude with increasing frequency.   Even in the case of a transmission network where no amplification is involved and in which the output is always smaller than the input, it is not uncommon to refer to the ratio V_o/V_{in}, for a sinusoidal signal, as the "amplification" or "gain" A of the circuit.   For the circuit of Fig. 2-1, the magnitude of the gain $|A|$ and the angle θ by which the output leads the input are given by

$$|A| = \frac{1}{[1 + (f_1/f)^2]^{\frac{1}{2}}} \qquad \text{and} \qquad \theta = \arctan \frac{f_1}{f} \tag{2-1}$$

where $f_1 = 1/2\pi RC$.   At this frequency f_1, *the magnitude of the capacitive reactance is equal to the resistance,* and the gain is 0.707.   This drop in signal level corresponds to a signal reduction of 3 decibels (dB), and accordingly f_1 is referred to as the *lower 3-dB frequency.*   The maximum possible value of the gain (unity) is approached asymptotically at high frequencies.

Step-voltage Input   A *step voltage* is one which maintains the value zero for all times $t < 0$ and maintains the value V for all times $t > 0$.   The transition between the two voltage levels takes place at $t = 0$ and is accomplished in an arbitrarily short time interval.   Thus in Fig. 2-2, $v_i = 0$ immediately before $t = 0$ (to be referred to as time $t = 0-$), and $v_i = V$ immediately after $t = 0$ (to be referred to as time $t = 0+$).

From elementary considerations, the response of the network is exponential, with a time constant $RC \equiv \tau$, and the output voltage is of the form

$$v_o = B_1 + B_2 \epsilon^{-t/\tau} \tag{2-2}$$

The constant B_1 is equal to the steady-state value of the output voltage because as $t \to \infty$, $v_o \to B_1$.   If this final value of output voltage is called V_f, then $B_1 = V_f$.   The constant B_2 is determined by the initial output voltage, say V_i, because at $t = 0$, $v_o = V_i = B_1 + B_2$ or $B_2 = V_i - V_f$.   Hence the general solution for a single-time-constant circuit having initial and final values

V_i and V_f, respectively, is

$$v_o = V_f + (V_i - V_f)\epsilon^{-t/\tau} \tag{2-3}$$

This basic equation will be used many times throughout this text.

The constants V_f and V_i must now be determined for the circuit of Fig. 2-1. We have already emphasized that the capacitor C blocks the d-c component of the input; since the input is a constant for $t > 0$, the final output voltage is zero, or $V_f = 0$. The value of V_i is determined from the following basic considerations.

If the instantaneous current through a capacitor is i, then the change in voltage across the capacitor in time t_1 is $(1/C) \int_0^{t_1} i\, dt$. If we restrict ourselves to those circuits in which the current is always of finite magnitude, then the above integral approaches zero as $t_1 \to 0$. Hence, it follows that *the voltage across a capacitor cannot change instantaneously*, provided that the current remains finite.

Applying the above principle to the network of Fig. 2-1, we must conclude that since at $t = 0$ the input voltage changes discontinuously by an amount V, the output must also change abruptly by this same amount. If we assume that the capacitor is initially uncharged, then the output at $t = 0+$ must jump to V. Hence, $V_i = V$ and since $V_f = 0$, Eq. (2-3) becomes

$$v_o = V\epsilon^{-t/\tau} \tag{2-4}$$

Input and output are shown in Fig. 2-2. Note that the output is 0.61 of its initial value at 0.5τ, 0.37 at 1τ, and 0.14 at 2τ. The output has completed

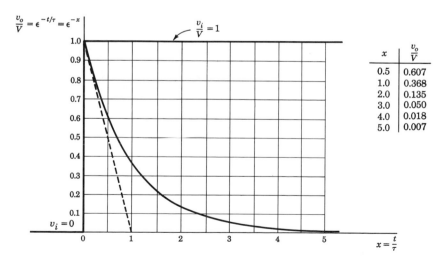

x	$\dfrac{v_o}{V}$
0.5	0.607
1.0	0.368
2.0	0.135
3.0	0.050
4.0	0.018
5.0	0.007

Fig. 2-2 Step-voltage response of the high-pass RC circuit. The dashed line is tangent to the exponential at $t = 0+$.

more than 95 percent of its total change after 3τ and more than 99 percent of its swing if $t > 5\tau$. Hence, although the steady state is approached asymptotically, we may assume for most applications that the final value has been reached after 5τ.

Pulse Input An ideal pulse has the waveform shown in Fig. 2-3a. The pulse *amplitude* is V and the pulse *duration* is t_p. It appears from Fig. 2-3a, b, and c that the pulse may be considered to be the sum of a step voltage $+V$ whose discontinuity occurs at $t = 0$ and a step voltage $-V$ whose discontinuity occurs at $t = t_p$.

If the pulse of Fig. 2-3a is applied to the circuit of Fig. 2-1, the response for times less than t_p is the same as that for the step-voltage input. Hence, the output at $t = t_p-$ is given by $v_o = V \exp(-t_p/RC) \equiv V_p$. At the end of the pulse, the input falls abruptly by the amount V, and, since the capacitor voltage cannot change instantaneously, the output must also drop by V. Hence, at $t = t_p+$, $v_o = V_p - V$. Since V_p is less than V, the voltage becomes negative and then decays exponentially to zero, as indicated in Fig. 2-3d.

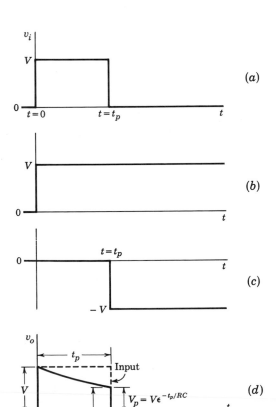

Fig. 2-3 (a) A pulse; (b, c) the step voltages which make up the pulse; (d) the pulse after transmission through the high-pass RC circuit.

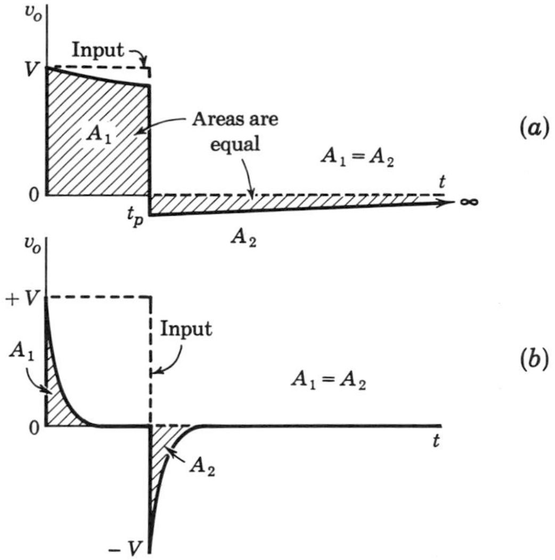

Fig. 2-4 (a) Response of high-pass circuit to a pulse if $RC/t_p \gg 1$; (b) pulse response if $RC/t_p \ll 1$.

For $t > t_p$, v_o is given by

$$v_o = V(\epsilon^{-t_p/RC} - 1)\epsilon^{-(t-t_p)/RC} \tag{2-5}$$

Note the distortion which has resulted from passing a pulse through an RC coupling network. There is a tilt to the top of the pulse and an undershoot at the end of the pulse. If these distortions are to be minimized, then the time constant RC must be very large compared with the width t_p. However, for all values of the ratio RC/t_p there must always be an undershoot, and *the area below the axis will always equal the area above.* The equality of areas can be verified by direct integration of the waveform in Fig. 2-3d. Because the input and output are separated by the blocking capacitor C, the d-c or average level of the output signal is zero for this linear circuit.

If the time constant is very large ($RC/t_p \gg 1$), there is only a slight tilt to the output pulse and the undershoot is very small. However, the negative portion decreases very slowly (as indicated in Fig. 2-4a), since its area must equal that of the positive portion. If the time constant is very small ($RC/t_p \ll 1$), the output consists of a positive *spike* or *pip* of amplitude V at the beginning of the pulse and a negative spike of the same size at the end of the pulse, as indicated in Fig. 2-4b. This process of converting pulses into pips by means of a circuit of short time constant is called *peaking*.

Square-wave Input A waveform which maintains itself at one constant level V' for a time T_1 and at another constant level V'' for a time T_2 and which is repetitive with a period $T = T_1 + T_2$, as indicated in Fig. 2-5a, is called a *square wave*. We are interested in the *steady-state* output waveform which results if this square wave is impressed on the circuit of Fig. 2-1.

First we shall prove that for *any periodic input waveform* the average level of the steady-state output signal from the circuit of Fig. 2-1 is always zero independently of the d-c level of the input. The network under consideration is governed by the equation

$$v_i = \frac{q}{C} + v_o \tag{2-6}$$

where q is the capacitor charge. Differentiating Eq. (2-6) gives

$$\frac{dv_i}{dt} = \frac{i}{C} + \frac{dv_o}{dt} \tag{2-7}$$

where $i = dq/dt$ is the mesh current. Since $v_o = iR$, this equation is equivalent to

$$\frac{dv_i}{dt} = \frac{v_o}{RC} + \frac{dv_o}{dt} \tag{2-8}$$

Multiplying by dt and integrating this equation over one period T we obtain

$$\int_{t=0}^{t=T} dv_i = v_i(T) - v_i(0) = \frac{1}{RC} \int_0^T v_o \, dt + v_o(T) - v_o(0) \tag{2-9}$$

Under steady-state conditions the output waveform (as well as the input signal) is repetitive with a period T, so that $v_o(T) = v_o(0)$ and $v_i(T) = v_i(0)$. Hence $\int_0^T v_o \, dt = 0$. Since this integral represents the area under the output waveform over one cycle, we have indeed verified that the average level of the steady-state output signal is always zero.

An alternative proof of this important principle, based upon a frequency-domain analysis, follows. The periodic input signal may be resolved into a Fourier series consisting of a constant term and an infinite number of sinusoidal components whose frequencies are multiples of $f = 1/T$. Since the blocking capacitor presents infinite impedance to the d-c input voltage, none of this d-c component reaches the output under steady-state conditions. Hence, the output signal is a sum of sinusoids whose frequencies are multiples of f. This waveform is therefore periodic with a fundamental period T but without a d-c component.

With respect to the circuit of Fig. 2-1 we have already established the following three points. First, the average level of the output signal is always zero independently of the average level of the input. The output must consequently extend in both the positive and negative direction with respect to the zero-voltage axis, and the area of the part of the waveform above the zero axis must equal the area which is below the zero axis. Second, when the input changes discontinuously by amount V, the output changes discontinuously by an equal amount and in the same direction. Third, during any finite time interval when the input maintains a constant level, the output decays exponentially toward zero voltage. In the limiting case where RC/T_1

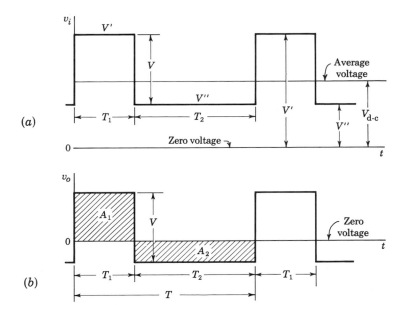

Fig. 2-5   (a) Square-wave input; (b) output voltage if the time constant is very large (compared with T).   The d-c component V_{d-c} of the output is always zero.   Area A_1 equals area A_2.

and RC/T_2 are both arbitrarily large in comparison with unity, the output waveform will be identical to the input except that the d-c component will be lacking.   Hence, the square wave of Fig. 2-5a, whose d-c level is different from zero, will appear after transmission with an average value zero, as in Fig. 2-5b.

At the other extreme, if RC/T_1 and RC/T_2 are both very small in comparison with unity, the output will consist of alternate positive and negative peaks, as in Fig. 2-6.   Observe in this case that the peak-to-peak amplitude of the output is twice the peak-to-peak amplitude of the input.

More generally, the response to a square wave must have the appearance shown in Fig. 2-7.   The equations from which to determine the four quantities V_1, V_1', V_2, and V_2', indicated in Fig. 2-7, are

$$V_1' = V_1 \epsilon^{-T_1/RC} \qquad V_1' - V_2 = V \tag{2-10a}$$

$$V_2' = V_2 \epsilon^{-T_2/RC} \qquad V_1 - V_2' = V \tag{2-10b}$$

A symmetrical square wave is one for which $T_1 = T_2 = T/2$.   Because of the symmetry, $V_1 = -V_2$ and $V_1' = -V_2'$.   Under this condition the equations in Eq. (2-10a) are identical with those in Eq. (2-10b).   Hence, the two equations in either line of Eqs. (2-10) suffice to determine the output.   We

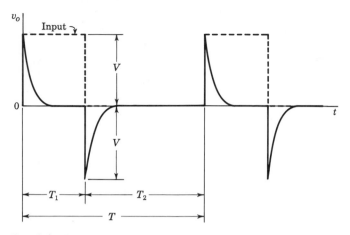

Fig. 2-6 Peaking of a square wave resulting from a time constant small compared with T.

find

$$V_1 = \frac{V}{1 + \epsilon^{-T/2RC}} \qquad V_1' = \frac{V}{1 + \epsilon^{+T/2RC}} \tag{2-11}$$

For $T/2RC \ll 1$ these reduce to

$$V_1 \approx \frac{V}{2}\left(1 + \frac{T}{4RC}\right) \qquad V_1' \approx \frac{V}{2}\left(1 - \frac{T}{4RC}\right) \tag{2-12}$$

The exponential portions of the output are now approximately linear, as shown in Fig. 2-8. The effect of the coupling network has been to introduce a tilt on the waveform. The percentage tilt P is defined by

$$P \equiv \frac{V_1 - V_1'}{V/2} \times 100 \approx \frac{T}{2RC} \times 100\% \tag{2-13}$$

Since the low-frequency 3-dB point is given by $f_1 = 1/2\pi RC$, we have the

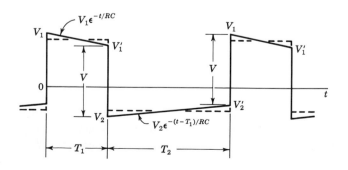

Fig. 2-7 The square-wave response of a high-pass RC circuit. The dashed curve would represent the output if $RC \gg T$.

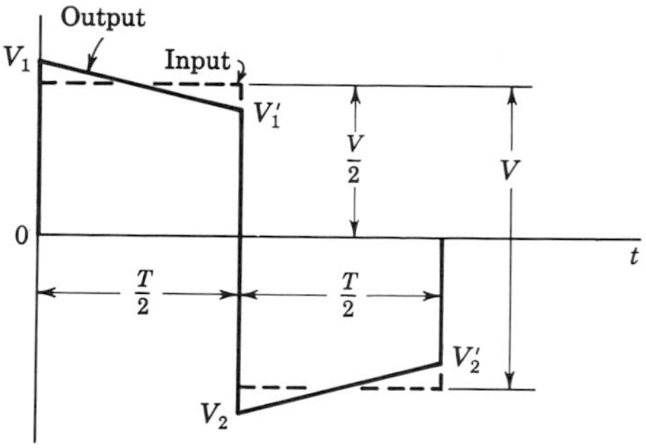

Fig. 2-8   Linear tilt of a square wave when $RC/T \gg 1$.

relationship

$$P \approx \pi \frac{f_1}{f} \times 100\%$$ (2-14)

in which $f = 1/T$ is the frequency of the applied square wave.

2-2   THE HIGH–PASS RC CIRCUIT (EXPONENTIAL AND RAMP INPUTS)

The response of the high-pass RC circuit to an input exponential waveform or to an input voltage which increases linearly with time is now to be considered.

Exponential Input   From the preceding discussion on peaking (see Fig. 2-6) we are led to conclude that, if the time constant of the circuit is decreased, the peaks obtained will be narrower, but the amplitude of the peak will remain equal to the discontinuity V of the input square wave.   This is true provided that the input has vertical sides, an impossibility in a physical waveform.   If RC is made extremely small, the finite rise time of the input waveform must be taken into account.

Consider a case in which the capacitor is initially uncharged and the input waveform rises rapidly but not discontinuously from zero to a level V, as shown in Fig. 2-9.   Since the initial capacitor voltage is assumed to be zero and since the input signal is zero at $t = 0$, then $v_o(0) = 0$ [see Eq. (2-6)].   It follows from Eq. (2-8) that

$$\left(\frac{dv_i}{dt}\right)_{\text{initial}} = \left(\frac{dv_o}{dt}\right)_{\text{initial}}$$ (2-15)

Since the initial rates of change of input and output are identical and both start from zero, we may anticipate that in the neighborhood of $t = 0$ the output will follow the input quite closely.   Furthermore, unless the time constant RC is very large in comparison with the time required for v_i to attain its final value, the capacitor will have acquired appreciable charge in this time.

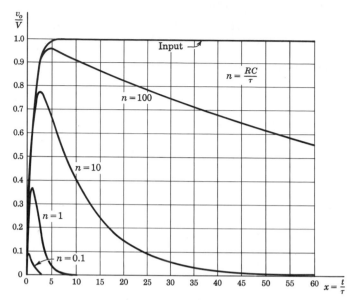

Fig. 2-9 Response of a high-pass RC circuit to an exponential input.

Hence it is apparent from Eq. (2-6) that v_o will fall short of attaining the voltage V. Eventually, of course, the output must decay exponentially to zero.

The above qualitative discussion will now be made quantitative by considering an exponential input waveform given by

$$v_i = V(1 - \epsilon^{-t/\tau}) \tag{2-16}$$

Equation (2-8) then becomes

$$\frac{V}{\tau} \epsilon^{-t/\tau} = \frac{v_o}{RC} + \frac{dv_o}{dt} \tag{2-17}$$

Defining x and n by

$$x \equiv \frac{t}{\tau} \quad \text{and} \quad n \equiv \frac{RC}{\tau} \tag{2-18}$$

the solution of Eq. (2-17), subject to the condition that initially the capacitor voltage is zero, is given by

$$v_o = \frac{Vn}{n-1} (\epsilon^{-x/n} - \epsilon^{-x}) \tag{2-19}$$

if $n \neq 1$ and by

$$v_o = Vx\epsilon^{-x} \tag{2-20}$$

if $n = 1$. These equations are plotted in Fig. 2-9, and it is seen that they have the shape predicted above. Note that if RC is much greater than $\tau(n \gg 1)$, the second term of Eq. (2-19) is negligible compared with the first except for very small values of time. Then

$$v_o \approx \frac{Vn}{n-1} \, \epsilon^{-x/n} \approx V\epsilon^{-t/RC} \tag{2-21}$$

This equation agrees with the way the circuit should behave for an ideal step voltage. Near the origin of time the output follows the input. Also, *the smaller the circuit time constant, the smaller will be the output peak.* For example, if RC just equals the time constant of the input wave ($n = 1$), the peak output will be only 37 percent of the peak input, but a very narrow pulse will result, as shown in Fig. 2-9. The larger RC is (relative to τ), the larger will be the peak output but also the wider will be the pulse. A value of RC is chosen to give the best compromise between these two conflicting characteristics for the particular application at hand. The choice is seldom critical.

Ramp Input   A waveform which is zero for $t < 0$ and which increases linearly with time for $t > 0$, $v = \alpha t$, is called a *ramp* or *sweep* voltage. Such a waveform is indicated as the "input" in Fig. 2-10a. If this waveform is applied to the circuit of Fig. 2-1, the output is governed by Eq. (2-8), which becomes

$$\alpha = \frac{v_o}{RC} + \frac{dv_o}{dt}$$

This equation has the solution, for $v_o = 0$ at $t = 0$,

$$v_o = \alpha RC(1 - \epsilon^{-t/RC}) \tag{2-22}$$

For times t which are very small in comparison with RC, we may replace the exponential in Eq. (2-22) by a series with the result

$$v_o = \alpha t \left(1 - \frac{t}{2RC} + \cdots\right) \tag{2-23}$$

The output signal falls away slightly from the input, as shown in Fig. 2-10a. As a measure of the departure from linearity, let us define the *transmission error* e_t as the difference between input and output divided by the input. The error at a time $t = T$ is then

$$e_t \equiv \frac{v_i - v_o}{v_i} \approx \frac{T}{2RC} = \pi f_1 T \tag{2-24}$$

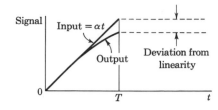

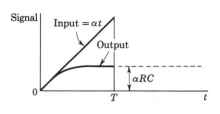

(a)

(b)

Fig. 2-10 (a) Response of a high-pass RC circuit to a ramp voltage for $RC/T \gg 1$; (b) response to a ramp voltage for $RC/T \ll 1$.

where $f_1 \equiv 1/2\pi RC$ is again the low-frequency 3-dB point. For example, if we desire to pass a 2-msec sweep with less than 0.1 percent deviation from linearity, the above equation yields

$$f_1 < 0.16 \text{ Hz}\dagger \quad \text{or} \quad RC > 1 \text{ sec}$$

For large values of t in comparison with RC, the output approaches the constant value αRC, as indicated in Fig 2-10b and Eq. (2-22).

2-3 THE HIGH–PASS RC CIRCUIT AS A DIFFERENTIATOR

If, in Fig. 2-1, the time constant is very small in comparison with the time required for the input signal to make an appreciable change, the circuit is called a *differentiator*. This name arises from the fact that under these circumstances the voltage drop across R will be very small in comparison with the drop across C. Hence we may consider that the total input v_i appears across C, so that the current is determined entirely by the capacitance. Then the current is $C \, dv_i/dt$, and the output signal across R is

$$v_o = RC \frac{dv_i}{dt}$$

Hence the output is proportional to the derivative of the input.

The derivative of a square wave is a waveform which is uniformly zero except at the points of discontinuity. At these points, precise differentiation would yield impulses of infinite amplitude, zero width, and alternating polarity. Referring to Fig. 2-6, we see that the RC differentiator provides, in the limit

† Hz = hertz = cycles per second. kHz = kilohertz. MHz = megahertz = megacycles per second.

of a very small time constant, a waveform which is correct except for the fact that the amplitude of the peaks never exceeds V. We may expect such an error since, at the time of the discontinuity, the voltage across R is not negligible compared with that across C.

For the ramp $v_i = \alpha t$, the value of $RC\, dv_i/dt$ is αRC. This result is verified in Fig. 2-10b except near the origin. The output approaches the proper derivative value only after a time has passed corresponding to several time constants. The error near $t = 0$ is again due to the fact that in this region the voltage across R is not negligible compared with that across C.

If we assume that the leading edge of a pulse can be approximated by a ramp, then we can measure the rate of rise of the pulse by using a differentiator. The peak output is measured on an oscilloscope, and from Fig. 2-10b we see that this voltage divided by the product RC gives the slope α. If R and C are not given to the desired accuracy, then the system must be calibrated by using a pulse of known rate of rise.

It is interesting to obtain a criterion for good differentiation in terms of steady-state sinusoidal analysis. If a sine wave is applied to the circuit of Fig. 2-1, the output will be a sine wave shifted by a leading angle θ such that

$$\tan \theta = \frac{X_C}{R} = \frac{1}{\omega RC} \tag{2-25}$$

and the output will be proportional to $\sin(\omega t + \theta)$. In order to have true differentiation we must obtain $\cos \omega t$. In other words, θ must equal 90°. This result can be obtained only if $R = 0$ or $C = 0$. However, if $\omega RC = 0.01$, then $1/\omega CR = 100$ and $\theta = 89.4°$, which is sufficiently close to 90° for most purposes. If $\omega RC = 0.1$, then $\theta = 84.3°$, and for some applications this may be close enough to 90°.

If the peak value of the input is V_m, the output is

$$\frac{V_m R}{\sqrt{R^2 + 1/\omega^2 C^2}} \sin(\omega t + \theta)$$

and if $\omega RC \ll 1$, then the output is approximately $V_m \omega RC \cos \omega t$. This result agrees with the expected value, $RC\, dv_i/dt$. If $\omega RC = 0.01$, then the output amplitude is 0.01 times the input amplitude.

Since it has been demonstrated that the output will be a small fraction of the input if the differentiation is satisfactory, then the output will frequently have to be followed by a high-gain amplifier. Any drift in amplifier gain will affect the level of the signal, and amplifier nonlinearity may affect the accuracy of differentiation. These difficulties are avoided by using the operational differentiator discussed in Sec. 1-8. This feedback amplifier does not suffer from the drifts just mentioned, the stability depending principally upon the constancy of R and C.

The operational-amplifier equivalent circuit for a differentiator is a capacitor C in series with a resistor $R/(1 - A)$, where A is the gain. The phase-shift angle θ between output and input for a frequency ω is given by

$$\tan \theta = \frac{1 - A}{\omega RC} \tag{2-26}$$

Comparing Eq. (2-26) with Eq. (2-25), we see that for the same values of R and C the frequency range of proper differentiation for the operational amplifier is $(1 - A)$ times that of the simple RC circuit and the output voltage has essentially the same magnitude for both circuits.

If the RC product for the operational amplifier is $(1 - A)$ times that of the simple circuit, then the output from the former will be $(1 - A)$ times that of the latter, whereas the quality of the differentiation is the same for both. The same result can be obtained by following the simple RC circuit by an amplifier of gain $(1 - A)$, but, as already emphasized, this arrangement will not have the stability and linearity of the operational system.

These considerations with respect to the conditions required for differentiation of sinusoidal waveforms suggest an alternative point of view in connection with the differentiation of an arbitrary waveform. Suppose we resolve an arbitrary signal into its Fourier components. If each of the components is shifted in phase by 90° and if the amplitude of each component is multiplied by a factor proportional to the frequency, then the Fourier series will have been effectively differentiated term by term. From this point of view the requirement for good differentiation is that the time constant RC shall be small in comparison with the period of the highest-frequency term of appreciable amplitude of the input signal.

2-4   DOUBLE DIFFERENTIATION

Figure 2-11 shows two RC coupling networks in cascade separated by an amplifier A. It is assumed that the amplifier operates linearly and that its output impedance is small relative to the impedance of R_2 and C_2, so that this combination does not load the amplifier. Let R_1 be the parallel combination of R and the input impedance of the amplifier. If the time constants R_1C_1 and R_2C_2 are small relative to the period of the input wave, then this circuit performs approximately a second-order differentiation.

If the input is a ramp ($v_i = \alpha t$) of long duration, the output v of the invert-

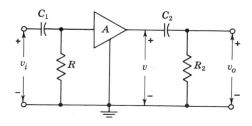

Fig. 2-11   A rate-of-rise amplifier.

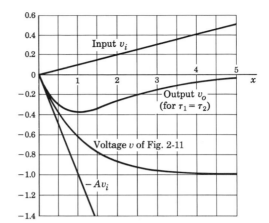

Fig. 2-12 Response of a rate-of-rise amplifier to a ramp input. $A = 10$, $\alpha\tau = 0.1$.

ing amplifier is the negative of the waveform in Fig. 2-10b and is given by [Eq. (2-22)]

$$v = -A\alpha\tau_1(1 - \epsilon^{-t/\tau_1}) \qquad (2\text{-}27)$$

where A is the magnitude of the amplifier gain and $\tau_1 \equiv R_1C_1$. This exponential input to the R_2C_2 network leads in turn to an output which is, as given in Eq. (2-19),

$$v_o = -A\alpha\tau_1 \frac{n}{n-1} (\epsilon^{-x/n} - \epsilon^{-x}) \qquad (2\text{-}28)$$

if $n \neq 1$, where $n \equiv \tau_2/\tau_1$, $\tau_2 \equiv R_2C_2$, and $x \equiv t/\tau_1$. Values of $-v_o/A\alpha\tau_1$ are plotted in Fig. 2-9 for values of n equal to 0.1, 1.0, 10, and 100. For $n = 1$, the output is given by

$$v_o = -A\alpha\tau x\epsilon^{-x} \qquad (2\text{-}29)$$

This special case is plotted in Fig. 2-12. It should be noted that a ramp voltage has been converted into a pulse. The *initial slope of the output wave is the initial slope of the input multiplied by the gain of the amplifier.* For this reason the stage in Fig. 2-11 is called a "rate-of-rise amplifier." For a single RC circuit, we demonstrate in Sec. 2-2 that the initial rate of change of output equals the initial rate of change of input independently of the time constant. Obviously, the same conclusion can be drawn for multiple differentiation. A direct check can be made from Eq. (2-28), where we find that at $t = 0$, $dv_o/dt = -A\alpha$.

As a second illustration of double differentiation consider the exponential waveform $v_i = V(1 - \epsilon^{-t/\tau})$ applied to the circuit of Fig. 2-11. If

$$R_1C_1 = R_2C_2 = \tau$$

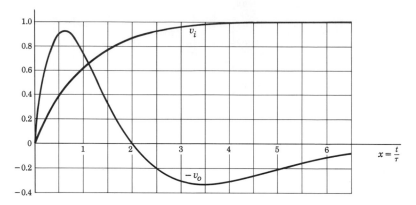

Fig. 2-13 Response of a double differentiator to an exponentially rising input. The numerical values correspond to an assumed amplifier-gain magnitude $A = 4$ and $V = 1$.

the output is found to be

$$v_o = -AVx\left(1 - \frac{x}{2}\right)\epsilon^{-x} \qquad (2\text{-}30)$$

where $x \equiv t/\tau$. This result is plotted in Fig. 2-13. The initial slope of v_o is $-AV/\tau$ since V/τ is the initial slope of v_i. Also, direct integration shows that the output waveform has as much area above the time axis as below, a fact of importance in some practical problems, such as in pulse spectrometry.[1]

2-5   THE LOW–PASS RC CIRCUIT

The circuit of Fig. 2-14 passes low frequencies readily, but attenuates high frequencies because the reactance of the capacitor C decreases with increasing frequency. At very high frequencies the capacitor acts as a virtual short circuit and the output falls to zero.

The importance of the basic circuit of Fig. 2-14 results from the fact that it may represent the situation which exists very frequently at the terminals of a signal source. The terminals of the source are 0-0'. Looking back into these terminals we may replace the source by a Thévenin's equivalent. The voltage v_i is the open-circuit voltage and R is the output impedance of the source, assumed purely resistive. The capacitance C represents all the capacitance which appears in shunt across 0-0'. This capacitance may arise from the wire used to couple terminals 0-0' to a load or may arise as a result of the capacitive component of admittance presented by the load or from stray capacitance across the terminals at the signal source itself. The fact is that almost invariably when we find ourselves pressed to extend the range of opera-

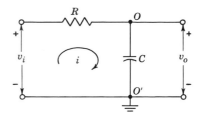

Fig. 2-14 The low-pass RC circuit.

tion of some electronic circuit to a higher frequency, it is because we have to contend with a low-pass RC circuit as shown in Fig. 2-14.

The network of Fig. 2-14 is identical with that of Fig. 2-1 except for the fact that the output is now taken across C instead of across R. Hence, the mathematical solution for the low-pass circuit can be obtained from the equations in Sec. 2-1. However, the physical behavior of the network of Fig. 2-14 is so different from that of the high-pass circuit of Fig. 2-1 that we shall give detailed consideration to the low-pass configuration.

Sinusoidal Input If the input voltage v_i is sinusoidal, the magnitude of the steady-state gain A and the angle θ by which the output leads the input are given by

$$|A| = \frac{1}{[1 + (f/f_2)^2]^{\frac{1}{2}}} \quad \text{and} \quad \theta = -\arctan\frac{f}{f_2} \tag{2-31}$$

where $f_2 \equiv 1/2\pi RC$. The gain falls to 0.707 of its low-frequency value at the frequency f_2. Hence, f_2 is called the *upper 3-dB frequency*.

Step-voltage Input The response of the circuit of Fig. 2-14 to a step input is exponential with a time constant RC. Since the capacitor voltage cannot change instantaneously, the output starts from zero and rises toward the steady-state value V, as shown in Fig. 2-15. The output is given by Eq. (2-3), or

$$v_o = V(1 - \epsilon^{-t/RC}) \tag{2-32}$$

Fig. 2-15 Step-voltage response of the low-pass RC circuit. The rise time t_r is indicated.

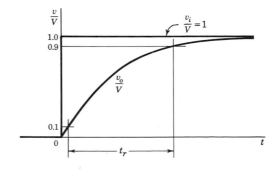

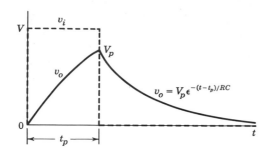

Fig. 2-16   Pulse response of the low-pass RC circuit.

The *rise time* t_r is defined as the time it takes the voltage to rise from 0.1 to 0.9 of its final value.   It gives an indication of how fast the circuit can respond to a discontinuity in voltage.   The time required for v_o to reach one-tenth its final value is readily found to be $0.1RC$ and the time to reach nine-tenths its final value is $2.3RC$.   The difference between these two values is the rise time t_r of the circuit and is given by

$$t_r = 2.2\tau = 2.2RC = \frac{2.2}{2\pi f_2} = \frac{0.35}{f_2} \tag{2-33}$$

Thus, the rise time is proportional to the time constant τ and inversely proportional to the upper 3-dB frequency.

Pulse Input   The response to a pulse, for times less than the pulse width t_p, is the same as that for a step input and is given by Eq. (2-32).   At the end of the pulse the voltage is V_p and the output must decrease to zero from this value with a time constant RC, as indicated in Fig. 2-16.   Note the waveform distortion that has resulted from passing a pulse through a low-pass RC circuit. In particular, it should be observed that the output will always extend beyond the pulse width t_p, because whatever charge has accumulated on the capacitor C during the pulse cannot leak off instantaneously.

If it is desired to minimize the distortion, then the rise time must be small compared with the pulse width.   If f_2 is chosen equal to $1/t_p$, then $t_r = 0.35t_p$. The output is as pictured in Fig. 2-17, which for many applications is a reasonable reproduction of the input.   We often use the rule of thumb that *a pulse*

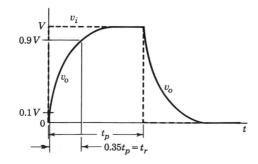

Fig. 2-17   Pulse response for the case $f_2 = 1/t_p$.

*shape will be preserved if the 3-dB frequency is approximately equal to the recip-
rocal of the pulse width.* Thus, to pass a 0.5-μsec pulse reasonably well requires
a circuit with an upper 3-dB frequency of the order of 2 MHz.

Square-wave Input Consider a periodic waveform whose instantaneous
value is constant at V' with respect to ground for a time T_1 and then
changes abruptly to V'' for a time T_2, the remainder of the cycle, as indicated
in Fig. 2-18a. As we have already observed above, a reasonable reproduction
of the input is obtained if the rise time t_r is small compared with the pulse
width. The steady-state response, in this case, is indicated in Fig. 2-18b.

If the time constant RC is comparable with the period of the input square
wave, the output will have the appearance shown in Fig. 2-18c. The equation
of the rising portion is determined by the fact that it must be an exponential

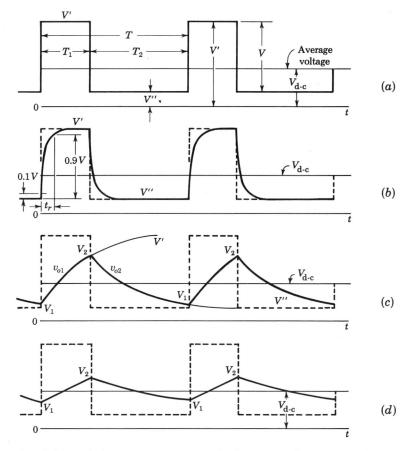

Fig. 2-18 (a) Square-wave input; (b–d) output of the low-pass RC
circuit. The time constant is smallest for (b) and largest for (d).

of time constant RC and that the voltage would rise to the steady-state value V' if the input remained at V'. If V_1 is the initial value of the output voltage, then from Eq. (2-3),

$$v_{o1} = V' + (V_1 - V')\epsilon^{-t/RC} \tag{2-34}$$

Similarly, the equation for the falling portion is

$$v_{o2} = V'' + (V_2 - V'')\epsilon^{-(t-T_1)/RC} \tag{2-35}$$

If we set $v_{o1} = V_2$ at $t = T_1$ and $v_{o2} = V_1$ at $t = T_1 + T_2$, the two resulting equations can be solved for the two unknowns V_1 and V_2.

If the time constant is very large compared with the period of the input square wave, the output consists of exponential sections which are essentially linear, as indicated in Fig. 2-18d.

Since the average voltage across R is zero (see the discussion on page 32), then the d-c voltage at the output is the same as that of the input. This average value is indicated as $V_{\text{d-c}}$ in all the waveforms of Fig. 2-18.

Consider a symmetrical square wave with zero average value, so that $T_1 = T_2 = T/2$ and $V' = -V'' = V/2$. For this case Eqs. (2-34) and (2-35) indicate that $V_1 = -V_2$, and we find that

$$V_2 = \frac{V}{2} \frac{\epsilon^{2x} - 1}{\epsilon^{2x} + 1} = \frac{V}{2} \tanh x \tag{2-36}$$

where T is the period of the square wave and $x \equiv T/4RC$.

2-6   THE LOW–PASS RC CIRCUIT (EXPONENTIAL AND RAMP INPUTS)

The response of the low-pass RC circuit to an input exponential waveform or to an input voltage which increases linearly with time is now to be considered.

Exponential Input   For an input of the form in Eq. (2-16),

$$v_i = V(1 - \epsilon^{-t/\tau})$$

the voltage across the resistor is given by Eq. (2-19) for $n \neq 1$. Hence, the voltage output across the capacitor is the difference between Eq. (2-16) and Eq. (2-19). Performing this subtraction, if $n \neq 1$,

$$\frac{v_o}{V} = 1 + \frac{1}{n-1}\epsilon^{-x} - \frac{n}{n-1}\epsilon^{-x/n} \tag{2-37}$$

and if $n = 1$,

$$\frac{v_o}{V} = 1 - (1 + x)\epsilon^{-x} \tag{2-38}$$

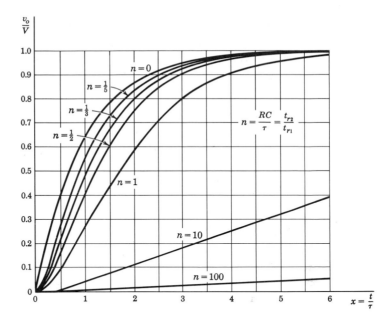

Fig. 2-19 Response of two isolated cascaded low-pass RC networks to a step input.

The parameters x and n are defined by $x \equiv t/\tau$ and $n \equiv RC/\tau$. Equations (2-37) and (2-38) give the response when an exponential of rise time t_{r1} ($= 2.2\tau$) is applied to a circuit of time constant RC (rise time $t_{r2} = 2.2RC$). The response, which has a rise time t_r, is plotted in Fig. 2-19 for various values of $n \equiv RC/\tau \equiv t_{r2}/t_{r1}$. An identical response results when a step is applied to a cascade of two circuits of rise times t_{r1} and t_{r2}, assuming that the second circuit does not load the first. Note, from Fig. 2-19, that, as n increases, a progressively longer time (called the *delay time*) is required for the response to attain 50 percent of its final value.

If two stages whose individual rise times are t_{r1} and t_{r2}, respectively, are cascaded and if the resultant rise time is t_r, then Fig. 2-20 is a plot of t_r/t_{r1}

Fig. 2-20 Relative rise time of two isolated cascaded low-pass RC networks. The ordinate is the rise time relative to $n = 0$ or to $t_{r2} = 0$. (If $t_{r2} = 0$, then $1/n = \infty$ and $t_r/t_{r1} = 1$.)

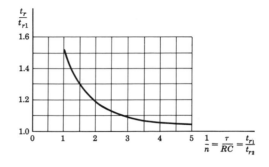

versus $t_{r1}/t_{r2} = 1/n$.   An excellent empirical relationship among the rise times is

$$t_r = 1.05 \sqrt{t_{r1}^2 + t_{r2}^2} \tag{2-39}$$

or

$$\frac{t_r}{t_{r1}} = 1.05 \sqrt{1 + n^2} \tag{2-40}$$

The ratio t_r/t_{r1} as given in Eq. (2-40) differs from the exact ratio as plotted in Fig. 2-20 by not more than 5 percent.

As an example of the usefulness of Fig. 2-20 consider that a cathode-ray oscilloscope of rise time t_{r2} is being used to observe and measure the rise time t_{r1} of an input waveform.   (We assume, for simplicity, that an oscilloscope may be represented by a single resistance-capacitance network.)   From Fig. 2-20 we see that if $t_{r2} = t_{r1}$ the observed rise time is 53 percent longer than the rise time of the input waveform.   On the other hand, if t_{r2} is less than $\frac{1}{3}t_{r1}$ the observed rise time differs from the input-signal rise time by less than 10 percent.   Hence an oscilloscope used to make a rise-time measurement should have a bandpass at least three times the bandpass of the circuit under test.

Ramp Input   For an input of the form $v_i = \alpha t$, the voltage v_R across the resistor is given by Eq. (2-22).   The voltage across the capacitor is $v_i - v_R$ or

$$v_o = \alpha(t - RC) + \alpha RC\epsilon^{-t/RC} \tag{2-41}$$

If it is desired to transmit the ramp with little distortion, then a small time constant must be used relative to the total ramp time T.   The output is given in Fig. 2-21a, where it is seen that the output follows the input but is delayed by one time constant RC from the input (except near the origin where there is distortion).   The transmission error e_t is defined as the difference between

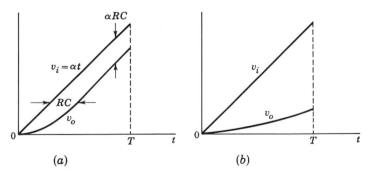

Fig. 2-21   Response of a low-pass RC circuit to a ramp voltage.   (a) $RC/T \ll 1$; (b) $RC/T \gg 1$.

input and output divided by the input at $t = T$. For $RC/T \ll 1$, we find

$$e_t \approx \frac{RC}{T} = \frac{1}{2\pi f_2 T} \tag{2-42}$$

where f_2 is the upper 3-dB frequency. For example, if we desire to pass a 2-msec sweep with less than 0.1 percent error, the above equation yields

$$f_2 > 80 \text{ kHz} \quad \text{and} \quad RC < 2 \text{ } \mu\text{sec}$$

If the time constant is large compared with the sweep duration, $RC/T \gg 1$, the output is very distorted, as it appears in Fig. 2-21b. By expanding the exponential in Eq. (2-41) in a power series in t/RC, we find

$$v_o \approx \frac{\alpha t^2}{2RC} \tag{2-43}$$

A quadratic response is obtained for a linear input, and hence the circuit acts as an integrator.

2-7   THE LOW–PASS RC CIRCUIT AS AN INTEGRATOR

If, in Fig. 2-14, the time constant is very large in comparison with the time required for the input signal to make an appreciable change, the circuit is called an *integrator*. This name arises from the fact that under these circumstances the voltage drop across C will be very small in comparison to the drop across R and we may consider that the total input v_i appears across R. Then the current is v_i/R and the output signal across C is

$$v_o = \frac{1}{C} \int i \, dt = \frac{1}{RC} \int v_i \, dt \tag{2-44}$$

Hence the output is proportional to the integral of the input.

If $v_i = \alpha t$, the result is $\alpha t^2/2RC$, as given by Eq. (2-43). As time increases, the drop across C will not remain negligible compared with that across R and the output will not remain the integral of the input. As a matter of fact, Fig. 2-21a shows that the output will change from a quadratic to a linear function of time.

The integral of a constant is a linear function, and this agrees with the curves of Fig. 2-18d which correspond to $RC/T \gg 1$. As the value of RC/T decreases, the departure from true integration increases, as indicated in Fig. 2-18c and b.

These examples show that the integrator must be used cautiously. We can obtain a criterion for good integration in terms of steady-state analysis by proceeding as in Sec. 2-3. If we define satisfactory integration as meaning that an input sinusoid has been shifted at least 89.4° (instead of the true value of 90°), then it is necessary that

$$RC > 15T$$

where T is the period of the sine wave.

Since the output is a small fraction of the input (because of the factor $1/RC$), amplification may be necessary. For the reasons given in Sec. 2-3, an operational amplifier may possibly be used to advantage.

Integrators are almost invariably preferred over differentiators in analog-computer applications for the following reasons. Since the gain of an integrator decreases with frequency whereas the gain of a differentiator increases nominally linearly with frequency, it is easier to stabilize the former than the latter with respect to spurious oscillations. As a result of its limited bandwidth an integrator is less sensitive to noise voltages than a differentiator. Further, if the input waveform changes very rapidly, the amplifier of a differentiator may overload. Finally, as a matter of practice, it is more convenient to introduce initial conditions in an integrator.

2-8   ATTENUATORS

We consider now the simple resistance attenuator which is used to reduce the amplitude of a signal waveform. We shall find the conditions under which it is possible to ensure no distortion even if shunt capacitance is taken into consideration. Also, we shall investigate the types of response which are obtained with a step voltage input if the circuit is improperly adjusted.

The simple resistor combination of Fig. 2-22a would multiply the input signal by the ratio $a = R_2/(R_1 + R_2)$ independently of the frequency, were it not for the inevitable stray capacitance C_2 which shunts R_2. The capacitance C_2 may be, for example, the input capacitance of a stage of amplification. Using Thévenin's theorem, the circuit in Fig. 2-22a may be replaced by its equivalent in Fig. 2-22b, in which R is equal to the parallel combination of R_1 and R_2. We ordinarily want both R_1 and R_2 to be large so that the nominal input impedance of the attenuator may be large enough to prevent loading down the input signal. If, say,

$$R_1 = R_2 = 1 \text{ M} \quad \text{and} \quad C_2 = 15 \text{ pF}$$

then the rise time in Fig. 2-22b is $2.2 \times 0.5 \times 15 \ \mu\text{sec} = 16.5 \ \mu\text{sec}$. So large a rise time is ordinarily entirely unacceptable.

The attenuator may be *compensated*, so that its attenuation is once again independent of the frequency, by shunting R_1 by a capacitance C_1, as indicated in Fig. 2-22c. The circuit has been redrawn in Fig. 2-22d to suggest that the two resistors and the two capacitors may be viewed as the four arms of a bridge. If $R_1C_1 = R_2C_2$, the bridge will be balanced, and no current

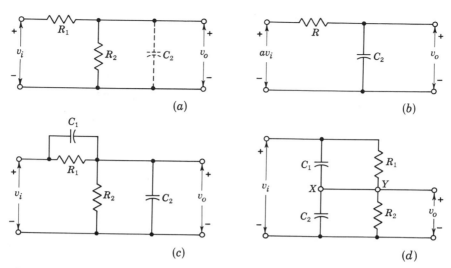

Fig. 2-22   An attenuator.   (a) Actual circuit; (b) equivalent circuit; (c) compensated attenuator; (d) compensated attenuator redrawn as a bridge.

will flow in the branch connecting the point X to the point Y.   For the purpose of computing the output, the branch X-Y may be omitted and the output is again equal to av_i independently of the frequency.   In practice, C_1 will ordinarily have to be made adjustable, and the final adjustment for compensation is made experimentally by the method of square-wave testing.   This procedure is necessary because the compensation is critically dependent on the condition $R_1C_1 = R_2C_2$ being satisfied precisely.

Let us consider the appearance of the output signal for a step-voltage input of magnitude V, if the compensation is incorrect.   Inasmuch as the input changes abruptly by V at $t = 0$ then the voltages across C_1 and C_2 must also change discontinuously.   On page 29 we demonstrated that the voltage across a capacitor cannot change instantaneously if the current remains finite.   Hence, we are led to the conclusion that an impulsive current must flow in the circuit of Fig. 2-22c.   An infinite current exists at $t = 0$ for an infinitesimal time, so that a finite charge $q = \int_{0-}^{0+} i \, dt$ is delivered to each capacitor.   At $t = 0+$, Kirchhoff's voltage law yields

$$V = \frac{q}{C_1} + \frac{q}{C_2} = \frac{(C_2 + C_1)q}{C_1 C_2} \tag{2-45}$$

or the output voltage at $t = 0+$ is

$$v_o(0+) = \frac{q}{C_2} = \frac{C_1}{C_1 + C_2} V \tag{2-46}$$

The initial output voltage is determined by the capacitors because they behave like short circuits for an instantaneous change. The final output voltage is determined by the resistors (because a capacitor acts as an open circuit under steady-state conditions for an applied d-c voltage). Hence,

$$v_o(\infty) = \frac{R_2}{R_1 + R_2} V \tag{2-47}$$

Looking back from the output terminals (with the input short-circuited) we see a resistor $R \equiv R_1R_2/(R_1 + R_2)$ in parallel with $C \equiv C_1 + C_2$. Hence, the decay of the output from initial to final value takes place exponentially with a time constant $\tau = RC$. The responses of an attenuator for C_1 equal to, greater than, and less than C_2R_2/R_1 are indicated in Fig. 2-23. Note that perfect compensation is obtained if $v_o(0+) = v_o(\infty)$ or from Eqs. (2-46) and (2-47)

$$\frac{C_1}{C_1 + C_2} V = \frac{R_2}{R_1 + R_2} V$$

This equation is equivalent to

$$R_1C_1 = R_2C_2 \tag{2-48}$$

which is the balanced-bridge condition obtained above for perfect compensation. It is also interesting to note that the extreme values of $v_o(0+)$ are 0 for $C_1 = 0$ and V for $C_1 = \infty$.

In practice we certainly cannot obtain infinite current. The reason we are led to the physically impossible impulsive response in the above analysis is that we have implicitly assumed a generator with zero source impedance. We shall now remove this restriction and show that even though the attenuator

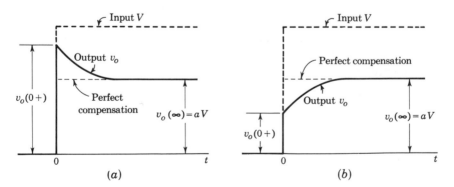

Fig. 2-23   Response of an attenuator to a step input.   For $C_1 = C_2R_2/R_1 \equiv C_p$ the compensation is perfect and the output is a step of magnitude $aV = R_2V/(R_1 + R_2)$.   (a) Overcompensation, $C_1 > C_p$; (b) undercompensation, $C_1 < C_p$.

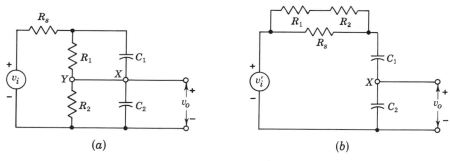

Fig. 2-24   (a) Compensated attenuator including impedance of source R_s;
(b) equivalent circuit: $v_i' = (R_1 + R_2)V/(R_1 + R_2 + R_s)$.

is compensated, the ideal step response can no longer be obtained. Nevertheless, an improvement in rise time does result if a compensated attenuator is used. For example, if the output is one-tenth the input, then the rise time of the output using the attenuator is one-tenth what it would be without the attenuator.

The compensated attenuator will reproduce faithfully the signal which appears at its input terminals. However, if the output impedance of the generator driving the attenuator is not zero, the signal will be distorted right at the input to the attenuator. This situation is illustrated in Fig. 2-24a, in which a generator of a step voltage V and of source resistance R_s is connected to the attenuator. Since, as was noted earlier, the lead which joins point X to point Y may be open-circuited, the circuit in Fig. 2-24a may be redrawn as in Fig. 2-24b. If $R_s \ll R_1 + R_2$, as is usually the case, the input to the attenuator will be an exponential of time constant R_sC', in which C' is the capacitance of the series combination of C_1 and C_2 or $C' = C_1C_2/(C_1 + C_2)$. It is this exponential waveform rather than the step which the attenuator will transmit faithfully.

If the generator terminals were connected to the terminals to which the attenuator output is connected, the generator would see a capacitance C_2. In this case the waveform at these terminals would be an exponential with time constant $\tau = R_sC_2$. When the attenuator is used, the time constant is $\tau' \approx R_sC'$. Since $\tau'/\tau \approx C'/C_2 = C_1/(C_1 + C_2) = a$, an improvement in waveform results. For example, if the attenuation is equal to 10 ($a = \frac{1}{10}$), then the rise time of the waveform will have been divided by a factor of 10. If we are able to afford a loss of signal level, this reduction of input capacitance may be used to advantage.

As an example of such an application, consider the problem associated with connecting the input terminals of an oscilloscope to a signal point in a circuit. If the point at which the signal is available is some distance from the oscilloscope terminals, and particularly if the signal appears at a high impedance level, we shall want to use shielded cable to connect the signal to the

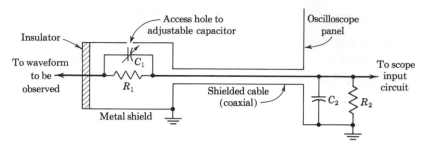

Fig. 2-25   A cathode-ray-oscilloscope probe.   (Not drawn to scale: the metal shield encasing R_1 and C_1 is a few inches long, whereas the coaxial cable is a few feet long.)

oscilloscope.   The shielding is necessary in this case to isolate the input lead from stray fields such as those of the ever-present power line.   The capacitance seen looking into several feet of cable may be as high as 100 to 150 pF.   This combination of high input capacitance together with the high output impedance (say, resistive) of the signal source will make it impossible to make faithful observations of fast waveforms.   A "probe" assembly which permits the use of shielded cable and still keeps the capacitance low is indicated in Fig. 2-25.   Typically, the attenuation introduced through use of the probe assembly is 10 or 20 and the input capacitance to the probe assembly is about 20 or 10 pF, respectively.   There are also units commercially available at present which consist of a probe assembly with an attenuation of 100 followed by an amplifier of gain 100.   The overall gain is 1, but the probe input capacitance may be as little as 2 or 3 pF.

The problem of providing continuously variable attenuation is not so easily solved.   In this case the resistors R_1 and R_2 must be replaced by a "potentiometer," and since the required compensating capacitance depends on the setting of the attenuator, the only practicable thing to do is to leave the attenuator uncompensated.   The bandpass is then a minimum when the potentiometer is set at its electrical midpoint and is given by

$$(f_2)_{min} = \frac{2}{\pi RC} \tag{2-49}$$

in which R is the total potentiometer resistance and C the total shunt capacitance between the potentiometer arm and ground.   If, say, $C = 20$ pF and $(f_2)_{min}$ is to be 10 MHz, $R = 3$ K.   The conflict between the necessary high potentiometer resistance to avoid loading down the signal source and the equally necessary low potentiometer resistance to maintain the bandpass suggests the use of a cathode follower with the potentiometer in the cathode circuit or an emitter follower with the potentiometer in the emitter leg.

2-9 RL CIRCUITS

Suppose the capacitor C and resistor R of the preceding sections in this chapter are replaced by a resistor R' and an inductor L, respectively. Then if the time constant L/R' equals the time constant RC, all the preceding results remain unchanged.

The inductor is seldom used if a large time constant is called for because a large value of inductance can be obtained only with an iron-core inductor which is physically large, heavy, and expensive relative to the cost of a capacitor for a similar application. Such an inductor will be shunted with a large amount of stray distributed capacitance. Furthermore, the nonlinear properties of the iron cause distortion, which may be undesirable. If it is required to pass a very low frequency through a circuit in which L is a shunt element, then the inductor may become prohibitively large. For example, with a lower 3-dB frequency of 10 Hz and for $R' = 100$ K, the inductance required is 1,600 H. Of course, in circuits where a small value of R' is tolerable, then a more reasonable value of inductance may be used.

The small, inexpensive, air-core inductor is used in low-time-constant applications. Figure 2-26a shows how a square wave may be converted into pulses by means of the peaking coil L. It is assumed that the bias voltage and the magnitude of the input are such that the tube operates linearly. The equivalent circuit is as indicated in Fig. 2-26b. The open-circuit voltage gain is the amplification factor μ of the tube and the output impedance R is the plate resistance r_p of the triode.

Since the instantaneous voltage $L\,di/dt$ across an inductor cannot be infinite, *the current through an inductor cannot change discontinuously.* Hence, an inductor acts as an open circuit at the time of an abrupt change in voltage. For a vacuum tube, with the output open-circuited, the change in plate voltage equals μ times the grid-voltage change. Hence, as indicated in Fig. 2-27, the peak of the output pulse (measured with respect to the quiescent voltage V_{PP}) equals μV, where V is the jump in voltage of the input signal (the peak-to-peak

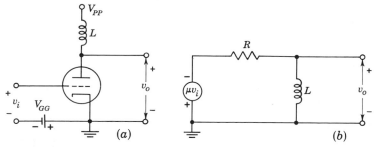

Fig. 2-26   (a) Peaking circuit using an inductor;
(b) linear equivalent circuit.

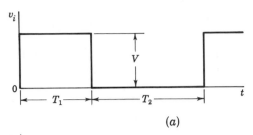

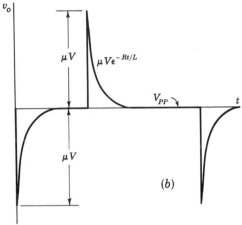

Fig. 2-27 Input v_i and output v_o for the circuit of Fig. 2-26, with $R = r_p$. It is assumed that the time constant L/R is much smaller than either T_1 or T_2.

voltage of the square wave). The output voltage falls or rises exponentially with a time constant L/R toward V_{PP}.

A peaking coil may also be used in the collector circuit of a transistor to obtain pulses. If the base input is a square wave of current whose peak-to-peak value is I, then the output voltage has the same waveform as in Fig. 2-27b. The peak voltage is now $h_{fe}I/h_{oe}$ and the time constant is $h_{oe}L$, assuming that the transistor can be represented by its low-frequency hybrid-parameter model of Fig. 1-3b.

The rate-of-rise amplifier of Fig. 2-11 often uses a peaking inductor in the output circuit instead of the R_2C_2 differentiating combination shown.

The situation where the square wave is large enough to cut the triode off, so that the circuit acts in a nonlinear manner, is considered in Sec. 8-10, where it will be found that the negative peaks are of smaller magnitude than the positive ones.

2-10 RLC CIRCUITS

In Fig. 2-26 there should be indicated a capacitor C across the output to include the effect of coil winding capacitance, output capacitance, and stray wiring capacitance to ground. This capacitance will modify the results of Sec. 2-9, as we shall now show.

Figure 2-28 shows a signal v_i applied through a resistor R to a parallel LC

circuit. From the differential equations for this network, and assuming a solution in the form ϵ^{st}, we find for the roots s of the characteristic equation[2] (or for the poles of the transfer function)

$$s = -\frac{1}{2RC} \pm \left[\left(\frac{1}{2RC}\right)^2 - \frac{1}{LC}\right]^{\frac{1}{2}}$$ (2-50)

Let us introduce the *damping constant* k and the *resonant* or *undamped period* T_o, defined by

$$k \equiv \frac{1}{2R}\sqrt{\frac{L}{C}} \qquad \text{and} \qquad T_o \equiv 2\pi\sqrt{LC}$$ (2-51)

in which case Eq. (2-50) can be put in the form

$$s = -\frac{2\pi k}{T_o} \pm j\frac{2\pi}{T_o}(1 - k^2)^{\frac{1}{2}}$$ (2-52)

If $k = 0$, we see that the roots are purely imaginary, $\pm j2\pi/T_o$, and hence that the response is an undamped sinusoid of period T_o. If $k = 1$, the two roots are equal, corresponding to the *critically damped* case. If $k > 1$, there are no oscillations in the output, and the response is said to be *overdamped*. If $k < 1$, the output will be a sinusoid whose amplitude decays with time, and the response is said to be *underdamped*.

The damping factor is inversely proportional to the Q of the circuit consisting of a parallel combination of R, L, and C. Thus

$$Q \equiv \omega_o RC = \frac{2\pi RC}{T_o} = \frac{RC}{\sqrt{LC}} = R\sqrt{\frac{C}{L}} = \frac{1}{2k}$$

If the input to Fig. 2-28 is a step voltage V and *if the initial current through the inductor is zero and the initial voltage across the capacitor is zero*, the response is given by the following equations, in which $x \equiv t/T_o$:

Critical Damping, $k = 1$ For the case of critical damping, we have

$$\frac{v_o}{V} = 4\pi x\epsilon^{-2\pi x}$$ (2-53)

If use is made of Eqs. (2-51), with $k = 1$, Eq. (2-53) can be put in the equivalent form

$$\frac{v_o}{V} = \frac{4Rt}{L}\epsilon^{-2Rt/L}$$ (2-54)

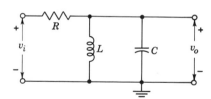

Fig. 2-28 A signal v_i is applied through a resistor R to a parallel LC circuit.

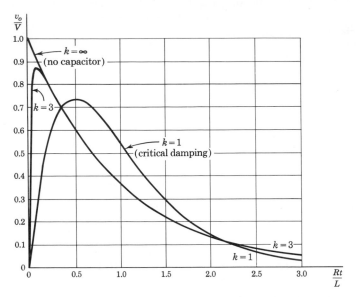

Fig. 2-29 Response of the circuit of Fig. 2-28 for the critically damped and overdamped cases for a fixed value of R and L. The parameter k is related to the capacitance C by $k \equiv (1/2R) \sqrt{L/C}$.

Overdamped, $k > 1$ In the overdamped case, it is convenient to rewrite Eq. (2-52) as

$$s = -\frac{2\pi k}{T_o} \pm \frac{2\pi k}{T_o} \sqrt{1 - \frac{1}{k^2}}$$

If we apply the binomial expansion to the radical and assume that k *is large enough so that* $4k^2 \gg 1$, we find for s the approximate values $-\pi/T_o k$ and $-4\pi k/T_o$. Subject to this restriction on the size of k, the response is

$$\frac{v_o}{V} \approx \epsilon^{-\pi x/k} - \epsilon^{-4\pi kx} \tag{2-55}$$

The first term is less than 1 everywhere except at $x = 0$. The second term is equal to the first term raised to the power $4k^2$. Hence, the second term is negligible compared with the first except near the origin. Thus Eq. (2-55) can be approximated by

$$\frac{v_o}{V} \approx \epsilon^{-\pi x/k} = \epsilon^{-\pi t/kT_o} = \epsilon^{-Rt/L} \tag{2-56}$$

in which we have made use of Eqs. (2-51). This result shows that the response approaches that for the zero-capacitance case (Fig. 2-26) as k becomes much greater than unity. Physically, this is just what we should expect, because

Eqs. (2-51) show that a large value of k means a small value of C for a given value of R and L.

Since the voltage across the capacitor cannot change instantaneously, Eq. (2-56) is in error at $t = 0$ and the more correct equation (2-55) must be used near the origin. The outputs for $k = 3$ and $k = 1$ are compared with that for $k = \infty$ $(C = 0)$ in Fig. 2-29.

Underdamped, $k < 1$ In the underdamped case, we have

$$\frac{v_o}{V} = \frac{2k}{\sqrt{1 - k^2}} \, \epsilon^{-2\pi k x} \sin 2\pi \sqrt{1 - k^2} \, x \tag{2-57}$$

where, as above, $x = t/T_o$. The damped period is seen to be $T_o/(1 - k^2)^{\frac{1}{2}}$ and hence is larger than the free period T_o.

Assume an amplifier with a coil in the output circuit so that L and C are fixed, but also assume that the damping can be varied by adjusting R. The response for several values of k is given in Fig. 2-30. The curves for k less

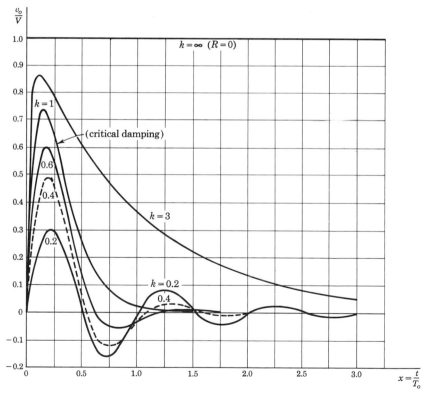

Fig. 2-30 Response of the circuit of Fig. 2-28 for fixed L and C. The parameter k is related to the resistance R by $k \equiv (1/2R)\sqrt{L/C}$.

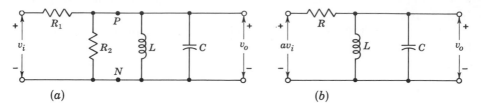

Fig. 2-31 (a) The circuit of Fig. 2-28 modified by the inclusion of a damping resistor R_2; (b) the equivalent circuit.

than unity are given by Eq. (2-57). The curve for $k = 1$ is plotted from Eq. (2-53) and the curve for $k = 3$ is given by Eq. (2-55). For $k = \infty$, $R = 0$ and the output equals the input as indicated. From Fig. 2-28 we note that the smaller the value of R the larger must be the source current. Note that *if the damping factor k is adjusted to be somewhat less than unity an excellent peaking circuit results.*

For a fixed L and C the damping may be increased by shunting the LC combination with an additional resistor, as indicated in Fig. 2-31a. If the circuit to the left of points P and N in this figure is replaced by its Thévenin's equivalent, the result is as in Fig. 2-31b. The resistor R represents R_1 and R_2 in parallel, and a is the amplification factor. Specifically,

$$R \equiv \frac{R_1 R_2}{R_1 + R_2} \quad \text{and} \quad a \equiv \frac{R_2}{R_1 + R_2} \tag{2-58}$$

Comparing Fig. 2-31b with Fig. 2-28, we see that the results obtained for the latter circuit are also valid for the former, provided that we multiply the output by the factor a.

2-11   RINGING CIRCUIT

In Sec. 2-10 we show that to obtain a pulse from a step voltage (peaking) the circuit should operate in the neighborhood of critical damping. In this section we are interested in having as nearly undamped oscillations as possible. Such a circuit is called a *ringing circuit.* If k is small, the circuit will ring for many cycles. It is often of interest to know the value required of the Q of a circuit which is to ring for a given number N of cycles before the amplitude decreases to $1/\epsilon$ of its initial value. From Eq. (2-57) we see that this decrement results when $2\pi kx = 1$. Since $x = t/T_o = NT_o/T_o = N$ and $k = 1/2Q$, we have

$$Q = \pi N \tag{2-59}$$

Thus a circuit with $Q = 12$ will ring for $Q/\pi \approx 4$ cycles before the amplitude of the oscillation decreases to 37 percent of its initial value.

Fig. 2-32   Ringing circuit with capacitor initially uncharged and with an initial inductor
current I.

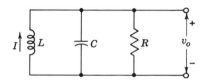

If the parallel LC combination is in series with a tube or transistor and
if the active device is cut off by means of a step voltage, then R_1 of Fig. 2-31a is
effectively infinite.   The equivalent circuit is given in Fig. 2-32.   For maximum ringing, no shunting resistor is added and R represents an effective
resistor to account for the losses in the coil.   The current I is the quiescent
current in the inductor before the step voltage is applied.

Outwardly, the circuits of Figs. 2-28 and 2-32 appear quite different.
When, however, the input to Fig. 2-28 is taken to be a step of amplitude V,
the output of the two circuits can be shown to be identical, provided only
that the initial inductor current I of Fig. 2-32 is taken to be V/R.   The two
circuits have the same characteristic roots given in Eq. (2-50).   And, under
the circumstance that $V = IR$, the conditions that apply in both cases to the
output voltage are that at $t = 0$, $v_o = 0$ and $dv_o/dt = I/C$.   Hence, provided
that we make the replacement of V for IR, all the equations from (2-53) to
(2-57) apply equally well to the circuit of Fig. 2-32.

If the damping is small enough the response approaches an undamped
sine wave.   We can easily find the amplitude of oscillation if we remember that
the initial magnetic energy stored in the inductor is converted into electric
energy in the capacitor at the end of one-quarter cycle.   Thus

$$\tfrac{1}{2}LI^2 = \tfrac{1}{2}CV^2_{\max} \qquad \text{or} \qquad V_{\max} = I\sqrt{\frac{L}{C}} \tag{2-60}$$

A ringing circuit may be used to generate a sequence of pulses regularly
spaced in time.   We shall see later how to obtain a pulse each time a sine wave
crosses the zero axis in the positive direction.   The sequence starts when the
device delivering the current I is cut off.   These pulses find application in
many timing operations.

If a pulse is applied to an active device which feeds a ringing LC combination, then at the end of the pulse there may be a voltage V_o across C as well as
a current I through L.   The possible responses of the circuit at the end of the
pulse subject to these initial conditions are discussed in Appendix A.

2-12   MEASUREMENT OF INDUCTANCE AND CAPACITANCE THROUGH CIRCUIT STEP RESPONSE

If a generator of output impedance R furnishes the voltage $v = V(1 - \epsilon^{-t/\tau})$
to an inductor of inductance L, the voltage v_L across the inductor is given

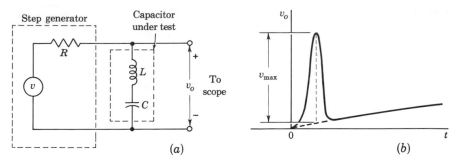

Fig. 2-33 A method for measuring the stray inductance L associated with a capacitance C. (a) The circuit; (b) the output waveform.

precisely by Eq. (2-19), in which $n \equiv (L/R)/\tau$. The waveform of v_L is a pulse, as shown in Fig. 2-9. The peak value of the pulse v_{max} is calculated (Prob. 2-49) to be

$$v_{max} = Vn^{1/(1-n)} \tag{2-61}$$

If $n \ll 1$ then $v_{max} \approx Vn = VL/R\tau$ and the peak value is proportional to the value of L. Accordingly, if the pulse is observed on a scope, an unknown inductance can be determined provided that the scope has first been calibrated by noting the pulse amplitude corresponding to a known value of inductance. It is to be noted that the pulse amplitude varies inversely with the time constant τ with which the generator waveform rises. Therefore this method of inductance measurement may be extended to the determination of smaller values of inductance as τ decreases. As an example, consider that $V = 10$ V, $R = 50 \ \Omega$, and $\tau = 1$ nsec (corresponding to a generator having a rise time of 2.2 nsec). Then an inductance of only $L = 2.5$ nH ($n = 0.05$) will give rise to a pulse of 0.5 V amplitude.

This method of inductance determination is well suited to the measurement[3] of stray or incidental inductance associated with a circuit component or device. As an example, in Fig. 2-33a an arrangement is shown by means of which we may determine the small inductance that is introduced into a circuit by a capacitor C. This inductance is often referred to as the "lead inductance" of the capacitor and has been represented in Fig. 2-33a by L. If the time constant $RC \gg \tau$ and if also $RC \gg L/R$, the output waveform v_o will appear as in Fig. 2-33b, that is, a pulse superimposed on a linear rise. The value of L is determined from v_{max}, as noted in Fig. 2-33b.

The measurement procedure indicated in Fig. 2-33 requires that all the generator current flow through L. Such would not be the case if the connection of the scope bridged a capacitance C' across the output terminals. Of course we might simply require that $RC' \ll L/R$. But for $L = 2.5$ nH and $R = 50 \ \Omega$ we would require $C' \ll 1$ pF, say $C' = 0.1$ pF. Scopes with conventional input connections have input capacitances of the order of 10 pF and would be completely unsuitable. Instead, matched delay-line inputs are required, as is discussed in the next chapter.

These measurement procedures may be adapted to measure not only stray inductance but also small incidental capacitances (Prob. 2-51).

REFERENCES

1. Chase, R. L.: "Nuclear Pulse Spectrometry," chap. 2, McGraw-Hill Book Company, New York, 1961.

2. Salvadori, M. G., and R. J. Schwarz: "Differential Equations in Engineering Problems," Prentice-Hall, Inc., Englewood Cliffs, N.J., 1954.

3. *Hewlett-Packard J.*, vol. 14, no. 2, October, 1962, Hewlett-Packard Company, Palo Alto, Calif.

3 / PULSE TRANSFORMERS AND DELAY LINES

This chapter continues the linear wave-shaping studies of the preceding chapter. First we shall consider what happens to a step or pulse waveform when it is transmitted through a transformer. In order to make such an analysis it is necessary to obtain a model or equivalent circuit of the pulse transformer. The distortions present in the output can then be calculated from this equivalent circuit.

The latter part of the chapter deals with transmission lines. A study is made of the waveshapes which are obtained under various terminating impedances and input excitations (a step, pulse, etc.).

3-1 PULSE–TRANSFORMER APPLICATIONS

Iron-cored transformers are used in the transmission and shaping of pulses which range in width from a fraction of a nanosecond to about 25 μsec. Among the extensive applications of pulse transformers are the following:

1. To change the amplitude and impedance level of a pulse
2. To invert the polarity of a pulse; also to provide, with the aid of a center-tapped winding, equal positive and negative pulses simultaneously
3. To produce a pulse in a circuit having negligible d-c resistance
4. To effect "d-c isolation" between a source and a load; in other words, to produce a pulse in a winding whose d-c voltage level may be arbitrarily selected
5. To couple between stages of pulse amplifiers
6. To differentiate a pulse
7. To act as a coupling element in certain pulse-generating circuits such as the *blocking oscillator* and the *multiar* (discussed in Chap. 16)

In many instances the functions listed above may be accomplished as well or better by transistor or vacuum-tube circuitry. But the transformer, being a completely passive circuit element, has none of the instability normally associated with tubes and transistors and in addition avoids the inconvenience of supplying the voltages required for the operation of these active devices.

3-2   TRANSFORMER MODELS[1]

The schematic diagram for a transformer is indicated in Fig. 3-1. The primary inductance is L_p, the secondary inductance is L_s, and the mutual inductance is M. The load resistance is R_L. In this section, we shall ignore the primary, secondary, and source resistances and also all capacitances. We shall also neglect core loss and the nonlinearity of the magnetic circuit. These parameters, however, will be added later to the equivalent circuit. The coefficient of coupling K between primary and secondary is defined by

$$K \equiv \frac{M}{\sqrt{L_p L_s}}$$

Under the circumstances specified above, an ideal transformer is one for which L_p is infinite and $K = 1$. In this case the output v_o is an exact replica of the input v_i and the transformation ratio n is independent of the load. For the ideal transformer,

$$\frac{v_o}{v_i} = \frac{i_p}{i_s} = \sqrt{\frac{L_s}{L_p}} = \frac{N_s}{N_p} = n \tag{3-1}$$

where i_p is the primary current, i_s is the secondary current, N_p is the primary number of turns, and N_s is the secondary number of turns.

An iron-cored transformer, such as a pulse transformer, behaves as a reasonable approximation to a perfect transformer *when used in connection with the fast waveforms it is intended to handle*. In such a case it is advantageous to replace the actual transformer by an ideal transformer together with additional circuit components which represent the departure of the real transformer from perfection. The reasons this procedure is useful and effective appear in Secs. 3-4 and 3-5. There it is shown how to determine the magnitudes of the circuit components which give rise to the departure of the transformer from ideal operation. These components may be calculated

Fig. 3-1   Schematic diagram of a transformer including source and load.

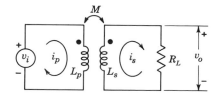

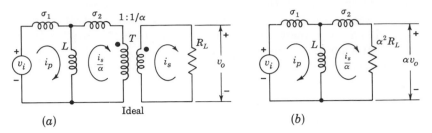

Fig. 3-2 (a) A circuit which is equivalent to that of Fig. 3-1, in which an ideal transformer T having a voltage step-up ratio $1/\alpha$ is introduced. (b) The same circuit with the transformer T eliminated by reflecting the secondary load into the primary.

from the geometry and mechanical construction of the transformer or measured experimentally. Alternatively, if a transformer response is specified, we may physically construct a transformer to meet the requirements.

We shall now show that the circuits of Fig. 3-2 are equivalent to the circuit of Fig. 3-1. Figure 3-2a includes an ideal transformer in cascade with a configuration of inductors which represent the departure of the actual transformer from perfection. The transformation ratio of the ideal transformer T is $1/\alpha \equiv$ secondary voltage/primary voltage, where α is a number which will be specified later. In Fig. 3-2 the load current has been reflected into the primary as i_s/α. The load resistance R_L also has been reflected to the primary side in Fig. 3-2b, where it appears as a resistance $\alpha^2 R_L$. We shall now find the values of the parameters σ_1, σ_2, and L in terms of α, L_p, L_s, and M.

The network of Fig. 3-2b is to be equivalent to the original transformer circuit in the sense that both are to draw the same current i_p from the source v_i and both are to furnish the same current i_s to the load R_L. In the circuit of Fig. 3-1 we may write the mesh equations

$$v_i = L_p \frac{di_p}{dt} - M \frac{di_s}{dt} \tag{3-2a}$$

$$0 = -M \frac{di_p}{dt} + L_s \frac{di_s}{dt} + i_s R_L \tag{3-2b}$$

The corresponding equations for the circuit of Fig. 3-2b are

$$v_i = (\sigma_1 + L) \frac{di_p}{dt} - L \frac{d}{dt}\left(\frac{i_s}{\alpha}\right) \tag{3-3a}$$

$$0 = -L \frac{di_p}{dt} + (\sigma_2 + L) \frac{d}{dt}\left(\frac{i_s}{\alpha}\right) + \alpha^2 R_L\left(\frac{i_s}{\alpha}\right) \tag{3-3b}$$

If Eq. (3-3b) is divided by α and then if Eqs. (3-2) and (3-3) are compared,

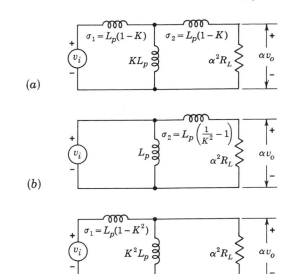

Fig. 3-3 The forms of the circuit of Fig. 3-2b for three particular values of α.

(a) $\alpha = \sqrt{L_p/L_s}$;

(b) $\alpha = (1/K)\sqrt{L_p/L_s}$;

(c) $\alpha = K\sqrt{L_p/L_s}$.

we find them to be identical provided that

$$L_p = \sigma_1 + L \qquad M = \frac{L}{\alpha} \qquad L_s = \frac{\sigma_2 + L}{\alpha^2} \qquad (3\text{-}4)$$

or

$$L = \alpha M \qquad \sigma_1 = L_p - \alpha M \qquad \sigma_2 = \alpha^2 L_s - \alpha M \qquad (3\text{-}5)$$

We are at liberty to select α. The circuits which result, together with the corresponding component values, are shown in Fig. 3-3a, b, and c for the choices $\alpha = \sqrt{L_p/L_s}$, $\alpha = (1/K)\sqrt{L_p/L_s}$, and $\alpha = K\sqrt{L_p/L_s}$. The results shown are calculated directly from Eq. (3-5) combined with the definition of the coefficient of coupling $K \equiv M/\sqrt{L_pL_s}$. There are an infinity of allowable values which may be selected for α, but the three indicated in Fig. 3-3 are those most commonly employed. Note that in Fig. 3-3a, $\sigma_1 = \sigma_2$; in Fig. 3-3b, $\sigma_1 = 0$; and in Fig. 3-3c, $\sigma_2 = 0$.

In a well-constructed pulse transformer the coefficient of coupling K differs from unity by less than 1 percent. Hence

$$1 - K^2 = (1 - K)(1 + K) \approx 2(1 - K)$$

And for such a transformer ($K \approx 1$), each of the circuits of Fig. 3-3 gives very nearly the same value for the total series inductance and the total shunt inductance. The total series inductance, called the *leakage inductance*, equals or has the approximate value

$$\sigma \approx 2L_p(1 - K) \qquad (3\text{-}6)$$

and the shunt inductance, called the *magnetizing inductance*, equals or has the approximate value

$$L \approx L_p \tag{3-7}$$

Further, the transformation ratio $1/\alpha$ of the ideal transformer equals or has the approximate value $1/\alpha \approx \sqrt{L_s/L_p}$. Since from Eq. (3-1) $\sqrt{L_s/L_p}$ is very nearly equal to the ratio n of the number of secondary turns to primary turns, we also have that

$$\frac{1}{\alpha} \approx n \tag{3-8}$$

We shall find the model of Fig. 3-3c most useful for studying the transmission of pulses through a transformer. This circuit is incomplete because we have neglected the capacitance and resistance associated with the device. We now introduce these parameters.

3-3   COMPLETE EQUIVALENT CIRCUIT

A pulse transformer, like any iron-cored transformer, consists of one or more layers of a primary winding on the core and one or more layers of a secondary winding wound over the primary.   Let us consider a very simple case in which the primary consists of a single layer of N_p turns wound in solenoidal form and the secondary consists also of a single layer separated somewhat from the primary but wound concentrically with it.   The secondary has N_s turns, and we shall assume that primary and secondary wire sizes are different so that the length of each winding is the same.   Now consider that the transformer is connected between a source and a load, as in Fig. 3-4.   Here we have connected opposite ends of the primary and secondary windings to a common ground.   Assuming that the windings are wound in the same direction on the core, the transformer must invert the input.   With respect to ground the output is $v_o = -(N_s/N_p)v_i = -nv_i$.

There is now a voltage nv_i between the bottom ends of the windings.   The voltage decreases linearly with distance along the windings and equals v_i at the top end.   As a consequence there exists an electric field in the space between the windings, and in this space electrostatic energy is being stored. The circuit element which stores energy electrostatically is capacitance. Therefore we must add to the model of the transformer in Fig. 3-3c a capaci-

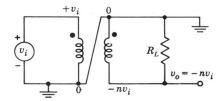

Fig. 3-4   An inverting transformer with turns ratio n.

Fig. 3-5 The equivalent circuit of a transformer, including resistances and the total shunt capacitance C.

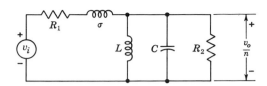

tance C which will give the proper electrostatic energy storage. This addition is made in Fig. 3-5.

We have taken the transformer capacitance into account by including a shunt capacitor C which is connected on the load side of the leakage inductance. Actually the transformer capacitance is a distributed element, and no matter what location is selected for a single lumped capacitance, the result is an approximate equivalent circuit. A somewhat better approximation would result if the capacitance C were split so that a part of it appeared on the generator side and a part appeared on the load side of the leakage inductance. Such a division, however, leads to an equivalent circuit whose extra complexity is not warranted, since a single capacitor in the equivalent circuit usually gives reasonably good agreement with experimental results.

Having decided to use a single lumped-capacitor representation of the distributed capacitance, we have located this capacitance at the load end of the leakage inductance for the following reasons. First, if C were located on the generator side, then if the generator had a nominally zero output impedance the effect of this capacitor would disappear—a result which does not agree with experimental findings. Second, the external shunt-loading capacitance C_L encountered with a pulse transformer is very frequently heavier on the output side of the leakage inductance, and this external capacitance, reflected into the primary side as $n^2 C_L$, may simply be added to the transformer capacitance. In Fig. 3-5 it is the total effective shunt capacitance that is represented by C.

In Fig. 3-5 we have included as well the resistance R_1, which represents the sum of the primary winding resistance and the generator impedance (assumed resistive). The resistance R_2 represents the combination of the load resistance R_L and the secondary winding resistance R'_2, so that

$$R_2 = \frac{R_L + R'_2}{n^2}$$

Before determining the pulse response of the circuit of Fig. 3-5 we shall, in the next two sections, learn how to calculate and measure the inductance and capacitance parameters of a transformer.

3-4   TRANSFORMER INDUCTANCES[2]

From the equivalent circuit of Fig. 3-3c it is clear that *the magnetizing inductance is the inductance presented at the input terminals when the secondary is open-*

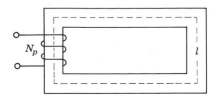

Fig. 3-6 A primary of N_p turns is wound on a magnetic core of mean magnetic path length l.

circuited. Or, more simply, the magnetizing inductance is the primary winding inductance. Similarly *the leakage inductance is the inductance presented at the terminals of the primary when the secondary is short-circuited.* These considerations may be used to estimate the magnetizing and leakage inductance from the transformer geometry and constructional features.

The primary inductance L_p may be calculated for the simple magnetic circuit of Fig. 3-6. If l is the mean length of the magnetic path, A the cross-sectional area of the core, N_p the number of primary turns, and μ the magnetic permeability,†

$$L_p = \frac{\mu A N_p{}^2}{l} \tag{3-9}$$

In order to see in a typical case how the leakage inductance σ depends on the geometry, consider the simple geometrical arrangement of Fig. 3-7a, where a single-layer secondary is wound over a one-layer primary. We have already noted that the secondary must be short-circuited in order to find σ. For this connection the output voltage is, of course, zero. Hence the net flux in the iron is zero, and the primary and secondary ampere-turns must be equal and oppositely directed, $N_p I_p = N_s I_s$. Almost all the flux appears in the space between the coils. For simplicity, we replace the coils by current sheets carrying the current $N_p I_p$ and $N_s I_s$ ($= N_p I_p$), respectively. A drawing of the concentric solenoidal windings is shown in Fig. 3-7b, and the magnetic field intensity H between windings is also indicated. The current sheets are the same length λ (in the direction perpendicular to the current flow) as the coils are long. We locate the current sheets at the point midway through the thickness of the wires of the coils. The value of H in the region between sheets is $H = N_p I_p/\lambda$. This result for H is computed by applying Ampère's circuital law to the path indicated in Fig. 3-7b. The energy density stored in the magnetic field is given by $\frac{1}{2}\mu H^2$. Accordingly the total energy W stored is $W = \frac{1}{2}\mu_o H^2 V$, where V is the volume between coils and where we have replaced μ by μ_o, the permeability of free space because the medium between the coils is air. The energy may also be calculated from $W = \frac{1}{2}\sigma I_p{}^2$ since this magnetic energy (with the secondary shorted) may be considered to reside in the leakage inductance σ. Equating the above two expressions for W,

---

† If the permeability of the iron relative to free space is μ_r, then $\mu = \mu_r \mu_o$, where $\mu_o = 4\pi \times 10^{-7}$ H/m is the permeability of free space.

we obtain

$$\sigma = \frac{\mu_o H^2 V}{I_p^2} = \frac{\mu_o N_p^2 V}{\lambda^2} \tag{3-10}$$

where all quantities are expressed in mks units. This calculation indicates clearly that σ is due to the leakage flux, that is, the flux which links one but not both windings. Hence, σ is essentially independent of the magnetic circuit of the transformer, since the leakage flux is almost entirely in air. Note that the ratio of magnetizing to leakage inductance $L_p/\sigma = \mu A \lambda^2/\mu_o V l$ is independent of the number of turns and is proportional to the permeability of the iron. One of the main reasons for using high-permeability cores in pulse transformers is to have a large ratio of magnetizing to leakage inductance.

The leakage inductance may be measured with a Q meter or an impedance bridge provided that the transformer secondary is shorted. Of course, the effect of the resistance which is in series with σ must be taken into account.

A second method of determining σ is to short the secondary, shunt the primary with a capacitance C_1, and measure the resonant frequency f_1. In order to eliminate the effect of the transformer and other unknown external capacitances which are in shunt with C_1, the above measurement is repeated with a second capacitor C_2. If the resonant frequency is now found to be f_2, we can show that

$$\sigma = \frac{f_1^2 - f_2^2}{(2\pi f_1 f_2)^2 (C_2 - C_1)} \tag{3-11}$$

A simple procedure for measuring the resonant frequencies f_1 and f_2 is the following: the transformer is placed in series with a tube or transistor and a steady current is established in the primary. Then a negative step cuts off the active device. The transformer will now ring, and the resonant period

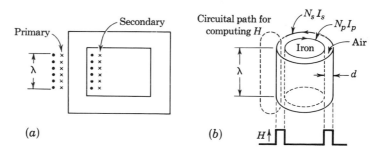

(a)   (b)

Fig. 3-7 (a) A one-layer secondary wound directly over a one-layer primary. A cross indicates current into the page, and a dot indicates current out of the page. (b) A schematic view of the windings considered as current sheets, and the magnetic-flux density between windings.

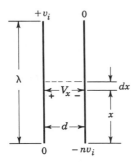

Fig. 3-8   The traces of the windings of Fig. 3-7.

may be measured from the waveform, as observed on a scope.   This method also allows the simultaneous measurement of the effective losses in the transformer.   Thus, as explained in Sec. 2-11, if the amplitude of the waveform falls to $1/\epsilon$ of its initial value in N cycles, the Q of the circuit is $N\pi$.

If the above experiments are repeated with the secondary open-circuited, then the magnetizing inductance L_p will be measured.

3-5   TRANSFORMER CAPACITANCES[2]

To illustrate how the transformer capacitance C in Fig. 3-5 depends on geometrical factors, we calculate C for the simple two-winding transformer represented geometrically in Fig. 3-7 whose circuit is given in Fig. 3-4.   The transformer is inverting, and opposite ends of each winding are connected.   The traces in the plane of the paper of the two windings of Fig. 3-7 are shown in Fig. 3-8.   At the bottom of the windings where $x = 0$ the voltage between windings is nv_i and at the top the voltage is v_i.   We assume that in between the voltage variation is linear with distance.   Therefore the voltage between windings at a distance x is

$$V_x = \left[n + (1 - n)\frac{x}{\lambda} \right] v_i \qquad (3\text{-}12)$$

The electric field is $E_x = V_x/d$.   The electrostatic energy stored per cubic meter is $\frac{1}{2}\epsilon E_x^2 = \frac{1}{2}\epsilon V_x^2/d^2$, where ϵ is the dielectric constant† of the medium separating the two windings.   If S is the mean circumference of the windings then the element of volume is $S d x$ and the total energy is

$$W = \int_0^\lambda \frac{1}{2}\frac{\epsilon V_x^2}{d^2} S \, dx \qquad (3\text{-}13)$$

If V_x from Eq. (3-12) is substituted into Eq. (3-13) and the integral is evalu-

† If the dielectric constant of the medium relative to free space is ϵ_r, then $\epsilon = \epsilon_r\epsilon_o$, where $\epsilon_o = (36\pi \times 10^9)^{-1}$ F/m is the permittivity of free space.

ated, the result is

$$W = \frac{1}{6}\frac{\epsilon S\lambda}{d}(n^2 + n + 1)v_i^2 \tag{3-14}$$

The voltage across the capacitor C introduced in shunt with the magnetizing inductance in Fig. 3-3 is approximately v_i because the drop across the leakage inductance will normally be small in comparison with v_i. The energy stored in the capacitor C is therefore $\frac{1}{2}Cv_i^2$. If this energy is equated to W we find

$$C = (n^2 + n + 1)\frac{C_o}{3} \tag{3-15}$$

where $C_o \equiv \epsilon S\lambda/d$. If the separation d between windings is small compared with the core thickness, then the two layers may be considered the plates of a parallel-plate capacitor whose capacitance is C_o. Note that for $n = 1$, $C = C_o$.

3-6   RISE–TIME RESPONSE OF A TRANSFORMER

The circuit of Fig. 3-5 is represented by a third-order equation whose solution would be quite involved. Furthermore, this complete solution would not clearly indicate the physical behavior of the circuit. Hence, if the input is a pulse, it is advantageous to divide the solution into three parts; the first gives the response near the front edge of the pulse, the second gives the response during the flat top, and the third gives the response after the termination of the pulse. In this section we consider the *rise-time response* and in the following two sections the remainder of the waveform is discussed.

The response near the front edge of the pulse is given by the high-frequency equivalent circuit of Fig. 3-9, which is obtained from Fig. 3-5 by neglecting the effect of L. (At high frequencies the reactance ωL of L is large compared with the parallel reactance $1/\omega C$ of C.) The magnitude of the input step is V. Writing down the differential equations for this network and assuming a solution in the form ϵ^{st}, we find for the roots s of the characteristic equation (or, equivalently, for the poles of the transfer function)

$$s = -\left(\frac{R_1}{2\sigma} + \frac{1}{2R_2C}\right) \pm \left[\left(\frac{R_1}{2\sigma} + \frac{1}{2R_2C}\right)^2 - \frac{R_1 + R_2}{\sigma C R_2}\right]^{\frac{1}{2}} \tag{3-16}$$

Let us introduce the *amplification factor* a, the *period* T, and the *damping*

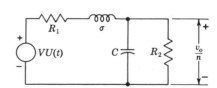

Fig. 3-9 The approximate equivalent circuit used to calculate the rise-time response.

constant k, defined by

$$a \equiv \frac{R_2}{R_1 + R_2} \qquad T \equiv 2\pi(\sigma C a)^{\frac{1}{2}} \qquad k \equiv \left(\frac{R_1}{\sigma} + \frac{1}{R_2 C}\right)\frac{T}{4\pi} \qquad (3\text{-}17)$$

in which case Eq. (3-16) can be put in the form

$$s = -\frac{2\pi}{T} k \pm j\frac{2\pi}{T}(1 - k^2)^{\frac{1}{2}} \qquad (3\text{-}18)$$

If $k = 0$, we see that the roots are purely imaginary, $\pm j 2\pi/T$, and hence the response is an undamped sinusoid of period T. In order for k to approach zero, we must have $R_1 \to 0$ and $R_2 \to \infty$, in which case $T = 2\pi\sqrt{\sigma C}$ is the free period of oscillations of the σC circuit. If $k = 1$, the two roots are equal, corresponding to the *critically damped* case. If $k > 1$, there are no oscillations in the output, and the response is said to be *overdamped*. If $k < 1$, the response will be a sinusoid whose amplitude decays with time, and the response is said to be *underdamped*.

If we introduce the parameters $x \equiv t/T$ and $y \equiv v_o/naV$, the response is given by the following:

Critical Damping, $k = 1$

$$y = 1 - (1 + 2\pi x)\epsilon^{-2\pi x} \qquad (3\text{-}19)$$

Overdamped, $k > 1$

$$y = 1 - \frac{4k^2}{4k^2 - 1}\epsilon^{-\pi x/k} + \frac{1}{4k^2 - 1}\epsilon^{-4\pi k x} \qquad (3\text{-}20)$$

If $4k^2 \gg 1$, the response may be approximated by

$$y \approx 1 - \epsilon^{-\pi x/k} \qquad (3\text{-}21)$$

Underdamped, $k < 1$

$$y = 1 - \left[\frac{k}{(1 - k^2)^{\frac{1}{2}}} \sin 2\pi(1 - k^2)^{\frac{1}{2}}x + \cos 2\pi(1 - k^2)^{\frac{1}{2}}x\right]\epsilon^{-2\pi k x} \qquad (3\text{-}22)$$

If the derivative of Eq. (3-22) is set equal to zero, the positions x_m and magnitudes y_m of the maxima and minima are obtained. The results are

$$x_m = \frac{m}{2(1 - k^2)^{\frac{1}{2}}} \qquad \text{and} \qquad y_m = 1 - (-1)^m \epsilon^{-2\pi k x_m} \qquad (3\text{-}23)$$

where m is an integer. The maxima occur for odd values of m and the minima are obtained for even values of m. By using Eq. (3-23) the waveshape of the underdamped output may be sketched very rapidly.

These responses are plotted in Fig. 3-10 for several values of k. If the *rise time* t_r is defined as the time interval required for the output to rise from 0.1 to 0.9 of its final value, we find, from Eq. (3-19) or Fig. 3-10, that for the

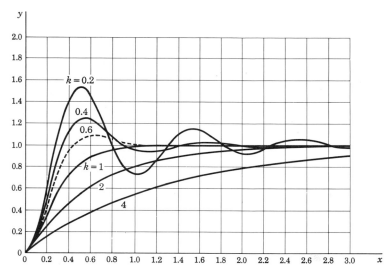

Fig. 3-10 The rise-time response of a pulse transformer.
$y \equiv v_o/naV$ and $x \equiv t/T$.

critically damped case

$$t_r = 0.53T = 3.35(\sigma Ca)^{\frac{1}{2}} \tag{3-24}$$

We note that in order for the output to rise rapidly, the leakage inductance and the shunt capacitance must be kept small.   The rise time may also be reduced by reducing a, but a small value of a will result in a highly attenuated output voltage.

For many applications an overshoot in the output of 5 or 10 percent is acceptable.   In such a case we may take advantage of the fact that permitting the overshoot will reduce the rise time.   For example, if $k = 0.6$, the overshoot is 9 percent (see Fig. 3-10) and $t_r = 0.27T$, whereas for the critically damped case $k = 1$ and $t_r = 0.53T$.   Since the value of T is itself a function of k, a detailed calculation must be made in each case before it is possible to state what the rise-time improvement will be.   As a simple example, however, consider the situation where $R_2 \gg R_1$, so that $a \approx 1$ and $T \approx 2\pi(\sigma C)^{\frac{1}{2}}$ is independent of the resistance values.   Then if R_2 is increased so that k decreases from 1 to 0.6, the rise time will be decreased to $(0.27/0.53) \times 100 \approx 50$ percent of what it was for the critically damped case.

Large step-up ratios are seldom used in pulse transformers, because the gain n can be obtained only at the price of increasing the rise time by the factor n.   This conclusion is easily verified.   If the step-up ratio is n, then load and interwinding capacitances are multiplied by approximately n^2 [see Eq. (3-15)].   Since the rise time t_r varies as $C^{\frac{1}{2}}$ [see Eq. (3-24)], t_r is proportional to n.   If, in order to accommodate additional secondary turns, the geometry is modified so that σ also increases, then the rise time deteriorates even further. Usually $n < 10$.

3-7   THE FLAT TOP OF THE PULSE

The response during the top of the pulse is obtained from the low-frequency equivalent circuit of Fig. 3-11a, which is obtained from Fig. 3-5 by neglecting the effect of the leakage inductance and shunt capacitance.   (At low frequencies ωL is small compared with $1/\omega C$, and $\omega\sigma$ is small compared with R_1.) Applying Thévenin's theorem, we obtain Fig. 3-11b, where $a \equiv R_2/(R_1 + R_2)$ and $R \equiv R_1 R_2/(R_1 + R_2)$.   The output is given by

$$y = \frac{v_o}{naV} = \epsilon^{-Rt/L} \tag{3-25}$$

For values of Rt/L much less than unity, the output is approximated by

$$y = 1 - \frac{Rt}{L} \tag{3-26}$$

Hence, the top of the output pulse will be tilted downward and the percent tilt P is given by

$$P = \frac{Rt_p}{L} \times 100\% \tag{3-27}$$

where t_p is the pulse width.   Near the beginning of the pulse there will be superimposed upon the linear fall the response pictured in Fig. 3-10.
   We have assumed that the inductance L is a constant.   This assumption is valid as long as the iron does not begin to saturate.   For a ferrite core discussed in Sec. 3-10 the permeability is fairly constant for flux densities B up to a maximum B_m, which is of the order of 1,500 to 5,000 G (0.15 to 0.5 Wb/m²). Saturation occurs if B exceeds the above value B_m.   Now

$$v_o = N_s \frac{d\phi}{dt} = nN_p A \frac{dB}{dt} \tag{3-28}$$

where N_s is the number of secondary turns, ϕ is the magnetic flux, n is the

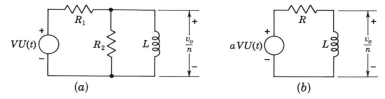

Fig. 3-11   (a) The equivalent circuit used to calculate the flat-top response of a transformer; (b) modified by Thévenin's theorem.

step-up ratio, N_p is the number of primary turns, and A is the cross-sectional area of the core. Assuming that the top of the pulse is flat and equals naV, the flux density at the end of the pulse is

$$B = \int_0^{t_p} \frac{v_o}{nN_pA} \, dt = \frac{aVt_p}{N_pA} \tag{3-29}$$

In any particular application we must be sure not to saturate the core. For example, consider that a pulse generator having an adjustable pulse width is applied to a transformer. The output pulse will be a reasonable reproduction of the input for small widths. When the input duration exceeds the value of t_p given by Eq. (3-29) with $B = B_m$, the output will drop rapidly. This behavior follows from the fact that, when the iron saturates, the magnetizing inductance drops to a very low value.

Note that it is the *volt-second product* which determines the maximum flux density. For example, if a given transformer saturates at a pulse width of 1 μsec and an amplitude of 10 V, then doubling the pulse height will cause saturation to take place in half the time, or at 0.5 μsec.

3-8   COMPLETE PULSE RESPONSE OF A TRANSFORMER

The waveform $y(t)$ is the composite of the *rise-time response* and the *flat-top response*, as found above. The composite waveform is obtained by first plotting the exponential (almost linear) portions at the top of the pulse. Then the positive and negative peak overshoots, given by $y_m - 1$ of Eq. (3-23), are superimposed upon the exponential.

The response beyond the pulse width, $t > t_p$, is obtained as follows. A pulse may be considered to be the sum of a step of voltage $+V$ whose discontinuity occurs at $t = 0$ and a step of voltage $-V$ whose discontinuity occurs at $t = t_p$ (see Fig. 2-3). Hence, if the transformer response to a step V at $t = 0$ is $y(t)$, then the output for $t > t_p$ is $y(t) - y(t - t_p)$. For the flat-top response, $y(t) = \epsilon^{-Rt/L}$ and hence

$$y(t) - y(t - t_p) = \epsilon^{-Rt/L} - \epsilon^{-R(t-t_p)/L}$$

$$= (1 - \epsilon^{Rt_p/L})\epsilon^{-Rt/L} \tag{3-30}$$

for $t > t_p$. Note that this response is an exponential with the same time constant as that of the top of the pulse. For the underdamped case, the trailing edge of the output waveform will contain the same high-frequency oscillations as are present on the leading edge. These are plotted by locating the maxima and minima with respect to the exponential of Eq. (3-30).

The response of a transformer of typical parameters is computed in the following illustrative example and shown in Fig. 3-12.

EXAMPLE A pulse transformer has the following parameters: $L = 5$ mH, $\sigma = 40$ μH, $C = 50$ pF, $R_1 = 200$ Ω, $R_2 = 2$ K, $n = 1$. Find the response to a 2-μsec 10-V pulse.

Solution For the rise-time response we have, from Eq. (3-17),

$$a = \frac{R_2}{R_1 + R_2} = \frac{2,000}{200 + 2,000} = 0.909$$

$$T = 2\pi(\sigma Ca)^{\frac{1}{2}} = 2\pi(40 \times 10^{-6} \times 50 \times 10^{-12} \times 0.909)^{\frac{1}{2}} = 0.267 \text{ μsec}$$

$$k = \left(\frac{R_1}{\sigma} + \frac{1}{R_2 C}\right)\frac{T}{4\pi} = \left(\frac{200}{40 \times 10^{-6}} + \frac{1}{2 \times 10^3 \times 50 \times 10^{-12}}\right)\frac{2.67 \times 10^{-7}}{4\pi}$$

$$= 0.318$$

Since $k < 1$, the response is underdamped and is given by Eq. (3-22), namely,

$$y = 1 - \left[\frac{k}{(1 - k^2)^{\frac{1}{2}}} \sin 2\pi(1 - k^2)^{\frac{1}{2}}\frac{t}{T} + \cos 2\pi(1 - k^2)^{\frac{1}{2}}\frac{t}{T}\right]\epsilon^{-2\pi kt/T}$$

Substituting numerical values into this equation, we obtain

$$y = 1 - (0.325 \sin 22.3t + \cos 22.3t)\epsilon^{-7.48t}$$

where t is expressed in microseconds.

From Eqs. (3-23) we find that the maxima and minima occur at

$$t_m = \frac{mT}{2(1 - k^2)^{\frac{1}{2}}} = 0.141m$$

where $m = 1, 2, 3, \ldots$, and that the magnitudes at t_m are

$$y_m = 1 - (-1)^m \epsilon^{-1.01m}$$

The flat-top response is given by Eq. (3-25), namely,

$$y = \epsilon^{-Rt/L} = \epsilon^{-0.0364t} \approx 1 - 0.0364t$$

where t is expressed in microseconds.   The percentage tilt of the top of the pulse is $3.64t_p = 7.28$ percent.

At the end of the pulse the response is given by Eq. (3-30), which for the given value of the transformer parameters reduces to

$$y = (1 - \epsilon^{0.0728})\epsilon^{-0.0364t} = -0.0758\epsilon^{-0.0364t}$$

At $t = t_p+ = 2$ μsec, a value of $y = -0.071$ is obtained.

The complete response (up to $t = 6$ μsec) is sketched in Fig. 3-12.   The long undershoot for $t > t_p$ should be noted.   This section of the response will slowly approach the zero axis so that the net area under the curve will equal zero, as we shall now demonstrate.

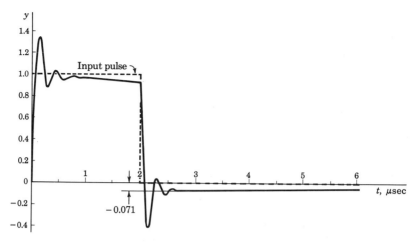

Fig. 3-12 The response of the transformer whose parameters are given in the illustrative example. $v_o = 9.09y$.

To show that the area under the plot of Fig. 3-12 is zero we write $v_o = N_s \, d\phi/dt$, where ϕ is the total flux in the magnetizing inductor. The total area under the output voltage waveform is

$$\int_0^\infty v_o \, dt = N_s \int_0^\infty \frac{d\phi}{dt} \, dt = N_s \phi \Big|_{t=0}^{t=\infty} = 0 \tag{3-31}$$

since $\phi = 0$ at $t = 0$ and at $t = \infty$. This proof does not assume that L is a constant. Even if saturation takes place, so that L is not constant, the area above the zero axis equals that below the zero axis and hence there is no shift in d-c level. However, if the iron core exhibits appreciable hysteresis, so that $\phi \neq 0$ at $t = \infty$, then the above theorem is not valid.

The high-frequency oscillations noted in Fig. 3-12 may be reduced to zero by increasing the loading on the transformer. Critical damping for the transformer of the above illustrative example is obtained when $R_2 = 400 \; \Omega$. This result is found by calculating the value of R_2 for $k = 1$. The attenuation factor is now

$$a = \frac{R_2}{R_1 + R_2} = \frac{400}{600} = 0.667$$

whereas the attenuation factor for $R_2 = 2 \text{ K}$ was 0.909. Thus, the oscillations have been removed at the expense of increased attenuation. Also, the output will rise somewhat more slowly toward its peak value. The rise time t_r calculated from Eq. (3-24) is 0.122 μsec. On the other hand, the tilt will now be smaller than it was for $R_2 = 2 \text{ K}$ because $R = R_1 a$ is reduced. With $R_2 = 400 \; \Omega$, the tilt is calculated to be 5.35 percent, which is to be compared with the value of 7.28 percent found above for $R_2 = 2 \text{ K}$.

It should be emphasized that the output impedance of the pulse generator may be different at the termination of the pulse than during the time of the pulse. If this condition exists, then a different value of R_1 must be used when calculating the response for $t > t_p$ than for $t < t_p$. For example, if $R_1(t > t_p)$ is much greater than $R_1(t < t_p)$, then the high-frequency oscillations at the trailing edge will be greatly damped. However, even when the pulse is terminated by opening a generator switch, in which case σ could be omitted, at least in Fig. 3-9, oscillations often persist. The reason for this behavior is that actually the capacitance is not properly included in the circuit as a lumped element, but should really be included as an element continuously distributed between the leakage inductance and ground.

3-9   PULSE–TRANSFORMER GENERAL CONSIDERATIONS

An ideal transformer, which would introduce none of the pulse distortion apparent in Fig. 3-12, would have an infinite L and zero σ and C. Actually, the magnetizing inductance determines the tilt during the pulse and the backswing at the end of the pulse. To minimize both tilt and backswing we require only, as appears in Eq. (3-27), that $L \gg Rt_p$. Accordingly, if, say, $t_p = 0.1$ μsec and $R = 200$ Ω, $Rt_p = 20$ μH and a magnetizing inductance $L = 1$ mH ($= 50 \times 20$ μH) is, for all practical purposes, infinite.

When a core has been selected, in both material and geometry, we may calculate the number of primary turns necessary to realize the required magnetizing inductance. Thereafter the number of secondary turns will be determined from the transformation ratio. The smallest core on which there is space available to place the windings is normally selected, and it is well to check with the aid of Eq. (3-29) that the core is not saturated at the peak of the pulse.

In a small pulse transformer, the preservation of the pulse shape is more important than efficiency of operation. The winding resistances may therefore be permitted to be quite large, often as large as 10 percent of the load or generator resistances. Small wire sizes may therefore be used, with a consequent reduction in capacitances. If the interwinding and interlayer distances are kept small, the leakage inductance will be small but the effective capacitance will increase. The reverse will be true if the interlayer distances are large. When the load and generator impedances are high, a large series leakage inductance may be much more readily tolerated than a large shunt capacitance. In this case the windings may be spaced far apart. If the load and generator resistances are very small, a close spacing may be preferred.

The finite rise time and ringing observed in the transformer response result from the leakage inductance σ and the capacitance C. Since the number of turns on the windings is fixed by the required magnetizing inductance, all we may do in connection with σ and C is decrease one at the expense of increasing the other. There is, however, in principle at least, one remedy

Fig. 3-13 A pulse-transformer core is made by winding a continuous strip of thin high-permeability alloy.

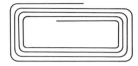

that will both minimize σ and C and increase L_p arbitrarily.   This method
consists in employing a core material whose magnetic permeability is infinite.
For in such a case, a one-turn primary would provide more than adequate
magnetizing inductance.   And since the turns are minimal we may shrink
the spacing between "windings" so that σ vanishes without introducing an
appreciable capacitance.   Pulse transformers employ as core material such
alloys as Hipersil (Westinghouse) $[\mu_r(\text{max}) \approx 12{,}000]$ or Permalloy (Western
Electric) $[\mu_r(\text{max}) \approx 80{,}000]$ or ferrites (Sec. 3-10).
     There is a second reason that great importance is attached to the perme-
ability of the core.   The permeability actually achieved in pulse transformers
is very much less than the maximum values indicated above.   When an abrupt
step of current is driven through the transformer winding, the magnetic flux
in the core is initially confined largely to the surface (the "skin effect") because
of the eddy currents that flow.   The effective cross section of the core is
thereby reduced.   As time passes, the flux penetrates deeper into the core
and eventually becomes uniform.   Accordingly, the effective permeability
of the core increases with increasing pulse duration.[1]   The effective permeability of Hipersil is of the order of 400 for microsecond pulses.   In order to reduce
eddy currents, to minimize both losses and the skin effect, it is important that
the core be laminated.   It has been found that Hipersil and Permalloy can be
rolled into strips as thin as 2 mils, and cores are often formed by winding a
continuous strip, as indicated in Fig. 3-13.

3-10   FERRITE TRANSFORMERS

Cores molded from a magnetic ceramic such as sintered manganese-zinc ferrite
are now available that are excellent for pulse transformers.   The maximum
permeability of this material is not very great, but its resistivity is at least
10 million times that of Hipersil or Permalloy.   This high resistivity means
that the skin effect due to eddy currents is very small and an effective permeability of the order of 1,000 is attained.   This value is larger than the effective
permeability of strip alloys.   Also, because of this high resistivity the core
loss is very small, and a Q of the order of 5 to 15 is obtained at a frequency
of 1 MHz.   One form in which ferrite cores are commercially available is
shown in Fig. 3-14a.   Because of its shape this element is called a "pot" or
"cup" core.   This type of core lends itself to "do-it-yourself" transformer
construction.   The windings are placed on a circular nylon or paper bobbin,
which is then inserted in the core.   An end view of the complete core, assem-

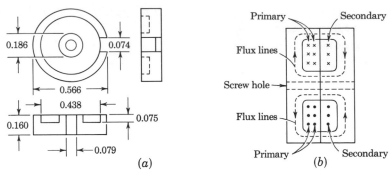

Fig. 3-14 (a) Three views of a small ferrite pot core. Dimensions are in inches. (Courtesy of Ferroxcube Corporation of America.) (b) The assembled transformer.

bled by putting two halves together, is indicated in Fig. 3-14b. The two sections are held together with a machine screw through a small hole in the center of the core, and the entire assembly is dipped into a hard-setting resin. The magnetic circuit thus completely encloses the windings. The primary inductance of a core whose dimensions are given in Fig. 3-14 can be calculated to be $L_p = 1.1N_p^2 \ \mu H$, to within 10 percent.

The windings in a pot core may be arranged in solenoidal layers as they are for a rectangular core or instead may be put side by side in slots in the bobbin. In the latter case, the turns pile up radially in the shape of a flat disk. For this arrangement the capacitance will be smaller but the leakage inductance larger than with the solenoidal winding.

Ferrite cores are also commercially available in the form of toroids (doughnuts) in very small sizes which make excellent pulse transformers for nanosecond applications.[3] For example, a transformer is found experimentally to have a rise time of 0.5 nsec and a tilt of 10 percent for a 50-nsec pulse if it is constructed according to the following specifications: Ferroxcube core, type 3B, 3C, or 102; outside diameter 0.2 in.; inside diameter 0.1 in.; height of toroid 0.125 in.; wound with 7 turns each of No. 27 Formvar bifilar wire for the primary and secondary. A primary and secondary wound in this manner constitute a "transmission-line" pulse transformer, as discussed in Sec. 3-20.

3-11   ELECTROMAGNETIC DELAY LINES[4]

Delay lines are passive four-terminal networks which have the property that a signal impressed at the input terminals appears at the output terminals at the end of a time interval t_d, called the *delay time*. Delays ranging from a few nanoseconds to hundreds of microseconds are obtainable with electromagnetic lines.

Fig. 3-15 A 1.0-μsec pulse
after passing through 1.0 μsec
of HH-2500 delay cable. The
rise time is 0.08 μsec. (Courtesy of Columbia Technical
Corporation.)

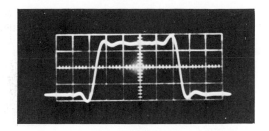

If a pulse is applied to a real (nonidealized) line, the signal will not only
be delayed but will also suffer attenuation and distortion, as indicated in
Fig. 3-15. In such a line, t_d is defined as the time interval between the
50 percent amplitude points on the rising edge of the incident and delayed
pulses. The important characteristics of delay lines are the following: the
time delay, the rise time, the attenuation, the distortion, the characteristic
impedance, the volume occupied by the line, the maximum voltage that may
be applied to the line, the stability of delay with temperature and time, the
ease and accuracy of adjusting the delay, and, finally, the cost.

The applications of delay lines are numerous. For example, a cathode-ray
oscilloscope which is to be used for observing fast waveforms has a built-in
delay line so that the input signal which also triggers the sweep is delayed
slightly before being applied to the vertical-deflection circuit. If the sweep
were not allowed to start before the signal was applied, then the first portion
of the waveform might not be visible on the scope face. Other applications
of delay lines occur in distributed amplifiers, in pulse coders and decoders, in
precise time measurement, in radar, in television, and in digital-computer
systems.

In many applications the type of distortion indicated in Fig. 3-15 (preshoot, overshoot, finite rise and fall times, and ringing) is acceptable. For
example, in computer circuits the occurrence or absence of a pulse is of more
importance than the exact form of the pulse. Moreover, where pulses have
become badly deteriorated in form, they may be reshaped. In other applications, notably in a scope, ringing is completely intolerable. As noted above,
a delay line is used in a scope to delay the signal until the sweep has started.
The remedy in this latter case is to construct a delay line whose ringing frequency is well beyond the bandpass of the system in which it is included.

3-12   TRANSMISSION–LINE CHARACTERISTICS

A uniform lossless transmission line, terminated in its characteristic impedance
Z_o, may be used as a delay line. If a sinusoidal voltage $V_s = A\,\epsilon^{j\omega t}$ is impressed
at the sending end of the line of Fig. 3-16, a traveling wave moves to the right
along the line. The voltage as a function of the distance x down the line is
given by $V_x = A\,\epsilon^{j(\omega t - \beta x)}$, and the voltage at the receiving end of the line is

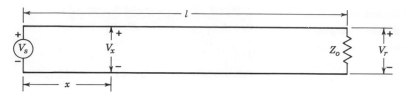

Fig. 3-16 A transmission line terminated in its characteristic impedance.

given by $V_r = A\epsilon^{j(\omega t - \beta l)}$. These facts follow from elementary transmission-line theory,[5,6] where it is shown that $\beta = \omega\sqrt{LC}$, ω being the angular frequency, L the inductance *per meter*, and C the capacitance *per meter*. Since the velocity with which the wave progresses is $u = (LC)^{-\frac{1}{2}}$, then $\beta = \omega/u$. Hence

$$V_r = A\epsilon^{j(\omega t - \beta l)} = A\epsilon^{j\omega(t - l/u)} = A\epsilon^{j\omega(t - t_d)} \tag{3-32}$$

where $t_d \equiv l/u$. From this equation we see that the voltage which appears at the receiving end is the same as that which was impressed on the sending end at a time t_d earlier. Since any waveform may be resolved into a Fourier spectrum and since the velocity u is independent of frequency, it follows from Eq. (3-32) that *for an ideal line, an arbitrary waveform impressed on the input terminals will appear without distortion at the output terminals after a delay time t_d.* It is also shown in transmission-line theory (Sec. 3-15) that if the line is terminated in a resistive impedance $Z_o = \sqrt{L/C}$, called the *characteristic impedance*, no reflection will take place when the signal reaches the end of the line.

Both L and C are functions of the geometry of the cross section of the line, but it turns out that for lines with a uniform cross section the product LC is independent of the geometry[5] and equals $\mu\epsilon$, where μ and ϵ are the magnetic permeability and the permittivity, respectively, of the medium between the conductors of the line. For a line whose conductors are in free space, $u = (LC)^{-\frac{1}{2}} = (\mu_o\epsilon_o)^{-\frac{1}{2}}$, where

$$\mu_o = 4\pi \times 10^{-7} \text{ H/m} \quad \text{and} \quad \epsilon_o = (36\pi \times 10^9)^{-1} \text{ F/m}$$

so that $u = 3 \times 10^8$ m/sec. This speed is the same as that with which a wave of electromagnetic radiation travels in free space, i.e., the velocity of light. The delay per meter T is given by $T = \sqrt{\mu\epsilon} = 1/u$ and, for air, $T = (3 \times 10^8)^{-1}$ sec/m = 3.3 nsec/m. For a medium of relative dielectric constant ϵ_r, the delay is $3.3\epsilon_r^{\frac{1}{2}}$ nsec/m. For the low-loss dielectric media which are available (polystyrene, polyethylene, or Teflon), $\epsilon_r \approx 2.3$ and $T \approx 5$ nsec/m. Such lines are useful in the nanosecond delay range, but the length of cable required is prohibitively long in the microsecond region. For example, a delay of 1 μsec requires a line 200 m long!

Delay-line Parameters   Consider the magnitudes of L, C, and Z_o for three types of transmission lines—the coaxial cable, the single-wire line over an infinite ground plane, and the line consisting of two parallel wires. These lines are shown in Fig. 3-17, together with the expressions for the corresponding capacitance C per unit length. We note the similarity in form in the three cases. Thus if $4h = D$ then C for the wire-over-ground case is the same as for the coaxial cable and the two-wire configuration has half as large a value of C. We note further that since it depends on the logarithm of the ratio of dimensions, the capacitance is extremely insensitive to changes in spacing. This similarity among lines and insensitivity to dimensional changes applies as well to the inductance L per unit length and the characteristic impedance $Z_o = \sqrt{L/C}$ of the line because

$$L = \frac{1}{u^2 C} \quad \text{and} \quad Z_o = \frac{1}{uC} \tag{3-33}$$

where u is the velocity of propagation. For a uniform line $u = \sqrt{\mu\epsilon}$.

For a lossless coaxial line in which the space between inner and outer conductor is filled with a material of dielectric constant ϵ_r we find

$$Z_o = 138\epsilon_r^{-\frac{1}{2}} \log \frac{D}{d} = 60\epsilon_r^{-\frac{1}{2}} \ln \frac{D}{d}$$

When attenuation in the line results principally from ohmic losses in the conductors, the loss (for a fixed D) is a minimum for $D/d = 3.6$. For this ratio and for $\epsilon_r = 2.3$, $Z_o = 51$ Ω. Most conventional and commercially available coaxial lines have impedances of this order of magnitude, i.e., from 50 to about 200 Ω. These lines have reasonable physical dimensions, whereas lines of appreciably higher characteristic impedance are not realistic. A line with $Z_o \approx 1{,}000$ Ω would require $\log (D/d) = 11$ or $D/d = 10^{11}$, which certainly is an impractical ratio.

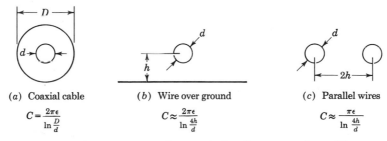

(a) Coaxial cable $\qquad$ (b) Wire over ground $\qquad$ (c) Parallel wires

$$C = \frac{2\pi\epsilon}{\ln \frac{D}{d}} \qquad\qquad C \approx \frac{2\pi\epsilon}{\ln \frac{4h}{d}} \qquad\qquad C \approx \frac{\pi\epsilon}{\ln \frac{4h}{d}}$$

Fig. 3-17   Several types of transmission lines, together with expressions for their capacitance per meter. The approximate expressions for the capacitance of the wire lines in (b) and (c) are correct to better than 5 percent if $h/d \geq 1$.

When we use the coaxial line we ordinarily do so through choice and employ for the purpose a cable of commercial manufacture. The coaxial line is important and useful especially because external points are shielded from the signal on the cable and because the cable is in turn shielded from neighboring signal sources. On the other hand a wire-over-ground line most often makes its appearance simply because a piece of wire has been used to connect together two points in an electrical circuit. Since the components are supported on a metal chassis this then constitutes the "infinite" ground plane. Such lines ordinarily operate with no dielectric or with a dielectric (the wire insulation) which occupies only a small part of the space between wire and ground. Accordingly $Z_o = 138 \log (4h/d)$, and if $h = d$, $Z_o \approx 83$ Ω. If $h = 2.5d$, $Z_o = 138$ Ω. We note again characteristic impedances in the neighborhood of 100 Ω and the same insensitivity of Z_o to spacing. Thus Z_o increases by 138 Ω each time $4h/d$ is multiplied by a factor of 10.

The two-wire line of Fig. 3-17c does not find wide application in fast circuitry and is included principally to suggest that the properties of the coaxial and wire-over-ground lines are characteristic of lines generally. A widely used two-wire line is the 300-Ω line used as an antenna leadin for television receivers.

The time delay per unit length T and the characteristic impedance Z_o depend on L and C, the inductance and capacitance per unit length, according to the formulas

$$T = \sqrt{LC} \qquad Z_o = \sqrt{\frac{L}{C}} \tag{3-34}$$

or

$$C = \frac{T}{Z_o} \qquad L = TZ_o \tag{3-35}$$

We have already noted that Z_o is usually limited to the range 50 to 200 Ω. Taking a nominal value $Z_o = 100$ Ω we have, *for an air dielectric,*

$$T \approx 1 \text{ nsec/ft} \quad \approx 3.3 \text{ nsec/m}$$

$$C \approx 10 \text{ pF/ft} \quad \approx 0.33 \text{ pF/cm} \tag{3-36}$$

$$L \approx 100 \text{ nH/ft} \approx 3.3 \text{ nH/cm}$$

3-13   THE TRANSMISSION LINE USED TO TRANSMIT A SIGNAL[7]

In Chap. 2 we find that in transmitting a signal from one point to another we must continually contend with shunt capacitance across the signal leads and inductance in series with the signal leads. Both of these ever-present circuit components limit the bandwidth of the transmission system and prevent the system from transmitting abrupt signal discontinuities. The consequent signal distortion is apparent in increased rise times, in ringing, etc.

If, however, the shunt capacitance and series inductance are distributed, as on a line, with capacitance C and inductance L per unit length, and if the line is terminated in a resistance $R = \sqrt{L/C}$, then such a line will convey a signal without distortion. Equally important is the fact that for a line so terminated the input impedance is also resistive and equal to R. Any actual line will produce some distortion and attenuation of a signal because of ohmic losses in the conductors and in the dielectric medium, both of which may be frequency-dependent. But with the lengths of lines normally employed in microsecond or nanosecond circuitry, the distortion so produced need not be serious.

It should now be apparent that when a fast waveform is to be transmitted for any appreciable distance, such transmission should take place over a matched coaxial cable so as to minimize distortion and coupling. Furthermore, the line should be matched not only at the receiving end but at the sending end as well. Any reflection which occurs on the line due to some line discontinuity or lack of perfect match at the receiving end will be absorbed at the sending end if the input is matched to the line.

To some extent in microsecond circuitry and to a much greater extent in nanosecond circuitry, commercial instruments (generators, scopes, etc.) have their output or input connections made through matched coaxial lines. In this way input and output impedances are made resistive, and it becomes feasible to make interconnections also through coaxial cables without introducing reflections. The characteristic impedances most common in commercial coaxial lines are nominally 50 Ω (as noted above) and nominally 72 Ω (to match the 72-Ω impedance of a dipole antenna).

Now let us turn our attention to the wire-over-ground line such as appears when a connection is made between components mounted over a metallic chassis. In Fig. 3-18 a source v_s of impedance R_s is connected through a line of characteristic impedance Z_o to a load of impedance $R = R_s$. Clearly, when possible, we would make $Z_o = R_s$. But, as we shall now see, it is possible to make an estimate of the distortion to be encountered if the condition $Z_o = R_s$ is not or cannot be fulfilled.

Let us assume, for the sake of being specific, that $Z_o > R_s$, as would be the case if the spacing of the line above the ground plane were too great. The capacitance per unit length of the line is $C = T/Z_o$ and $L = CZ_o^2$. If the line were matched to the terminations, the inductance would be $L' = CR_s^2$. Accordingly, there is an excess of inductance ΔL in a line of length l, given by

$$\Delta L = l(L - L') = lCZ_o^2 \left(1 - \frac{R_s^2}{Z_o^2}\right) = lTZ_o \left(1 - \frac{R_s^2}{Z_o^2}\right) \tag{3-37}$$

We now assume that the configuration of Fig. 3-18a may be replaced as in Fig. 3-18b. Here we consider that the inductance $L' = CR_s^2$ has "canceled"

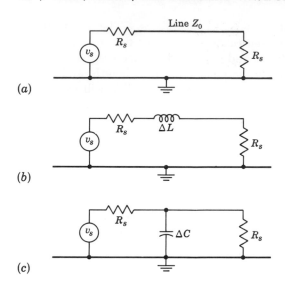

(a)

(b)

(c)

Fig. 3-18 (a) A line of characteristic impedance Z_o connects
impedances R_s. (b) If $Z_o > R_s$
the line is replaced by a lumped
excess inductance. (c) If
$Z_o < R_s$ the line is replaced by
a lumped excess capacitance.

the line capacitance. The line is now replaced by an inductor ΔL and the
transmission-line properties of the line may be disregarded. It is now clear
that if the source at the input end of the line generated a step of amplitude V,
the waveform at the load would be $v = (V/2)(1 - \epsilon^{-t/\tau})$, with $\tau = \Delta L/2R_s$.
As an example, consider a signal being transmitted over a 50-Ω cable to a 50-Ω
termination. Let us assume that a 3-cm lead is used to make the connection
from the cable to the termination and that this constitutes a 100-Ω wire-over-
ground line. Since $R_s = 50$ Ω, $Z_o = 100$ Ω, and $l = 3$ cm, then $\Delta L = 7.5$ nH,
and the rise time is, from Eq. (2-33), $2.2\tau = 0.17$ nsec. This time is proportional to the wire length. If a 1-cm lead were used it would be reduced to
one-third the above value, or 0.06 nsec (60 psec).

If $Z_o < R_s$, as when the wire-to-ground spacing is too small, then there
will be an excess of capacitance, which is given by

$$\Delta C = \frac{lT}{Z_o}\left(1 - \frac{Z_o^2}{R_s^2}\right) \tag{3-38}$$

This capacitance is shown in the circuit of Fig. 3-18c, from which it appears
that the corresponding time constant is $\tau = (R_s/2)\,\Delta C$.

Altogether, it appears that when $Z_o > R_s$ there will be an excess of inductance and when $Z_o < R_s$ an excess of capacitance. Since line impedances are
in the neighborhood of 100 Ω, we shall have excess inductance or capacitance
depending on whether R_s is smaller or larger than about 100 Ω. Suppose that
we are required by certain constraints to use terminations of 1,000 Ω. We
might then be inclined to increase the Z_o of the line by increasing the line-to-

Fig. 3-19 Illustrating how the
presence of a ground plane reduces
the coupling between conductors
mounted above a metal chassis.

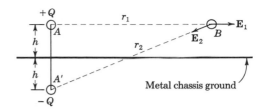

ground spacing, since such increase of spacing increases L and decreases C. Such a maneuver is both ineffective and inadvisable. It is ineffective because, as we have seen, the line parameters are remarkably insensitive to spacing and the separation corresponding to $Z_o = 1,000$ Ω is unrealistically large. It is also inadvisable because raising a lead well above the metal chassis reduces the shielding effect between leads which results from the presence of the chassis metal. We now look into this matter of shielding.

In Fig. 3-19 A and B are two wires running above a ground plane. It is intended that these wires should have no coupling one to the other. But of course there is some unavoidable capacitive coupling between them simply because of their proximity to one another. To examine this coupling, let us consider that a charge $+Q$ per unit length is placed on A. Then the magnitude of the electric field at B due to Q would be $E_1 = Q/2\pi\epsilon r_1$ if the ground plane were not present. The net effect of the induced charges on the ground plane is replaceable by an *image* wire of charge $-Q$ located as shown, and this charge causes at B in Fig. 3-19 an electric field E_2 whose magnitude is $Q/2\pi\epsilon r_2$. The net field at B depends now on the vector difference $\mathbf{E}_1 - \mathbf{E}_2$. The magnitude of this resultant field intensity decreases as h decreases. Thus even though the metal chassis does not project upward between wires A and B, its presence in their neighborhood furnishes a measure of shielding provided that the wires are *close* to the chassis. In the same way, if a current flows in wire A, an image current flows in the opposite direction at A', and the net magnetic coupling with the circuit of which B is a part is thereby reduced. As a matter of fact, it is this shielding ability of the presence of a large mass of metal that constitutes a most important reason why circuits are assembled on a metallic chassis. And while it now appears that we may deliberately choose to mount a wire close to a chassis, it hardly need be emphasized that generally every advantage lies with making leads as short as possible.

3-14   NONUNIFORM LINES[4,6]

The low values of characteristic impedance obtainable in lines of uniform cross section are often as much an inconvenience as is the short delay per meter. For example, consider that we are required to transmit a 10-V pulse along a line. If $Z_o = 50$ Ω, the generator must supply a 200-mA peak current, whereas if $Z_o = 1,000\ \Omega$ only 10 mA is required. Accordingly, in such applications, the higher-impedance line has a distinct advantage over the lower-impedance cable.

From Eqs. (3-34) it follows that both T and Z_o can be increased if a constructional change is made which increases L. One such modification is to replace the straight center conductor by a continuous coil of wire in the form of a helix. In a cable with this configuration the cross section is not uniform and the product LC is no longer equal to $\mu\epsilon$. The inductance may be further increased if the inner conductor is wound upon a ferromagnetic core. Such delay lines are described in Appendix B, which also lists commercially available cables with values of T and Z_o as large as 1 μsec/ft and 4,000 Ω, respectively. We note that the delay has been increased by a factor of 1,000 over that of an air-dielectric cable and the impedance has been multiplied by about 100.

A given delay can often be obtained with less attenuation and in a smaller volume with a lumped-parameter line than with a distributed-parameter line. The design of such lumped delay networks is given in Appendix C. Several manufacturers supply physically small lumped-parameter lines having fixed delays in standard values up to about 20 μsec, with impedances in the range from 50 Ω to 10 K, and with the ratio of delay to rise time of the order of 10. Larger physical units are available with delays up to 200 msec and with rise times which are less than 3 percent of the delay.[8]

3-15   REFLECTIONS ON TRANSMISSION LINES

It will be recalled[5,6] that the general solution for the voltage v and current i on an ideal (lossless) transmission line is given by

$$v = f_1\left(t - \frac{x}{u}\right) + f_2\left(t + \frac{x}{u}\right) \tag{3-39}$$

and

$$i = \frac{1}{R_o}\left[f_1\left(t - \frac{x}{u}\right) - f_2\left(t + \frac{x}{u}\right)\right] \tag{3-40}$$

The positive assumed directions of v and i are indicated in Fig. 3-20. The characteristic impedance of the line is R_o, and u is the propagation velocity. The function f_1 is an arbitrary function of the argument $t - x/u$ and represents a wave traveling to the right (in the positive x direction) with velocity u. Similarly, f_2 represents a wave traveling to the left. For a wave traveling to the right, $v/i = R_o$, whereas for a wave moving to the left, $v/i = -R_o$. This difference in sign results simply from the fact that in both cases the assumed positive current direction is as shown in Fig. 3-20. The general solution for wave propagation on a transmission line consists in combining a wave traveling to the right with a wave traveling to the left in such a way that

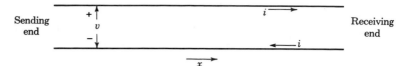

Fig. 3-20   Sign conventions for current and voltage on a transmission line.

the boundary conditions at the sending and receiving ends are satisfied (at each
end of the line the ratio v/i must equal the terminating resistance).   We shall
now illustrate this principle by applying it to a number of important special
cases.

Infinite Line   Assume that a unit step $U(t)$ is applied to the sending end
of a line which is arbitrarily long, so that the conditions at the receiving end
need never be considered.   Then the boundary conditions are obviously satisfied by taking

$$v = U\left(t - \frac{x}{u}\right) \qquad i = \frac{1}{R_o} U\left(t - \frac{x}{u}\right) \qquad\qquad (3\text{-}41)$$

It is understood, from the definition of the unit step, that $U(t - x/u)$ is zero
whenever the argument is negative.   The voltage distributions along the line
at two successive times are shown in Fig. 3-21.   The abrupt discontinuity in
voltage travels down the line with velocity u.

Finite Line Terminated in Its Characteristic Impedance   An additional
boundary condition now must be satisfied at the termination where v/i must
equal R_o.   But the solution given by Eqs. (3-41) already satisfies this additional condition, so that the voltage and current on the line remain as before
(for an applied unit step).   In general, a line terminated in its characteristic
impedance behaves as an infinitely long line.

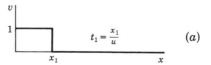

Fig. 3-21   The voltage distribution along
an infinite line at two particular instances
of time t_1 and t_2, with $t_2 > t_1$.

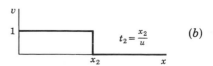

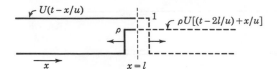

Fig. 3-22   Incident and reflected waves at a termination with $R > R_o$ for $t > l/u$.

Finite Line Terminated in $R \neq R_o$ The boundary condition at the termination is no longer satisfied by Eqs. (3-41). It is now required that at the termination the ratio v/i equal R rather than R_o. Hence, we must now find a combination of waves traveling to the right and to the left which will satisfy the boundary condition. The circumstances which exist at the termination of the line ($x = l$) for the case of a resistive termination $R > R_o$ are shown at a time $t > l/u$ in Fig. 3-22. The incident wave of voltage $U(t - x/u)$ has progressed to the point where the discontinuity has passed beyond the end of the line. The second or reflected wave is represented by $\rho U(t - 2l/u + x/u)$ and is one which travels from right to left and whose discontinuity passes $x = l$ at $t = l/u$. (Of course, it is understood that the dashed portions of the waves to the right of $x = l$ do not actually exist, because the line ends at $x = l$.) The constant ρ is called the *reflection factor*. For times $t \geq l/u$, the net voltage at the termination is $1 + \rho$. The current associated with the original wave is $1/R_o$ flowing to the right. The current associated with the reflected wave is ρ/R_o flowing to the left. The net current is $(1 - \rho)/R_o$ flowing to the right. If the termination is R, then it is required that

$$\frac{1 + \rho}{(1 - \rho)/R_o} = R$$

or

$$\rho = \frac{R/R_o - 1}{R/R_o + 1} \tag{3-42}$$

This result for ρ, which measures the ratio of the amplitudes of the two voltage waves, is consistent with our expectation that $\rho = 0$ if $R = R_o$. We also note that ρ is positive if $R > R_o$, whereas the reflected voltage wave is inverted (ρ is negative) if the terminating resistance is less than the characteristic

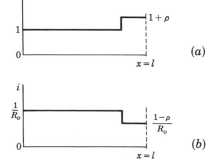

(a)

(b)

Fig. 3-23   The voltage (a) and current (b) distributions along a line with a termination $R > R_o$ for $t > l/u$.

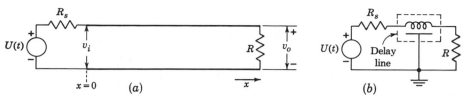

Fig. 3-24 (a) A generator with an output impedance R_s at the sending end of a line. (b) The same circuit drawn using the standard symbol for a delay line.

resistance $(R < R_o)$. The voltage and current distributions along the line for a particular instant of time, $t > l/u$, are shown in Fig. 3-23. The net voltage and current at any point on the line result from the simultaneous existence of the incident and reflected waves. The time $l/u \equiv t_d$ which it takes the wave to travel down the entire length of the line is called the *one-way delay time.*

To summarize, for voltage waves,

$$-1 \leq \rho \leq +1 \qquad \text{for any } R$$

$$\rho = +1 \qquad \text{if } R = \infty \text{ (an open-circuited line)}$$

$$\rho = -1 \qquad \text{if } R = 0 \text{ (a short-circuited line)}$$

$$\rho > 0 \qquad \text{if } R > R_o$$

$$\rho < 0 \qquad \text{if } R < R_o$$

$$\rho = 0 \qquad \text{if } R = R_o$$

If one end of the line is terminated in R_o, then when the discontinuity reaches this termination it is completely "absorbed" and no additional reflections result.

In the case where the generator at the sending end provides a voltage $v_s(t)$ and has a source impedance R_s, the amplitude of the wave which starts down the line is easily calculated. In Fig. 3-24, the ratio of voltage to current on the line is R_o until the discontinuity of the first wave reaches the termination R. Hence, at time $t = 0$, the impedance seen looking to the right is R_o and the amplitude of the wave is

$$v_1(t) = \frac{R_o}{R_o + R_s} v_s(t)$$

3-16   A SHORTED OR AN OPEN LINE

Consider a generator of a step voltage V and impedance $R_s = R_o$ connected to a line which is short-circuited at the receiving end, as indicated in Fig. 3-25a. What is the appearance of the voltage waveform at the sending end? At $t = 0$

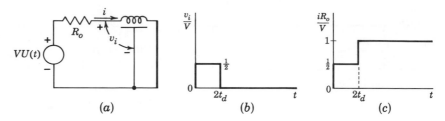

Fig. 3-25 (a) A step voltage applied to a short-circuited line from a generator whose impedance matches that of the line; (b) the resulting voltage v_i and (c) the resulting current i at the input of the line.

a step $VR_o/(R_o + R_s) = V/2$ appears at $x = 0$. This discontinuity travels to the shorted end, where a second discontinuity $-V/2$ (since $\rho = -1$) will start toward the left. When this second edge reaches the input end, it will add a voltage $-V/2$ to the voltage $+V/2$ established previously. The resultant waveform will be a pulse of amplitude $V/2$ and duration $2t_d$, as indicated in Fig. 3-25b. The advantage of producing a pulse in this manner is that the duration depends only on passive elements (the L and C of the line) and thus may have a stability not shared by pulse generators which depend upon active elements. The initial current is $V/2R_o$. This current discontinuity is reflected as $-\rho V/2R_o = +V/2R_o$, so that at time $t \geq 2t_d$ the input current is $V/2R_o + V/2R_o = V/R_o$, as it should be, since the steady-state voltage at the input to the line is zero. The current waveform is indicated in Fig. 3-25c.

Attenuation In the above discussion we have neglected the attenuation of the line, which we shall now take into account. Consider the circuit of Fig. 3-25 again. The initial discontinuity at the input end will arrive at the shorted end as $\frac{1}{2}V\epsilon^{-a}$, where $a = \zeta l$, ζ is the attenuation factor, and l the length of line. At $t = 2t_d$, a negative step of amplitude $\frac{1}{2}V\epsilon^{-2a}$ will appear at the input end and the resultant wave will be as in Fig. 3-26a. We see that a small step voltage or "pedestal" v' remains after the pulse.

The above result will yield an expression for ζ in terms of the d-c input resistance $R_\text{d-c}$ of the shorted distortionless line. Since for $t > 2t_d$ there are no further discontinuities, v' may be calculated from

$$v' = \frac{VR_\text{d-c}}{R_\text{d-c} + R_o} \approx V\frac{R_\text{d-c}}{R_o}$$

in which we have taken into account the fact that on any practically useful line $R_\text{d-c} \ll R_o$. From Fig. 3-26a we see that

$$V\frac{R_\text{d-c}}{R_o} + \frac{V}{2}\epsilon^{-2a} = \frac{V}{2}$$

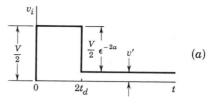

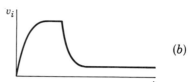

(a)

Fig. 3-26 The voltage at the input to the line in Fig. 3-25 when attenuation is taken into consideration if the input waveform is (a) an abrupt step and (b) an exponential rise. (c) An exponential input step shaped by a double delay-line differentiator (if an amplifier with an odd number of stages is used, then the waveform should be inverted).

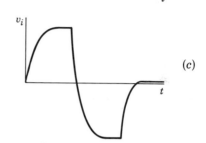

(b)

(c)

Assuming small attenuation, so that $\epsilon^{-2a} \approx 1 - 2a$, we find

$$a = \zeta l = \frac{R_{\text{d-c}}}{R_o} \tag{3-43}$$

Delay-line Differentiation If the generator in Fig. 3-25 supplies an exponential rise of voltage, the waveform at the input to the line is as pictured in Fig. 3-26b. It is assumed that the exponential time constant τ is short compared with the delay time t_d of the line. The waveform in Fig. 3-26b follows from that of Fig. 3-26a, which gives the response for an ideal step applied to a shorted line. If an exponential waveform is applied to an RC differentiating circuit ($RC < \tau$), then the output (Fig. 2-9) resembles that of Fig. 3-26 in the sense that a step voltage has been converted into a pulse. Hence, a shorted transmission line is referred to as a *delay-line differentiator*. This terminology is further explained below.

If the waveform of Fig. 3-26b is applied through a buffer amplifier to a second shorted line, the result is that indicated in Fig. 3-26c, which should be compared with Fig. 2-13. Note that this double-delay-line differentiator almost eliminates the pedestal due to the cable attenuation and gives a waveform whose average value is approximately zero. Double-delay-line shaping is an important technique used in nuclear pulse spectrometry.[9]

We shall now explain how a length of delay line, shorted at the receiving end, may be used to accomplish much the same function as is served by a differentiating circuit.

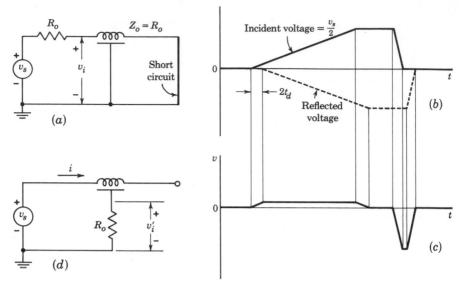

Fig. 3-27   (a) A delay-line differentiator using a short-circuited line; (b) the incident and reflected waveforms; (c) the resultant waveform at the input to the line; (d) a delay-line differentiator using an open-circuited line.

Consider the situation pictured in Fig. 3-27a, which consists of the same circuit as in Fig. 3-25a but excited by a signal waveform v_s. To illustrate the point we intend to make, a signal waveform has been selected which consists of a portion where the voltage changes slowly with time and another portion where the voltage changes rapidly with time.   The waveform has the shape shown by the solid curve in Fig. 3-27b.   Initially the line behaves like a resistance equal to its characteristic impedance Z_o.   The voltage v_i which travels down the line is $v_s/2$, which has the same shape as the input waveform. However, the short circuit at the line end causes the appearance at the line input of a delayed and inverted voltage waveform, shown by the dashed curve in Fig. 3-27b.   We shall assume no attenuation, so that the dashed reflected waveform is equal in amplitude to the incident signal.   If the line introduces a delay t_d, the dashed waveform is delayed by $2t_d$.   The resultant waveform of v_i is given by the sum of the two waveforms in Fig. 3-27b and is shown in Fig. 3-27c.

We now observe a marked similarity between this resultant waveform and the waveform that would result from transmitting the input signal through a differentiating network.   The principal difference in the two cases is that, with the differentiating circuit, portions of the waveform of Fig. 3-27c that appear as linear rises and falls would be replaced by rising and falling exponentials.   Like the differentiating circuit, the delay-line circuit gives a larger response when the input changes rapidly and a smaller response when the input changes slowly. When a ramp signal of slope α is applied to a differentiating network, the response, after an initial transient, is constant and equal to $\tau\alpha$, where τ is the circuit time constant ($\tau = RC$ or $\tau = L/R$, depending on the circuit). The corresponding response for the delay-line differentiator is $2t_d\alpha$. Qualitatively, we may say that a differentiating network is one which will not transmit a signal unless it is changing rapidly. Quantitatively we have that, for a ramp, the output signal is equal to the magnitude of the input change in one time constant. Qualitatively, the shorted delay line acts as a short circuit except when the signal changes rapidly. Quantitatively we have the result that, for a ramp input, the output is equal to the magnitude of the input change in the time $2t_d$.

A type of waveform which is frequently encountered is one in which there occurs periodically an abrupt change in an otherwise slowly varying shape. If this waveform is passed through a differentiating circuit, the waveform generally will be "washed out" and only the response to the abrupt change will stand out. In this way, the signal may be converted to a waveform that is essentially a train of pulses. If the differentiation is achieved in a lumped single-time-constant circuit, the pulse train will consist of a series of spikes, as in Fig. 2-6, with a sharp leading edge, an exponentially falling trailing edge, and a top which is pointed rather than flat. In such an application the use of a delay-line differentiator has obvious merit because the pulses will be much more nearly rectangular and will have a duration which is controllable by an adjustment of the length of the delay line. Nanosecond pulse generators making use of delay-line differentiation are discussed in Secs. 13-17 and 20-8.

Another form of delay-line differentiator using an open-circuited line is indicated in Fig. 3-27d. If the signal voltage v_s has the waveform in Fig. 3-27b, then the voltage $v_i' = iR_o$ across the input resistance R_o has the form in Fig. 3-27c. This statement follows from the fact that in Fig. 3-27d the current wave i is inverted at the open-circuited end, just as the voltage wave in Fig. 3-27a is inverted at the short-circuited end. Clearly, the circuit in Fig. 3-27d is the dual of that in Fig. 3-27a.

3-17   MULTIPLE REFLECTIONS

Let $v_1(t)$ be the voltage wave which starts down the line at $t = 0+$. At $t = t_d$ this incident wave reaches the end of the line $x = l$ and a reflected voltage wave $v_2 = \rho v_1$ (and a reflected current wave $-\rho v_1/R_o$) start back along the line. At $t = 2t_d$ this first reflection reaches the beginning of the line $x = 0$. The condition for no reflection at the sending end is that the ratio of voltage to current at this termination be R_o. If the generator impedance does not equal the characteristic impedance of the line, then this boundary condition is not satisfied. We must therefore postulate, for any other termination, the existence of a third voltage wave $v_3 = \rho' v_2$ which starts to the right

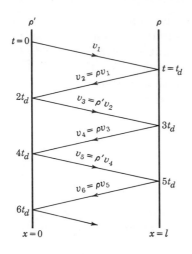

Fig. 3-28 Reflection chart of voltage waves traveling between sending end at $x = 0$ with reflection coefficient ρ' and receiving end at $x = l$ with reflection coefficient ρ. The same chart may be used for current waves provided that each voltage is divided by R_o and the signs of the reflection coefficients ρ and ρ' are changed.

from $x = 0$ at $t = 2t_d+$. The reflection factor ρ' for the input-end termination is given by Eq. (3-42), where $R = R_s$ is the sending-end resistance. This third wave is the reflection of the second wave and will in turn produce a reflection v_4 at the receiving end, and so on indefinitely. This sequence of events is conveniently represented in the reflection chart of Fig. 3-28. *The resultant voltage is the algebraic sum of all the individual reflection components.* For example, the voltage at $x = 0$ for $0 < t < 4t_d$ is the sum of v_1 (starting at $t = 0$), v_2 (starting at $t = 2t_d$), and v_3 (starting at $t = 2t_d$). If the signal source is a ramp generator αt of internal resistance R_s, then with $k = \alpha R_o/(R_o + R_s)$, $v_1 = kt$. To this must be added v_2, which is a ramp of slope ρk starting at $t = 2t_d$ and v_3, another ramp, also passing through $t = 2t_d$ and of slope $\rho' \rho k$. The individual components and the resultant are indicated in Fig. 3-29 for the special case where $\rho = -\frac{1}{2}$ and $\rho' = +\frac{1}{3}$. The complete waveform at $x = 0$ is seen to be a broken line function of time, with the changes in slope occurring at $t = 2t_d,\ 4t_d,\ 6t_d,\ \ldots\ .$

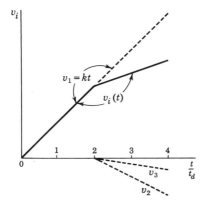

Fig. 3-29 The voltage $v_i(t)$ at the sending end $(x = 0)$ of a line for $0 < t < 4t_d$ if the source is a ramp generator and $\rho = -\frac{1}{2}$ and $\rho' = +\frac{1}{3}$.

$$v_2 = \rho v_1(t - 2t_d)U(t - 2t_d)$$
$$= -\tfrac{1}{2}k(t - 2t_d)U(t - 2t_d)$$
$$v_3 = \rho'v_2(t - 2t_d)U(t - 2t_d)$$
$$= -\tfrac{1}{6}k(t - 2t_d)U(t - 2t_d)$$

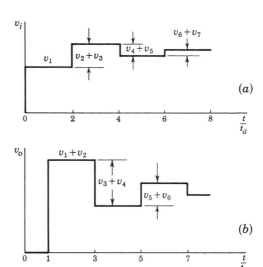

Fig. 3-30 The input to a line is a step voltage. The waveforms at the input (a) and output (b). The figure is drawn for $\rho = +1$ and $\rho' = -\frac{1}{2}$.

If the signal source is a step generator, then the resultant waveform is a staircase function of time with steps up and down. At the sending end the steps start at $t = 0$, $2t_d$, $4t_d$, $6t_d$, . . . , and at the receiving end at $t = t_d$, $3t_d$, $5t_d$, . . . , as indicated in Fig. 3-30. For example, the resultant constant voltage v_i for $4t_d < t < 6t_d$ is seen from Fig. 3-30a to be the algebraic sum $v_1 + v_2 + v_3 + v_4 + v_5$. Some of the v's may be negative.

If the input is a pulse (whose width t_p is less than $2t_d$), then the output will be a train of pulses. At $x = 0$ the pulse beginning at $t = 4t_d$ will be equal to $v_4 + v_5$. This result follows from the fact that the waveforms v_1, v_2, and v_3 are all zero for $t > 4t_d$, since $t_p < 2t_d$. It is possible for some of the pulses to be inverted, as in the following illustration.

EXAMPLE A pulse generator whose impedance is $R_s = 600$ Ω delivers a 2-μsec 8-V pulse to a 1,000-Ω line terminated in 9,000 Ω. The one-way delay of the cable is 5 μsec. Find the voltage waveforms at the input and output of the line.

Solution The magnitude of the pulse applied to the line is

$$V_1 = \frac{8R_o}{R_o + R_s} = \frac{8 \times 1,000}{1,000 + 600} = 5.00 \text{ V}$$

From Eq. (3-42)

$$\rho = \frac{R/R_o - 1}{R/R_o + 1} = \frac{9 - 1}{9 + 1} = +0.80$$

and

$$\rho' = \frac{0.6 - 1}{0.6 + 1} = -0.25$$

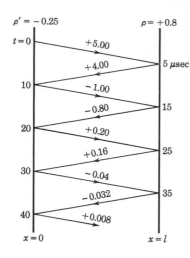

Fig. 3-31   Reflection chart for illustrative example.

The reflection chart for these specific values (corresponding to Fig. 3-28) is given in Fig. 3-31.   From this chart and the above theoretical discussion it follows that the pulse trains at the sending and receiving ends of the line are as pictured in Fig. 3-32.

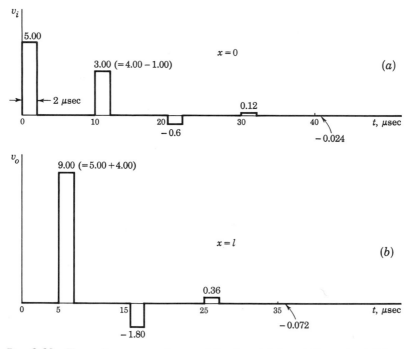

Fig. 3-32   The voltage waveforms at the input (a) and the output (b) of the illustrative example.

If the attenuation ζ of the line is appreciable, then it is taken into account in the reflection chart by multiplying each pulse amplitude by $\epsilon^{-\zeta l}$ when the pulse travels a distance l.   If we are interested in the current, rather than the voltage, waveform, then we must remember that a current pulse i is reflected at a termination as a pulse $-\rho i$.

3-18   DISCHARGE OF AN INITIALLY CHARGED LINE

Figure 3-33 indicates a line charged to a voltage V with the switch S closed. If S is opened at $t = 0$, what is the voltage (or current) waveform at the beginning or the end of the line?   To answer this question we must first find the magnitude of the initial voltage step traveling down the line.   Then we can draw a reflection chart and proceed as before.

At $t = 0-$ the line voltage is V and the line current is $i(0-) = V/R_1$. At $t = 0+$ the voltage at the input to the line will change to another value $v_i = V'$, which can be determined as follows.   The current in R_2 at $t = 0+$ is $i' = V'/R_2$, and from Fig. 3-33 this must equal the negative of the input line current, or $i(0+) = -i' = -V'/R_2$.

The first voltage step V_1 which must travel down the line equals the change in input voltage, or $V_1 = V' - V$.   The corresponding input-current step is $i(0+) - i(0-) = -(V'/R_2) - (V/R_1)$.   Since the ratio of voltage to current for a traveling wave must equal the characteristic impedance R_o,

$$V' - V = R_o\left(-\frac{V'}{R_2} - \frac{V}{R_1}\right) \tag{3-44}$$

If this equation is solved for V' and then $V_1 = V' - V$ is evaluated, we find

$$\frac{V_1}{V} = -\frac{R_2/R_1 + 1}{R_2/R_o + 1} \tag{3-45}$$

Note that, independently of R_2, if $R_1 = R_o$, then $V_1 = -V$ and

$$V' = V_1 + V = 0$$

This result is consistent with the fact that there can be no reflection if a line is terminated in its characteristic impedance.   Hence the wave traveling down the line must wipe out the voltage on the line.   Since *the net voltage at any distance along a line is the sum of the initial voltage V plus all the traveling waves which have reached that point, then, for $0 < t < t_d$, $V + V_1 = 0$ or $V_1 = -V$*

Fig. 3-33 With the switch S closed the line is charged to a voltage V. Then S is opened at $t = 0$.

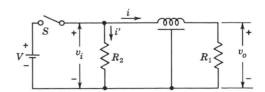

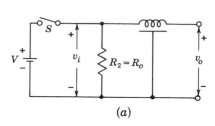 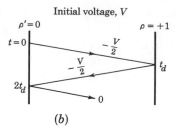

Fig. 3-34   (a) Open-circuited line charged to an initial voltage V with input end terminated in its characteristic resistance; (b) corresponding reflection chart.

At $t = t_d$ the wave reaches the end of the line, the line is completely discharged, and there is no reflected wave. However, if $R_1 \neq R_o$, then there will be a reflection at $t = t_d$.

EXAMPLE   A line charged to a voltage V is terminated at its input end in its characteristic impedance and at its output end is open-circuited as in Fig. 3-34a. When the voltage source V is removed find the voltage waveforms at both ends of the line.

Solution   Since $R_1 = \infty$ and $R_2 = R_o$, then $\rho = +1$, $\rho' = 0$, and from Eq. (3-45)

$$\frac{V_1}{V} = -\frac{0+1}{1+1} = -\frac{1}{2}$$

The reflection chart is given in Fig. 3-34b and the waveforms in Fig. 3-35. The initial voltage at $t = 0+$ is $V + V_1 = V - V/2 = V/2$. The discontinuity

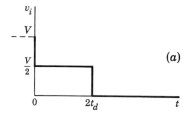

(a)

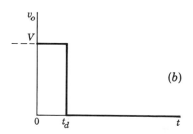

(b)

Fig. 3-35   Waveforms at input (a) and output (b) of the line of Fig. 3-34, which is charged to a voltage V for $t < 0$.

$V_1 = -V/2$ travels down the line, discharging it to half voltage as it progresses. At the end of the line at $t = t_d$ the discontinuity is reflected as $-V/2$ (since $\rho = 1$), and it discharges the line to zero as it moves toward the beginning of the line. At $t = 2t_d$ the line is completely discharged. The resultant voltage across R_o is a pulse of amplitude $V/2$ and duration $2t_d$, as indicated in Fig. 3-35a.

For service as the switch S in Fig. 3-34 special types of relays are commercially available. The contacts of these relays are hermetically sealed in a container which also contains some liquid mercury and a gas such as hydrogen at a pressure as high as 250 atm. The mercury is drawn to the contact points by capillary action and the contact is made through the mercury. Since the metal contacts, which are wetted by the mercury, never actually touch, the relay is capable of millions of operations without wear on the contact points. The very high gas pressure prevents arcing at the contacts, both on making and on breaking. Accordingly, the time interval which elapses between the moment when the contact just begins to open and the time when it is completely open may be a small fraction of a nanosecond. The repetition rate of the closing or opening operation is limited, as in any mechanical relay. Repetition rates up to several hundred per second are possible. A number of commercial nanosecond pulse generators[6] are available in which the waveforms are generated as in Fig. 3-34 by using a mercury relay in connection with a delay line. The polarity and amplitude of the pulse depend upon the charging voltage, and the pulse width is determined by the line length. Repetition rates as high as several megacycles per second are possible if the mechanical relay is replaced by an avalanche transistor (Sec. 13-17).

A charged line may be used as a *pulse stretcher*.[10] Consider, for example, a lumped line open-circuited at the output and terminated in R_o at the input. Each capacitor is charged simultaneously from a pulse of width t_p through an emitter follower and buffer diodes. At the end of the input pulse the line starts to discharge, but the output will remain constant for a time t_d, as indicated in Fig. 3-35b. Hence, the input width t_p has been *stretched* to t_d.

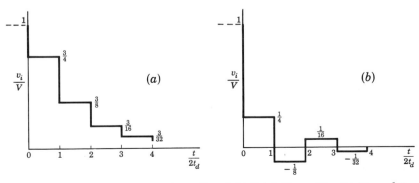

Fig. 3-36 The voltage across R_2 in Fig. 3-34 if (a) $R_2 = 3R_o$, (b) $R_2 = \frac{1}{3}R_o$.

If neither R_1 nor R_2 in Fig. 3-33 equals R_o, then there will be multiple reflections from each end of the line. The voltages across R_2 for the two special cases $R_2 = 3R_o$ ($\rho = +\frac{1}{2}$) and $R_2 = \frac{1}{3}R_o$ ($\rho = -\frac{1}{2}$) are indicated in Fig. 3-36 for $R_1 = \infty$.

3-19   REFLECTIONS FROM REACTIVE TERMINATIONS[6]

If the line ends in an impedance Z consisting of some combination of reactive and resistive elements, then the reflection coefficient is given by

$$\rho(s) = \frac{Z(s)/R_o - 1}{Z(s)/R_o + 1} \tag{3-46}$$

where $Z(s)$ is the terminating impedance, with the reactance of each inductor L represented by Ls and that of each capacitor by $1/Cs$. If $v_1(s)$ is the Laplace transform of the first wave traveling down the line, then the wave reflected from the termination and starting at $t = t_d$ is the inverse Laplace transform of $\rho(s)v_1(s)$. In other words, the reflection chart of Fig. 3-28 continues to be valid, provided that ρ, ρ', and the v's are written as functions of s and the inverse Laplace transform is taken to obtain the time waveforms.

EXAMPLE A unit step generator is matched to a line of characteristic impedance R_o which is terminated in a capacitor C, as in Fig. 3-37. Find the voltage v_i at the input to the line and v_o at the termination.

Solution The incident wave is $\frac{1}{2}U(t) = v_1(t)$ and $v_1(s) = 1/2s$. The reflection coefficient in the s plane is

$$\rho(s) = \frac{1/sCR_o - 1}{1/sCR_o + 1} = \frac{-s + 1/CR_o}{s + 1/CR_o} \tag{3-47}$$

The reflected wave is

$$v_2(s) = \rho(s)v_1(s) = \frac{-s + 1/CR_o}{2s(s + 1/CR_o)} \tag{3-48}$$

Taking the inverse Laplace transform yields

$$v_2(t) = \frac{1}{2} - \epsilon^{-t'/R_oC} \tag{3-49}$$

where we have written $t' \equiv t - t_d$ to remind us that this reflection starts at $t = t_d$ or $t' = 0$. Since $R_s = R_o$, $\rho' = 0$ and hence $v_3 = 0$. In other words, there are no reflections beyond the first originating at the end of the line.

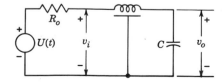

Fig. 3-37 The termination of the line is a capacitor C.

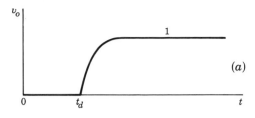

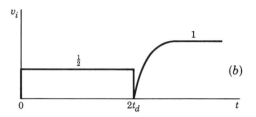

Fig. 3-38 The waveforms (a) at the output and (b) at the input of a line excited from a matched step generator if the line terminates in a capacitor.

The voltage v_o at the end of the line is zero until $t = t_d$ and then is $v_1 + v_2 = \frac{1}{2} + v_2 = 1 - \epsilon^{-t'/RC}$. This waveform is plotted in Fig. 3-38a. The waveform at the input to the line is v_1, a constant of magnitude $\frac{1}{2}$ until $t = 2t_d$. From then on $v_i = v_1 + v_2 = 1 - \epsilon^{-t''/RC}$, where $t'' \equiv t - 2t_d$. The waveform is plotted in Fig. 3-38b. Note that the capacitor voltage starts at zero and ends with a magnitude equal to that of the generator (unity).

If the termination in Fig. 3-37 consists of a resistor R in parallel with C, then the waveforms in Fig. 3-38 remain valid with the following modifications. The steady-state voltage is $R/(R + R_o)$ instead of unity and the time constant is $CRR_o/(R + R_o)$ instead of CR_o (Prob. 3-67).

If the termination of the line in Fig. 3-37 is an inductor in series with a resistor R, the waveforms can be calculated to be those indicated in Fig. 3-39

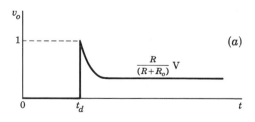

Fig. 3-39 The waveforms of the line of Fig. 3-37 terminated in a series combination of an inductor L and resistor R. The time constant of the exponential portions is $L/(R + R_o)$.

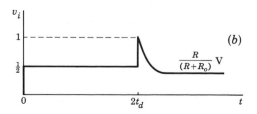

(Prob. 3-66). Note that if $R = 0$, the steady-state voltage is zero, which is obviously correct since an inductor cannot sustain a constant voltage. If $R = R_o$, then the steady-state voltage is $\frac{1}{2}$ V.

3-20   "TRANSMISSION–LINE" PULSE TRANSFORMERS

When a signal is transmitted from a generator to a load over a pair of wires, the signal waveform is distorted by the combination of the distributed shunt capacitance between wires and inductance in series with the wires. We find, however (Secs. 3-12 and 3-13), that if the inductance and capacitance are "balanced" against one another for a particular load, as in the case of a uniform transmission line, the signal may be transmitted to the load without distortion and without reflection at the load if the line is properly terminated. In being transmitted through a pulse transformer, fast waveforms are distorted by the shunt capacitance and leakage inductance of the windings. We shall now show that, in special cases, it is possible to arrange the transformer windings mechanically in such a manner that the signal is transferred from generator to load as if over a transmission line. Under these circumstances distortion and reflections are minimized.

1 : 1 Inverting Transformer In the circuit of Fig. 3-40a a generator is supplying power to a load through a transmission line. In Fig. 3-40b the line is bent (with a radius which is large compared with the spacing between wires), so that the two points P' and S are brought close to one another. We now join P' and S and connect this common point to ground. This connection between P' and S perturbs the power flow from source to load only slightly since the traveling wave carrying the power is guided by the wires and travels in the direction indicated by the arrow.

Grounding of P' and S, though it does not affect appreciably the power delivered to the load, does introduce one complication. For example, consider in Fig. 3-40b that v is a low-frequency source. Then the complete low-impedance loop $PP'SP$ around the outside coil will require of the source that it provide a large current which will then circulate around this loop. To minimize this circulating current we may wind the transmission line around a leg of a magnetic circuit. The inductance of the outer coil will be increased and the circulating current decreased thereby. If the line is long enough, we may wind it many times around a magnetic toroid, as shown in Fig. 3-40c. Note in Fig. 3-40c that P' and S are both grounded, making these points common, as in Fig. 3-40b. Also, as in Fig. 3-40b, the source is connected between P and S $(= P')$ and the load between S' and P' $(= S)$. But in Fig. 3-40c it is apparent that the source is connected to a primary winding P-P', whereas the load is connected to a secondary S-S'. In this transformer the windings of the primary and the secondary lie side by side and may even be twisted together. The entire arrangement has finally been redrawn in Fig. 3-40d,

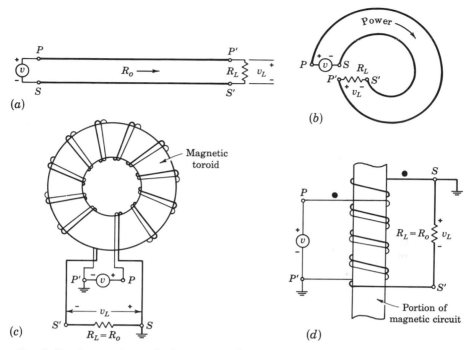

Fig. 3-40 Concerning a 1:1 inverting "transmission-line" transformer, (a) a source supplies power to a load over a delay line; (b) the line is curved to place the source and load close together; (c) the line is wrapped around a magnetic core; (d) the transformer in (c) drawn in conventional form.

where it looks more like a conventional representation of an inverting transformer. The dots assigned to the windings are consistent with the fact that the windings encircle the magnetic core in the same direction.

In summary, we have established that a 1:1 inverting transformer constructed in the manner described above actually conveys power from source to load along a transmission line. If the line is matched, the transformer capacitance and leakage inductance do not affect the transmission, and the rise time of the transformer is now determined by secondary effects not considered in the above discussion. The flat-top response is determined, as in a conventional transformer (Sec. 3-7), in part by the magnetizing inductance, which in turn depends upon the number of times the line is wrapped around the toroid and upon the permeability of the iron. The constructional details of a nanosecond "transmission-line" pulse transformer are given in Sec. 3-10.

Assume now that we have no interest in polarity inversion. Then, returning to Fig. 3-40a, we would ground S and S', but we would not be able to operate source and load at different d-c voltages. Accordingly we now

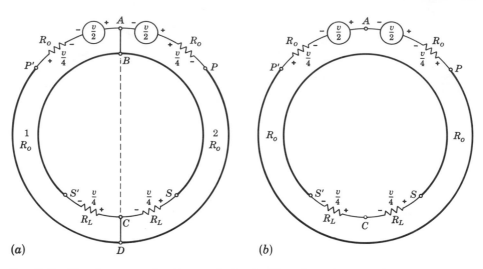

Fig. 3-41 Development of a "transmission-line" noninverting transformer.

need a noninverting transformer with d-c isolation between primary and secondary.

1:1 Noninverting Transformer Starting with a set of transmission lines, we shall now show how to construct a 1:1 noninverting transformer with d-c isolation. We might imagine that if we devise a transmission-line transformer with d-c isolation we might make of it either an inverting or a noninverting transformer simply by moving one ground lead. We shall see that such is not the case.

In Fig. 3-41a are drawn two independent transmission lines, 1 and 2, which have been curved so that they may meet at sending and receiving ends. If an equipotential plane is placed perpendicular to the plane of the page and along the line $ABCD$, then the electric and magnetic field lines will remain unchanged throughout space. This statement follows from the fact that system 2 is the electromagnetic image of system 1 in the equipotential (ground) plane. We may note now, for future reference, that points A and C are at the same potential.

Each line has a source $v/2$ and is matched, $R_L = R_o$, at both sending and receiving ends. Because of the generator polarities and the symmetry of the arrangement it is apparent that the leads AB and CD carry no current. The circuit has been redrawn in Fig. 3-41b with the leads AB and CD deleted.

We see that we have here a 1:1 transformer in which the primary and secondary are isolated. Observe the most interesting conclusion which must be drawn from Fig. 3-41b, namely, that although each line has a characteristic impedance R_o, proper matching requires a source and load resistor equal to $2R_L = 2R_o$.

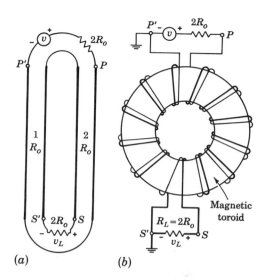

Fig. 3-42 (a) Two lines connected in series at the sending and receiving ends constitute a transformer; (b) physical realization of the "transmission-line" 1 : 1 noninverting transformer.

We noted earlier that there is zero voltage between A and C. Hence there is also zero potential between P' and S' because the voltage drop from A to P' is the same as the drop from C to S'. Therefore, we are free to make a common ground connection at P' and S'. In this case the voltage $v_{PP'} = v_{SS'}$ and the transformer is a 1:1 noninverting transformer. We are not at liberty to make an inverting transformer by grounding, say, P' and S because there is a voltage difference between these two points.

An alternative manner of arriving at the configuration of Fig. 3-41b is shown in Fig. 3-42. Here two lines, 1 and 2, each of characteristic impedance R_o, are placed side by side. The inputs to the lines are placed in series, and the input signal is applied to this combination. Similarly the receiving ends are placed in series, and the load is connected to this combination. The matching resistance must equal $2R_o$, which is the sum of the characteristic impedances of the two lines.

In order to limit the current in the outer loops of Figs. 3-41b and 3-42a, it is necessary to wind the lines around a magnetic core. A physical realization of a 1:1 noninverting transformer is shown in Fig. 3-42b. Other transmission-line transformers are possible[3] in which there is an integral relationship in the voltage transformation ratio between primary and secondary.

REFERENCES

1. Lee, Reuben: "Electronic Transformers and Circuits," John Wiley & Sons, Inc., New York, 1947.

2. Glasoe, G. N., and J. V. Lebacqz: "Pulse Generators," Massachusetts Institute of Technology Radiation Laboratory Series, vol. 5, McGraw-Hill Book Company, New York, 1948.

3. Winningstad, C. N.: Nanosecond Pulse Transformers, *IRE Trans. Nucl. Sci.*, vol. NS-6, pp. 26–31, March, 1959.
   Ruthroff, C. L.: Some Broad-band Transformers, *Proc. IRE*, vol. 47, pp. 1337–1342, August, 1959.

4. Blackburn, J. F. (ed.): "Components Handbook," Massachusetts Institute of Technology Radiation Laboratory Series, vol. 17, sec. 1–12 and chap. 6, McGraw-Hill Book Company, New York, 1948.
   Kallmann, H. E.: High-impedance Cable, *Proc. IRE*, vol. 34, pp. 348–351, June, 1946.
   Anderson, J. R.: Electrical Delay Lines for Digital Computer Applications, *IRE Trans. Electron. Computers*, vol. EC-2, no. 2, pp. 5–13, June, 1953.

5. Johnson, W. C.: "Transmission Lines and Networks," chap. 1, McGraw-Hill Book Company, New York, 1950.

6. Lewis, I. A. D., and F. H. Wells: "Millimicrosecond Pulse Techniques," 2d ed., Pergamon Press, New York, 1959.

7. Winningstad, C. N.: Nanosecond Pulse Measurements, *IRE WESCON Conv. Record*, 1961.

8. AD-YU Electronics Laboratory Inc., Passaic, N.J.

9. Chase, R. L.: "Nuclear Pulse Spectrometry," chap. 2, McGraw-Hill Book Company, New York, 1961.

10. Craib, J. F.: Improved Pulse Stretcher, *Electronics*, vol. 24, no. 6, pp. 129–131, June, 1951.

4 / WIDEBAND AMPLIFIERS (UNCOMPENSATED)

Frequently the need arises in pulse systems for amplifying a signal with a minimum of distortion. Under these circumstances the active devices involved must operate linearly. In the analysis of such circuits the first step is the replacement of the actual circuit by a linear model. Thereafter it becomes a matter of circuit analysis to determine the distortion produced by the transmission characteristics of the linear network. In this sense the present discussion is an extension of the material of Chaps. 2 and 3 with the following important differences. Previously we were satisfied with simply observing the distortion introduced by various simple transmission networks for several representative waveforms. Now we shall be concerned with the problem of how the distortion may be minimized and how the signal may be amplified.

The frequency range of the amplifiers discussed in this chapter extends from a few cycles per second (hertz), or possibly from zero, up to hundreds of megahertz. The original impetus for the study of such wideband amplifiers was supplied because they were needed to amplify the pulses occurring in a television signal. Therefore, such amplifiers are often referred to as *video amplifiers*.

Basic amplifier circuits are discussed here. Modifications of these configurations to give improved characteristics are considered in the following chapter.

4-1 FREQUENCY RESPONSE OF AN AMPLIFIER[1]

A criterion which may be used to compare one amplifier with another with respect to fidelity of reproduction of the input signal is suggested by the following considerations. Any arbitrary waveform of engineering importance may be resolved into a Fourier spectrum. If the waveform is periodic, the Fourier spectrum will consist of a series of sines and cosines whose frequencies are all integral multiples of a fundamental frequency. The fundamental frequency is the reciprocal of the time which must elapse before the waveform repeats itself. If the waveform is not periodic, the fundamental period extends in a sense from a time $-\infty$ to a time $+\infty$. The fundamental frequency is then infinitesimally small, the frequencies of successive terms in the Fourier series differ by an infinitesimal amount rather than by a finite amount, and the Fourier series becomes instead a Fourier integral. In either case the spectrum includes terms whose frequencies extend, in the general case, from zero frequency to infinity.

Consider a sinusoidal signal of angular frequency ω represented by $V_m \sin (\omega t + \phi)$. If the gain of the amplifier has a magnitude A and if the signal suffers a phase lag θ, then the output will be

$$A V_m \sin (\omega t + \phi - \theta) = A V_m \sin \left[\omega \left(t - \frac{\theta}{\omega} \right) + \phi \right]$$

Therefore, *if the amplification A is independent of frequency and if the phase shift θ is proportional to frequency (or is zero) then the amplifier will preserve the form of the input signal, although the signal will be delayed in time by an amount $D = \theta/\omega$.*

This discussion suggests that the extent to which an amplifier's amplitude response is not uniform and the extent to which its time delay is not constant with frequency may serve as a measure of the lack of fidelity to be anticipated in it. In principle, it is really not necessary to specify both amplitude and delay response since, for most practical circuits, the two are related and, one having been specified, the other is uniquely determined. However, in particular cases, it may well be that either the time-delay response or amplitude response is the more sensitive indicator of frequency distortion.

Video amplifiers of either the transistor or tube variety are almost invariably of the RC-coupled type. For such a stage the frequency characteristics may be divided into three regions. There exists a range, called the *midband frequencies*, over which the amplification is reasonably constant and equal to A_o and over which the delay is also quite constant. In the second (low-frequency) region, below the midband, an amplifier stage will be shown to behave like a simple high-pass circuit of time constant τ_1 of the type depicted in Fig. 2-1. We have, then, that the ratio of the gain A_1 at low frequency to the midband gain A_o is [see Eq. (2-1)]

$$\left| \frac{A_1}{A_o} \right| = \frac{1}{\sqrt{1 + (f_1/f)^2}} \tag{4-1}$$

where $f_1 = 1/2\pi\tau_1$ is the lower 3-dB frequency, or half-power frequency. In the third (high-frequency) region, above the midband, the amplifier stage has as its equivalent circuit the low-pass combination of Fig. 2-14 with a time constant τ_2.   Hence the ratio of the gain A_2 at high frequency to the midband gain A_o is [Eq. (2-31)]

$$\left|\frac{A_2}{A_o}\right| = \frac{1}{\sqrt{1 + (f/f_2)^2}} \tag{4-2}$$

where $f_2 = 1/2\pi\tau_2$ is the upper 3-dB frequency.   The normalized time delays, D_1 and D_2, for the low- and high-frequency ranges, respectively, are given by

$$f_1 D_1 = \frac{f_1\theta_1}{\omega} = -\frac{1}{2\pi}\frac{f_1}{f}\arctan\frac{f_1}{f} \tag{4-3}$$

and

$$f_2 D_2 = \frac{f_2\theta_2}{\omega} = \frac{1}{2\pi}\frac{f_2}{f}\arctan\frac{f}{f_2} \tag{4-4}$$

In the above expressions θ_1 and θ_2 represent the angle by which the output lags the input, neglecting the initial 180° phase shift through the amplifier. The frequency dependence of the gains in the high- and low-frequency range is to be seen in Fig. 4-1.

The frequency range from f_1 to f_2 is called the *bandwidth* of the amplifier stage.   We may anticipate in a general way that a signal, all of whose Fourier components of appreciable amplitude lie well within the range f_1 to f_2, will

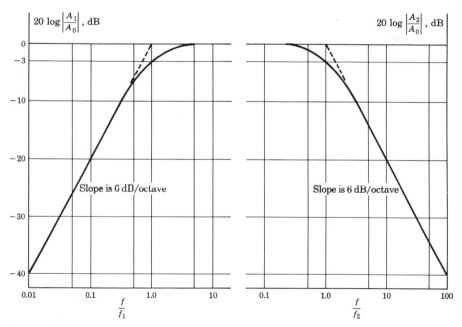

Fig. 4-1   A log-log plot of the gain characteristics of an RC-coupled amplifier.

pass through the stage without excessive distortion. This criterion must be applied, however, with extreme caution, as will be indicated later.

4-2   STEP RESPONSE OF AN AMPLIFIER

An alternative criterion of amplifier fidelity is the response of the amplifier to a particular input waveform. Of all possible available waveforms, the most generally useful is the step voltage. In terms of a circuit's response to a step, the response to an arbitrary waveform may be written in the form of the super-position integral. Another feature which recommends the step voltage is the fact that this waveform is one which permits small distortions to stand out clearly. Additionally, from an experimental viewpoint, we note that excellent pulse (a short step) and square-wave (a repeated step) generators are available commercially.

As long as an amplifier can be represented by a single-time-constant circuit, then the correlation between its frequency response and the output waveshape for a step input is that noted in Chap. 2. The nonzero rise time t_r and percentage sag or tilt P introduced by the amplifier are related to the high and low 3-dB frequencies, respectively, by Eqs. (2-33) and (2-14):

$$f_2 t_r = 0.35 \tag{4-5}$$

and

$$P \approx \pi \frac{f_1}{f} \times 100\% \tag{4-6}$$

where f is the frequency of the testing square wave.

Quite generally, even for more complicated amplifier circuits than the uncompensated single stage, there continues to be an intimate relationship between the distortion of the leading edge of a step and the high-frequency response. Similarly, there is a close relationship between the low-frequency response and the distortion of the flat portion of the step. We should, of course, expect such a relationship, since the high-frequency response measures essentially the ability of the amplifier to respond faithfully to rapid variations in signal, while the low-frequency response measures the fidelity of the amplifier for slowly varying signals. An important feature of a step is that it is a combination of the most abrupt voltage change possible and of the slowest possible voltage variation.

An important experimental procedure (called *square-wave testing*) is to observe with an oscilloscope the output of an amplifier excited by a square-wave generator. We shall see later that it is possible to improve the response of an amplifier by adding to it certain circuit elements which then must be adjusted with precision. It is a great convenience to be able to adjust these elements and to see simultaneously the effect of such an adjustment on the amplifier output waveform. The alternative is to take data, after each successive adjustment, from which to plot the amplitude and phase responses.

Aside from the extra time consumed in this latter procedure we have the problem that it is usually not obvious which of the attainable amplitude and phase responses corresponds to optimum fidelity. On the other hand, the step response gives immediately useful information.

It is possible, by judicious selection of two square-wave frequencies, to examine individually the high-frequency and low-frequency distortion. For example, consider an amplifier which has a high-frequency time constant of 1 μsec and a low-frequency time constant of 0.1 sec. A square wave of half period equal to several microseconds, on an appropriately fast oscilloscope sweep, will display the rounding of the leading edge of the waveform and will not display the tilt. At the other extreme, a square wave of half period approximately 0.01 sec on an appropriately slow sweep will display the tilt and not the distortion of the leading edge.

It should *not* be inferred from the above comparison between steady-state and transient response that the phase and amplitude responses are of no importance at all in the study of amplifiers. The frequency characteristics are useful for the following reasons. In the first place, much more is known generally about the analysis and synthesis of circuits in the frequency domain than in the time domain, and for this reason the design of coupling networks is often done on a frequency-response basis. Second, it is often possible to arrive at least at a qualitative understanding of the properties of a circuit from a study of the steady-state response in circumstances where transient calculations are extremely cumbersome. Finally, it happens occasionally that an amplifier is required whose characteristics are specified on a frequency basis, the principal emphasis being to amplify a sine wave.

4-3   THE *RC*–COUPLED AMPLIFIER

A cascaded arrangement of common-cathode (CK) vacuum-tube stages is shown in Fig. 4-2a and of common-emitter (CE) transistor stages in Fig. 4-2b. The output Y_1 of one stage is coupled to the input X_2 of the next stage via a blocking capacitor C_b which is used to keep the d-c component of the output voltage at Y_1 from reaching the input X_2. The resistor R_g is the grid leak and the plate (collector) circuit resistor is R_p (R_c). The cathode resistor R_k, the emitter resistor R_e, the screen resistor R_{sc}, and the resistors R_1 and R_2 are used to establish the bias. The bypass capacitors, used to prevent loss of amplification due to negative feedback, are C_k in the cathode, C_z in the emitter, and C_{sc} in the screen circuit. There are also present interelectrode capacitances in the case of a tube and junction capacitances if a transistor is used. These will be taken into account when we consider the high-frequency response, which is limited by their presence. In any practical mechanical arrangement of the amplifier components there are also capacitances associated with tube sockets and the proximity to the chassis of components (for example, the body of C_b) and signal leads. These stray capacitances will also be considered

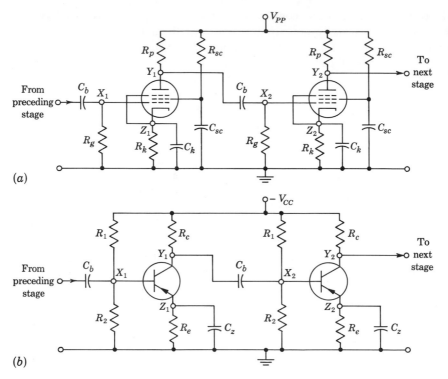

(a)

(b)

Fig. 4-2 A cascade of (a) common-cathode (CK) vacuum-tube stages or (b) common-emitter (CE) transistor stages.

later.   We shall assume that the active device operates linearly, so that small-signal models will be used throughout this chapter.

4-4   LOW–FREQUENCY RESPONSE OF AN RC–COUPLED STAGE[1]

The effect of the bypass capacitors C_k, C_z, and C_{sc} on the low-frequency characteristics will be discussed later.   For the present we shall assume that these capacitances are arbitrarily large and act as a-c short circuits across R_k, R_e, and R_{sc}, respectively.   A single intermediate stage of either of the cascades

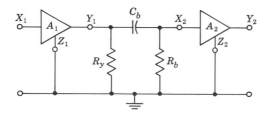

Fig. 4-3 A schematic representation of either a tube or transistor stage. Biasing arrangements and supply voltages are not indicated.

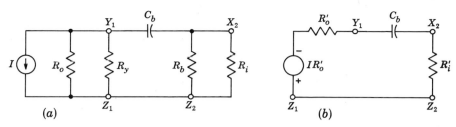

Fig. 4-4 (a) The low-frequency model of an RC-coupled amplifier; (b) an equivalent representation. For a tube: $I = g_m V_i$, $R_o = r_p$, $R_y = R_p$, $R_b = R_g$, and $R_i = \infty$. For a transistor: $I = h_{fe} I_b$, $R_o \approx 1/h_{oe}$, $R_b = R_1 R_2/(R_1 + R_2)$, $R_y = R_c$, $R_i \approx h_{ie}$, $R'_i = R_i R_b/(R_i + R_b)$, and $R'_o = R_o R_y/(R_o + R_y)$.

in Fig. 4-2 may be represented schematically as in Fig. 4-3. The resistor R_b represents the grid-leak resistor for a tube and equals R_1 in parallel with R_2 if a transistor stage is under consideration. The resistor R_y represents R_p for a tube and R_c for a transistor.

The low-frequency equivalent circuit is obtained by neglecting all shunting capacitances and all junction capacitances, by replacing A_2 by its input impedance R_i, and by replacing A_1 by its Norton's equivalent, as indicated in Fig. 4-4a. For a vacuum tube $R_i = \infty$; the output impedance is $R_o = r_p$ (the plate resistance); and $I = g_m V_i$ (transconductance times grid signal voltage). For a transistor these quantities may be expressed in terms of the CE hybrid parameters as in Sec. 1-3; $R_i \approx h_{ie}$ (for small values of R_c), $R_o = 1/h_{oe}$ (for a current drive), and $I = h_{fe} I_b$, where I_b is the base signal current. Let R'_o represent R_o in parallel with R_y and let R'_i be R_i in parallel with R_b. Then replacing I and R'_o by the Thévenin's equivalent, the single-time-constant high-pass circuit of Fig. 4-4b results. Hence the lower 3-dB frequency is

$$f_1 = \frac{1}{2\pi (R'_o + R'_i) C_b} \tag{4-7}$$

This result is easy to remember since the time constant equals C_b multiplied by the sum of the effective resistances R'_o to the left of the blocking capacitor and R'_i to the right of C_b. For a vacuum-tube amplifier $R'_i = R_g \gg R_p$. Since $R'_o < R_p$ because R'_o is R_p in parallel with R_o, then $R_g \gg R'_o$ and $f_1 \approx 1/2\pi C_b R_g$.

EXAMPLE It is desired to have a sag or tilt of no more than 10 percent when a 50-Hz square wave is impressed upon an amplifier stage. The output-circuit resistance is $R_y = 1$ K. What minimum value of coupling capacitance is required if

  a. Vacuum tubes with $R_g = 1$ M are used?
  b. Transistors with $R_i = 1$ K are used?

Solution   *a.* From Eqs. (4-6) and (4-7), with $f = 50$ Hz and $P \leq 10$ percent,

$$P \approx \frac{100\pi f_1}{f} = \frac{1}{(R_o' + R_i')C_b} \leq 10$$

or

$$C_b \geq \frac{1}{10(R_o' + R_i')}$$

Since $R_i' = 1$ M and $R_o' < R_p = 1$ K, then $C_b \geq (1/10^7)$ F $= 0.1\ \mu$F.

b. From Prob. 1-29 we find that for a transistor $R_o \geq 1/h_{oe} \approx 40$ K and hence $R_o' \approx R_c = 1$ K. If we assume that $R_b \gg R_i = 1$ K, then $R_i' \approx 1$ K. Hence

$$C_b \geq \frac{1}{(10)(2,000)}\ F = 50\ \mu\text{F}$$

Note that because the input impedance of a transistor is much smaller than that of a tube, a coupling capacitor is required with the transistor which is 500 times larger than that required with the tube. Fortunately it is possible to obtain physically small electrolytic capacitors having such high capacitance values at the low voltages at which transistors operate.

The value $R_i = 1$ K is reasonable for a grounded emitter transistor. If, however, an unbypassed emitter resistor R_e is used, then the input impedance is increased [Eq. (1-50)] by approximately $h_{fe}R_e$, to the order of 10 K or higher. Under these circumstances C_b can be reduced greatly, perhaps by a factor of 10. Now, however, because of the degeneration introduced by R_e, the magnitude of the midband gain is much smaller.

Phase Distortion   In the above illustration $f_1 = 1/2\pi(0.1) = 1.6$ Hz. Since the fundamental frequency is $f = 50$ Hz, then $f/f_1 = 50/1.6 = 31$, and every frequency in the Fourier series of the input signal is at least 31 times the 3-dB frequency. Since, from Eq. (4-1), $|A_1/A_o| = (1 + 1/31^2)^{-\frac{1}{2}} = 0.9995$, then the amplification of all harmonics may be taken identical. Hence the signal lies well within the passband of the amplifier as far as the low frequencies which determine the flat top of the wave are concerned. And yet the horizontal portions of the input are distorted into a 10 percent tilt in the output waveform. We shall now demonstrate that the reason for this apparently anomalous situation is to be found in the extreme sensitivity of the shape of the output to a shift in phase of the fundamental frequency component.

A symmetrical square wave of unity amplitude and of fundamental frequency f has a Fourier series,

$$v = \frac{4}{\pi}(\sin \alpha + \tfrac{1}{3}\sin 3\alpha + \tfrac{1}{5}\sin 5\alpha + \cdot \cdot \cdot) \tag{4-8}$$

in which $\alpha \equiv 2\pi ft$. Consider first only the influence on the square wave of

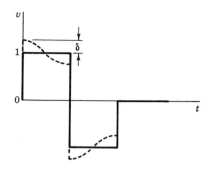

Fig. 4-5 Modification of a square wave because of the phase shift of the fundamental.

the phase shift of the fundamental. The phase shift is, from Eq. (4-3),

$$\theta_1 = \arctan\frac{f_1}{f} \approx \frac{f_1}{f}$$

for small angles. The output is obtained by replacing α in Eq. (4-8) by $\alpha + \theta_1$. The waveform is then modified by

$$\Delta v = \frac{4}{\pi}\left[\sin\left(\alpha + \frac{f_1}{f}\right) - \sin\alpha\right]$$

Since, for small angles, $\cos(f_1/f) \approx 1$ and $\sin(f_1/f) \approx f_1/f$, this equation reduces to

$$\Delta v \approx \frac{4}{\pi}\frac{f_1}{f}\cos\alpha = \delta\cos\alpha \tag{4-9}$$

where $\delta \equiv 4f_1/\pi f$. The waveform, modified by the addition of Δv, is shown in Fig. 4-5. The percentage tilt is

$$P = 2\delta \times 100\% = \frac{8}{\pi}\frac{f_1}{f} \times 100\% \tag{4-10}$$

For $f_1/f = \frac{1}{31}$, $P = 8.1$ percent. To take into account the effect of the phase shift of the remaining harmonics (which will, incidentally, change the cosine tilt into a linear tilt), we need but note that the nth harmonic is of relative amplitude $1/n$ and is shifted in phase $1/n$th as much as the fundamental. Therefore the above result may be corrected by writing

$$P = 8.1\left(1 + \frac{1}{3^2} + \frac{1}{5^2} + \frac{1}{7^2} + \cdots\right) = 8.1 \times 1.23 = 10\%$$

This result agrees with the value $P = 10$ percent given above.

4-5 HIGH–FREQUENCY RESPONSE OF A VACUUM–TUBE RC–COUPLED AMPLIFIER STAGE[1]

For frequencies above the midband range we may neglect the reactance of the large series capacitance C_b. However, we must now include in Fig. 4-2

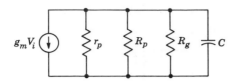

$g_m V_i$

r_p R_p R_g C

Fig. 4-6 The high-frequency model of an RC-coupled stage using a pentode.

the output capacitance C_o from Y_1 to ground and the input capacitance C_i from X_2 to ground. To these capacitances must also be added the stray capacitance to ground. If the sum of all these shunt capacitances is called C, then the high-frequency model of Fig. 4-6 can be drawn. In order to keep the input capacitance C_i as small as possible, a pentode, rather than a triode, is used for the tube. (See Sec. 1-7.) Hence, r_p is of the order of magnitude of a megohm, as is also R_g, whereas R_p is at most a few kilohms. Therefore the parallel combination of these three resistors can be approximated by R_p without introducing appreciable error. As predicted above, the amplifier stage at high frequencies behaves like a single-time-constant low-pass circuit. Hence, the upper 3-dB frequency f_2 and the rise time t_r are given by

$$f_2 = \frac{1}{2\pi R_p C} \qquad t_r = 2.2 R_p C \tag{4-11}$$

In the midband region, where the shunting effect of C can be neglected $(X_c \gg R_p)$, the output voltage is $V_o = -g_m R_p V_i$, and hence the midband gain $A_o \equiv V_o/V_i$ (for $R_p \ll r_p$ and $R_p \ll R_g$) is given by

$$A_o = -g_m R_p \tag{4-12}$$

The rise time of the amplifier may be improved by reducing the product $R_p C$. Every attempt should be made to reduce C by careful mechanical arrangement to decrease the shunt capacitance. The rise time may also be lessened by reducing R_p, but this reduces simultaneously the nominal amplifier gain. A figure of merit which is very useful in comparing tube types is obtained by computing the ratio of the nominal gain to the rise time in the limiting case where stray capacitance is considered to have been reduced to zero. Alternatively we may define the figure of merit F as the product of A_o and f_2. From Eqs. (4-11) and (4-12) we have, since $C = C_i + C_o$,

$$F \equiv |A_o| f_2 = \frac{g_m}{2\pi(C_o + C_i)} \tag{4-13}$$

Since $f_2 \gg f_1$, the bandwidth $f_2 - f_1 \approx f_2$ and $|A_o| f_2 = F$ is called the *gain-band-width product*. It should be noted that f_2 varies inversely with plate-circuit resistance whereas A_o is proportional to R_p, so that the gain-bandwidth product is a constant independent of R_p. It is possible to reduce R_p to such a low value that a midband gain $|A_o| = 1$ is obtained. Hence the figure of merit F may be interpreted as giving the maximum possible bandwidth obtainable with a given tube if R_p is adjusted for unity gain. For video pentodes such as the 6AK5, 6BH6, 6AU6, 6BC5, and 6CL6, values of g_m ranging from 5 to 11 millimhos (mA/V) and values of $C_o + C_i$ from 7 to

20 pF are obtainable. The value of F for all these tubes lies between 80 and 120 MHz, which is attained with the 6AK5.

An amplifier with a gain of unity is not very useful. Hence, let us assume that $|A_o|$ is at least 2. Then $f_2 = F/|A_o| = 60$ MHz for the 6AK5 tube. The corresponding rise time is $t_r = 0.35/60$ μsec $= 6$ nsec. In a practical circuit, the inevitable extra stray capacitance might easily reduce the bandwidth by a factor of 2. Hence we may probably take a rise time of 12 nsec or a bandwidth of 30 MHz as a reasonable estimate of a practical upper limit for an uncompensated tube amplifier using lumped parameters. If the desired gain is 10 instead of 2, the maximum 3-dB frequency is about 6 MHz.

The highest transconductance available in tubes is about 50 millimhos and is obtained with frame grid pentodes having very close (0.05 mm) grid-to-cathode spacing. For example, the Amperex type 7788 pentode has $g_m = 50$ mA/V and $C_o + C_i \approx 20$ pF, corresponding to $F = 400$ MHz. With this tube a 3-dB frequency of about 20 MHz is possible with a gain of 10. If more bandwidth is needed, then distributed amplifiers are used (Sec. 5-8).

The above discussion is valid for any stage of a tube amplifier, including the output stage. For this last stage C_i, representing the input capacitance to the following stage, is missing and its place is taken by any shunt capacitance of the device being driven (say a cathode-ray tube). Since the input impedance of a transistor cannot be represented by a parallel resistance-capacitance combination, the analysis of an internal stage differs from the final stage. In the next section we consider the output stage and then we shall proceed to the analysis of an internal stage of a cascade of transistors.

4-6 FREQUENCY RESPONSE OF A TRANSISTOR STAGE—THE SHORT–CIRCUIT CURRENT GAIN

Consider a single-stage CE transistor amplifier—or the last stage of a cascade. The load R_L on this stage is the collector-circuit resistor, so that $R_c = R_L$.

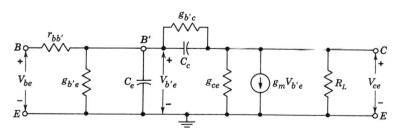

Fig. 4-7 The hybrid-Π circuit for a single transistor with a resistive load R_L. Typical values for a high-frequency transistor are $r_{bb'} = 100$ Ω, $r_{b'e} \equiv 1/g_{b'e} = 1$ K, $r_{b'c} \equiv 1/g_{b'c} = 4$ M, $r_{ce} \equiv 1/g_{ce} = 80$ K, $C_{b'e} \equiv C_e = 100$ pF, $C_{b'c} \equiv C_c = 3$ pF, and $g_m = 50$ millimhos (at $I_C = 1.30$ mA and $V_{CE} = 6$ V).

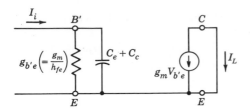

Fig. 4-8 Approximate equivalent circuit for the calculation of the short-circuit CE current gain.

To analyze the frequency response of the transistor amplifier we use the hybrid-Π model of Sec. 1-4.   The equivalent circuit of the transistor with a resistive load R_L is shown in Fig. 4-7.   In the caption, representative values of the circuit components are specified for a transistor intended for use at high frequencies.   We shall use these values as a guide in making simplifying assumptions.

The approximate equivalent circuit from which to calculate the short-circuit current gain is shown in Fig. 4-8.   A current source furnishes a sinusoidal input current of magnitude I_i, and the load current is I_L.   We have neglected $g_{b'c}$, which should appear across terminals $B'E$, because $g_{b'c} \ll g_{b'e}$.   And of course g_{ce} disappears because it is in shunt with a short circuit.   There is an additional approximation involved in that we have neglected the current delivered directly to the output through $g_{b'c}$ and C_c.   We shall see shortly that this approximation is justified.

The load current is $I_L = -g_m V_{b'e}$, where

$$V_{b'e} = \frac{I_i}{g_{b'e} + j\omega(C_e + C_c)} \tag{4-14}$$

The current amplification under short-circuited conditions is

$$A_i = \frac{I_L}{I_i} = \frac{-g_m}{g_{b'e} + j\omega(C_e + C_c)} \tag{4-15}$$

Using the result given in Eq. (1-11), also noted in Fig. 4-8, that $g_{b'e} = g_m/h_{fe}$, we have

$$A_i = \frac{-h_{fe}}{1 + j(f/f_\beta)} \tag{4-16}$$

where the frequency at which the CE short-circuit current gain falls by 3 dB is given by

$$f_\beta = \frac{g_{b'e}}{2\pi(C_e + C_c)} = \frac{1}{h_{fe}} \frac{g_m}{2\pi(C_e + C_c)} \tag{4-17}$$

This frequency is also represented by the symbols f_{hfe} and f_{ae}.   The frequency range up to f_β is referred to as the *bandwidth* of the circuit.   Note that the value of A_i at $\omega = 0$ is $-h_{fe}$, in agreement with the definition of $-h_{fe}$ as the low-frequency short-circuit CE current gain.

Since for a single-time-constant circuit the 3-dB frequency f_2 is given by $f_2 = 1/2\pi RC$, where R is the resistance in parallel with the capacitance, we could have written f_β by inspection as

$$f_\beta = \frac{1}{2\pi r_{b'e}(C_e + C_c)}$$

in agreement with Eq. (4-17).

The Parameter f_T We introduce now f_T, which is defined as *the frequency at which the short-circuit common-emitter current gain attains unit magnitude.* Since $h_{fe} \gg 1$ we have, from Eqs. (4-16) and (4-17), that f_T is given by

$$f_T \approx h_{fe}f_\beta = \frac{g_m}{2\pi(C_e + C_c)} \approx \frac{g_m}{2\pi C_e} \qquad (4\text{-}18)$$

since $C_e \gg C_c$.   Hence, from Eq. (4-16),

$$A_i \approx \frac{-h_{fe}}{1 + jh_{fe}(f/f_T)} \qquad (4\text{-}19)$$

The parameter f_T is an important high-frequency characteristic of a transistor. Like other transistor parameters its value depends on the operating conditions of the device.   Typically the dependence of f_T on collector-to-emitter voltage and emitter current is as shown in Fig. 4-9.   Note that it does not require a very large change in operating conditions to change f_T by a factor of 2.

Since $f_T \approx h_{fe}f_\beta$ this parameter may be given a second interpretation.   It represents the *short-circuit current-gain–bandwidth product;* that is, for the CE configuration with the output shorted, f_T is the product of the low-frequency

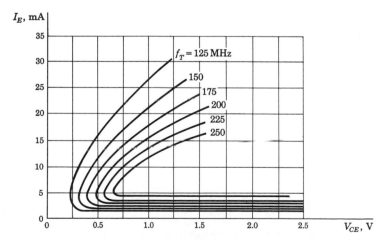

Fig. 4-9   Contours of constant f_T versus emitter current and collector voltage for type 2N501A p-n-p germanium MADT transistor.   (Courtesy of General Instrument Corporation.)

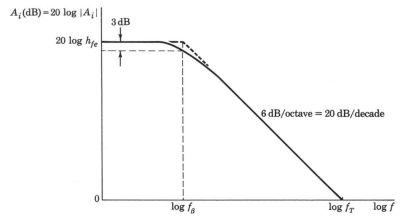

Fig. 4-10 The short-circuit CE current gain versus frequency (plotted on a log-log scale).

current gain and the upper 3-dB frequency. For our typical transistor (Fig. 4-7) $f_T = 80$ MHz, and $f_\beta = 1.6$ MHz. It is to be noted from Eq. (4-18) that there is a sense in which gain may be sacrificed for bandwidth and vice versa. Thus, if two transistors are available with equal f_T, the transistor with lower h_{fe} will have a correspondingly larger bandwidth.

In Fig. 4-10, A_i expressed in decibels (i.e., 20 log $|A_i|$) is plotted against frequency on a logarithmic frequency scale. When $f \ll f_\beta$, $A_i \approx -h_{fe}$ and A_i (dB) approaches asymptotically the horizontal line A_i (dB) $= 20$ log h_{fe}. When $f \gg f_\beta$, $|A_i| \approx h_{fe}f_\beta/f = f_T/f$, so that A_i (dB) $= 20$ log $f_T - 20$ log f. Accordingly, A_i (dB) $= 0$ dB at $f = f_T$. And for $f \gg f_\beta$ the plot approaches as an asymptote a straight line passing through the point $(f_T, 0)$ and having a slope which causes a decrease in A_i (dB) of 6 dB per octave or 20 dB per decade. The intersection of the two asymptotes occurs at the "corner" frequency $f = f_\beta$ where A_i is down by 3 dB.

Earlier we neglected the current delivered directly to the output through $g_{b'c}$ and C_c. Now we may see that this approximation is justified. Consider, say, the current through C_c. The magnitude of this current is $\omega C_c V_{b'e}$ whereas the current due to the controlled generator is $g_m V_{b'e}$. The ratio of currents is $\omega C_c/g_m$. At the highest frequency of interest f_T we have, from Eq. (4-18), using the typical values of Fig. 4-7,

$$\frac{\omega C_c}{g_m} = \frac{2\pi f_T C_c}{g_m} = \frac{C_c}{C_e + C_c} \approx 0.03$$

In a similar way the current delivered to the output through $g_{b'c}$ may be shown to be negligible.

The frequency f_T is often inconveniently high to allow a direct experimental determination of f_T. However, a procedure is available which allows a measurement of f_T at an appreciably lower frequency. We note from Eq. (4-16) that for $f \gg f_\beta$ we may neglect the unity in the denominator and write $|A_i|f \approx h_{fe}f_\beta = f_T$ from Eq. (4-18). Accordingly, at some particular frequency f_1 (say f_1 is five or ten times f_β) we measure the gain $|A_{i1}|$. The parameter f_T may be calculated now from $f_T = f_1|A_{i1}|$. In the case of our typical transistor, for which $f_T = 80$ MHz and $f_\beta = 1.6$ MHz, the frequency f_1 may be $f_1 = 5 \times 1.6 = 8.0$ MHz, a much more convenient frequency than 80 MHz.

The experimentally determined value of f_T is used to calculate the value of C_e in the hybrid-Π circuit. From Eq. (4-18)

$$C_e = \frac{g_m}{2\pi f_T} \qquad (4\text{-}20)$$

The Parameter f_α In addition to f_T and f_β, still another parameter f_α is used to characterize the high-frequency performance of a transistor. This parameter f_α is called the *short-circuit common-base cutoff frequency* and is the frequency at which the short-circuit CB current gain α has fallen by 3 dB from its low-frequency value. The current gain α is given by

$$\alpha = \frac{\alpha_o}{1 + j(f/f_\alpha)} \qquad (4\text{-}21)$$

in which α_o is the low-frequency value of α and f_α is the common-base cutoff frequency. We find in Prob. 4-6 that $f_\alpha = f_\beta h_{fe} \approx f_T$. Although it is expected that f_α and f_T should be very close in value, experimentally it is found that in diffusion transistors $f_\alpha \approx 1.2f_T$, whereas in drift transistors $f_\alpha \approx 2f_T$. Diffusion transistors are transistors in which the base doping is uniform, so that minority carriers cross the base entirely through diffusion. In drift transistors the doping is nonuniform, and an electric field exists in the base that causes a drift of minority carriers which aids the mechanism of diffusion.

The reason for the discrepancy is to be found in the fact that our lumped-circuit equivalent representation of the transistor is simply not accurate enough. By way of example, consider Eq. (4-21), which predicts that at $f = f_\alpha$, $|\alpha| = \alpha_o/\sqrt{2}$ and predicts also that α has undergone a 45° phase shift in comparison with its low-frequency value. This calculated amplitude response is in close agreement with experiment, but the phase-shift calculation may well be far off. It is found, empirically, that the discrepancy between calculation and experiment can be very substantially reduced by introducing an "excess-phase" factor[2] in the expression for α, so that Eq. (4-21) becomes

$$\alpha = \frac{\alpha_o}{1 + j(f/f_\alpha)} \, \epsilon^{-jmf/f_\alpha} \qquad (4\text{-}22)$$

In this equation m is an adjustable parameter that ranges from about 0.2 for a diffusion transistor to about unity for a drift transistor.

Another parameter related to the high-frequency operation of a transistor is f_{max}. This parameter gives the frequency at which the power gain (under matched conditions of generator and load impedances) is reduced to unity. It is the maximum frequency of operation of a transistor oscillator and is given by

$$f_{\max} = \sqrt{\frac{f_T}{8\pi r_{bb'}C_c}}$$

(4-23)

The parameter $f_{\max}$ is not of immediate applicability in amplifier circuits.

4-7   CURRENT GAIN WITH RESISTIVE LOAD

To minimize the complications which result when the load resistor R_L in Fig. 4-7 is not zero, we shall find it convenient to deal with the parallel combination of $g_{b'c}$ and C_c using Miller's theorem of Sec. 1-7. We identify $V_{b'e}$ with V_1 in Fig. 1-7 and V_{ce} with V_2. On this basis the circuit of Fig. 4-7 may be replaced by the circuit of Fig. 1-9, which is repeated in Fig. 4-11a for convenience. This circuit is still rather complicated because it has two independent time constants, one associated with the input circuit and one associated with the output. We shall now show that in a practical situation the output time constant is negligible in comparison with the input time constant and may therefore be ignored. Let us therefore delete the output capacitance $C_c(K - 1)/K$, consider the resultant circuit, and then show that the reintroduction of the output capacitance makes no significant change in the performance of the circuit.

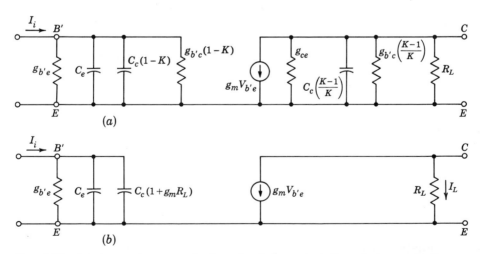

Fig. 4-11   (a) Approximate equivalent circuit for calculation of response of a transistor amplifier stage with a resistive load; (b) further simplification of the equivalent circuit.

Since $K \equiv V_{ce}/V_{b'e}$ is the stage amplification, we shall normally have $|K| \gg 1$. Hence $g_{b'c}(K-1)/K \approx g_{b'c}$. Since $g_{b'c} \ll g_{ce}$ ($r_{b'c} \approx 4$ M and $r_{ce} \approx 80$ K), we may omit $g_{b'c}$ from Fig. 4-11a. In a wideband amplifier R_L seldom exceeds 2 K. The conductance g_{ce} may be neglected compared with R_L and the output circuit consists of the current generator $g_m V_{b'e}$ feeding the load R_L, as indicated in Fig. 4-11b. Even if the above approximations were not valid for some particular transistor or load, the analysis to follow is still valid provided that R_L is interpreted as the parallel combination of the collector-circuit resistor, r_{ce} and $r_{b'c}$.

By inspection, $K = V_{ce}/V_{b'e} = -g_m R_L$, and with $g_m = 50$ millimhos and $R_L = 2,000$ Ω, $K = -100$. For this maximum value of K the conductance $g_{b'c}(1-K) \approx 0.025$ millimho is negligible compared with $g_{b'e} \approx 1$ millimho. Hence the circuit of Fig. 4-11a is reduced to that shown in Fig. 4-11b. The load resistance R_L has been restricted to a maximum value of 2 K because at values of R_L much above 2,000 Ω the capacitance $C_c(1 + g_m R_L)$ becomes excessively large and the bandpass correspondingly small.

Now let us return to the capacitance $C_c(K-1)/K \approx C_c$, which we neglected above. For $R_L = 2,000$ Ω,

$$R_L C_c = 2 \times 10^3 \times 3 \times 10^{-12} = 6 \times 10^{-9} \text{ sec}$$

The input time constant is

$$r_{b'e}[C_e + C_c(1 + g_m R_L)] = 10^3(100 + 3 \times 101)10^{-12} = 403 \times 10^{-9} \text{ sec}$$

It is therefore apparent that the bandpass of the amplifier will be determined by the time constant of the input circuit and that in the useful frequency range of the stage the capacitance C_c in the output circuit will not make itself felt. Of course, if the transistor works into a highly capacitive load, this capacitance would have to be taken into account and it then might happen that the output time constant would predominate.

The circuit of Fig. 4-11b is different from the circuit of Fig. 4-8 only in that a load R_L has been included and that C_c has been augmented by $g_m R_L C_c$. To the accuracy of our approximations the low-frequency current gain A_{Io} under load is the same as the low-frequency gain A_{io} with output shorted. Therefore

$$A_{Io} = -h_{fe}$$

However, the 3-dB frequency is now f_2 (rather than f_β), where

$$f_2 = \frac{1}{2\pi r_{b'e}C} = \frac{g_{b'e}}{2\pi C} \tag{4-24}$$

where

$$C \equiv C_e + C_c(1 + g_m R_L) \tag{4-25}$$

4-8   TRANSISTOR AMPLIFIER RESPONSE TAKING SOURCE IMPEDANCE INTO ACCOUNT

In the preceding discussions we assumed that the transistor stage was driven from an ideal current source, that is, a source of infinite resistance. We now remove that restriction and consider that the source has a resistive impedance R_s. We may represent the source by its Norton's equivalent, as in Fig. 4-12a, or by its Thévenin's equivalent, as in Fig. 4-12b. At low frequencies the current gain $A_{Io} \equiv I_L/I_i = -g_m V_{b'e}/g_{b'e} V_{b'e} = -h_{fe}$, from Eq. (1-11). Therefore the low-frequency current gain,† taking the load and source impedances into account, is

$$A_{Iso} \equiv \frac{I_L}{I} = \frac{I_L}{I_i}\frac{I_i}{I} = -h_{fe}\frac{R_s}{R_s + r_{bb'} + r_{b'e}} = \frac{-h_{fe}R_s}{R_s + h_{ie}} \tag{4-26}$$

since $h_{ie} = r_{bb'} + r_{b'e}$. Note that A_{Iso} is independent of R_L. The 3-dB frequency is determined by the time constant consisting of C and the equivalent resistance R shunted across C. Accordingly,

$$f_2 = \frac{1}{2\pi RC} \tag{4-27}$$

in which C is given by Eq. (4-25) and R is the parallel combination of $R_s + r_{bb'}$ and $r_{b'e}$, namely,

$$R \equiv \frac{(R_s + r_{bb'})r_{b'e}}{R_s + h_{ie}} \tag{4-28}$$

From Eq. (1-21) and the results of Sec. 1-6 we have that the voltage gain A_{Vso} at low frequency, taking load and source impedances into account, is

$$A_{Vso} = A_{Iso}\frac{R_L}{R_s} = \frac{-h_{fe}R_L}{R_s + h_{ie}} \tag{4-29}$$

Note that A_{Vso} increases linearly with R_L. The 3-dB frequency for voltage gain A_{Vs} is also given by Eq. (4-27). Note that f_2 increases as the load

† See Sec. 1-6 for the definitions of the various current and voltage gains.

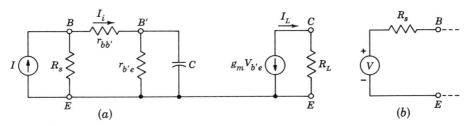

Fig. 4-12   (a) A transistor is driven by a generator of impedance R_s which is represented by its Norton's equivalent circuit.   (b) The generator is represented by its Thévenin's equivalent.

resistance is decreased because C is a linear function of R_L. At $R_L = 0$ the 3-dB frequency is finite (unlike the vacuum-tube amplifier, which has infinite bandpass for zero plate-circuit resistance) and, from Eq. (4-20), is given by

$$f_2 = \frac{1}{2\pi R(C_e + C_c)} = \frac{f_T}{g_m R} = \frac{f_\beta}{g_{b'e} R} \qquad R_L = 0 \qquad (4\text{-}30)$$

For $R_s = 0$ this quantity is of the order of $f_T/5 \approx 10 f_\beta$ and for $R_s = 1$ K (and $R_L = 0$), $f_2 \approx f_T/25 \approx 2 f_\beta$. Of course, for $R_L = 0$ the voltage gain is zero. In practice, when $R_L \neq 0$, much lower 3-dB frequencies than those indicated above will be obtained.

As noted in Sec. 1-6, the equality in 3-dB frequencies for current and voltage gains applies only in the case of a fixed source resistance. The voltage gain A_V (for the case of an ideal voltage source) and the current gain A_I (for the case of an ideal current source) do not have the same value of f_2. In the former case $R_s = 0$ and in the latter case $R_s = \infty$. Equation (4-27) applies in both cases provided that for A_V we use $R = R_V$, where, from Eq. (4-28),

$$R_V = \frac{r_{bb'} r_{b'e}}{r_{bb'} + r_{b'e}} = \frac{r_{bb'} r_{b'e}}{h_{ie}} \qquad (4\text{-}31)$$

and for A_I we use $R = R_I$, where

$$R_I = r_{b'e} \qquad (4\text{-}32)$$

Since $R_V \ll R_I$, the 3-dB frequency f_{2V} for an ideal voltage source is higher than f_{2I} for an ideal current source.

The gain-bandwidth product is found in Prob. 4-9 to be

$$|A_{Vso} f_2| = \frac{g_m}{2\pi C} \frac{R_L}{R_s + r_{bb'}} = \frac{f_T}{1 + 2\pi f_T C_c R_L} \frac{R_L}{R_s + r_{bb'}} \qquad (4\text{-}33)$$

$$|A_{Iso} f_2| = \frac{f_T}{1 + 2\pi f_T C_c R_L} \frac{R_s}{R_s + r_{bb'}} \qquad (4\text{-}34)$$

The quantities f_2, A_{Iso}, and A_{Vso} which characterize the transistor stage depend on both R_L and R_s. The form of this dependence, as well as the order of magnitude of these quantities, may be seen in Fig. 4-13. Here f_2 has been plotted as a function of R_L, up to $R_L = 2,000 \ \Omega$, for several values of R_s. The topmost f_2 curve in Fig. 4-13 for $R_s - 0$ corresponds to ideal voltage source drive. The current gain is zero and the voltage gain ranges from zero at $R_L = 0$ to 90.9 at $R_L = 2,000 \ \Omega$. Note that a source impedance of only 100 Ω reduces the bandwidth by a factor of about 1.8. The bottom curve has $R_s = \infty$ and corresponds to the ideal current source. The voltage gain is zero for all R_L if $R_s = \infty$. For any R_L the bandwidth is highest for lowest R_s.

In the case of a vacuum-tube stage of amplification, the gain-bandwidth product is a useful number. It does not depend on the plate-circuit resistance and no driving generator impedance enters into the discussion. The gain-bandwidth product depends only on the tube parameters and serves as a figure

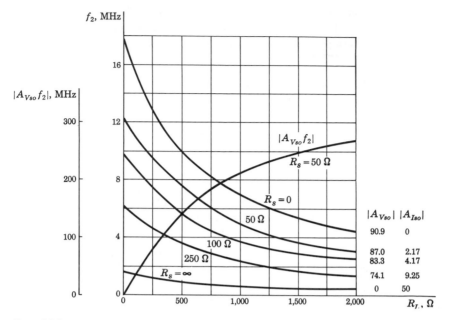

Fig. 4-13 Bandwidth f_2 as a function of R_L with source impedance as a parameter for an amplifier consisting of one CE transistor whose parameters are given in Fig. 4-7. Also the gain-bandwidth product for a 50-Ω source is plotted. The tabulated values of $|A_{Vso}|$ correspond to $R_L = 2,000\ \Omega$ and to the values of R_s on the curves. The values of $|A_{Iso}|$ are independent of R_L.

of merit for the tube. The constancy of the gain-bandwidth product is useful in assuring us that the extent to which we sacrifice gain is exactly the extent to which we improve bandwidth, and vice versa. In the case of a transistor amplifier consisting of a single stage, the gain-bandwidth product is ordinarily not a useful parameter. It is not independent of R_s and R_L and varies widely with both. The current-gain–bandwidth product decreases with increasing R_L and increases with increasing R_s. The voltage-gain–bandwidth product increases with increasing R_L and decreases with increasing R_s. Even if we know the gain-bandwidth product at a particular R_s and R_L we cannot use the product to determine the improvement, say, in bandwidth corresponding to a sacrifice in gain. For if we change the gain by changing R_s or R_L or both, generally the gain-bandwidth product will no longer be the same as it had been.

The important formulas for a single-stage CE amplifier are summarized in Table 4-1. Also included in the table are the results (obtained in Sec. 4-10) for an internal stage of a chain of cascaded stages.

TABLE 4-1   CE transistor amplifier summary

$$f_T = \frac{g_m}{2\pi(C_e + C_c)} = h_{fe}f_\beta \qquad h_{ie} = r_{bb'} + r_{b'e}$$

$$f_2 = \frac{1}{2\pi RC} \qquad\qquad t_r = 2.2RC$$

Single stage: $A_{Vs} = A_{Is}R_L/R_s$. The collector-circuit resistance is designated by R_L.

$$A_{Iso} = \frac{-h_{fe}R_s}{R_s + h_{ie}} \qquad A_{Vso} = \frac{-h_{fe}R_L}{R_s + h_{ie}}$$

$$R = \frac{(R_s + r_{bb'})r_{b'e}}{R_s + h_{ie}} \qquad C = C_e + C_c(1 + g_mR_L)$$

$$|A_{Vso}f_2| = \frac{f_T}{1 + 2\pi f_T C_c R_L} \frac{R_L}{R_s + r_{bb'}}$$

Cascaded stages (internal stage): $A_V = A_{Is} = A$. The parallel combination of the collector-circuit resistance and the effective resistance at the base is designated by R_c.

$$A_o = \frac{-h_{fe}R_c}{R_c + h_{ie}} \qquad R_L = \frac{h_{ie}R_c}{R_c + h_{ie}}$$

$$R = \frac{(R_c + r_{bb'})r_{b'e}}{R_c + h_{ie}} \qquad C = C_e + C_c(1 + g_mR_L)$$

$$|A_o f_2| = \frac{f_T}{1 + 2\pi f_T C_c R_L} \frac{R_c}{R_c + r_{bb'}}$$

$$(R_c)_{opt} = \frac{h_{ie}}{\sqrt{x} - 1} \qquad x = \frac{h_{fe}C_c}{C_e + C_c} \frac{h_{ie}}{r_{bb'}}$$

$$(f_2)_{max} = \frac{f_T h_{ie}}{h_{fe}r_{bb'}} \qquad \text{for } R_c = 0$$

4-9   TRANSIENT RESPONSE OF A TRANSISTOR STAGE

From the preceding analysis we see that the frequency response of the transistor stage driven from a resistive source and working into a resistive load is that of a low-pass circuit with a single time constant τ. For such a network, the response to a unit input step is given by $A_o(1 - \epsilon^{-t/\tau})$, where A_o is the steady-state voltage. The rise time t_r of this exponential waveform is given by Eq. (2-33),

$$t_r = 2.2\tau = \frac{0.35}{f_2}$$

If a current generator of impedance R_s applies a step of current I at the transistor input, then

$$i_L = -\frac{h_{fe}R_s I}{R_s + h_{ie}}(1 - \epsilon^{-2\pi f_2 t}) \qquad (4\text{-}35)$$

If the source is a voltage generator which supplies a step of voltage V, then the output voltage is

$$v_o = -\frac{h_{fe}R_L V}{R_s + h_{ie}} (1 - \epsilon^{-2\pi f_2 t}) \tag{4-36}$$

In each case f_2 depends on R_s and R_L in accordance with Eqs. (4-27), (4-28), and (4-25). The rise time increases linearly with R_L since t_r is proportional to $C = C_e + C_c(1 + g_m R_L)$.

For an ideal current source ($R_s = \infty$) the current gain is $|A_{Iso}| = h_{fe}$, and it follows from Eq. (4-34) with $\omega_T = 2\pi f_T$ that

$$\tau = \frac{1}{2\pi f_2} = \frac{h_{fe}}{\omega_T} (1 + \omega_T C_c R_L) \tag{4-37}$$

4-10   CASCADED CE TRANSISTOR STAGES[3]

We consider now the operation of one transistor amplifier stage in a cascade of many stages. Such a cascade is shown in Fig. 4-14. We have omitted from this diagram all supply voltages and components, such as coupling capacitors, which serve only to establish proper bias and do not affect the high-frequency response. The collector-circuit resistor R_c is included, however, since this resistor has an effect on both the gain and frequency response. The base-biasing resistors R_1 and R_2 in Fig. 4-2b are assumed to be large compared with R_c. If this condition is not satisfied, then the symbol R_c represents the parallel combination of R_1, R_2, and the collector-circuit resistance. A complete stage from collector to collector is included in the dashed box. We define the current gain of the stage to be $A_{Is} \equiv I_2/I_1$. A comparison of Fig. 4-14 with Fig. 4-12 reveals that each stage behaves like a current generator of impedance $R_s = R_c$ delivering current to the following stage. We define the voltage gain to be $A_V \equiv V_2/V_1$. Since we have specified V_1 as the voltage precisely at the stage input, then A_V is the gain for an ideal voltage source. We shall now prove that $A_{Is} = A_V$ for an infinite cascade of similar stages.

In a long chain of stages the input impedance Z_i between base and emitter of each stage is identical. Let Z_i' represent Z_i in parallel with R_c. Accordingly

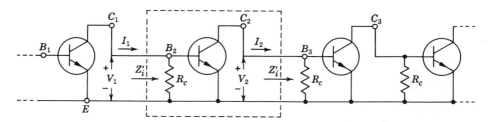

Fig. 4-14 An infinite cascade of CE stages. The dashed box encloses one stage.

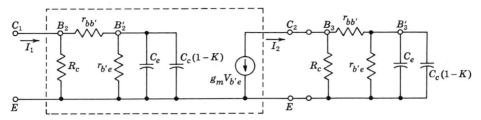

Fig. 4-15   The equivalent circuit of the enclosed stage of Fig. 4-14.

$Z_i' = V_1/I_1 = V_2/I_2$, so that $I_2/I_1 = A_{Is} = V_2/V_1 = A_V$ in this special case. We shall now calculate this gain $A_{Is} = A_V \equiv A$.   For this purpose Fig. 4-15 shows the circuit details of the stage in the dashed box in Fig. 4-14.   Also shown is the input portion of the next stage so that we may take account of its loading effect on the stage of interest.   The symbol K used in the expression $C_c(1 - K)$ for one of the capacitors is $K \equiv V_{ce}/V_{b'e}$.   The elements involving $g_{b'c}$ have been omitted since, as above, their omission can be shown to introduce little error.

The gain $A_o = I_2/I_1$ at low frequencies is given by Eq. (4-26) except with R_s replaced by R_c, and we have

$$A_o = \frac{-h_{fe}R_c}{R_c + h_{ie}} \tag{4-38}$$

To calculate the bandwidth we must evaluate K.   From Fig. 4-15 we obtain for K an unwieldy expression.   Since K is a function of frequency, the element marked $C_c(1 - K)$ is not a true capacitor but rather is a complex network.   Thus in order to proceed with a simple solution which will give reasonable accuracy we shall use the zero-frequency value of K.   We shall show below that the response obtained experimentally will be somewhat better than that predicted by this analysis and hence that we are erring in the conservative direction.   At zero frequency $K = K_o = -g_m R_L$, in which R_L is the resistive load on the transistor from C to E and consists of R_c in parallel with $r_{bb'} + r_{b'e} = h_{ie}$.   Therefore

$$R_L = \frac{R_c h_{ie}}{R_c + h_{ie}} \tag{4-39}$$

and the capacitance is

$$C = C_e + C_c(1 + g_m R_L) \tag{4-40}$$

The gain is $A = I_2/I_1 = -g_m V_{b'e}/I_1$, where $V_{b'e} = V_{b'2e}$ represents the voltage across C.   Instead of calculating $V_{b'e}$ directly from the input network of Fig. 4-15 we again make the observation that this is a single-time-constant circuit.   Hence we can calculate the 3-dB frequency f_2 by inspection.   Since the capacitance C is charged through a resistance R consisting of $r_{b'e}$ in parallel with $R_c + r_{bb'}$, or

$$R = \frac{(R_c + r_{bb'})r_{b'e}}{R_c + h_{ie}} \tag{4-41}$$

the 3-dB frequency is

$$f_2 = \frac{1}{2\pi RC} \tag{4-42}$$

This half-power frequency is the same for the current gain and voltage gain.

In using the approximation $K = K_o = -g_m R_L$ we are making a conservative error, since K_o is the maximum magnitude of K and is attained only at zero frequency. Using K_o leads to the largest value of shunt capacitance C and consequently to an overly low estimate of the bandwidth f_2.

From the above equations the gain-bandwidth product is found to be

$$|A_o f_2| = \frac{g_m}{2\pi C} \frac{R_c}{R_c + r_{bb'}} = \frac{f_T}{1 + 2\pi f_T C_c R_L} \frac{R_c}{R_c + r_{bb'}} \tag{4-43}$$

where R_L depends upon R_c as indicated in Eq. (4-39).

Gain and Bandwidth Considerations   Our only adjustable parameter is R_c and we now discuss its selection. At one extreme, if we set $R_c = 0$ we would simply shunt all output current away from the following transistor. As a matter of fact, it seems initially not unreasonable to set R_c arbitrarily high so as to avoid this shunting effect. However, as we reduce R_c and thereby lose gain, a compensating advantage appears. A reduction of R_c reduces R_L in Eq. (4-39) and also reduces R in Eq. (4-41). The reduction in R_L reduces $C = C_e + C_c(1 + g_m R_L)$, and this reduction together with the reduction in R increases f_2, as is seen in Eq. (4-42). It may be that a decrease in gain may be more than compensated for by an increase in f_2. To investigate this point we differentiate the gain-bandwidth product $|A_o f_2|$ with respect to R_c. Setting the derivative equal to zero we find that a maximum does occur. The value of R_c for which this optimum gain-bandwidth product is obtained is designated by $(R_c)_{\text{opt}}$ and is given by

$$(R_c)_{\text{opt}} = \frac{h_{ie}}{\sqrt{x} - 1} \tag{4-44}$$

with

$$x = \frac{h_{fe} C_c}{C_e + C_c} \frac{h_{ie}}{r_{bb'}} \tag{4-45}$$

In Fig. 4-16 we have plotted the gain, the bandwidth, and the gain-bandwidth product. The maximum which is apparent [at $R_c = 360\ \Omega$, as found from Eq. (4-44)] is not particularly pronounced.[3] Nevertheless there is enough of a falling off at values of R_c above or below $(R_c)_{\text{opt}}$ so that it may be worthwhile to operate near the maximum. It is important to bias the transistor so that at the quiescent point a large value of f_T is obtained (Fig. 4-9).

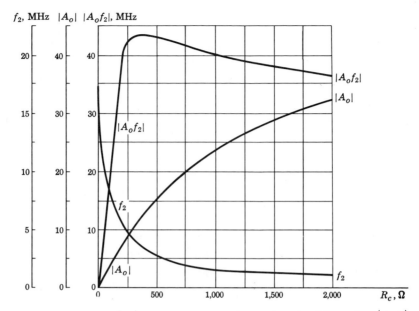

Fig. 4-16 Gain $|A_o|$, bandwidth f_2, and gain-bandwidth product $|A_o f_2|$ as a function of R_c for one stage of a CE cascade. The transistor parameters are given in Fig. 4-7.

Note in Fig. 4-16 that $|A_o f_2|$ remains roughly constant for values of R_c in the neighborhood of $(R_c)_{\text{opt}}$ or for larger values of R_c. Hence, for a cascade of stages (as distinct from the single stage considered in Sec. 4-8) the gain-bandwidth product takes on some importance as a figure of merit. For our typical transistor $f_T = 80$ MHz, whereas the constant value of $|A_o f_2|$ in Fig. 4-16 is approximately 40 MHz, or $0.5 f_T$. The authors have found that a good general rule in choosing a transistor as a broadband amplifier is to assume $A_o f_2 \approx 0.6 f_T$. This conclusion is based upon calculations on more than twenty transistors for which the hybrid-Π parameters were known. These had values of f_T ranging from 700 kHz to 700 MHz. In each case $(R_c)_{\text{opt}}$ was found and the value of $A_o f_2$ at this optimum resistance was calculated. All values of gain-bandwidth product were in the range between $0.4 f_T$ and $0.8 f_T$. The values of $A_o f_2$ were also calculated for several values of R_c besides $(R_c)_{\text{opt}}$, and it was confirmed that the gain-bandwidth product remained constant over a wide range of values of R_c.

It must be remembered that bandwidth cannot be exchanged for gain at low values of gain because $A_o f_2$ is not constant for small values of R_c or A_o. The maximum value of f_2, which occurs at $R_c = 0$, is given by

$$(f_2)_{\text{max}} = \frac{f_T}{g_m R} = \frac{f_T h_{ie}}{h_{fe} r_{bb'}} \tag{4-46}$$

For our typical transistor $(f_2)_{max} = 17$ MHz, as indicated in Fig. 4-16, and this value of bandwidth is obtained only at zero voltage gain.

The design of the pulse amplifier represents, as usual, a compromise between gain and bandwidth. If A_o is specified, then the load R_c which must be used is found from Eq. (4-38). Then the bandwidth which will be obtained is found from Eq. (4-42). On the other hand, if the desired rise time t_r is specified, then $f_2 = 0.35/t_r$ substituted into Eq. (4-42) will not allow a direct calculation of R_c. The reason for the difficulty is that R depends upon R_c and that $C = C_e + C_c(1 + g_m R_L)$ is also a function of R_c through R_L, as given in Eq. (4-39). Under these circumstances an arbitrary value of R_c, say 1,000 Ω, is chosen and f_2 is calculated. If this value is larger (smaller) than the desired value of f_2, then the next approximation to R_c must be larger (smaller) than 1,000 Ω. By plotting f_2 versus R_c the desired value of R_c can be found by interpolation.

The approximations which we have made in this analysis are valid if R_L is less than 2,000 Ω. Since R_L is the parallel combination of R_c and $h_{ie} \approx 1,100\,\Omega$, there are no restrictions on the magnitude of R_c. As $R_c \to \infty$, $R_L = h_{ie}$ and $A_o = -h_{fe}$. The asymptotic limits in Fig. 4-16 are found to be $|A_o| = 50$, $f_2 = 0.59$ MHz, and $|A_o f_2| = 29.5$ MHz.

The important results obtained above for an internal stage of a cascade are summarized in Table 4-1 on page 131. These formulas are not valid for the first or last stage. For the first stage the equations in Table 4-1 for a single stage apply, provided that the load R_L is taken as the collector-circuit resistance in parallel with the input resistance of the second stage,

$$R_L = \frac{R_c h_{ie}}{R_c + h_{ie}}$$

For the last stage in a cascade use the formulas for a single stage, with R_s equal to the collector-circuit resistance R_c of the preceding stage and with R_L equal to the R_c of the last stage.

4-11   RISE–TIME RESPONSE OF CASCADED STAGES[4]

We have seen that one stage of an amplifier, whether of the tube or transistor type, behaves like a circuit with a single time constant τ. If the midband gain is A_o the high-frequency transfer function is $A_o/(1 + j\omega\tau)$ for a single stage. It is convenient to carry out a transient analysis in the complex-frequency s plane (or the Laplace domain). Using this notation and with $V_i(s) = $ the input voltage and $V_o(s) = $ the output voltage, then

$$\frac{V_o(s)}{V_i(s)} = \frac{A_o}{1 + s\tau}$$

Since the gains of cascaded stages are multiplicative, the transfer function for n identical stages is

$$\frac{I_o(s)}{I(s)} = \frac{V_o(s)}{V(s)} = \frac{A_o{}^n}{(1 + s\tau)^n} \tag{4-47}$$

If the input is a step V, then $V(s) = V/s$. The output function has an nth-order pole at $s = -1/\tau$ and a first-order pole at $s = 0$. Defining the normalized response y and the normalized time x by

$$y \equiv \frac{v_o(t)}{A_o{}^n V} \qquad \text{and} \qquad x \equiv \frac{t}{\tau}$$

we find from the inverse Laplace transform that

$$y = 1 - \left[1 + x + \frac{x^2}{2!} + \frac{x^3}{3!} + \cdots + \frac{x^{n-1}}{(n-1)!} \right] \epsilon^{-x} \tag{4-48}$$

If it is observed that the polynomial in the brackets is the first n terms of the power-series expansion of ϵ^{+x}, then this equation can be written in the form

$$y = \left[\frac{x^n}{n!} + \frac{x^{n+1}}{(n+1)!} + \cdots \right] \epsilon^{-x} \tag{4-49}$$

Near the origin ($x = 0$) the curves start out like x^n. Therefore the first $n - 1$ derivatives are zero. As n increases there is introduced into the response a progressively larger delay. This is observed in Fig. 4-17, where the response

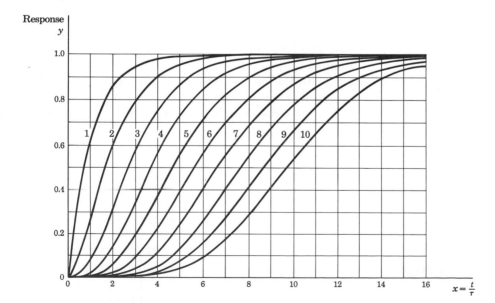

Fig. 4-17 Response $[y \equiv v_o(t)/A_o{}^n V]$ of cascaded identical uncompensated stages. (Adapted from Ref. 4, fig. 1-25.)

for the cases $n = 1$ to $n = 10$ is shown. We also note that the rise time $t_r^{(n)}$ (for n stages) increases with n.

By determining the time between 0.1 and 0.9 values from the graph in Fig. 4-17 we can calculate $t_r^{(n)}$. For a single stage $(n = 1)$, $t_r = 2.2\tau$. Values of $t_r^{(n)}/t_r$ have been tabulated in Table 4-2. Also listed are values of $1.1\sqrt{n}$, which we see is an excellent empirical approximation for $t_r^{(n)}/t_r$.

From Table 4-2 we see that if we wish to have an overall bandpass of 1 MHz with a two-stage amplifier, then the upper 3-dB frequency of each must be 1.5 MHz. If 1 MHz is desired with a three-stage amplifier, then each (identical) stage must be 1.9 MHz wide, etc.

It will be shown in the next chapter that even for a circuit which is more complicated than a single RC combination the rise-time–bandwidth product remains approximately constant at 0.35. This result suggests that we try to calculate the rise time from the bandwidth. The upper 3-dB frequency for n cascaded amplifiers is $f_2^{(n)}$ and may be computed from

$$\left[\frac{1}{\sqrt{1 + (f_2^{(n)}/f_2)^2}}\right]^n = \frac{1}{\sqrt{2}} \tag{4-50}$$

so that

$$\frac{f_2}{f_2^{(n)}} = \frac{1}{\sqrt{2^{1/n} - 1}} \tag{4-51}$$

Therefore, if we assume that $t_r^{(n)} f_2^{(n)} = t_r f_2$, we have

$$\frac{t_r^{(n)}}{t_r} = \frac{1}{\sqrt{2^{1/n} - 1}} \tag{4-52}$$

Values of $(2^{1/n} - 1)^{-\frac{1}{2}}$ are also listed in Table 4-2. Note the very good agreement with the correct values of rise-time ratios.

The delay associated with the curves of Fig. 4-17 may be specified by the time required for the response to go from 0 to 0.5. We see that each stage beyond the first introduces the same amount of delay. For n amplifiers, the

TABLE 4-2   Comparison of exact rise-time ratios with ratios given by square-root rule and bandwidth rule

	n								
	2	3	4	5	6	7	8	9	10
$\dfrac{t_r^{(n)}}{t_r}$ [Eq. (4-49)]...........	1.5	1.9	2.2	2.5	2.8	3.0	3.3	3.45	3.6
$1.1\sqrt{n}$...........	1.55	1.9	2.2	2.5	2.7	2.9	3.1	3.3	3.5
$(2^{1/n} - 1)^{-\frac{1}{2}}$ [Eq. (4-52)]........	1.55	2.0	2.3	2.6	2.85	3.1	3.3	3.5	3.7

delay is given approximately by $(n - 0.3)RC$. This delay is not ordinarily considered as a distortion.

From the above discussion we may expect that, at least approximately, the following rule applies with respect to the overall rise time t_r of n nonidentical stages. If the rise time of the individual amplifiers is $t_{r1}, t_{r2}, \ldots, t_{rn}$ and of the input waveform rise time is t_{ro}, then the output-signal rise time is given (to within 10 percent) by

$$t_r \approx 1.1 \sqrt{t_{ro}^2 + t_{r1}^2 + t_{r2}^2 + \cdots + t_{rn}^2} \tag{4-53}$$

Having considered the rise-time response of cascaded stages we shall now inquire into the flat-top response of such cascaded stages. Before doing so, however, let us investigate all possible causes for a tilt in the step response of an amplifier. We already know that the blocking capacitor between stages is one contributor to such a tilt. In the next two sections it is shown that the screen, cathode, and emitter bypass capacitors also affect the tilt in the amplifier output waveform.

4-12   EFFECT OF SCREEN BYPASS ON LOW–FREQUENCY RESPONSE

The screen-grid circuit consists of a voltage-dropping resistor R_{sc} and a capacitor C_{sc} from screen to ground, as in Fig. 4-18. If a positive step voltage is applied to the grid, the plate current increases and hence so does the screen current. At $t = 0$, the screen voltage V_{sc} is at its quiescent value. As time passes, the capacitor must discharge to a steady-state voltage equal to the plate-supply voltage minus the new value of the screen current times R_{sc}. Therefore there is a droop in screen voltage with time and a corresponding tilt in the output plate voltage. The waveshape is similar to that encountered if too small a blocking capacitor C_b is used.

The method of calculating the size of the screen capacitor to keep the tilt below a certain value is best illustrated by a numerical example. Consider a 6CL6 with a quiescent current of 20 mA and $V_{sc} = 150$ V. Because of a step input to the grid, the plate current increases to 30 mA. What is the

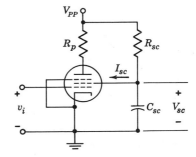

Fig. 4-18 Use of a screen-dropping resistor and a bypass capacitor to supply screen bias voltage.

minimum value of C_{sc} if the tilt is to be less than 10 percent for a 50-Hz square wave?

The screen current for a 6CL6 is approximately one-fourth the plate current. Hence, the screen quiescent current is 5 mA and the screen current under signal conditions is 7.5 mA. The difference, or 2.5 mA, must come from the screen capacitor, and this current will discharge the capacitor. If we assume that the plate current is approximately proportional to the screen voltage, then we can allow only 10 percent drop in V_{sc}, or 15 V. Thus

$$\Delta V_{sc} = \frac{(\Delta I_{sc})t}{C_{sc}} \tag{4-54}$$

where t is the time for half a cycle = 0.01 sec. Thus

$$C_{sc} = \frac{(\Delta I_{sc})t}{\Delta V_{sc}} = \frac{2.5 \times 10^{-3}}{15} \times 10^{-2} \text{ F} = 1.7 \ \mu\text{F}$$

This is a reasonable value, and hence screen grids are usually bypassed.

When it is desired to decrease even the small tilt which might be introduced by the screen circuit and when an appropriate low-impedance screen power supply is not available, the screen may be fed from a cathode follower whose grid is at a fixed voltage with respect to ground. If the screen current changes by 2.5 mA, as in the above illustration, and if the cathode-follower output impedance is, say, 400 Ω, then the change in screen voltage is only $2.5 \times 10^{-3} \times 400 = 1$ V.

4-13   EFFECT OF AN EMITTER (OR A CATHODE) BYPASS CAPACITOR ON LOW–FREQUENCY RESPONSE

If an emitter resistor R_e is used for self-bias in an amplifier and if it is desired to avoid the degeneration and hence the loss of gain due to R_e, then we might attempt to bypass this resistor with a very large capacitor C_z. The circuit is indicated in Fig. 4-2. It will be shown that if the input is a square wave, the output is a square wave with a tilt similar to that due to the coupling capacitor between stages.

In the analysis it is convenient to use the hybrid-Π model of Fig. 4-7. At low frequencies we may neglect the capacitances C_e and C_c. We have already demonstrated in Sec. 4-6 that for the usual parameters encountered in a pulse amplifier the conductances $g_{b'c}$ and g_{ce} may also be taken as zero. The equivalent circuit, subject to these approximations, is given in Fig. 4-19a.

A current i_b causes a drop $r_{b'e}i_b$ across $r_{b'e}$ and, as a consequence, the collector current is $i_c = g_m r_{b'e}i_b = h_{fe}i_b$. The current $i_z = i_b + i_c = (h_{fe} + 1)i_b$. If the impedance in the emitter lead is $Z_e(s)$, then, as a consequence of the current $i_b(s)$, the drop across Z_e is $(h_{fe} + 1)Z_e(s)i_b(s)$. Therefore, looking into terminals EN from the input side we see an impedance $(h_{fe} + 1)Z_e$. The current i_b may be determined, accordingly, from the equivalent circuit of Fig. 4-19b.

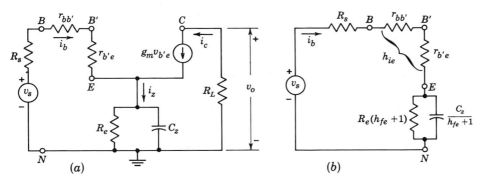

Fig. 4-19 (a) The low-frequency hybrid-II equivalent circuit of a transistor with an emitter impedance; (b) the equivalent input circuit.

The time constant τ of this circuit is given by the product of $C_z/(h_{fe} + 1)$ and a resistance given by the parallel combination of R' and R, where

$$R' \equiv R_e(h_{fe} + 1) \qquad R \equiv R_s + r_{bb'} + r_{b'e} = R_s + h_{ie} \qquad (4\text{-}55)$$

so that

$$\tau = \frac{RR'}{R + R'} \frac{C_z}{h_{fe} + 1} \qquad (4\text{-}56)$$

If the input signal v_s is a step of magnitude V, then $i_b = V/R$ at $t = 0+$ (because the capacitor acts as a short circuit) and $i_b = V/(R + R')$ at $t = \infty$ (because the capacitor behaves as an open circuit). Since the output voltage is

$$v_o = -i_c R_L = -g_m R_L v_{b'e} = -g_m R_L r_{b'e} i_b = -h_{fe} R_L i_b \qquad (4\text{-}57)$$

then, using Eq. (2-3),

$$v_o = -\frac{h_{fe} R_L V}{R + R'} \left(1 + \frac{R'}{R} \epsilon^{-t/\tau} \right) \qquad (4\text{-}58)$$

If $t \ll \tau$, then we may expand the exponential and, retaining only the linear term, we obtain

$$v_o = -\frac{h_{fe} R_L V}{R} \left(1 - \frac{R'}{R + R'} \frac{t}{\tau} \right) \qquad (4\text{-}59)$$

The percentage tilt in the output is

$$P = \frac{R'}{R + R'} \frac{t}{\tau} \times 100\% = \frac{(h_{fe} + 1)t}{RC_z} \times 100\% \qquad (4\text{-}60)$$

Note that the tilt is independent of R_e and R_L provided that the tilt is small.

Assuming $R_s = 0$, let us calculate the size of C_z so that we may reproduce a 50-Hz square wave with a tilt of less than 10 percent. Using the parameters of our typical transistor, namely, $h_{fe} = 50$, $r_{b'e} = 1$ K, $r_{bb'} = 100\ \Omega$, and with

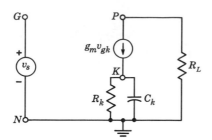

Fig. 4-20 The equivalent circuit of a pentode stage $(r_p \gg R_L + R_k)$ with a cathode impedance.

$t = 0.01$ sec for one-half a cycle,

$$C_z = \frac{(51)(0.01)(100)}{(1,100)(10)} \text{ F} = 4,600 \ \mu\text{F}$$

For a 1 percent tilt, C_z would have to be at least 46,000 μF! If $R_s = h_{ie} = 1,100 \ \Omega$ then R is doubled and C_z is cut in half. Such large values of capacitance are impractical and it must be concluded that if accurate reproduction of the flat top of a square wave of low frequency is desired the emitter bias resistor must be unbypassed. The flatness will then be obtained at the sacrifice of gain because of the degeneration caused by R_e. If the loss in amplification cannot be tolerated, R_e cannot be used.

If the active device is a pentode instead of a transistor, the equivalent circuit of Fig. 4-20 must be used. From a comparison with Fig. 4-19a it is clear that the above analysis remains valid for a tube, provided that we take $r_{b'e} = \infty$, $r_{bb'} = 0$, $R_s = 0$, $C_z = C_k$, and $R_e = R_k$. Since, as $r_{b'e}$ approaches infinity,

$$\frac{h_{fe} + 1}{R} = \frac{g_m r_{b'e} + 1}{r_{b'e}} = g_m$$

the percentage tilt is, from Eq. (4-60),

$$P = \frac{g_m t}{C_k} \times 100 \% \tag{4-61}$$

If g_m for the pentode is 5 millimhos (one-tenth that of the transistor), then for a 1 percent output tilt with a 50-Hz square-wave input the capacitor C_k must be at least 5,000 μF, still an unreasonable value.

4-14   FLAT–TOP RESPONSE OF CASCADED STAGES[4]

If, upon application of a voltage step, one resistance-capacitance coupling circuit produces a tilt of P_1 percent and if a second circuit produces a tilt of P_2 percent, the effect of cascading these circuits is to produce a tilt of $P_1 + P_2$ percent. This result applies only if the individual tilts and combined tilt are small enough so that in each case the voltage falls approximately linearly with time. We shall now prove these statements.

For a step input of amplitude V, the output of the first circuit is $V\epsilon^{-t/\tau_1} \approx V(1 - t/\tau_1)$, in which τ_1 is the time constant. If this signal is applied to the second circuit, of time constant $\tau_2 = R_2C_2$, then, neglecting the possible gain of the active device, the result may be computed from the equation

$$R_2 i + \frac{q}{C_2} = V\left(1 - \frac{t}{\tau_1}\right)$$

Differentiating this equation with respect to t (remembering that $v_o = R_2 i$) yields

$$\frac{dv_o}{dt} + \frac{v_o}{\tau_2} = -\frac{V}{\tau_1}$$

The solution for the output voltage v_o, subject to $v_o = V$ at $t = 0$, is

$$v_o = -V\frac{\tau_2}{\tau_1} + V\left(1 + \frac{\tau_2}{\tau_1}\right)\epsilon^{-t/\tau_2}$$

$$\approx V\left(1 - \frac{t}{\tau_1} - \frac{t}{\tau_2}\right) \tag{4-62}$$

Since t/τ_1 is the tilt due to the first network and t/τ_2 is the tilt due to the second network, Eq. (4-62) verifies the rule started above: the resultant tilt caused by two RC circuits in cascade is the sum of the tilts due to each network. Since the output again has a linear tilt, we may extend the result to an arbitrary number of stages, provided only that the net tilt remains small enough to be represented by a linear fall.

It was noted earlier that, within a single amplifier stage, tilt may be introduced by the coupling circuit, the screen circuit, and the cathode or emitter circuit. Since each of these produces its tilt by a mechanism which is independent of the others, the net tilt produced by an individual stage may be computed again by simply adding the individual tilts.

We have seen that one stage of an RC-coupled amplifier, whether of the tube or transistor type, behaves like a circuit with a single time constant τ. If the midband gain is A_o, then the low-frequency transfer function is $A_o/(1 - j/\omega\tau)$, which, in the Laplace domain, may be written for a single stage as

$$\frac{A_o}{1 + 1/s\tau} = \frac{A_o s\tau}{1 + s\tau}$$

For n identical stages with a step voltage of magnitude V, the Laplace transform of the output is

$$V_o(s) = \frac{V}{s}\left(\frac{A_o s\tau}{1 + s\tau}\right)^n = \frac{V A_o{}^n s^{n-1}}{(s + 1/\tau)^n} \tag{4-63}$$

The output function has an nth-order pole at $s = -1/\tau$. Taking the inverse transform we obtain for $y \equiv v_o/A_o{}^n V$ versus $x \equiv t/\tau$

$$y = \frac{1}{(n-1)!}\frac{d^{n-1}}{dx^{n-1}}\left(x^{n-1}\epsilon^{-x}\right) \tag{4-64}$$

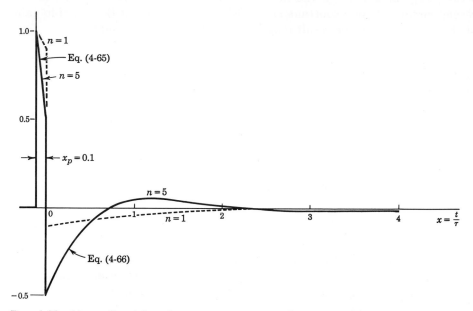

Fig. 4-21 Normalized low-frequency response of one and five identical cascaded stages to an input pulse (whose trailing edge occurs at $x = 0$).

For small values of x, we may write $\epsilon^{-x} = 1 - x$ in Eq. (4-64), which then reduces to

$$y = 1 - nx = 1 - \frac{nt}{\tau} \tag{4-65}$$

This equation again verifies the fact that the tilt of n identical stages is n times the tilt of a single stage.

A pulse of width t_p may be considered to be the result of the superposition of a positive step which occurs at $t = 0$ and a negative step at $t = t_p$. Suppose that a pulse of width t_p is applied to a cascade of a number of amplifiers with identical low-frequency time constants. The response at the end of the pulse (for $t \geq t_p$) may be obtained from Eq. (4-64) by forming the difference

$$v_d \equiv v_o(t) - v_o(t - t_p)$$

For very small values of t_p (more specifically if $t \gg t_p$), it follows from the definition of the derivative that

$$v_d \approx t_p \frac{dv_o}{dt} = x_p \frac{dv_o}{dx}$$

where $x_p \equiv t_p/\tau$. Using Eq. (4-64), we have for $y_d \equiv v_d/A_o{}^n V$

$$y_d = \frac{x_p}{(n-1)!} \frac{d^n}{dx^n}(x^{n-1}\epsilon^{-x}) \qquad\qquad (4\text{-}66)$$

Let us assume $x_p = 0.1 = t_p/\tau$, so that for a single stage there is a tilt of 10 percent during the pulse. The response during and after the pulse is indicated by the dashed curve marked $n = 1$ in Fig. 4-21. For five cascaded stages the tilt will be about 50 percent during the pulse, and the response is indicated by the solid curve ($n = 5$) in Fig. 4-21. Theoretically for n stages there should be $n - 1$ crossings of the zero voltage axis, but because the attenuation is so great not all of these are clearly visible in Fig. 4-21. It is of practical importance to note that the response to a pulse persists for a very long time relative to the pulse itself.

4-15   CATHODE INTERFACE RESISTANCE[5]

In many vacuum tubes there develops with use a *cathode interface layer* between the base metal of the cathode and the active emitting surface of the cathode, as shown in Fig. 4-22. The interface compound is a semiconductor compound formed as a result of the chemical interaction between the oxide-emitting material and the base metal or with some reducing constituent of the base metal. The resistance of the interface layer may lie in the range from several ohms to several hundred ohms and may therefore have an appreciable influence on tube operation. Additionally, the emitting surface and the cathode base metal serve as the electrodes of a capacitor, the cathode interface layer acting as a leaky dielectric between these electrodes. The overall effect of the interface layer is to introduce into the cathode a parallel resistance-capacitance combination whose time constant, it is found experimentally, normally lies in the approximate range 0.2 to 2.0 μsec.

In video amplifiers the effect of cathode interface resistance may well be serious. For a signal whose period is very large in comparison with the interface time constant, the principal effect is a loss in gain since the effective transconductance of the tube will be reduced from g_m to $g_m/(1 + g_m R_i)$, R_i being the

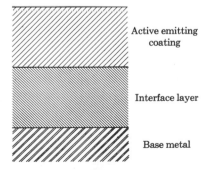

Active emitting
coating

Fig. 4-22   Cross section of cathode, showing interface layer.

Interface layer

Base metal

interface resistance. An abrupt discontinuity applied to the tube grid will appear at the output similarly reduced in amplitude but accompanied by an overshoot at the leading edge of the pulse.

Interface resistance is present to some extent in all tubes with oxide-coated cathodes but is usually particularly pronounced in tubes whose cathode base material contains a large amount of silicon. Interface resistance is inversely proportional to cathode area and is therefore more serious in tubes with small cathode areas. Also, since the effect of interface resistance is to reduce the effective transconductance by the factor $1 + g_m R_i$, high-g_m tubes are particularly sensitive to interface effects. Interface resistance increases with the total number of hours that the cathode has been heated, and the end of the useful life of a tube may be the result of interface resistance rather than loss in cathode emission.

A second disease which is often characteristic of video amplifier tubes has the popular designation "slump." The term is applied to a tube which behaves as though there were present in the cathode a parallel resistance-capacitance combination with a time constant in the range of several seconds. The response of such a tube to an input negative step is an output positive step which gradually slumps to a lower voltage level. The origin of "slump" is not well understood. The effect is often a source of difficulty in the design of d-c amplifiers for cathode-ray oscilloscopes.

REFERENCES

1. Millman, J.: "Vacuum-tube and Semiconductor Electronics," chap. 15, McGraw-Hill Book Company, New York, 1958.

2. Thomas, D. E., and J. L. Moll: Junction Transistor Short-circuit Current Gain and Phase Determination, *Proc. IRE*, vol. 46, no. 6, pp. 1177–1184, June, 1958.
   Kressel, H., H. S. Veloric, and A. Blicher: Design Consideration for Double-diffused Silicon Switching Transistors, *RCA Rev.*, vol. 23, no. 4, pp. 587–616, December, 1962.

3. Johannes, V.: "Transient Response of Transistor Video Amplifiers," sec. 1, University Microfilms, Ann Arbor, Mich., 1961.

4. Valley, G. E., Jr., and H. Wallman: "Vacuum Tube Amplifiers," Massachusetts Institute of Technology Radiation Laboratory Series, vol. 18, chap. 2, McGraw-Hill Book Company, New York, 1948.

5. Dukat, F. M., and I. E. Levy: Cathode-interface Effects in TV Receiver Design, *Electronics*, vol. 26, no. 4, pp. 169–171, April, 1953.
   Eisenstein, A.: The Leaky-condenser Oxide Cathode Interface, *J. Appl. Phys.*, vol. 22, no. 2, pp. 138–148, February, 1951.

5 / WIDEBAND AMPLIFIERS (COMPENSATED)

By adding a few passive circuit elements to the basic amplifier configurations discussed in the preceding chapter improved characteristics may be obtained. The rise time may be shortened and the tilt may be decreased. These compensated amplifiers are considered in detail in the present chapter.

By employing transmission-line coupling between active elements it is possible to extend the bandpass of a vacuum-tube amplifier into the hundreds-of-megahertz region. These so-called "distributed amplifiers" are also discussed in this chapter.

5-1   SHUNT COMPENSATION OF A VACUUM–TUBE STAGE[1–4]

One of the simplest methods available for improving the rise-time (or high-frequency) response of an amplifier without loss of gain is to add an inductor L in series with the plate-circuit resistor R_p, as in Fig. 5-1. This arrangement places L in parallel with the capacitance which shunts the stage, and hence the circuit is called a "shunt-compensated" or "shunt-peaked" amplifier.

We can readily see, qualitatively, why the inductor improves the high-frequency response. The plate-circuit impedance is now $Z_p = R_p + j\omega L$ and increases with frequency, so that the gain is larger than it would be if L were absent. This increase in amplification tends to offset the decrease in gain due to the shunting capacitance C, whose reactance decreases with increasing frequency.

The analysis of the uncompensated amplifier discussed in the preceding chapter was made in the frequency domain. The gain function turned out to be that corresponding to a single-time-constant

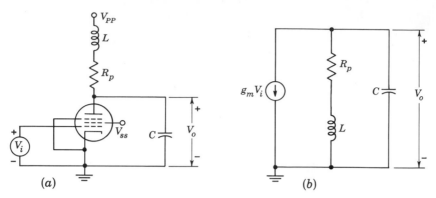

Fig. 5-1 (a) A shunt-compensated stage; (b) the equivalent circuit if $r_p \gg R_p$.

circuit or, equivalently, the transfer function contained a single pole. Under these circumstances we see in Sec. 4-9 that the transient response can be written down immediately. For the shunt-compensated stage we shall find a more complicated transfer function (one containing two poles and one zero). For this case a knowledge of the frequency response (the amplitude and phase versus frequency) is of very little practical help in finding the transient response. Hence, the present analysis will be made in the complex-frequency s plane (the Laplace domain) so that the rise-time response can be obtained directly.

Since the input circuit for a stage containing a tube is different from that using a transistor, each analysis will be considered separately, the former in this section and the latter in the next. Since for a pentode $r_p \gg R_p$, the equivalent circuit of a shunt-compensated stage is that given in Fig. 5-1b. The output voltage equals the short-circuit current times the impedance (or divided by the admittance) across the output port [Eqs. (1-1)], or

$$V_o(s) = \frac{-g_m V_i(s)}{1/(R_p + sL) + sC} \tag{5-1}$$

The transfer function $A(s)$ is the ratio of output to input voltage, or

$$A(s) \equiv \frac{V_o(s)}{V_i(s)} = \frac{-g_m(sL + R_p)}{s^2 CL + sCR_p + 1} \tag{5-2}$$

This function has two poles and one zero and may be written in the form

$$\frac{V_o(s)}{V_i(s)} = \frac{-(g_m/C)(s + s_o)}{(s + s_1)(s + s_2)} \tag{5-3}$$

It is convenient to introduce the parameters K and f_2, defined by

$$K \equiv R_p \sqrt{\frac{C}{L}} \qquad f_2 \equiv \frac{1}{2\pi R_p C} \tag{5-4}$$

in which case

$$s_o = \frac{R_p}{L} \qquad \left.\begin{matrix} s_1 \\ s_2 \end{matrix}\right\} = \pi f_2 K^2 \pm \pi f_2 K^2 \sqrt{1 - \frac{4}{K^2}} \qquad (5\text{-}5)$$

The parameter f_2 is, of course, the upper 3-dB frequency of the uncompensated amplifier $(L = 0)$, and $K = 1/Q_o$. Here Q_o is the Q at the resonant frequency $(\omega_o = 1/\sqrt{LC})$ of the series combination of R_p, L, and C, so that $Q_o = \omega_o L/R_p$.

Let us consider the transient response to an applied step V so that

$$V_i(s) = \frac{V}{s}$$

Depending upon whether K is equal to, smaller than, or larger than 2, the response will be critically damped, underdamped (oscillatory), or overdamped. The results of taking the inverse Laplace transform of $V_o(s)$ are given below for the various special cases. For convenience we introduce the normalized time x and normalized response y given by

$$x \equiv \frac{t}{2\pi R_p C} = f_2 t \qquad y \equiv -\frac{v_o}{g_m R_p V} \qquad (5\text{-}6)$$

and we have

Critical Damping, $K = 2$

$$y = 1 - \epsilon^{-4\pi x} - 2\pi x \epsilon^{-4\pi x} \qquad (5\text{-}7)$$

Underdamped, $K < 2$

$$y = 1 + \epsilon^{-\pi K^2 x}\left[\frac{2 - K^2}{K\sqrt{4 - K^2}} \sin\left(\pi K \sqrt{4 - K^2}\, x\right)\right.$$
$$\left. - \cos\left(\pi K \sqrt{4 - K^2}\, x\right)\right] \qquad (5\text{-}8)$$

Overdamped, $K \gg 2$ In this case of large K, $s_1 \approx 2\pi f_2 K^2$ and $s_2 \approx 2\pi f_2$. The term in the solution associated with the first of these roots will decay very rapidly; we may therefore neglect it and write

$$y \approx 1 - \epsilon^{-2\pi f_2 t} = 1 - \epsilon^{-t/R_p C} \qquad (5\text{-}9)$$

as is to have been expected.

From Eq. (5-4) the inductance L is given by

$$L = m R_p^2 C \qquad (5\text{-}10)$$

where $m \equiv 1/K^2$. (Both parameters, m and K, are used in the literature.)

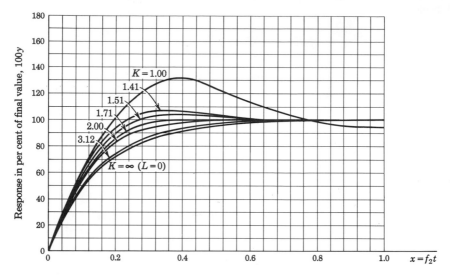

Fig. 5-2   Response of a shunt-compensated vacuum-tube stage to a unit voltage step. (Adapted from Ref. 2.)

The unit step response for several values of K is shown in Fig. 5-2. As the peaking inductor is increased in value, there is a progressive improvement in rise time without accompanying overshoot up to the point of critical damping. Beyond this point the amplifier response exhibits a progressively larger overshoot. The factor by which the rise time is improved (divided) by compensation is $\rho \equiv t_r/t_r'$, in which t_r and t_r' are, respectively, the rise time of the amplifier before and after compensation. The parameter ρ and the percentage overshoot γ are plotted in Fig. 5-3. For the case of critical damping, $m = 0.25$, the rise time is improved by the factor 1.43.

If the frequency response is determined by replacing s by $j\omega$ in Eq. (5-2) it is found that the curve having the most uniform amplitude response (maximum flatness) corresponds to $K = 1.54$. The curve having the most constant time delay is given by $K = 1.71$. The curve for which $|A_2/A_o| = 1$ at $f/f_2 = 1$ is given by $K = 1.41$. The overshoot γ and rise-time improvement ρ for these special cases are summarized in Table 5-1.

TABLE 5-1   Overshoot and rise-time improvement

K	$m = 1/K^2$	γ, %	ρ	Characteristic
2	0.25	0	1.43	Critical damping
1.71	0.34	1.0	1.70	Most constant delay
1.54	0.42	3.8	1.90	Maximum flatness
1.41	0.50	6.5	2.00	$|A_2/A_o| = 1$ at $f/f_2 = 1$

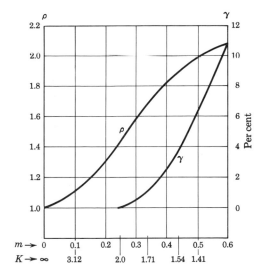

Fig. 5-3 Percent overshoot γ and rise-time improvement ρ of a shunt-compensated vacuum-tube stage.

$m \equiv 1/K^2$;

$K \equiv R_p \sqrt{C/L}$.

The amount of overshoot which is tolerable is very largely a function of the application of the amplifier. For example, for an amplifier to be used in oscillography, any visible overshoot would be objectionable. On the other hand, in television amplifiers, overshoots as large as 5 percent not only may be acceptable but may actually improve the quality of the resultant picture. If the number of stages used is large, then the overshoot should be kept below about 2 percent (Sec. 5-4).

In the case of no compensation it will be recalled [Eq. (2-33)] that the product $t_r f_2 = 0.35$. It is of interest to note that the same rule also applies quite well in the present case of shunt compensation. For example, we may calculate that for critical damping the amplitude response falls by 3 dB at $f/f_2 = \sqrt{2}$. Since we estimated above that in this case the rise time was divided by the factor 1.43, we have that

$$f_2' t_r' = \sqrt{2}\, f_2 \frac{t_r}{1.43} \approx f_2 t_r = 0.35$$

where, here, t_r' and f_2' are the rise time and bandwidth for critical damping.

An initial estimate of the peaking inductance required may be made by estimating the total shunt capacitance. The required inductance is usually in the range 1 to 100 μH. Adjustable coils are available for this application, and the final adjustment is made experimentally by the method of square-wave testing. The inductance is changed by varying the depth of insertion into the coil form of a powdered-iron slug. The square-wave frequency is set so that the half period of the square wave is several times the rise time and the inductance is adjusted to give the type of response most suitable for the application for which the amplifier stage is intended.

5-2   SHUNT COMPENSATION OF A TRANSISTOR STAGE IN A CASCADE[1,5]

The circuit in the dashed box of Fig. 4-15 is a complete stage, collector to collector, in a cascade of stages. This complete stage is redrawn in Fig. 5-4 with the addition of the shunt-compensating inductor L. The capacitor $C = C_e + C_c(1 - K)$. As before, we use the low-frequency gain $K = K_o$ in the calculation of C. Since at low frequencies the inductor acts as a short circuit, K_o is unaffected by L. Hence, C is given by Table 4-1 (and is repeated in Table 5-2, on page 156, for convenience).

The transfer function (current gain or voltage gain, since in Sec. 4-10 we show that the two are identical) has the same form as Eq. (5-3), namely,

$$A(s) = \frac{I_2(s)}{I_1(s)} = \frac{-(g_m/C)(s + s_o)}{(s + s_1)(s + s_2)} \tag{5-11}$$

where

$$s_o = \frac{R_c}{L} \tag{5-12}$$

$$\left. \begin{array}{c} s_1 \\ s_2 \end{array} \right\} = \frac{1}{2}\left(\frac{R_c + r_{bb'}}{L} + \frac{1}{r_{b'e}C} \right) \pm \frac{1}{2}\left[\left(\frac{R_c + r_{bb'}}{L} + \frac{1}{r_{b'e}C} \right)^2 \right. $$
$$\left. - \frac{4}{LC}\frac{R_c + h_{ie}}{r_{b'e}} \right]^{\frac{1}{2}} \tag{5-13}$$

We seek to adjust the circuit so that, for an applied step input waveform, the output response will have the shortest rise time consistent with no overshoot. The transfer function of Eq. (5-11) has a single zero and two poles, just as does the transfer function of Eq. (5-3) for the shunt-compensated vacuum-tube stage. In correspondence with that case we might have expected that an adjustment which satisfies the condition $s_1 = s_2$ would ensure no overshoot. However, we shall now find that an additional condition (besides $s_1 = s_2$) must be satisfied if the response is to be monotonic.

Identical Poles   For an input current step $I_1(s) = I/s$. When s_1 and s_2 are identical, so that $s_1 = s_2 = s_i$, we have from Eq. (5-11) that

$$I_2(s) = \frac{-(g_m/C)(s + s_o)I}{s(s + s_i)^2} \tag{5-14}$$

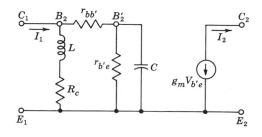

Fig. 5-4   The equivalent circuit of one stage of a cascade of shunt-compensated transistor stages.

$h_{ie} = r_{bb'} + r_{b'e}$.

The final-value theorem of Laplace transform theory is $i_2(\infty) = \lim\limits_{s\to 0} sI_2(s)$. Hence the steady-state value of the current, the value attained by $i_2(t)$ at $t = \infty$, is $i_2(\infty) = -g_m s_o I/C s_i{}^2$. Taking the inverse transform of Eq. (5-14), the output current, normalized with respect to this final value, is

$$y \equiv \frac{-C s_i{}^2 i_2(t)}{g_m s_o I} = 1 - \left[1 + s_i t \left(1 - \frac{s_i}{s_o}\right)\right] \epsilon^{-s_i t} \tag{5-15}$$

If the response is to be monotonic, then the slope dy/dt must not be zero except, of course, at $t = \infty$. We find for the derivative

$$\frac{dy}{dt} = (s_i{}^2 \epsilon^{-s_i t}) \left[t\left(1 - \frac{s_i}{s_o}\right) + \frac{1}{s_o}\right] \tag{5-16}$$

from which it is clear that if $s_o \geq s_i$, then $1 - s_i/s_o \geq 0$, so that there will be no value of t for which $dy/dt = 0$. *The magnitude of the zero must equal or exceed that of the pole in order to ensure a response without overshoot.* In the vacuum-tube case it turns out that this condition is automatically satisfied since there $s_o = R_p/L$ and $s_i = R_p/2L$.

From Eq. (5-13) we have, since the quantity within the square root must be zero for equal poles, that

$$s_i = \frac{1}{2}\left(\frac{R_c + r_{bb'}}{L} + \frac{1}{r_{b'e}C}\right) = \sqrt{\frac{1}{LC} \frac{R_c + h_{ie}}{r_{b'e}}} \tag{5-17}$$

Since Eq. (5-17) is quadratic in L, the equation is satisfied by either of two inductance values. Correspondingly s_i may assume one of two values. The larger s_i, the smaller is the rise time, since the exponent in Eq. (5-15) is proportional to s_i. The smaller L, the larger is s_i. After some algebraic manipulation we find that the smaller value L' of L that satisfies Eq. (5-17) is

$$L' = r_{b'e}^2 C \left(1 - \sqrt{\frac{R_c + h_{ie}}{r_{b'e}}}\right)^2 \tag{5-18}$$

Using this value of L' we find that the condition $s_o \geq s_i$ imposes on R_c the restriction

$$R_c \geq \frac{r_{bb'} h_{ie}}{r_{b'o}} \tag{5-19}$$

The rise time may be determined by using numerical or graphical means to calculate from Eq. (5-15) the value of $s_i t$ for which $y = 0.1$ and again the value of $s_i t$ for which $y = 0.9$. The difference between the two is $s_i t'_r$, where t'_r is the compensated rise time. From Eq. (5-15) we see that $s_i t'_r$ is a function of $s_i/s_o \equiv d$. Of more interest than the compensated rise time is the rise-time improvement due to compensation, $\rho \equiv t_r/t'_r$. We have, using t_r from Table 4-1, that

$$\rho \equiv \frac{t_r}{t'_r} = \frac{2.2 RC}{t'_r} = \frac{2.2}{(s_i/s_o)s_i t'_r} \frac{s_i{}^2}{s_o} RC \tag{5-20}$$

Using Eqs. (4-41), (5-17), and (5-12) we find that $s_i{}^2 RC/s_o = (R_c + r_{bb'})/R_c$. Further, the quantity $(s_i/s_o)s_i t_r'$ is a function only of the ratio $s_i/s_o = d$. Therefore

$$\rho = \frac{R_c + r_{bb'}}{R_c} B(d) \qquad (5\text{-}21)$$

in which[5]

$$B(d) \equiv \frac{2.2}{(s_i t_r')d} \qquad (5\text{-}22)$$

The quantity $B(d)$ is plotted in Fig. 5-5 as a function of d. We have already proved that $s_o \geq s_i$, and hence the maximum value of $d = s_i/s_o$ is unity. From Eqs. (5-12) and (5-17), with $L = L'$

$$d \equiv \frac{s_i}{s_o} = \frac{1}{2} + \frac{1}{2}\left(\frac{r_{bb'}}{R_c} + \frac{L'}{r_{b'e}R_cC}\right) \qquad (5\text{-}23)$$

Since all the parameters in this equation are positive, the minimum value of d is 0.5. Hence the abscissa in Fig. 5-5 extends only from 0.5 to 1.0.

Substituting for L' from Eq. (5-18) into Eq. (5-23) we obtain, after some algebraic manipulation,

$$d = \frac{r_{b'e}}{R_c}\left(\frac{R_c + h_{ie}}{r_{b'e}} - \sqrt{\frac{R_c + h_{ie}}{r_{b'e}}}\right) \qquad (5\text{-}24)$$

Since at low frequencies the inductance acts as a short circuit, the compensated midband gain is the same as that of the uncompensated stage.

We have now completed the analysis on the basis of which we may select the proper value of inductance and calculate the rise-time improvement for the case where we have initially selected R_c so that it satisfies the condition given in Eq. (5-19). We shall now show how an overshoot may be avoided even when R_c is too small to satisfy Eq. (5-19).

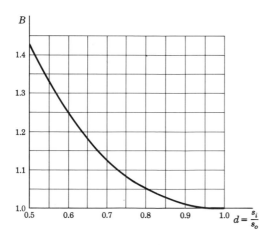

Fig. 5-5   The function $B(d)$.

Pole-Zero Cancellation A response without overshoot may also be obtained if we arrange for a pole-zero cancellation. For example, if $s_o = s_2$, then for an applied step $I_1(s) = I/s$, Eq. (5-11) yields

$$I_2(s) = \frac{-g_m I}{Cs(s + s_1)} \tag{5-25}$$

Taking the inverse transform, we find for the normalized response

$$y \equiv -\frac{Cs_1 i_2(t)}{g_m I} = 1 - \epsilon^{-s_1 t} \tag{5-26}$$

which increases monotonically with time.

Using Eqs. (5-12) and (5-13) we find that either of the conditions $s_o = s_1$ or $s_o = s_2$ leads to the same value L'' of inductance, which is given by

$$L'' = \frac{r_{b'e} r_{bb'} R_c C}{h_{ie}} \tag{5-27}$$

The rise time corresponding to the solution given in Eq. (5-26) is $2.2/s_1$. Had we canceled the other pole, so that $s_o = s_1$, the rise time would have been $2.2/s_2$. Since we are interested in obtaining the fastest possible response, we must cancel the smaller pole. This smaller pole is that at s_2, where s_2 corresponds to the minus sign in Eq. (5-13). Accordingly

$$\frac{R_c}{L''} \leq \frac{1}{2}\left(\frac{R_c + r_{bb'}}{L''} + \frac{1}{r_{b'e}C}\right) \tag{5-28}$$

Combining Eqs. (5-27) and (5-28), we have the condition on R_c that

$$R_c \leq \frac{r_{bb'} h_{ie}}{r_{b'e}} \tag{5-29}$$

Since from Eq. (5-13)

$$s_1 + s_2 = \frac{R_c + r_{bb'}}{L''} + \frac{1}{r_{b'e}C} \qquad \text{and} \qquad s_2 = s_o = \frac{R_c}{L''}$$

then

$$s_1 = \frac{r_{bb'}}{L''} + \frac{1}{r_{b'e}C} = \frac{R_c + h_{ie}}{R_c r_{b'e} C} \tag{5-30}$$

where we have used the value L'' from Eq. (5-27). The compensated rise time t_r' is now given by

$$t_r' = \frac{2.2}{s_1} = \frac{2.2 R_c r_{b'e} C}{R_c + h_{ie}} \tag{5-31}$$

From Table 4-1 and Eq. (5-31) we find that

$$\rho \equiv \frac{t_r}{t_r'} = \frac{R(R_c + h_{ie})}{R_c r_{b'e}} = \frac{R_c + r_{bb'}}{R_c} \tag{5-32}$$

Since $s_1 > s_2 = s_o$, $s_1/s_o > 1$. If we define $B(d) \equiv 1$ for $d \equiv s_1/s_o > 1$, then Eq. (5-21) for ρ may be used for all values of R_c.

Design Considerations The results derived in this section are summarized in Table 5-2, which is convenient for reference when designing a shunt-compensated stage. The midband gain A_o is the same with compensation (L in series with R_c) as without compensation because at low frequencies the reactance of L may be neglected compared with the resistance R_c.

It is interesting to compare the performance of the shunt-compensated transistor stage in a cascade with that of a similar tube amplifier. The rise-time improvement for the tube (under conditions of critical damping) is 1.43 regardless of the parameters. For the transistor stage $\rho = [(R_c + r_{bb'})/R_c]B$,

---

TABLE 5-2   Shunt-compensated CE transistor stage in a cascade

---

Notation

$$R = \frac{(R_c + r_{bb'})r_{b'e}}{R_c + h_{ie}} \qquad R_L = \frac{R_c h_{ie}}{R_c + h_{ie}} \qquad C = C_e + C_c(1 + g_m R_L)$$

The low-frequency gain (compensated or uncompensated) or collector resistance

$$A_o = \frac{-h_{fe}R_c}{R_c + h_{ie}} \qquad \text{or} \qquad R_c = \frac{-A_o h_{ie}}{A_o + h_{fe}}$$

Inductance for $R_c \geq r_{bb'}h_{ie}/r_{b'e}$ (identical poles)

$$L' = r_{b'e}^2 C\left(1 - \sqrt{\frac{R_c + h_{ie}}{r_{b'e}}}\right)^2$$

Inductance for $R_c \leq r_{bb'}h_{ie}/r_{b'e}$ (pole-zero cancellation)

$$L'' = \frac{r_{b'e}r_{bb'}R_c C}{h_{ie}}$$

Uncompensated rise time

$$t_r = 2.2RC$$

Compensated rise time

$$t_r' = \frac{2.2C}{B(d)}\frac{r_{b'e}R_c}{R_c + h_{ie}}$$

Rise-time improvement factor $\rho \equiv t_r/t_r'$

$$\rho = \frac{R_c + r_{bb'}}{R_c}B(d)$$

where

$$d = \frac{r_{b'e}}{R_c}\left(\frac{R_c + h_{ie}}{r_{b'e}} - \sqrt{\frac{R_c + h_{ie}}{r_{b'e}}}\right)$$

and $B(d)$ is plotted in Fig. 5-5 and is given in the following table:

d	0.5	0.55	0.6	0.65	0.7	0.8	0.9	≥ 1.0
B	1.43	1.33	1.25	1.18	1.13	1.05	1.01	1.00

where $B(d)$ is between 1 and 1.43. It is thus seen that for low values of R_c shunt compensation is more effective with transistors than with tubes; an improvement considerably greater than 1.43 may be attained. With values of R_c much greater than $r_{bb'}$, the parameters d, B, and ρ each approach unity and shunt compensation is less effective with transistor than with tube amplifiers.

The design of a stage for a specified gain A_o is straightforward. From Table 5-2 we can solve directly for the resistance R_c in the collector circuit and for the compensating inductance L' or L''. Corresponding to these values we can calculate the compensated rise time t_r' by first finding C and B from the formulas in Table 5-2. However, if the desired rise time t_r' is specified, then the solution is more complicated because of the dependence of t_r' upon resistance R_c implicitly through C and B. A method of successive approximations can be used,[5] or we may proceed by assuming various values of R_c and obtaining the corresponding rise time t_r'. From a plot of R_c versus t_r' we can find the value of R_c which yields the desired rise time. This method will now be illustrated.

EXAMPLE Design a shunt-compensated stage for the transistor whose parameters are given in Fig. 4-7 so that a rise time of 40 nsec is obtained. Find R_c, L, and A_o. What is the rise-time improvement due to the compensation?

Solution The parameters are $g_m = 5 \times 10^{-2}$ mho, $r_{b'e} = 1,000 \ \Omega$, $r_{bb'} = 100 \ \Omega$, $h_{ie} = 1,100 \ \Omega$, $C_e = 100$ pF, and $C_c = 3$ pF. We shall now find t_r' as a function of R_c. Start with $R_c = 1,000 \ \Omega$ as a first try. Using the formulas in Table 5-2,

$$R_L = \frac{R_c h_{ie}}{R_c + h_{ie}} = \frac{(1,000)(1,100)}{2,100} = 524 \ \Omega$$

$$C = C_e + C_c(1 + g_m R_L) = 100 + (3)(1 + 5 \times 10^{-2} \times 524) = 182 \text{ pF}$$

$$d = \frac{r_{b'e}}{R_c}\left(\frac{R_c + h_{ie}}{r_{b'e}} - \sqrt{\frac{R_c + h_{ie}}{r_{b'e}}}\right)$$

$$= \left(\frac{1,000}{1,000}\right)\left(\frac{2,100}{1,000} - \sqrt{2.1}\right) = 0.65$$

$$B(0.65) = 1.18 \qquad \text{from Fig. 5-5}$$

$$t_r' = \frac{2.2C}{B(d)} \frac{r_{b'e} R_c}{R_c + h_{ie}} = \frac{(2.2)(182 \times 10^{-12})(1,000)(1,000)}{(1.18)(2,100)} = 161 \text{ nsec}$$

This value is very much larger than the desired 40 nsec rise time. Hence, our first guess of 1,000 Ω for R_c was much too large. Let us therefore now try $R_c = 250$ Ω. Proceeding as above, we find

$$R_L = 204 \ \Omega \qquad C = 134 \text{ pF} \qquad d = 0.76 \qquad B = 1.08 \qquad t_r' = 50.5 \text{ nsec}$$

This rise time is still too large, and hence R_c must be somewhat smaller. Assuming $R_c = 200 \ \Omega$, we find

$$R_L = 167 \ \Omega \qquad C = 128 \text{ pF} \qquad d = 0.80 \qquad B = 1.05 \qquad t_r' = 41.2 \text{ nsec}$$

Extrapolating from 250 Ω and 50.5 nsec through 200 Ω and 41.2 nsec we obtain $R_c = 195\ \Omega$ for 40 nsec. Corresponding to $R_c = 195$ we find $C = 128$ pF and $B = 1.05$. Since

$$R_c > \frac{r_{bb'}h_{ie}}{r_{b'e}} = \frac{(100)(1,100)}{1,000} = 110\ \Omega$$

then we must use the value L' and not L'':

$$L' = r_{b'e}^2 C \left(1 - \sqrt{\frac{R_c + h_{ie}}{r_{b'e}}}\right)^2 = (10^6)(128 \times 10^{-12})(1 - \sqrt{1.295})^2\ \text{H}$$

$$= 2.44\ \mu\text{H}$$

$$A_o = \frac{-h_{fe}R_c}{R_c + h_{ie}} = \frac{-(50)(195)}{1,295} = -7.53$$

$$\rho = \frac{R_c + r_{bb'}}{R_c} B(d) = \frac{(295)(1.05)}{195} = 1.59$$

The above analysis gives the inductance value for the response without overshoot. Using an adjustable choke of this nominal value the method of square-wave testing is used to obtain experimentally the desired type of response (perhaps a few percent overshoot) for the application under consideration.

5-3   ADDITIONAL METHODS OF RISE–TIME COMPENSATION[6]

Various networks of the type shown in Fig. 5-6 have been suggested for coupling the output Y_1 of one tube stage to the input X_2 of the next in an attempt to improve the rise-time response of a video amplifier. The following terminology is common in the literature:

$C_1 = 0$	$L_1 \neq 0$	$L_2 = 0$	*shunt compensation*
$C_1 \neq 0$	$L_1 \neq 0$	$L_2 = 0$	*tuned-shunt compensation*
$C_1 = 0$	$L_1 = 0$	$L_2 \neq 0$	*series compensation*
$C_1 = 0$	$L_1 \neq 0$	$L_2 \neq 0$	*shunt-series compensation*
$C_1 \neq 0$	$L_1 \neq 0$	$L_2 \neq 0$	*Dietzold compensation*

The detailed analysis of any of these circuits except shunt peaking is very involved and will not be considered here. Optimum values of C_1, L_1, and L_2 are best determined experimentally. The process of adjustment for best response is usually quite complicated because the various parameters interact with one another. Furthermore, a rise-time improvement of only about 1.5 is obtained over that for shunt peaking even with the most complicated circuit

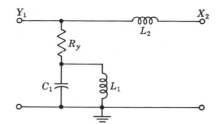

Fig. 5-6 Four-terminal coupling networks.

of Fig. 5-6. For these reasons the four terminal coupling networks are not as popular as the simple shunt-compensated network. Occasionally one of the more complicated networks is used because it makes possible the elimination of one stage in a multistage tube amplifier and thus effects a worthwhile economy of parts. Similar arrangements are possible also with transistor amplifiers, but these more complicated networks have found practically no applications.

In passing, it is interesting to note that the product of bandwidth and rise time for any of the networks of Fig. 5-6 remains approximately equal to that of the uncompensated stage, or

$$f_2 t_r \approx f_2' t_r' \approx 0.35 \tag{5-33}$$

5-4   RISE TIME OF CASCADED COMPENSATED STAGES[7]

When identical stages which individually have overshoot are cascaded, it is still possible to make some general rules concerning the overall response. These rules apply only very roughly but are nevertheless of some value.

When the individual stages have very small overshoot, of the order of 1 or 2 percent, the overshoot increases very slowly with the number of stages or may even fail entirely to increase. For example, if the overshoot for a single stage is about 1 percent, at 16 stages it has grown to only about 4 percent and is still about the same at 64 stages. If the rise time and bandwidth for n stages are $t_r^{(n)}$ and $f_2^{(n)}$, respectively, then it is found that the equations $t_r^{(n)} = 1.1 \sqrt{n} \, t_r$ and $f_2^{(n)} t_r^{(n)} = f_2 t_r$ also hold reasonably well in this case.

Circumstances are different when the overshoot is in the range 5 to 10 percent. In this case the rise time increases appreciably more slowly than $\sqrt{n}$, whereas the overshoot instead grows approximately as $\sqrt{n}$. If, therefore, an amplifier is to have a fairly large number of stages, it is clear that the individual stages must be adjusted for very slight or no overshoot.

5-5   LOW–FREQUENCY COMPENSATION[8]

In Sec. 4-4 it is seen that for a step input the amplifier response is not flat-topped but exhibits a downward tilt because of the coupling or blocking

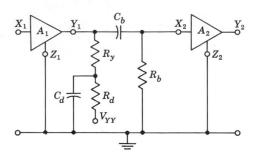

Fig. 5-7 The decoupling filter R_d and C_d is used for low-frequency compensation. For a tube $R_b = R_g$ and $R_y = R_p$. For a transistor, R_b is R_1 in parallel with R_2, the base bias resistors of Fig. 4-2b, and $R_y = R_c$.

capacitor C_b. If, in Fig. 4-2, we add a high resistance R_d in series with R_p or R_c which connects to the supply voltage and bypass this resistor to ground with a large capacitance C_d we may compensate for the tilt in the output waveform. The added elements are indicated in Fig. 5-7, which should be compared with Fig. 4-3. These additional components R_d and C_d are often used with a multistage amplifier as a decoupling filter to minimize the interactions between stages which result from the use of a common power supply. This same decoupling network compensates for the low-frequency distortion introduced by C_b. Thus at high frequencies (in the midband region) C_d acts as a short circuit across R_d, and the gain of the amplifier is determined by R_y. At low frequencies, however, C_d becomes a large reactance and the effective output-circuit resistance increases toward $R_y + R_d$. This increase in amplification tends to compensate for the loss in output due to the attenuation caused by the reactance of C_b, which increases as the frequency is reduced.

It is clear from the above qualitative discussion that R_d should be as large as possible. Therefore, let us begin the quantitative analysis by assuming R_d to be infinite. Furthermore, let us consider that the active device—pentode or transistor—is a perfect current source, so that its output impedance may be taken as infinite. For a negative step voltage applied to the input a constant current I_o is delivered, as indicated in Fig. 5-8. The resistance R_i' is R_b in parallel with the input resistance R_i of the following stage.

Since the current I_o divides inversely as the parallel impedances,

$$Z_1 \equiv R_y + \frac{1}{sC_d} \quad \text{and} \quad Z_2 \equiv R_i' + \frac{1}{sC_b}$$

the Laplace transform of the current in R_i' is $(I_o/s)[Z_1/(Z_1 + Z_2)]$, or

$$V_o(s) = \frac{R_i'I_o}{s} \frac{R_y + 1/sC_d}{R_y + R_i' + (1/s)(1/C_d + 1/C_b)} \tag{5-34}$$

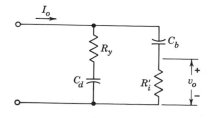

Fig. 5-8 The equivalent circuit of Fig. 5-7 for the flat-top response, assuming the active device behaves as an ideal current generator. For a tube $R_i' = R_g$. For a transistor R_i' is R_b in parallel with R_i, the input impedance of the next stage.

Let R_σ be the series combination of R_y and R_i' and let C_σ represent C_d in series with C_b, so that

$$R_\sigma \equiv R_y + R_i' \qquad C_\sigma \equiv \frac{C_d C_b}{C_d + C_b} \qquad (5\text{-}35)$$

Then

$$V_o(s) = \frac{R_y R_i' I_o (s + 1/R_y C_d)}{R_\sigma s (s + 1/R_\sigma C_\sigma)} \qquad (5\text{-}36)$$

Taking the inverse Laplace transform we obtain

$$y(t) = \frac{1}{\lambda} [(\lambda - 1)\epsilon^{-x} + 1] \qquad (5\text{-}37)$$

where

$$y \equiv \frac{v_o(t) R_\sigma}{I_o R_y R_i'} \qquad x \equiv \frac{t}{R_\sigma C_\sigma} \qquad (5\text{-}38)$$

$$\lambda \equiv \frac{R_y C_d}{R_\sigma C_\sigma} = \frac{1 + C_d/C_b}{1 + R_i'/R_y} \qquad (5\text{-}39)$$

Let us inquire about the output waveform as a function of C_d or λ. The results are shown in Fig. 5-9.

Perfect Compensation, $\lambda = 1$

$$y = 1 \qquad (5\text{-}40)$$

This case corresponds to a flat-topped output, or to perfect compensation. Note that $\lambda = 1$ or $R_y C_d = R_\sigma C_\sigma$ results in a pole-zero cancellation in Eq. (5-36). From Eq. (5-39) we see that this special case of $\lambda = 1$ is equivalent to

$$\frac{C_d}{C_b} = \frac{R_i'}{R_y}$$

or to

$$C_d = \frac{C_b R_i'}{R_y} \equiv C_f \qquad (5\text{-}41)$$

Fig. 5-9 Normalized output voltage response to a step input of a stage compensated for low-frequency tilt. The parameter λ given in Eq. **(5-39)** is adjusted by means of the compensating capacitance C_d.

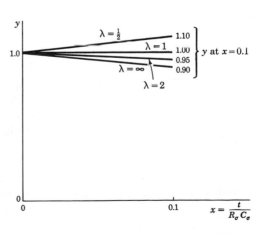

We see that perfect compensation corresponds to the case where the two branches of the circuit in Fig. 5-8 have the same time constants. The capacitance C_f for perfect compensation may be quite large. For example, for a tube circuit with $C_b = 0.1 \ \mu\text{F}$, $R_i' = R_g = 1 \ \text{M}$, and $R_y = R_p = 1 \ \text{K}$ we find $C_f = 100 \ \mu\text{F}$. For a transistor for which R_i' is of the same order of magnitude as $R_y = R_c$, $C_f \approx C_b \approx 50 \ \mu\text{F}$.

No Compensation, $\lambda \gg 1$

$$y \approx \epsilon^{-x} \tag{5-42}$$

This case corresponds to no compensation at all (C_d is very large). For times small compared with the time constant $R_\sigma C_\sigma$, so that $x \ll 1$, we have approximately

$$y \approx 1 - x \tag{5-43}$$

Undercompensation, $\lambda = 2$

$$y = \tfrac{1}{2}(1 + \epsilon^{-x}) \tag{5-44}$$

and for $x \ll 1$

$$y \approx 1 - \frac{x}{2} \tag{5-45}$$

We see that even if C_d is twice as large as the value C_f required for exact compensation, the percentage tilt is one-half the tilt of the uncompensated case, as may be seen by comparing Eq. (5-45) and Eq. (5-43).

Overcompensation, $\lambda = \tfrac{1}{2}$

$$y = 2 - \epsilon^{-x} \tag{5-46}$$

and for $x \ll 1$,

$$y \approx 1 + x \tag{5-47}$$

In this case the positive tilt is equal to the negative tilt of the uncompensated amplifier. Note that if "overcompensation" (an upward tilt) is observed in square-wave testing of an amplifier, it is to be corrected by *increasing* the size of C_d.

The curves of Fig. 5-9 are unrealistic because of the assumption that R_d is infinite. For large values of t, the capacitors C_b and C_d act as open circuits and hence eventually all the current must flow through R_d and not R_i'. Thus, all curves of Fig. 5-9 must eventually drop to zero. For example, the curve marked $\lambda = \tfrac{1}{2}$ would then have a rounded top. It is to be noted that if a square wave were to be applied to the amplifier for testing purposes, we would normally select the half period of the square wave to be approximately $0.1 R_\sigma C_\sigma$. In this case the input square wave would, after transmission through the amplifier, have the appearance of one of the curves of Fig. 5-10.

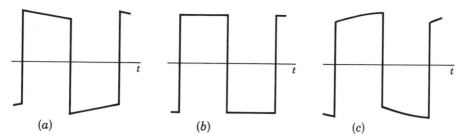

Fig. 5-10   Low-frequency square-wave response of an amplifier.   (a) C_d too large; (b) perfect compensation, $C_d = C_b R_i'/R_y$; (c) C_d too small.

Finite R_d   In the above analysis R_d was infinite.   Let us remove this restriction and examine the effect of a finite value of R_d.   It is desired that the output start out as flat as possible, or that $dy/dt = 0$ at $t = 0$.   This condition was satisfied for $R_d = \infty$ by choosing C_d in accord with Eq. (5-41). Since at $t = 0$ the capacitor C_d acts as a short circuit across R_d, then regardless of the value of R_d the output will begin with a horizontal slope, provided that $C_d = C_b R_i'/R_y$.   For this particular value of capacitance C_d we find in Prob. 5-17 that the first two terms of the power-series expansion of the normalized output voltage are

$$y = 1 - \frac{R_i'}{R_i' + R_y} \frac{x^2}{2n} \tag{5-48}$$

where

$$x \equiv \frac{t}{R_\sigma C_\sigma} \qquad n \equiv \frac{R_d}{R_y} \tag{5-49}$$

It is now seen that the tilt is parabolic and not linear and that the initial slope is indeed zero.   The amount of tilt to be expected is given by the following example.   If the uncompensated tilt is, say, 10 percent, so that $x = 0.1$, then Eq. (5-48) gives a compensated tilt of less than $(0.1)^2/2n \times 100$ percent $= 0.5/n$ percent because $R_i'/(R_i' + R_y) \leq 1$.   Hence, even if $n = 1$ or R_d is just equal to R_y (instead of $R_d \gg R_y$), the tilt is at most 0.5 percent. For a pentode $R_i' = R_g \gg R_y$ and $R_i'/(R_i' + R_y) \approx 1$, whereas for a transistor R_i' may be of the same order of magnitude as R_y and the tilt is reduced even further.

The above theory indicates that the proper procedure for low-frequency compensation is to choose R_d as large as possible and then to choose C_d to satisfy

$$R_y C_d = R_i' C_b$$

The upper limit on R_d is determined by the fact that the quiescent device current passes through R_d and that the power supply must be able to furnish this voltage drop.

Tilt Due to Several Sources   In Secs. 4-12 and 4-13 it is shown that the tilt may be produced not only by the coupling circuit but also by the screen and cathode or emitter circuits.   Further, the overall tilt is the sum of the tilts due to each of these three causes, provided that each sag is small.   Let the resultant tilt be P percent in the time $T/2$ of half a cycle.   If the screen and cathode or emitter bypass capacitors were arbitrarily large, then this tilt would be obtained if the blocking capacitance had an effective value C_b given by $P = 100T/2R_\sigma C_b$.   Hence, it is possible to compensate for the total tilt by choosing C_d to balance this effective C_b, or from Eq. (5-41)

$$C_d = \frac{C_b R_i'}{R_y} = \frac{100T}{2PR_\sigma}\frac{R_i'}{R_y} = \frac{50}{fPR_y}\frac{R_i'}{R_y + R_i'} \tag{5-50}$$

where f is the frequency of the testing square wave.

5-6   RISE–TIME COMPENSATION IN THE CATHODE CIRCUIT

In Sec. 4-13 we see that an unreasonably large bypass capacitor C_k is needed across the cathode self-biasing resistor R_k in order to prevent low-frequency distortion.   If no capacitor C_k is used, the midband gain is divided by $1 + g_m R_k$ because of the feedback in the cathode resistor, but there will be no tilt in the output for a step input.   Suppose we now add a small capacitor C_k in an attempt to improve the high-frequency response.   Since the voltage across a capacitor cannot change abruptly, C_k acts initially as a short circuit across R_k, so that there is no degeneration at the instant that the step is applied, and therefore the rise time should be improved.   It will indeed be shown below that if we adjust $R_k C_k = R_p C$, the bandwidth will be multiplied by the factor $1 + g_m R_k$.   However, since at the same time the nominal gain will be divided by the same factor, then unlike the compensation methods described above (shunt peaking, etc.) the gain-bandwidth product will remain unaltered.   If the circuit served no other purpose than to extend the bandwidth at the expense of gain it would be of little interest, since the same end may be achieved by the much simpler expedient of reducing the plate-circuit resistance R_p.   However, since we have here a case of current feedback, this circuit has better stability of gain and more linearity of operation.

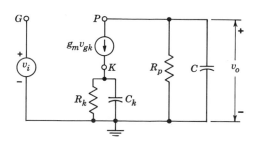

Fig. 5-11   The equivalent circuit of a pentode with a cathode impedance.

To consider the effect of the cathode impedance on the gain and rise time we draw the equivalent circuit in Fig. 5-11. Since for a pentode $r_p \gg R_p + R_k$, no plate resistance appears in this circuit. A straightforward analysis yields for the transfer function

$$\frac{V_o(s)}{V_i(s)} = \frac{-g_m Z_L}{1 + g_m Z_k} \tag{5-51}$$

where Z_L represents R_p and C in parallel, and Z_k is the parallel combination of R_k and C_k:

$$Z_L \equiv \frac{R_p}{1 + sCR_p} \qquad Z_k \equiv \frac{R_k}{1 + sC_k R_k}$$

Combining these three equations gives

$$\frac{V_o(s)}{V_i(s)} = \frac{-g_m R_p}{1 + sCR_p} \frac{1 + sC_k R_k}{1 + g_m R_k + sC_k R_k} \tag{5-52}$$

This function has two poles and one zero, and there is the possibility of overshoot in the response to a step input. However, if we choose C_k so that the zero cancels one pole, then a single time constant remains and a response without overshoot is assured. Thus, let

$$C_k R_k = CR_p \tag{5-53}$$

For a step input, so that $V_i(s) = V/s$,

$$V_o(s) = \frac{A_o V}{s(1 + s\tau_c)} \tag{5-54}$$

where the midband gain A_o and the compensated time constant τ_c are given by

$$A_o \equiv \frac{-g_m R_p}{1 + g_m R_k} \qquad \tau_c \equiv \frac{CR_p}{1 + g_m R_k} \tag{5-55}$$

The output is

$$v_o = \frac{-g_m R_p V}{1 + g_m R_k} (1 - \epsilon^{-t/\tau_c}) \tag{5-56}$$

If the cathode were connected directly to ground, the output would be

$$v_o = -g_m R_p V (1 - \epsilon^{-t/\tau_o}) \tag{5-57}$$

in which $\tau_o = R_p C$. In either case the ratio of gain to rise time is the same and equals $g_m/2.2C$.

In order to see some of the useful features of cathode compensation, let us compare two amplifiers, one without and one with cathode compensation. The first amplifier, without compensation, has a plate resistor R_p. The second amplifier has a plate resistor αR_p and a cathode resistor selected to make

$$1 + g_m R_k = \alpha$$

The quiescent tube current and voltage are to remain as before, which means that the plate resistor αR_p must be returned to a higher supply voltage. It may happen that the cathode resistor selected will furnish the bias required for optimum linearity of tube operation. More generally, however, some additional external bias will be required. The capacitance C shunting the plate to ground is to be the same in both cases. These two amplifiers now have the *same gain* and the *same bandwidth*.

One advantage of the compensated amplifier that is readily apparent is its greater stability of nominal gain with respect to variation of tube parameters. In the case where the nominal gain is given by $-g_m \alpha R_p / (1 + g_m R_k)$ the gain will be a less sensitive function of g_m than in the case where the gain is given by $-g_m R_p$. In the limiting case in which $g_m R_k \gg 1$, the gain for the compensated case is simply $-\alpha R_p / R_k$, independently of g_m.

A second advantage of the compensated amplifier is an improvement in linearity of operation. The nonlinearity of a pentode amplifier results from the variation of transconductance g_m with tube current. The effective transconductance of the compensated stage is $g_m / (1 + g_m R_k)$. For large values of $g_m R_k$ the effective transconductance becomes quite insensitive to variations in the g_m of the tube. Additionally, since the load resistor is α times as large in the compensated as in the uncompensated stage, the current swing in the compensated amplifier will be $1/\alpha$ times the current swing in the uncompensated amplifier for the same output signal. Hence for the same output signal from the two amplifiers the response of the compensated amplifier will be more linear. For comparable linearity in the two cases the compensated stage can provide a larger output signal.

If the capacitor C_k were not present, then because of cathode-follower action the amplifier would handle a peak-to-peak input signal larger than the grid base of the tube. However, it must be emphasized that because of the presence of C_k the input signal must be restricted in amplitude to the grid base. Otherwise the operation of the circuit will be highly nonlinear, as explained in Sec. 8-15.

5-7   HIGH–FREQUENCY RESPONSE OF A STAGE WITH AN EMITTER IMPEDANCE

In Fig. 5-12 there are indicated circuits with emitter impedances. If an amplifier is under consideration the output is taken at the collector. Under these circumstances we shall show that it is advantageous to bypass the emitter resistance, as in Fig. 5-12a (or at least a portion of it as in Fig. 5-12b), in order to improve the transient response. On the other hand, the output may be taken across R_e if an emitter follower is desired. In this case C_z, in

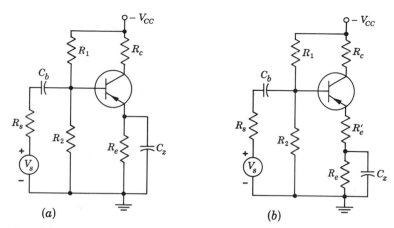

Fig. 5-12 Stages with an emitter impedance. These circuits may be used either as emitter followers or to provide gain.

Fig. 5-12a, represents the inevitable shunt capacitance of the load and the stray capacitance. We wish to study the high-frequency behavior of both these circuits. We shall, for simplicity, assume that R_1, R_2, and C_b are so large that they have negligible effect on the circuit operation.

We shall first investigate the effect of an unbypassed emitter resistor R_e on the high-frequency response of a CE amplifier. Such a circuit is indicated in Fig. 1-19a. The equivalent circuit of Fig. 1-19b applies independently of the particular model chosen for the transistor and hence may be used at high frequencies in connection with the hybrid-Π model. Assuming that the current gain A_I is large compared with unity over the entire passband, then, from Fig. 1-19b, we see that the effective collector-circuit resistance is increased from R_c to R_c', where, if $|A_I| \gg 1$, then

$$R_c' = R_c + \frac{A_I - 1}{A_I} R_e \approx R_c + R_e \qquad (5\text{-}58)$$

However, the principal effect of the emitter resistor is to increase the input impedance by $(1 - A_I)R_e$. Hence, subject to the same reasonable assumptions made in Sec. 4-7, the equivalent circuit is that given in Fig. 4-12 with $(1 - A_I)R_e$ added to the input and R_e added to the output, as indicated in Fig. 5-13. The capacitor C is, from Fig. 4-11b,

$$C = C_e + C_c(1 + g_m R_c') \qquad (5\text{-}59)$$

The current gain A_I is, from Fig. 5-13, given by

$$A_I \equiv \frac{I_L}{I_i} = \frac{-g_m V_{b'e}}{(V_{b'e}/r_{b'e}) + j\omega C V_{b'e}} = \frac{-h_{fe}}{1 + j\omega C r_{b'e}} \qquad (5\text{-}60)$$

since, from Eq. (1-11), $h_{fe} = g_m r_{b'e}$. Note that the current gain at low frequencies is $-h_{fe}$, the short-circuit current amplification, independently of the

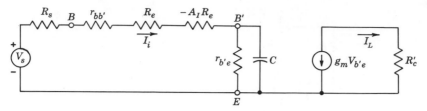

Fig. 5-13 The equivalent high-frequency circuit of a CE amplifier with an emitter resistor R_e. The effective load is $R_c' \approx R_c + R_e$.

emitter resistance. However, the upper 3-dB frequency for current gain does depend somewhat on R_e because C is a function of R_c' and hence R_e.

We shall now demonstrate that $-A_I R_e$ in Fig. 5-13 is equivalent to a resistance in parallel with a capacitance. The admittance Y is

$$Y = \frac{1}{-A_I R_e} = \frac{1 + j\omega C r_{b'e}}{h_{fe} R_e} = \frac{1}{h_{fe} R_e} + j\omega \frac{C}{g_m R_e} \tag{5-61}$$

Hence, $-A_I R_e$ may be represented by a resistance $h_{fe} R_e$ in parallel with a capacitance $C/g_m R_e$, as indicated in Fig. 5-14a, where $R_1 \equiv R_s + r_{bb'} + R_e$. The time constant of the parallel combination is

$$h_{fe} R_e \frac{C}{g_m R_e} = \frac{h_{fe} C}{g_m} = r_{b'e} C$$

and this is identical with the time constant between B' and E. Hence the circuit may be redrawn as in Fig. 5-14b. Points B'' and B' are at the same potential because of the equality of the time constants of the parallel branches in Fig. 5-14a (Sec. 2-8). It is now clear that a single-time-constant circuit is under consideration. The low-frequency voltage gain A_{Vso}, taking the source impedance into account, and the 3-dB frequency for voltage gain are found from Fig. 5-14b to be

$$A_{Vso} \equiv \frac{V_o}{V_s} = -\frac{h_{fe} R_c}{R_s + R_i} \tag{5-62}$$

and

$$f_2 = \frac{1}{2\pi C r_{b'e}} \frac{R_s + R_i}{R_s + r_{bb'} + R_e} \tag{5-63}$$

where R_i is the input resistance given by

$$R_i = r_{bb'} + r_{b'e} + R_e(1 + h_{fe}) = h_{ie} + R_e(1 + h_{fe}) \tag{5-64}$$

The gain-bandwidth product is

$$|A_{Vso} f_2| = \frac{g_m}{2\pi C} \frac{R_c}{R_s + r_{bb'} + R_e} \tag{5-65}$$

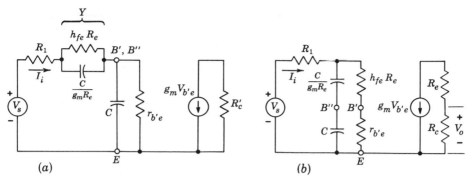

Fig. 5-14 (a) The impedance $-A_I R_e$ in Fig. 5-13 is replaced by a parallel resistor-capacitor combination. Also, $R_1 \equiv R_s + r_{bb'} + R_e$. (b) A circuit equivalent to that in (a).

[compare with Eq. (4-33)]. The effect of adding an emitter resistance is to greatly increase the input impedance R_i, because $h_{fe} \gg 1$. If $R_s \gg R_i$, then the circuit behavior is unaffected by the addition of R_e. Usually this is an unrealistic inequality, and more often the consequences of inserting an unbypassed emitter resistor are the following: the voltage gain is greatly decreased, the bandpass is greatly increased, but the gain-bandwidth product is decreased only slightly (if $R_s + r_{bb'} > R_e$). (An unbypassed cathode resistor decreases the gain of a tube amplifier but does not improve the bandpass.)

The physical reason that the emitter resistor R_e improves the bandwidth is that, as the frequency increases, the capacitor $C/g_m R_e$ in Fig. 5-14 bypasses the resistor $h_{fe} R_e$ to a greater extent. Hence with increasing frequency the current I_i increases and the voltage $V_{b'e}$ increases correspondingly. And, finally, the output increases as well, since the output voltage is proportional to $V_{b'e}$. It is therefore reasonable to expect that a small capacitance C_z placed across R_e, as in Fig. 5-12a, will further extend the bandwidth, since this capacitance will cause the input impedance to decrease even faster with frequency. The calculation of the value of C_z required to give, for a step input, an output having the fastest rise time without overshoot is quite formidable[5] algebraically because we no longer are dealing with a single-time-constant circuit. The proper value of C_z is best found experimentally. A reasonable initial choice for C_z is based upon the consideration that the parallel impedance of R_e and C_z should start to decrease significantly at the 3-dB frequency f_2 of the stage with $C_z = 0$, or $C_z \approx 1/2\pi R_e f_2$. Incidentally, it is interesting to note[5] that an emitter-compensated stage will generally have a smaller bandwidth than a shunt-compensated stage for the same gain.

Oscillations A practical difficulty with an emitter-compensated stage is that it may break into oscillations. We shall now explain why this possibility exists and give several methods for minimizing this effect.

If we were to attempt a solution for the transient response of Fig. 5-12a, we would use the equivalent circuit of Fig. 5-14a with R_e replaced by Z_e, the impedance consisting of R_e in parallel with C_z. Hence, Eq. (5-61) is now modified as follows:

$$Y = \frac{1}{-A_I Z_e} = \frac{1 + j\omega C r_{b'e}}{h_{fe}} \left(\frac{1}{R_e} + j\omega C_z \right)$$

$$= \frac{1}{h_{fe} R_e} - \frac{\omega^2 C C_z}{g_m} + j\omega \left(\frac{C}{g_m R_e} + \frac{C_z}{h_{fe}} \right) \tag{5-66}$$

where again use was made of the relationship $h_{fe} = g_m r_{b'e}$. As might have been anticipated, a capacitance C_z/h_{fe} appears in parallel with $C/g_m R_e$. However, it follows from Eq. (5-66) that these capacitances no longer shunt $h_{fe} R_e$ but instead are in parallel with a resistance whose value is

$$\frac{1}{(1/h_{fe} R_e) - (\omega^2 C C_z/g_m)} = \frac{h_{fe} R_e}{1 - \omega^2 C C_z R_e r_{b'e}} \tag{5-67}$$

For sufficiently high frequencies this resistance will be negative. Hence, the possibility exists that the input impedance Z_i to the amplifier will consist of a capacitance shunted by a negative resistance. Under this circumstance, where Z_i has a negative real part, the transient response may consist of large overshoots or, if the stage is driven from an inductive source, oscillations may result.

Incidentally, it should be noted that not only must Y in Fig. 5-14a be modified, as indicated by Eq. (5-66), but that R_1 and R'_c contain R_e, which must now be changed to Z_e. Clearly, to obtain the transient response of such a network is a complicated chore.

The instability discussed above is accentuated if the stage in Fig. 5-12 is used as an emitter follower. In this case the output is taken from the emitter, and hence the inevitable shunt capacitance of the load or interconnecting wiring capacitance appears directly across R_e. The tendency for an emitter follower to oscillate may be minimized in a number of ways. The obvious first step is to keep the driving-source inductance as low as possible. A resistance R_b may be added to the base so that the net input resistance remains positive over the passband of the stage. Alternatively, a resistance R'_e may be added in series with the emitter, as indicated in Fig. 5-12b. Since this resistance will be reflected into the input circuit as $(1 + h_{fe})R'_e$, then a few tens of ohms in the emitter are as effective as a few kilohms added to the base circuit. In either case, the voltage gain is reduced somewhat and there is a d-c level shift between input and output. A third way[9] to improve the stability is to use a coil on a lossy ferrite core in series with the emitter. The core may be regarded as a resistor in parallel with a small inductor. At high frequencies the losses in the core are reflected as a resistance R'_e in series with the emitter and, as noted above, this tends to suppress the possibility of oscillations. Since this technique does not introduce a d-c level shift it is particularly advantageous with emitter followers used in low-level logic circuits (Chap. 9).

The above discussion refers only to the small-signal behavior of the emitter follower. For large-signal excitation the transistor acts as a switch, and the response under these conditions is discussed in Sec. 8-16.

5-8   DISTRIBUTED AMPLIFIERS

Lumped-circuit delay lines (Appendix C) are essential elements in a type of pulse amplifier which is referred to as a *distributed amplifier*. Distributed amplifiers are principally effective with vacuum tubes and provide worthwhile gain over a bandwidth which exceeds appreciably the bandwidth attainable with conventional tube amplifiers.

In a conventional uncompensated amplifier stage, increased bandwidth may be achieved only at the expense of gain. When the stage gain has been reduced to unity or less, the stage is no longer useful for the purpose of amplification. Furthermore, since the gain of a number of stages in cascade is the product of the individual gains, no advantage accrues from cascading such stages of unity gain or less. To pursue the matter further, let us compute approximately the relationship between the gain, the bandwidth, and the number of stages in a conventional vacuum-tube uncompensated amplifier. The magnitude of the gain of n stages is

$$A = (g_m R_p)^n \qquad (5\text{-}68)$$

in which g_m is the transconductance and R_p the plate-circuit resistor of a stage. In Sec. 4-11 we see that the result of cascading n amplifiers is to divide the bandwidth approximately by the factor $\sqrt{n}$. The upper 3-dB frequency of an n-stage amplifier is therefore approximately

$$f_2 = \frac{1}{2\pi R_L C_s \sqrt{n}} \qquad (5\text{-}69)$$

in which C_s is the sum of the input and output capacitance in a stage. The *figure of merit F* of a tube may be defined as the product of gain and bandwidth and is given by

$$F = \frac{g_m}{2\pi C_s} \qquad (5\text{-}70)$$

From Eqs. (5-68) to (5-70) we have

$$f_2 A^{1/n} \sqrt{n} = F \qquad (5\text{-}71)$$

Now as an example of the limitations of the bandwidth capabilities of a conventional amplifier let us compute the bandwidth possible in an amplifier where the gain is required to be, say, $\epsilon^2 = (2.72)^2 = 7.4$. We shall use the type 6AK5 vacuum tube, for which $g_m = 5.1$ millimhos and the sum of the input and output capacitance is 6.8 pF. We shall assume (unrealistically) that we may neglect all additional stray capacitance. The figure of merit $F = (5.1 \times 10^{-3})/(2\pi \times 6.8 \times 10^{-12}) = 120 \times 10^6 \text{ sec}^{-1}$. We may compute from Eq. (5-71) that f_2 will be a maximum for a given value of gain if

$$n = 2 \ln A \qquad (5\text{-}72)$$

For a gain $A = \epsilon^2$, $n = 4$ and, from Eq. (5-71), we have $f_2 = 36.4$ MHz. We have the result, then, that even using a tube of high figure of merit and assuming that every conceivable precaution is taken to reduce shunt capacitance, it is not possible to build a conventional amplifier of gain 7.4 with a bandwidth in excess of 36.4 MHz. This situation may, of course, be remedied somewhat by the use of some form of high-frequency compensation. But, as we noted in Sec. 5-3, a really worthwhile improvement in bandpass is achieved only with a circuit of considerable complexity, with its attendant difficulties of adjustment, particularly in a multistage amplifier.

The basic limitation of the conventional cascade of amplifier stages is overcome by combining amplifier tubes in the manner indicated in Fig. 5-15. Such an arrangement is called a *distributed amplifier*.[10] The capacitances C_g and C_p represent, respectively, the input and output capacitances of the tubes together with the stray or any other capacitances present. A signal applied at the input travels down the grid transmission line, reaches each grid in turn, and is finally absorbed in the matched termination. Each pentode delivers current to the plate line, which is matched at both ends. One-half the tube current flows to each plate-line termination. The delays per section of the plate and grid lines are adjusted to be identical. Then all the current which reaches the plate-line output termination, in response to a given input voltage, will arrive at this termination at the same time. If the characteristic impedance of the plate line is R_{op}, then it follows that the

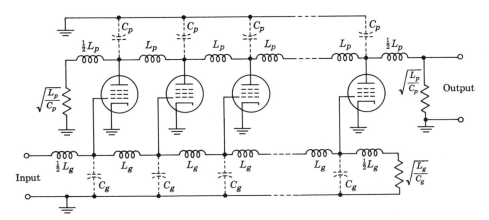

Fig. 5-15 A stage of distributed amplification consisting of n sections.

gain of an amplifier having n sections is

$$A = \tfrac{1}{2}ng_mR_{op} \tag{5-73}$$

The upper frequency limit of the amplifier may be considered to be determined essentially by the cutoff frequency of the delay lines. The cutoff frequencies of the plate and grid lines are the same since the delay per section is the same for both lines.

We may observe the following distinctive features. The gain of a distributed amplifier is computed by *adding* the gains provided by each tube individually. Hence, even if each tube provides a gain less than unity, the overall gain still increases with increasing number of tubes. Further, since the cutoff frequency of a delay line is not a function of the number of sections, the upper frequency limit of the amplifier is not decreased as more tubes are added to increase the gain.

At low frequencies, where the reactances of the elements of the transmission lines are negligibly small, the amplifier of Fig. 5-15 may be viewed simply as n parallel pentodes feeding a plate load resistor $R_{op}/2$. Hence the gain is given again by Eq. (5-73). However, a simple parallel connection of n tubes would not serve a useful purpose, since in such a case the effective g_m and shunt capacitance would increase in the same proportion. The figure of merit F for a parallel combination of tubes is the same as for a single tube. The distributed amplifier arrangement, however, effectively parallels the tubes so far as transconductance is concerned, but manages to keep the capacitances separate.

The delay lines indicated in Fig. 5-15 are constructed of prototype sections, and the terminations are simple resistors. Actually, of course, any of the other types of lines described in Appendix C may be used instead, and m-derived half-section terminations may be used to advantage. Any improvement of the delay lines will improve the performance of the amplifier.

5-9   DISTRIBUTED AMPLIFIERS IN CASCADE

In Fig. 5-15 each tube with its portion of transmission line is called a *section*. The combination of n such sections is called a *stage*. Distributed amplifier stages may be cascaded in the conventional sense. Suppose that we consider a cascade of m such stages. Then the total number of tubes involved in such an amplifier is $N = nm$. We may now show that, for a fixed gain, there is an optimum arrangement of tubes which reduces to a minimum the number of tubes required.

When stages are cascaded, the output end of the plate line of one amplifier must be coupled into the input end of the grid line of the succeeding amplifier. Since generally the plate and grid lines will have different characteristic impedances, an impedance transformer must be interposed between the lines. An impedance-transforming device which matches the grid-line impedance R_{og}

to the plate impedance R_{op} will simultaneously produce a voltage transformation in the ratio $(R_{og}/R_{op})^{\frac{1}{2}}$. Thus, from Eq. (5-73), we have the result that the gain from one grid line to the next is

$$A = \frac{ng_m}{2} R_{op} \sqrt{\frac{R_{og}}{R_{op}}} = \frac{ng_m}{2} \sqrt{R_{op}R_{og}} \tag{5-74}$$

If the gain of the entire amplifier of m stages is G, then $G = A^m$, or $G^{1/m} = A$. Using Eq. (5-74) to determine n we have

$$N = nm = \frac{2mA}{g_m \sqrt{R_{op}R_{og}}} = \frac{2mG^{1/m}}{g_m \sqrt{R_{op}R_{og}}} \tag{5-75}$$

The minimum number of tubes for a fixed gain and fixed line impedances is found by setting the derivative $dN/dm = 0$ in Eq. (5-75). The result is $m = \ln G$. Thus $G = A^m = \epsilon^m$, and when N is adjusted for a minimum,

$$A = \epsilon \tag{5-76}$$

in which $\epsilon = 2.72$. Hence the tubes are used in optimum fashion when each stage produces a gain ϵ corresponding to 8.68 dB. From Eq. (5-75) the number of tubes per stage is

$$n = \frac{2\epsilon}{g_m \sqrt{R_{op}R_{og}}} \tag{5-77}$$

We shall now calculate the bandpass of the distributed amplifier in terms of the tube parameters at the specified gain. The characteristic impedances of the plate and grid lines are

$$R_{op} = \sqrt{\frac{L_p}{C_p}} \qquad R_{og} = \sqrt{\frac{L_g}{C_g}} \tag{5-78}$$

Since the delays of the plate and grid lines are the same, we have from Eq. (C-8) that

$$L_pC_p = L_gC_g \tag{5-79}$$

and from Eq. (C-5) that the cutoff frequency f_c of the lines is

$$f_c = \frac{1}{\pi \sqrt{L_pC_p}} = \frac{1}{\pi \sqrt{L_gC_g}} \tag{5-80}$$

Using Eqs. (5-80), (5-79), (5-78), and (5-74) with $A = \epsilon$ we find that

$$f_c = \frac{ng_m}{2\pi\epsilon \sqrt{C_pC_g}} \tag{5-81}$$

The bandwidth of one stage is given by f_c in Eq. (5-81). When m stages are cascaded the overall bandwidth will be reduced. We shall assume that we may reasonably apply to a cascade of identical distributed-amplifier stages the same rule that applies in the case of conventional stages, namely, that the bandwidth $f_c^{(m)}$ of m stages is $1/\sqrt{m}$ times the bandwidth of one stage. In this case we find that the bandwidth $f_c^{(m)}$ is

$$f_c^{(m)} = \frac{n}{\sqrt{m}} \frac{g_m}{2\pi\epsilon \sqrt{C_p C_g}} \tag{5-82}$$

The bandwidth will be a maximum when C_p and C_g are reduced to a minimum, that is, when no additional capacitance is introduced and $C_p = C_o$ is the tube output capacitance and $C_g = C_i$ is the tube input capacitance. In this case Eq. (5-82) may be written

$$f_c^{(m)} = \frac{n}{\sqrt{m}} \frac{F'}{\epsilon} \tag{5-83}$$

with

$$F' = \frac{g_m}{2\pi \sqrt{C_o C_i}} \tag{5-84}$$

The parameter F' is a figure of merit for the tube for distributed-amplifier service, just as F in Eq. (5-70) is a figure of merit for a conventional amplifier.

In Sec. 5-8 we find that, in a conventional cascade when a gain ϵ^2 is specified, four tubes should be used and that with 6AK5 tubes a bandwidth of 36.4 MHz results. Let us now use the same four tubes in a distributed amplifier. We set $A = \epsilon$, as in Eq. (5-76). Hence, for a gain of ϵ^2, two stages are required, leaving two tubes per stage. We now find from Eqs. (5-83) and (5-84) that for the 6AK5 with $C_i = 4.0$ pF and $C_o = 2.8$ pF,

$$f_c^{(m)} = \frac{2}{\sqrt{2}\, 2.72} \frac{5.1 \times 10^{-3}}{2\pi \sqrt{4.0 \times 2.8 \times 10^{-12}}} \text{ Hz} = 126 \text{ MHz}$$

Hence, the distributed amplifier bandwidth is approximately four times that of the conventional amplifier using the same number of tubes and the same gain.

In the analysis above, we specified the overall gain G and found that tubes were used most economically when $A = \epsilon$. It can be shown (Prob. 5-26) that if we specify both the overall gain G and the overall bandwidth $f_c^{(m)}$, then the tubes will be most economically used if $A = \epsilon^{\frac{3}{2}}$.

Practical Considerations in Distributed Amplifiers The discussion, so far, of distributed amplifiers has been unrealistic in that it has been assumed that the frequency range of the amplifier is limited only by the cutoff frequency of the delay lines. The fact is, however, that as the frequency increases, the impedance seen looking into the grid of the tubes exhibits not only a capacitive reactance but a resistive loading as well. The resistive loading at the grid has two sources. The first of these is due to the presence of inductance in the cathode-to-ground lead of the tube. The presence of such a cathode-lead inductance results in a conductive component of admittance at the grid. For this reason, good construction practice requires that the cathode connection

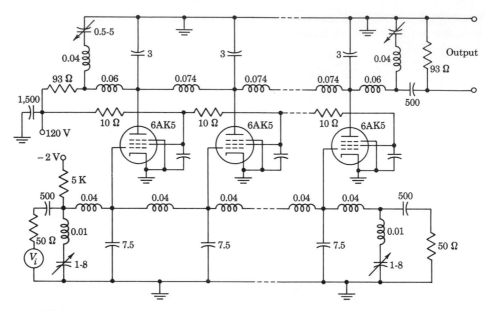

Fig. 5-16 A practical form of a distributed amplifier. Inductances are given in microhenrys, and capacitances in picofarads.

to ground be made as short and direct as possible. But, of course, some residual conductive component at the grid will always remain.

A second and much more important source of conductive loading at the grid results from effects due to the finite time of transit of an electron across the tube. For a sinusoidal signal, the loading due to transit-time effects[11] begins to make itself felt when the period of the signal becomes small enough to be comparable to the transit time. Each of these components of conductance at the grid is proportional to the square of the frequency. The severity of this loading with increasing frequency may be noted by observing that at 400 MHz the input resistance of a 6AK5 is only 250 Ω.

An example of a stage of distributed amplification[12] is shown in Fig. 5-16. The plate and grid lines are constructed of prototype sections. They are terminated in m-derived half sections (Fig. C-9), which serve to improve the match between the lines and the terminating resistors. The lines are designed to have a cutoff frequency of 400 MHz. The grid line has a characteristic impedance of 50 Ω. This low impedance has been selected to minimize the effect of the loading of the grid line due to transit-time effects. The general formulas

$$R_o = \sqrt{\frac{L}{C}} \quad \text{and} \quad f_c = \frac{1}{\pi \sqrt{LC}} \tag{5-85}$$

may be solved for L and C with the result

$$L = \frac{R_o}{\pi f_c} \tag{5-86}$$

$$C = \frac{1}{\pi R_o f_c} \tag{5-87}$$

The inductance per section of the grid line is calculated from Eq. (5-86) with the result $L_g = 0.04$ μH. The grid-line capacitance is calculated from Eq. (5-87) with the result $C_g = 16$ pF. After the 6AK5 input capacitance, the tube-socket capacitance, and other stray capacitances have been taken into account; it is found that an additional 7.5 pF must be added in the grid circuit to bring the total to the required 16 pF.

It is advantageous to make the impedance of the plate line as large as possible since the amplifier gain increases with plate-line impedance. The impedance would be a maximum if the capacitance per section were kept at a minimum. Actually 3 pF of capacitance has been added to each section of the line to bring the line impedance down to 93 Ω. A 93-Ω impedance is particularly convenient since there is available a commercial coaxial cable (RG-62/U) whose impedance is 93 Ω.

The amplifier uses a total of nine 6AK5 tubes. The transit-time loading of the grid line is therefore quite heavy, and actually the signal level on the grid line falls appreciably as the upper frequency limit of the amplifier is approached. This effect, however, is counterbalanced by the fact that the impedance of the plate line (as viewed at the point where the tube plates are connected) increases substantially as the line cutoff frequency is approached. The result is that the gain remains reasonably uniform up to a frequency nearly equal to the cutoff frequency.

REFERENCES

1. Pettit, J. M., and M. M. McWhorter: "Electronic Amplifier Circuits," McGraw-Hill Book Company, New York, 1961.

2. Bedford, A. V., and G. L. Fredendall: Transient Response of Multistage Video-frequency Amplifiers, *Proc. IRE*, vol. 27, pp. 277–284, 1939.

3. Kallmann, H. E., R. E. Spencer, and C. P. Singer: Transient Response, *Proc. IRE*, vol. 33, pp. 169–195, 1945.

4. Palmer, R. C., and L. Mautner: A New Figure of Merit for the Transient Response of Video Amplifiers, *Proc. IRE*, vol. 37, pp. 1073–1077, 1949.

5. Johannes, V.: "Transient Response of Transistor Video Amplifiers," sec. 1, University Microfilms, Ann Arbor, Mich., 1961.

6. Muller, F. A.: High-frequency Compensation of *RC* Amplifiers, *Proc. IRE*, vol. 42, no. 8, pp. 1271–1276, August, 1954.

Seeley, S. W., and C. N. Kimball: Analysis and Design of Video Amplifiers, *RCA Rev.*, vol. 3, no. 3, pp. 290–308, January, 1939.

Design of Video-amplifier Peaking Circuits for Optimum Transient Response, *RCA Ind. Serv. Lab. Rept.* LB-930.

7. Valley, G. E., Jr., and H. Wallman: "Vacuum Tube Amplifiers," chap. 2, Massachusetts Institute of Technology Radiation Laboratory Series, vol. 18, McGraw-Hill Book Company, New York, 1948.

8. Larsen, M. J.: Low-frequency Compensation of Video-frequency Amplifiers, *Proc. IRE*, vol. 33, pp. 666–670, 1945.

9. Hunter, L. P. (ed.): "Handbook of Semiconductor Electronics," 2d ed., p. 15-21, McGraw-Hill Book Company, New York, 1962.

10. Ref. 1, chap. 6.

11. Spangenberg, K. R.: "Vacuum Tubes," 1st ed., chap. 16, McGraw-Hill Book Company, New York, 1948.

12. Scharfman, H.: Distributed Amplifier Covers 10 to 360 Mc, *Electronics*, vol. 25, no. 7, pp. 113–115, July, 1952.

6 / STEADY–STATE SWITCHING CHARACTERISTICS OF DEVICES

Junction diodes, thermionic diodes, transistors, and vacuum tubes all have extreme regions of operation in which they nominally do not conduct even when large voltages are applied and regions in which they conduct heavily even when relatively small voltages are applied. In the first of these regions the device is described as being "off," "open," or "nonconducting." In the latter region the device is said to be "on," "closed," or "conducting." When the device is driven from one extreme condition to the other, it operates much like a switch. In the present chapter we shall be interested only in the steady-state characteristics of these extreme end states of the switching operation.

There is a limit to the speed with which the switching transition between the extreme states can be made. This switching speed, together with other matters relating to the transition between end states, is discussed in detail in Chap. 20.

6-1 THE SEMICONDUCTOR DIODE

For an ideal p-n junction the current I is related[1] to the voltage V by the equation

$$I = I_o(\epsilon^{V/\eta V_T} - 1) \tag{6-1}$$

A positive value of I indicates that the current flows from the p to the n side. The diode is forward-biased if V is positive, indicating that the p side of the junction is positive with respect to the n side. The symbol η is unity for germanium and is approximately 2 for silicon. The parameter η takes into account the recombination of carriers in the junction transition region.[2] Such recombination is negligible in germanium but not in silicon.

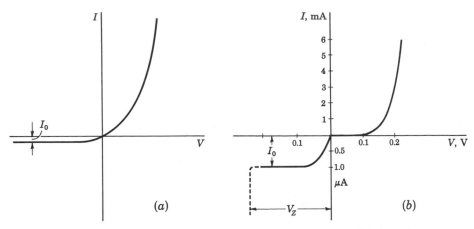

Fig. 6-1 (a) The volt-ampere characteristics of an ideal diode; (b) the volt-ampere characteristic for a germanium diode redrawn to show the order of magnitude of currents. Note the expanded scale for reverse currents. The dashed portion indicates breakdown at V_Z.

The symbol V_T stands for the electron-volt equivalent of temperature and is given by

$$V_T \equiv \frac{kT}{e} \tag{6-2}$$

in which k is the Boltzmann constant ($k = 1.380 \times 10^{-23}$ J/°K), e is the electronic charge ($e = 1.602 \times 10^{-19}$ C), and T is the absolute temperature. Substituting, we find that $V_T = (T/11,600)$ V and that at room temperature ($T = 300°$K) $V_T = 0.026$ V $= 26$ mV.

The form of the volt-ampere characteristic described by Eq. (6-1) is shown in Fig. 6-1a. When the voltage V is positive and several times V_T, the unity in the parentheses of Eq. (6-1) may be neglected. Accordingly, except for a small range in the neighborhood of the origin the current increases exponentially with voltage. When the diode is reverse-biased, and $|V|$ is several times V_T, $I \approx -I_o$. The reverse current is therefore constant, independently of the applied reverse bias. Consequently I_o is referred to as the *reverse saturation current.*

For the sake of clarity, the current I_o in Fig. 6-1a has been greatly exaggerated in magnitude. Ordinarily the range of forward currents over which a diode is operated is many orders of magnitude larger than the reverse saturation current. In order to display forward and reverse characteristics conveniently it is necessary, as in Fig. 6-1b, to use two different current scales. The volt-ampere characteristic shown in that figure has a forward current scale in milliamperes and a reverse scale in microamperes.

The dashed portion of the curve of Fig. 6-1*b* indicates that at a reverse-biasing voltage V_Z the diode characteristic exhibits an abrupt and marked departure from Eq. (6-1).   At this critical voltage a large reverse current flows, and the diode is said to be in the *breakdown* region, discussed in Sec. 6-4.

The Cutin Voltage V_γ   Both silicon and germanium diodes are commercially available.   There are a number of differences between these two types that are relevant in circuit design.   The difference in volt-ampere characteristics is brought out in Fig. 6-2.   Here are plotted the forward characteristics at room temperature of a general-purpose germanium switching diode and a general-purpose silicon diode, the 1N270 and 1N3605, respectively.   The diodes have comparable current ratings.   A noteworthy feature in Fig. 6-2 is that there exists a *cutin, offset, break-point,* or *threshold* voltage V_γ below which the current is very small (say less than 1 percent of maximum rated value).   Beyond V_γ the current rises very rapidly.   From Fig. 6-2 we see that V_γ is approximately 0.2 V for germanium and 0.6 V for silicon.

Note that the break in the silicon diode characteristic is offset about 0.4 V with respect to the break in the germanium diode characteristic.   The reason for this difference is to be found, in part, in the fact that the reverse saturation current in a germanium diode is normally larger by a factor of about 1,000 than the reverse saturation current in a silicon diode of comparable ratings.   Thus if I_o is in the range of microamperes for a germanium diode, I_o will be in the range of nanoamperes for a silicon diode.

Since $\eta = 2$ for small currents in silicon, the current increases as $\epsilon^{V/2V_T}$ for the first several tenths of a volt and increases as ϵ^{V/V_T} only at higher voltages.   This initial smaller dependence of the current on voltage accounts for the further delay in the rise of the silicon characteristic.

Fig. 6-2   The forward volt-ampere characteristics of a germanium (1N270) and a silicon (1N3605) diode at 25°C.

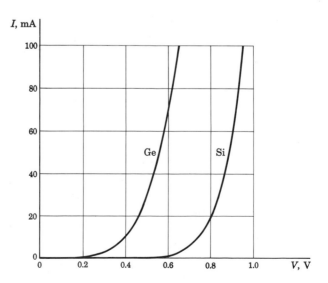

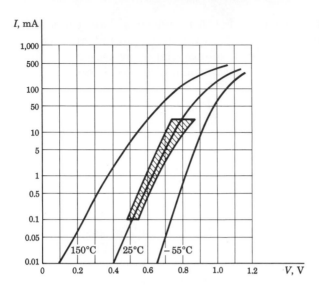

Fig. 6-3 Volt-ampere characteristics at three different
temperatures for a silicon
diode. (Planar epitaxial
passivated types 1N3605,
1N3606, 1N3608, and
1N3609.) The shaded area
indicates 25°C limits of controlled conductance. Note
that the vertical scale is
logarithmic and encom-
passes a current range of
50,000. (Courtesy of
General Electric Company.)

Logarithmic Characteristic   It is instructive to examine the family of curves for the silicon diodes shown in Fig. 6-3. A family for a germanium diode of comparable current rating is quite similar, with the exception that corresponding currents are attained at lower voltage.

From Eq. (6-1), assuming that V is several times V_T so that we may drop the unity, we have $\log I = \log I_o + 0.434V/\eta V_T$. We therefore expect in Fig. 6-3, where $\log I$ is plotted against V, that the plots will be straight lines. We do indeed find that at low currents the plots are linear and correspond to $\eta \approx 2$. At large currents an increment of voltage does not yield as large an increase of current as at low currents. The reason for this behavior is to be found in the ohmic resistance of the diode. At low currents, the ohmic drop is negligible and the externally impressed voltage simply decreases the potential barrier at the p-n junction. At high currents the externally impressed voltage is called upon principally to establish an electric field to overcome the ohmic resistance of the semiconductor material. Therefore at high currents the diode behaves more like a resistor than a diode and the current increases linearly rather than exponentially with applied voltage.

6-2   THE TEMPERATURE DEPENDENCE OF p-n CHARACTERISTICS

Let us inquire into the diode voltage variation with temperature at fixed current. This variation may be calculated from Eq. (6-1), where the temperature is contained implicitly in V_T and also in the reverse saturation current. The dependence of I_o on temperature T is given approximately by[3]

$$I_o = KT^{2/\eta}\epsilon^{-V_o/\eta V_T} \tag{6-3}$$

where K is a constant and eV_g (e the electronic charge) is the energy required to break a covalent bond in the semiconductor. For germanium $\eta = 1$ and $V_g = 0.75$ V and for silicon $\eta \approx 2$ and $V_g = 1.12$ V. Taking the derivative of the logarithm of Eq. (6-3) we find

$$\frac{1}{I_o}\frac{dI_o}{dT} = \frac{d(\ln I_o)}{dT} = \frac{2}{\eta T} + \frac{V_g}{\eta T V_T} \approx \frac{V_g}{\eta T V_T} \tag{6-4}$$

since $V_g/V_T \gg 2$. At room temperature, we deduce from Eq. (6-4) that $d(\ln I_o)/dT = 0.075/°C$ for Si and $0.10/°C$ for Ge. The performance of commercial diodes is only approximately consistent with these results. The reason for the discrepancy is that in a physical diode there is a component of the reverse saturation current due to leakage over the surface that is not taken into account in Eq. (6-3). Since this leakage component is independent of temperature, we may expect to find a smaller rate of change of I_o with temperature than that predicted above. From the data presented in Sec. 6-8, we find that the reverse saturation current increases approximately 7 percent/°C for both silicon and germanium. Since $(1.07)^{10} \approx 2.0$, we conclude that the *reverse saturation current approximately doubles for every 10°C rise in temperature.*

From Eq. (6-1), dropping the unity in comparison with the exponential, we find, for constant I,

$$\frac{dV}{dT} = \frac{V}{T} - \eta V_T \left(\frac{1}{I_o}\frac{dI_o}{dT}\right) \approx \frac{V - V_g}{T} \tag{6-5}$$

where use has been made of Eq. (6-4). Consider a diode operating at room temperature (300°K) and just beyond the threshold voltage V_γ (say at 0.2 V for Ge and 0.6 for Si). Then we find from Eq. (6-5)

$$\frac{dV}{dT} = \begin{cases} -1.8 \text{ mV/°C} & \text{for Ge } (V_g = 0.75 \text{ V}) \\ -1.7 \text{ mV/°C} & \text{for Si } (V_g = 1.12 \text{ V}) \end{cases} \tag{6-6}$$

Since these data are based on "average characteristics" it might be well for conservative design to assume a value of

$$\frac{dV}{dT} = -2 \text{ mV/°C} \tag{6-7}$$

for either Ge or Si at room temperature. Note from Eq. (6-5) that $|dV/dT|$ decreases with increasing T.

The temperature dependence of forward voltage is given in Eq. (6-5) as the difference between two terms. The positive term V/T on the right-hand side results from the temperature dependence of V_T. The negative term results from the temperature dependence of I_o and does not depend on the voltage V across the diode. The equation predicts, accordingly, that at increasing V, dV/dT should become less negative, reach zero at $V = V_g$, and thereafter reverse sign and go positive. This behavior is regularly

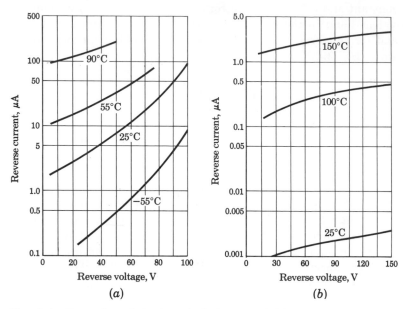

Fig. 6-4 Examples of diodes which do not exhibit a constant reverse saturation current. (a) Germanium diode 1N270; (b) silicon 1N461. (Courtesy of Raytheon Company.)

exhibited by diodes. Normally, however, the reversal takes place at a current which is higher than the maximum rated current. The curves of Fig. 6-3 also suggest this behavior. At higher voltages the separation between curves of different temperatures is smaller than at low voltages.

Typical reverse characteristics of germanium and silicon diodes are given in Fig. 6-4a and b. Observe the very pronounced dependence of current on reverse voltage, a result which is not consistent with our expectation of a constant saturated reverse current. This increase in I_o results from leakage across the surface of the diode and also from the additional reason that new current carriers may be generated by collision in the transition region at the junction (Sec. 6-8). On the other hand there are many commercially available diodes, both germanium and silicon, that do exhibit a fairly constant reverse current with increasing voltage. The much larger value of I_o for a germanium than for a silicon diode, to which we have previously referred, is apparent in comparing Fig. 6-4a and b. Since the temperature dependence is approximately the same in both types of diodes, at elevated temperatures the germanium diode will develop an excessively large reverse current, whereas for silicon I_o will be quite modest. Thus we see that for Ge in Fig. 6-4 an increase in temperature from room temperature (25°C) to 90°C increases the reverse current to hundreds of microamperes, although in silicon at 100°C the reverse current has increased only to some tenths of a microampere.

6-3   DIODE TRANSITION CAPACITANCE

A diode is driven to the reverse-biased condition when it is desired to turn off a current or prevent the transmission of a signal. When diodes are used for such purposes in circuits which handle fast waveforms or high frequencies, we must take account of the capacitance which appears across a reverse-biased junction. This capacitance is called the *barrier* or *transition capacitance* C_T. If this capacitance is large enough, the current which is to be restrained by the low conductance of the reverse-biased diode will flow through the capacitor.

Diodes intended for service with fast waveforms have transition capacitances of the order of 1 to 10 pF. The barrier capacitance decreases with increasing reverse voltage, as is illustrated for two typical diodes in Fig. 6-5. For certain assumed simple junction geometries it is calculated[4] that the transition capacitance is given by

$$C_T = \frac{\lambda}{(V_B)^n} \approx \frac{\lambda}{V^n} \tag{6-8}$$

Here λ is a constant, and V_B is the voltage across the junction and is equal to the externally impressed voltage V, except for quite small voltages. The exponent n is $\frac{1}{2}$ or $\frac{1}{3}$ for an abrupt or gradual junction, respectively.

6-4   AVALANCHE DIODE

The reverse voltage characteristic of the diode, including the breakdown region, is redrawn in Fig. 6-6a. Diodes which are designed with adequate power-dissipation capabilities to operate in the breakdown region may be employed as voltage-reference or constant-voltage sources. Such diodes are known as

Fig. 6-5 Typical barrier-capacitance variation, with reverse voltage, of silicon diodes 1N914 and 1N916. (Courtesy of Fairchild Semiconductor Corporation.)

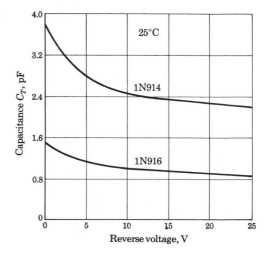

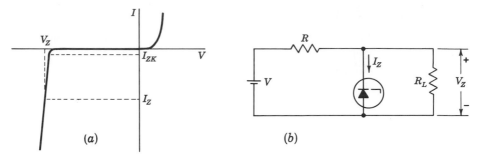

Fig. 6-6 (a) The volt-ampere characteristic of an avalanche or Zener diode; (b) a circuit in which such a diode is used to regulate the voltage across R_L against changes due to variations in load current and supply voltage.

avalanche, breakdown, or *Zener diodes.* They are used characteristically in the manner indicated in Fig. 6-6b and are replacing the gaseous glow tubes previously employed in this circuit. The source V and resistor R are selected so that initially the diode is operating in the breakdown region. Here the diode voltage, which is also the voltage across the load R_L, is V_Z, as in Fig. 6-6a, and the diode current is I_Z. The diode will now regulate the load voltage against variations in load current and against variations in supply voltage V because in the breakdown region large changes in diode current produce only small changes in diode voltage. Moreover, as load current or supply voltage changes, the diode current will accommodate itself to these changes to maintain a nearly constant load voltage. The diode will continue to regulate until the circuit operation requires the diode current to fall to I_{ZK}, in the neighborhood of the knee of the diode volt-ampere curve. The upper limit on diode current is determined by the power-dissipation rating of the diode.

Two mechanisms of diode breakdown for increasing reverse voltage are recognized. In one mechanism, the thermally generated electrons and holes acquire sufficient energy from the applied potential to produce new carriers by removing valence electrons from their bonds. These new carriers, in turn, produce additional carriers again through the process of disrupting bonds. This cumulative process, which is analogous to the Townsend discharge[1] in a gas diode, is referred to as *avalanche multiplication.* It results in the flow of large reverse currents, and the diode finds itself in the region of *avalanche breakdown.* Even if the initially available carriers do not acquire sufficient energy to disrupt bonds, it is possible to initiate breakdown through a direct rupture of the bonds because of the existence of the strong electric field. Under these circumstances the breakdown is referred to as *Zener breakdown.* This Zener effect is now known to play an important role only in diodes with breakdown voltages below about 6 V. Nevertheless, the term *Zener* is commonly used for the *avalanche* or *breakdown diode* even at higher voltages. Silicon diodes operated in avalanche breakdown are available with maintaining voltages from several volts to several hundred volts and with power ratings up to 50 W.

A matter of interest in connection with Zener diodes, as with semiconductor devices generally, is their temperature sensitivity. The temperature dependence of the reference voltage, which is indicated in Fig. 6-7a and b, is typical of what may be expected generally. In Fig. 6-7a the temperature coefficient of the reference voltage is plotted as a function of the operating current through the diode for various different diodes whose reference voltage at 5 mA is specified. The temperature coefficient is given as percentage change in reference voltage per centigrade degree change in diode temperature. In Fig. 6-7b has been plotted the temperature coefficient at a fixed diode current of 5 mA as a function of Zener voltage. The data which are used to plot this curve are taken from a series of different diodes of different Zener voltages but of fixed dissipation rating. From the curves in Fig. 6-7a and b we note that the temperature coefficients may be positive or negative and will normally be in the range ± 0.1 percent/°C. Note that if the reference voltage is above 6 V, where the physical mechanism involved is avalanche multiplication, the temperature coefficient is positive. However, below 6 V, where true Zener breakdown is involved, the temperature coefficient is negative.

A second matter of importance in connection with Zener diodes is the slope of the diode volt-ampere curve in the operating range. If the reciprocal slope $\Delta V_Z/\Delta I_Z$, called the *dynamic resistance*, is r, then a change ΔI_Z in the

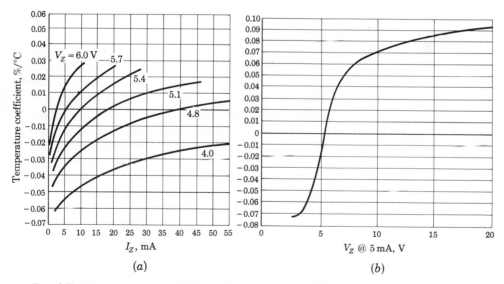

(a)   (b)

Fig. 6-7 Temperature coefficients for a number of Zener diodes having different operating voltages (a) as a function of operating current, (b) as a function of operating voltage. The voltage V_Z is measured at $I_Z = 5$ mA (from 25 to 100°C). (Courtesy of Pacific Semiconductors, Inc.)

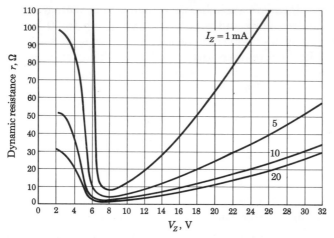

Fig. 6-8 The dynamic resistance at a number of currents for Zener diodes of different operating voltages at 25°C. The measurements are made with a 60-Hz current at 10 percent of the d-c current.   (Courtesy of Pacific Semiconductors, Inc.)

operating current of the diode produces a change $\Delta V_Z = r \, \Delta I_Z$ in the operating voltage.   Ideally $r = 0$, corresponding to a volt-ampere curve which, in the breakdown region, is precisely vertical.   The variation of r at various currents for a series of avalanche diodes of fixed power-dissipation rating and various voltages is shown in Fig. 6-8.   Note the rather broad minimum which occurs in the range 6 to 10 V and note that at large V_Z and small I_Z, the dynamic resistance r may become quite large.   Thus we find that a TI 3051 (Texas Instruments Company) 200-V Zener diode operating at 1.2 mA has an r of 1,500 Ω.   Finally we observe that to the left of the minimum, at low Zener voltages, the dynamic resistance rapidly becomes quite large.   Some manufacturers specify the minimum current I_{ZK} (Fig. 6-6a) below which the diode should not be used.   Since this current is on the knee of the curve, where the dynamic resistance is large, then for currents lower than I_{ZK} the regulation will be poor.

The capacitance across a breakdown diode is the transition capacitance and hence varies inversely as some power of the voltage, as in Eq. (6-8). Since C_T is proportional to the cross-sectional area of the diode, high-power avalanche diodes have very large capacitances.   Values of C_T from 10 to 10,000 pF are common.

The performance and reliability of Zeners have been greatly improved with the oxide-passivation process.   They may be used over the range ~200°C. Also, some diodes exhibit a very sharp knee even down into the microampere region.

Zener diodes are available with voltages as low as about 2 V. Below this voltage it is customary, for reference and regulating purposes, to use diodes in the *forward* direction. As appears in Fig. 6-2, the volt-ampere characteristic of a forward-biased diode (sometimes called a *stabistor*) is not unlike the reverse characteristic with the exception that in the forward direction the knee of the characteristic occurs at lower voltage. A number of forward-biased diodes may be operated in series to reach higher voltages. Such series combinations, packaged as single units, are available with voltages up to about 5 V and may be preferred to reverse-biased Zener diodes, which at low voltages, as seen in Fig. 6-8, have very large values of dynamic resistance.

When it is important that a Zener diode operate with a low temperature coefficient, it may be feasible to operate an appropriate diode at a current where the temperature coefficient is at or near zero. Quite frequently such operation is not convenient, particularly at higher voltages and when the diode must operate over a range of currents. Under these circumstances *temperature-compensated* avalanche diodes find application. Such diodes consist of a reverse-biased Zener diode with a positive temperature coefficient combined in a single package with a forward-biased diode whose temperature coefficient is negative. As an example, the Transitron SV3176 silicon 8-V reference diode has a temperature coefficient of ± 0.001 percent/°C at 10 mA over the range -55 to $+100$°C. The dynamic resistance is only 15 Ω. The temperature coefficient remains below 0.002 percent/°C for currents in the range 8 to 12 mA. The voltage stability with time of some of these reference diodes is comparable to that of conventional standard cells.

When a high-voltage reference is required it is usually advantageous (except of course with respect to economy) to use two or more diodes in series rather than a single diode. This combination will allow higher voltage, higher dissipation, lower temperature coefficient, and lower dynamic resistance.

6-5 THE VACUUM–TUBE DIODE

For an ideal diode the current I is related to the voltage V by the Langmuir-Child space-charge equation[1]

$$I = GV^{\frac{3}{2}} \tag{6-9}$$

The specific value of the parameter G, called the *perveance*, depends upon the geometry of the system. The derivation of this *three-halves-power equation* assumes an equipotential cathode surface, zero contact potential between electrodes, no trace of gas in the tube, zero initial velocity of the emitted electrons, and a heater temperature sufficiently high to supply the desired plate current. A plot of Eq. (6-9) is given in Fig. 6-9a.

The diode characteristic does not follow Eq. (6-9) for small currents or voltages because the initial velocities of the electrons and the contact potential cannot be neglected in this region. An expanded view of the volt-ampere

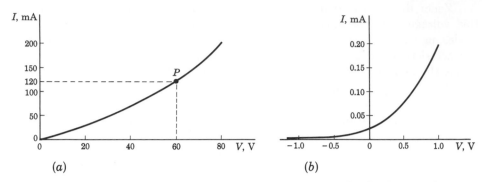

Fig. 6-9 The volt-ampere characteristic of a vacuum diode (a) for large voltages and (b) for small voltages.

curve near the origin is given in Fig. 6-9b.   Space charge is negligible at these small currents, and the current is given by

$$I = I_o \epsilon^{V/V_T} \tag{6-10}$$

where I_o is the plate current at zero applied voltage V and $V_T = T/11{,}600$ is the electron-volt equivalent of temperature given in Eq. (6-2).   Note that the curve does *not* pass through the origin and that the *cutin* or *break point* V_γ is quite indefinite.   Whereas V_γ for a semiconductor diode is positive (approximately 0.2 V for Ge and 0.6 V for Si), for most vacuum-tube diodes the break point is negative and is located at a voltage between -0.25 and -0.75 V.

There is a shift in the characteristic with filament temperature.   Experiment reveals this drift to be about 0.1 V *for a* 10 *percent change in heater voltage*. The higher the filament voltage, the more the curve shifts to the left, because the increase in the initial velocities of the electrons with increase in temperature results in higher currents at a given voltage.   The displacement with tube aging or tube replacement is found in practice to be of the order of ± 0.25 V.

Since a diode consists of two metallic electrodes separated by a dielectric— a vacuum—this device constitutes a capacitor.   The order of magnitude of the capacitance in receiving-type tubes is 5 pF and is nominally constant, independent of voltage (unlike the semiconductor diode which has a nonlinear junction capacitance).

Commercial vacuum diodes are available for rectifying very high voltages, up to about 200,000 V.

Because of their small size and long life and because no filament power is required, semiconductor diodes are replacing vacuum rectifiers in many applications.   The tube must be used, however, if very high voltage or power is involved, if extremely low reverse currents are necessary, or if the diode is located in an unusual environment (high nuclear radiation or very high ambient temperature).

6-6   DIODE RESISTANCE

The static resistance R of a diode is defined as the ratio V/I of the voltage to the current. At any point on the volt-ampere characteristic of the diode (Figs. 6-1 and 6-9) the resistance R is equal to the reciprocal of the slope of a line joining the operating point to the origin. The static resistance varies widely with V and I and is not a useful parameter. The rectification property of a diode is indicated on the manufacturer's specification sheet by giving the maximum forward voltage V_F required to attain a given forward current I_F and also the maximum reverse current I_R at a given reverse voltage V_R. Typical values for a silicon planar epitaxial diode are $V_F = 0.8$ V at $I_F = 10$ mA (corresponding to $R_F = 80\ \Omega$) and $I_R = 0.1\ \mu\text{A}$ at $V_R = 50$ V (corresponding to $R_R = 500$ M).

For small-signal operation the *dynamic* or *incremental resistance* r is an important parameter and is defined as the reciprocal of the slope of the volt-ampere characteristic, $r \equiv dV/dI$. The dynamic resistance is not a constant but depends upon the operating voltage. For example, for a semiconductor diode we find from Eq. (6-1) that the dynamic conductance $g \equiv 1/r$ is

$$g \equiv \frac{dI}{dV} = \frac{I_o \epsilon^{V/\eta V_T}}{\eta V_T} = \frac{I + I_o}{\eta V_T} \tag{6-11}$$

For a reverse bias greater than a few tenths of a volt (so that $|V/\eta V_T| \gg 1$) g is extremely small and r is very large. On the other hand, for a forward bias greater than a few tenths of a volt $I \gg I_o$, and r is given approximately by

$$r \approx \frac{\eta V_T}{I} \tag{6-12}$$

The dynamic resistance varies inversely with current; at room temperature and for $\eta = 1$, $r = 26/I$, where I is in mA and r is in ohms. For a forward current of 26 mA, the dynamic resistance is 1 Ω. The ohmic body resistance of the semiconductor may be of the same order of magnitude or even much higher than this value. Although r varies with current, in a small-signal model, it is reasonable to use the parameter r as a constant.

A Piecewise Linear Diode Characteristic   A large-signal approximation which often leads to a sufficiently accurate engineering solution is the *piecewise linear* representation. For example, the three-halves-power curve of Fig. 6-9a may be replaced by a straight line through the origin and the point P ($V = 60$ V, $I = 120$ mA), as pictured in Fig. 6-10a. This piecewise linear model replaces the diode by an open circuit (infinite back resistance) if $V < 0$ and a constant forward resistance of $60/0.12 = 500\ \Omega$ for $V > 0$. In this case, where the characteristic is a straight line passing through the origin, the dynamic resistance dV/dI equals the static resistance V/I and the break-point or cutin voltage V_γ is zero.

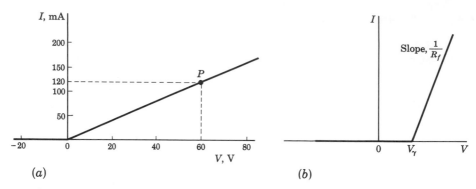

Fig. 6-10   The piecewise linear characterization of a diode.   (a) A vacuum tube;
(b) a semiconductor.

The piecewise linear approximation for a semiconductor characteristic is indicated in Fig. 6-10b.   The break point is not at the origin and hence V_γ is also called the *offset* or *threshold voltage*.   The diode behaves like an open circuit if $V < V_\gamma$ and has a constant incremental resistance $r = dV/dI$ if $V > V_\gamma$.   Note that the resistance r (also designated as R_f and called the *forward resistance*) takes on added physical significance even for this large-signal model, whereas the static resistance $R = V/I$ is not constant and is not useful.

The numerical values V_γ and R_f to be used depend upon the type of diode and the contemplated voltage and current swings.   For example, from Fig. 6-2 we find that for a current swing from cutoff to 10 mA with a germanium diode reasonable values are $V_\gamma = 0.2$ V and $R_f = 20$ Ω, and for a silicon diode $V_\gamma = 0.6$ V and $R_f = 15$ Ω.   On the other hand, a better approximation for current swings up to 50 mA leads to the following values—germanium: $V_\gamma = 0.3$ V, $R_f = 6$ Ω; silicon: $V_\gamma = 0.65$ V, $R_f = 5.5$ Ω.   For an avalanche diode $V_\gamma = V_Z$, and R_f is the dynamic resistance in the breakdown region.

6-7   THE TRANSISTOR AS A SWITCH

The transistor Q in Fig. 6-11a is being used as a switch to connect and disconnect the load R_L from the source V_{CC}.   Except that the transistor may be operated electrically and may be made to respond more rapidly, it serves the same function as that of the mechanical switch in Fig. 6-11b.   The mechanical switch arrangement allows no current to flow when the switch is open, but when the switch is closed, all of the voltage V_{CC} appears across the load R_L. Ideally, the transistor switch should have these same properties.   In this section we discuss the steady-state characteristics of the circuit of Fig. 6-11a corresponding to the cases when the transistor switch is open and when it is

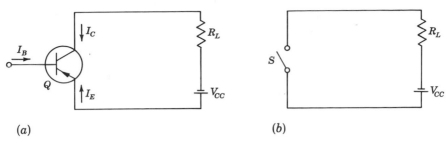

(a)   (b)

Fig. 6-11 The transistor in (*a*) is being used as a switch. It serves the same function as the switch S in (*b*). The positive reference direction for each current is as shown. The symbol V_{CC} is a positive number representing the magnitude of the supply voltage.

closed. Chapter 20 contains a detailed discussion of the speed with which a transition between these two states may be made.

When a transistor is used as a switch it is useful to divide its range of operation into three regions: the *cutoff*, the *active*, and the *saturation* regions. These regions are easily identified on the common-base characteristics of the transistor, as in Fig. 6-12. In the cutoff region both the emitter junction and the collector junction are reverse-biased, and only very small reverse saturation currents flow across the junctions. The transistor operates in the region below the characteristic for $I_E = 0$. This characteristic corresponds to a collector current I_{CO}, the reverse collector saturation current. It is almost

Fig. 6-12 Typical common-base characteristics of a *p-n-p* transistor. The cutoff, active, and saturation regions are indicated. Note the expanded voltage scale in the saturation region.

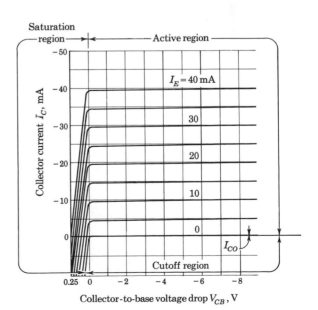

but not precisely coincident with the axis $I_C = 0$. It is required that the transistor must be in the cutoff region at times when it is to behave as an open (nonconducting) switch.

When the emitter junction is forward-biased and the collector junction is reverse-biased, the transistor output current responds most linearly to an input signal. The pulse amplifiers of Chaps. 4 and 5 were operated in this active region. In switching operations, this region is not of great interest because the transistor switches abruptly from the cutoff region to the saturation region (or in the reverse direction) and spends, ideally, a relatively insignificant time in the active region.

The region to the left of the ordinate $V_{CB} = 0$ and above $I_E = 0$ is the saturation region. Here the emitter junction and the collector junction are both forward-biased. The voltages across the individual junctions or across the combination of junctions are small (in the millivolt range). Accordingly, when a transistor switch is required to be in the closed (conducting) condition it is driven into saturation.

When a transistor is used as a switch in the common-base configuration, the input emitter current required to operate the switch is nominally as large as the collector current being switched. In the common-collector configuration, the input voltage required to operate the switch is nominally as large as the supply voltage.

In the common-emitter configuration the input switching signal, current or voltage, is small in comparison with the switched output current or voltage. Hence, the common-emitter configuration is the most generally useful for a transistor switch.

Return now to the common-emitter switching circuit of Fig. 6-11a and consider that the transistor is a type 2N404 p-n-p alloy-junction germanium transistor with CE characteristics shown in Fig. 6-13. This transistor type is fairly representative of alloy-junction germanium transistors capable of

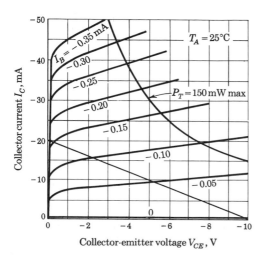

Fig. 6-13 Common-emitter characteristics of the 2N404 alloy-junction germanium transistor. Maximum allowable dissipation $P_T = 150$ mW. Load line corresponding to $V_{CC} = 10$ V, $R_L = 500\ \Omega$. (Courtesy of Texas Instruments, Inc.)

dissipating about 150 mW at room temperature (25°C). We have selected a load resistor $R_L = 500\ \Omega$ and a supply voltage $V_{CC} = 10$ V and have super-imposed the corresponding load line on the characteristics. The cutoff and saturation regions are not as clearly shown in Fig. 6-13 as in Fig. 6-12. We shall, in the next several sections, discuss details of importance in connection with the transistor operation at cutoff and operation in saturation.

6-8   THE TRANSISTOR AT CUTOFF

Cutoff in a transistor is defined by the condition $I_E = 0$. A good approximation to the common-base characteristics, even to the point of cutoff, is given by the equation

$$I_C = -\alpha I_E + I_{CO} \tag{6-13}$$

in which α is the common-base short-circuit forward current gain. From Eq. (6-13) we see that the cutoff condition $I_E = 0$ implies that $I_C = I_{CO}$.

It is important to note that in the common-emitter configuration the transistor will not be at cutoff if the base is open-circuited. For, from Fig. 6-11a, if $I_B = 0$, then $I_E = -I_C$ and from Eq. (6-13) we have

$$I_C = -I_E = \frac{I_{CO}}{1 - \alpha} \tag{6-14}$$

In germanium, even near cutoff, α may be as large as 0.9 and $I_C \approx 10 I_{CO}$. Therefore the transistor is not at cutoff. In Sec. 6-17 we find that $I_B = 0$ corresponds to a small forward bias and that to bring the germanium transistor to cutoff we need to establish a reverse-biasing voltage between base and emitter of about 0.1 V. In silicon, at collector currents of the order of I_{CO}, it is found[2,3] that α is very nearly zero because of recombination in the junction transition region. Hence, even with $I_B = 0$, we find from Eq. (6-14) that $I_C = I_{CO} = -I_E$, so that the transistor is still very close to cutoff. We verify in Sec. 6-17 that, in silicon, cutoff ($I_E = 0$) occurs at $V_{BE} \approx 0$ V, corresponding to a short-circuited base.

The Reverse Collector Saturation Current I_{CBO} The collector current when the emitter current is zero is designated by the symbol I_{CBO}. Two factors cooperate to make $|I_{CBO}|$ larger than $|I_{CO}|$. First, there exists a leakage current which flows not through the junction but around it and across the surfaces. The leakage current is proportional to the voltage across the junction. The second reason why $|I_{CBO}|$ exceeds $|I_{CO}|$ is that new carriers may be generated by collision in the junction transition region, leading to avalanche multiplication of current and eventual breakdown, as discussed in Sec. 6-9. But even before breakdown is approached, this *multiplication* component of current may attain considerable proportions.

At 25°C, I_{CBO} for a germanium transistor, whose power dissipation is in the range of some hundreds of milliwatts, is of the order of microamperes.

Under similar conditions a silicon transistor has an I_{CBO} in the range of nano-amperes. The temperature sensitivity of I_{CBO} in silicon is approximately the same as that of germanium. However, because of the lower absolute value of I_{CBO} in silicon, these transistors may be used up to about 200°C, whereas germanium transistors are limited to about 100°C.

Plots showing typical variations of I_{CBO} with temperature for a germanium and for a silicon transistor are given in Fig. 6-14. These plots indicate an almost linear relationship between the logarithm of I_{CBO} and the temperature. In this respect the plots are in only fair agreement with Eq. (6-3). That equation predicts a noticeable decrease in slope with increasing temperature, particularly over the large temperature range, 235°C, in the case of silicon. A more pronounced failing of Eq. (6-3) is that it predicts a larger factor of growth for I_{CBO} over the temperature range contemplated in Fig. 6-14 than is actually observed. The discrepancy arises from the fact that a part of the current that is measured as reverse saturation current actually results from surface leakage and is independent of temperature. Hence, the rate of change of I_{CBO} with T should be less than that predicted by Eq. (6-3), which assumes that all the current is temperature-sensitive. By examining a large number of commercially available plots like those in Fig. 6-14 we find an average value for $(1/I_{CBO})(dI_{CBO}/dT)$ of 8 percent/°C for Ge and 6 percent/°C for Si (as

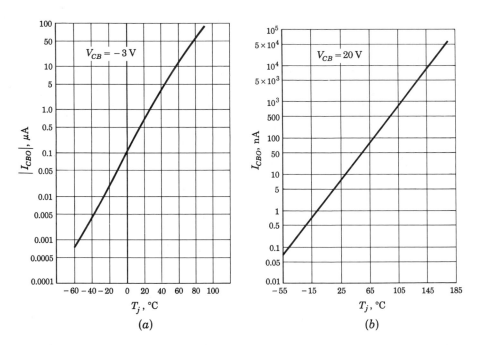

Fig. 6-14 Plot of I_{CBO} versus the junction temperature T_j for (a) germanium alloy type 2N1175 transistor and (b) silicon planar epitaxial type 2N914 transistor. (Courtesy of General Electric Company.)

Fig. 6-15 The source V_{BB} applies a reverse-biasing voltage to the n-p-n transistor base through R_B and maintains the transistor in cutoff in the presence of the reverse saturation current I_{CBO}.

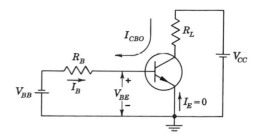

against the theoretical values of 10 and 7.5 percent/°C, respectively). Since $(1.08)^{10} = 2.2$ and $(1.06)^{10} = 1.8$, then *for every 10°C rise in temperature I_{CBO} is multiplied by 2.2 for Ge and by 1.8 for Si.* If not too large a temperature range is contemplated, then we can say that I_{CBO} approximately doubles for every 10°C increase for either Ge or Si.

In addition to the variability of reverse saturation current with temperature there is also a wide variability of reverse current with particular samples of a given transistor type. For example, the specification sheet for a Texas Instruments type 2N337 grown diffused silicon switching transistor indicates that this type number includes units with values of I_{CBO} extending over the extremely large range from 0.2 nA to 0.3 μA. Accordingly, any particular transistor may have an I_{CBO} which differs very considerably from the average characteristic for the type.

Circuit Considerations at Cutoff Because of temperature effects, avalanche multiplication, and the wide variability encountered from sample to sample of a particular transistor type, even silicon may have values of I_{CBO} of the order of many tens of microamperes. Consider the circuit configuration of Fig. 6-15, where V_{BB} represents a biasing voltage applied to keep the transistor cut off. We have considered that the transistor is just at the point of cutoff, with $I_E = 0$, so that $I_B = -I_{CBO}$. Since at cutoff $V_{BE} \leq -0.1$ V, then the condition of cutoff requires that

$$V_{BE} = -V_{BB} + R_B I_{CBO} \leq -0.1 \text{ V} \qquad (6\text{-}15)$$

As an extreme example assume that R_B is as large as 100 K and assume that we want to allow for the contingency that I_{CBO} may become as large as 100 μA. Then V_{BB} must be at least 10.1 V. When I_{CBO} is small the voltage across the base-to-emitter junction will be 10.1 V. Hence we must use a transistor whose maximum allowable reverse base-to-emitter junction voltage before breakdown exceeds 10 V. It is with this contingency in mind that a manufacturer supplies a rating for the reverse *breakdown voltage* between emitter and base, represented by the symbol BV_{EBO}. The subscript O indicates that BV_{EBO} is measured under the condition that the collector current is zero. Breakdown voltages BV_{EBO} may be as high as some tens of volts or as low as 0.5 V. If $BV_{EBO} = 1$ V, then V_{BB} must be chosen to have a maximum value of 1 V. For $V_{BB} = 1$ V and for $I_{CBO} = 0.1$ mA maximum, R_B cannot exceed 9 K. For example, if $R_B = 8$ K, then $-V_{BB} + I_{CBO}R_B = -1 + 0.8 = -0.2$ V, so that the transistor is indeed cut off.

We consider now the unhappy consequences which might result if, for any one of many reasons, Eq. (6-15) is not satisfied. Suppose, in Fig. 6-15, that $-I_B$ falls short of the value I_{CBO}, necessary to maintain cutoff, by the current ΔI_B. Then the change in collector current above the value I_{CBO} will be very nearly $\Delta I_C = h_{fe} \Delta I_B$, in which h_{fe} is the common-emitter forward-current gain. The increased collector current will carry the transistor somewhat into the active region, where the collector-junction power dissipation is increased. The dissipation at the collector junction will increase its temperature and thereby increase I_{CBO}. The increase in I_{CBO} is aided further by the fact that at low collector currents an increase in collector current increases h_{fe}. The possibility exists that the process whereby I_C increases may become regenerative and a runaway condition may occur.

6-9   BREAKDOWN VOLTAGES[5]

In a transistor switch, the voltage change which occurs at the collector with switching is nominally equal to the collector supply voltage V_{CC}. Since this voltage change will be used to operate other circuits and devices, then for the sake of reliability of operation, V_{CC} should be made as large as possible. The maximum allowable voltage depends, as we shall see, not only on the characteristics of the transistor but also on the associated transistor base circuitry.

The maximum reverse-biasing voltage which may be applied before breakdown between the collector and base terminals of the transistor, under the condition that the emitter lead be open-circuited, is represented by the symbol BV_{CBO}. This breakdown voltage is a characteristic of the transistor alone. Breakdown occurs because of avalanche multiplication of the current I_{CO} that crosses the collector junction. As a result of this multiplication the current becomes MI_{CO}, in which M is the factor by which the original current I_{CO} is multiplied by the avalanche effect. (We neglect leakage current, which does not flow through the junction and is therefore not subject to avalanche multiplication.) At a high enough voltage, namely BV_{CBO}, the multiplication factor M becomes nominally infinite and the region of breakdown is then attained. Here the current rises abruptly, and large changes in current accompany small changes in applied voltage.

The avalanche multiplication factor depends on the voltage V_{CB} between collector and base. We shall consider that

$$M \equiv \frac{1}{1 - (V_{CB}/BV_{CBO})^n} \tag{6-16}$$

Equation (6-16) is employed because it is a simple expression which gives a good empirical fit to the breakdown characteristics of many transistor types.

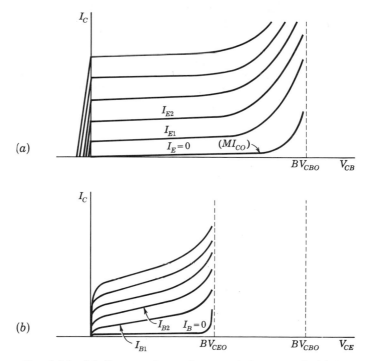

Fig. 6-16   (a) Common-base characteristics extended into breakdown region.   (b) Idealized common-emitter characteristics extended into breakdown region.

The parameter n is found to be in the range of about 2 to 10 and controls the sharpness of the onset of breakdown.   When n is large, M continues at nearly unity until V_{CB} approaches very close to BV_{CBO}, at which point M soars upward abruptly.   When n is small the onset of breakdown is more gradual.

In Fig. 6-16a the common-base characteristics of Fig. 6-12 have been extended into the breakdown region.   The curve for $I_E = 0$ is a plot, as a function of V_{CB}, of the product of the reverse collector current I_{CO} and the avalanche multiplication factor M.   The abrupt growth in I_C as BV_{CBO} is approached is shown, along with the slower increase of I_C over the active region that results from the small but not negligible avalanche multiplication.

If a current I_E is caused to flow across the emitter junction, then, neglecting the avalanche effect, a fraction αI_E, where α is the common-base current gain, reaches the collector junction.   Taking multiplication into account, I_C has the magnitude $M\alpha I_E$.   Consequently, it appears that, in the presence of avalanche multiplication, the transistor behaves as though its common-base current gain were α^*, where

$$\alpha^* = M\alpha \tag{6-17}$$

The CE Configuration   Since $h_{FE} = \alpha/(1 - \alpha)$ it follows that in the presence of avalanche multiplication the CE current gain is h_{FE}^*, where

$$h_{FE}^* = \frac{\alpha^*}{1 - \alpha^*} = \frac{M\alpha}{1 - M\alpha} \tag{6-18}$$

Now α is a positive number with a maximum magnitude less than unity, but $M\alpha$ may equal unity in magnitude, at which point h_{FE}^* becomes infinite. Accordingly, any base current, no matter how small, will give rise to an arbitrarily large collector current whenever $M\alpha = 1$. This situation is, of course, to be interpreted to mean that breakdown has occurred. Therefore, whenever the *base current is kept fixed*, breakdown occurs at the voltage V_{CB} which satisfies the equation

$$M \equiv \frac{1}{1 - (V_{CB}/BV_{CBO})^n} = \frac{1}{\alpha} \tag{6-19}$$

or at the voltage V_{CB} given by

$$V_{CB} = BV_{CBO} \sqrt[n]{1 - \alpha} \tag{6-20}$$

Since V_{CB} at breakdown is much larger than the small forward base-to-emitter voltage V_{BE}, we may replace V_{CB} by V_{CE} in Eq. (6-20). Also

$$1 - \alpha = \frac{\alpha}{h_{FE}} \approx \frac{1}{h_{FE}}$$

since α is very close to unity. Finally we note that the condition imposed above that the base current be fixed implies a current generator at the base, that is, a source of infinite impedance. Equation (6-20) consequently gives the breakdown voltage under the condition that the base is open-circuited (with respect to signal or a-c variations). Altogether, the collector-to-emitter breakdown voltage *with open-circuited base*, designated BV_{CEO}, is

$$BV_{CEO} = BV_{CBO} \sqrt[n]{\frac{1}{h_{FE}}} \tag{6-21}$$

For an *n-p-n* germanium transistor a reasonable value for n, as determined experimentally, is $n = 6$. If we now take $h_{FE} = 50$ we find that

$$BV_{CEO} = 0.52 BV_{CBO}$$

so that if $BV_{CBO} = 40$ V, BV_{CEO} is about half as much, or about 20 V. Idealized common-emitter characteristics extended into the breakdown region are shown in Fig. 6-16b.

The Breakdown Voltage with Base Not Open-circuited   Assume that the base is returned to the emitter through a resistor R_B, as in Fig. 6-17. We may expect that the breakdown voltage, designated BV_{CER}, will lie between BV_{CEO} and BV_{CBO}. To estimate BV_{CER} we shall make some simplifying assumptions concerning the emitter-junction diode. We had noted in Fig. 6-2

Fig. 6-17   The breakdown voltage between
collector and emitter is increased above
$|BV_{CEO}|$ by returning the base to the emitter
through R_B.

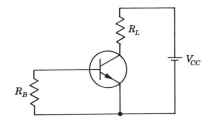

that the semiconductor junction diode exhibits a threshold voltage V_γ in the
forward direction.   That is, until the forward voltage attains about 0.2 V in
germanium or 0.6 V in silicon the forward current is very small.   We shall
assume that, until the threshold voltage has been reached, the collector current
will flow entirely to the base and hence through R_B.   We also assume that once
the threshold voltage is exceeded nearly all the additional collector current will
flow through the emitter junction and the corresponding breakdown voltage is
BV_{CEO}.   Therefore   when   the   collector-to-emitter   voltage   is   larger   than
BV_{CEO} and the threshold voltage of the emitter junction is reached, breakdown will occur.   On this basis, we expect breakdown when the collector
current MI_{CO} satisfies

$$MI_{CO}R_B = V_\gamma \tag{6-22}$$

Proceeding as above, we now calculate that

$$BV_{CER} = BV_{CBO} \sqrt[n]{1 - \frac{I_{CO}R_B}{V_\gamma}} \tag{6-23}$$

The value of BV_{CER} for $R_B = 0$, that is, when the base is short-circuited to
the emitter, is denoted by the symbol BV_{CES}.   Equation (6-23) suggests
that $BV_{CES} = BV_{CBO}$.   However, we must recognize the presence of the
base-spreading resistance $r_{bb'}$ and R_B should properly be replaced by $R_B + r_{bb'}$.
Accordingly, even when $R_B = 0$, BV_{CES} is lower in magnitude than BV_{CBO}.

   Equation (6-23) was derived on the basis of the assumption that, until
the occurrence of breakdown, the current through R_B was very large in comparison with the emitter-junction current.   If R_B becomes so large that this
condition is not satisfied, then Eq. (6-23) is not applicable.   Finally we should
note that, after breakdown has occurred, the collector and the emitter currents will become very large in comparison with the base current.   Therefore
at large currents the presence of R_B makes no difference, and the voltage across
the transistor will drop from BV_{CER} to BV_{CEO}, as noted in Fig. 6-18.

   The breakdown voltage may also be increased by returning the resistor
R_B to a voltage V_{BB}, as in Fig. 6-15, which provides some back bias for the
emitter junction.   In this case the condition which determines the onset of
breakdown is

$$MI_{CO}(R_B + r_{bb'}) = V_\gamma + V_{BB} \tag{6-24}$$

and the breakdown voltage, now represented by the symbol BV_{CEX}, is given approximately by

$$BV_{CEX} = BV_{CBO} \sqrt[n]{1 - \frac{I_{CO}(R_B + r_{bb'})}{V_\gamma + V_{BB}}} \tag{6-25}$$

Figure 6-18 shows plots of the collector current against the collector-to-emitter voltage extending into the breakdown region. The dependence of the breakdown voltage on the base return circuit is indicated. Also indicated is the fact that, after breakdown, the voltage across the transistor at large currents drops to a lower *sustaining* voltage. The curves are not drawn to scale vertically. Before breakdown the currents are very small, being of the order of microamperes, and the negative-resistance region of the curves may be attained after the currents have risen to several milliamperes, whereas after breakdown the currents may rise to many tens of milliamperes.

Note that in the breakdown region the characteristic of Fig. 6-18 gives more than one value of current for a given voltage but has a single value of voltage for each current. The transistor possesses a negative-resistance characteristic of the current-controlled type as defined in Sec. 13-1. A device used in this breakdown region is called an *avalanche transistor*. Because both the current and voltage may be large the power dissipation may be exceeded unless pulsed operation at a low duty cycle is employed (Sec. 13-16).

For the sake of avoiding complexity at the outset of our discussion concerning breakdown, we have allowed an error which we shall now correct. Figure 6-18 makes it appear that, unlike the other plots, the curve corresponding to "open base" approaches BV_{CEO} monotonically. The fact is that some transistors have an "open-base" characteristic that displays a breakdown voltage which is larger than the sustaining voltage and is hence of the same

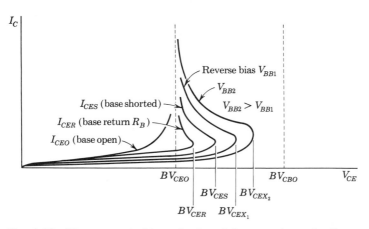

Fig. 6-18 Plot, extended into the breakdown region, of collector current against V_{CE} for various connections to the base. The sustaining voltage is BV_{CEO}.

Fig. 6-19 A possible set of break-down characteristics. The curve marked O is for an open-circuited base, R for the base connected to the emitter through a resistor, S for a short-circuited base, and X for a reverse-biased base. The break-down-voltage curve between collector and base with the emitter open is marked V_{CBO}. (Courtesy of Silicon Transistor Corporation.)

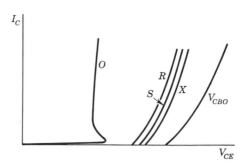

general form as all the other plots in Fig. 6-18. Qualitatively the reason for such anomalous behavior, as well as certain other features in breakdown characteristics, is explained by the following. We note from Eq. (6-21) that BV_{CEO} is a function of h_{FE}. The parameter h_{FE}, however, is a function of the current. It is small at small currents, increases with increasing current to a maximum, and then falls off again. In Fig. 6-18 the dependence of BV_{CEO} on h_{FE} and consequently on collector current has not been taken into account. Because the variation of h_{FE} with I_C depends on the transistor construction (alloy, planar epitaxial, etc.), the breakdown characteristics may take on a variety of shapes. One possibility is indicated in Fig. 6-19.

6-10   LATCHING VOLTAGES

A possibility for which we must check is illustrated in Fig. 6-20. Here we have selected a supply voltage V_{CC} which is less than the breakdown voltage BV. But the load selected is one which causes the load line to intersect the transistor breakdown characteristic at three points. Suppose, then, that the transistor is initially at saturation at S, which is the point of intersection of the load line and a common-emitter characteristic corresponding to some base current I_B adequate to drive the transistor to saturation. Next suppose that the transistor is driven off by applying a reverse voltage at the base corresponding to which condition the transistor characteristic is the breakdown characteristic shown. The transistor will leave the point S and move along the load line toward cutoff, where the collector-to-emitter voltage is nominally V_{CC}. Cutoff, however, will not be attained; instead, a stable state of appreciable current will be reached at L. This point of intersection of the load line with the breakdown characteristic is called the *latching* point.

Suppose that the transistor is originally at cutoff with $V_{CE} \approx V_{CC}$ and that the base current then is driven to I_B. Under these conditions the transistor will be driven to saturation at the point S. Thus the latching point is

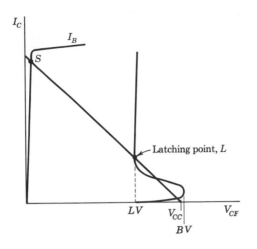

Fig. 6-20 Illustrating the phenomenon of "latching" in a transistor switch.

properly named. Like the latch on a door, it restrains transfer of the operating point in one direction but not the other. The latching voltage is ordinarily identified by the same subscripts used for the corresponding breakdown voltage. Thus, in Fig. 6-20, if it is appropriate to label the breakdown voltage BV_{CEX} the latching voltage would be designated LV_{CEX}.

Figure 6-18 suggests that, if the latching point occurs at high enough currents, then the latching voltage will be very close to BV_{CEO} and that the latching voltage is largely independent of the type of base return. Figure 6-18 was drawn on the basis of the assumption that at high current very little current would flow in the base circuit of the transistor. If this condition is not satisfied, then the lower voltage to which the breakdown characteristic drops after the breakdown voltage has been passed is a function of the type of base return. This lower voltage will also be the latching voltage corresponding to each case, and we may therefore continue to designate it by LV. When the transistor itself is being referred to rather than the proposed application of the transistor as a switch, the latching voltage is referred to as the *lower limiting voltage* and the symbol LV is still appropriate. Just as $BV_{CEX} > BV_{CES} > BV_{CER} > BV_{CEO}$, so too is $LV_{CEX} > LV_{CES} > LV_{CER} > LV_{CEO}$. These lower limiting or latching voltages are also referred to as *sustaining* voltages and designated by V (sustaining). Thus $LV_{CEX} \equiv V_{CEX}$ (sustaining), etc.

6-11 REACH–THROUGH

There is a second mechanism by which a transistor's usefulness may be terminated as the collector-to-base voltage is increased. This mechanism is called *punch-through* or *reach-through* and results from the increased width of the collector-junction transition region that results from increased collector-junction voltage (the Early effect).

The transition region at a junction is the region of uncovered charges on both sides of the junction at the positions occupied by the impurity atoms. As the voltage applied across the junction increases, the transition region penetrates deeper into the collector and base. Because neutrality of charge must be maintained, the number of uncovered charges on each side remains equal. Since the doping of the base is ordinarily substantially smaller than that of the collector, the penetration of the transition region into the base is larger than into the collector. Since the base is very thin, it is possible that, at moderate voltages, the transition region will have spread completely across the base to reach the emitter junction. At this point normal transistor action ceases, since emitter and collector are effectively shorted. Punch-through differs from avalanche breakdown in that it takes place at a fixed voltage between collector and base, and is not dependent on circuit configuration.

In a particular transistor, the voltage limit is determined by punch-through or breakdown, whichever occurs at the lower voltage. If reach-through occurs first (as it usually does for an alloy transistor), this voltage is easily determined in the following manner. An adjustable reverse-biasing voltage is connected between collector and base, and at the same time the voltage between emitter and base is monitored with a high-impedance volt-meter. As the collector-to-base voltage is increased the emitter voltage will respond very little until the punch-through voltage is reached. At this point an increment in collector-to-base voltage will cause an equal change in the emitter voltage.

6-12   THE TRANSISTOR SWITCH IN SATURATION

When a transistor switch is driven from saturation to cutoff, one of the factors which has an important effect on the speed of response is the time required to charge the capacitance which appears in shunt across the output terminals of the transistor. This capacitance must charge through the load resistance R_L, and for this reason, in fast switching circuits, R_L must be kept small. In saturation, the transistor current is nominally V_{CC}/R_L, and since R_L is small, it may well be necessary to keep V_{CC} correspondingly small in order to stay within the limitations imposed by the transistor on maximum current and dissipation. The total voltage swing at the transistor switch is $V_{CC} - V_{CE}(\text{sat})$. The symbol $V_{CE}(\text{sat})$ stands for the collector-to-emitter voltage of the transistor when it is in saturation. If V_{CC} is fixed, then in order to make the output swing as large as possible it is necessary that $V_{CE}(\text{sat})$ be as small as possible. The largest possible output swing is desirable in order to reduce the sensitivity of the switching circuit to noise, supply voltage fluctuations, transistor aging and replacement, etc. There are also occasions when the output of one switch is d-c-coupled to the input of other switches, and a small change in $V_{CE}(\text{sat})$ will determine whether the succeeding switches are cut off or driven to saturation. For both these reasons, a manufac-

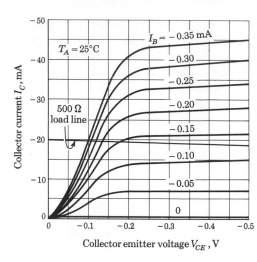

Fig. 6-21 Saturation-region common-emitter characteristics of the type 2N404. A load line corresponding to $V_{CC} = 10$ V, $R_L = 500$ Ω, is superimposed. (Courtesy of Texas Instruments, Inc.)

turer of a switching transistor invariably specifies $V_{CE}(\text{sat})$ under particular operating conditions or furnishes information from which $V_{CE}(\text{sat})$ may be determined.

For the transistor switch of Fig. 6-11a we are not able to read $V_{CE}(\text{sat})$ with any precision from the plots of Fig. 6-13. Accordingly, we refer instead to the characteristics shown in Fig. 6-21. In these characteristics the 0- to -0.5-V region of Fig. 6-13 has been expanded, and we have superimposed the same load line as before, corresponding to $R_L = 500$ Ω. We observe from Figs. 6-13 and 6-21 that V_{CE} and I_C no longer respond appreciably to base current I_B after the base current has attained the value -0.15 mA. At this point the transistor is in saturation and $|V_{CE}| \approx 175$ mV. At $I_B = -0.35$, $|V_{CE}|$ has dropped to $|V_{CE}| \approx 100$ mV. Larger magnitudes of I_B will of course decrease $|V_{CE}|$ slightly further.

For a transistor operating in the saturation region, a quantity of interest is the ratio $V_{CE}(\text{sat})/I_C$. This parameter is called the *common-emitter saturation resistance* and variously abbreviated R_{CS}, R_{CES}, or $R_{CE}(\text{sat})$. To specify R_{CS} properly we must indicate the operating point at which it was determined. For example, from Fig. 6-21 we find that at $I_C = -20$ mA and $I_B = -0.35$ mA, $R_{CS} = -0.1/(-20 \times 10^{-3}) = 5$ Ω. The usefulness of R_{CS} stems from the fact, as appears in Fig. 6-21, that to the left of the knee each of the plots, for fixed I_B, may be approximated, at least roughly, by a straight line. In instances where the manufacturer is not inclined to make available saturation-region collector characteristics as in Fig. 6-21, he may instead specify values of R_{CS} for one or more values of I_B. Still another way to supply the necessary information is indicated in Fig. 6-22. Here $V_{CE}(\text{sat})$ is plotted as a function of I_C and I_C/I_B for the 2N914 silicon transistor.

The saturation voltage $V_{CE}(\text{sat})$ depends not only on the operating point but also on the semiconductor material (germanium or silicon) and on the type of transistor construction. Alloy-junction transistors and epitaxial transistors give the lowest values for $V_{CE}(\text{sat})$ (corresponding to about 1 Ω satura-

Fig. 6-22   The collector-to-emitter saturation voltage $V_{CE}(\text{sat})$ as a function of the ratio I_C/I_B with I_C as a parameter for a 2N914 planar epitaxial passivated silicon transistor. (Courtesy of General Electric Company.)

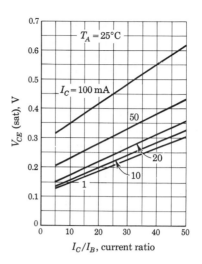

tion resistance), whereas grown-junction transistors yield the highest.   Germanium transistors have lower values for $V_{CE}(\text{sat})$ than do silicon.   For example, we see in Fig. 6-21 that an alloy-junction Ge transistor may allow, with adequate base currents, values for $V_{CE}(\text{sat})$ as low as tens of millivolts at collector currents which are some tens of milliamperes.   Similarly, epitaxial silicon transistors may yield saturation voltages as low as 0.2 V with collector currents as high as an ampere.   On the other hand, grown-junction germanium transistors have saturation voltages which are several tenths of a volt, and silicon transistors of this type may have saturation voltages as high as several volts.

The D-C Current Gain   A parameter of interest in connection with the transistor is the ratio I_C/I_B, where I_C is the collector current and I_B is the base current.   This ratio is known as the *d-c forward-current transfer ratio*, the *d-c current gain*, or sometimes simply as the *d-c beta* and is represented by the symbol h_{FE}.   The symbol h_{FE} is not to be confused with h_{fe}.   The latter, with the lowercase subscripts, equals the incremental ratio $\Delta I_C/\Delta I_B$.   We note from Fig. 6-21 that at $V_{CE} = -0.25$ V and at $I_C = -30$ mA, $h_{FE} = -30/-0.23 = 130$.   At the same operating point $h_{fe} \approx 6/0.05 = 120$.

In the saturation region, particularly, the parameter h_{FE} is a useful number and one which is usually supplied by the manufacturer when a switching transistor is involved.   We know I_C, which is given approximately by V_{CC}/R_L, and a knowledge of h_{FE} tells us how much input base current (I_C/h_{FE}) will be needed to saturate the transistor.   For the type 2N404, the variation of h_{FE} with collector current at a low value of V_{CE} is as given in Fig. 6-23.   Note the wide spread (a ratio of 3:1) in the value which may be obtained for h_{FE} even for a transistor of a particular type.   Commercially available transistors have values of h_{FE} that cover the range from 10 to 150 at currents as small as 5 mA and as large as 30 A.

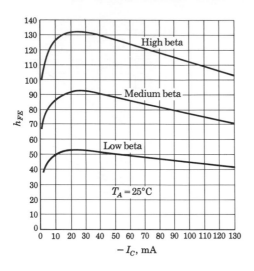

Fig. 6-23 Plots of d-c current gain h_{FE} (at $V_{CE} = -0.25$ V) versus collector current for three samples of the type 2N404 germanium transistor. (Courtesy of General Electric Company.)

From Fig. 6-21 we see that $h_{fe} \to 0$ in the saturation region because $\Delta I_C \to 0$ for a finite ΔI_B. However, from Fig. 6-13 we find that for saturation at $I_C = -30$ mA the base current required is $I_B = -0.25$ mA, so that $h_{FE} = -30/(-0.25) = 120$. We note that the value of h_{FE} (at the edge of saturation) is of the same order of magnitude as h_{fe} (in the active region).

6-13   INPUT CHARACTERISTICS

Knowing the base driving current required, we next naturally wish to learn how large a voltage must be applied at the input base to establish this base current. Information on this point is to be found in the plot of Fig. 6-24. Here the base-to-emitter voltage V_{BE} is plotted against the base current. The values of V_{CE} for which plots are given are low values such as are normally encountered in a transistor switch in saturation. A qualitative understanding of the form of the curves is not difficult. If we were dealing with the emitter junction alone there would be just a single plot, for the emitter-junction current would depend only on the voltage applied between base and emitter. In the present case, however, a collector current flows and the principal part of this collector current continues through the base to cross the emitter junction as well. Accordingly, for a fixed base-to-emitter voltage, the base current varies with collector current and therefore varies also with collector-to-emitter voltage *so long as collector-to-emitter voltage affects the collector current.* Or, for fixed base current, the base-to-emitter voltage varies only so long as this same condition applies. Thus, considering any one of the plots for fixed base current in Fig. 6-21, we expect that V_{BE} will depend on V_{CE} only to the left of the knee and not above the knee, where collector current is insensitive to collector-to-emitter voltage.

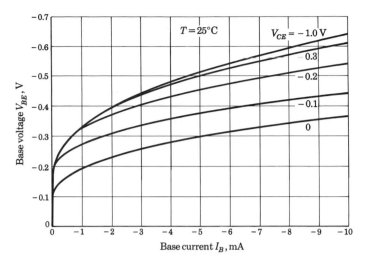

Fig. 6-24 Input characteristics for type 2N404 Ge transistor in the CE configuration.

We now determine from Fig. 6-21 for the given load line and $I_B = -0.35$ mA that $I_C = -20$ mA and $V_{CE}(\text{sat}) = -0.1$ V. From Fig. 6-24 we find $V_{BE}(\text{sat}) = -0.23$ V. As noted in Fig. 6-25, we calculate that

$$V_{BC}(\text{sat}) = -0.23 + 0.10 = -0.13 \text{ V}$$

We observe that the transistor is indeed in saturation since the collector junction is forward-biased rather than reverse-biased. (The p-type collector is positive with respect to the n-type base.) We observe further that the voltage between collector and emitter is smaller than the input voltage between base and emitter. This circumstance arises because, in saturation, the voltage drops across the two junctions are in opposition.

The input characteristics for silicon transistors are similar in form to those in Fig. 6-24. The only notable difference in the case of silicon is that the curves break away from zero current at $V_{BE}(\text{sat})$ in the range 0.5 to 0.6 V rather than in the range 0.1 to 0.2 V as for germanium. Manufacturers give input characteristics in a variety of ways in addition to the form in Fig. 6-24.

Fig. 6-25 Calculated values of terminal voltages to illustrate that $|V_{CE}| < |V_{BE}|$. (a) and (b) are equivalent representations. Note that a p-n-p transistor is involved and that both junctions are forward-biased.

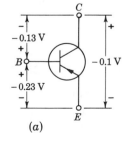

(a)

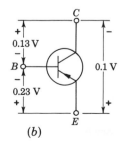

(b)

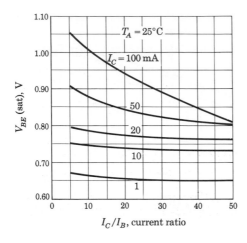

Fig. 6-26 The base-to-emitter saturation voltage V_{BE}(sat) as a function of the ratio I_C/I_B with I_C as a parameter for a 2N914 silicon transistor. (Courtesy of General Electric Company.)

Sometimes V_{BE}(sat) is plotted against I_B (or the ratio I_C/I_B) for various collector currents I_C, as in Fig. 6-26. Another method of presenting the data is to plot I_C against V_{BE}(sat) for various values of I_B.

We note in Fig. 6-24 that beyond the break, a quite small change in input voltage causes a large change in base current. For this reason it is desirable in transistor switches to avoid "voltage drive," that is, applying a voltage from a low-impedance voltage source to drive the transistor to saturation. Rather it is advantageous to use "current drive," that is, to furnish a signal from a high-impedance source which consequently behaves nominally as a current generator.

Tests for Saturation   It is often important to know whether or not a transistor is in saturation. We have already given two methods for making such a determination. These may be summarized as follows:

1. *If I_C and I_B can be determined independently from the circuit under consideration, then the transistor is in saturation if $I_B \geq I_C/h_{FE}$.*
2. *If V_{CB} is determined from the circuit configuration and if this quantity is positive for a p-n-p transistor (or negative for an n-p-n), then the transistor is in saturation.* Of course, the emitter junction must be simultaneously forward-biased, but then we would not be testing for saturation if this condition were not satisfied. We shall make applications of these rules in the transistor-circuit analyses to follow in later chapters.

6-14   TEMPERATURE VARIATION OF SATURATION PARAMETERS

The variation with temperature of the saturation parameters of the transistor switch is of interest. At constant base and collector currents we find that the forward base-to-emitter voltage $|V_{BE}|$ has a typical temperature sensitivity

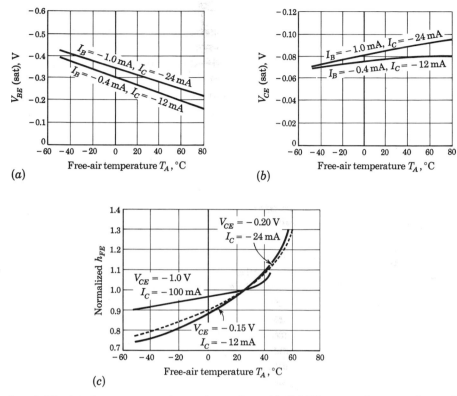

Fig. 6-27 (a) Temperature dependence for type 2N404 germanium transistor of the base-to-emitter voltage in the saturation region at constant base and collector current; (b) temperature dependence of collector-to-emitter saturation voltage; (c) temperature dependence of h_{FE} for various collector currents normalized to 25°C.   (Courtesy of Texas Instruments, Inc.)

in the range -1.5 to -2.0 mV/°C.   This figure applies both to germanium and to silicon units and agrees reasonably well with the calculation in Sec. 6-2. A plot for germanium of V_{BE} against ambient temperature is shown in Fig. 6-27a.   A similar characteristic for silicon has approximately the same slope.

In saturation, the transistor consists of two forward-biased junctions back to back, series opposing.   It is consequently to be anticipated that the temperature-induced voltage change in one junction will be canceled in some measure by the change in the other junction.   We do indeed find such to be the case for V_{CE}(sat).   At small and moderate transistor currents the compensation may be very good, as is borne out by the plots of Fig. 6-27b.   At high currents the voltage drops across the body resistance of the emitter and collector may become comparable to or even larger than the drops across the junctions.   Since the ohmic drops are additive, the cancellation is not possible at these high currents.

The temperature dependence of h_{FE} is shown in Fig. 6-27c.   At small and moderate currents h_{FE} increases substantially with temperature.   At high currents h_{FE} may well become rather insensitive to temperature.

6-15   VARIABILITY OF PARAMETERS

In spite of the great strides that have been made in the manufacturing technology of semiconductor devices, transistors of any particular type still come out of production with a wide spread in the values of some parameters.   We have already noted the wide variability that may appear in I_{CBO}.   Fortunately, other parameters are less difficult to control.   Nevertheless, a variation by a factor of 3 or 4 in a parameter as important as h_{FE} is not unusual (see Fig. 6-23).

To provide information about this variability a transistor data sheet, in tabulating parameter values, invariably provides columns headed *minimum*, *typical* (or *average*), and *maximum*.   Not all columns will necessarily have entries.   Thus, if a particular parameter is not especially important in the application for which the transistor is intended, a listing may appear only under *typical*.   Or, in the case of an important parameter, it may be that the extreme corresponding to the worst case is given and possibly the typical value as well.   By way of example, we find that the Texas Instruments Company data sheet on the 2N404 gives only the minimum value of BV_{CBO}, whereas for h_{FE} it gives minimum and typical values (30 and 100, respectively).   Also, the data sheet gives only a typical value (135) for the small-signal CE short-circuit current gain h_{fe}, whereas for $V_{CE}(\text{sat})$ it gives maximum and typical values.   Any design calculated to ensure that the circuit will operate as predicted with nearly all transistors of a particular type must take account of the most unfavorable possible value of each parameter.   Design on this basis is called *worst-case* design and is employed invariably when reliability is important.

6-16   ANALYTIC EXPRESSIONS FOR TRANSISTOR CHARACTERISTICS

We have found that the transistor characteristics which we require for switching-circuit design are normally furnished for each transistor type by the manufacturer.   It is nonetheless important that we should have some analytic procedure for determining the operating states of a transistor.   In the first place, the availability of such a theoretical analysis will relieve us of complete dependence on published specifications and plots for each transistor type we may plan to use.   Second, it will permit us to arrive at general principles concerning transistor operation.   Third, we shall be able to obtain some numerical values (such as the cutoff voltage) not usually supplied by the manufacturer.

The basic equation of transistor operation in the active region is

$$I_C = -\alpha I_E + I_{CO} \tag{6-26}$$

In Eq. (6-26) I_C and I_E are the collector and emitter currents, respectively, and are positive when currents flow into the corresponding transistor terminal. The parameter α is known as the *short-circuit common-base current gain*, with a value close to but always less than unity. The parameter I_{CO} is the reverse saturation current discussed in Sec. 6-8. (In an *n-p-n* transistor I_{CO} is a positive number, and in a *p-n-p* transistor it is a negative number.)

Suppose now that we seek to generalize Eq. (6-26) so that it will apply not only when the collector junction is substantially reverse-biased but also for any voltage across the collector junction. To do this we need but to replace the term I_{CO} by the more complete expression from which the I_{CO} term is derived as a special case. We recall that the volt-ampere characteristic of any junction is given by Eq. (6-1):

$$I = I_o(\epsilon^{V/\eta V_T} - 1) \tag{6-27}$$

where $\eta = 1$ for germanium and $\eta \approx 2$ for silicon. We apply Eq. (6-27) to the collector junction by replacing I_o by $-I_{CO}$ and V by V_C. The symbol V_C represents the drop across the junction, as in Fig. 6-28, from the p side to the n side and not necessarily the voltage V_{CB} between the corresponding transistor terminals. The difference between V_{CB} and V_C is due to the ohmic drop across the body resistances of the transistor, particularly the base-spreading resistance $r_{bb'}$. A positive value of V_C means that the collector is

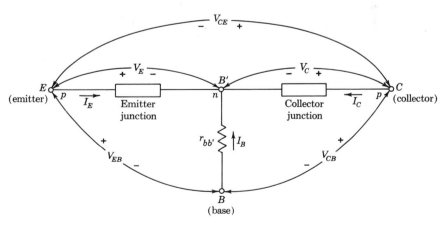

Fig. 6-28 Defining the voltages and currents in the Ebers and Moll equations. For either a *p-n-p* or an *n-p-n* transistor a positive value of current means that positive charge flows into the corresponding junction, and a positive V_E (V_C) means that the emitter (collector) is forward-biased (the p side positive with respect to the n side).

forward-biased with respect to the base.   Equation (6-26) now becomes

$$I_C = -\alpha_N I_E + I_{CO}(1 - \epsilon^{V_C/\eta V_T}) \tag{6-28}$$

We have added a subscript N to α for a reason that will be apparent shortly.
Equation (6-28) reduces again to Eq. (6-26) if V_C is negative (a reverse-biased
junction) and large in magnitude in comparison with ηV_T.   Equation (6-28)
represents the analytical expression for the common-base characteristics of
Fig. 6-12.

To generalize further, we must recognize that there is no essential reason
which constrains us from using a transistor in an *inverted* fashion, that is,
interchanging the roles of the emitter junction and the collector junction and
returning the base current to the collector instead of the emitter.   From a
practical point of view such an arrangement might not be as effective as use
in the *normal* fashion, but this matter does not concern us now.   With this
inverted mode of operation in mind we may now write, in correspondence with
Eq. (6-28),

$$I_E = -\alpha_I I_C + I_{EO}(1 - \epsilon^{V_E/\eta V_T}) \tag{6-29}$$

Here α_I is the *inverted* common-base current gain, just as α_N in Eq. (6-28)
is the current gain in *normal* operation. I_{EO} is the emitter-junction reverse
saturation current and V_E is the voltage drop from p side to n side at the
emitter junction and is positive for a forward-biased emitter.   In the literature
α_R (*reversed* alpha) and α_F (*forward* alpha) are sometimes used in place of
α_I and α_N, respectively.

If we use these equations simultaneously we shall be assuming tacitly
that the response of the transistor to a current injected at the collector junction
is independent of any possible current injected at the emitter junction and
vice versa.   Thus we shall be assuming that *superposition* applies to the
transistor currents, and this assumption implies, in turn, that the transistor
is a linear device.   This linearity does not apply, of course, to the volt-ampere
characteristic of the junctions but only to relationships of the junction currents
to one another.   If the doping of the base region is uniform such linearity
will indeed prevail.   If the doping is not uniform, linearity will not apply.
However, the range of applicability of these equations is very wide and covers
many transistor types.

We may use Eqs. (6-28) and (6-29) to solve explicitly for the junction
currents in terms of the junction voltages as defined in Fig. 6-28, with the
result that

$$I_E = \frac{\alpha_I I_{CO}}{1 - \alpha_N \alpha_I}(\epsilon^{V_C/\eta V_T} - 1) - \frac{I_{EO}}{1 - \alpha_N \alpha_I}(\epsilon^{V_E/\eta V_T} - 1) \tag{6-30}$$

$$I_C = \frac{\alpha_N I_{EO}}{1 - \alpha_N \alpha_I}(\epsilon^{V_E/\eta V_T} - 1) - \frac{I_{CO}}{1 - \alpha_N \alpha_I}(\epsilon^{V_C/\eta V_T} - 1) \tag{6-31}$$

These two equations were first presented by Ebers and Moll in a now classic
paper.[6]

The third current I_B is determined from the condition

$$I_B + I_E + I_C = 0 \tag{6-32}$$

We may solve explicitly for the junction voltages in terms of the currents from Eqs. (6-29) and (6-28), with the result that

$$V_E = \eta V_T \ln \left(1 - \frac{I_E + \alpha_I I_C}{I_{EO}} \right) \tag{6-33}$$

$$V_C = \eta V_T \ln \left(1 - \frac{I_C + \alpha_N I_E}{I_{CO}} \right) \tag{6-34}$$

The parameters α_N, α_I, I_{CO}, and I_{EO} are not independent but are related by the condition[6]

$$\alpha_N I_{EO} = \alpha_I I_{CO} \tag{6-35}$$

Manufacturers' data sheets often provide information about α_N, I_{CO}, and I_{EO}, so that α_I may be determined. For many of the transistor types such as alloy junction and grown junction, for which the present analysis is appropriate, I_{EO} lies in the range $0.5I_{CO}$ to I_{CO}.

6-17   ANALYSIS OF CUTOFF AND SATURATION REGIONS

Let us now apply the above equations to find the d-c currents and voltages in the grounded-emitter transistor switch.

The Cutoff Region   If we define *cutoff* as we did in Sec. 6-8 to mean zero emitter current and reverse saturation current in the collector, what emitter-junction voltage is required for cutoff? Equation (6-33) with $I_E = 0$ and $I_C = I_{CO}$ becomes

$$V_E = \eta V_T \ln \left(1 - \frac{\alpha_I I_{CO}}{I_{EO}} \right) = \eta V_T \ln (1 - \alpha_N) \tag{6-36}$$

where use was made of Eq. (6-35). At 25°C, $V_T = 26 \text{ mV}$, and for $\alpha_N = 0.98$, $V_E = -100 \text{ mV}$ for germanium ($\eta = 1$). Near cutoff we may expect that α may be smaller than the nominal value of 0.98. With $\alpha = 0.9$ for germanium, we find that $V_E = -60 \text{ mV}$. For silicon near cutoff, $\alpha \approx 0$ (page 195), and from Eq. (6-36), $V_E \approx \eta V_T \ln 1 = 0 \text{ V}$. The voltage V_E is the drop from the p to the n side of the emitter junction. To find the voltage which must be applied between base and emitter terminals we must in principle take account of the drop across the base-spreading resistance $r_{bb'}$ in Fig. 6-28. If $r_{bb'} = 100 \ \Omega$ and $I_{CO} = 2 \ \mu\text{A}$, then $I_{CO}r_{bb'} = 0.2 \text{ mV}$, which is negligible. Since the emitter current is zero, the potential V_E is called *the floating emitter potential*.

The above analysis indicates that a reverse bias of approximately 0.1 V (0 V) will cut off a germanium (silicon) transistor. It is interesting to determine what currents will flow if a larger reverse input voltage is applied.

Assuming that both V_E and V_C are negative and much larger than ηV_T, so that the exponentials may be neglected in comparison with unity, Eqs. (6-28) and (6-29) become

$$I_C = -\alpha_N I_E + I_{CO} \qquad I_E = -\alpha_I I_C + I_{EO} \qquad (6\text{-}37)$$

Solving these equations and using Eq. (6-35), we obtain

$$I_C = \frac{I_{CO}(1 - \alpha_I)}{1 - \alpha_N \alpha_I} \qquad I_E = \frac{I_{EO}(1 - \alpha_N)}{1 - \alpha_N \alpha_I} \qquad (6\text{-}38)$$

Since (for Ge) $\alpha_N \approx 1$, $I_C \approx I_{CO}$ and $I_E \approx 0$. Using $\alpha_N = 0.9$ and $\alpha_I = 0.5$, then $I_C = I_{CO}(0.50/0.55) = 0.91 I_{CO}$ and $I_E = I_{EO}(0.10/0.55) = 0.18 I_{EO}$ and represents a very small *reverse* current. Using $\alpha_I \approx 0$ and $\alpha_N \approx 0$ (for Si), we have that $I_C \approx I_{CO}$ and $I_E \approx I_{EO}$. Hence, increasing the magnitude of the reverse base-to-emitter bias beyond cutoff has very little effect on the very small transistor currents.

Short-circuited Base Suppose that instead of reverse-biasing the emitter junction we simply short the base to the emitter terminal. The currents which now flow are found by setting $V_E = 0$ and by neglecting exp $(V_C/\eta V_T)$ in the Ebers and Moll equations. The results are

$$I_C = \frac{I_{CO}}{1 - \alpha_N \alpha_I} \equiv I_{CES} \qquad \text{and} \qquad I_E = -\alpha_I I_{CES} \qquad (6\text{-}39)$$

where we have used the notation of Sec. 6-9; namely, I_{CES} represents the collector current in the common-emitter configuration with a short-circuited base. If (for Ge) $\alpha_N = 0.9$ and $\alpha_I = 0.5$, then I_{CES} is about $1.8 I_{CO}$ and $I_E = -0.9 I_{CO}$. If (for Si) $\alpha_N \approx 0$ and $\alpha_I \approx 0$, then $I_{CES} \approx I_{CO}$ and $I_E \approx 0$. Hence, even with a short-circuited emitter junction the transistor is virtually at cutoff.

Open-circuited Base If instead of a shorted base we allowed the base to "float" so that $I_B = 0$, we see, in Sec. 6-8, that the cutoff condition is not reached. The collector current under this condition is called I_{CEO} and is given by

$$I_{CEO} = \frac{I_{CO}}{1 - \alpha_N} \qquad (6\text{-}40)$$

It is interesting to find the emitter-junction voltage under this condition of a floating base. From Eq. (6-33) with $I_E = -I_C$, and using Eq. (6-35),

$$V_E = \eta V_T \ln \left[1 + \frac{\alpha_N(1 - \alpha_I)}{\alpha_I(1 - \alpha_N)} \right] \qquad (6\text{-}41)$$

For $\alpha_N = 0.9$ and $\alpha_I = 0.5$ (for Ge) we find $V_E = +60$ mV. For $\alpha_N \approx 2\alpha_I \approx 0$ and $\eta = 2$ (for Si) we have $V_E \approx 2V_T \ln 3 = +57$ mV. Hence an open-circuited base represents a slight *forward* bias.

The Cutin Voltage   The volt-ampere characteristic between base and emitter at constant collector-to-emitter voltage is not unlike the volt-ampere characteristic of a simple junction diode.   When the emitter junction is reverse-biased the base current is very small, being of the order of nanoamperes or microamperes for silicon and germanium, respectively.   When the emitter junction is forward-biased, again as in the simple diode, no appreciable base current flows until the emitter junction has been forward-biased to the extent where $|V_{BE}| \geq |V_\gamma|$, where V_γ is called the *cutin voltage*.   Since the collector current is nominally proportional to the base current, no appreciable collector current will flow until an appreciable base current flows.   Therefore, a plot of collector current against base-to-emitter voltage will exhibit a cutin voltage, just as does the simple diode.

In principle, a transistor is in its active region whenever the base-to-emitter voltage is on the forward-biasing side of the cutoff voltage, which occurs at a reverse voltage of 0.1 V for germanium and 0 V for silicon.   In effect, however, a transistor enters its active region when $V_{BE} > V_\gamma$.

We may estimate the cutin voltage V_γ in a typical case in the following manner.   Assume that we are using a transistor as a switch so that when the switch is ON it will carry a current of 2 mA.   We may then consider that the cutin point has been reached when, say, the collector current equals 1 percent of the maximum current or a collector current $I_C = 20\ \mu\text{A}$.   Hence, V_γ is the value of V_E given in Eq. (6-33), with $I_E = -(I_C + I_B) \approx -I_C = -20\ \mu\text{A}$.   Assume a silicon transistor with $\alpha_I = 0.5$ and $I_{EO} = 1$ nA.   Since $\eta = 2$ and at room temperature $V_T = 0.026$ V, we obtain from Eq. (6-33)

$$V_\gamma = 2(0.026)(2.30)\log_{10}\left[1 + \frac{20 \times 10^{-6}(1 - 0.5)}{10^{-9}}\right] = 0.48\ \text{V}$$

With a germanium transistor ($\eta = 1$ and $I_{EO} = 1\ \mu\text{A}$) and again using $\alpha_I = 0.5$ and $I_E = -I_C = -20\ \mu\text{A}$, Eq. (6-33) yields $V_\gamma = 0.06$ V.   If the switch had been called upon to carry 20 mA rather than 2 mA, then cutin voltages of 0.60 and 0.12 V for Si and Ge, respectively, would have been obtained.

Figure 6-29 shows plots, for several temperatures, of the collector current as a function of the base-to-emitter voltage at constant collector-to-emitter voltage for a typical silicon transistor.   We see that the value calculated above for V_γ (of the order of 0.5 V) is entirely reasonable.   The temperature dependence results from the temperature variation in the forward direction of the emitter-junction diode.   Therefore the lateral shift of the plots with change in temperature and the change with temperature of the cutin voltage V_γ are given by Eq. (6-6), that is, approximately -2 mV/°C.

The Saturation Region   Let us consider the 2N404 *p-n-p* germanium transistor operated as in Fig. 6-21 with $I_C = -20$ mA and $I_B = -0.35$, so that $I_E = +20.35$ mA.   Assume the following reasonable values: $I_{CO} = -2.0\ \mu\text{A}$, $I_{EO} = -1.0\ \mu\text{A}$, and $\alpha_N = 0.99$.   From Eq. (6-35) $\alpha_I = 0.50$.

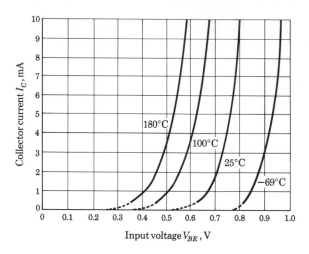

Fig. 6-29 Plot of collector current against base-to-emitter voltage for various temperatures for the type 2N337 silicon transistor. (Courtesy of Transitron Electronic Corporation.)

From Eqs. (6-33) and (6-34) we calculate that at room temperature

$$V_E = (0.026)(2.30) \log_{10}\left[1 - \frac{20.35 - (0.50)(20)}{-10^{-3}}\right] = 0.24 \text{ V}$$

and

$$V_C = (0.026)(2.30) \log_{10}\left[1 - \frac{-20 + 0.99(20.35)}{-(2)(10^{-3})}\right] = 0.11 \text{ V}$$

For a *p-n-p* transistor

$$V_{CE} = V_C - V_E = 0.11 - 0.24 \approx -0.13 \text{ V}$$

Taking the voltage drop across $r_{bb'}$ ($\approx 100 \ \Omega$) into account,

$$V_{CB} = V_C - I_B r_{bb'} = 0.11 + 0.035 \approx 0.15 \text{ V}$$

and

$$V_{BE} = I_B r_{bb'} - V_E = -0.035 - 0.24 \approx -0.28 \text{ V}$$

These are in fair agreement with the values in Fig. 6-25, which were obtained from the published saturation curves ($V_{CE} = -0.10$, $V_{CB} = 0.13$, and $V_{BE} = -0.23$).

Note that the base-spreading resistance does not enter into the calculation of the collector-to-emitter voltage. For a diffused-junction transistor the voltage drop resulting from the collector-spreading resistance may be significant for saturation currents. If so, this ohmic drop can no longer be neglected as we have done above. For example, if the collector resistance is 5 Ω, then with a collector current of 20 mA the ohmic drop is 0.10 V and $|V_{CE}|$ increases from 0.13 to 0.23 V.

6-18   TYPICAL TRANSISTOR–JUNCTION VOLTAGE VALUES

Quite often, in making a transistor-circuit calculation, we are beset by a complication when we seek to determine the transistor currents. These currents are influenced by the transistor-junction voltages. However, to determine these junction voltages we would first have to know the very currents we seek to determine. A commonly employed and very effective procedure to overcome this problem arises from the recognition that certain of the transistor-junction voltages are ordinarily small in comparison with externally impressed voltages, the junction voltages being in the range of only tenths of volts. We may therefore start the calculation by making the first-order approximation that these junction voltages are all zero. On this basis we calculate a first-order approximation of the current. These first-order currents are now used to determine the junction voltages either from transistor characteristics or, if they are applicable, from the Ebers-Moll equations. The junction voltages so calculated are used to determine a second-order approximation of the currents, etc. As a matter of practice it ordinarily turns out that not many orders are called for, since the successive approximations converge to a limit very rapidly. Furthermore, a precise calculation is not justifiable because of the variability from sample to sample of transistors of a given type.

The required number of successive approximations may be reduced, or, more importantly, the need to make successive approximations may usually be eliminated completely by recognizing that for many low- and medium-power transistors, over a wide range of operating conditions, certain transistor-junction voltages lie in a rather narrow range and may be approximated by the entries in Table 6-1. Table 6-1 lists the collector-to-emitter saturation voltage $[V_{CE}(\text{sat})]$, the base-to-emitter saturation voltage $[V_{BE}(\text{sat}) \equiv V_\sigma]$, the base-to-emitter voltage in the active region $[V_{BE}(\text{active})]$, at cutin $[V_{BE}(\text{cutin}) \equiv V_\gamma]$, and at cutoff $[V_{BE}(\text{cutoff})]$. The entries in the table are appropriate for an n-p-n transistor. For a p-n-p transistor the signs of all entries should be reversed. Observe that the total range of V_{BE} between cutin and saturation is rather small, being only 0.2 V. The voltage $V_{BE}(\text{active})$ has been located somewhat arbitrarily but nonetheless reasonably at the midpoint of the range.

Of course, particular cases will depart from the estimates of Table 6-1. But it is unlikely that the larger of the numbers will be found in error by more than about 0.1 V or that the smaller entries will be wrong by more than about

TABLE 6-1   **Typical n-p-n transistor-junction voltages at 25°C†**

	$V_{CE}(\text{sat})$	$V_{BE}(\text{sat}) \equiv V_\sigma$	$V_{BE}(\text{active})$	$V_{BE}(\text{cutin}) \equiv V_\gamma$	$V_{BE}(\text{cutoff})$
Si	0.3	0.7	0.6	0.5	0.0
Ge	0.1	0.3	0.2	0.1	−0.1

† The temperature variation of these voltages is discussed in Secs. 6-14 and 6-17.

0.05 V. In any event, starting a calculation with the values of Table 6-1 may well make further approximations unnecessary.

Finally, it should be noted that the values in Table 6-1 apply to the intrinsic junctions. The base terminal-to-emitter voltage includes the drop across the base-spreading resistance $r_{bb'}$. Ordinarily the drop $r_{bb'}I_B$ is small enough to be neglected. If, however, the transistor is driven very deeply into saturation the base current I_B may not be negligible, and we must take

$$V_{BE} = V_\sigma + I_B r_{bb'}$$

6-19   THE VACUUM–TUBE TRIODE[1]

Typical triodes used in pulse applications, as well as in other types of circuits, are the 6CG7, the 12AU7 (or its equivalent the 5963), the 12AT7, the 12AX7, and the 5965. These are miniature tubes and each contains two triode sections in one envelope. The 6SN7 is a nonminiaturized tube similar to the type 6CG7 and was the tube most commonly used in pulse-type equipment during World War II. The 5963 and 5965 were designed for use in high-speed digital computers. The volt-ampere characteristics of the above tubes are given in Appendix D or in this section. The curves for the 5965 are given in Figs. 6-30 and 6-31. In these latter characteristics, curves for positive grid voltages have been included because, as we shall see, the grid of a tube

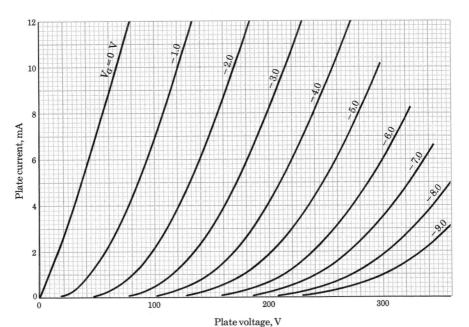

Fig. 6-30 Negative-grid plate characteristics of 5965 tube.

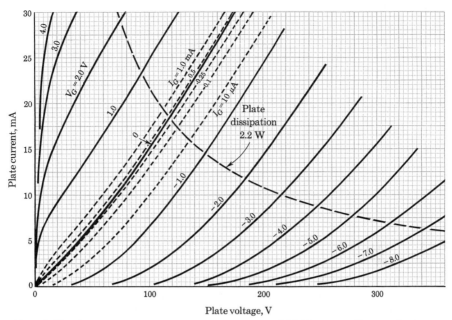

Fig. 6-31   Positive-grid plate characteristics of 5965 tube.   The dashed lines are loci of constant grid current.

is often driven positive in pulse circuits.   If the region near small plate voltages is ignored, then the positive-grid curves are very similar in shape and spacing to those for negative-grid values.   Hence, if the grid signal is supplied from a source of low impedance, so that the loading effect on the source due to the flow of grid current may be ignored, the tube will continue to operate linearly even if the grid signal makes an excursion into the positive-grid region.   This linearity will continue so long as the grid current is a small fraction of the total cathode current.

In pulse applications, large voltage swings are often encountered, and the small-signal equivalent circuit of Sec. 1-5 is meaningless because the tube parameters μ, r_p, and g_m are *not* constant.   The variation of these parameters with plate current is given in Fig. 6-32.

The grid volt-ampere characteristics of the 5965 tube are given in Fig. 6-33. At a given plate voltage the grid circuit behaves as a diode.   By analogy with the definition of the dynamic plate resistance, the *dynamic grid resistance* r_g is given by dV_G/dI_G, where V_G and I_G are the instantaneous values of grid voltage and current, respectively.   The *static grid resistance* r_G is defined as the ratio V_G/I_G.   From Fig. 6-33 it appears that the difference in values between the static and dynamic resistances is not great, except possibly for small grid voltages.   Furthermore, the value of the grid resistance r_G is not a sensitive function of plate voltage.   From Fig. 6-33 we find that for the 5965

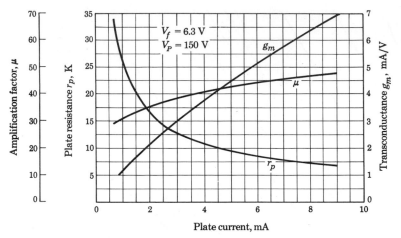

Fig. 6-32   Average small-signal parameters of 5965 tube.

tube, 250 Ω is a reasonable value for r_G.   For other tubes, the **grid resistance** may be much more variable than indicated above.   For example, for a 12AU7 the static r_G has values ranging from about 500 to 1,500 Ω, depending upon the values of grid and plate voltages (see the grid current curves of Fig. 6-33).

The grid current at zero grid-to-cathode voltage, and even for slightly negative grid voltages, is often large enough to have an appreciable effect on the operation of a circuit.   We estimate from Fig. 6-31 that the grid current is

$$I_G = 400 \ \mu A \qquad \text{at } V_G = 0 \text{ V}$$

$$I_G = 10 \ \mu A \qquad \text{at } V_G = -0.5 \text{ V}$$

$$I_G = 0.25 \ \mu A \qquad \text{at } V_G = -1 \text{ V}$$

Consider, then, that a grid-leak resistor is connected from grid to cathode of the 5965.   If this grid-leak resistor is $R_g = (1/0.25) \text{ M} = 4 \text{ M}$, then a negative bias of 1 V will be developed.   At a plate voltage of 100 V, the plate current corresponding to the 1-V bias is seen from Fig. 6-30 to be 7 mA.   If we were to neglect the effect of the grid current and assume that $V_G = 0$, we would expect a current of 15 mA, or more than twice the value actually obtained. Even if the grid leak were reduced to 50 K, the bias due to grid current would be -0.5 V, since $(5 \times 10^4)(10 \times 10^{-6}) = 0.5$, and the plate current would be about 11 mA, which is still much less than the zero grid-bias value of 15 mA.

A Clamped Grid   If, as in Fig. 6-34, the grid leak is tied to the V_{PP} supply instead of to the cathode, then the grid-to-cathode voltage will approach nominal zero for values of R_g which are large compared with r_G.   For example, if $R_g = 1 \text{ M}$ and $V_{PP} = 300 \text{ V}$, then the grid current will be approximately

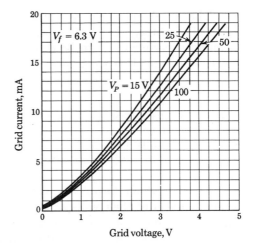

Fig. 6-33 Average volt-ampere grid characteristics of the 5965 tube.

300 μA.   From Fig. 6-31, we find that the grid voltage corresponding to this grid current is about -0.05 V.   (If we assume that the value of $r_G = 250$ Ω is valid at low grid voltages, then the calculated value of V_G is $0.3 \times 0.25 = +0.075$ V.)   In many pulse circuits it is common to use this connection of the grid leak to a high positive voltage.   Under such circumstances, where the grid is held at the cathode voltage because of the flow of grid current, we shall refer to the grid as being *clamped* to the cathode.   Alternatively, the tube is said to be *in clamp*.   Clamping is discussed in detail in Chap. 8.

Variability of Characteristics   If the grid voltage is made a few volts negative, then the grid current reverses.[7]   This negative current is caused by the positive ions which are attracted to the grid.   Since the positive-ion current comes from the residual gas in the "vacuum" tube, it is very variable from tube to tube, and is usually a small fraction of a microampere.   Negative grid current can also result from thermionic or photoelectric emission from the grid.

The characteristics given in Figs. 6-30 to 6-33 are average values as supplied by the manufacturer, and the curves for a specific tube may differ appreciably from these published values.   The Military Specification, MIL-E-1, for Electron Tubes gives the limits of variability which may be expected.

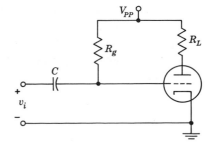

Fig. 6-34 A triode with the grid leak R_g connected to the V_{PP} supply.

The volt-ampere characteristics vary with heater temperature and with aging of the tube. In multielement tubes as in diodes, the temperature effect is found experimentally to be equivalent to a 0.1-V shift in cathode voltage (relative to the other electrodes) for each 10 percent change in heater voltage.

6-20   ADDITIONAL DEVICES

A family of plate characteristics for the type 6AU6 pentode is given in Fig. D-1. The switching characteristics of a pentode are quite similar to those of a triode.

Devices which, between a pair of terminals, exhibit a negative resistance over a portion of their volt-ampere characteristic are considered in Chap. 12. Included among these are the *tunnel diode*, the *unijunction transistor*, the *p-n-p-n diode*, the *silicon controlled switch*, and the *avalanche transistor*.

The use of the *field-effect transistor* in chopper circuit applications is discussed in Chap. 17.

REFERENCES

1. Millman, J., and C. C. Halkias: "Electronic Devices and Circuits," McGraw-Hill Book Company, New York, 1967.

2. Sah, C. T., R. N. Noyce, and W. Shockley: Carrier-generation and Recombination in P-N Junctions and P-N Junction Characteristics, *Proc. IRE*, vol. 45, pp. 1228–1243, September, 1957.
   Moll, J. L.: The Evolution of the Theory for the Voltage-Current Characteristic of P-N Junctions, *Proc. IRE*, vol. 46, pp. 1076–1082, June, 1958.

3. Phillips, A. B.: "Transistor Engineering," chap. 6, McGraw-Hill Book Company, New York, 1962.

4. Ref. 3, chap. 5.

5. Turner, C. R.: Interpretation of Voltage Ratings for Transistors, *Appl. Note* SMA-2, Radio Corporation of America, November, 1961.

6. Ebers, J. J., and J. L. Moll: Large Signal Behavior of Junction Transistors, *Proc. IRE*, vol. 42, pp. 1761–1772, December, 1954.

7. Valley, G. E., Jr., and H. Wallman: "Vacuum Tube Amplifiers," p. 418, Massachusetts Institute of Technology Radiation Laboratory Series, vol. 18, McGraw-Hill Book Company, New York, 1948.

7/CLIPPING AND COMPARATOR CIRCUITS

In Chaps. 2 through 5 we consider the behavior of linear circuits and devices. With such circuits a sinusoidal excitation results in a sinusoidal steady-state response of the same frequency but of a different amplitude and phase. Any other excitation results in an output which has suffered a change in waveshape.

In the preceding chapter we discuss the nonlinear properties of devices. In electronic systems many useful and interesting functions can be performed by taking advantage of the nonlinearity of semiconductor or thermionic devices. In this chapter we describe and analyze the manner in which these nonlinear elements, in combination with resistors, can function as clipper or as comparator circuits.

Energy-storage circuit components are not required in the basic process of clipping or comparison. Any reactive elements introduced in this chapter represent spurious unavoidable stray components (such as shunt capacitance). In the following chapters we consider some combinations of nonlinear devices, resistors, and energy-storage elements which perform other basic functions such as clamping, switching, and the generation of square-wave or pulse waveforms.

7-1 CLIPPING (LIMITING) CIRCUITS

Clipping circuits are used to select for transmission that part of an arbitrary waveform which lies above or below some particular reference voltage level. Clipping circuits are also referred to as voltage (or current) *limiters, amplitude selectors,* or *slicers*. Limiting circuits usually fall into one of the following configurations: (1) a series combination of a diode, resistor, and reference supply (the "diode" may be the input circuit of a vacuum-tube triode, pentode, or transistor);

225

(2) a network consisting of several diodes, resistors, and reference voltages; (3) two emitter-coupled or cathode-coupled triodes operating as an overdriven difference amplifier. Some of the more commonly employed clipping circuits are now to be described.

7-2   DIODE CLIPPERS

In Chap. 6, it is suggested that a diode volt-ampere characteristic may be approximated as shown in Fig. 6-10 (and repeated for easy reference in Fig. 7-1a) by a curve which is *piecewise linear* and *continuous*. This idealized diode characteristic exhibits a discontinuity in slope at the voltage V_γ, and this point of slope discontinuity is called a *break point*. The break point occurs at $V_\gamma \approx 0.2$ V for Ge, at $V_\gamma \approx 0.6$ V for Si, at $V_\gamma \approx -0.5$ V for a thermionic diode, and for a "backward" diode (Sec. 12-2), V_γ is very nearly zero. Using this piecewise linear model of the diode, we find that the clipping circuit of Fig. 7-1b has the transmission characteristic shown in Fig. 7-1c. The transmission characteristic, which is a plot of the output voltage v_o as a function of the input voltage v_i, also exhibits piecewise linear continuity. A break point occurs at the voltage $V_R + V_\gamma$. To the left of the break point (for $v_i < V_R + V_\gamma$) the diode is reverse-biased (OFF). In this region, the signal v_i may be transmitted directly to the output, since there is no load across the output to cause a drop across the series resistor R. To the right of the break point, increments Δv_i in the input are attenuated and appear at the output as increments $\Delta v_o = \Delta v_i R_f / (R_f + R)$, in which R_f is the diode forward resistance.

Figure 7-1c shows a sinusoidal input signal of amplitude large enough so that the signal makes excursions past the break point. The corresponding output exhibits a suppression of the positive peak of the signal. If $R_f \ll R$, then this suppression will be very pronounced, and the positive excursion of the output will be sharply limited at the voltage $V_R + V_\gamma$. The output will appear as though the positive peak had been "clipped off" or "sliced off." Often it turns out that $V_R \gg V_\gamma$, in which case one may consider that V_R itself is the limiting reference voltage.

In Fig. 7-2a the clipping circuit has been modified in that the diode in Fig. 7-1b has been reversed. The corresponding piecewise linear representation of the transfer characteristic is shown in Fig. 7-2b. In this circuit, the portion of the waveform more positive than $V_R - V_\gamma$ is transmitted without attenuation, but the less positive portion is greatly suppressed.

The Break Region The piecewise linear approximation given in Fig. 7-1a indicates an abrupt discontinuity in slope at V_γ. Actually, the transition of the diode from the OFF condition to the ON condition is not abrupt. Therefore, a waveform which is transmitted through a clipper will not show an abrupt onset of clipping at a break point but will instead exhibit a *break region*, that is, a region of transition from unattenuated to attenuated transmission. We shall now estimate the range of this break region.

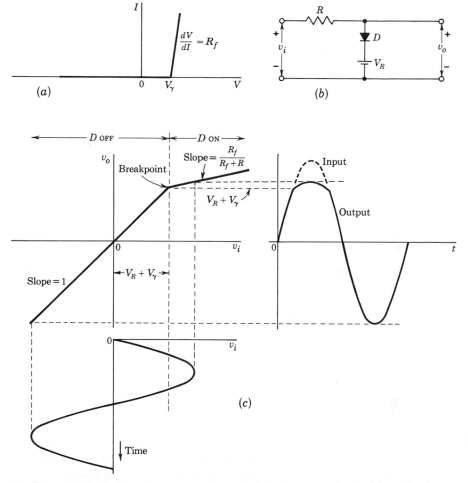

Fig. 7-1 (a) A diode volt-ampere characteristic is approximated by a broken straight line having piecewise linear continuity. (b) A diode clipping circuit which transmits that part of the waveform more negative than $V_R + V_\gamma$. (c) The piecewise linear transmission characteristic of the circuit. A sinusoidal input and the clipped output are shown.

The semiconductor diode equation [Eq. (6-1)] is

$$I = I_o(\epsilon^{V/\eta V_T} - 1)$$

Beyond the diode break point, for currents I which are large compared with I_o we may write

$$I = I_o \epsilon^{V/\eta V_T} \tag{7-1}$$

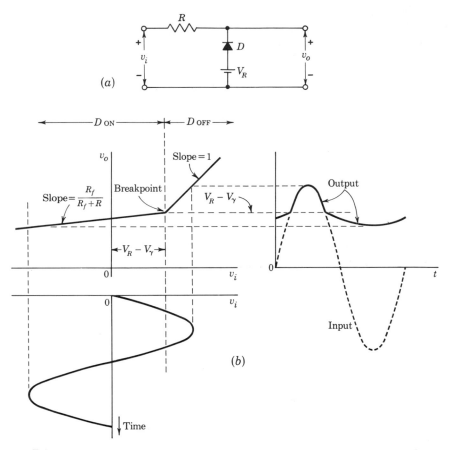

Fig. 7-2 (a) A diode clipping circuit which transmits that part of the waveform more positive than $V_R - V_\gamma$. (b) The piecewise linear transmission characteristic of the circuit. A sinusoidal input and the clipped output are shown.

From the discussion in Sec. 6-5 it follows that Eq. (7-1) applies equally to the thermionic diode in the region of small currents, where space-charge effects are negligible, provided that I_o is interpreted to be the current at zero voltage and $\eta = 1$. We must, of course, use a value of V_T appropriate to the cathode temperature. For an oxide-coated cathode, a reasonable value of the temperature is $T = 1000°\text{K}$ and correspondingly $V_T = 0.086$ V. Since the limiting circuit clips or does not clip depending on whether the diode incremental resistance r is very large or very small in comparison with the circuit resistance R, let us arbitrarily define the break region as the range over which the diode resistance is multiplied by some large factor, say 100. The incremental diode resistance $r \equiv dV/dI$ and is, from Eq. (7-1),

$$r = \frac{\eta V_T}{I_o} \epsilon^{-V/\eta V_T} = \frac{\eta V_T}{I} \tag{7-2}$$

Note that r varies inversely with the quiescent current and directly with the absolute temperature.

The resistance will be multiplied by a factor 100 over the voltage range ΔV provided that $\epsilon^{\Delta V/\eta V_T} = 10^2$. We have then, with $\eta = 1$ for Ge, $\eta = 2$ for Si, $V_T = 0.026$ V at room temperature, and $V_T = 0.086$ V at $T = 1000°$K,

$$\Delta V = 2\eta V_T \ln 10$$

$$= 0.12 \text{ V (Ge)}$$

$$= 0.24 \text{ V (Si)}$$

$$= 0.40 \text{ V (thermionic)} \tag{7-3}$$

Note that the break region is independent of the quiescent current. Thus, at any current $I \gg I_o$ (at an arbitrary value of resistance) the dynamic resistance is multiplied by a factor of 100 if the voltage is decreased by the value of ΔV given in Eq. (7-3) and divided by 100 if V is increased by ΔV.

If a diode is to be effective in a clipping circuit, the signal applied to the circuit must carry the diode from a point well to one side of the break region to a point well to the other side. If the signal is only of the order of magnitude of the extent of the break region, the output will not display sharp limiting.

Reverse Characteristic We now consider the diode reverse-bias characteristic. In thermionic diodes the magnitude of the diode current decreases monotonically with increasing negative plate-to-cathode voltage. At voltages in excess of 0.5 to 0.75 V the current will have fallen to the point where it ordinarily need no longer be taken into account. In an ideal semiconductor diode, at reverse voltages the reverse current is constant. In a circuit such as in Fig. 7-1b this constant current will give rise to a fixed voltage difference across R, but this fixed voltage will not prevent the output signal from following the input signal without attenuation. However, in many semiconductor diodes (Fig. 6-4) the magnitude of the diode reverse current increases with reverse voltage. A piecewise linear and continuous volt-ampere model which represents such diodes better than does the characteristic of Fig. 7-1a is shown in Fig. 7-3. To the right of the break point, in the forward-biased region, the diode resistance is R_f. To the left of the break point, in the back or reverse-biased region, the diode resistance is R_r. (Actually, of course, in the region from the origin to the break point the diode is forward-biased.)

Fig. 7-3 A piecewise linear and continuous approximation to the volt-ampere characteristic of a semiconductor diode. The break point is at $V = V_\gamma$. To the right of V_γ the forward resistance R_f is small, and to the left the reverse resistance R_r is large.

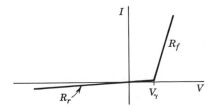

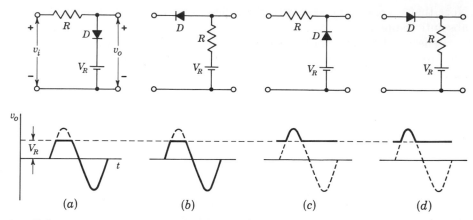

Fig. 7-4 Four diode clipping circuits. In (a) and (c) the diode appears as a shunt element. In (b) and (d) the diode appears as a series element. Under each circuit appears the output waveform (solid) for a sinusoidal input. The clipped portion of the input is shown dashed.

In Figs. 7-1c and 7-2b we have assumed R_r arbitrarily large in comparison with R. If this condition does not apply, the transmission characteristics must be modified. The portions of these curves which are indicated as having unity slope must instead be considered to have a slope $R_r/(R_r + R)$.

In a transmission region of a diode clipping circuit we require that $R_r \gg R$, for example, that $R_r = kR$, where k is a large number. In the attenuation region, we require that $R \gg R_f$, for example, that $R = kR_f$. From these two equations we deduce that $R = \sqrt{R_f R_r}$ and that $k = \sqrt{R_r/R_f}$. On this basis we conclude that it is reasonable to select R as the geometrical mean of R_r and R_f. And we note that the ratio R_r/R_f may well serve as a figure of merit for diodes used in the present application.

Clipping Circuits Figures 7-1 and 7-2 appear again in Fig. 7-4 together with variations in which the diodes appear as series elements. If in each case a sinusoid is applied at the input, the waveforms at the output will appear as shown by the heavy lines. In these output waveforms we have neglected V_γ in comparison with V_R and we have assumed that the break region is negligible in comparison with the amplitude of the waveforms. We have also assumed that $R_r \gg R \gg R_f$. In two of these circuits the portion of the waveform transmitted is that part which lies below V_R; in the other two the portion above V_R is transmitted. In two the diode appears as an element in series with the signal lead; in two it appears as a shunt element. The use of the diode as a series element has the disadvantage that when the diode is OFF and it is intended that there be no transmission, fast signals or high-frequency waveforms may be transmitted to the output through the diode

capacitance.   The use of the diode as a shunt element has the disadvantage
that when the diode is open (back-biased) and it is intended that there be
transmission, the diode capacitance, together with all other capacitance in
shunt with the output terminals, will round sharp edges of input waveforms
and attenuate high-frequency signals.   A second disadvantage of the use
of the diode as a shunt element is that in such circuits the impedance R_s of the
source which supplies V_R must be kept low.   This requirement does not arise
in circuits where V_R is in series with R, which is normally large compared with
R_s.

When a diode clipper is used with fast waveforms, then, as the example
to follow will show, the stray capacitances associated with the circuit may
not be neglected.

EXAMPLE   The clipper of Fig. 7-5a is to be used with the input waveform indicated.   This input may represent a pulse or half a cycle of a square wave.   The
capacitance C_1 is the total effective capacitance shunting the diode (for which 5 pF
is a reasonable value), and C_2 is the total capacitance shunting the output load
resistor R ($\gg R_f$).   The value $C_2 = 20$ pF is nominally the input capacitance of an
oscilloscope probe which we might be using.   Find the output waveform, assuming
that the back resistance is infinite.

Solution   If the diode were perfect and the capacitances were neglected, the output
waveform would be as shown in Fig. 7-5b.

Assume that a steady-state condition has been reached in which the input is
-5 V and the output is 0 V.   Now let the input rise abruptly by 10 V.   If the
source impedance is negligible, an impulsive current results and the initial output
voltage rise is determined entirely by the capacitors.   Since $C_2 = 4C_1$, only one-
fifth of the input rise will appear across C_2; hence the output will jump abruptly
by 2 V.   The voltage across the diode is now 3 V and in the direction to make the
diode conduct.   The output v_o will rise to its final value of 5 V with a time constant
$\tau_1 = (C_1 + C_2)R_f$, where R_f is the forward resistance of the diode.   Similarly,
when the input voltage drops by 10 V, the output voltage will drop abruptly by

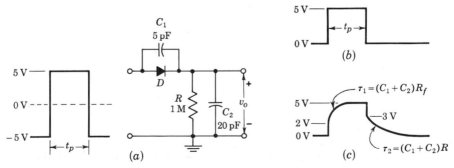

Fig. 7-5   (a) A diode clipper circuit with a pulse input.   The output (b) neglecting capacitances, (c) taking capacitances into account.

2 V.   The cathode of the diode is now at $+3$ V, and the anode is at -5 V.   The diode will not conduct, and the decay of the output signal to zero will take place with a time constant $\tau_2 = (C_1 + C_2)R$.   The resultant waveform is shown in Fig. 7-5c.   If $R_f = 100$ Ω, then $\tau_1 = 2.5$ nsec.   On the other hand, $\tau_2 = 25$ μsec.   If, say, $t_p \approx 50$ μsec, the slow decay on the trailing edge of the signal will be very apparent, whereas the rise time of the front edge is negligible.

If in the above example the capacitance C_1 across the diode were larger than C_2 across the load, then the output waveform would not have rounded rising and falling sides, as indicated in Fig. 7-5.   Instead there would be a spike overshoot at the front and rear edges (Prob. 7-6).

7-3   VACUUM–TUBE CLIPPER AT CUTOFF

A vacuum-tube circuit such as that shown in Fig. 7-6a will limit a signal when the grid is driven beyond cutoff.   A triode vacuum tube is indicated but a pentode would serve as well.   In Fig. 7-6b a plot of a typical dynamic transfer characteristic of the tube is indicated.   The sinusoidal input signal is of such magnitude and so biased that the signal makes excursions above and below the cutin point, which occurs at the voltage V_γ (also called the *cutoff voltage*).   As appears in the output current waveform, only that part of the input is transmitted which is above this cutoff voltage.   Note that since the transfer characteristic is not linear near cutin, some distortion appears in the output waveform.

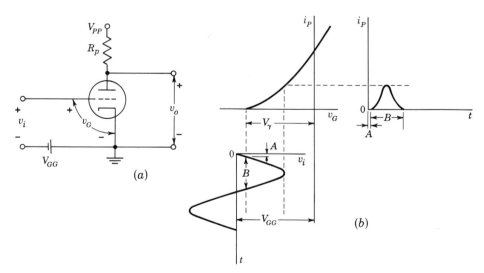

Fig. 7-6   (a) A vacuum-tube clipper; (b) the dynamic transfer curve.   A sinusoidal input signal and the clipped output are shown.   The time durations marked A are equal, and those marked B are equal.

The cutoff point in a vacuum tube is not sharply defined. The abrupt-ness with which the plate current increases as the grid voltage rises depends upon the tube type and upon the plate voltage at which the tube is operating. High-μ triodes give sharper cutin break regions than do low-μ triodes, and the break region becomes less sharp as the plate voltage increases. Except for very low plate voltages, the break region at cutoff is less sharp than that of either a semiconductor or thermionic diode. Further, the break region in a diode is appreciably more stable with respect to device replacement than is the case with triodes or pentodes. On the other hand, the nominally infinite input impedance of a triode or pentode and the fact that the tube will provide gain are both important advantages.

7-4   CLIPPER USING GRID–CATHODE "DIODE"

We observe in Fig. 6-33 that the grid-voltage–grid-current characteristic of a multielectrode tube has much the same form as the volt-ampere characteristic of a simple thermionic diode. In the case of a triode, the characteristic depends somewhat on plate voltage, but this dependence is small enough to be neglected in our present considerations. It then appears, as in Fig. 7-7a, that a triode may be viewed as a combination of a diode and an ideal triode which draws no grid current. A clipping circuit using this grid-cathode "diode," called a *grid-current limiter*, is shown in Fig. 7-7b. The operation of the circuit is illustrated in Fig. 7-8. In Fig. 7-8a the piecewise linear and continuous transmission characteristic of the resistor R and diode combination has been drawn, assuming for simplicity that the diode break is abrupt, that the break occurs at zero voltage, and that $R_r \gg R \gg R_f$. A sinusoidal input is indi-cated and, as appears, the grid-to-cathode signal v_G displays a clipped positive

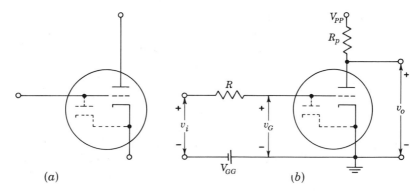

Fig. 7-7   (a) A representation of a triode taking into account the fact that the grid and cathode are the terminals of a diode; (b) a clipping circuit which uses the grid-cathode diode.

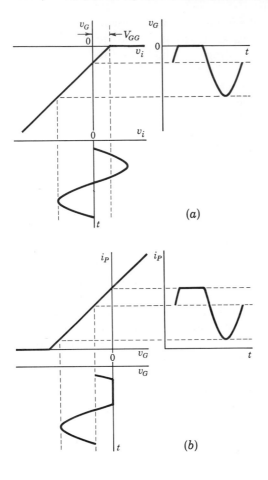

Fig. 7-8 (*a*) The grid-cathode diode clipper of Fig. 7-7 is represented by a piecewise linear and continuous transfer curve. A sinusoidal input gives a clipped grid signal. (*b*) The clipped grid signal is faithfully reproduced in the plate current if the dynamic transfer characteristic is linear.

peak. As noted in Sec. 6-19, under these circumstances the grid is said to be *clamped* to the cathode. In Fig. 7-8*b* a piecewise linear representation is shown for the dynamic transfer characteristic of the triode. The grid signal v_G is applied to the characteristic. Since the grid signal does not make excursions which cross a break point of this transfer characteristic, the output current waveform has the same form as the grid signal. It is important to recognize that the clipping which appears in the output current waveform is *not* due to the failure of the plate current to respond to the grid voltage but rather to the failure of the grid voltage to respond to the applied signal. As appears in Fig. 7-8*b* and as is pointed out in Sec. 6-19, the plate current will respond almost linearly to grid voltage, even for positive grid voltages, up to the point where the grid current becomes an appreciable part of the total cathode current.

We have already noted that the diode break is sharper than the cutoff break in a triode. Therefore, if the series resistor R in Fig. 7-7*b* is large enough, the clipping which takes place in a triode at the occurrence of grid current may well be sharper than the clipping at cutoff. Further advantages of grid-current limiting over cutoff limiting are the following. Assume that the cathode temperature of the triode has increased. If the grid-to-cathode voltage remained the same, the plate current would increase. However, if the grid is drawing current, this current simultaneously increases and, because of the resistance R, the grid-to-cathode voltage decreases. As a consequence, the plate current remains more nearly constant. The resultant stability of the break associated with grid current permits this break to be used in precision circuits. Furthermore, when the tube is conducting, the output impedance is the parallel combination of the plate resistance r_p and R_p, whereas in the case of cutoff clipping the output impedance is R_p. The effect of capacitive loading on fast waveforms is therefore less for grid-current limiting than for cutoff limiting. On the other hand, the input capacitance is greater when the tube conducts and amplifies than when it is cut off and the gain is zero. Hence, the capacitive input impedance of a grid-current limiter will produce more distortion in a fast signal than will a cutoff limiter because this capacitance must charge through R.

7-5   LIMITING BY BOTTOMING

There is a third type of limiting possible with a triode. Consider the circuit of Fig. 7-7b, but without the series grid resistor which is necessary for grid-current limiting. The largest possible plate current is V_{PP}/R_p. If we apply to the grid, from a low-impedance source, a signal large enough to make the plate current nearly equal to V_{PP}/R_p, limiting will take place. For example, if the tube is a type 5965 with $V_{PP} = 300$ and $R_p = 30$ K, the current will be about 10 mA and the break will be at $+2$ V at the grid instead of 0 V (see Fig. 6-31). Such clipping is sometimes referred to as *plate-current saturation*, but it is not to be confused with any effect associated with maximum cathode emission. This type of limiting is also referred to as *bottoming*, since it results when the plate voltage has gone as low as it can go and yet still leave some tube voltage to supply the tube current. This type of limiting is not particularly stable, but it is useful where precision is not required.

7-6   PENTODE CLIPPERS

Pentodes may be used as grid-current, cutoff, or bottoming limiters. If a high value of plate-circuit resistance is used, then the bottoming takes place while the grid voltage is still negative. For example, for a 6AU6 tube with $V_{PP} = 300$, $R_p = 100$ K, and a screen voltage of 150 V, the plate characteristics in Fig. D-1 show that the limiting takes place at -2 V on the grid.

As a cutoff limiter, a pentode with a fixed screen voltage may provide a sharper break than a triode. When the grid voltage in a triode changes from a value just below cutoff to a value just above cutoff, the plate voltage drops because of the plate resistor. The change in plate current is therefore smaller than in the case of a pentode with fixed screen voltage.

7-7   THE TRANSISTOR CLIPPER

The transistor has two pronounced nonlinearities which may be used for clipping purposes. One occurs as the transistor crosses from cutin into the active region, and the second occurs when the transistor crosses from the active region to saturation. Therefore, if an input signal waveform makes excursions which carry the transistor across the boundary between cutin and active operation or across the boundary between the active and saturation regions, a portion of the input waveform will be clipped off. Presumably we require that the portion of the input waveform which keeps the transistor in the active region shall appear at the output without distortion. If such be the case, we require that the input *current* rather than the input voltage have the waveform of the signal of interest. The reason for this requirement is that over a large signal excursion in the active region the transistor output current responds nominally linearly to the input current but is related in a quite nonlinear manner (exponentially) to the input voltage. Therefore, in a transistor clipper as in other *large-signal* transistor circuits, we are led to the use of a current drive, as indicated in Fig. 7-9. The resistor R, which represents either the signal source impedance or a resistor deliberately introduced, must be large in comparison with the input resistance of the transistor in the active region. Under these circumstances the input base current will very nearly have the waveform of the input voltage, for we shall have $i_B = (v_i - V_\gamma)/R$, where V_γ is the base-to-emitter cutin voltage.

The Cutin Region   We wish to know at what input voltage the circuit of Fig. 7-9 enters the active region. We shall now discuss this matter. In Sec. 6-17 we find that a transistor comes out of cutoff in the neighborhood of 0.1-V reverse bias for germanium or 0 V for silicon. More precisely, we find that at this back-biasing voltage the emitter current is zero and that the collector current, which equals the reverse-saturation current I_{CBO}, flows also in the

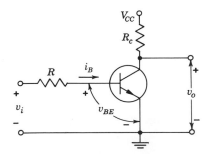

Fig. 7-9   A transistor clipper.

base lead.   The collector current starts to increase above $|I_{CBO}|$ as we begin to forward-bias the transistor.   However, the reverse saturation current is so small (nanoamperes for Si and microamperes for Ge) that a substantial *forward bias* V_γ is required before the collector current becomes a small fraction (say 1 percent) of its rated value.   In Table 6-1 (page 219) we find that this *cutin voltage* V_γ is of the order of 0.5 V for Si and 0.1 V for Ge.

The cutin values just quoted are to be taken only as reasonable estimates and are not intended to apply precisely to any particular transistor type. Our definition of the point at which the transistor comes into conduction depends on the transistor and the circuit application.   In one instance we may consider that a collector-current change of only 10 μA brings the transistor into the active region, whereas in another instance we may require 100 μA or more.   In any event, we find experimentally that germanium transistors, in general, cut in at v_{BE} some tens of millivolts either side of 0.1 V, and in silicon v_{BE} will be some tens of millivolts either side of 0.5 V.   These considerations also make it clear that to keep a silicon transistor in cutoff it may be quite adequate at room temperature to return its base to the emitter through a resistor and thus not to use a reverse-biasing voltage.   In the case of a germanium transistor, however, such an arrangement would almost never be suitable.   At elevated temperatures, even with silicon a reverse-bias supply will be necessary, as indicated in Sec. 6-8 in connection with Fig. 6-15.

Input Resistance   Another parameter of interest for the transistor clipper of Fig. 7-9 is the incremental resistance seen looking into the terminals between base and emitter.   When the switch is in the cutoff condition the incremental resistance may well be many tens of megohms in magnitude.   Its precise value is of no great importance so long as it is large in comparison with R in Fig. 7-9.   As noted in Sec. 1-12, when the transistor collector-circuit resistance is small (say below about 5 K) and the transistor enters its active region, the input impedance is approximately the short-circuit input resistance h_{ie}.   From Eqs. (1-12) and (1-11)

$$h_{ie} = r_{bb'} + r_{b'e} = r_{bb'} + \frac{h_{fe}}{g_m} \tag{7-4}$$

in which $r_{bb'}$ is the base-spreading resistance, h_{fe} is the short-circuit common-emitter current gain, and g_m the transconductance.   If the value of g_m from Eq. (1-9) is substituted into Eq. (7-4), we obtain

$$h_{ie} = r_{bb'} + (h_{fe} + 1)\frac{\eta V_T}{|I_E|} \tag{7-5}$$

Note that the input resistance varies inversely with emitter current.   (This result remains valid even for large values of R_c.   The general expression for the input impedance is given in Prob. 7-10.)   Accordingly, as the transistor comes out of cutoff and moves further into the active region its input resistance decreases.   As an example, in a germanium transistor for which $\eta = 1$, $r_{bb'} = 100\ \Omega$, and $h_{fe} = 49$, at an emitter current I_E of 100 μA the input resistance is $h_{ie} = 100 + 50 \times 260 = 13,100$ Ω.   At $I_E = 1$ mA, $h_{ie} = 1,400$ Ω, and at $I_E = 10$ mA, $h_{ie} = 230$ Ω.   We have neglected the variation of the parameter h_{fe} with transistor current because this variation is small in comparison with the variation in g_m.   Accordingly, it is quite customary, in driving a transistor switch from cutin well into the active region, to encounter changes in input resistance by a factor of 100 or more with resistance variation from tens or hundreds of thousands of ohms to a few hundred ohms.

Waveforms   Waveforms for the clipper in Fig. 7-9 are shown in Fig. 7-10. Here we consider that the input signal v_i is a ramp which starts at a voltage below cutoff and carries the transistor into saturation.   The voltage scale in Fig. 7-10a is appropriate for an *n-p-n* germanium transistor.   The slope dv_{BE}/dt of the base waveform is related to the slope dv_i/dt of the input by

$$\frac{dv_{BE}}{dt} = \frac{h_{ie}}{R + h_{ie}} \frac{dv_i}{dt} \tag{7-6}$$

The input impedance h_{ie} decreases as the transistor goes further into the active region, and consequently the slope of dv_{BE}/dt decreases also.   In saturation the current gain $h_{fe} = 0$ and h_{ie} falls to the low value $r_{bb'} + \eta V_T/|I_E|$.   For example, if the saturation current is 1 mA, then the minimum input resistance in the active region, just before saturation, is 1,400 Ω, whereas when the transistor enters saturation, the input impedance drops to about 126 Ω.   This abrupt lowering of the input resistance results in a sharp limiting of the voltage

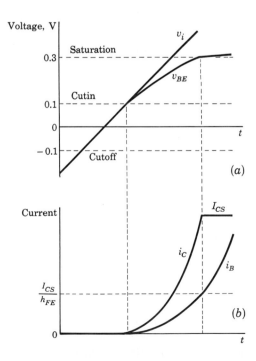

Fig. 7-10 Waveforms of the transistor clipper of Fig. 7-9. (a) The voltage v_{BE} which results when a ramp input drives the switch from cutoff into saturation. (b) The base and collector currents.

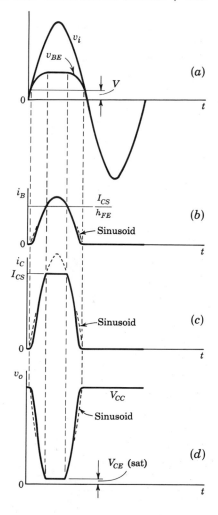

Fig. 7-11 Waveforms for the transistor clipper of Fig. 7-9. The input v_i is sinusoidal and large enough to carry the transistor both into saturation and below cutoff. The base is biased so that cutin occurs at the voltage V. (a) The input voltage v_i and the base-to-emitter voltage v_{BE}; (b) the base current; (c) the collector current and (d) the output (collector) voltage.

v_{BE}, and the waveform remains nominally constant at the base-to-emitter voltage corresponding to saturation.

The slope of the base current i_B is given by

$$\frac{di_B}{dt} = \frac{1}{R + h_{ie}} \frac{dv_i}{dt}$$

and hence increases as the transistor goes further into the active region and eventually into saturation. In the active region, as shown in Fig. 7-10b, the collector current will have the same form as the base current. In saturation, however, the collector current will remain constant at

$$i_C = \frac{V_{CC} - V_{CE}(\text{sat})}{R_c} \equiv I_{CS}$$

This limiting occurs when $i_B > I_{CS}/h_{FE}$.

The waveforms which result when a sinusoidal voltage v_i carries the transistor from cutoff to saturation are shown in Fig. 7-11. The base circuit is biased so that cutin occurs when v_{BE} reaches the voltage V.

7-8   CLIPPING AT TWO INDEPENDENT LEVELS

The waveform i_C in Fig. 7-11 exhibits clipping at two currents. In the waveform of Fig. 7-11, however, the levels at which clipping occurs are not independently adjustable but are separated by the current I_{CS}. Similarly, a vacuum tube may be used to clip both the positive and negative extremities of a signal by adjusting the tube bias and signal amplitude so that clipping occurs both at cutoff and as the tube is driven to clamp. In this case also the clipping levels

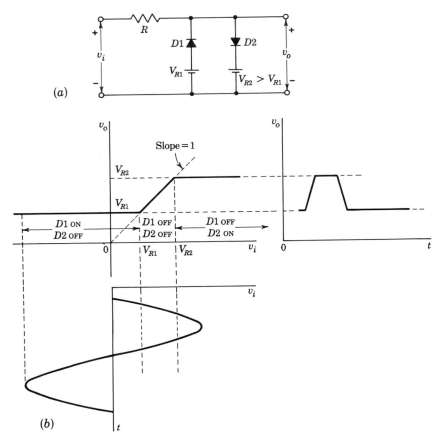

Fig. 7-12 (a) A double-diode clipper which limits at two independent levels. (b) The piecewise linear transfer curve for the circuit in (a). The doubly clipped output for a sinusoidal input is shown.

are not independently adjustable. The region between cutoff and zero grid voltage is called the *grid base* of the tube.

Diode clippers may be used in pairs to perform double-ended limiting at independent levels. A parallel, a series, or a series-parallel arrangement may be used. A parallel arrangement is shown in Fig. 7-12a. Figure 7-12b shows the piecewise linear and continuous input-output voltage curve for the circuit in Fig. 7-12a. The transfer curve has two break points, one at $v_o = v_i = V_{R1}$ and a second at $v_o = v_i = V_{R2}$, and has the following characteristics (assuming $V_{R2} > V_{R1}$):

Input v_i	*Output v_o*	*Diode states*
$v_i \leq V_{R1}$	$v_o = V_{R1}$	D1 ON, D2 OFF
$V_{R1} < v_i < V_{R2}$	$v_o = v_i$	D1 OFF, D2 OFF
$v_i \geq V_{R2}$	$v_o = V_{R2}$	D1 OFF, D2 ON

The circuit of Fig. 7-12a is referred to as a *slicer* because the output contains a slice of the input between the two reference levels V_{R1} and V_{R2}.

The circuit is used as a means of converting a sinusoidal waveform into a square wave. In this application, to generate a symmetrical square wave, V_{R1} and V_{R2} are adjusted to be numerically equal but of opposite sign. The transfer characteristic passes through the origin under these conditions, and the waveform is clipped symmetrically top and bottom. If the amplitude of the sinusoidal waveform is very large in comparison with the difference in the reference levels, then the output waveform will have been *squared*.

Two avalanche diodes in series opposing, as indicated in Fig. 7-13a, constitute another form of double-ended clipper. If the diodes have identical characteristics then a symmetrical limiter is obtained. If the breakdown (Zener) voltage is V_Z and if the cutin voltage in the forward direction is V_γ (≈ 0.5 V for Si), then the transfer characteristic of Fig. 7-13b is obtained.

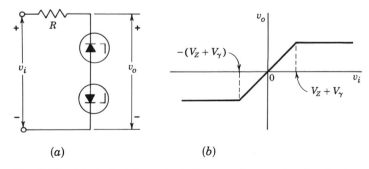

(a)   (b)

Fig. 7-13 (a) A double-ended clipper using avalanche diodes; (b) the transfer characteristic.

7-9   CATHODE–COUPLED AND EMITTER–COUPLED CLIPPER

The cathode-coupled and emitter-coupled circuits of Fig. 7-14a and b may be used as double-ended clippers. Qualitatively the operation of these circuits is the following. Consider initially that the input voltage v_i is negative enough to ensure that $V1$ (or $Q1$) is in cutoff. Then only $V2$ ($Q2$) is carrying current. We consider that V_{GG2} (V_{BB2}) has been adjusted so that $V2$ is within its grid base ($Q2$ is in its active region). As v_i increases, $V1$ ($Q1$) will eventually come out of cutoff, both tubes (transistors) will be carrying current, and the input signal will appear at the output, amplified but not inverted. As v_i continues its excursion in the positive direction the common cathode (common emitter) will follow the grid of $V1$ (the base of $Q1$). Since the grid of $V2$ (the base of $Q2$) is fixed, a point will be reached when the rising cathode (emitter) cuts off $V2$ ($Q2$). In summary, the input signal is amplified but twice limited, once by the cutoff of $V1$ ($Q1$) and once by the onset of cutoff in $V2$ ($Q2$).

To consider these circuits more quantitatively, let us make the simplifying assumption that the current I in Fig. 7-14 is constant. Such would be the case if $|v_{GK2}|$ ($|v_{BE2}|$) were small compared with $V_{GG2} + V_{KK}$ ($V_{BB2} + V_{EE}$). In making this assumption we lose no essential feature of the operation of the circuit, and we may now describe the operation of the circuit in the following way. When v_i is below the cutoff point of $V1$ ($Q1$), all the current I flows through $V2$ ($Q2$). As v_i carries $V1$ ($Q1$) above cutoff, the current in $V2$ ($Q2$) decreases while the current in $V1$ ($Q1$) increases, the sum of the currents in the two active devices remaining constant and equal to I. The total range Δv_o over which the output can follow the input is $R_p I$ ($R_c I$) and is therefore adjustable through an adjustment of I. The absolute voltage of the portion of the input waveform selected for transmission may be selected through an adjust-

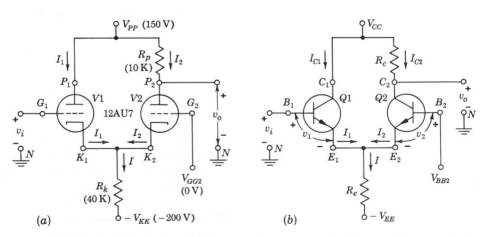

Fig. 7-14 (a) A cathode-coupled two-level clipping circuit; (b) an emitter-coupled clipping circuit.

ment of a biasing voltage on which v_i is superimposed or through an adjustment of V_{GG2} (V_{BB2}).   The total range of input voltage Δv_i between clipping limits is $\Delta v_o/A$, in which A is the gain through the amplifier stage.

The vacuum-tube circuit of Fig. 7-14a has the merit of offering a high input impedance, since the tube grids need never draw grid current.   The illustrative example which follows shows how the piecewise linear transmission characteristic of the vacuum-tube circuit may be determined.

EXAMPLE   The circuit of Fig. 7-14a has $V_{PP} = 150$ V, $V_{KK} = 200$ V, $V_{GG2} = 0$, $R_p = 10$ K, and $R_k = 40$ K.   Find the values of input voltage v_i at which limiting occurs and find the corresponding limited voltages v_o at the output.   The tube is a 12AU7 (Fig. D-5).

Solution   Consider initially that $V1$ is OFF.   Using the procedure of Sec. 1-10 we calculate that the grid-to-cathode voltage drop of $V2$ is $V_{GK2} = -3$ V.   The current in $V2$ is $I_2 = (200 + 3)/40 = 5.08$ mA.   The lower limiting voltage, from the plate of $V2$ to the ground terminal N, is $V_{PN2} = V_{PP} - I_2R_p = 150 - 5.08 \times 10 = 99$ V $\equiv V_{oL}$.   Since the common-cathode voltage with respect to ground is $V_{KN} = 3$ V, then $V_{PK1} = V_{PP} - V_{KN} = 150 - 3 = 147$ V.   At this voltage, grid cutin occurs at $V_{GK1} = -11$ V $= V_{\gamma1}$.   Therefore the lower of the limiting voltages at the grid of $V1$ is $V_{GN1} = V_{GK1} + V_{KN} = -11 + 3 = -8$ V $\equiv V_{iL}$.

As the grid of $V1$ rises, so also will the common cathode.   We need to calculate the cathode voltage at which $V2$ cuts off.   Here we encounter a minor complication.   For, to determine this cutin voltage $V_{\gamma2}$ we need to know V_{PK2} and hence need to know the very cathode voltage we seek to determine.   This complication, which occurs frequently in the analysis of tube (and transistor) circuits, is easily resolved by using a series of successive approximations.   Thus, in the present instance, we estimate that at cutin $V_{PK2} = 150$ V.   Correspondingly, from Fig. D-5 we find that $V_{\gamma2} = -11$ V.   On this basis, however, $V_{PK2} = 139$ V, and the corresponding $V_{\gamma2}$ is closer to -10 V.   Now $V_{PK2} = 140$ V and, in principle, we might make a further correction in cutin voltage, etc.   But the accuracy with which published average tube characteristics represent a particular sample of a tube type hardly warrants further extension of the approximation.

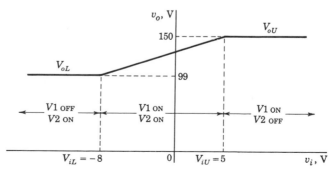

Fig. 7-15   The piecewise linear transfer characteristic of the cathode-coupled clipper specified in the example.

We have found that $V2$ cuts off when the cathode voltage is $V_{KN} = +10$ V. Hence, $I_1 = (200 + 10)/40 = 5.25$ mA. From the tube characteristic we read (at $V_{PK1} = 140$ V, $I_1 = 5.25$ mA) that $V_{GK1} \approx -5$ V. Thus the upper limiting voltage at the grid of $V1$ is $V_{GN1} = V_{GK1} + V_{KN} = -5 + 10 = +5$ V $\equiv V_{iU}$. The corresponding limit at the output plate is $V_{oU} = V_{pp} = 150$ V. The change in total cathode current from one limit to the other is $5.25 - 5.08 \approx 0.2$ mA, a change of less than 4 percent. In passing, we may note that, having determined the lower limiting point, we might have determined the upper limiting point by calculating the gain A as indicated earlier. We are, however, reluctant to do this because the tubes operate from cutoff to well into their grid bases and it is uncertain what values are to be assigned to the tube parameters. Finally, we may see in Fig. 7-15 a piecewise linear and continuous transfer characteristic of the circuit. To draw this figure we have idealized the triodes by considering that cutin is sharp and that within the grid base the tubes operate linearly. The average gain is now seen to be

$$A = \frac{V_{oU} - V_{oL}}{V_{iU} - V_{iL}} = \frac{150 - 99}{5 + 8} = 3.9$$

Transfer Characteristic of the Emitter-coupled Clipper   We shall find the transfer curve of the circuit of Fig. 7-14b. The emitter current is

$$I = I_1 + I_2 = \frac{V_{BB2} + V_{EE} - v_2}{R_e} \tag{7-7}$$

In Table 6-1 (page 219) we see that the base-to-emitter voltage changes by only 0.2 V from cutin to saturation. Hence, as long as $V_{BB2} + V_{EE} \gg v_2$, I remains essentially constant. The upper output limited level is $V_{oU} = V_{CC}$ and corresponds to $I_2 = 0$ and $I_1 = I$. The lower output limited level is

$$V_{oL} = V_{CC} - I_{C2}R_c \tag{7-8}$$

and corresponds to $I_1 = 0$ and $I_2 = I$. Note that

$$I_2 = I_{C2} + I_{B2} = I_{C2}\left(1 + \frac{1}{h_{FE}}\right)$$

Hence, in Eq. (7-8), $I_{C2} = I/(1 + 1/h_{FE})$, where I is obtained from Eq. (7-7). From Fig. 7-14b we find

$$v_i = V_{BB2} + v_1 - v_2 \tag{7-9}$$

Since the current varies exponentially with base-to-emitter voltage (v_1 or v_2), the cutoff (and hence the clipped) levels are approached asymptotically. Let us therefore define (as indicated in Fig. 7-16) the upper input level V_{iU} to correspond to $I_2 = 0.1I$ and $I_1 = 0.9I$. Similarly, we define the lower input level V_{iL} to correspond to $I_1 = 0.1I$ and $I_2 = 0.9I$. From Eq. (6-33) the base-to-emitter voltage is

$$V_E = \eta V_T \ln\left(1 - \frac{I_E + \alpha_I I_C}{I_{EO}}\right)$$

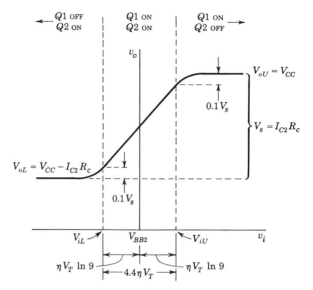

Fig. 7-16 The transfer characteristic of the emitter-coupled clipper.

Neglecting the small base current in the active region, $I_C \approx -I_E$, and since the second term in the parentheses is large compared with the first,

$$V_E \approx \eta V_T \ln \frac{-(1-\alpha_I)I_E}{I_{EO}} \tag{7-10}$$

Since $v_1 = V_E$ if $I_1 = -I_E$ and $v_2 = V_E$ if $I_2 = -I_E$, then from Eq. (7-9)

$$v_i = V_{BB2} + \eta V_T \ln \frac{(1-\alpha_I)I_1}{I_{EO}} - \eta V_T \ln \frac{(1-\alpha_I)I_2}{I_{EO}}$$

$$= V_{BB2} + \eta V_T \ln \frac{I_1}{I_2} \tag{7-11}$$

Hence

$$V_{iU} = V_{BB2} + \eta V_T \ln \frac{0.9I}{0.1I} = V_{BB2} + \eta V_T \ln 9 \tag{7-12}$$

and

$$V_{iL} = V_{BB2} + \eta V_T \ln \frac{0.1I}{0.9I} = V_{BB2} - \eta V_T \ln 9 \tag{7-13}$$

These results are indicated in Fig. 7-16. Since at room temperature $V_T = 26$ mV and since $\ln 9 = 2.2$, then the input swing is ± 57 mV for Ge and ± 114 mV for Si, centered about the reference voltage V_{BB2}. In general, the total input voltage swing Δv_i to carry the output through its entire swing $V_s = I_{C2}R_c$ is

$$\Delta v_i = 2\eta V_T \ln 9 = 4.4\eta V_T \tag{7-14}$$

This increment is proportional to the junction temperature. These results have been verified experimentally.[1] If the transistor parameters are not identical for the two sections, then the transfer curve will not be symmetrical with respect to V_{BB2}.

Since the current switches from $Q2$ to $Q1$, with the total emitter current remaining constant, this circuit is often referred to as a "current-mode switch." Note that neither transistor need be in saturation at any point over the entire range of operating voltages.

7-10   COMPENSATION FOR TEMPERATURE CHANGES IN DIODES

The break point of a semiconductor diode depends on the junction temperature, and the break point of a thermionic diode depends on the cathode temperature. From Eq. (6-7) we note that for semiconductor diodes the break point decreases by about 2 mV/°C with increase in temperature. Experimentally it is found that in a thermionic diode the break point becomes more negative by about 100 mV for a 10 percent increase in heater voltage.

It is convenient in the discussion to follow to represent the actual diode as an ideal diode in which the break point occurs at zero voltage and to add in series with the diode a voltage source equal to the offset voltage V_γ. A simple clipping circuit in which the diode is represented in this manner is shown in Fig. 7-17a. The polarity of the source V_γ is appropriate for semiconductor diodes. For thermionic diodes the polarity must be reversed. It is apparent from Fig. 7-17a, as well as from earlier discussions, that since V_γ depends on temperature, so also must the point on the input waveform at which clipping takes place.

A scheme for temperature compensation is shown in Fig. 7-17b. The second diode $D2$ introduces a second source V_γ, and we may expect that if both diodes experience the same change in junction temperature or heater voltage the change in V_γ in $D2$ will compensate for the change in $D1$. The break point occurs at $v_i = V_R = v_o$. In practice we find that, because of differences in diodes even of a particular type, compensation is not perfect.

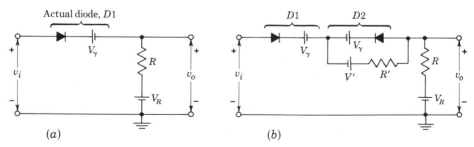

Fig. 7-17 (a) Illustrating that the point of clipping is a function of the diode temperature, because V_γ depends upon T; (b) a scheme for temperature compensation.

With some selection of diodes, however, an improvement by a factor of 5 or even better is not difficult to achieve.

The auxiliary supply V' and resistor R' are necessary to keep diode $D2$ conducting at all times. Otherwise, if $D1$ were conducting, but if $D2$ were back-biased, there would be no transmission of signal to the output. This condition—that $D2$ must conduct—establishes an upper limit to the allowable magnitude of v_i before a second limiting (not intended) occurs as $D2$ goes OFF. From the condition that the current in the diode $D2$ must always be in the forward direction we find

$$v_i(\text{maximum}) = V_R + \frac{R}{R'}(V' - V_\gamma) \qquad (7\text{-}15)$$

An alternative compensating circuit which avoids the need for a separate ungrounded source V' is shown in Fig. 7-18a. Here the source V_R serves simultaneously as the reference voltage and as the source that causes current to flow through $D2$. As in the previous circuit, the input must be restrained from becoming so large that it reduces to zero the current through $D2$. In this circuit, a change in V_γ occasioned by a change in temperature will cause a change in the d-c level of the output voltage since (if $R_r \gg R \gg R_f$)

$$v_o = V_R - V_\gamma \qquad \text{if } v_i \le V_R$$
$$v_o = v_i - V_\gamma \qquad \text{if } v_i \ge V_R \qquad (7\text{-}16)$$

We observe, however, that independently of V_γ, just before diode $D1$ conducts, the voltage at the cathode of the ideal diode in $D1$ is V_R. Accordingly, a change in temperature will not change the point $(v_i = V_R)$ *on the input waveform* at which clipping occurs.

Still another temperature-compensated clipping circuit is shown in Fig. 7-18b. This circuit has the transfer characteristic indicated in Eq. (7-16). When $v_i < V_R$, $D1$ is OFF, $D2$ is ON, and $v_o = V_R - V_\gamma$. When $v_i > V_R$, $D1$ is ON, $v_o = v_i - V_\gamma$, which is greater than $V_R - V_\gamma$, and therefore $D2$ is OFF. Thus, at $v_i = V_R$ the diodes switch simultaneously, $D1$ from OFF to ON

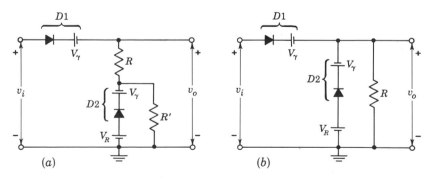

Fig. 7-18   **Alternative schemes for temperature compensation of clippers.**

and $D2$ from ON to OFF. The clipping level is V_R, independently of the voltage V_γ and hence of the temperature.

This circuit does not have the inconvenient requirement that a current be maintained in the compensating diode, as is the case with the previous circuits. It does have the disadvantage, however, that the sharpness of the break region may be adversely affected. For when the input signal v_i rises to the point where $D1$ starts to conduct, all the current initially flowing through $D2$ must transfer to $D1$ before the output v_o can respond fully to the input signal. Because of the finite forward resistances of the diodes this transfer does not take place abruptly but extends over a finite range of the input voltage, thus extending the break region (Prob. 7-25).

7-11   COMPARATORS

The nonlinear circuits which we have used to perform the operation of clipping may also be used to perform the operation of *comparison*. In this case the circuits become elements of a *comparator* system and are usually referred to simply as *comparators*. A comparator circuit is one which may be used to mark the instant when an arbitrary waveform attains some reference level. The distinction between comparator circuits and the clipping circuits considered earlier is that in a comparator there is no interest in reproducing any part of the signal waveform. For example, the comparator output may consist of an abrupt departure from some quiescent level which occurs at the time the signal attains the reference level but is otherwise independent of the signal. Or the comparator output may be a sharp pulse which occurs when signal and reference are equal.

The diode circuit of Fig. 7-19 which we encountered earlier as a clipping circuit is used here in a comparator operation. For the sake of illustration the input signal is taken as a ramp. This input crosses the voltage level $v_i = V_R$ at time $t = t_1$. The output remains quiescent at $v_o = V_R$ until $t = t_1$, after which it rises with the input signal. The output waveform is drawn on the basis of the assumption that the diode has a break point at zero voltage and that the diode resistance changes abruptly at the break point from infinite reverse resistance to a finite forward resistance.

The device to which the comparator output is applied will respond when the comparator voltage has risen to some level V_o above V_R. However, the precise voltage at which this device responds is subject to some variability Δv_o because of gradual changes which result from aging of components, temperature changes, etc. As a consequence (as shown in Fig. 7-19) there will be a variability Δt in the precise moment at which this device responds and an uncertainty Δv_i in the input voltage corresponding to Δt. Furthermore, if the device responds in the range Δv_o, the device will respond not at $t = t_1$ but at some later time t_2. The situation may be improved by increasing the slope of the rising portion of the output waveform v_o. If the diode were indeed ideal, it would be advantageous to follow the comparator of Fig. 7-19

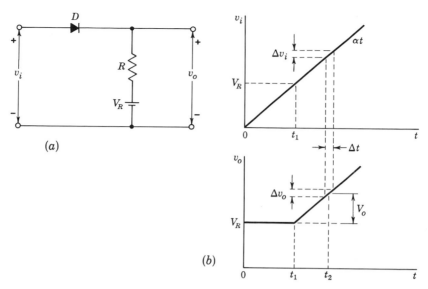

Fig. 7-19 (a) A diode comparator; (b) the comparison operation is illustrated with a ramp input signal v_i and the corresponding output waveform is indicated.

by an amplifier. For then, if the amplifier had a gain A, the output v_o would pass through the range Δv_o in an interval $\Delta t/A$ and the delay in response would be reduced to $(t_2 - t_1)/A$. We shall, however, now show that because of the characteristics of an actual diode such an anticipated advantage is not realized.

In Chap. 6 we define the diode break point to be the point in the neighborhood of the diode voltage which yields a diode current approximately 1 percent of the rated diode current. This definition does not define a break point for a diode-resistor series combination, as in Fig. 7-19. For such a combination we may arrive at a definition of a break point in the following way. To the left of the break point the diode incremental resistance r should be very much larger than the resistance R. To the right of the break point the diode resistance r should be very much smaller than R. It seems reasonable, then, to consider that the break point is located at about the place where $r = R$. At this point the transmission gain $\Delta v_o/\Delta v_i = R/(R + r) = \frac{1}{2}$. Now let us suppose that we have connected the comparator of Fig. 7-19 directly to the device it is to actuate. The device will respond when the diode current is, say, I, and the drop across R is RI. Now let us interpose an amplifier of gain A between the comparator and the device to be actuated. It is to be understood that the amplifier is coupled to the comparator in such a way that it amplifies only the changes in comparator output voltage and not the reference voltage V_R. Then the comparator will cause a response when the amplifier output is RI or when the drop across R is RI/A, corresponding to a diode current I/A. From Eq. (6-12) we have the result that the diode incremental resistance is inversely proportional to I. Therefore, if an amplifier is used, the device actuated by the comparator will respond at a current such that $r = RA$. The transmission gain between amplifier output and comparator input is A times that of the diode-resistor combination with $r = RA$, or with the amplifier in place,

$$\frac{\Delta v_o}{\Delta v_i} = A\,\frac{R}{R + r} = \frac{AR}{R + RA} = \frac{A}{1 + A} \tag{7-17}$$

Even if A were to become arbitrarily large, this last ratio will attain a maximum value of unity. Since we found above that the transmission gain without the amplifier was $\frac{1}{2}$, then an infinite gain device gives an improvement by only a factor of 2. This discussion concerning a diode-resistor comparator break point is summarized by noting that the voltage of the input signal at which the comparator yields an output depends not only on the diode but also on the value selected for R, that the comparator break point depends also on the gain of the amplifier following the comparator, and finally that the sharpness of the comparator break is not materially improved by an amplifier following the comparator. All these features are to be seen in the diode comparator input-output curves shown in Fig. 7-20.

In addition to the lack of sharpness in the comparator-element break region, there is a second source of difficulty in comparators. This difficulty results from the instability of the components constituting the comparator circuit. For example, as is pointed out in Sec. 7-10, there is a shift of about -2 mV/°C in the voltage at which a constant current is obtained in a semiconductor diode. Such a shift will have a corresponding effect on the effective

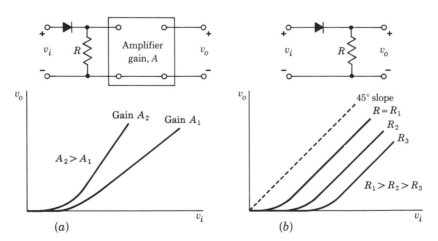

(a)   (b)

Fig. 7-20 (a) Illustrating that an amplifier following a diode comparator does not improve the comparator sharpness but does move the break point; (b) illustrating that variation of R does not change comparator sharpness but does move the break point.

reference point of a diode comparator.   To minimize instability of this type, it
may be necessary to use some one of the compensating schemes described in
Sec. 7-10, or for a thermionic diode it may be necessary to regulate the heater
voltage.

There is also a third type of error encountered in comparators which
results from the presence of reactive elements in the circuit.   Such reactive
components are energy-storing elements, and therefore the effective reference
point of a comparator may depend somewhat on the past history of the
circuit.   The reference point will then be a function of the nature of the input
signal, i.e., its amplitude, repetition rate, etc.   With the simple diode comparator considered thus far, the principal reactive elements are the capacitances
across the diode and across the load.

A metal-semiconductor diode [also called a *hot-carrier diode* (Sec. 20-9)],
with the very sharp break in its volt-ampere characteristic, its very low capacitance, and negligible storage time, makes an excellent comparator element.

7-12   BREAKAWAY DIODE AND AMPLIFIER

If, in a series combination of resistor and diode, the output is taken across the
diode, the circuit will continue to operate as a comparator.   A comparator
of this type with a ramp input is shown in Fig. 7-21.   The diode in Fig. 7-19
is often referred to as a *pick-off diode*, while the diode in Fig. 7-21 is called a
breakaway diode.   The stray capacitance shunting the output will limit the

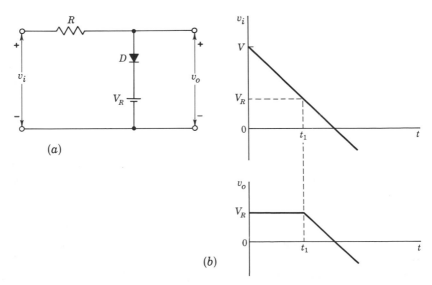

Fig. 7-21   (a) A comparator using a "breakaway" diode; (b) the output
waveform for a ramp input signal given by $v_i = V - \alpha t$, where $V > V_R$.

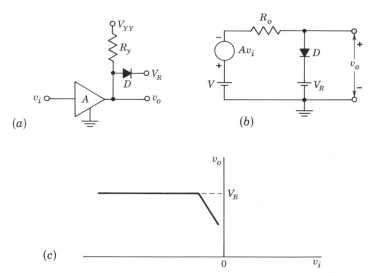

Fig. 7-22 (a) An amplifier precedes the diode comparator; (b) equivalent circuit where A is the magnitude of the amplifier gain; (c) transfer characteristic.

abruptness of change in the output waveform more seriously in the circuit of Fig. 7-21 than in the circuit of Fig. 7-19. For this reason the former circuit is not frequently used where fast waveforms are encountered. On the other hand, in the series diode circuit of Fig. 7-19 some of the input voltage, for fast waveforms, will appear at the output even for voltages less than V_R because of the diode shunt capacitance.

Although an amplifier which follows a diode-resistor comparator does not improve the sharpness of the comparator break, an amplifier *preceding* the comparator will do so. Thus, suppose that the input signal to a diode comparator must go through a range Δv_i to carry the comparator through its uncertainty region. Then, if the amplifier has a gain A, the input signal need only go through the range $\Delta v_i/A$ to carry the comparator output through the same voltage range. A simple example of a comparator using this principle is shown in Fig. 7-22a. Replacing the amplifier by its Thévenin's equivalent, we have the circuit shown in Fig. 7-22b, in which A is the amplifier gain and R_o is the output impedance of the amplifier including R_y. The battery V represents the quiescent output voltage. The circuit is of the type shown in Fig. 7-21, in which the output is taken across the diode. A plot of output against input voltage exhibits the break shown in Fig. 7-22c.

A disadvantage normally encountered when an amplifier precedes a comparator results from the fact that, as in Fig. 7-22a, the amplifier must be direct-coupled to the comparator so as to avoid energy-storage elements. If, therefore, there should be a drift in the output voltage of the amplifier corresponding to a fixed input voltage, then the comparator reference point would shift correspondingly. Accordingly, unless care is taken to stabilize the amplifier against such drift the amplifier-comparator combination may lose more in accuracy through the use of the amplifier than it gains in precision.

7-13   DIODE–DIFFERENTIATOR COMPARATOR

The diode circuit of Fig. 7-19 is not a complete comparator system by itself. When the input signal rises past V_R the output voltage will depart from V_R, and this change in output voltage will be applied to a device which is to be actuated. The precise moment at which the device will respond will depend on its own characteristics. Why, then, use the diode comparator at all? The answer is that the diode network lends convenience to the operation. The output device will not have to contend with the input signal until shortly before the comparison point is reached. The diode circuit is especially convenient when the signal is passed through an amplifier in order to improve the precision of the comparison.

Just as it is convenient to restrain the signal from reaching the actuated device until shortly before the moment of comparison, so also it is convenient to remove the signal after the comparison has been completed. A signal appropriate for this purpose can be generated by a double differentiation of the output signal of Fig. 7-19. Such a circuit is shown in Fig. 7-23. The amplifier is used to prevent the second differentiator from loading the first and also serves to increase the amplitude of the output to a convenient level. Waveforms for the circuit of Fig. 7-23 are shown in Fig. 7-24 for the special case where $R_1C_1 = R_2C_2 = \tau$, where R_1 is the parallel combination of R_1' and the input impedance of the amplifier. Beginning at $t = t'$, at which time $v_i = V_R$, the amplifier input voltage v_A rises from zero and approaches $\alpha\tau$ exponentially, α being the slope of the input ramp. The second differentiator converts the exponential to a pulse, as shown in Fig. 7-24c. The pulse waveform is given by Eq. (2-20) as

$$v_o = A\alpha\tau x\epsilon^{-x} \tag{7-18}$$

where now $x \equiv (t - t')/\tau$.

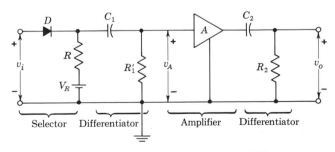

Fig. 7-23   A comparator followed by two differentiators to provide a pulse-type comparator output.

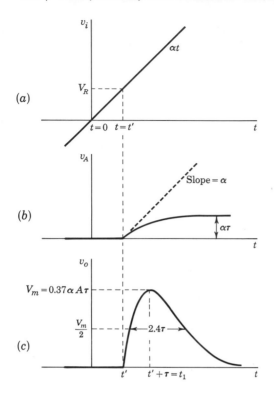

Fig. 7-24 Waveforms of the circuit of Fig. 7-23. (a) Ramp input; (b) waveform after first differentiation; (c) pulse-type output.

From Eq. (7-18) we deduce that the pulse amplitude V_m is $0.37\alpha A\tau$, the pulse width at half maximum is 2.4τ, and the pulse peak occurs $x = 1$ or at $t = t' + \tau \equiv t_1$. In arriving at the pulse waveform of Fig. 7-24 we have again assumed an ideal diode. The actual form of the pulse will appear more nearly as shown in Fig. 7-25, where two different amplifier gains A_1 and A_2 $(A_2 > A_1)$ are considered. The pulse does not rise abruptly, and conse-

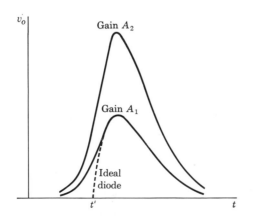

Fig. 7-25 The waveform of the output pulse of Fig. 7-23 for two gains, taking into account the lack of abruptness of the diode break.

quently when the gain is increased the pulse becomes not only larger but also broader. Accordingly, we see that the pulse is not made sharper with increased gain. Of course, as before, if the gain were introduced before the diode the output pulse would indeed be sharper.

7-14   ACCURATE TIME DELAYS

To see one way in which the circuit of Fig. 7-23 may be usefully employed, consider that the slope α of the input ramp is known. If the comparison or reference voltage V_R is set at some value V_{R1}, a pulse will be generated with a peak at $t = t_1$. If in a second comparator $V_R = V_{R2}$, then a pulse will occur at $t = t_2$. We need then but measure the d-c voltage difference $V_{R2} - V_{R1}$ in order to establish the time difference between the two pulses, since

$$t_2 - t_1 = \frac{V_{R2} - V_{R1}}{\alpha}$$

Alternatively, if in a single comparator circuit we change V_R by ΔV_R, then we also have changed the time of occurrence of the pulse by $\Delta t = \Delta V_R/\alpha$. If it should happen that the ramp is being used as the time base of a CRO display, the pulse may then be used as a timing marker. Comparators are used extensively in this manner for timing purposes in radar systems. Altogether, we see that the combination of a ramp of known slope used in connection with a comparator allows us to establish or measure a time interval simply by measuring a voltage. We have here a *voltage-to-time converter*.

We have defined the operation of comparison as one in which a determination is made of the precise moment at which a signal voltage attains some reference voltage. On this basis, the pulse-generating circuit of Fig. 7-23 may or may not be a comparator, depending on the exact manner in which its signal is used. On the one hand, suppose that the pulse signal is applied to some device which responds at some particular voltage on the rising edge of the pulse. Then the device would respond as well when this particular voltage is attained even if the diode and double differentiation were not used. It is, then, the device that performs the operation of comparison, and the circuit of Fig. 7-23 is used as a matter of convenience. Such might be the case if we used the pulse to modulate the intensity of a CRO and take note of the position where the trace intensity just dropped to zero. In this case the comparator operation would actually be performed by the combination of the nonlinearity of the CRO screen (intensity versus beam current) and the eye of the observer. On the other hand, if we superimposed the pulse on the deflection signal and noted, with whatever precision is possible (taking account of the pulse width), the position of the pulse peak, then the circuit of Fig. 7-23 is itself performing the operation of comparison.

7-15　　AN AMPLIFIER FOR A COMPARATOR

We have already noted the advantage of placing an amplifier before a comparator. An amplifier which finds frequent use in this application is the emitter-coupled or the cathode-coupled amplifier of Fig. 7-14. When so used, the signal is applied to one base (or grid), say $B1$ (or $G1$), and a referencing voltage V_R is applied to the other base (grid), say $B2$ ($G2$). The output signal at the collector (plate) is applied to the nonlinear device (diode, tube, or transistor at cutoff, etc.). To change the point of comparison we change the referencing voltage.

Let us assume that the comparator element has been coupled to the amplifier output in such a way that the comparator element will just respond when the signal v_i equals V_i. We require that, when V_R is fixed, the output voltage v_o should respond sensitively to changes Δv_i in input voltage. The comparator precision is improved by the ratio $\Delta v_o / \Delta v_i$. On the other hand, suppose that we change V_R by an amount ΔV_R. Then we require that the comparator shall respond when the input voltage attains the value $v_i = V_i + \Delta V_R$. Since the comparator which is coupled to the output responds at some fixed output voltage, then we also require that when both bases (or grids) change by the same increment, there shall be *no* corresponding change in the output.

The above requirements may be described in terms of the *common-mode gain* A_c and the *difference-mode gain* A_d of the amplifier. The common-mode gain is defined by $A_c \equiv \Delta v_o / \Delta v_c$ under the circumstances that when an input voltage increment $\Delta v_i = \Delta v_c$ is applied to one base (grid), an identical signal change Δv_c is simultaneously applied to the other base (grid). The difference-mode gain is defined by $A_d \equiv \Delta v_o / \Delta v_d$ under the circumstances that when a signal increment $\Delta v_i = \Delta v_d / 2$ is applied to one base (grid), a signal change $-\Delta v_d / 2$ is simultaneously applied to the other base (grid). When the common-mode gain is very small the difference-mode gain may also be defined as $A_d \equiv \Delta v_o / \Delta v_d$ under the circumstances that the full signal change Δv_d is applied to one base (grid) while no signal is applied to the other base (grid). We may now specify, in connection with the amplifier of Fig. 7-14, that we require that A_d be as large and A_c as small as feasible. The *common-mode rejection ratio* A_d / A_c may well be taken as a figure of merit in the present application. Expressions for A_c and A_d in terms of the circuit and device parameters are given in Prob. 7-30 for the emitter-coupled comparator and in Prob. 7-32 for the cathode-coupled circuit.

We may now see that if the common emitter (cathode) in the circuits of Fig. 7-14 is returned to a constant-current source, the common-mode gain will be reduced to zero. For let each base, say, in Fig. 7-14b be increased by the same voltage. Then, from symmetry, the current in each transistor must increase by the same amount ΔI, and the emitter current must increase by $2 \Delta I$. However, if the emitter current is being supplied by a constant-current source, then $2 \Delta I$ must be zero. Consequently the change in each collector current must also be zero. Hence, in response to the simultaneous change in voltage at both bases, no change at all will take place in the output voltage. A constant-current source is approximated through the use of a large emitter resistor R_e returned to a correspondingly large source V_{EE}. A more nearly perfect current source is obtained if, as in Fig. 7-26, the emitter resistor is replaced by a transistor $Q4$ with its own emitter resistor. The incremental resistance seen looking into the collector of this added transistor may be extremely large (Prob. 1-32).

We have described how the suppression of the common-mode gain makes the amplifier stages of Fig. 7-14 suitable for use as a comparator amplifier. We may now see that this same stage is itself useful as a comparator. Assume that we propose to use for comparison purposes the cutoff of transistor $Q2$. We saw that a simultaneous variation at both inputs produced negligible change in the load current through $Q2$. In the present instance we need but apply this result to the special case in which the collector current has just been set at zero. Accordingly, if for some reference voltage V_R a voltage v_i just brings $Q2$ into conduction, a change ΔV_R will require a change $\Delta v_i = \Delta V_R$ to bring $Q2$ again just to the point of cutin. For example, in Fig. 7-16 (with $V_{BB2} = V_R$) note that approximate cutin of $Q2$ occurs at

$$V_{iU} = V_R + 2.2\eta V_T$$

so that, indeed, if V_R changes by ΔV_R, then ΔV_{iU} changes by the same amount.

A comparator circuit which uses the difference amplifier twice is shown in Fig. 7-26.[1,2] The first stage ($Q1a$ and $Q1b$) is an amplifier, and the second

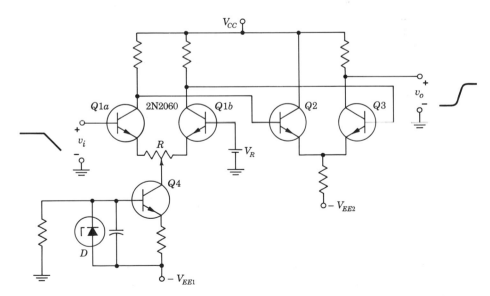

Fig. 7-26 A symmetrical transistor amplifier followed by a difference-amplifier comparator.

stage ($Q2$ and $Q3$) is the comparator. The transistor $Q4$ in the emitter of the first stage provides a large, effective emitter resistance. Hence, in effect, the emitter current is being supplied by a current source. The avalanche (Zener) breakdown diode D is used here to establish a fixed voltage between the base of $Q4$ and the $-V_{EE1}$ supply.

The operation of the circuit is as follows. If the amplifier is completely symmetrical, then when the signal voltage v_i is equal to the reference voltage V_R the voltages at the two collectors of $Q1$ will be identical. The resistor R may be adjusted so that under these conditions the currents through $Q1a$ and $Q1b$ are nominally equal. The difference amplifier can handle linearly only a limited range of input signal. Hence, as the voltage difference between the collectors of $Q1$ departs from zero, a break occurs in the comparator output as $Q2$ or $Q3$ passes through cutoff. (Note the emitter-coupled clipper waveform of Fig. 7-16.) For a negative-going sweep input signal, as indicated in Fig. 7-26, the response at the collector of $Q3$ will be a positive-going step, as shown. The step will begin when the input signal v_i is different from the reference V_R by some fixed voltage. Alternatively, the resistor R may be adjusted so that the step forms at the moment $v_i = V_R$. We may now discuss some of the advantageous features of the circuit of Fig. 7-26.

The gain of the amplifier $Q1$, defined as the ratio of the collector-to-collector output voltage to the input signal v_i, may be quite high, perhaps 50 to

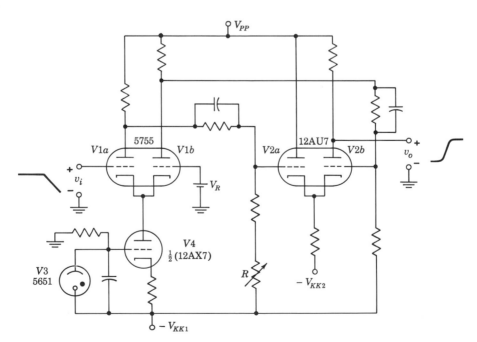

Fig. 7-27 A symmetrical tube amplifier followed by a difference-amplifier comparator.

500.   Hence there is an improvement by this gain factor in the precision of the comparator as compared with the precision which would result without the amplifier.   This circuit has a very high common-mode rejection ratio because of the extremely large effective dynamic resistance in the emitter of $Q1$.

The amplifier is highly stable.   This stability results not from any particular circuit features but rather from the nature of the devices employed. The type 2N2060 was designed especially for difference-amplifier applications. It consists of two high-gain n-p-n silicon planar transistors in the same hermetically sealed enclosure.   The manufacturer guarantees that for equality of collector currents the maximum difference in base voltages is 5 mV, that the base voltage differential change at fixed collector current will not exceed 10 μV/°C, and that h_{fe} for one transistor will not differ from h_{fe} for the other by more than 10 percent.

A tube circuit which is analogous to the transistor configuration of Fig. 7-26 is given in Fig. 7-27.   The avalanche diode is now replaced by a glow-tube regulator type 5651.   The input stage $V1$ is a type 5755 tube, especially designed for balanced d-c amplifier service.   The manufacturer guarantees that, for equality of tube currents, the maximum difference in grid voltage is 0.3 V.   More importantly, it is claimed that over a 7-hr period a maximum change of no more than 5 mV need be made at one grid to keep the currents balanced.   Similarly, the 5651 tube is a very stable reference-voltage regulator.   Over the normal operating range of 1.5 to 3.5 mA, its maintaining voltage of 87 V will not drift with age by more than about 0.3 V under continuous operation.

7-16   APPLICATIONS OF VOLTAGE COMPARATORS

The gain of the difference amplifier of Fig. 7-26 may be increased by adding more stages.   The number of stages which may be used is limited by the d-c stability (drift).   With a very high gain, the output will approach a step waveform, which begins at the instant when the input differs from the reference voltage V_R by a fixed amount.   In this chapter we consider only nonregenerative comparators, but by using circuits with positive feedback it is possible to obtain infinite forward gain (unity loop gain).   Such regenerative comparators (for example, the Schmitt circuit of Sec. 10-12) do indeed approximate a step output.   Other types of regenerative comparators (for example, the blocking oscillator of Chap. 16) generate a pulse rather than a step output waveform.   Most applications of comparators make use of the step or pulse nature of the output.

Accurate Time Measurements   This application is discussed in Sec. 7-14.

Pulse-time Modulation   If a periodic sweep waveform is applied to a comparator whose reference voltage is modulated by some information, it is possible to obtain a succession of pulses whose relative spacing reflects the input information. The result is a *time-modulation* system of communication.

Timing Markers Generated from a Sine Wave　If a comparator is used to detect the instant of equality of the instantaneous value of a sine wave with a d-c reference voltage, pulses will be obtained which are synchronized with the sine wave. Thus, a sine wave is converted into a series of pulses. There are many important applications of this type.

Phasemeter　When a comparator is used with sinusoidal signals, as above, then it is advantageous that the reference voltage be zero. Under these circumstances the pulses are locked in phase with the zero of the sine wave and, ideally, are independent of the amplitude of the sinusoidal voltage. The phase angle between two voltages can be measured by a method based upon this principle. Both voltages are converted into pulses, and the time interval between the pulse of one wave and that obtained from the second sine wave is measured. This time interval is proportional to the phase difference. Such a phasemeter can measure angles from 0 to 360°.

Square Waves from a Sine Wave　If the comparator output is a signal which assumes either one of two levels (a step output), then a sine-wave input will result in a square-wave output. If the reference voltage V_R is set equal to zero, then a symmetrical square wave results.

Amplitude-distribution Analyzer　A comparator is a basic building block in a system used to analyze the amplitude distribution of the noise generated in an active device or the voltage spectrum of the pulses developed by a nuclear-radiation detector,[3] etc. To be more specific, suppose that the output of the comparator is 100 V if $v_i > V_R$ and 0 V if $v_i < V_R$. Let the input to the comparator be tube noise. A d-c meter is used to measure the average value of the output square wave. For example, if V_R is set at zero, the meter will read 100 V, which is interpreted to mean that the probability that the amplitude is greater than zero is 100 percent. If V_R is set at some value V_R' and the meter reads 70 V, this is interpreted to mean that the probability that the amplitude of the noise is greater than V_R' is 70 percent, etc. In this way the cumulative amplitude probability distribution of the noise is obtained by recording meter readings as a function of V_R.

Analog-to-Digital Converter　It is often required that data taken in a physical system be converted into digital form. Such data would normally appear in electrical analog form. For example, a temperature difference would be represented by the output of a thermocouple, the strain of a mechanical member would be represented by the electrical unbalance of a strain-gauge bridge, etc. The need therefore arises for a device that converts analog information into digital form. A very large number of such devices have been invented.[4] We shall consider below one such system which involves a time-modulation scheme of high precision.

In this system[5] a continuous sequence of equally spaced pulses is passed through a gate. The gate is normally closed and is opened at the instant of the beginning of a linear ramp. The gate remains open until the linear sweep voltage attains the reference level of a comparator, the level of which is set equal to the analog voltage to be converted. The number of pulses in the train that passes through the gate is therefore proportional to the analog voltage. If the analog voltage varies with time, it will, of course, not be possible to convert the analog data continuously, but it will be required that the analog data be sampled at intervals. The maximum value of the analog voltage will be represented by a number of pulses n. It is clear that n should be made as large as possible consistent with the requirement that the time interval between two successive pulses shall be larger than the timing error of the time modulator. The recurrence frequency of the pulses is equal, at a minimum, to the product of n and the sampling rate. Actually, the recurrence rate will be larger in order to allow time for the circuit to recover between samplings.

One form of digital voltmeter uses the above-described analog-to-digital converter. The number of pulses which pass through the gate is proportional to the voltage being measured. These pulses go to a counter whose reading is indicated visually by means of some form of luminous display (for example, the Nixie tube, which is a cold-cathode glow tube described in Sec. 18-13). Another system for analog-to-digital conversion is given in Sec. 18-3.

REFERENCES

1. De Matteis, W. M.: Current Switch Comparator for Analog to Digital Conversion, *Appl. Lab. Rept.* 760, Philco Corporation, Lansdale, Pa.

2. Middlebrook, R. D.: "Differential Amplifiers," John Wiley & Sons, Inc., New York, 1963.

3. Van Rennes, A. B.: Pulse-amplitude Analysis in Nuclear Research, *Nucleonics*, vol. 10, no. 7, pp. 20–27, July, 1952; no. 10, pp. 50–56, October, 1952.

4. Susskind, A. K.: "Notes on Analog-Digital Conversion Techniques," John Wiley & Sons, Inc., New York, 1957.

5. Slaughter, D. W.: An Analog-to-Digital Converter with an Improved Linear-sweep Generator, *IRE Natl. Conv. Record—Electron. Computers*, 1953, pp. 7–12.

8 / CLAMPING AND SWITCHING CIRCUITS

In this chapter we study a number of circuits in which reactive elements (particularly capacitors) appear in combination with nonlinear devices. In *clamping circuits* the capacitors are essential; in *switching circuits* the capacitors are often unavoidable.

8-1 THE CLAMPING OPERATION

A function that must frequently be accomplished with a periodic waveform is the establishment of the recurrent positive or negative extremity at some constant reference level V_R. Since, in the steady state, the circuits used to perform this function restrain the extremity of the waveform from going beyond V_R, the circuits are referred to as *clamping circuits*. However, the term "clamping" is rather widely used for a variety of related but not identical operations. For example, in Sec. 6-19 we encountered in Fig. 6-34 a "clamped" grid. Here the grid was constrained to remain in the neighborhood of 0 V because of the flow of grid current, and this restraint was in no way related to a signal. Generally, whenever a circuit point becomes connected through a low impedance (as through a conducting diode) to some reference voltage V_R, we say that the point has been clamped to V_R since the voltage at the point will not be able to depart appreciably from V_R. In this sense, the diode limiting circuit of Fig. 7-2 is an example of such a clamping circuit, since the output is clamped to V_R whenever the input voltage exceeds V_R. Since, in these applications, a voltage change in only one direction is restrained, the circuits are called *one-way clamps*. Two diodes may be used to establish a *two-way clamp*.

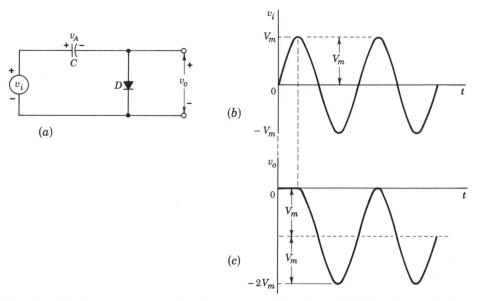

Fig. 8-1   (a) The basic circuit of a d-c restorer; (b) a sinusoidal signal is applied at $t = 0$; (c) the output waveform.

The need to establish the extremity of the positive or negative signal excursion at some reference level often appears in connection with a signal which has passed through a capacitive coupling network. Such a signal has lost its d-c component, and the clamping circuit introduces a d-c component. For this reason the circuit is often referred to as a *d-c restorer* or *d-c reinserter*. These terms are somewhat misleading, because the prefix "re" suggests that the d-c component so introduced is identical with the d-c component lost in transmission. Such is normally not the case. A term which avoids this objection is *d-c inserter*. In practice all the terms mentioned above are used interchangeably.

Waveforms   We turn our attention now to Fig. 8-1, which consists of a signal source v_i of negligible output impedance, a capacitor, and the diode D. We assume that the diode is ideal in that it exhibits an arbitrarily sharp break at 0 V and that its forward resistance is zero.   The input signal is a sinusoid, as shown, which begins at $t = 0$.   The capacitor C is uncharged at $t = 0$. Our interest is in finding the waveform of the voltage v_o across the diode.

During the first quarter cycle the input signal rises from zero to the maximum value V_m.   The diode being ideal, no forward voltage may appear across it.   Accordingly, during this first quarter cycle the capacitor voltage $v_A = v_i$.   The voltage across C rises sinusoidally, the capacitor being charged through the series combination of the signal source and the diode.   Throughout this first quarter cycle the output v_o has remained zero. At the end of the quarter cycle there exists across the capacitor a voltage $v_A = V_m$.

When, after the first quarter cycle, the peak has been passed and the input signal begins to fall, the voltage v_A across the capacitor is no longer able to follow the input voltage. For in order to do so, it would be required that the capacitor discharge, and because of the diode, such a discharge is not possible. Accordingly, the capacitor remains charged to the voltage $v_A = V_m$, and, after the first quarter cycle, the output voltage is $v_o = v_i - V_m$. During succeeding cycles the positive excursion of the signal just barely reaches zero. The diode need never again conduct, and the positive extremity of the signal has been "clamped" or "restored" to zero.

Suppose that after the steady-state situation has been attained, the amplitude of the input signal is increased. Then there will again be an interval of one quarter cycle, at most, during which the diode will conduct. The d-c voltage across C will increase as required, again to restore the positive excursions of the signal to zero. But what if the amplitude of the input signal should decrease? In this case, it is required that the d-c voltage across the capacitor decrease. But in the circuit of Fig. 8-1 there is no mechanism to allow such a discharge. To permit a decrease in capacitor voltage it is necessary to shunt a resistor across C or equivalently to shunt a resistor across the diode. In this latter case the capacitor will discharge through the series combination of the resistor shunted across the diode and the resistance of the source.

A circuit with such a resistor R is shown in Fig. 8-2a. Figure 8-2b shows the output waveform for a case where an abrupt decrease of the amplitude of the sinusoidal input signal has taken place. The first two complete cycles correspond to the steady-state condition after the large-amplitude signal had been applied for a long time. At the time $t = t_1$ the amplitude is abruptly reduced in magnitude. Since the capacitor cannot discharge rapidly, the positive peaks of the sinusoid fall short of attaining 0 V. The voltage across C falls exponentially as C discharges, and after some cycles, the positive peaks again reach zero. In this case, unlike the case where R was absent, some diode current will flow at each positive peak. For now, even after the voltage across C has dropped to the point where the positive peaks reach zero voltage, the capacitor continues to discharge through R and the generator resistance. Therefore the diode must now supply to the capacitor, at the time of this positive excursion of the signal, the charge lost by the capacitor when the diode is not conducting. Fortunately, in most applications there is no great need to reestablish clamping immediately after a reduction in signal amplitude. Many cycles may be permitted to elapse before clamping is again obtained. Therefore, since the capacitor discharges slowly, the diode will not have to conduct for any but a small part of the cycle to supply the lost charge.

It is of interest to examine the waveform in the neighborhood of a positive peak where the diode conducts. This portion of the waveform is shown on an expanded scale in Fig. 8-2c. If it were not for the diode the signal would

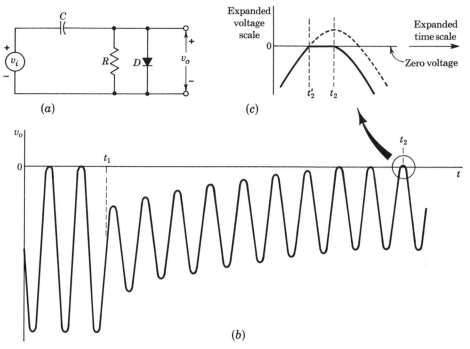

(a)

(c)

(b)

Fig. 8-2 (a) The clamping circuit is completed through the inclusion of a resistor which permits the capacitor to discharge. (b) At time $t = t_1$ the amplitude of the input signal is abruptly reduced. Output waveform shows approach to steady state. (c) The details of the waveform v_o at the first positive peak which would cross the zero-voltage axis at t_2' if the diode were absent.

follow the dashed sinusoidal waveform, with peak at $t = t_2$. Because of the diode the part of the waveform between t_2' and t_2 is clamped at 0 V. The time t_2' is the time at which the rising sinusoidal waveform just reaches the zero axis. To the right of t_2 the waveform is sinusoidal, with maximum at $l = t_2$. That is, to the right of $t = t_2$ the actual waveform is the dashed waveform lowered so that the peak occurs at 0 V. If the distortion of the waveform is to be kept to a minimum the capacitor must lose only a very small fraction of its total charge during any one cycle. This condition requires that the time constant RC be very large in comparison with the period of the signal.

8-2   CLAMPING CIRCUIT TAKING SOURCE AND DIODE RESISTANCES INTO ACCOUNT

We turn now to the more realistic circuit shown in Fig. 8-3. Here we have included a source resistance R_s and a diode forward resistance R_f. The

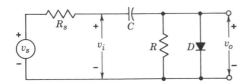

resistance R_f will lie in the range tens to hundreds of ohms, depending on the type of diode used. The source resistance may be negligible or may range up to many thousands of ohms, depending on the source. We shall assume for the present that the diode break point V_γ occurs at zero voltage. This restriction is removed in Sec. 8-4.

As already suggested, and as we shall shortly see more exactly, the precision of operation of the circuit depends on the condition that $R \gg R_f$. To analyze the circuit we need only to recognize that the equivalent circuits for the purpose of calculating the output voltage v_o are as shown in Fig. 8-4a and b. The circuit in Fig. 8-4a applies when the diode is conducting and Fig. 8-4b applies when the diode is not conducting. If the inequality $R \gg R_f$ is not valid, then in Fig. 8-4a, R_f must be replaced by the parallel combination of R and R_f. In connection with Fig. 8-4b it may also turn out that the back resistance of the diode is not very large in comparison with R. In this case R must be replaced by the parallel combination of the two. In some instances it may be that no physical resistor R is placed in the circuit and that R may actually represent the diode reverse resistance R_r.

The Transient Waveform We wish to follow the waveform which results after a signal is suddenly applied to the circuit and to see how the steady state is reached. After a number of cycles have passed we expect to approach a steady-state condition in which the positive peaks have been clamped to zero. For this purpose we need but make use of the equivalent circuits of Fig. 8-4 and proceed as in the following illustrative example.

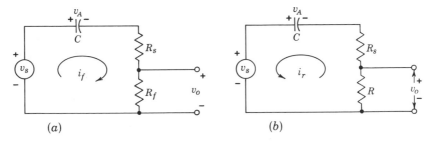

(a)   (b)

Fig. 8-4 Circuits equivalent to the circuit of Fig. 8-2 for the purpose of calculating v_o (a) when the diode is conducting, (b) when the diode is not conducting. It is assumed that $R_r \gg R \gg R_f$.

EXAMPLE   In the circuit of Fig. 8-3, $R_s = R_f = 100\ \Omega$, $R = 10$ K, and $C = 1.0$ μF.   At $t = 0$ there is applied a symmetrical square-wave signal of amplitude 10 V and frequency 5 kHz.   As indicated in Fig. 8-5, the signal v_s extends from 0 to $+10$ V.   Draw the first several cycles of the output waveform.

Solution   Assume that the capacitor C is initially uncharged.   Using the equivalent circuit of Fig. 8-4a we find that, at the first 10-V jump of the input signal, the output jumps to $+5$ V.   The output now decays toward zero exponentially with a time constant

$$\tau = (R_s + R_f)C = 200\ \mu\text{sec}$$

Since the period $T = (5,000)^{-1}$ sec $= 200$ μsec, then at the end of a half cycle of the square wave, at $t = T/2$, the output, indicated by the solid line, has fallen to

$$v_o\left(t = \frac{T}{2}\right) = 5\epsilon^{-T/2\tau} = 5\epsilon^{-\frac{1}{2}} = 3.0\ \text{V}$$

At this time, since the voltage across R_f is 3.0 V, so also is the voltage across R_s, leaving the capacitor voltage $v_A = 4$ V.   When at $t = (T/2)+$ the input drops back to zero, the diode cuts off, and we now use Fig. 8-4b.   In this circuit, $v_A = 4.0$ V and $v_s = 0$, so that, neglecting R_s compared with R, $v_o = -4.0$ V, as shown in Fig. 8-5.   The output now again starts to decay toward zero.   However, the time constant now is $RC = 10$ K $\times$ 1.0 μF $= 10{,}000$ μsec, or 100 times larger than the time $T/2 = 100$ μsec.   Therefore the decay is negligible and is not indicated in the figure.

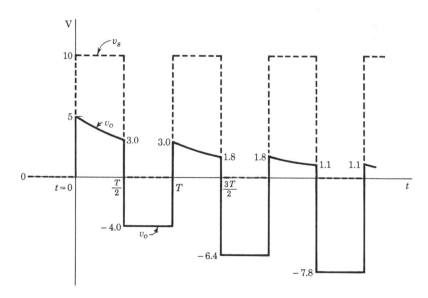

Fig. 8-5   Example of the transient approach to the steady state in a clamping circuit.

Since in the interval $t = T/2$ to $t = T$ the voltage across the capacitor has not changed, then at $t = T+$ the output returns to $+3.0$ V. Again the output decays toward zero. The portion of the exponential decay in the interval $t = T$ to $t = 3T/2$ is a continuation of the portion in the interval $t = 0$ to $t = T/2$. If all of the decays indicated were moved together so that they just joined, they would form one continuous exponential decay from 5.0 V toward zero.

At $t = 3T/2$, $v_o = 3\epsilon^{-\frac{1}{2}} = 1.8$. The remaining calculations are repetitions of those above and the results are shown on the figure. We observe that cycle by cycle the output waveform is approaching the steady-state case, where the positive excursion of the waveform is clamped approximately to zero. (The calculation of the clamped level is given in Sec. 8-3.)

The Steady-state Output Waveform for a Square-wave Input   Henceforth we shall concentrate exclusively on the *steady-state response*. Consider the square wave of Fig. 8-6a applied to the clamping circuit of Fig. 8-3. The general form of the output waveform is indicated in Fig. 8-6b and is determined by the four voltages V_1, V_1', V_2, and V_2'. We shall now indicate how to calculate these voltage values.

Consider conditions at $t = 0-$ when $v_s = V''$ and $v_o = V_2'$. Since the diode is reverse-biased at this time, Fig. 8-4b is applicable, and the capacitor voltage is

$$v_A = V'' - \frac{R + R_s}{R} V_2' \tag{8-1}$$

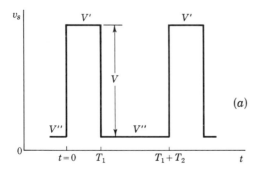

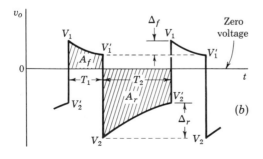

Fig. 8-6   (a) A square-wave input signal of peak-to-peak amplitude V; (b) the general form of the steady-state output of a clamping circuit with the input in (a).

At $t = 0+$ the input signal jumps to V', the output to V_1, the diode conducts, and Fig. 8-4a is applicable. Since the voltage across the capacitor cannot change instantaneously, it remains at the value given in Eq. (8-1). From Fig. 8-4a

$$v_A = v_s - \frac{R_f + R_s}{R_f} v_o \tag{8-2a}$$

or, with $v_s = V'$,

$$V'' - \frac{R + R_s}{R} V_2' = V' - \frac{R_f + R_s}{R_f} V_1 \tag{8-2b}$$

Since the peak-to-peak input amplitude is $V = V' - V''$, Eq. (8-2b) becomes

$$V = \frac{R_f + R_s}{R_f} V_1 - \frac{R + R_s}{R} V_2' \tag{8-3}$$

In a similar manner, by considering conditions at $t = T_1-$ and T_1+ we obtain

$$V = \frac{R_f + R_s}{R_f} V_1' - \frac{R + R_s}{R} V_2 \tag{8-4}$$

Since in the interval T_1 the diode is conducting, the output decays with a time constant $(R_f + R_s)C$. Hence

$$V_1' = V_1 \epsilon^{-T_1/(R_f + R_s)C} \tag{8-5}$$

Similarly, during the interval T_2 the diode is reverse-biased and the circuit time constant is $(R + R_s)C$, so that

$$V_2' = V_2 \epsilon^{-T_2/(R + R_s)C} \tag{8-6}$$

Equations (8-3), (8-4), (8-5), and (8-6) suffice to determine the four voltages V_1, V_1', V_2, and V_2'. We have in effect used these same equations in calculating the voltages in Fig. 8-5. Note that if, as in the present case, the source impedance is taken into account the output voltage jumps are smaller than the abrupt discontinuity V in the input. Only if $R_s = 0$ are the jumps in input and output voltages alike. Thus, from Eqs. (8-3) and (8-4) with $R_s = 0$,

$$V = V_1 - V_2' = V_1' - V_2 \tag{8-7}$$

Observe, also, that the response is independent of the absolute levels V' and V'' of the input signal and is determined only by the amplitude V. It is possible, for example, for V'' to be negative or even for both V' and V'' to be negative. *The average level of the input plays no part in determining the steady-state output waveform.*

An interesting result is obtained if Eq. (8-4) is subtracted from Eq. (8-3):

$$\frac{R_f + R_s}{R_f} (V_1 - V_1') - \frac{R + R_s}{R} (V_2' - V_2) = 0$$

or using the notation in Fig. 8-6, where

$$\Delta_r \equiv V_2' - V_2 \qquad \Delta_f \equiv V_1 - V_1' \tag{8-8}$$

we obtain

$$\Delta_f = \frac{R_f}{R_f + R_s} \frac{R + R_s}{R} \Delta_r \tag{8-9}$$

Since R_s is usually much smaller than R, then the tilt Δ_f in the forward direction is almost always less than the tilt Δ_r when the diode is reverse-biased. Only when $R_s \ll R_f$ are the two tilts almost equal.

8-3   A CLAMPING–CIRCUIT THEOREM

We shall now demonstrate that, in the steady state, the area A_f under the output voltage waveform in the forward direction (when the diode conducts) is related to the area A_r in the reverse-biased direction (when the diode does not conduct) by the relationship

$$\frac{A_f}{A_r} = \frac{R_f}{R} \tag{8-10}$$

The areas referred to are indicated in Fig. 8-6b, for which case the input signal was a square wave. However, the result applies quite generally, independently of the input waveform and of the magnitude of the source resistance R_s.

If $v_f(t)$ is the output waveform in the forward direction, then, from Fig. 8-4a, the capacitor charging current is $i_f = v_f/R_f$. Therefore the charge acquired by the capacitor during the forward interval is

$$\int_0^{T_1} i_f \, dt = \frac{1}{R_f} \int_0^{T_1} v_f \, dt = \frac{A_f}{R_f} \tag{8-11}$$

Similarly, if $v_r(t)$ is the output voltage in the reverse direction, then the current which discharges the capacitor is $i_r = v_r/R$, and the total charge lost is

$$\int_{T_1}^{T_1+T_2} i_r \, dt = \frac{1}{R} \int_{T_1}^{T_1+T_2} v_r \, dt = \frac{A_r}{R} \tag{8-12}$$

In the steady state, the net charge acquired by the capacitor must be zero. Therefore we equate Eqs. (8-11) and (8-12) and arrive at Eq. (8-10). This equation says that *for any input waveform the ratio of the area under the output-voltage curve in the forward direction to that in the reverse direction is equal to the ratio R_f/R.* An application of this principle is given in the following illustrative example.

EXAMPLE   (a) An unsymmetrical square wave with $T_1 = 1$ msec and $T_2 = 1$ μsec has an amplitude of 10 V. This signal is applied to the restorer circuit of Fig. 8-3, in which $R_f = 50$ Ω, $R = 50$ K, and $R_s = 0$. Assume that the capacitor C is arbitrarily large, so that the output is a square wave without tilt. Find where, on the waveform, the zero level is located. (b) If the waveform is inverted so that $T_1 = 1$ μsec and $T_2 = 1$ msec, find the location of the zero level. (c) If the diode is inverted, but the input is as in part b, locate the zero level.

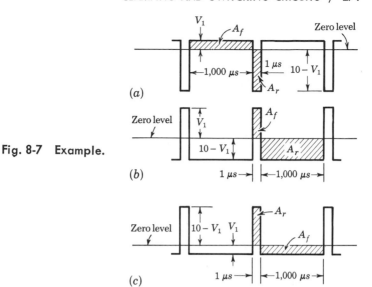

Fig. 8-7   **Example.**

Solution   *a.* The output waveform and zero level are shown in Fig. 8-7*a.* In the figure we have already taken into account the fact that, since $R_s = 0$, the peak-to-peak amplitude of the output signal must be the same as that for the input signal, namely 10 V. We have $A_f = 1{,}000V_1$ and $A_r = 10 - V_1$, where the areas are in units of microsecond-volts. From Eq. (8-10) we have

$$\frac{A_f}{A_r} = \frac{1{,}000V_1}{10 - V_1} = \frac{R_f}{R} = \frac{50}{50 \times 10^3} = 10^{-3}$$

and we find that $V_1 = 10^{-5}$ V. This example illustrates that clamping of the broad base line of the waveform is quite precise, since only one-millionth of the input waveform is above the zero level.

    b. In Fig. 8-7*b,* $A_f = V_1$ and $A_r = (1{,}000)(10 - V_1)$. Again from Eq. (8-10) we have

$$\frac{V_1}{1{,}000(10 - V_1)} = 10^{-3}$$

or $V_1 = 5$ V. The zero level is now not near the positive peak but is halfway down the waveform, and the circuit has done very poorly as a clamp. This example illustrates that it is not advisable to attempt to clamp the peak of narrow pulses. To yield good precision of clamping in this case it would be necessary to increase R by a large factor. Such an increase in R would greatly lengthen the circuit recovery time for a signal decrease. It may even be that when semiconductor diodes are used, such an increase in R would be ineffective because of the finite diode back resistance.

    c. Positive voltages now back-bias the diode and negative voltages forward-bias the diode, because the diode has been inverted. Comparing Fig. 8-7*c* and *a* we see that one is inverted with respect to the other, and hence $V_1 = 10^{-5}$ V, as in part *a.*

We note that the d-c level of the input did not enter these calculations.   This observation confirms our previous assertion that the average value of the input has no effect on the output level in the steady state.

We can summarize the results of this example by observing that it is very difficult to achieve d-c restoration to the peak (either positive or negative) of a narrow pulse, but we can very effectively clamp to the broad base line.

If the clamping-circuit theorem is applied to the waveform in Fig. 8-5, it is found that the peak of v_o is at 0.1 V.

8-4   PRACTICAL CLAMPING CIRCUITS

To maintain perfect flatness of the positive and negative peaks of a square wave would require that C be arbitrarily large, as was assumed to be the case in the problem above. In a practical situation it will normally turn out that $(R_f + R_s)C \ll T_1$ and $(R + R_s)C \gg T_2$.   A square-wave or pulse-type waveform, after restoration, typically appears as in Fig. 8-8.   During the interval T_2 there is a small tilt Δ_r, and at the beginning of the interval T_1 a sharp spike of magnitude Δ_f appears.   The capacitor recharges through the diode in a very short time, and during the remainder of the time T_1 no appreciable diode current flows.   From Eq. (8-9) the overshoot Δ_f will usually be smaller than the tilt Δ_r.

For the important special case depicted in Fig. 8-8, the voltage values are found from Eqs. (8-5), (8-4), (8-6), (8-9), and (8-8) to be

$$V_1' = 0 \qquad V_2 = -\frac{R}{R + R_s} V \tag{8-13}$$

$$V_2' = V_2 \epsilon^{-T_2/(R+R_s)C} \tag{8-14}$$

$$V_1 = \Delta_f = \frac{R_f}{R_f + R_s} \frac{R + R_s}{R} (V_2' - V_2) \tag{8-15}$$

We have already observed that even if we assume C arbitrarily large, unless the source resistance R_s is zero, the part of the input signal which

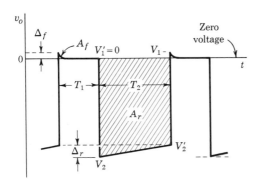

Fig. 8-8   The form of the output of a clamping circuit with a square-wave input for $(R_f + R_s)C \ll T_1$ and $(R + R_s)C \gg T_2$.

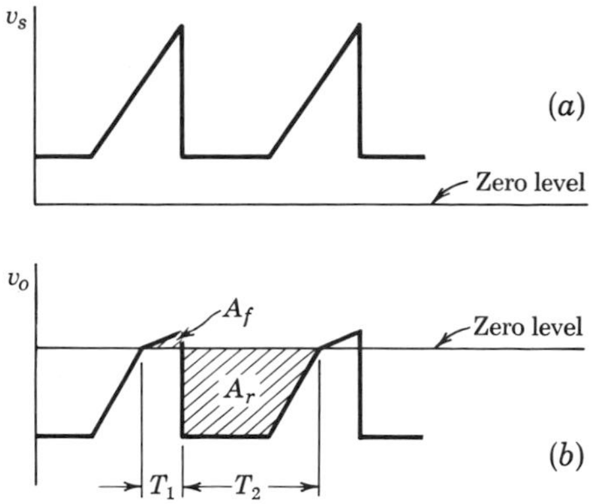

Fig. 8-9 (a) A ramp input to a restorer and (b) the output taking into account the finite source impedance.

occurs when the diode is conducting appears at the output multiplied by the factor $R_f/(R_f + R_s) \equiv F_f$. That part of the signal which occurs when the diode is not conducting is transmitted to the output multiplied by $R/(R + R_s) \equiv F_r$. Usually F_r is much closer to unity than is F_f. Such selective attenuation obviously produces distortion in the output signal by blunting or flattening that part of the signal which drives the diode to conduction. This distortion is not readily apparent in the case of a square wave, which already has a flat top. Nor is it even especially apparent with a sinusoidal signal, whose peak is already rounded. The distortion is much more easily observed in the case of a signal which comes to a sharp peak, such as is the case with the ramp type of waveform of Fig. 8-9. In that figure it is assumed that the capacitor is large enough so that the only distortion is due to the attenuation. We emphasize again that quite independently of the distortion the relationship $A_f/A_r = R_f/R$ is valid.

If the diode in Fig. 8-3 is reversed, it was shown in the illustrative example above that the negative rather than the positive extremity of the signal will be established at zero. If the circuit is modified to include a fixed voltage V_R, as in Fig. 8-10, the positive extremity (or negative extremity, if the diode is reversed) of the output will be established at V_R. That such is the case may be seen in the following manner. Assume first that the output signal is taken across the diode and that the voltage V_R is part of the input signal. We have already seen that, in the steady state, the output voltage across the diode is unaffected by the direct component of the input. Accordingly, the waveform which appears across the diode is the signal v_i with its positive extremity clamped to zero. The voltage v_o is this same voltage translated upward by the amount V_R, so that the positive extremity of v_o is clamped to V_R. The clamping theorem, Eq. (8-10), remains valid provided that the areas A_f and A_r are measured with respect to the level V_R rather than with respect to ground.

Sometimes the device which receives the output signal of the restorer of Fig. 8-10 bridges a resistance across the output, and it is necessary to operate

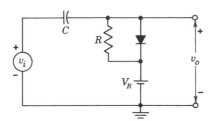

Fig. 8-10   A circuit which clamps
to the voltage V_R.

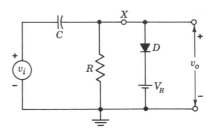

Fig. 8-11   A modification of the
circuit of Fig. 8-10 in which the
resistor R is bridged across both
the diode and the reference
voltage V_R.

the circuit as in Fig. 8-11.   Here the resistor R is bridged across the combination of both the diode and the reference source.   This arrangement will
operate perfectly well, provided that the amplitude of the input signal is
adequately large.   To see how this restriction arises, we consider that the
lead marked X in Fig. 8-11 has been opened, separating the RC circuit from the
diode-reference source combination.   The input signal v_i makes excursions in
the positive and negative direction with respect to its average value.   The
signal between X and ground makes equal excursions with respect to its average value, which is zero.   If the diode is to conduct, so that clamping may
take place, the positive excursions at X must at least equal V_R.   Accordingly,
the condition to be imposed on the signal to ensure proper circuit operation is
that *the positive excursion of the signal with respect to its average value must be
larger than V_R.*
   We shall now remove the assumption we have made thus far in this
chapter that the diode break point occurs at zero voltage.   The diode piecewise linear model is as indicated in Fig. 7-3, with the break point at a voltage
V_γ.   Hence, the clamping circuit of Fig. 8-2 is equivalent to that of Fig. 8-11
with $V_R = V_\gamma$.   Therefore, if a silicon diode ($V_\gamma \approx 0.6$ V) is used, the circuit
will function properly only if the positive excursion of the signal above its
average value exceeds 0.6 V.   Moreover, the peak of the output will be
clamped not to zero but to a voltage of 0.6 V above ground.   If an external
reference voltage V_R is added to the diode, then the clamping level is $V_R + V_\gamma$.
   By following the procedure outlined in Sec. 8-3 the clamping-circuit
theorem expressed by Eq. (8-10) can be generalized (for Fig. 8-11) to

$$\frac{A_f - (V_R + V_\gamma)T_1}{A_r} = \frac{R_f}{R} \tag{8-16}$$

where T_1 is the interval over which the diode is forward-biased, and $R \gg R_f$.
If in the illustrative example in Sec. 8-3 a silicon diode is used, then an application of the above equation yields the following results.   In parts a and c

$V_1 = 0.6 + 10^{-5} \approx 0.6$ V, so that the broad base line is now clamped to $V_\gamma = 0.6$ V rather than to ground. In part b, $V_1 = 5.3$ V instead of 5.0 V, which was obtained on the assumption that $V_\gamma = 0$.

8-5   EFFECT OF DIODE CHARACTERISTICS ON CLAMPING VOLTAGE

Up to this point in our discussion of restorers, we have represented the diode as an ideal diode in series with a battery V_γ and a resistor R_f. We shall now take explicit account of the diode volt-ampere characteristic to calculate, in a simple case, the effect of the diode characteristic on the clamping voltage. Let us consider that the input to the clamping circuit of Fig. 8-3 is the square wave of Fig. 8-6a with a peak-to-peak amplitude V. Assume that the capacitor C is arbitrarily large, so that the output waveform v_o across the diode is similarly a square wave. We shall again consider, as in Sec. 7-2, that the relation $I = I_o \epsilon^{V/\eta V_T}$ may be used for both semiconductor and thermionic diodes. As noted before, this assumption is valid for semiconductor diodes whenever the diode operates at a point where $\epsilon^{V/\eta V_T} \gg 1$. Finally, for the sake of simplicity, we shall assume that $R_s = 0$ in this discussion, which takes the diode volt-ampere characteristic into account.

During the interval when the input signal is at its positive extremity V' the diode clamps the output at some clamping voltage V_{cl}, and the corresponding diode current is

$$I_{cl} = I_o \epsilon^{V_{cl}/\eta V_T}$$

This current charges the capacitor C so that $v_A = V' - V_{cl}$. During the negative interval of the input signal, the diode is not conducting and the voltage across R in Fig. 8-4b has the magnitude $v_A - V'' = V - V_{cl}$. We expect that V_{cl} will be of the order of tenths of a volt and we shall consider that V is of the order of tens of volts, so that we may set $V - V_{cl} \approx V$. Then the discharge current of the capacitor is V/R. Since the square wave is symmetrical in the sense that the two portions of the cycle are of equal duration and the net accumulation of charge must be zero, we must set the charging current equal to the discharging current, or

$$I_{cl} = \frac{V}{R} = I_o \epsilon^{V_{cl}/\eta V_T} \tag{8-17}$$

Solving for V_{cl} we find

$$V_{cl} = \eta V_T \ln \frac{V}{R I_o} \tag{8-18}$$

and taking the differential we obtain

$$dV_{cl} = \eta V_T \frac{dV}{V} \tag{8-19}$$

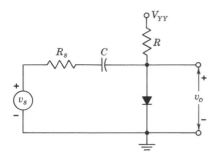

Fig. 8-12   A clamping circuit in which the diode is biased to operate higher on its characteristic.

The clamping voltage may be obtained from Eq. (8-18), and from Eq. (8-19) we may determine how the clamping voltage varies with the amplitude of the input signal.   As is to be expected, we find that in typical cases the clamping voltages for the various diodes are in the neighborhood of the break point for the diode ($V_{cl} \approx V_\gamma$), that is, some tenths of a volt negative for a thermionic diode and about 0.2 and 0.6 V positive, respectively, for Ge and Si.

From Eq. (8-19) we find for Ge, with $V = 10$ V and $\eta = 1$, that a 1-V change in input amplitude yields a change $dV_{cl} = 2.6$ mV.   We observe that Eq. (8-19) indicates a reduced dependence of V_{cl} on V with increased V.   The reason for this feature is that at increased V the diode clamps higher on its volt-ampere characteristic, where a smaller voltage change is required for a fixed current change.   To take advantage of this feature it is useful to bias the diode in its forward direction, as is done in Fig. 8-12.   Here the resistor R is returned to a supply V_{YY}, which causes a current V_{YY}/R to flow through the diode even in the absence of a signal.   It may be proved (Prob. 8-18) that in the presence of a symmetrical square wave of peak-to-peak value V the diode current during conduction (when clamping takes place) is

$$I_{cl} = \frac{2V_{YY} + V}{R} \tag{8-20}$$

and that

$$dV_{cl} = \eta V_T \frac{dV}{2V_{YY} + V} \tag{8-21}$$

In the illustration above, with $V = 10$ V and $dV = 1$ V, if we assume $V_{YY} = 50$ V then $dV_{cl} = 0.24$ mV, which is a value approximately one-tenth that obtained with $V_{YY} = 0$.   This example indicates the advantage with respect to the stability of the clamping level of using a biased diode.

If the diode is represented by a piecewise linear model, then the clamping-circuit theorem of Sec. 8-3 when applied to the biased diode of Fig. 8-12 must be generalized as follows:

$$\frac{A_f - V_\gamma T_1}{A_r + V_{YY} T} = \frac{R_f}{R} \tag{8-22}$$

where T is the period of the input square wave, T_1 is the interval over which the diode conducts, and V_γ is the diode break-point voltage.

If the input signal is too small the circuit of Fig. 8-12 will not operate properly, but instead the diode will remain conducting continuously. The minimum signal amplitude required is obtained as follows. Assume that when the input is at its upper level, the output is constant at V_γ, so that the current in R is $I_R = (V_{YY} - V_\gamma)/R$. When the input drops by V then the current through R_s is $I_S = -V/R_s$. If I_R always exceeds I_S in magnitude, then current flows continuously in the diode. Hence, in order to ensure proper circuit operation with an applied square waveform, *the peak-to-peak magnitude V of the signal must be larger than* $(V_{YY} - V_\gamma)(R_s/R)$. Note that if the signal-source impedance is negligible the circuit always functions correctly (the diode becomes nonconducting during the lower level of the input signal waveform).

8-6   CLAMPING IN BASE OR GRID CIRCUITS

Clamping may be accomplished in the base circuit of a transistor or the grid circuit of a triode (or pentode). The operation is identical with that described above. In the case of a tube the grid and cathode serve as the electrodes of the diode, whereas in a transistor the emitter-junction diode provides the necessary nonlinearity. Restorer action in the input circuit of a multielectrode device has interesting applications, as we shall now indicate.

Consider a transistor amplifier in which fixed-current bias is obtained by connecting a resistor R from the base to the supply voltage, as indicated in Fig. 8-13a. The behavior of the base circuit is identical with the diode clamp of Fig. 8-12. If the quiescent base current $(V_{CC} - V_\gamma)/R$ places the transistor in the active region and if the signal swing is small enough, then, as noted in the preceding section, the base-to-emitter diode always conducts. The emitter-junction voltage remains forward-biased, and the transistor behaves as a small-signal class A amplifier. On the other hand, if a large-amplitude signal is employed, clamping will take place at the positive extremities of the

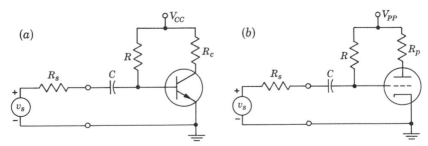

Fig. 8-13   Clamping in the input circuit of an amplifier using (a) a transistor and (b) a tube.

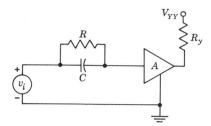

Fig. 8-14 Self-bias is obtained at the input of an oscillator because of the flow of grid or base current at the peaks of the signal.

waveform and the transistor will be below cutoff during a portion of each cycle. Such an overdriven amplifier or switch is treated in detail in Sec. 8-8.

If the transistor in Fig. 8-13a is replaced by a tube, as indicated in Fig. 8-13b, then there is a fundamental difference in behavior. For no input signal the grid must be clamped to the cathode ($V_{GK} \approx 0$), whereas for the circuit of Fig. 8-13a the transistor could be within its active region in the quiescent state. If the signal is supplied by a voltage source ($R_s \approx 0$) and if the peak-to-peak signal amplitude does not exceed the grid base, then essentially linear operation is obtained, for the following reason. The clamping action at the grid will adjust the d-c level of the signal between grid and cathode to be negative, so that no grid current is drawn except for a small portion of the cycle at its positive extremity. The signal thus provides its own bias. As long as $R_s \ll R$, the signal distortion during the small part of the cycle when grid current is drawn will be negligible. If, on the other hand, a large input signal is employed, then the tube will be cut off during a portion of each cycle, and an overdriven amplifier results.

For large-signal operation the resistor R in Fig. 8-13 may also be connected to ground instead of to the supply voltage. Alternatively R may be placed directly across C. This latter connection is indicated in Fig. 8-14 and is commonly used to provide self-bias in a resonant-circuit oscillator which operates class C.

In all of the preceding discussion concerning clamping we have assumed a periodic signal. It is, however, to be noted that clamping may be accomplished with a nonperiodic signal, provided only that the signal has a periodic positive or negative extremity. We consider next how clamping may be accomplished with a signal (not necessarily periodic) in which the level to be clamped occurs neither as the positive nor as the negative extremity but at some level in between.

8-7   SYNCHRONIZED CLAMPING

The d-c restorers discussed above are examples of clamping circuits in which the time during which the clamping is effective is controlled by the signal itself. Useful features result when the time of clamping is not determined directly by the signal but is determined rather by an auxiliary voltage, called a *control signal*, which occurs synchronously with the signal. For example, suppose the waveform of Fig. 8-15a is to be used to displace the beam of a cathode-ray tube linearly with time, first in one direction and then in the other direction from some fixed initial point. If the signal is transmitted through an a-c coupling network whose low-frequency time constant is not very large in comparison with the interval T_1, the signal will distort into the form shown in Fig. 8-15b. The principal defect in the waveform is that the two displacements will start from different places (A and B). In addition the d-c level V_R has been lost. If, however, the signal is passed through the circuit of Fig. 8-16 and if switch S is closed during time T_2 and is open during time T_1, the waveform will appear as in Fig. 8-15c. The pips which appear when the voltage returns to the level V_R will be reduced to infinitesimally narrow spikes as the resistance of the switch (R_f) approaches zero.

It is, of course, required that the switch S be open throughout the time interval T_1, but it is not necessary that the switch be closed for the entire interval T_2. It is only required that the switch be closed for a period long enough to allow the capacitor C to acquire or lose enough charge to bring the output terminal to the reference level V_R.

It is not possible to use synchronized clamping with a signal of arbitrary waveform. For example, if the waveform were sinusoidal, it would necessarily

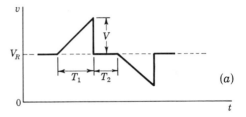

(a)

Fig. 8-15 Illustrating the necessity for synchronized clamping for a signal which may vary in both directions from a voltage which is to be established at some reference level.

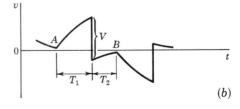

(b)

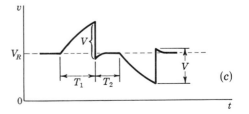

(c)

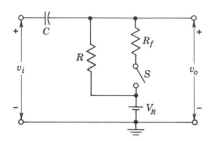

Fig. 8-16 Switch S closes in synchronism with the signal during those intervals when it is desired that the output be clamped to V_R.

be distorted every time the switch S closed. Synchronous clamping may be used whenever the signal has periodically occurring intervals during which the input waveform is quiescent. Where synchronized clamping is feasible, it may be used to provide d-c restoration even when the positive and negative excursions of the signal fluctuate from cycle to cycle.

A synchronous clamping circuit is shown in Fig. 8-17. The signal is transmitted from input to output through the capacitor C_s. The two-diode circuit which is bridged between signal lead and ground serves the function of the switch S in Fig. 8-16. Two control-signal pulse trains v_1 and v_2 are required. These waveforms are identical in all respects except that one is the inverse of the other. The d-c levels of the waveforms are of no consequence.

This circuit may be analyzed in much the same manner as the d-c-restorer circuit, for the two circuits have many features in common. Such an analysis[1] reveals that, in the steady state, during the interval T_c the diodes are brought to conduction and that the voltage at the point A is the same as at point B. As in the d-c restorer, the diodes conduct briefly to supply to the capacitors C the charge lost through the resistors R during the nonconducting interval T_n. During T_n both diodes are back-biased, and the output signal lead is entirely free to follow the input signal. Suppose that at the end of an interval T_n the voltage at A is not equal to V_R. Then when the diodes are brought to the point of conduction, if it should happen that $V_A > V_R$, diode $D1$ will conduct, discharging capacitor C_s into capacitor C until $V_A = V_R$. If $V_A < V_R$, diode $D2$ will conduct until $V_A = V_R$.

For proper operation of the circuit it is required that $C \gg C_s$ and that $RC \gg T_n$. There is, however, an upper limit on R which results from the fact that during the interval T_n the capacitor must be able to discharge through R the charge it may have acquired from C_s during the interval T_c. When C_s discharges it does so through a diode forward resistance R_f and through the output impedance R_s of one of the generators that furnish the control signal. Hence it is also required that $C_s(R_f + R_s) \ll T_c$. The required minimum amplitude V_P of the clamping pulse is determined by the condition that neither diode be brought into conduction during the interval T_n by the signal. This condition leads to the restriction that $V_P > V_S$, where V_S is the peak signal excursion above or below the average value of the signal. Finally, we may note that if the clamping-pulse amplitudes are not equal or if the resistors

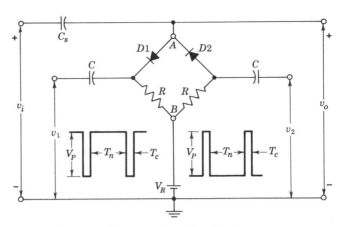

Fig. 8-17 A synchronous clamping circuit.

R are not equal, the circuit will not clamp to V_R but to a somewhat different voltage. All these matters are discussed in detail in Ref. 1.

8-8   THE TRANSISTOR OR TUBE AS A SWITCH

We have already noted that amplifier tubes (vacuum triodes and pentodes) and transistors are widely used in applications where they operate as switches. In these circuits, the tube or transistor is placed in series with the load and a supply voltage.   The tube is then driven by an external signal between cutoff and clamp, or the transistor is driven between cutoff and saturation.   The waveforms generated in such nonlinear switching circuits are of interest and are now to be considered.   The principles already established in connection with clipping and clamping circuits are useful in studying these switching circuits.

A transistor switching circuit is shown in Fig. 8-18a.   In the absence of a signal, the transistor would be held in saturation by virtue of the connection of the base to the supply voltage through R.   The signal v_s in Fig. 8-18c is applied to the base through the capacitor C from a source of resistance R_s. This signal, of peak-to-peak amplitude V, may be described variously as a square wave, a train of negative pulses (of width T_2 separated by intervals T_1), or a train of positive pulses (of width T_1 separated by intervals T_2).   In the present instance, because of the application involved, the signal is commonly referred to as a *gating waveform* or as a train of negative gating pulses.   The transistor is viewed as a "gate" which opens and closes, and the input waveform is the agency through which the transistor is so operated.   The tube circuit of Fig. 8-18b is similar and similarly operated.   In the absence of a signal the tube would be maintained in clamp by virtue of the connection of the

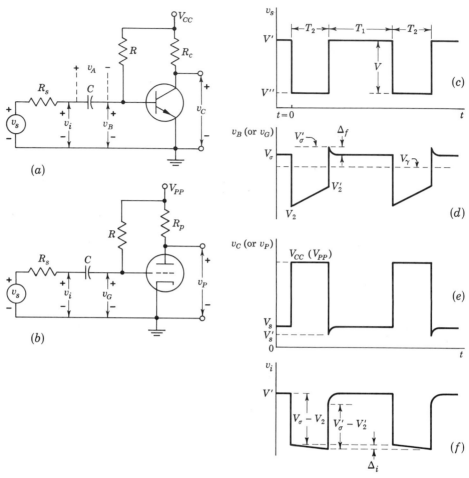

Fig. 8-18 (a) Transistor switch; (b) vacuum-tube switch; (c) input waveform; (d) base (or grid) waveforms; (e) collector (or plate) waveforms; (f) waveform v_i at junction of R_s and C.

grid to the supply voltage through the resistor R. The input gating waveform drives the tube from clamp to cutoff and back again.

Base (Grid) Waveform   If we view the base (grid) and emitter (cathode) terminals of the transistor (tube) as the terminals of a diode, then it is evident that the base (grid) circuit is precisely the clamping circuit of Fig. 8-12. The waveform at the base (grid) will appear as in Fig. 8-18d, which should be compared with Fig. 8-8. In these two figures we have considered that the time constant with which the capacitor C charges at the positive extremity of the signal is small in comparison with the interval T_1 and that the base

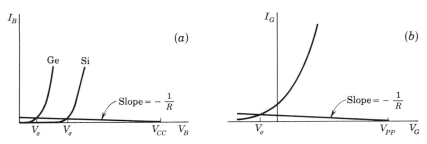

Fig. 8-19 Load-line construction to find quiescent clamping voltage (a) at base of Ge or Si transistor, (b) at grid of triode.

(grid) voltage returns to the level at which it would remain constantly in the absence of an input signal. This saturation (clamping) level V_σ will ordinarily be some tenths of a volt negative in the case of a thermionic tube and will be approximately 0.3 or 0.7 V for a germanium or silicon transistor (page 219). If there were available the base-emitter or the grid-cathode volt-ampere characteristics, and there were an interest in so doing, we might find the clamping voltage V_σ more precisely by the load-line construction of Fig. 8-19.

We shall now calculate the voltage levels of the base waveform v_B of Fig. 8-18d. Since the signal v_s is at the level V' just before the onset of the interval T_2 (at $t = 0-$) and the base is at the clamped or saturation level V_σ, then the voltage across the capacitor is $V' - V_\sigma$. At $t = 0+$, immediately after the abrupt negative drop, $v_s = V''$, and the capacitor voltage remains $V' - V_\sigma$ because the voltage across a capacitor cannot change instantaneously. We assume that the abrupt change V in v_s is large enough to drive the transistor below the cutin level V_γ. We shall neglect the small base current at cutoff. The equivalent circuit from which to calculate v_B is indicated in Fig. 8-20, where at $t = 0+$, $v_A = V' - V_\sigma$. The generator source impedance may be different during the two portions of the square wave. Hence we have indicated in Fig. 8-20 the symbol R_{s2} to represent the value of the source resistance during the interval T_2. By using the principle of superposition and remembering that $V = V' - V''$, we may calculate that the base drops to the voltage

$$v_B(0+) = (V_\sigma - V)\frac{R}{R + R_{s2}} + V_{CC}\frac{R_{s2}}{R + R_{s2}} \equiv V_2 \qquad (8\text{-}23)$$

Note that this result is independent of the d-c level of the input signal. Equa-

Fig. 8-20 The input equivalent circuit for the transistor base circuit of Fig. 8-18a during the time that the transistor is cut off. At $t = 0+$, $v_A = V' - V_\sigma$.

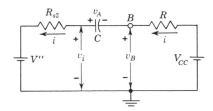

tion (8-23) is valid only if the base is driven below the cutin level, that is, if $V_2 < V_\gamma$.

If it should happen that $R_{s2} \ll R$, then $V_2 \approx V_\sigma - V$, and the change in voltage at the base is $V_\sigma - V_2 = V$, which just equals the abrupt drop in the gating waveform v_s. At the other extreme consider the case in which, say, $R_{s2} = R$, so that

$$V_2 = \tfrac{1}{2}(V_\sigma - V + V_{CC}) \tag{8-24}$$

The condition $V_2 < V_\gamma$ requires that

$$V > V_{CC} + V_\sigma - 2V_\gamma \tag{8-25}$$

Since $V_\sigma - 2V_\gamma$ is usually small compared with the supply voltage V_{CC}, then, approximately, $V > V_{CC}$. In other words, for $R_{s2} = R$, the peak-to-peak voltage must be at least as large as the supply voltage in order to bring the transistor just below the cutin point. This same conclusion is reached in Sec. 8-5. It is also to be noted, in the above situation, that any drop in input voltage beyond $V = V_{CC}$ will appear at the base attenuated by a factor of 2.

During the interval T_2 the voltage at the base rises asymptotically toward V_{CC}. Using Eq. (2-3) we may calculate V_2' from

$$V_2' = V_{CC} - (V_{CC} - V_2)\epsilon^{-T_2/\tau} \tag{8-26}$$

in which $\tau = C(R + R_{s2})$. We have assumed in Fig. 8-18 that the voltage V_2' is below the cutin level V_γ.

During the interval T_2 the capacitor C charges and the voltage v_B rises, bringing v_B closer to its initial level V_σ. Therefore, when at $t = T_2$ the input rises abruptly by amount V, it carries the base above its initial level V_σ. Consequently, as shown in Fig. 8-18d, an overshoot appears in the v_B waveform. We shall now calculate the amplitude Δ_f of this overshoot. To make this calculation we first find the change Δv_A in the voltage across the capacitor C in the interval T_2.

During the interval T_2 the base side of the capacitor C in Fig. 8-20 *rises* in voltage by the amount $V_2' - V_2$. Consequently, the current i through R falls by the amount $(V_2' - V_2)/R$, as does also the current through R_{s2}. Hence the decrease in voltage drop across R_{s2} is $R_{s2}(V_2' - V_2)/R$, and the voltage v_i at the generator side of the capacitor *falls* by this same amount. The total change in voltage across C is

$$-\Delta v_A = (V_2' - V_2) + \frac{R_{s2}}{R}(V_2' - V_2) = (V_2' - V_2)\left(1 + \frac{R_{s2}}{R}\right) \tag{8-27}$$

At $t = T_2+$, immediately after the input v_s has returned to its initial level, the circuit is different from its initial state only in that the voltage Δv_A exists across the capacitor C and that the source resistance is now R_{s1}. The *change* from the initial state which results from this voltage change Δv_A may be calculated from the circuit of Fig. 8-21. We have assumed that the incremental input impedance of the transistor is the base-spreading resistance $r_{bb'}$. The

Fig. 8-21 An equivalent circuit from which
to calculate the overshoot Δ_f at $t = T_2+$.

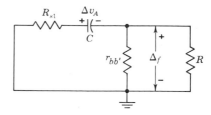

change in voltage across $r_{bb'}$ is the overshoot. From Fig. 8-21 and using Eq.
(8-27) we find, subject to the condition $R \gg r_{bb'}$, that

$$
\Delta_f \approx -\Delta v_A \frac{r_{bb'}}{R_{s1} + r_{bb'}}
$$

$$
= (V_2' - V_2) \frac{R + R_{s2}}{R} \frac{r_{bb'}}{R_{s1} + r_{bb'}} \tag{8-28}
$$

Very commonly $R_{s2} \ll R$ and $r_{bb'} \ll R_{s1}$. In this case we have

$$
\Delta_f \approx (V_2' - V_2) \frac{r_{bb'}}{R_{s1}} \tag{8-29}
$$

The peak value of v_B is $V_\sigma' = V_\sigma + \Delta_f$. The overshoot decays with a time
constant

$$
\tau' = (R_{s1} + r_{bb'})C \tag{8-30}
$$

In the case of the vacuum tube, $r_{bb'}$ is replaced by r_G, the grid resistance.

Collector (Plate) Waveform Consider the waveform of Fig. 8-18e. During the interval T_2, when the transistor is below cutoff, its collector voltage
is $v_C = V_{CC}$ (neglecting the small reverse saturation current). At $t = 0-$,
when the transistor is in saturation, $v_C \equiv V_s = V_{CE}(\text{sat})$. This saturation voltage is obtained by drawing a load line corresponding to V_{CC} and
R_c on the collector characteristics and locating the intersection for a base
current $I_B = (V_{CC} - V_\sigma)/R$. A good approximation for $V_{CE}(\text{sat})$ is given
in Table 6-1 and is 0.3 V for a silicon and 0.1 V for a germanium transistor.
At $t = T_2+$, when the base overshoots, there will be a small undershoot in
collector voltage. Since v_C cannot go negative, V_s' differs from V_s by a maximum of 0.1 V for germanium and 0.3 V for silicon. In many cases when the
collector waveform is observed on a cathode-ray oscilloscope the undershoot
will be small enough so that it will not be apparent unless special pains are
taken to observe it.

For a tube, the plate voltage equals V_{PP} during the interval T_2, when it is
cut off. At $t = 0-$, when the tube is in clamp, its grid voltage is $V_\sigma \approx 0$.
The plate voltage V_s is found by drawing on the plate characteristics the load
line for R_p and V_{PP} and locating the intersection for $V_G = 0$. Corresponding
to the grid overshoot at $t = T_2+$ there is a pronounced plate undershoot.
This value of plate voltage V_s' is now found by drawing a load line on the plate

characteristics and locating the intersection at $V_G = \Delta_f$. Since the grid over-shoot carries the grid into the region of positive grid voltages, a set of characteristics for such positive voltages is required. The plate undershoot will be larger than the grid overshoot by the gain of the tube. And the tube gain will be even larger in the neighborhood of clamp than it is in the tube's grid base. This increased gain results from the fact that with increasing current the tube transconductance increases and the plate resistance decreases. Since Δ_f is of the order of a few volts, then $V_s - V'_s$ is of the order of tens of volts.

Waveform v_i Consider the waveform v_i, which is the voltage to ground of the junction of R_s and the capacitor. The upper level of this waveform is identical with the upper level V' of v_s, since at $t = 0-$ there is no current in C. At the beginning of the interval T_2 the voltage v_B drops abruptly by $V_\sigma - V_2$, as does the voltage v_i on the other side of the capacitor. The voltage v_i drops less than does the input voltage v_s because of the voltage drop across R_{s2} that results from the current which is charging the capacitor C. As the capacitor charges, the capacitor current decreases, the drop across R_{s2} becomes smaller, and therefore during the interval T_2 the voltage v_i falls. As we saw earlier, in the discussion leading to Eq. (8-27), the change Δ_i in v_i is

$$\Delta_i = (V'_2 - V_2)\frac{R_{s2}}{R} \qquad (8\text{-}31)$$

At the end of the interval T_2 there is an abrupt jump in v_i by $V'_\sigma - V'_2$ to match the abrupt jump in v_B on the other side of the capacitor. This jump in v_i leaves v_i short of returning to its starting point on account of the fact that during T_2 the capacitor has been charging through R_{s2} and R. For $t > T_2$ the capacitor discharges through R_{s1} and the input diode. As the capacitor discharges, the collector (plate) undershoot decays with the same time constant τ' [Eq. (8-30)] as does the base (grid) overshoot and the voltage v_i rises to its starting point asymptotically with this same time constant.

EXAMPLE   In the circuit of Fig. 8-18a the transistor is an n-p-n germanium triode with characteristics similar to those of the p-n-p type 2N404, $V_{CC} = 10$ V, $R_c = 500\ \Omega$, $R = 40$ K, $C = 0.1\ \mu\text{F}$, $R_{s1} = R_{s2} = 10$ K, $V = 10$ V, and $T_2 = 1.0$ msec. Find the voltage levels of all the waveforms of Fig. 8-18, assuming $r_{bb'} = 100\ \Omega$.

Solution   In the quiescent state the transistor is in saturation. Assuming the transistor in saturation, we estimate from Table 6-1 that $V_{BE}(\text{sat}) = V_\sigma = 0.3$ V. Then the base current is $I_B = (10 - 0.3)/40 = 0.24$ mA. We observe from Fig. 6-21 that such a base current is adequate to keep the transistor in saturation. From Eq. (8-23) we find

$$V_2 = (0.3 - 10)\frac{40}{40 + 10} + 10\frac{10}{40 + 10} = -5.8\ \text{V}$$

From Eq. (8-26)

$$V_2' = 10 - (10 + 5.8)\epsilon^{-1.0/5.0} = -2.9 \text{ V}$$

From Eq. (8-28), the base overshoot is

$$\Delta_f = (-2.9 + 5.8) \frac{40 + 10}{40} \frac{0.1}{10.1} = 0.04 \text{ V}$$

and

$$V_\sigma' = V_\sigma + \Delta_f = 0.3 + 0.04 \approx 0.3 \text{ V}$$

In the waveform for v_C we find the upper level at 10 V. From Table 6-1 the lower level is $V_{CE}(\text{sat}) = 0.1$ V. Or if we refer to Fig. 6-21 we read $V_{CE}(\text{sat}) \approx 0.12$ V. Considering, however, that Fig. 6-21 gives *average* characteristics, in any particular case the former estimate from Table 6-1 is probably as reliable as the number read from the characteristic curves. The level V_s' lies somewhere between 0 and 0.1 V and cannot be determined with any precision from average characteristics.

The changes in level in v_i are

$$V_\sigma - V_2 = 0.3 + 5.8 = 6.1 \text{ V}$$

$$V_\sigma' - V_2' = 0.3 + 2.9 = 3.2 \text{ V}$$

and from Eq. (8-31)

$$\Delta_i = (-2.9 + 5.8)\tfrac{10}{40} = 0.7 \text{ V}$$

The recovery time constant is

$$\tau' = (R_{s1} + r_{bb'})C = (10.1 \times 10^3)(0.1 \times 10^{-6}) \text{ sec} \approx 1 \text{ msec}$$

In order for the transient for $t > T_2$ to die down in the interval T_1, as indicated in Fig. 8-18, we must have $T_1 \gg \tau'$. In the present example, since $T_2 \approx \tau'$, then it is necessary that $T_1 \gg T_2$.

EXAMPLE In the circuit of Fig. 8-18b the tube is a type 5965, $V_{PP} = 200$ V, $R = 0.5$ M, $R_{s1} = R_{s2} = 20$ K, $R_p = 10$ K, $C = 0.01$ μF, $T_2 = 1.0$ msec, and $V = 100$ V. Find the voltage levels of all the waveforms of Fig. 8-18.

Solution The grid current when the grid rests quiescently at clamp is $200/0.5 = 400$ μA. From Fig. 6-31 and from the discussion of Sec. 6-19 we find that at this grid current the grid is close to 0 V. In a tube circuit the voltages encountered are of the order of hundreds of volts. Therefore the 0.1- or 0.2-V departure of the clamping voltage in either direction from zero normally encountered is of no great consequence and is often neglected. From Eq. (8-23) we see that

$$V_2 = -100 \frac{500}{500 + 20} + 200 \frac{20}{500 + 20} = -88 \text{ V}$$

and from Eq. (8-26)

$$V_2' = 200 - (200 + 88)\epsilon^{-(1.0/5.2)} = -38 \text{ V}$$

which is more than adequate to keep the tube cut off.

From Fig. 6-33 we estimate $r_G = 0.25$ K.   From Eq. (8-28), with $r_{bb'}$ replaced by r_G,

$$\Delta_f = (-38 + 88) \frac{500 + 20}{500} \frac{0.25}{20 + 0.25} = 0.7 \text{ V}$$

Turning now to the waveform for v_P, we have an upper level $V_{PP} = 200$ V.   To find the lower level we draw the load line for 200 V and $R_p = 10$ K on the plate characteristics of Fig. 6-31, and we find that $V_s = 80$ V at $V_G = 0$ and that the minimum plate voltage at $V_G = 0.7$ V is $V_s' \approx 60$ V.   Thus a 0.7-V grid overshoot has given rise to a 20-V undershoot, or a gain of 30 in the positive grid region. On the waveform for v_i we find that the drop is

$$V_\sigma - V_2 = 88 \text{ V}$$

Since $V_\sigma \approx 0$, then $V_\sigma' \approx \Delta_f$ and at $t = T_2+$ the rise is

$$V_\sigma' - V_2' = 0.7 + 38 = 38.7 \text{ V}$$

and the tilt of the bottom is, from Eq. (8-31),

$$\Delta_i = (-38 + 88)\tfrac{20}{500} = 2.0 \text{ V}$$

The time constant with which the overshoots decay is

$$\tau' = (20.3 \times 10^3)(0.01 \times 10^{-6}) \text{ sec} = 0.20 \text{ msec}$$

8-9   TWO–STAGE OVERDRIVEN AMPLIFIER

A circuit configuration which is frequently encountered is shown in Fig. 8-22. Here the transistor switch $Q2$ is driven not directly by a gating signal but rather by the output of a preceding transistor $Q1$.   The transistor $Q1$ has applied to it a gating signal which drives this transistor alternately from cutoff to clamp and back again.   Accordingly, in the absence of the loading effect of transistor $Q2$, the signal at the collector of $Q1$ would be a gating wave-

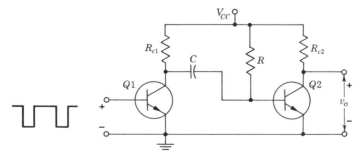

Fig. 8-22 A transistor switch capacitively coupled to and driven by a preceding transistor switch.

form of the same form as the input signal except inverted in polarity. This circuit may now be analyzed in the same manner as we analyzed the circuit of Fig. 8-18. We need only take into account that the second transistor $Q2$ is being driven by a gating source whose output impedance is not constant. That is, when $Q1$ is cut off, the output impedance of $Q1$ is equal to the collector-circuit resistance R_{c1}. When transistor $Q1$ is in saturation the output impedance is the parallel combination of R_{c1} and the collector saturation resistance R_{CE} and hence nominally equal to R_{CE}. Therefore one may repeat all the calculations performed above, taking into account that $R_{s1} = R_{c1}$ and $R_{s2} = R_{CE}$. The waveform v_i in Fig. 8-18 corresponds in Fig. 8-22 to the waveform to be found at the collector of $Q1$. Of course, a circuit involving two vacuum tubes in the manner that Fig. 8-22 involves two transistors may be similarly analyzed. For the tube circuit, $R_{s1} = R_{p1}$ and R_{s2} is the parallel combination of R_{p1} and r_{p1}.

EXAMPLE In the circuit of Fig. 8-22, both transistors are germanium, $V_{CC} = 10$ V, $R_{c1} = 10$ K, $R_{c2} = 500$ Ω, $R = 40$ K, and $C = 0.1$ μF. The square wave which is applied to the base of $Q1$ makes excursions between such levels that transistor $Q1$ is turned ON and OFF. The ON period of $Q1$ is $T_2 = 1.0$ msec. Find the voltage levels of the waveforms at the collector of $Q1$ and at the base and collector of $Q2$.

Solution The waveform v_{C1} at the collector of $Q1$ is similar to the waveform v_i in Fig. 8-18f, and the waveforms at the base and collector of $Q2$ are similar to the waveforms v_B and v_C in Fig. 8-18d and e. When $Q1$ goes ON, v_{C1} drops from $V_{CC} = 10$ V to $v_{C1} = 0.1$ V, the collector saturation voltage. Therefore using, in Eq. (8-23), $V = 10 - 0.1 = 9.9$ V and setting $R_{s2} = R_{CE} \approx 0$, we find, with $V_\sigma = 0.3$ V,

$$V_2 = (0.3 - 9.9) = -9.6 \text{ V}$$

From Eq. (8-26)

$$V_2' = 10 - (10 + 9.6)\epsilon^{-(1.0/4.0)} = -5.3 \text{ V}$$

At the time of the generation of the overshoot, $Q1$ is OFF and $R_{s1} = R_{c1} = 10$ K. From Eq. (8-28), using $r_{bb'} = 100$ Ω,

$$\Delta_f = (-5.3 + 9.6)\frac{0.1}{10.1} = 0.04 \text{ V}$$

and

$$V_\sigma' = V_\sigma + \Delta_f \approx 0.3 \text{ V}$$

The waveform v_{C1} at the collector of $Q1$ starts at $V_{CC} = 10$ V and drops to 0.1 V when $Q1$ goes ON. Since $R_{s2} \approx 0$, we have from Eq. (8-31) that $\Delta_i \approx 0$, so that v_{C1} exhibits no tilt but is constant at $v_{C1} = 0.1$ V in the interval T_2 during which $Q1$ is ON. At the end of this interval v_{C1} jumps by the amount

$$V_\sigma' - V_2' = 0.3 + 5.3 = 5.6 \text{ V}$$

The waveform at the collector of $Q2$ has the form of Fig. 8-18e, with $V_{CC} = 10$ V, $V_s = 0.1$, and V_s' somewhere between 0 and 0.1 V. The overshoot in v_{B2}, the undershoot in v_{C2}, and the approach of v_{C1} to its initial level all take place with a time constant given in Eq. (8-30) as $\tau' = (10.1 \times 10^3)(0.1 \times 10^{-6}) \approx 1$ msec.

8-10   SWITCH WITH INDUCTIVE LOAD

In Fig. 8-23a and b are shown a transistor and a tube switch in which the load is an inductor L shunted by a resistor R. A gating waveform of arbitrary d-c level (Fig. 8-24a) is applied, and, because of the clamping in the base and grid circuits, the transistor is carried from saturation to cutoff and the tube from clamp to cutoff. An additional resistor R_c has been included in the collector circuit, since otherwise when the transistor is in saturation the collector current will become intolerably large.

After the transistor and tube have been in saturation or clamp for a long enough time so that all transients have decayed, each inductor will

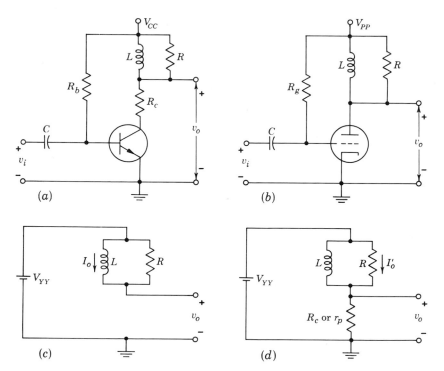

Fig. 8-23 (a) A transistor switch driving an inductor; (b) a tube switch driving an inductor; (c) equivalent circuit for finding the output when the switch is open (V_{YY} represents either V_{CC} or V_{PP}); (d) equivalent circuit when the switch is closed.

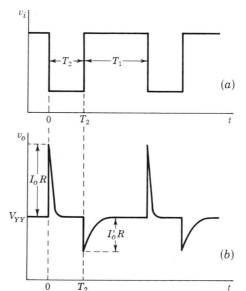

Fig. 8-24 (a) The input to the switches of Fig. 8-23; (b) the output voltage.

be carrying a current I_o. In the case of the transistor the current will be $[V_{CC} - V_{CE}(\text{sat})]/R_c$. In the case of the tube the current may be found from the plate characteristics as the current corresponding to a plate voltage V_{PP} and a grid voltage nominally equal to zero. At the moment the switches are driven to cutoff, the equivalent circuit from which to calculate the output voltage is given in Fig. 8-23c, in which the initial condition that the inductor current is I_o has been noted. We have neglected the small leakage current of the transistor when it is in the cutoff condition. The inductor current, which must flow through R, decays from I_o to zero with a time constant L/R. The output voltage is as shown in Fig. 8-24b and is given by

$$v_o = V_{YY} + I_o R \epsilon^{-Rt/L} \tag{8-32}$$

Thus, driving the switch to cutoff develops at the collector or plate a positive spike of amplitude $I_o R$ superimposed on the supply voltage. This spike may become very large for large R, and this possibility is a cause for concern, particularly in the case of transistors, where the maximum collector breakdown voltage must not be exceeded. In practice, the peak voltage may be limited by stray capacitance across the inductor, but even in this case a peak value several times the supply voltage may be obtained. A difficulty may arise from a rather unexpected source. Suppose, for example, that a d-c milliammeter has been introduced into the collector circuit of a transistor resistive switch to monitor the collector current. Then, because of the inductance of the meter coil, the simple act of abruptly turning off the collector current may damage the transistor.

When the switches are returned to saturation or clamp again at the time $t = T_2$, the inductor acts initially as an open circuit. The transistor current and hence the current through R is $[V_{CC} - V_{CE}(\text{sat})]/(R + R_c)$. The tube current may be found by drawing on the tube characteristics a load line passing through V_{PP} corresponding to a resistance R and noting the point of intersection of this load line with the tube characteristic for a grid-to-cathode voltage of Δ_f.

Let I'_o represent the transistor or tube current which flows at the moment the switches are turned ON. The current through R will decay from I'_o to zero with a time constant L/R', in which R' is the parallel combination of R and R_c (or of R and the tube plate resistance r_p). The output for $t > T_2$ is

$$v_o = V_{YY} - I'_o R \epsilon^{-R'(t-T_2)/L} \tag{8-33}$$

and has the form of the negative spike shown in Fig. 8-24b superimposed on the supply voltage. The negative spike is always smaller than the supply voltage because the collector or plate voltage cannot reverse in polarity. The time constant of the negative spike is larger than that of the positive spike and therefore the negative spike decays more slowly. The output waveform of Fig. 8-24 should be compared with the waveform of the peaking circuit of Fig. 2-6, where the active device operates linearly.

The above discussion has neglected the capacitance shunting the output terminals. If this capacitance is taken into account, we may have responses similar to those discussed in Sec. 2-10. For example, if the circuit is underdamped, each pulse in Fig. 8-24 will be converted into a train of damped oscillations. Because the damping is greater when the tube or transistor conducts than when they are not conducting, it is possible to have oscillations near $t = 0$ instead of a single positive peak and yet have a single negative spike at $t = T_2$.

8-11   DAMPER DIODES

If, in the peaking circuit just discussed, it is desired to have only positive output pulses, we may connect a diode across the coil as indicated in Fig. 8-25a. If the output voltage falls below V_{YY}, the diode conducts and the small forward resistance of the diode quickly damps out this portion of the waveform. If in Fig. 8-24 there were oscillations in the vicinity of $t = 0$, the diode of

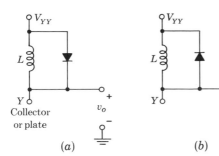

Fig. 8-25 (a) The damper diode allows the output to go only positive with respect to V_{YY}. (b) The diode allows the output to go only negative with respect to V_{YY}.

Fig. 8-25a would allow only the *first* positive peak to appear in the output because of the heavy damping which it imposes on the ringing circuit.   This action accounts for the name *damper diode*.

If the damper diode is inserted across the peaking coil with the polarity indicated in Fig. 8-25b, then the output will contain a *single* negative peak in the vicinity of $t = T_2$.   In this case the diode conducts whenever the output rises above V_{YY}.

8-12   SWITCH WITH CAPACITIVE LOAD

A switch with a capacitive load behaves somewhat differently depending upon whether the active device is a tube or transistor.   We shall first consider the vacuum-tube circuit of Fig. 8-26a, including the shunt capacitance C_s.

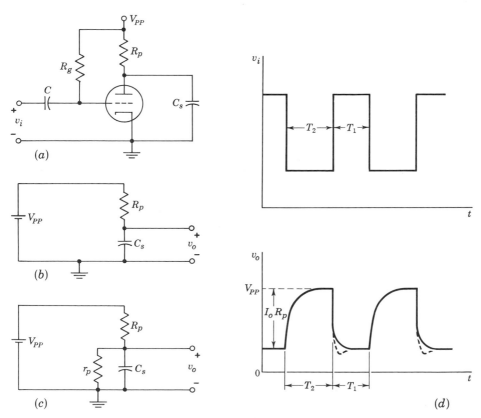

Fig. 8-26 (a) A tube switch with a capacitive load; (b) equivalent circuit for calculating output when the tube is OFF; (c) equivalent circuit when the tube is ON; (d) the input and output waveforms.

The Capacitively Loaded Tube Switch   At $t = 0-$, when the tube is in clamp, the plate current I_o is found from the intersection of the load line (corresponding to V_{PP} and R_p) and the plate characteristic for $V_G = 0$. The plate voltage at this time is therefore $V_{PP} - I_o R_p$, as indicated in Fig. 8-26d. At $t = 0+$, when the switch is driven to cutoff, the equivalent circuit from which to determine the plate waveform is given in Fig. 8-26b. The plate rises toward V_{PP} exponentially with a time constant $R_p C_s$. When, on the other hand, the switch has been driven into clamp during the interval T_1, the equivalent circuit from which to calculate the fall of the plate voltage is shown in Fig. 8-26c. The output falls with a time constant $R'C_s$, where R' is the parallel combination of R_p and the dynamic plate resistance r_p. From the waveform of Fig. 8-26d it is apparent that the waveform falls more rapidly than it rises, since $R' < R_p$.

If the grid resistance r_G is taken into account and if the input time constant is not large enough compared with T_2, then there will be an overshoot at the grid at $t = T_2+$ and an undershoot at the plate (as indicated in Fig. 8-18e). The dashed portion in Fig. 8-26d corresponds to this situation.

The Capacitively Loaded Transistor Switch   Consider the circuit indicated in Fig. 8-27a. At $t = 0-$, when the transistor is in saturation, the output voltage v_o and the collector current i_C are given by

$$v_o = V_{CE}(\text{sat}) \qquad \text{and} \qquad i_C = \frac{V_{CC} - V_{CE}(\text{sat})}{R_c} \equiv I_o$$

These quantities are indicated in Fig. 8-27d. At $t \geq 0+$, when the switch is driven to cutoff, the equivalent circuit from which to determine the collector voltage is given in Fig. 8-27b. The collector rises toward V_{CC} with a time constant $R_c C_s$.

At $t = T_2+$ the input rises abruptly, the base-emitter diode is forward-biased by V_σ (a few tenths of a volt), and a base current $I_B = (V_{CC} - V_\sigma)/R_b$ flows. The collector voltage at this time is the voltage V_{CC} across C_s. Since this voltage reverse-biases the collector junction, the transistor remains in its active region and a collector current $I_C \approx h_{FE}I_B \equiv I'_o$ results. The equivalent circuit from which to calculate the discharge of the capacitor is indicated in Fig. 8-27c, where I'_o is a constant. The output voltage starts at a value V_{CC} and falls exponentially with a time constant $R_c C_s$ toward a steady-state value of $V_{CC} - I'_o R_c$. For $t \geq T_2$ and until the transistor returns to saturation, the output is given by

$$v_o = V_{CC} - I'_o R_c + I'_o R_c \epsilon^{-(t-T_2)/R_c C_s} \tag{8-34}$$

These conditions are pictured in Fig. 8-27d, where we have indicated that $I'_o > I_o$. This condition follows from the fact that I_o is the saturation collector current and hence that $I_o < h_{FE}I_B$, whereas $I'_o = h_{FE}I_B$. Therefore $I'_o > I_o$.

When the collector falls to $V_{CE}(\text{sat})$ the transistor enters into saturation, the collector current drops to its saturation value I_o, and v_o remains at

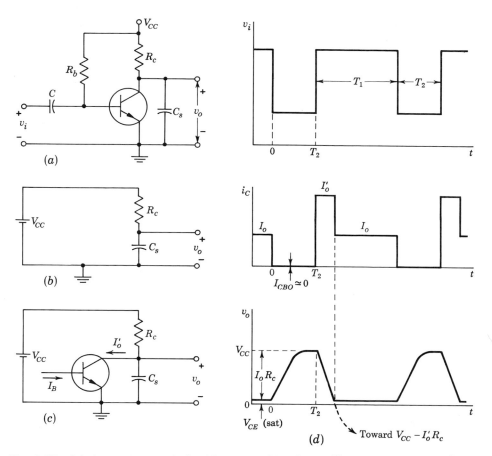

Fig. 8-27 (a) A transistor switch with a capacitive load; (b) equivalent circuit for calculating the output when the transistor is OFF; (c) equivalent circuit when the base circuit is driven into clamp but the collector circuit is still in its active region; (d) the input voltage v_i, the collector current i_C, and the output voltage v_o waveforms. The dashed portion of v_o is an exponential which asymptotically approaches $V_{CC} - I_o'R_c$.

$V_{CE}(\text{sat})$, as indicated in Fig. 8-27d. If $I_o' \gg I_o$, then v_o falls almost linearly with time and the fall time is much smaller than the rise time.

In drawing the waveforms in Figs. 8-26d and 8-27d we have implicitly assumed that the time constants were small compared with T_2 or T_1. If, instead, we assume that R_pC_s (or R_cC_s) is much greater than T_2, then the exponential rise can be approximated by a linear increase. The circuit behaves as an integrator and a step input is converted into a ramp output. This ramp-generating circuit is discussed in detail in Sec. 14-5.

The effect of a **capacitance shunting across a** set of terminals of resistive output impedance is to cause a rounding of what would otherwise be an abrupt jump in voltage. We have already observed in Sec. 2-5 that such is the case. In the present circuits we note the additional feature that because of the switch nonlinearity the rounding is more pronounced on the rising edge of the waveform than on the falling edge. We should now also observe that even when shunt capacitances are not deliberately introduced, such stray shunt and wiring capacitance is always present. Therefore, in retrospect, all of the voltage changes that have been indicated as abrupt and discontinuous (for example, in Fig. 8-18) will in practice be observed to be somewhat rounded. If, however, the time constants of the rise and fall are small in comparison with the intervals T_1 and T_2, the rounding will be little noticed.

If a *p-n-p* transistor is used in Fig. 8-27, a negative supply is required and the waveforms in Fig. 8-27d are inverted. In this case the falling edge is slower (more rounded) than the rising edge.

8-13   PLATE AND COLLECTOR CATCHING DIODES

Diodes used at the plate or collector of a tube or transistor to limit the plate or collector excursion are called *catching diodes*. One application of catching diodes is shown in Fig. 8-28. A tube switch has a capacitive load as in Fig. 8-28a and for this reason, in the absence of the diodes, the output would have the waveform shown dashed in Fig. 8-28b. However, diode $D1$ restrains v_o from rising above V_1, and diode $D2$ prevents v_o from dropping below V_2. The output waveform is as shown by the solid curve. The waveform is now reasonably square, the rise and fall times having been reduced appreciably. For example, assume that diode $D1$ was used alone and that V_1 was adjusted to reduce the output voltage amplitude by only 20 percent. In this case a simple calculation shows that the rise time would be reduced to approximately one-half its original value. Finally we note that in the absence of the catching

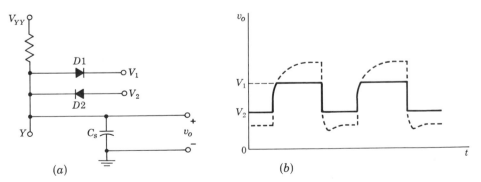

Fig. 8-28   (a) Plate or collector catching diodes; (b) squaring of waveform of Fig. 8-26d due to diodes.

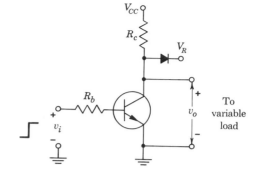

Fig. 8-29   A collector catching diode
is used to maintain constant output
swing in the presence of a variable
load.

diodes the limits on the waveform are approached asymptotically (except for
the fall in Fig. 8-27d), so that, in principle, the limits are never really attained.
With the catching diodes the limits are reached in a finite time.

A further application of a catching diode is shown in Fig. 8-29.   A transistor switch is to be called upon to provide a signal for a variable resistive load.
Such a situation is of interest in digital-computer circuitry, where the output
of one switch must provide the input signal to a variable number of succeeding
switches.   As is ordinarily the case in computers, the coupling from stage to
stage is direct (d-c), such as has been indicated in the figure.   The levels of the
input signal v_i are assumed appropriate to carry the transistor from cutoff to
saturation.   In the absence of the diode, the output voltage at cutoff and
hence the swing at the collector would depend on the load.   For example, if
there is no load the swing will be nominally equal to V_{CC}.   If, on the other hand,
the load resistor equals R_c, then the swing is only one-half V_{CC}.   In the presence
of the diode, however, the situation is different.   In the absence of a load the
collector will rise to V_R when the transistor stops conducting, and a current
$(V_{CC} - V_R)/R_c$ will flow through the diode.   As load current is now drawn,
the diode current will reduce correspondingly, but the collector voltage at
cutoff will remain constant at V_R.   The output voltage will remain constant
until the load current has become so large that the diode current has dropped
to zero.   The net result is that, up to a limit, the output swing at the collector
is independent of the number of other switches to which it must "fan out"
and that this swing is $V_R - V_{CE}(\text{sat})$.

8-14   NONSATURATING SWITCHES

The limitation on speed imposed by the storage time of a saturated transistor
prompts us (Chap. 20) to inquire about the operation of a switch which is not
permitted to attain saturation.   We consider now a number of such nonsaturating transistor switches.

An obvious scheme for avoiding saturation is indicated in Fig. 8-30.
When the input is at its lower voltage the transistor is cut off, the collector

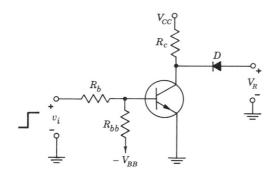

Fig. 8-30   The diode D is used to prevent the transistor from entering the saturation region. Note that $V_{CC} \gg V_R \approx 1$ V.

is at V_{CC}, and diode D does not conduct if $V_R < V_{CC}$. When the input v_i rises, the collector may fall only to the voltage V_R. At that voltage, diode D conducts and restrains the collector from falling further. The voltage V_R is small in comparison with V_{CC} but large enough (say ~ 1 V) to ensure that the collector junction will never become forward-biased. The usefulness of the circuit depends, of course, on the selection of a diode D which has a recovery time appreciably smaller than the recovery time of the transistor.

As noted earlier, the current gain of a transistor h_{FE} is a function of the particular sample of a transistor type as well as a function of temperature. The base current must be at least large enough so that even in the worst case of lowest h_{FE} the collector current is adequate to drop the collector voltage to V_R. The transistor is in the active region. Therefore when either the base current i_B increases above this minimum or, because of a temperature change or transistor replacement, h_{FE} increases, the collector current will increase even though the collector voltage does not. This rise in collector current increases the collector dissipation and the diode dissipation as well, because all the excess collector current passes through the diode. Altogether it is found that the method of Fig. 8-30 is not very reliable; it leads frequently to the destruction of both diode and transistor.

The principle of an improved scheme is presented in Fig. 8-31. The diode, which in Fig. 8-30 was returned to a fixed reference voltage, is here returned to the base through a voltage V_B. The collector voltage may drop freely until the diode $D1$ conducts. From this point the drop from collector to base will be

$$v_{CB} = -V_{D1} + V_B \tag{8-35}$$

in which V_{D1} is the diode voltage drop from p side to n side. Since we require that v_{CB} be positive (for an n-p-n transistor) to avoid saturation we require that

$$V_B > V_{D1} \tag{8-36}$$

even at the largest voltage across diode $D1$ corresponding to the largest current through it.

The circuit operates by virtue of the nonlinear feedback which exists between the collector and base. Suppose that the diode is conducting and

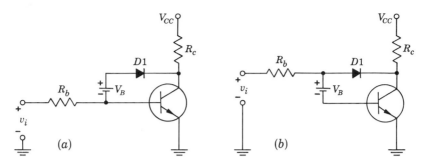

Fig. 8-31 (a) A nonlinear feedback method to restrain the transistor from entering saturation; (b) an alternative connection.

that, as a consequence of a change in h_{FE}, the collector current increases. This collector-current increment will flow in part through R_c and in part through the diode and then through R_b. The result will be a lowering of the collector voltage which will be transmitted to the base. The lowered base voltage will undo much of the effect of the increased collector current. Alternatively, we may describe the operation by saying that when the collector current increases because of an increase in h_{FE} the current through the diode increases. Therefore some of the base current is shunted away from the base.

　　As noted in Fig. 8-31, the driving current may be applied either directly to the base, as in Fig. 8-31a, or to the junction of the diode and voltage source, as in Fig. 8-31b. If the diode $D1$ is germanium, the voltage source V_B may be realized through the use of a forward-biased silicon diode. If a silicon diode is used for $D1$, two silicon diodes in series or an avalanche diode may be used for V_B.

　　One of a number of possible nonsaturating switches is shown in Fig. 8-32. Here diode $D1$ transmits the feedback signal from collector to base. Diode $D2$ supplies the voltage V_B. The additional diode $D3$ performs a function which is not restricted to nonsaturating switches and is often used in saturating switches as well. This diode ensures that initially, before the transistor is

Fig. 8-32 An example of a non-saturating switch.

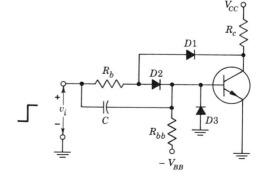

driven to conduction, the base of the transistor is not more negative than the forward voltage drop across $D3$. In this way, the delay time at the beginning of the response is kept small because the transistor is only slightly below cutoff. Also, $D3$ ensures that the breakdown voltage BV_{EB} is not exceeded. And finally, as we see in greater detail in Chap. 20, the capacitor C assists in ensuring that the switch responds rapidly to the input signal.

8-15   THE CATHODE FOLLOWER WITH A CAPACITIVE LOAD

One of the useful properties of a cathode follower is its ability to handle a large input signal. As an example, consider a cathode follower, as in Fig. 8-33a, using a type 12AT7 vacuum tube with a cathode resistance $R_k = 20$ K.   From the plate characteristics of Fig. D-4 it may be determined that the tube will operate within its grid base for the range of voltage of v_i from -107 to $+92$ V. At $v_i = -107$ V the tube is just at cutoff, and at $v_i = +92$ V, $V_{GK} = 0$. Further increase in v_i will result in the flow of large grid current. However, over this allowable range of grid voltage (two-thirds as large as the total supply voltage of 300 V), the cathode follower possesses high input impedance, low output impedance, excellent linearity, and stability. If the tube were used as a conventional amplifier, with the load in the plate circuit the allowable grid swing would be only 7 V, because with a supply voltage of 300 V the tube cuts off at $V_{GK} = -7$ V.

The output impedance R_o of a cathode follower (seen looking between cathode and ground) is equal to the parallel combination of the cathode resistor and a resistor $1/g_m$, where g_m is the tube transconductance. Rather commonly $R_k \gg 1/g_m$, so that $R_o \approx 1/g_m$. The output resistance so attained is rather small in comparison with resistances normally encountered in vacuum-tube circuits. For the 12AT7, $g_m \approx 3$ mA/V and $R_o = 330$ Ω. Other small tubes are available with $g_m = 10$ mA/V or higher, yielding $R_o = 100$ Ω or less. This low output resistance is advantageous when, as in Fig. 8-33a, the cathode

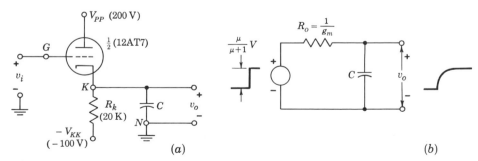

Fig. 8-33   (a) A cathode follower with capacitive load; (b) small-signal equivalent circuit for the purpose of finding the rise time of the output waveform.

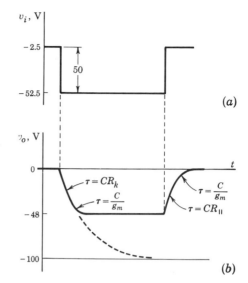

Fig. 8-34 Waveforms of (a) the input and (b) the output of the circuit of Fig. 8-33.

follower is called upon to drive a capacitive load and a fast signal must be handled. If the input v_i is a small step voltage of amplitude V, the equivalent circuit for calculating the output voltage is as shown in Fig. 8-33b. The output rises with the time constant $CR_o = C/g_m$ and for a fixed C is minimized by using a tube with the largest possible g_m.

We shall now see, however, that when the cathode follower is loaded by a capacitor, it may handle a signal which is large but slow or fast but small. It is not able to handle a signal which is simultaneously larger than the grid base and possesses a short rise-time edge. Assume, for example, that in Fig. 8-33a, $v_o = 0$, in which case we find that $v_i \approx -2.5$ V. Now let the input signal (Fig. 8-34a) consist of a negative step of magnitude 50 V, so that the grid drops to -52.5 V. Because of the capacitor C the cathode cannot respond immediately, and it remains at 0 V. For $V_{PK} = 200$ V, cutoff occurs at $V_{GK} = -6$ V. Hence the tube will be driven to cutoff. Since the tube is now not conducting, the output v_o will start to fall exponentially with a time constant CR_k rather than with a time constant C/g_m. We find, again from the plate characteristics of Fig. D-4, that the steady-state cathode voltage corresponding to $V_{GN} = -52.5$ V is $V_{KN} = -48$ V. The waveform of the falling edge of the cathode waveform, $v_o = v_{KN}$, is now as shown in Fig. 8-34b. The voltage starts asymptotically toward -100 V but is stopped at the -48-V level. Just slightly (about 6 V) before the -48-V level is reached the tube again enters its grid base, and the time constant associated with the fall changes from CR_k to C/g_m.

Now let us assume that after the output has attained its steady-state value $v_o = -48$ V, the input is returned abruptly from -52.5 V back to its -2.5-V level, so that the cathode is now to return to 0 V. Again, because the cathode voltage cannot change abruptly, the grid will now be driven into the positive grid region.   The output now rises as the capacitor C is charged both by the tube current and by current delivered directly to the output from the input signal source through the grid-to-cathode resistance r_g.   If $r_g \approx 1/g_m$ (an entirely reasonable possibility), then, of the current which charges the output capacitor, one-half will be supplied by the tube and one-half by the signal source.   This situation will result in a rapid charging of the capacitor but certainly defeats the purpose for which the cathode follower is being employed, since it no longer is a high-input-impedance device isolating source from load.   The rise of the waveform is shown in Fig. 8-34b.   The time constant is initially $CR_{\parallel}$, where $R_{\parallel}$ is the parallel combination of r_g and $1/g_m$. Near the end of the rise the tube enters its grid base, and the time constant becomes again C/g_m.

We have so far neglected the effect of the resistance R_s of the source which provides the input grid signal.   If this resistance is appreciable, then on the rising edge of the waveform the extent to which the grid will go positive with respect to the cathode will be limited and so also will be the amount of current which the tube will supply.   For this condition of very large R_s the grid will be clamped to the cathode and the output impedance of the tube will be its plate resistance r_p.   When R_s is taken into account, the output impedance of the cathode follower on the rising edge and until the tube returns to its grid base is (Prob. 8-39)

$$R_o = \frac{r_p(R_s + r_g)}{R_s + r_p + (\mu + 1)r_g} \tag{8-37}$$

In the limit when $R_s \to \infty$, $R_o = r_p$, and when $R_s = 0$, and for $\mu \gg 1$, $R_o \approx R_{\parallel}$.

8-16   THE EMITTER FOLLOWER WITH A CAPACITIVE LOAD

An emitter follower with capacitive load is shown in Fig. 8-35.   The resistor R_s represents the sum of the source resistance, the base-spreading resistance, and any other resistance added deliberately in series with the base.   In the active region the input and output are coupled by the forward-biased emitter junction.   The usefulness of the emitter follower results from the impedance transformation effected by the transistor.   As noted in Sec. 1-12, the input impedance is high and is given by Eq. (1-50) and the output impedance is low and is given by Eq. (1-54).

Neglecting the effect of the capacitor C, an emitter follower can handle a signal which approximates much more closely the supply voltage than is the case with a cathode follower.   With reference to Fig. 8-35 we note that one limit on the input is approximately 0 V, at which point the transistor would effectively be cut off.   The other limit is saturation.   When a silicon transistor is in saturation the collector-to-emitter voltage is about 0.3 V and the base-to-emitter voltage about 0.7 V, leaving 0.4 V between collector and base.   Therefore saturation will occur when the base-to-ground voltage is about 0.4 V

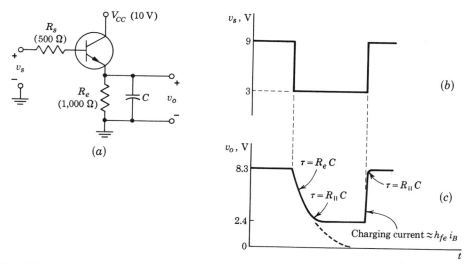

Fig. 8-35 (a) A capacitively loaded emitter follower; (b) the input and (c) the output waveforms.

larger in magnitude than V_{CC}. The input v_s becomes somewhat larger still because of the voltage drop in R_s. This result is to be compared with the situation in the cathode follower, where an input signal of the order of two-thirds of the supply voltage is obtained.

We consider now the output waveform of the circuit of Fig. 8-35 in response to the input signal shown, which makes abrupt transitions between the level 9 and 3 V. The waveforms are shown in Fig. 8-35b and c. Allowing a 0.6-V drop across the emitter junction in the active region and assuming $h_{fe} = 50$, so that the emitter current is 51 times the base current, we find that when $v_s = 9$ V, the drop across R_s is about 0.1 V, and $v_o = 8.3$ V. When the input drops abruptly to 3 V, the transistor is driven off because the emitter voltage is 8.3 V and hence the emitter junction is reverse-biased. The capacitor now discharges asymptotically toward 0 V with a time constant R_eC. The discharge toward zero is halted at $v_o = 2.4$ V, which is the steady-state output voltage corresponding to $v_s = 3$ V. As the voltage v_o drops to about 0.5 V below the 3.0-V base level, at which time v_o is about 0.1 V above the 2.4-V level, the transistor enters its active region and the time constant changes to $R_{\parallel}C$, where $R_{\parallel}$ is the parallel combination of R_e with the emitter-follower output impedance $\approx (R_s + h_{ie})/(h_{fe} + 1) = 30\ \Omega$, with $h_{ie} = 1$ K.

At the rising edge of the input waveform the emitter junction is driven to clamp ($V_{BE} \approx V_{\sigma}$), so that the base current is initially

$$i_B = \frac{v_s - V_{\sigma} - v_o}{R_s} = \frac{9 - 0.7 - 2.4}{0.5} = 11.8\ \text{mA}$$

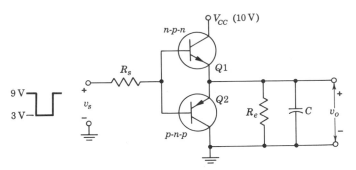

Fig. 8-36 An emitter-follower circuit using complementary transistors.

Since the voltage difference between the supply and the capacitor voltage $(10 - 2.4 = 7.6$ V) appears from collector to emitter and is of such polarity as to reverse-bias the collector junction, the transistor is therefore in its active region. A very large collector current should flow, since

$$i_C \approx h_{FE}i_B = (50)(11.8) = 590 \text{ mA}$$

The actual collector current may be considerably smaller because of the collector body resistance and the power-supply internal resistance. The large collector current charges the capacitor rapidly.

We observe in both Fig. 8-34 for the cathode follower and Fig. 8-35 for the emitter follower that the rates of fall and rise may be very unequal. In the case of the vacuum-tube circuit this situation is not correctable by any simple means. In the case of the transistor there is a simple remedy because of the availability of *complementary* transistor pairs. A complementary transistor pair is a set, one of which is an n-p-n unit and the other is p-n-p and which have similar current and voltage ratings. An emitter-follower circuit using such a pair which serves to give the falling edge a speed comparable to that of the rising edge is shown in Fig. 8-36. Let us assume that v_s is at its upper level. Then $Q1$ will be conducting and $v_o = 8.4$ V if the drop across R_s is negligible. The voltage v_o appears across $Q2$, but since $Q2$ is a p-n-p transistor and the base voltage of $Q2$ is 9.0 V, this transistor will be well beyond cutoff. At the moment that v_s makes its abrupt negative transition, $Q1$ will be driven to cutoff, but $Q2$ will be in its active region. The collector current of $Q2$ will become very large so that C may discharge abruptly through $Q2$. Altogether, C will charge rapidly through $Q1$ and discharge rapidly through $Q2$, yielding an output waveform v_o in which both edges are fast.

In Sec. 5-7 it is pointed out that a small-signal capacitively loaded emitter follower may break into oscillations. The same difficulty may arise with the overdriven emitter follower as it passes through the active region. The several methods indicated in Sec. 5-7 for minimizing the possibility of oscillations are equally effective for the large-signal transistor circuits considered here.

REFERENCE

1. Millman, J., and H. Taub: "Pulse and Digital Circuits," pp. 447–453, McGraw-Hill Book Company, New York, 1956.

9 / LOGIC CIRCUITS

Even in a large-scale digital system, such as in a computer, or in a data-processing, control, or digital-communication system, there are only a few basic operations which must be performed. These operations, to be sure, may be repeated very many times. The four circuits most commonly employed in such systems are known as the OR, AND, NOT, and FLIP-FLOP. These are called *logic* gates or circuits because they are used to implement Boolean algebraic equations (as we shall soon demonstrate). This algebra was invented by G. Boole in the middle of the nineteenth century as a system for the mathematical analysis of logic.

This chapter discusses in detail the four basic logic circuits mentioned above. Auxiliary circuits such as delay devices, amplifiers, wave-shaping circuits, etc., which are also used in digital systems are treated elsewhere in this text.

Since logic gates are used extensively in digital computers we shall take our illustrations of these switching gates from this field.

9-1   SOME FEATURES OF A DIGITAL COMPUTER[1-4]

An electronic digital computer is a system which processes and stores very large amounts of data and which solves scientific problems of numerical computation of such complexity and with such speed that solution by human calculators is not feasible. We may get some sense of the basic processes involved if we think of the computer as a system which is able to perform numerical computation and to follow instructions with extreme rapidity but which is not able to program itself. The numbers and the instructions which form the *program* the com-

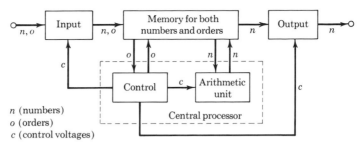

Fig. 9-1 Showing interrelationship of basic elements of a digital computer (Ref. 1).

puter is to follow are stored in an essential part of the computer called the *memory*. A second important portion of the computer is the *control*, whose function it is to interpret orders. The control must convert the order into an appropriate set of voltages to operate switches, etc., and thereby carry out the instructions conveyed by the order. A third basic element of a computer is the *arithmetic unit*, which contains the circuits which actually perform the arithmetic computations: addition, subtraction, etc. The control and arithmetic components combined are called the *central processor*. Finally, a computer requires appropriate input-output devices for inserting numbers and orders into the memory and for reading the final result. The input-output components may be punched cards or paper tape, magnetic tape, photographic film, typewriters, printers, etc., and the study of these devices is outside the scope of this book. Our principal interest is in the logic gates of the central processor and in some types of memory circuits.

Suppose we consider that, as part of a larger routine, an order to perform an addition or division, etc., has been transmitted to the central processor. In response to this order the control must select the correct operands from the memory, it must transmit these operands to the correct arithmetic unit, and it must return to the memory in some previously designated place the result of this computation. The memory serves, then, to store not only the original input data but also the partial results which will have to be used again as the computation proceeds. Lastly, if the computation is not to cease with the execution of this instruction and the storage of the partial result, the control unit must automatically sequence to the next instruction.

In terms of this crude representation of the functioning of a digital computer the interrelationship of the various components is as indicated in Fig. 9-1. The connection of the control unit back to the input is to permit insertion of more data when room becomes available in the memory.

9-2   DIGITAL (BINARY) OPERATION OF A SYSTEM[2-4]

A digital system functions in a binary manner. It employs devices which are permitted to exist in only two possible states. A transistor is allowed to

TABLE 9-1   Binary-state terminology

	1	2	3	4	5	6	7	8	9	10	11
One of the states.....	True	High	1	Up	Pulse	Excited	Off	Hot	Closed	North	Yes
The other state......	False	Low	0	Down	No pulse	Non-excited	On	Cold	Open	South	No

operate at cutoff or in saturation, but not in its active region.   A node may be at a high voltage of, say, 12 ± 2 V or at a low voltage of, say, 0 ± 0.2 V, but no other values are allowed.   Various designations are used for these two quantized states, and the most common of these are listed in Table 9-1. In logic, a statement is characterized as *true* or *false*, and this is the first binary classification listed in the table.   A switch may be *closed* or *open*, which is the notation under 9, etc.   Binary arithmetic and mathematical manipulation of switching or logic functions are best carried out with classification 3, which involves two symbols, 0 (zero) and 1 (one).

The binary system of representing numbers will now be explained by making reference to the familiar *decimal system*.   In the latter the base is 10 (ten), and ten numerals, 0, 1, 2, 3, . . . , 9, are required to express an arbitrary number.   To write numbers larger than 9, we assign a meaning to the *position* of a numeral in an array of numerals.   For example, the number 1264 (one thousand two hundred sixty four) has the meaning

$$1264 \equiv 1 \times 10^3 + 2 \times 10^2 + 6 \times 10^1 + 4 \times 10^0 \tag{9-1}$$

Thus the individual digits in a number represent the coefficients in an expansion of the number in powers of 10.   The digit which is furthest to the right is the coefficient of the zeroth power, the next is the coefficient of the first power, and so on.

In the *binary system* of representation the base is 2, and only the two numerals 0 and 1 are required to represent a number.   The numerals 0 and 1 have the same meaning as in the decimal system, but a different interpretation is placed on the position occupied by a digit.   In the binary system the individual digits represent the coefficients of powers of *two* rather than *ten* as in the decimal system.   For example, the decimal number 19 is written in the binary representation as 10011 since

$$10011 \equiv 1 \times 2^4 + 0 \times 2^3 + 0 \times 2^2 + 1 \times 2^1 + 1 \times 2^0$$
$$= \quad 16 \quad + \quad 0 \quad + \quad 0 \quad + \quad 2 \quad + \quad 1 \quad = 19 \tag{9-2}$$

A short list of equivalent numbers in decimal and binary notation is given in Table 9-2.

TABLE 9-2   Equivalent numbers in decimal and binary notation

Decimal notation	Binary notation	Decimal notation	Binary notation
0	00000	11	01011
1	00001	12	01100
2	00010	13	01101
3	00011	14	01110
4	00100	15	01111
5	00101	16	10000
6	00110	17	10001
7	00111	18	10010
8	01000	19	10011
9	01001	20	10100
10	01010	21	10101

A general method for converting from a decimal to a binary number is indicated in Table 9-3. The procedure is the following. Place the decimal number (in this illustration, 19) on the extreme right. Next divide by 2 and place the quotient (9) to the left and indicate the remainder (1) directly below it. Repeat this process (for the next column $9 \div 2 = 4$ and a remainder of 1) until a quotient of 0 is obtained. The array of 1's and 0's in the second row is the binary representation of the original decimal number. In this example, decimal 19 = 10011 binary.

A binary digit (a 1 or a 0) is called a *bit*. A group of bits having a significance is a *bite, word,* or *code*. For example, to represent the 10 numerals (0, 1, 2, . . . , 9) and the 26 letters of the English alphabet would require 36 different combinations of 1's and 0's. Since $2^5 < 36 < 2^6$, then a minimum of 6 bits per bite are required in order to accommodate all the alphanumeric characters. In this sense a bite is sometimes referred to as a *character* and a group of one or more characters as a *word*.

Logic Systems In a *d-c* or *level-logic* system a bit is implemented as one of two voltage levels. If, as in Fig. 9-2a, the more positive voltage is the 1 level and the other is the 0 level, the system is said to employ d-c *positive* logic. On the other hand, a d-c *negative*-logic system, as in Fig. 9-2b, is one which designates the more negative voltage state of the bit as the 1 level and the more positive as the 0 level. It should be emphasized that the absolute values of the two voltages are of no significance in these definitions. In

TABLE 9-3   Decimal-to-binary conversion

Divide by 2..........	0	1	2	4	9	19 decimal
Remainder..........	1	0	0	1	1	Binary

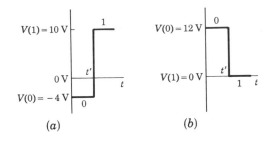

Fig. 9-2 Illustrating the definitions of (a) positive and (b) negative logic. The numerical value of the voltage $V(1)$ of the 1 state and of the voltage $V(0)$ of the 0 state is arbitrary. A transition from one state to the other occurs at $t = t'$.

particular, the 0 state need not represent a zero voltage level (although in some systems it might). In Fig. 9-2b we have intentionally illustrated the case where the value of the 1 state is 0 V.

In a *dynamic* or *pulse logic* system a bit is recognized by the presence or absence of a pulse. A 1 signifies the existence of a positive pulse in a dynamic positive-logic system; a negative pulse denotes a 1 in a dynamic negative-logic system. In either system a 0 at a particular input (or output) at a given instant of time designates that no pulse is present at that particular moment. In a "double-rail" system the variable appears on two leads. A pulse on one lead indicates that the variable has the 0 value, whereas a pulse on the other lead signifies a 1. A system employing pulses for bits may be constructed with capacitive or transformer (a-c) coupling between stages, although d-c coupling may also be used with pulses.

Most computers using pulses operate as *synchronous systems* since all operations are performed during definite constant intervals of time. There is available in a computer, to achieve this synchronism, a continuous sequence of pulses of good waveshape, whose frequency is usually established by a crystal oscillator. This stable oscillator determines the basic rate at which the computer operates and for this reason is referred to as the *master clock*. These clock pulses are distributed to all parts of the computer, where they are used to maintain the timing of the system.

In a synchronous dynamic system a number is represented in serial form by a train of pulses. A 1 is implemented by a pulse occurring at the same time as a clock pulse, whereas for a 0 a signal pulse is absent at a particular clock-pulse time. For example, the pulse train in Fig. 9-3a represents the binary number 11010111 (decimal 215) in a positive "single-rail" logic system. Such a signal is called a *binary-coded pulse train*. Note that since time increases from left to right, the least-significant pulse occurs at the extreme left (at $t = 0$), whereas in representing a binary number the least-significant bit is placed at the extreme right. In a similar way, the instructions which must be conveyed from place to place are also transmitted in the form of a train of pulses. Actually, then, a waveform representing a number is indistinguishable from a waveform representing an instruction.

A d-c system may function synchronously in the sense that all operations are performed sequentially and in coincidence with a timing pulse from the clock. In such a computer the clock interval must be longer than the minimum time of any operation. On the other hand, a d-c system may also be used in an asynchronous manner. In this mode, there is no clock, but instead each completed logic operation generates a timing pulse to start the next operation. The overall speed of an asynchronous system is higher than that of the synchronous machine since the former is determined by the average speed of an operation whereas the latter is established by the maximum time of any operation. However, because of the need for additional circuitry to designate the end of a logical operation, an asynchronous system is more complex than one run synchronously.

The signal waveform of Fig. 9-3*a* consists of a time sequence of pulses. The pulses (or absence of pulses) occur serially, one after another, and the information (number or instruction) conveyed by this pulse sequence may be transmitted from one place to another over a single communication link (i.e., in the simplest case, a pair of wires). This mode of representing information is described by the word *serial*. Alternatively, we may devise that each of the pulses (or absence of pulses) needed to represent the information occurs simultaneously on a separate channel (i.e., in the simplest case, on a separate wire, there being, say, a common ground). This mode of operation is described by the word *parallel*. In the serial mode the time required to transmit the information is the duration T of a pulse interval multiplied by the number of bits in the character. In the parallel mode, the information is transmitted in one pulse interval but we require as many channels as there are bits in the character. The serial mode is slower but cheaper; the parallel mode is faster but more expensive.

With the above brief description of digital-computer fundamentals, the binary method of coding information, and the common modes of operation we are now ready to study the logic circuits (OR, AND, NOT, and FLIP-FLOP) which form the basic building blocks of the machine. We shall first define each of these logic gates and then show how to implement the desired function in hardware. We shall then consider the sequential operation of various combinations of these switching gates to perform more advanced logic, such

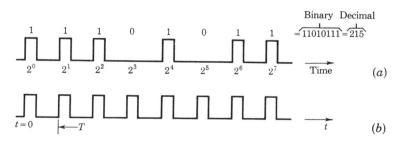

Fig. 9-3 (*a*) A pulse train representing a number (or an order) in a synchronous positive-logic digital system; (*b*) the clock pulses.

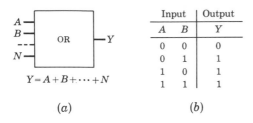

	Input		Output
	A	B	Y
	0	0	0
	0	1	1
	1	0	1
	1	1	1

$Y = A + B + \cdots + N$

(a)

(b)

Fig. 9-4 (a) The IEEE standard for an OR gate and its Boolean expression; (b) the truth table for a two-input OR gate.

as is needed in the control, the memory, or the arithmetic sections of the computer.

9-3   THE OR GATE[2-6]

An OR gate has two or more inputs and a single output, and it operates in accordance with the following definition:[5] *The output of an* OR *assumes the* 1 *state if one or more inputs assume the* 1 *state.* The n inputs to a logic circuit will be designated by $A, B, \ldots, N$ and the output by Y. It is to be understood that each of these symbols may assume one of two possible values, either 0 or 1. The IEEE standard symbol for the OR circuit is given in Fig. 9-4a, together with the Boolean expression for this gate. The equation is to be read "Y equals A or B or $\cdots$ or N." Instead of defining a logical operation in words, an alternative method is to give a *truth table* which contains a tabulation of all possible input values and their corresponding outputs. It should be clear that the two-input truth table of Fig. 9-4b is equivalent to the above definition of the OR operation.

In a *diode-logic* (DL) system the logical gates are implemented by using diodes. A diode OR for negative logic is shown in Fig. 9-5, where the symbol D is used to represent either a thermionic or a semiconductor diode (although the former would be used only in those infrequent circumstances when very high voltages are required). The generator source resistance is designated by R_s. We consider first the case where the supply voltage V_R has a value equal to the voltage $V(0)$ of the 0 state for d-c logic.

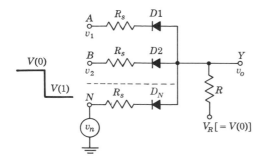

Fig. 9-5 A diode OR circuit for negative logic. [It is also possible to choose the supply voltage such that $V_R > V(0)$, but that arrangement has the disadvantage of drawing stand-by current when all inputs are in the 0 state.]

If all inputs are in the 0 state, then the voltage across each diode is $V(0) - V(0) = 0$. Since in order for a diode to conduct it must be forward-biased by at least the cutin voltage V_γ (Fig. 7-3), then none of the diodes conducts. Hence the output voltage is $v_o = V(0)$, and Y is in the 0 state.

If now input A is changed to the 1 state, which for negative logic is at the potential $V(1)$, less positive than the 0 state, then $D1$ will conduct. The output becomes

$$v_o = V(0) - [V(0) - V(1) - V_\gamma] \frac{R}{R + R_s + R_f}$$

where R_f is the diode forward resistance. Usually R is chosen much larger than $R_s + R_f$. Under this restriction

$$v_o \approx V(1) + V_\gamma$$

Hence, the output voltage exceeds the more negative level $V(1)$ by V_γ (approximately 0.2 V for germanium or 0.6 V for silicon). Furthermore, the step in output voltage is *smaller* by V_γ than the change in input voltage.

From now on, unless explicitly stated otherwise, we shall assume $R \gg R_s$ and ideal diodes with $R_f = 0$ and $V_\gamma = 0$. The output, for input A excited, is then $v_o = V(1)$ and the circuit has performed the following logic: if $A = 1$, $B = 0, \ldots, N = 0$, then $Y = 1$, which is consistent with the OR operation.

For the above excitation, the output is at $V(1)$ and each diode, except $D1$, is back-biased. Hence, the presence of signal sources at $B, C, \ldots, N$ does not result in an additional load on generator A. Since the OR configuration minimizes the interaction of the sources on one another, this gate is sometimes referred to as a *buffer* circuit. Since it allows several independent sources to be applied at a given node it is also called a (nonlinear) *mixing* gate.

If two or more inputs are in the 1 state, then the diodes connected to these inputs conduct and all other diodes remain reverse-biased. The output is $V(1)$ and again the OR function is satisfied. If for any reason the level $V(1)$ is not identical for all inputs then *the most negative value of $V(1)$ (for negative logic) appears at the output*, and all diodes except one are nonconducting.

A positive-logic OR gate uses the same configuration as that in Fig. 9-5 except that all diodes must be reversed. *The output now is equal to the most positive level $V(1)$* [or more precisely is smaller than the most positive value of $V(1)$ by V_γ]. If a dynamic logic system is under consideration, then *the output-pulse magnitude is* (approximately) *equal to the largest input pulse* (regardless of whether the system uses positive or negative logic).

Dynamic Systems   The influence of shunt capacitance and diode capacitance on the output pulse is easily seen. Assume, for simplicity, that the level $V(0)$ is at ground potential and that only one generator is furnishing an input pulse. Since, therefore, all diodes but one are back-biased during the

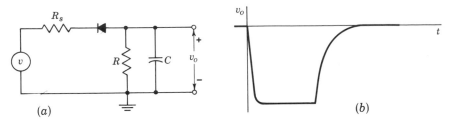

Fig. 9-6 (a) Equivalent circuit for an OR circuit with one input excited; (b) output waveform in response to input pulse.

input pulse, the capacitance shunted across the output is $C = C_o + (n - 1)C_d$, in which C_o is the capacitance across R, and C_d is the diode capacitance. We neglect here the impedance of all generators connected to the back-biased diodes and also assume that $R \gg R_s$, where R_s is the output impedance of the generator supplying the pulse. This input pulse will appear at the output with a rounded leading edge whose time constant is R_sC. The equivalent circuit and waveform are shown in Fig. 9-6a and b. When the input voltage rises at the end of the pulse, the output capacitor will maintain the output voltage and every diode will be back-biased. The capacitor C (whose capacitance is now equal to $C_o + nC_d$) must discharge through R. The trailing-edge rise time will therefore be very much longer than the leading-edge rise time since $R \gg R_s$. The number of input circuits which may be used is determined by the required transient response of the network. There will also be a small amount of coupling between generators because of the diode capacitances, but this effect need not be serious.

The above discussion has neglected entirely the diode forward and reverse recovery times. If the rise and fall times in Fig. 9-6 due to shunt capacitance are small compared with the diode transient times due to minority-carrier storage, then the recovery-time considerations given in Secs. 20-1 to 20-6 become important. For very-high-speed circuits, the diode internal characteristics rather than the shunt capacitance may be the limiting factor in the transient response.

A second mode of operation of the OR circuit of Fig. 9-5 is possible if V_R is set equal to a voltage more positive than $V(0)$ by at least V_γ. For this condition *all diodes conduct in the 0 state*, and $v_o \approx V(0)$ if $R \gg R_s + R_f$. If one or more inputs are excited then the diode connected to the most negative $V(1)$ conducts, the output equals this value of $V(1)$, and all other diodes are back-biased. Clearly, the OR function has been satisfied. The output waveform for a pulse input is the same as in Fig. 9-6 on the leading edge. However, on the trailing edge the output rises toward V_R but is clamped when it reaches $V(0)$ (Fig. 9-12).

A third mode of operation of the circuit of Fig. 9-5 results if we select $V_R < V(0)$. This arrangement has the disadvantage that the output will not

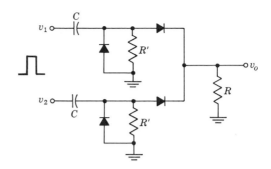

Fig. 9-7 A two-input OR gate for positive pulses using capacitive coupling.

respond until the input falls enough to overcome the initial reverse bias of the diodes.

In a dynamic logic system, if the pulses are at an inconvenient average-voltage level, then capacitive coupling is employed as indicated in Fig. 9-7. The diode polarities shown are appropriate for positive pulses. Note that because of the presence of the blocking capacitors C, the d-c level of the signal is immaterial, and hence, for convenience, the resistor R may be connected to ground instead of to the voltage V_R, as in Fig. 9-5. Now, however, d-c restoration is required, as indicated by the shunt diodes in Fig. 9-7. The resistors R' are large and may possibly be omitted altogether with semiconductor diodes. When an input is excited with a pulse, the corresponding input capacitor will acquire a charge because of the current which flows through the series diode. At the termination of the pulse the capacitor will quickly discharge through the shunt diode. Hence the next pulse in the train will be transmitted with full amplitude to the output. In general, the complication of a-c coupling with restorers should be avoided.

Emitter-follower Logic (EFL)   A second implementation of the OR circuit, using an *emitter-coupled* OR gate, is indicated in Fig. 9-8, where an emitter follower is used for each input and where the output is taken from the common

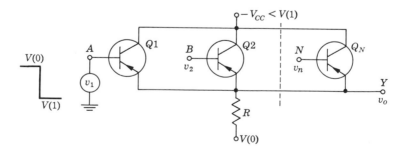

Fig. 9-8 An OR circuit for negative logic using transistors in an emitter-follower configuration. (This same circuit is also a positive AND gate, as is shown in Sec. 9-4.)

emitter resistor R. The bottom of this resistor goes to a supply voltage equal to $V(0)$. For negative logic the transistors must be of the p-n-p type as shown. If all inputs are at the 0 level $V(0)$ then each base-to-emitter voltage is at

$$V(0) - V(0) = 0 \text{ V}$$

and each transistor is virtually at cutoff. The output is then $v_o = V(0)$, or $Y = 0$. On the other hand, if any input is at the 1 level, then, because of emitter-follower action, the output is $v_o = V(1)$ (neglecting the small base-to-emitter voltage), or $Y = 1$. The base-to-emitter voltage of all transistors other than the one which is excited is $V(0) - V(1)$. Since this is a positive voltage (for negative logic), these p-n-p transistors are OFF. Here, then, is a circuit obeying the OR logic and having a higher input impedance than the diode-logic gate. If more than one input is excited simultaneously, then the output follows the most negative value of $V(1)$. Note that the collector supply voltage $-V_{CC}$ must be more negative than $V(1)$ if the excited transistor is to be driven into its active region. (If the transistor went into saturation then the input impedance would be shunted by the low-saturation base-to-collector resistance.) In practice a resistor would probably be added in series with the supply voltage $-V_{CC}$ to protect the transistors in case of an accidental short circuit to ground at the output.

For positive logic, n-p-n transistors must be used and V_{CC} must be more positive than $V(1)$. If high voltages are required, then vacuum tubes may be used as cathode followers for positive logic. Since there is no tube equivalent of a p-n-p transistor, a vacuum-tube negative-logic OR circuit analogous to that in Fig. 9-8 is not possible.

The general appearance of an output pulse from the circuit of Fig. 9-8 will be as shown in Fig. 9-6. That is, the leading edge will have a shorter rise time than the trailing edge. The reason is again that the total shunt capacitance across the output will charge through an impedance which is relatively low. In the active region the output resistance of an emitter follower is approximately $(R_s + h_{ie})/(1 + h_{fe})$, where R_s is the source impedance. At the end of the pulse, however, the capacitor must discharge through R alone.

Boolean Identities If it is remembered that A, B, and C can take on only the value 0 or 1, then the following equations from Boolean algebra pertaining to the OR ($+$) operation are easily verified:

$$A + B + C = (A + B) + C = A + (B + C) \tag{9-3}$$

$$A + B = B + A \tag{9-4}$$

$$A + A = A \tag{9-5}$$

$$A + 1 = 1 \tag{9-6}$$

$$A + 0 = A \tag{9-7}$$

These equations may be justified by referring to the definition of the OR operation, to a truth table, or to the action of the OR circuits discussed above.

9-4   THE AND GATE[2-6]

An AND gate has two or more inputs and a single output, and it operates in accordance with the following definition: *The output of an AND assumes the 1 state if and only if all the inputs assume the 1 state.* The IEEE standard for the AND circuit is given in Fig. 9-9a, together with the Boolean expression for this gate. The equation is to be read "Y equals A *and* B *and* $\cdots$ *and* N." [Sometimes a dot ($\cdot$) or a cross ($\times$) is placed between symbols to indicate the AND operation.] It may be verified that the two-input truth table of Fig. 9-9b is consistent with the above definition of the AND operation.

A diode-logic (DL) configuration for a negative AND gate is given in Fig. 9-10a. To understand the operation of the circuit, assume initially that all source resistances R_s are zero and that the diodes are ideal. If *any* input is at the 0 level $V(0)$, the diode connected to this input conducts and the output is clamped at the voltage $V(0)$, or $Y = 0$. However, if *all* inputs are at the 1 level $V(1)$, then all diodes are reverse-biased and $v_o = V(1)$, or $Y = 1$. Clearly, the AND operation has been implemented. The AND gate is also called a *coincidence circuit.*

A positive-logic AND gate uses the same configuration as that in Fig. 9-10a except that all diodes are reversed. This circuit is indicated in Fig. 9-10b and should be compared with Fig. 9-5. It is to be noted that the symbol $V(0)$ in Fig. 9-5 designates the same voltage as $V(1)$ in Fig. 9-10b because each represents the upper binary level. Similarly, $V(1)$ in Fig. 9-5 equals $V(0)$ in Fig. 9-10b since both represent the lower binary level. Hence, these two circuits are identical, and we conclude that *a negative OR gate is the same circuit as a positive AND gate.* This result is not restricted to diode logic, and, by using Boolean algebra, we show in Sec. 9-8 that it is valid independently of the hardware used to implement the circuit.

In Fig. 9-10b it is possible to choose V_R to be more positive than $V(1)$. If this condition is met, then all diodes will conduct upon a coincidence (all inputs in the 1 state) and the output will be clamped to $V(1)$. The output impedance is low in this mode of operation, being equal to $(R_s + R_f)/n$ in parallel with R. On the other hand, if $V_R = V(1)$ then all diodes are cut off

Fig. 9-9   (a) The IEEE standard for an AND gate and its Boolean expression; (b) the truth table for a two-input AND gate.

Input		Output
A	B	Y
0	0	0
0	1	0
1	0	0
1	1	1

$Y = AB \cdots N$

(a)   (b)

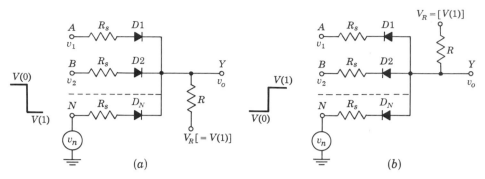

Fig. 9-10   A diode-logic AND circuit for (a) negative logic and (b) positive logic.

at a coincidence, and the output impedance is high (equal to R). If for any reason not all inputs have the same upper level $V(1)$, then the output of the positive AND gate of Fig. 9-10b will equal $V(1)_{\min}$, the *least* positive value of $V(1)$. Note that the diode connected to $V(1)_{\min}$ conducts, clamping the output to this minimum value of $V(1)$ and maintaining all other diodes in the reverse-biased condition. If, on the other hand, V_R is smaller than all inputs $V(1)$, then all diodes will be cut off upon coincidence and the output will rise to the voltage V_R. Similarly, if the inputs are pulses, then *the output pulse will have an amplitude equal to the smallest input amplitude* [provided that V_R is greater than $V(1)_{\min}$].

Let us now examine the positive-logic circuit of Fig. 9-10b more carefully, taking into account the source resistance R_s, the diode forward resistance R_f, and the diode break-point voltage V_γ. Assume that m inputs are at $V(1)$ and hence that m diodes are reverse-biased. The remaining $n - m$ diodes conduct, and hence the effective circuit of these diodes in parallel consists of a resistance $(R_s + R_f)/(n - m)$ in series with a voltage V_γ. For this excitation the output is

$$v_o = V(1) - [V(1) - V(0) - V_\gamma]\frac{R}{R + (R_s + R_f)/(n - m)} \qquad (9\text{-}8)$$

Note that if all inputs are excited, $m = n$ and $v_o = V(1)$, which is the expected output voltage for a coincidence. Also, if we neglect $R_s + R_f$ compared with R, then if $m \neq n$,

$$v_o = V(0) + V_\gamma \qquad (9\text{-}9)$$

and the output is clamped at a value V_γ above the $V(0)$ level. However, if we take the nonzero value of $R_s + R_f$ into account, then we see from Eq. (9-8) that the output will respond to the number m of excited inputs. The output increases by small steps as m increases from 0 to $n - 1$. This variation in level is called *logical noise*. If $R \gg R_s + R_f$, then the response at a coincidence will be very much larger than the response resulting even when all but one of the diodes are caused to stop conducting ($m = n - 1$). However, in an AND circuit even the slight response (the noise) to something less than a complete coincidence is often undesirable.   To reduce this effect a shunt diode D is added to Fig. 9-10b, converting it to Fig. 9-11 in order to clamp the output to a fixed voltage V' until all inputs are excited.

Use of Clamping Diode   In Fig. 9-11, V' must be adjusted so that the individual diode currents I_1, I_2, . . . , I_n (which are nominally equal to one another) are each larger than I.   If we assume, for the moment, perfect diodes, then the restriction $I_1 > I$ means

$$I_1 = \frac{V' - V(0)}{R_s} > \frac{V_R - V'}{R} = I \qquad (9\text{-}10)$$

In this case, even if all but one diode is back-biased by input signals, the diode D will be required to continue to conduct, and the output will remain clamped to V'.   If, however, all the input diodes stop conducting, the diode D must also become nonconducting and the output must rise toward V_R.   If $V_R \geq V(1)$, then the input diode connected to $V(1)_{\min}$ will clamp the output to this lowest input level, as already noted above.   A limitation on the number of input circuits which may be employed is the current-carrying capacity of the diode D. The diode D keeps the output close to V' for anything less than a complete coincidence but, of course, will not act as a perfect clamp because of the finite forward resistance R_f of D.   If, say, some number $m < n$ diodes are cut off, the current in D must change by an amount $\Delta I_o \approx m I_1$ and the output will change by $(\Delta I_o) R_f$.   The magnitude of this change is usually quite small in comparison to the output which results at a coincidence.

The circuit of Fig. 9-11 is an AND circuit since its output remains in the 0 state unless all inputs are in the 1 state, in which case the output goes to the 1 state.   However, there has been a *d-c level shift* because the 0 state $V(0)$ of the input is not the same as the 0 state V' of the output.   From Eq. (9-10) we must choose $V' > V(0)$ in order to have a positive value for the current I_1. We could have emphasized earlier that a d-c level shift takes place to some extent in all gates using real (nonidealized) elements.   For example, in the OR circuit of Fig. 9-5 having an input $V(1)$ in the 1 state, the output is also at the 1 level but the voltage of the output 1 state is given by $V(1) + V_\gamma$, which differs from the input by V_γ.   If in passing through a logic gate the level shift, or the delay, rise, and fall times, or the attenuation is too large then it may be necessary to follow this stage with a circuit designed to reshape, to retime, or to amplify the signal.

Fig. 9-11   A positive AND gate with a clamping diode D shunting the output.

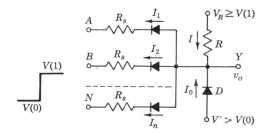

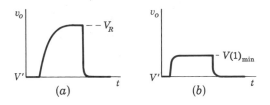

Fig. 9-12 Effect of capacitance on waveform of output of an AND circuit. (a) $V(1)_{min} > V_R$; (b) $V(1)_{min} < V_R$.

A Dynamic System   The waveform of an output pulse is easily calculated if we neglect the capacitances across the diodes.   When the output pulse is formed, all the diodes are reverse-biased and the output capacitance C_o must charge through R.   The output will therefore rise from V' to V_R with a time constant RC_o.   At the termination of the input pulses the series diodes conduct and hence introduce a resistance $R' \equiv (R_s + R_f)/n$.   The output capacitance now discharges with a time constant equal to the product of C_o and the parallel combination of R and R'.   The trailing edge of the output pulse will therefore decay much more rapidly than the leading edge rises.   The waveform is indicated in Fig. 9-12a, provided that the peak value of the pulse exceeds V_R. This result is to be compared with the corresponding waveform in Fig. 9-6 for the OR circuit.   The rise time is improved if the peak value of the pulse is smaller than V_R.   Under these circumstances the output rises toward V_R but is clamped at the peak of the smallest pulse by the diode connected to the generator supplying this pulse.   The waveform is indicated in Fig. 9-12b. If the diode capacitances are taken into account the output waveforms are modified only slightly from those indicated in Fig. 9-12.

The circuit of Fig. 9-11 may be modified for capacitive coupling by inserting a series blocking capacitor C before each input and a resistor from each input to ground or to some reference voltage.   No additional diodes are needed for restoration since at the end of each pulse C may discharge rapidly through the combination of its own diode and the shunt diode D.   As with the OR gate, here also direct coupling is preferable to capacitive coupling.

Transistors in *emitter-follower logic* (EFL) can implement the AND function. For example, the circuit of Fig. 9-8 *without modification* functions as a positive AND gate.   A negative AND is obtained by using n-p-n transistors or tubes.

Boolean Identities   Since A, B, and C can only have the value 0 or 1, then the following expressions involving the AND operation may be verified:

$$ABC = (AB)C = A(BC) \tag{9-11}$$

$$AB = BA \tag{9-12}$$

$$AA = A \tag{9-13}$$

$$A1 = A \tag{9-14}$$

$$A0 = 0 \tag{9-15}$$

$$A(B + C) = AB + AC \tag{9-16}$$

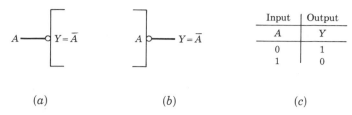

Input	Output
A	Y
0	1
1	0

(a)   (b)   (c)

Fig. 9-13 Logic negation at the input (a) and output (b) of a logic block and the Boolean equation for negation; (c) the truth table.

These equations may be proved by reference to the definition of the AND operation, to a truth table, or to the behavior of the AND circuits discussed above. Also, by using Eqs. (9-14), (9-16), and (9-6) it can be shown that

$$A + AB = A \tag{9-17}$$

Similarly, it follows from Eqs. (9-16), (9-13), and (9-6) that

$$A + BC = (A + B)(A + C) \tag{9-18}$$

We shall have occasion to refer to these last two equations later.

9-5   THE NOT OR INVERTER CIRCUIT

The NOT circuit has a single input and a single output and performs the operation of LOGIC NEGATION in accordance with the following definition:[5] *The output of a* NOT *circuit takes on the 1 state if and only if the input does* not *take on the 1 state.* The IEEE standard to indicate a LOGIC NEGATION is a small circle drawn at the point where a signal line joins a logic symbol. Negation at the input of a logic block is indicated in Fig. 9-13a and at the output in Fig. 9-13b. The truth table and the Boolean expression for negation are given in Fig. 9-13c. The equation is to be read "Y equals NOT A" or "Y is the complement of A." [Sometimes a prime (′) is used instead of the bar (‾) to indicate the NOT operation.]

A circuit which accomplishes a logic negation is called a NOT circuit, or, since it inverts the sense of the output with respect to the input, it is also known as an *inverter*. The output of an INVERTER is relatively more positive

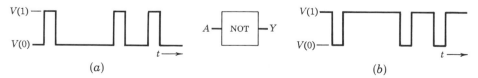

(a)   (b)

Fig. 9-14 (a) The input A and (b) the output Y of a NOT circuit.

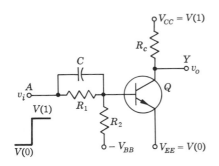

Fig. 9-15 An INVERTER for positive logic. A similar circuit using a *p-n-p* transistor is used for a negative-logic NOT circuit.

if and only if the input is relatively less positive. In a truly binary system only two levels $V(0)$ and $V(1)$ are recognized, and the output, as well as the input, of an inverter must operate between these two voltages. When the input is at $V(0)$, the output must be at $V(1)$, and vice versa. Ideally, then, a NOT circuit inverts a signal while preserving its shape and the binary levels between which the signal operates, as indicated in Fig. 9-14.

The transistor circuit of Fig. 9-15 implements an inverter for positive logic having a 0 state of $V(0) = V_{EE}$ and a 1 state of $V(1) = V_{CC}$. If the input is low, $v_i = V(0)$, then the parameters are chosen so that the Q is OFF and hence $v_o = V_{CC} = V(1)$. On the other hand, if the input is high, $v_i = V(1)$, then the circuit parameters are picked so that Q is in saturation and then $v_o = V_{EE} = V(0)$, if we neglect the collector-to-emitter saturation voltage $V_{CE}(\text{sat})$. A detailed calculation of quiescent conditions is made in the following example.

EXAMPLE If the silicon transistor in Fig. 9-16 has a minimum value of h_{FE} of 30, find the output levels for input levels of 0 and 12 V.

Solution For $v_i = V(0) = 0$ the open-circuited base voltage V_B is

$$V_B = -12 \frac{15}{100 + 15} = -1.56 \text{ V}$$

Since a bias of about 0 V is adequate to cut off a silicon emitter junction (Table 6-1, page 219), then Q is indeed cut off. Hence, $v_o = 12$ V for $v_i = 0$.

For $v_i = V(1) = 12$ V let us verify the assumption that Q is in saturation. Assume for the moment that the transistor saturation voltages are zero. The minimum base current required for saturation is

$$(I_B)_{\min} = \frac{I_C}{h_{FE}}$$

We have

$$I_C = \frac{12}{2.2} = 5.45 \text{ mA} \qquad (I_B)_{\min} = \frac{5.45}{30} = 0.18 \text{ mA}$$

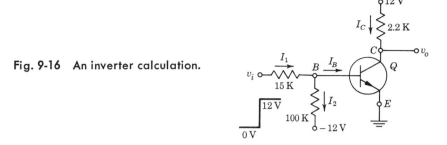

Fig. 9-16   An inverter calculation.

From Fig. 9-16 we can find the base current. We find

$$I_1 = \frac{12}{15} = 0.80 \text{ mA} \qquad I_2 = \frac{12}{100} = 0.12 \text{ mA}$$

and

$$I_B = I_1 - I_2 = 0.80 - 0.12 = 0.68 \text{ mA}$$

Since this value exceeds $(I_B)_{\min}$, Q is indeed in saturation and the drop across the transistor is zero. Hence, $v_o = 0$ for $v_i = 12$ V, and the circuit has performed the NOT operation.

The assumptions made in this example that $V_{BE}(\text{sat}) = 0$ and $V_{CE}(\text{sat}) = 0$ may be discarded by referring to the manufacturers' curves. Corresponding to $I_C = 5.45$ mA and $I_B = 0.68$ mA we can find the saturation junction voltages. It is usually sufficiently accurate to use the approximate values given in Table 6-1, which for silicon are $V_{BE}(\text{sat}) = 0.7$ V and $V_{CE}(\text{sat}) = 0.3$ V. With these values

$$I_C = \frac{12 - 0.3}{2.2} = 5.31 \text{ mA} \qquad (I_B)_{\min} = \frac{5.31}{30} = 0.18 \text{ mA}$$

$$I_1 = \frac{12 - 0.7}{15} = 0.75 \text{ mA} \qquad I_2 = \frac{0.7 - (-12)}{100} = 0.13 \text{ mA}$$

and

$$I_B = 0.75 - 0.13 = 0.62 > (I_B)_{\min} = 0.18 \text{ mA}$$

Note that the values obtained from this more exact calculation do not differ greatly from those obtained by assuming that the transistor in saturation is an ideal short circuit.

If the input to the inverter is obtained from the output of a similar gate, then the input levels are $V(0) = V_{CE}(\text{sat}) = 0.3$ V and $V(1) = 12$ V. The corresponding output levels are 12 and 0.3 V, respectively.

The capacitor C across R_1 in Fig. 9-15 is added to improve the transient response of the inverter. This capacitor aids in the removal of the minority-carrier charge stored in the base when the signal changes abruptly between logic states. A detailed discussion of this phenomenon, including rise time, fall time, and propagation delay time, is given in Chap. 20. The order of magnitude of C is 100 pF, but its exact value depends upon the transistor.

Transistor Limitations   There are certain transistor characteristics as well as certain circuit features which must particularly be taken into account in designing transistor inverters.

1. *The back-bias emitter-junction voltage V_{EB}.* This voltage must not exceed the emitter-to-base breakdown voltage BV_{EBO} specified by the manufacturer.   For the type 2N914, $BV_{EBO} = 5$ V, and for the 2N1304, $BV_{EBO} = 25$ V.   However, for some (diffused-base) transistors BV_{EBO} may be quite small (less than 1 V).

2. *The d-c current gain h_{FE}.* Since h_{FE} decreases with decreasing temperature (Fig. 6-27c), the circuit must be designed so that at the lowest expected temperature the transistor will remain in saturation.   The maximum value of R_1 is determined principally by this condition.

3. *The reverse collector saturation current I_{CBO}.* Since $|I_{CBO}|$ increases about 7 percent/°C (doubles every 10°C for either germanium or silicon), then we cannot continue to neglect the effect of I_{CBO} at high temperatures.   At cutoff the emitter current is zero and the base current is I_{CBO} (in a direction opposite to that indicated as I_B in Fig. 9-16).   Let us calculate the value of I_{CBO} which just brings the transistor to the point of cutoff.   If we assume, as in Table 6-1, that at cutoff $V_{BE} = 0$ V, then $I_1 = 0$ and the drop across the 100-K resistor is

$$100 \, I_{CBO} = 12 \text{ V} \qquad \text{or} \qquad I_{CBO} = 0.12 \text{ mA}$$

The ambient temperature at which $I_{CBO} = 0.12$ mA $= 120 \, \mu$A is the maximum temperature at which the inverter will operate satisfactorily.   From Fig. 6-14b we see that the 2N914 silicon transistor could be operated at temperatures in excess of 185°C.   If a germanium transistor were under consideration then the peak temperature would be about 100°C (Fig. 6-14a).

Other Inverter Gates   A plate-loaded tube amplifier is also an inverter for positive logic, but in general there will be a large shift in the binary levels. However, a cathode follower may be used in cascade with a d-c amplifier in

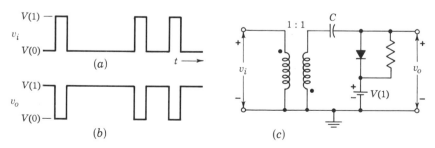

Fig. 9-17   (a) The input and (b) the output of (c) the transformer INVERTER circuit.

order to obtain the ideal NOT operation between two fixed levels, as indicated in Fig. 9-14 (Prob. 9-16).

For pulse logic an inverting transformer may be used to perform the NOT function. As indicated in Fig. 9-17, a d-c restorer is added to establish the absolute levels of the two states at the output.

Boolean Identities   From the basic definition of the NOT, AND, and OR connectives we can verify the following Boolean identities:

$$\bar{\bar{A}} = A \tag{9-19}$$

$$\bar{A} + A = 1 \tag{9-20}$$

$$\bar{A} A = 0 \tag{9-21}$$

$$A + \bar{A}B = A + B \tag{9-22}$$

9-6   THE INHIBIT OPERATION

A NOT circuit preceding one terminal (N) of an AND gate acts as an INHIBITOR. This modified AND circuit implements the logical statement: *If $A = 1, B = 1,$ . . . , $M = 1$, then $Y = 1$ provided that $N = 0$. However, if $N = 1$, then the coincidence of $A, B, \ldots , M$ is inhibited, and $Y = 0$.* Such a configuration is also called an *anticoincidence* circuit. The logical block symbol is drawn in Fig. 9-18a, together with its Boolean equation. The equation is to be read "Y equals A and B and $\cdots$ and M and not N." The truth table for a three-input AND gate with one inhibitor terminal (C) is given in Fig. 9-18b.

A combination of the AND circuit of Fig. 9-10b and the INVERTER of Fig. 9-16 satisfying the logic given in the truth table (Fig. 9-18b) is indicated in Fig. 9-19. If either input A or B or both are in the 0 state, $V(0) = 0$ V, then at least one of the diodes $D1$ or $D2$ conducts and clamps the output to 0 V or $Y = 0$. This argument verifies all items in the truth table except lines 4 and 8. Consider now the situation where a coincidence occurs at A and B. If C is in the 0 state, then Q is cut off, and the output of the NOT circuit is $\bar{C} = 1$ (12 V). Hence, all three diodes are reverse-biased and the output rises to

Fig. 9-18 (a) The logic block and Boolean expression for an AND with an inhibitor terminal N. (b) The truth table for $Y = AB\bar{C}$. The column on the left numbers the eight possible input combinations.

	Input	Output
	A B C	Y
1	0 0 0	0
2	0 1 0	0
3	1 0 0	0
4	1 1 0	1
5	0 0 1	0
6	0 1 1	0
7	1 0 1	0
8	1 1 1	0

$Y = AB \cdots M\bar{N}$

(a)   (b)

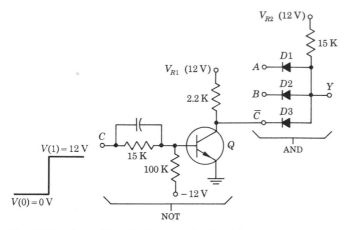

Fig. 9-19 A positive-logic AND circuit with a negation input terminal.

12 V, or $Y = 1$, which verifies line 4 of the truth table.   (If A, B, V_{R1}, and V_{R2} are not all equal to the same voltage, then the output will rise to the smallest of these values.)   Finally, consider the condition in line 8 of Fig. 9-18b. If C is in the 1 state, then Q is driven into saturation, and the output of the transistor drops to 0 V (ideally).   Hence, $\bar{C} = 0$, $D3$ conducts, and $Y = 0$, which indeed is the logic in the last row of the truth table.

It is possible to have a two-input AND, one terminal of which is inhibiting. This circuit satisfies the logic: "The output is true (1) if input A is true (1) provided that B is not true (0) [or equivalently, provided that B is false (0)]." Another possible configuration is an AND with more than one inhibit terminal.

In a dynamic system, if an inhibit pulse is to allow none of the signal to be transmitted through the gate, it is necessary that the inhibit pulse begin earlier and last longer than the signal pulses.   A method of effectively stretching the inhibit pulse is indicated in Prob. 9-17.

9-7   THE EXCLUSIVE OR CIRCUIT

An EXCLUSIVE OR gate obeys the definition:[5]   *The output of a two-input* EXCLU-*SIVE OR assumes the 1 state if one and only one input assumes the 1 state.*   The

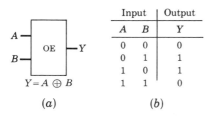

Input		Output
A	B	Y
0	0	0
0	1	1
1	0	1
1	1	0

$Y = A \oplus B$

(a)

(b)

Fig. 9-20   (a) The IEEE standard for an EXCLUSIVE OR symbol and its Boolean expression; (b) the truth table.

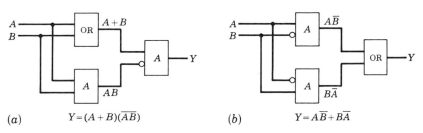

Fig. 9-21 Two logical blocks for the EXCLUSIVE OR (OE) gate.

IEEE standard symbol for an EXCLUSIVE OR is given in Fig. 9-20a and the truth table in Fig. 9-20b. The circuit of Sec. 9-3 is referred to as an INCLUSIVE OR if it is desired to distinguish it from the EXCLUSIVE OR (OE).

The above definition is equivalent to the statement: "If $A = 1$ or $B = 1$ but not simultaneously, then $Y = 1$." In Boolean notation,

$$Y = (A + B)(\overline{AB}) \tag{9-23}$$

This function is implemented in logic diagram form in Fig. 9-21a.

A second logical statement equivalent to the definition of the OE is the following: "If $A = 1$ and $B = 0$, or if $B = 1$ and $A = 0$, then $Y = 1$." The Boolean expression is

$$Y = A\bar{B} + B\bar{A} \tag{9-24}$$

The block diagram which satisfies this logic is indicated in Fig. 9-21b.

An EXCLUSIVE OR is employed within the arithmetic section of a computer (Sec. 9-12). Another application is as an *inequality comparator, matching circuit,* or *detector* because, as can be seen from the truth table, $Y = 1$ only if $A \neq B$. This property is used to check for the inequality of two bits. If bit A is not identical with bit B then an output is obtained. Equivalently, "If A and B are both 1 or if A and B are both 0, then no output is obtained, and $Y = 0$." This latter statement may be put into Boolean form as

$$Y = \overline{AB + \bar{A}\bar{B}} \tag{9-25}$$

This equation leads to a third implementation for the OE block, which is indicated by the logic diagram of Fig. 9-22a. An *equality detector* gives an

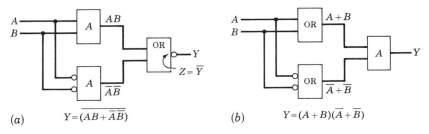

Fig. 9-22 Two additional logic diagrams for the EXCLUSIVE OR (OE) gate.

output $Z = 1$ if A and B are both 1 or if A and B are both 0, and hence

$$Z = \bar{Y} = AB + \bar{A}\bar{B} \tag{9-26}$$

where use was made of Eq. (9-19). If the output Z is desired, the negation in Fig. 9-22a may be omitted or an additional inverter may be cascaded with the output of the OE.

A fourth possibility for the OE is

$$Y = (A + B)(\bar{A} + \bar{B}) \tag{9-27}$$

which may be verified from the definition or from the truth table. This logic is depicted in Fig. 9-22b.

We have demonstrated that there often are several ways to implement a logical circuit. In practice one of these may be realized more advantageously than the others. Boolean algebra is sometimes employed for manipulating a logic equation so as to transform it into a form which is better from the point of view of implementation in hardware. In the next section we shall verify through the use of Boolean algebra that the four expressions given above for the EXCLUSIVE OR are equivalent.

9-8   DE MORGAN'S LAWS

The statement "If and only if all inputs are true (1), then the output is true (1)" is logically equivalent to the statement "If at least one input is false (0), then the output is false (0)." In Boolean notation this equivalence is written

$$ABC \cdots = \overline{\bar{A} + \bar{B} + \bar{C} + \cdots} \tag{9-28}$$

If we take the complement of both sides of this equation and use Eq. (9-19) we obtain

$$\overline{ABC \cdots} = \bar{A} + \bar{B} + \bar{C} + \cdots \tag{9-29}$$

This equation and its dual (which may be proved in a similar manner)

$$\overline{A + B + C + \cdots} = \bar{A}\bar{B}\bar{C} \cdots \tag{9-30}$$

are known as De Morgan's laws. These complete the list of basic Boolean identities. For easy future reference, all these relationships are summarized in Table 9-4.

With the aid of Boolean algebra we shall now demonstrate the equivalence of the four EXCLUSIVE OR circuits of the preceding section. Using Eq. (9-29) it is immediately clear that Eq. (9-23) is equivalent to Eq. (9-27). Now the latter equation can be expanded with the aid of Table 9-4 as follows:

$$(A + B)(\bar{A} + \bar{B}) = A\bar{A} + B\bar{A} + A\bar{B} + B\bar{B} = B\bar{A} + A\bar{B} \tag{9-31}$$

Table 9-4   Summary of basic Boolean identities

Fundamental laws

OR	AND	NOT
$A + 0 = A$	$A0 = 0$	$A + \bar{A} = 1$
$A + 1 = 1$	$A1 = A$	$A\bar{A} = 0$
$A + A = A$	$AA = A$	$\bar{\bar{A}} = A$
$A + \bar{A} = 1$	$A\bar{A} = 0$	

Associative laws

$$(A + B) + C = A + (B + C) \qquad (AB)C = A(BC)$$

Commutative laws

$$A + B = B + A \qquad AB = BA$$

Distributive law

$$A(B + C) = AB + AC$$

De Morgan's laws

$$\overline{AB \cdots} = \bar{A} + \bar{B} + \cdots \quad \overline{A + B + \cdots} = \bar{A}\bar{B} \cdots$$

Auxiliary identities

$$A + AB = A \qquad A + \bar{A}B = A + B$$

$$(A + B)(A + C) = A + BC$$

This result shows that the EXCLUSIVE OR of Eq. (9-24) is equivalent to that of Eq. (9-27).   Finally, applying Eq. (9-30) to Eq. (9-25) gives

$$\overline{AB + \bar{A}\bar{B}} = (\overline{AB})(\overline{\bar{A}\bar{B}}) \tag{9-32}$$

Using Eq. (9-29), we have

$$(\overline{AB})(\overline{\bar{A}\bar{B}}) = (\bar{A} + \bar{B})(\bar{\bar{A}} + \bar{\bar{B}}) = (\bar{A} + \bar{B})(A + B) \tag{9-33}$$

where use is made of the identity $\bar{\bar{A}} = A$.   Comparing Eqs. (9-32) and (9-33) shows that the EXCLUSIVE OR of Eq. (9-25) is equivalent to that of Eq. (9-27).

With the aid of De Morgan's law we can show that *an* AND *circuit for positive logic also functions as an* OR *gate for negative logic.*   Let Y be the output and $A, B, \ldots, N$ be the inputs to a positive AND so that

$$Y = AB \cdots N \tag{9-34}$$

Then, by Eq. (9-29),

$$\bar{Y} = \bar{A} + \bar{B} + \cdots + \bar{N} \tag{9-35}$$

If the output and all inputs of a circuit are complemented so that a 1 becomes a 0 and vice versa, then positive logic is changed to negative logic (refer to

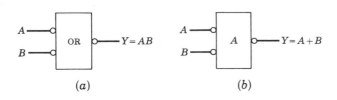

Fig. 9-23 (a) An AND can be implemented with OR and NOT gates. (b) An OR can be constructed from AND and NOT gates.

Fig. 9-2). Since Y and $\bar{Y}$ represent the *same* output terminal, A and $\bar{A}$ the *same* input terminal, etc., then the circuit which performs the positive AND logic in Eq. (9-34) also operates as the negative OR gate of Eq. (9-35). Similar reasoning is used to verify that the same circuit is either a negative AND or a positive OR, depending upon how the binary levels are defined. We verified this result for diode logic in Sec. 9-4, but the present proof is independent of how the circuit is implemented.

It should now be clear that it is really not necessary to use all three connectives, OR, AND, and NOT. The OR and the NOT are sufficient because from the De Morgan law of Eq. (9-28) the AND can be obtained from the OR and the NOT, as is indicated in Fig. 9-23a. Similarly the AND and the NOT may be chosen as the basic logic circuits, and from the De Morgan law of Eq. (9-30) the OR may be constructed as shown in Fig. 9-23b. This figure makes clear once again that an OR (AND) circuit negated at input and output performs the AND (OR) logic.

9-9   THE NAND AND NOR GATES (DTL LOGIC)[7]

In Fig. 9-21a the negation before the second AND could equally well be put at the output of the first AND without changing the logic. Such an AND-NOT sequence is also present in Fig. 9-23b and in many other logic operations. This negated AND is called a NOT-AND or a NAND gate. The logic symbol, Boolean expression, and truth table for the NAND are given in Fig. 9-24. The NAND may be implemented by placing a transistor NOT circuit *after* the diode AND in Fig. 9-19. Such a transposition is shown in Fig. 9-25. Circuits involving diodes and transistors as in Fig. 9-25 are called *diode-transistor logic* (DTL) *gates*.

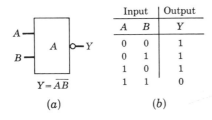

Input		Output
A	B	Y
0	0	1
0	1	1
1	0	1
1	1	0

$Y = \overline{AB}$

(a)   (b)

Fig. 9-24 (a) The logic symbol and Boolean expression for a two-input NAND gate; (b) the truth table.

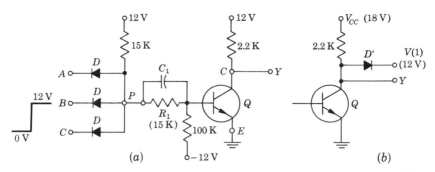

Fig. 9-25 (a) A three-input positive NAND (or negative NOR) gate; (b) a collector clamping diode may be used to improve the characteristics of the output circuit.

EXAMPLE (a) Verify that the circuit of Fig. 9-25 is a positive NAND for the binary levels 0 and 12 V. Neglect source impedance and junction saturation voltages and diode voltages in the forward direction. Find the minimum h_{FE}. (b) If the drop across a conducting diode is 0.6 V and if the sum of source and diode resistances is 1 K, is a clamping diode (Fig. 9-11) required at the output of the AND circuit? (c) Will the circuit operate properly if the inputs are obtained from the outputs of similar NAND gates? Assume silicon transistors and diodes and neglect collector saturation resistance and diode forward resistance.

Solution a. If any input is at 0 V, then the junction point P of the diodes is at 0 V because a diode conducts and clamps this point to $V(0) = 0$. The base voltage of the transistor is then

$$V_B = -(12)\left(\frac{15}{115}\right) = -1.56 \text{ V}$$

Hence, Q is cut off and Y is at 12 V, or $Y = 1$. This result confirms the first three rows of the truth table of Fig. 9-24b.

If all inputs are at $V(1) = 12$ V, assume that all diodes are reverse-biased and that the transistor is in saturation. We shall now verify that these assumptions are indeed correct. If Q is in saturation, then with $V_{BE} = 0$ the voltage at P is $(12)(\frac{15}{30}) = 6$ V. Hence, with 12 V at each input all diodes are reverse-biased by 6 V. Since the diodes are nonconducting, then the two 15-K resistors are in series and the base current of Q is

$$\frac{12}{30} - \frac{12}{100} = 0.40 - 0.12 = 0.28 \text{ mA}$$

Since the collector current is

$$I_C = \frac{12}{2.2} = 5.45 \text{ mA} \quad \text{and} \quad (h_{FE})_{\min} = \frac{5.45}{0.28} = 19$$

then Q will indeed be in saturation if $h_{FE} \geq 19$. Under these circumstances the output is at ground, or $Y = 0$. This result confirms the last row of the truth table.

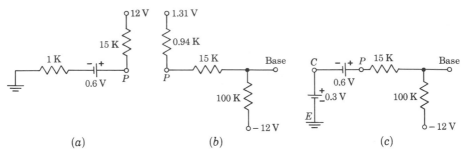

Fig. 9-26   Relating to the calculation of the base voltage of the transistor in the circuit of Fig. 9-25.

b. The transistor must be OFF if at least one input is at 0. The worst case occurs when all diodes except one are reverse-biased, because then the voltage at P is a maximum. For this case the Thévenin's equivalent from P to ground is, from Fig. 9-26a, a voltage of

$$(12)\left(\frac{1}{16}\right) + (0.6)\left(\frac{15}{16}\right) = 0.75 + 0.56 = 1.3 \text{ V}$$

in series with a resistance of $(1)(\frac{15}{16}) = 0.94$ K. The open-circuit voltage at the base of the transistor is, from Fig. 9-26b,

$$V_B = -(12)\left(\frac{15.9}{115.9}\right) + (1.31)\left(\frac{100}{115.9}\right) = -1.65 + 1.13 = -0.52 \text{ V}$$

This voltage is more than adequate to reverse-bias Q, and hence no clamping diode is required.

c. If the inputs are high the situation is exactly as in part a. With respect to keeping the base node at a low voltage when there is no coincidence, the worst situation occurs when all but one input are high. The low input now comes from a transistor in saturation, and $V_{CE}(\text{sat}) \approx 0.3$ V. The open-circuit voltage at the base of Q is, from Fig. 9-26c,

$$V_B = -12\left(\frac{15}{115}\right) + 0.9\left(\frac{100}{115}\right) = -0.78 \text{ V}$$

which cuts off Q and $Y = 1$, as it should.

Neglecting the inherent speed limitations of the transistor (Chap. 20), the rise time of the output, when Q is cut off, depends upon the shunt capacitance C_s and the collector resistance R_c. If V_{CC} is increased, then for fixed values of C_s and R_c, less time is required to reach the particular voltage at which the next stage switches (is driven into saturation). If such an increased value of V_{CC} [$> V(1)$] is used, then a collector clamping diode is often added as indicated in Fig. 9-25b. This diode limits the collector voltage of Q to $V(1)$

Fig. 9-27 (a) The logic symbol and
Boolean expression for a two-input NOR gate;
(b) the truth table.

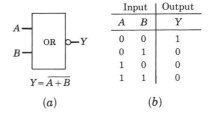

Input		Output
A	B	Y
0	0	1
0	1	0
1	0	0
1	1	0

$Y = \overline{A + B}$

(a)   (b)

and also prevents the following stage from being driven too heavily into saturation.   Secondarily, the diode helps to reduce the time to discharge C_s when Q
is driven into saturation by providing a lower collector starting voltage.

A NOR Gate   A negation following an OR is called a NOT-OR or a NOR gate.
The logic symbol, Boolean expression, and truth table for the NOR are given
in Fig. 9-27.   A positive NOR circuit is implemented in Fig. 9-28a by a cascade
of a diode OR and a transistor INVERTER.

In Fig. 9-28a the base supply $-V_{BB}$ may also be used as the reference voltage V_R for the OR, and hence V_R and R may be omitted from the circuit.   Such
a simplified configuration is indicated in Fig. 9-28b for the specific binary levels
$V(0) = 0$ and $V(1) = 12$ V.   We can readily confirm that this circuit obeys
NOR logic.   If all inputs are in the 0 state, all diodes conduct and the input
to the inverter is 0 V.   If any input is high the diode connected to this input
conducts and all other diodes are reverse-biased.   The voltage at the diode
node P is now $V(1) = 12$ V.   Hence, from input to point P the OR function
has been satisfied.   Since from P to the output we have an inverter identical to
that in Fig. 9-16, no further calculations are necessary to justify NOR operation.
The direct connection from the junction of the OR diodes to the external pin C
is convenient for expanding the number of inputs by adding more diodes as

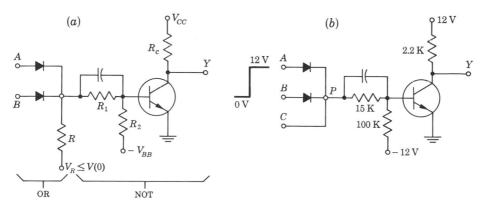

Fig. 9-28 (a) A positive OR in cascade with a NOT to form a NOR gate.   (b) A
more practical form of positive NOR (or negative NAND) gate.

needed.   It will also turn out to be important for combining two NOT gates into another basic circuit (the FLIP-FLOP, described in Sec. 9-13).

The circuits of Figs. 9-25 and 9-28 employ *diode-transistor logic* (DTL), or *transistor-diode logic* (TDL).   The NAND and NOR may also be implemented in configurations which do not use both diodes and transistors, as is indicated in Secs. 9-16 through 9-19.   With the aid of De Morgan's laws it can be shown that, regardless of the hardware involved, a positive NAND is also a negative NOR, whereas a negative NAND may equally well be considered a positive NOR.

It is clear that a single input NAND is a NOT.   Also, a NAND followed by a NOT is an AND.   In Sec. 9-8 it is pointed out that all logic can be performed by using only the two connectives AND and NOT.   Therefore, we now conclude that by repeated use of the NAND circuit alone, any logical function can be carried out.   A similar argument leads equally well to the result that all logic can be performed by using only the NOR circuit.

9-10   TWO LOGIC–CIRCUIT CONVERSION THEOREMS

We have seen that a particular gating circuit may perform one or another function, depending on whether positive or negative logic is used.   Thus a positive AND gate is a negative OR gate, etc.   Suppose, however, our interest is in preserving the type of logic performed by the gate (i.e., an AND gate is to remain an AND gate) but we wish to reverse the logic from positive to negative or the other way around.   Then the following generalization provides a method for achieving this modification.

THEOREM I   A circuit using positive (negative) logic can be converted into a configuration performing the *same* logic function but with negative (positive) logic, provided that all supply voltages are reversed in polarity, the input voltage levels are reversed in polarity, all diodes are reversed, and all transistors are changed from *p-n-p* to *n-p-n* and vice versa.

Proof   If all voltages in a circuit are reversed, then the current in any branch is also reversed.   If all diodes and transistors are reversed ($p\text{-}n\text{-}p \rightleftharpoons n\text{-}p\text{-}n$), then a diode or transistor which was reverse-biased (conducting) in the original circuit remains reverse-biased (conducting) in the converted network.   Hence, the two circuits perform the same logic.   However, if the original level $V(1)$ was more positive than $V(0)$, then $-V(1)$ is now more negative than $-V(0)$ and positive logic has been changed to negative logic.

As an application of this theorem, suppose it were desired to build a positive NAND with the binary levels $V(0) = -12$ V and $V(1) = 0$ V.   Since the circuit of Fig. 9-28b is a negative NAND, it can be converted by the theorem into the configuration of Fig. 9-29, which is now a positive NAND.

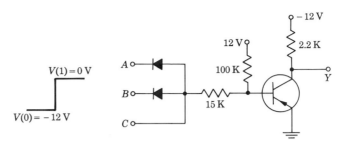

Fig. 9-29   A positive NAND **gate.**

The following theorem is useful in converting a logic circuit with one set of binary voltages to an equivalent circuit with another set of logical levels (in which, perhaps, neither is 0 V).

THEOREM II   If the same voltage is added to all supply voltages, to any leads that are grounded, and to both binary levels, then the logic function performed by the circuit remains unchanged.

Proof   The voltage difference between any two nodes is invariant under the above procedure, which is equivalent simply to shifting the zero reference of voltage. Hence, the same currents flow in the modified circuit under the identical logical conditions as in the original circuit.   Clearly, the logic performed is unchanged.

As an application of this theorem, consider a positive NAND with logic levels of -9 and $+3$ V.   This logic is performed by the circuit of Fig. 9-29 if the $+12$-V supply is changed to $+15$ V, the -12-V supply to -9 V, and if the emitter is connected to a $+3$-V supply.

9-11   PACKAGING OF LOGIC CIRCUITS

A digital computer uses a large number of switching circuits, but, as we have already emphasized, the variety of different types of gates is quite small. Hence, the fundamental circuits, which are used over and over again, are mounted on a number of plug-in units called *logic cards*.   The advantage with respect to manufacturing, replacement, trouble shooting, convenience, etc., is apparent.   Several manufacturers[8] market such logic cards, consisting of a glass epoxy printed-circuit board with the individual components mounted to the board through funnel eyelets.   Also, a number of vendors[9] have a line of micrologic gates manufactured by integrated-circuit techniques.   These will undoubtedly replace the conventional lumped-component circuits within a few years.   Building a digital system consists principally in interconnecting these packages to perform the desired logic.   A card or integrated-circuit

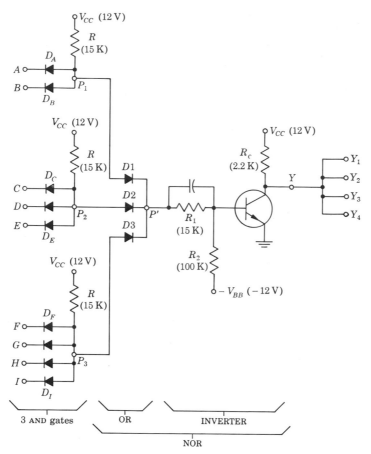

Fig. 9-30 A two-level AND-NOR (AOI) package.

package might consist of ten 2-input NAND gates, or eight NOT circuits (inverting amplifiers), or three 5-input OR gates, etc.

Often the logical design calls for an AND followed by an OR or vice versa. Such a configuration is known as *two-level logic*. One of the most useful logic packages[10] for a large-scale computer is the AND-OR-INVERT (AOI) or AND-NOR configuration. If a two-input AND is fed to terminal A of Fig. 9-28b, a three-input AND to terminal B, and a four-input AND to the third terminal (through a diode) we obtain the AOI circuit of Fig. 9-30. The number of inputs, called the *fan-in* of each AND, is not critical and neither is the number of AND clusters which feed the OR gate. In this illustration the fan-in for the OR circuit is 3. The number of outputs from a logic circuit is called the *fan-out*. Figure 9-30 indicates a fan-out of 4, so that this AOI block may feed four other logic circuits. For the particular inputs indicated the logic is

$$Y = \overline{AB + CDE + FGHI} \qquad (9\text{-}36)$$

We shall now proceed to verify the above logic. If at least one input to each AND is at 0, then the corresponding AND diodes conduct and the output of each AND is 0. All junction voltages are neglected for the conducting devices. The OR diodes in series with these outputs conduct, the transistor is cut off by -1.56 V at the base, and the output is $V(1) = 12$ V (refer to the NAND calculation in Sec. 9-9). If all inputs in one AND (say A and B) are excited to $V(1)$, then D_A and D_B are reverse-biased and P_1 rises toward $+12$ V. Hence $D1$ conducts more heavily, and Q goes into saturation with a base current of 0.28 mA. The output of the OR, point P', is at 6 V, and since the inputs to $D2$ and $D3$ are zero, these diodes are reverse-biased. Since $Y = V(0) = 0$ V, then the circuit has functioned as a simple two-input NAND.

Consider the case where a coincidence occurs simultaneously in more than one AND. For example, let us assume that all inputs are at the high level $V(1) = 12$ V. Then all AND diodes are reverse-biased, all OR diodes are forward-biased, and Q is in saturation, as we shall now prove. If the AND diodes were cut off, then P_1, P_2, and P_3 would rise toward 12 V and $D1$, $D2$, and $D3$ would conduct. Under these circumstances the three resistors R are in parallel, as are also the three OR diodes, and the equivalent circuit is that given in Fig. 9-31. The base current is now

$$I_B = \frac{12}{5 + 15} - \frac{12}{100} = 0.48 \text{ mA}$$

This is an increase over 0.28 mA which was obtained when there was a coincidence at only one AND gate. Since the smaller value was sufficient to drive Q into saturation (provided that h_{FE} exceeded 19), then certainly Q is now also in saturation. Note that the voltage at P (P_1, P_2, or P_3) is $(12)(\frac{15}{20}) = 9$ V, whereas the voltage at each input is 12 V. All input diodes are reverse-biased by 3 V, thus verifying our original assumption.

In summary, the AOI circuit behaves like independent NAND circuits with no interaction of one AND upon the other, because the OR gate acts as

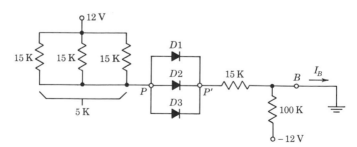

Fig. 9-31 The equivalent circuit of Fig. 9-30 when there is a simultaneous coincidence at all inputs. The point P represents either P_1, P_2, or P_3 since these three nodes are at the same potential.

a buffer. However, the larger the number of AND circuits which are excited simultaneously, the further into saturation is the transistor driven.

The AOI function or any other logic block may be implemented in many ways besides using diode-transistor logic. The most frequently used alternative configurations are discussed in Secs. 9-16 through 9-19. We shall first, however, show how these logic blocks may be combined to perform arithmetical or control functions.

9-12   BINARY ADDITION[2,4,11]

A digital computer must obviously contain circuits which will perform arithmetic operations, i.e., addition, subtraction, multiplication, and division. The basic operations are addition and subtraction, since multiplication is essentially repeated addition, and division is essentially repeated subtraction. It is entirely possible to build a computer in which an *adder-subtractor* is the only arithmetic unit present. Multiplication, for example, may then be performed by *programming;* that is, the computer may be given instructions telling it how to use the adder repeatedly to find the product of two numbers.

Suppose we wish to sum two numbers in decimal arithmetic and obtain, say, the hundreds digit. We must add together not only the hundreds digit of each number but also a carry from the tens digit (if one exists). Similarly in binary arithmetic we must add not only the digit of like significance of the two numbers to be summed but also the carry bit (should one be present) of the next lower significant digit. This operation may be carried out in two steps: first, add the two bits corresponding to the 2^k digit, and then add the resultant to the carry from the 2^{k-1} bit. A two-input adder is called a *half adder*, because to complete an addition requires two such half adders.

We shall first show how a *half adder-subtractor* is constructed from the basic logic gates and then indicate how the *full* or *complete adder-subtractor* is assembled. A half adder-subtractor has two inputs—A and B—representing the bits to be added, and three outputs—D (for the digit of the same significance as A and B represent), C (for the carry bit), and P (for the borrow bit). In a half adder D and C are used, while in a half subtractor D and P are used.

The symbol for a half adder-subtractor is given in Fig. 9-32a and the truth table in Fig. 9-32b. Note that the D column gives the sum of A and B

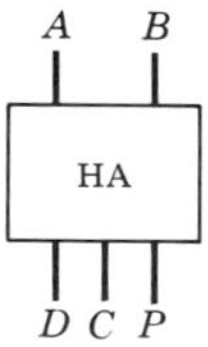

Input		Output		
A	B	D	C	P
0	0	0	0	0
0	1	1	0	1
1	0	1	0	0
1	1	0	1	0

(a)   (b)

Fig. 9-32 (a) The symbol for a half adder-subtractor; (b) the truth table for the digit (D), carry (C), and borrow (P).

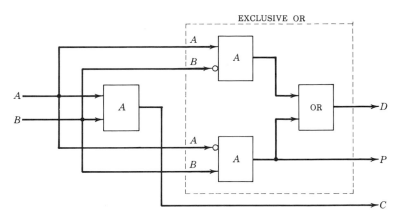

Fig. 9-33   Block diagram of a half adder-subtractor.

as long as the sum can be represented by a single digit. When, however, the sum is larger than can be represented by a single digit, then D gives the digit in the result which is of the same significance as the individual digits being added. Thus, in the first three rows of the truth table D gives the sum of A and B directly. Since the decimal equation "1 plus 1 equals 2" is written in binary form as "01 plus 01 equals 10," then in the last row $D = 0$. Because a 1 must now be carried to the place of next higher significance, $C = 1$. Finally, where subtraction of B from A is contemplated, the P (borrow) column gives the digit which must be borrowed from the place of next higher significance when B is larger than A, as in the second row of Fig. 9-32.

From Fig. 9-32b we see that D obeys the EXCLUSIVE-OR (OE) function, C follows the logic of an AND gate, and P obeys the logic "B and not A." Figure 9-33 shows a configuration which satisfies this half adder-subtractor logic based upon the OE circuit of Fig. 9-21b. Any of the other implementations given in Sec. 9-7 for the OE may be used in the half adder-subtractor. A half adder constructed by using only NOR circuits is given in Prob. 9-36.

Parallel Operation   Two multidigit numbers may be added serially (one column at a time) or in parallel (all columns simultaneously). Consider parallel operation first. For an n-digit binary number there are (in addition to a common ground) n signal leads in the computer for each number. The kth line for number A (or B) is excited by A_k (or B_k), the bit for the 2^k digit ($k = 0$, 1, . . . , n). A parallel binary adder is indicated in Fig. 9-34. Each digit except the least-significant one (2^0) requires a complete adder consisting of two half adders in cascade. The sum digit for the 2^0 bit is $S_o = D_o$ of a half adder because there is no carry to be added to A_o plus B_o. The sum S_k ($k \neq 0$) of A_k plus B_k is made in two steps. First the digit D_k is obtained from one half adder, and then D_k is summed with the carry C_{k-1} which may have resulted from the next lower place. As an example, consider $k = 2$ in Fig. 9-34. There the carry bit C_1 may be the result of the direct sum of A_1 plus B_1 if

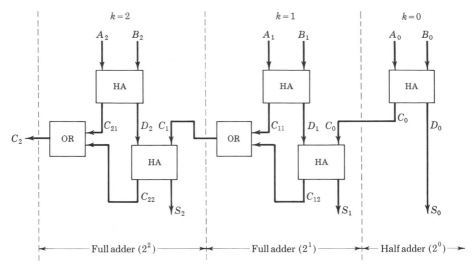

Fig. 9-34   A parallel binary adder consisting of half adders.

each of these is 1.   This first carry is called C_{11} in Fig. 9-34.   A second possibility is that $A_1 = 1$ and $B_1 = 0$ (or vice versa), so that $D_1 = 1$, but that there is a carry C_o from the next lower significant bit.   The sum of $D_1 = 1$ and $C_o = 1$ gives rise to the carry bit designated C_{12}.   It should be clear that C_{11} and C_{12} cannot both be 1, although they will both be 0 if $A_1 = 0$ and $B_1 = 0$.   Since either C_{11} or C_{12} must be transmitted to the next stage, an OR gate must be interposed between stages, as indicated in Fig. 9-34.   This circuit is equally effective for subtraction, provided that the borrow bit P is used in place of the carry C.

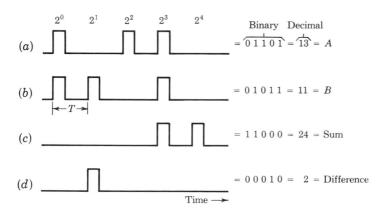

Fig. 9-35   (a, b) Pulse waveforms representing numbers A and B; (c, d) waveforms representing sum and difference.

Fig. 9-36 A serial binary adder consisting of two half adders.

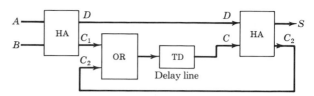

Serial Operation   In a serial adder the inputs A and B are synchronous pulse trains on two lines in the computer.   Figure 9-35a and b show typical pulse trains representing, respectively, the decimal numbers 13 and 11. Pulse trains representing the sum (24) and difference (2) are shown in Fig. 9-35c and d, respectively.   A serial *adder* is a device which will take as inputs the two waveforms of Fig. 9-35a and b and deliver the output waveform in Fig. 9-35c.   Similarly, a *subtractor* will yield the output shown in Fig. 9-35d.

We have already emphasized that the sum of two multidigit numbers may be formed by adding to the sum of the digits of like significance the carry (if any) which may have resulted from the next lower place.   With respect to the pulse trains of Fig. 9-35, the above statement is equivalent to saying that, at any instant of time, we must add (in binary form) to the pulses A and B the carry pulse (if any) which comes from the resultant formed one period T earlier.   The carry pulse may arise from the direct sum of two digits (each 1) or from the addition of digits 1 and 0 and a carry 1 from the preceding interval.   The logic outlined above is performed by the full adder circuit of Fig. 9-36, which consists essentially of two half adders in cascade.   This circuit differs from the configuration in the parallel adder of Fig. 9-34 by the inclusion of an electromagnetic delay line.   The time delay TD of this line is equal to the time T between pulses.   Hence, the carry pulse (from either of the two sources mentioned above) is delayed a time T and added to the digit pulses in A and B, exactly as it should be.

Flow Chart   It is instructive to construct the "flow chart" in Fig. 9-37 for the addition of two binary numbers A and B, in either the parallel method

Fig. 9-37 Relating to a full adder consisting of two half adders.   The arrows show the origin of the carry bit C.   The bits correspond to those in Fig. 9-35 and the symbols to those in Fig. 9-36.

	2^4	2^3	2^2	2^1	2^0
A	0	1	1	0	1
B	0	1	0	1	1
D	0	0	1	1	0
C_1	0	1	0	0	1
C	1	1	1	1	0
C_2	0	0	1	1	0
S	1	1	0	0	0

Input			Output	
A	B	C	S	C'
0	0	0	0	0
0	0	1	1	0
0	1	0	1	0
0	1	1	0	1
1	0	0	1	0
1	0	1	0	1
1	1	0	0	1
1	1	1	1	1

Fig. 9-38 Truth table for a three-input adder.

of Fig. 9-34 or the serial arrangement of Fig. 9-36. In applying the table to the parallel method of Fig. 9-34 it is to be kept in mind that C_{11} in Fig. 9-34 corresponds to C_1 in Fig. 9-36 and C_{12} corresponds to C_2. That is, in the double subscript attached to the C's in Fig. 9-34 the first indicates the significance of the digit and the second indicates whether the carry comes from the first or second half adder. The illustration is taken for numbers A and B pictured in Fig. 9-35. In any particular case the chart is filled out as follows: A and B are the given numbers. D and C_1 are obtained from the truth table of Fig. 9-32b for the addition of two single-digit numbers. Then C is obtained from C_1 by shifting each 1 to the next higher significant place, as indicated by the downward arrows in the table. Now, remembering that D and C are the inputs to a half adder, we obtain the outputs S and C_2 following the rules of Fig. 9-32b. We must, however, start with the right-hand column (2^0) and work toward the left. Each time $C_2 = 1$ this bit is shifted into C, as indicated by the upward arrows. Note that in this particular example two of the carries come from C_1 (corresponding to $A = B = 1$) and two from C_2 (corresponding to the addition of 1, 0, and a carry 1 from the next lower place). From the chart of Fig. 9-37 we can construct the waveforms existing at the input or output of any block in Fig. 9-36.

A comparison of Figs. 9-34 and 9-36 indicates that parallel addition is faster than serial because all digits are added simultaneously in the former but in sequence in the latter. However, whereas only one full adder is needed for serial arithmetic we must use a full adder for each bit in parallel addition (except the 2^0 bit, which requires a half adder). Hence, parallel addition is much more expensive than serial operation.

The Three-input Adder   It is possible to construct a complete adder without the use of half adders. The circuit has three inputs: A, B, and the carry C. A truth table for such an adder is given in Fig. 9-38. The output carry C' in a serial system is delayed one synchronizing interval T and then becomes the input carry C. From the truth table we can verify that the Boolean expressions for the sum S and the carry C' are given by

$$S = \bar{A}\bar{B}C + \bar{A}B\bar{C} + A\bar{B}\bar{C} + ABC \qquad (9\text{-}37)$$

$$C' = \bar{A}BC + A\bar{B}C + AB\bar{C} + ABC \qquad (9\text{-}38)$$

By algebraic manipulation these expressions may be transformed into a number of different forms. In particular it turns out that

$$S = (A + B + C)(\bar{C}') + ABC \tag{9-39}$$

$$C' = AB + BC + CA \tag{9-40}$$

These expressions can also be verified from the truth table. The three-input adder is considered in Prob. 9-37.

9-13   REGISTERS

In addition to the AND, OR, and NOT logic gates a fourth important basic circuit, called the FLIP-FLOP, is required in many digital systems. A FLIP-FLOP consists of two NOT circuits interconnected in the manner shown in Fig. 9-39a. Each NOT could be, for example, the transistor INVERTER of Fig. 9-15. The FLIP-FLOP configuration is studied in detail in Chap. 10. For the present we are interested only in certain external characteristics which are relevant in digital systems. The most important property of the FLIP-FLOP is that, on account of the interconnection, the circuit may persist indefinitely in a state in which one device (say $Q1$) is ON while the other ($Q2$) is OFF. A second stable state of the FLIP-FLOP is one in which the roles of the two devices are interchanged so that $Q1$ is OFF and $Q2$ is ON. Since the FLIP-FLOP has two stable states it may be used to *store* one bit of information. For these reasons the FLIP-FLOP is also called a BINARY.

An output, designated as Y in Fig. 9-39, may be taken from a collector. This output may take on two voltage levels, corresponding to either $Y = 1$ or $Y = 0$. If we designate the output at the other collector as $\bar{Y}$, then the

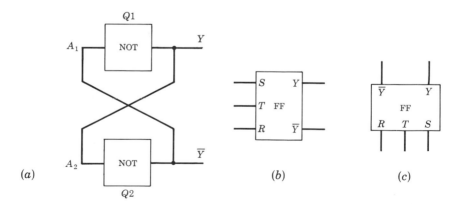

Fig. 9-39 (a) A FLIP-FLOP assembled from two NOT circuits; (b or c) the logic symbol. An input to T effectively applies excitation to S and R simultaneously.

FLIP-FLOP has two stable states, one in which $Y = 1$ and $\bar{Y} = 0$ and the other in which $Y = 0$ and $\bar{Y} = 1$. The existence of these stable states is consistent with the interconnection shown in Fig. 9-39a. For example, if the output Y of one NOT circuit is 1 then so also is the input A_2 to the second NOT circuit. The second INVERTER then has the state 0 at its output $\bar{Y}$ and at the input A_1 to the first gate. This result is consistent with our original assumption that the first NOT gate had a 1 at its output. It is readily verified that the situation in which both outputs are in the same state is not consistent with the interconnection.

A FLIP-FLOP is represented in block form as in Fig. 9-39b or c, where three input terminals are indicated—S (*set*), R (*reset*), and T (*trigger*). The points of connection of these terminals are not shown in Fig. 9-39a but are described in the next chapter (Fig. 10-14). An excitation of the *set* input causes the FLIP-FLOP to establish itself in the state $Y = 1$. If the binary is already in that state the excitation has no effect. A signal at the *reset* input causes the FLIP-FLOP to establish itself in the state $Y = 0$. If the binary is already in that state the excitation has no effect. The waveform of the input signal (a pulse, a step, etc.), auxiliary circuits through which the excitation is applied to the binary, and related matters are discussed in the next chapter.

A triggering signal applied to the T input causes the FLIP-FLOP to change its state regardless of the existing state of the binary. Thus each successive excitation applied to T causes a transfer, and T is referred to as the *toggle* or *complementing input*. This type of excitation is called *symmetrical triggering* (Sec. 10-9) and is used in binary counters (Chap. 18) and in other applications. *Unsymmetrical triggering* (Sec. 10-7) through the S or R input is most useful in logic applications, as we demonstrate below.

A One-word Memory Suppose that it is required to carry out the addition of two numbers which are stored in the main computer memory. Now, ordinarily, it will not be possible to abstract both numbers from the memory simultaneously. Since in the adders of the preceding section both numbers are applied simultaneously, it will generally be required that at least one of the numbers be stored, temporarily, in a one-word memory device. Similarly, it may not be feasible to return the arithmetic-unit output immediately to the main memory. In this case, again, a one-word *memory*[3,4,12] or *storage device*, which is called a *register*, is needed.

A set of n flip-flop circuits may clearly be used to store an n-digit binary number, since we have but to set the states of the binaries at 0 or 1, depending on the value of the digit which the FLIP-FLOP is to represent. The binary number may appear in serial form as a train of pulses, and one method for inserting the number into the register is as shown in Fig. 9-40. The input pulse train is applied to a delay line which is tapped at time-delay intervals TD equal to the basic pulse separation time (a one-bit delay T). Hence, at the moment the last pulse (2^3) of the train appears at the input of the delay line, the earlier pulses will appear at the delay-line taps. If, at this moment, the *register line* is pulsed, then the AND circuits will transmit to each binary the

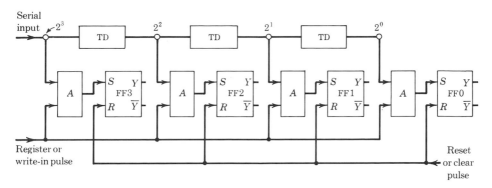

Fig. 9-40 A register for converting a 4-bit code from serial to parallel form. The time delay TD equals the interval between bits.

pulse (or lack of pulse) at the corresponding delay-time taps. The output of each AND circuit is coupled to the set input of a FLIP-FLOP so that the AND-circuit pulse (if one is present) will leave the corresponding binary in state 1. Thus, the 2^3 bit is registered in $FF3$, the 2^2 bit in $FF2$, etc. The register may be cleared by a pulse on the reset line. This pulse will cause each binary to remain in or return to state 0. The circuit of Fig. 9-40 is a *serial-to-parallel converter*, since each bit of information in a pulse train is now available in a separate FLIP-FLOP. A *temporal code* (a time arrangement of bits) has been changed to a *spatial code* (information stored in a static memory).

Consider that the outputs of successive binaries are coupled through a second set of AND circuits to corresponding points of an additional delay line which is a duplicate of the input line. Then the reset pulse, which is also applied to the second set of AND circuits, will not only clear the register but also establish on this output line the initial temporal arrangement of the pulses. Hence, the clearing operation will also regenerate the original pulse train. Here, then, is a *parallel-to-serial converter*—or a *spatial-to-temporal converter*—since each bit of information stored statically in individual FLIP-FLOPS has been converted into a pulse train on a single line.

The process outlined in the preceding paragraph is called *destructive read-out* because the information is removed from the register when the pulse train is formed. Consider now that one input of the AND circuit is again excited by the d-c level of a FLIP-FLOP in the register and that the second input is a pulse from an external source. If this pulse is *not* used to reset the register, then nondestructive readout is possible; that is, a pulse-train output is obtained while retaining the digital information in the register. An alternative form of coincidence gate having a binary input and a pulse input is given in Sec. 17-2.

A Shift Register   For low repetition frequencies (say below 200 kHz) the time delay TD (> 5 μsec) may require an impractically long delay line. A circuit which avoids this difficulty (and also has other favorable character-

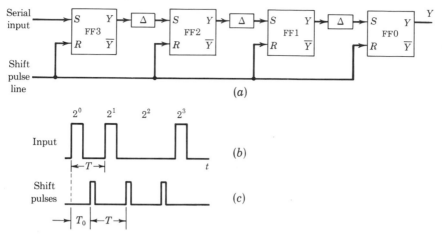

Fig. 9-41 (a) A shift register. The time delay Δ is very much smaller than the interval between pulses. (b) A typical input pulse train. (c) The shift-pulse waveform.

istics) is indicated in Fig. 9-41a; it is called a *shift register*. The input consists again of the train of pulses which is to be stored in the register. The reset or *shift* pulse line is excited now, not by a single pulse, but rather by a continuous train of pulses (Fig. 9-41c) which are timed to occur nominally *midway* between the pulses of the input number. The delay sections have a delay Δ much smaller than the time interval between pulses and are required to ensure that an individual binary shall not receive a triggering signal simultaneously from the shift line and from a preceding FLIP-FLOP. The shift pulses always drive the binaries to state 0. The coupling between binaries is such that a *succeeding* FLIP-FLOP *will respond only if the preceding binary goes from state* 1 *to state* 0. The pulse which results from this transition will drive the succeeding FLIP-FLOP to state 1.

Now suppose we want to register the number 1011. The pulse pattern is as indicated in Fig. 9-41b. The first pulse (2^0) drives $FF3$ to state 1. The shift pulse now returns $FF3$ to 0, and a short time later (depending on the delay Δ) FLIP-FLOP $FF2$ is driven to state 1 by the pulse received from the previous binary. The first digit (2^0), which was initially registered in $FF3$, has been shifted to $FF2$, and $FF3$ has been cleared (returned to 0) so that it may now register the next pulse (2^1). We may now easily follow the procedure from this point and see that, by this process of registering and shifting to make room for the next pulse, the input number will eventually become installed in the register. Of course, the shift pulses must cease at the moment the number has been registered.

This register may be read in parallel (each FLIP-FLOP output going to a separate line), if desired, and hence the shift register is also a *serial-to-parallel converter.* Or we may take the output at Y of $FF0$ in Fig. 9-41 and obtain a pulse train. To read the register serially in this fashion it will be necessary to apply four shift pulses. In response to these shift pulses the original character will appear at the output Y of $FF0$. Note that the shift-out rate may be greater or smaller than the original pulse repetition rate. Hence, here is a method for effectively changing the spacing of a pulse sequence, a process referred to as *buffering*.

Assume that a binary number is stored in a shift register, with the least-significant bit stored in $FF0$. Now apply one shift pulse. Each bit then moves to the next lower significant place and hence is divided by 2. The number now held in the register is half the original number, provided that $FF0$ was originally 0. Since the 2^0 bit is lost in the shift to the right, then if $FF0$ was originally in the 1 state, corresponding to the decimal number 1, after the shift the register is in error by the decimal number 0.5. If the circuit is wired so that each shift pulse causes a shift to the left, then each bit moves to the next higher significant place and the number stored is multiplied by 2. A computer uses shift registers in this way for performing the multiplication of two binary numbers.

A shift register may function as a *digital delay device*. Thus, the input pulse train appears at the output of an n-stage register delayed by a time equal to $T_o + (n - 1)T$, where T is the interval between shift pulses and T_o is the time between the first bit and the first shift pulse (Fig. 9-41).

An important application of a shift register is as a character generator.[13] The FLIP-FLOPS may be preset individually with push-button switches at S and R to give the desired code (the pattern of 1's and 0's making up the desired character). Then a pulse generator is used to apply shift pulses and the output of the shift register gives the temporal pattern corresponding to the character. For test purposes it is often important that the code be repeated continuously. This mode of operation is easily obtained by feeding the output of the register back into the input to form a "reentrant shift register." Such an arrangement is called a *dynamic* or *circulating memory*.

9-14   DYNAMIC REGISTERS

If a pulse train already available in a computer is to be stored, then a more economical form of dynamic memory is obtained by replacing the shift register of the preceding section with a delay line. The word (pulse train) is introduced at one end of the delay line, whose time delay TD is equal to the time duration for the word, and the output signal is returned to the delay-line input so that the word continues to circulate around a closed path. A *dynamic register* of this type is indicated in Fig. 9-42. Synchronization between the circulating pulse train and the pulses in other parts of the computer is required since the word may take many trips around the circuit, and if the total loop delay is even slightly incorrect a large error may accumulate. Therefore means are usually

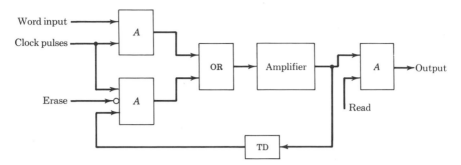

Fig. 9-42 A dynamic register using a delay line whose time delay TD equals a word length. The clock pulses are used for synchronization.

provided[10] (but this circuit is not shown in Fig. 9-42) to ensure the synchronization and also to reshape the pulses which suffer attenuation and distortion in each round trip.

The pulse train circulates and reappears continuously at the output. The register may be read (nondestructively) by exciting the *read* lead of the output AND circuit by a number of pulses equal to the digits in one word. The register may be cleared by exciting the *erase* lead of the inhibitor in the circulation path. Note that to erase an n-digit word, n consecutive erase pulses are required, which pulses coincide with the clock pulses. Alternatively, a single erase pulse may be used whose duration is sufficient to encompass the entire word.

In order to represent a 10-digit number in the decimal system, 34 binary bits are required. If the clock-pulse repetition rate is 2 MHz, then one word will require 17 μsec. If the total storage capacity of the computer is 1,000 words, then a single delay line 17 msec long could supply all the storage. However, a prohibitively long waiting time would then be required before a word is available. As a compromise between speed and equipment some 50 lines, each 20 words (or 340 μsec) long, might be used. The attenuation of electrical delay lines (about 6 dB/μsec delay at a frequency of a few megacycles per second) is excessive for the present application. An improvement results if the block marked TD in Fig. 9-42 is an acoustical delay line. The pulse train representing a word is used to modulate a carrier frequency of 30 MHz. These 30-MHz pulses in turn drive a quartz-crystal transducer which generates waves in a mercury column. A receiving crystal at the other end of the line detects the pulses after they have traveled down the column.

Many different memory systems,[12] besides the mercury line, are now in use or have been suggested for future computers. These memory devices include a magnetic-core matrix, solid acoustical lines, magnetostrictive delay lines, magnetic drums, and others.

The dynamic-register circuit of Fig. 9-42 suggests an interesting special case. Suppose that the input were to consist of a single pulse instead of a pulse train and that the total circuital delay were adjusted to be equal to the time interval between pulses.   In this case the circuit could exist in either of two possible states, i.e., a state 0 in which there is no circulating pulse or a state 1 in which there is a circulating pulse.   This one-digit (or one-bit) dynamic storage circuit then shares the basic digital property of a binary and is called a *dynamic* FLIP-FLOP.

9-15   DIODE MATRICES OR CODE–OPERATED SWITCHES[3,4,14]

As noted earlier, in a digital computer, instructions as well as numbers are conveyed by means of binary levels or pulse trains.   If, say, 4 bits of a character are set aside to convey instructions, then 16 different instructions are possible.   This information is *coded* in binary form.   There arises frequently a need for a multiposition switch, also called a *translational network*, which may be operated in accordance with this code.   In other words, for each of the 16 codes, one and only one line is to be excited.   This process of identifying a particular code is called *decoding*.

Consider that the codes are stored in a 4-bit register.   The four FLIP-FLOPS, the 16 output lines, and a decoding network made up of diodes is indicated in Fig. 9-43.   Because of the schematic arrangement of wires into (16)

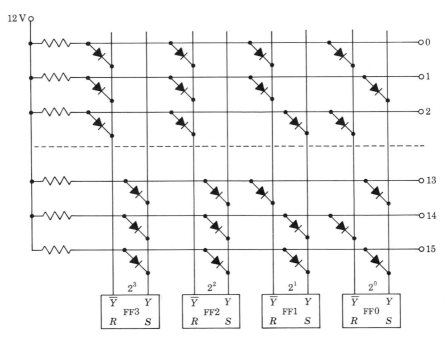

Fig. 9-43   A 16-position code-operated switch (lines 3 through 12 are not indicated).

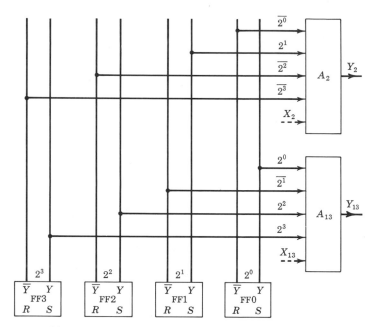

Fig. 9-44 Decoding lines 2 and 13 of a 16-position switch. If the AND gate uses diodes, then this circuit is identical with that of Fig. 9-43. An external signal X_2 will be gated onto control line Y_2 by code 0010.

rows and (8) columns this array is called a *rectangular diode matrix*. Assume that the binary outputs are $Y = 0$ V for binary 0 and $Y = 12$ V for binary 1. If a FLIP-FLOP is set (storing a 1) then the right-hand output is at 12 V, but if it stores a 0 then the left-hand output is at 12 V. The left-hand output $\bar{Y}$ is, of course, the complement of the right-hand output Y. Let us now confirm that the diodes are correctly positioned to perform the desired decoding. Consider, for example, line 13. Note that the diodes are connected to the code 1101, which is the binary equivalent of the decimal number 13. For this 1101 code all the cathodes to the diodes are at 12 V and the output on line 13 rises to 12 V. In other words, this line is excited—it is a binary 1. For any other code, at least one of the diodes connected to line 13 is at 0 V and the output is clamped to 0 V. The reader should confirm that all the diodes follow this logical pattern. For example, line 2 is excited if and only if the code corresponds to the binary representation of decimal 2, namely, 0010.

It should be clear that the decoder is simply a diode AND gate and that an equivalent representation for Fig. 9-43 is given in Fig. 9-44. Only the AND gates for channels 2 and 13 are indicated but, of course, there are a total of 16 four-input AND gates. If only the first 10 lines are used, namely those labeled 0, 1, 2, . . . , 9, then this translational network converts from a *binary-coded decimal* (BCD) to a decimal number.

If in the diode matrix of Fig. 9-43 n flip-flops are used, the number of switch combinations, instructions, or positions available will be 2^n, and in this case $(n)(2^n)$ diodes will be required in a rectangular array. For example, a 256-position switch will require 2,048 diodes. It turns out[14] that where the number of switch positions is 16 or more, it is possible to rearrange the network into other than a rectangular array, with an attendant saving in diodes. For example, the most economical 256-position switch requires only 608 diodes, against 2,048 for the rectangular array. The effect of the finite back resistance of semiconductor diodes used in a multiposition switch is considered in Ref. 14.

Decoder Applications Suppose that for a specific code (say 1101) it is desired that a signal X_{13} (in the form of a pulse train) appear on line Y_{13}. This is accomplished, as indicated in Fig. 9-44, by adding a fifth input X_{13} and a fifth diode to the AND gate which controls output Y_{13}.

In the above application each code can, of course, control a different signal X_k on the kth line. If instead of this parallel operation it is desired that all signals appear serially on one line—but each signal in response to its own control code—then the outputs Y_k in Fig. 9-44 are applied to an OR circuit. This configuration is an example of two-level AND-OR logic. An alternative, slightly simpler arrangement, but one which has the disadvantage that it does not use standard logic packages, is indicated in Fig. 9-45.

The decoder circuit may also be used as an electronic distributor or commutator. Suppose that in Fig. 9-43 the binaries are coupled together in such a way that they may be pulsed from one code to the next in succession: 0000, 0001, 0010, . . . , 1111, 0000, 0001, etc. (such an arrangement is called

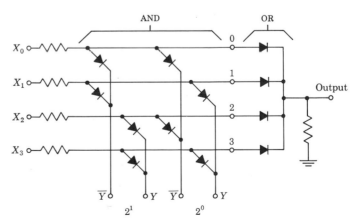

Fig. 9-45 A code-operated switch used for direct gating of four signals onto the same line.

Input	Outputs			
	Y_1	Y_2	Y_3	Y_4
A	1	1	1	0
B	0	0	1	1
C	1	0	1	1

(a)   (b)

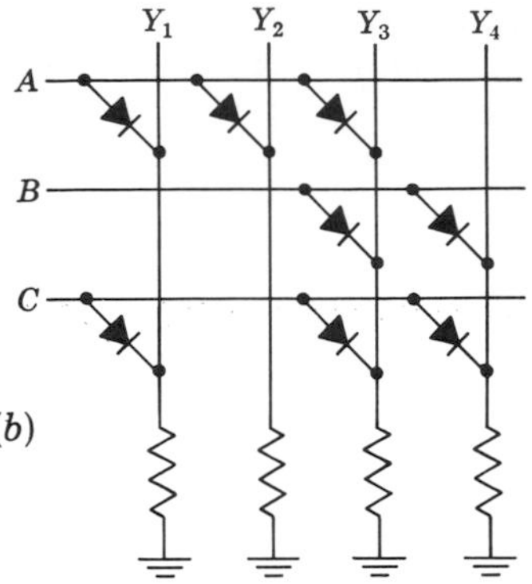

Fig. 9-46 An encoder. The diode matrix in (b) follows the logic in (a).

a scale-of-16 counter and is discussed in Sec. 18-1). The circuit will then commutate from one channel to the next at the occurrence of each input pulse.

The Encoding Process The decoding process is one in which a binary code establishes the state 1 on one (and only one) of a number of output lines. For example, in Fig. 9-43 the code stored in the form of FLIP-FLOPS selects any one of 16 channels. The inverse process is called *encoding*. In this process there are a number of input lines and a code is generated depending on which of the input channels is in the binary state 1. For example, consider the situation in which there are three inputs and the code is to have four binary digits. Such a situation is represented in the table of Fig. 9-46a. This table is to be interpreted to mean that if $A = 1$ then

$$Y_1 = 1 \qquad Y_2 = 1 \qquad Y_3 = 1 \qquad Y_4 = 0 \qquad \text{etc.}$$

Clearly, the OR connective is invoked here at each output line because

$$Y_1 = A + C \qquad Y_2 = A \qquad Y_3 = A + B + C \qquad Y_4 = B + C$$

The diode matrix for this encoder is indicated in Fig. 9-46b.

A binary-coded decimal (BCD) encoder which converts the decimal numbers 0, 1, 2, . . . , 9 into the binary code is indicated in Fig. 9-47. Note, for example, that since it is necessary to excite the 2^2 channel for any decimal number whose binary code contains 2^2, then OR diodes will be connected to this line from switches 4, 5, 6, and 7. For permanent storage the set terminal S of a flip-flop would be connected to each output channel. The reset terminal R would be excited from the 0 line.

Finally, the problem of converting one code into another is a combination of the two situations treated above. The first code is decoded as in Fig. 9-43 and then encoded as in Fig. 9-46.

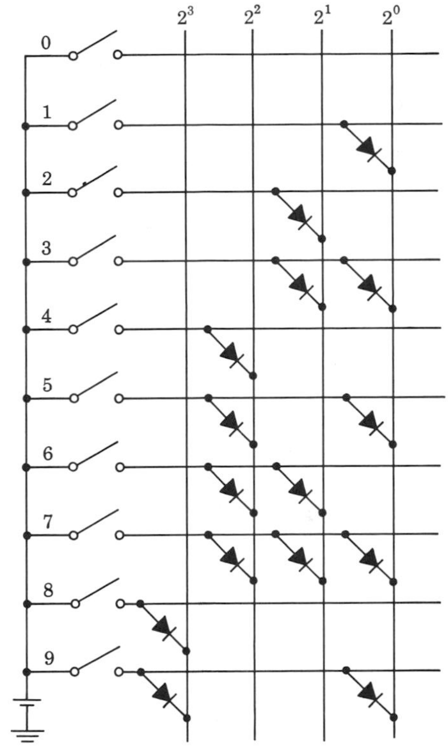

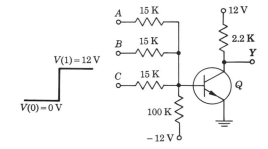

Fig. 9-47 An encoding matrix to transform a decimal number into a binary code (BCD).

9-16 RESISTOR–TRANSISTOR LOGIC[2] (RTL AND RCTL)

In addition to the logic configurations we have already considered—diode logic, diode-transistor logic, and emitter-follower logic—there are several other possible systems.[15] We shall now show how the NOR (NAND) gate may be implemented in some of these other configurations.

If in the DTL circuit of Fig. 9-28b we omit the diodes and fan in through 15-K resistors, as in Fig. 9-48, then we have the *resistor-transistor logic* NOR gate, RTL (or *transistor-resistor logic*, TRL).

Fig. 9-48 Resistor-transistor positive NOR (negative NAND) logic gate.

EXAMPLE Verify that the RTL circuit of Fig. 9-48 obeys NOR logic. Neglect junction voltages and source resistance.

Solution If all inputs are in the zero-voltage state, then the base is connected through the three 15-K resistances in parallel, or through 5 K to ground. The open-circuit base voltage is

$$V_B = -(12)\left(\frac{5}{105}\right) = -0.57 \text{ V}$$

which is sufficient to reverse-bias Q. Hence, the output is at 12 V, or $Y = 1$.

Assume that with one input in the 1 state, Q is in saturation. The equivalent circuit is indicated in Fig. 9-49a. The base current is

$$I_B = \frac{12}{15} - \frac{12}{100} = 0.68 \text{ mA}$$

Since the collector current is $I_C = 12/2.2 = 5.5$ mA, then Q is indeed in saturation if $h_{FE} > 5.5/0.68 = 8$.

If more than one input is excited, then the transistor is driven further into saturation. For example, if all three inputs are in state 1, then, from Fig. 9-49b,

$$I_B = \frac{12}{5} - \frac{12}{100} = 2.28 \text{ mA}$$

Hence, if $h_{FE} > 8$ and if one or more inputs are at 12 V, the output is at V_{CE}(sat), or $Y = 0$, and NOR logic has been performed.

The RTL gate has a relatively slow transient response, particularly if all inputs are excited so that the transistor is driven heavily into saturation. By using capacitors across the input resistors the minority-carrier storage time (Sec. 20-21) can be reduced. Such a configuration is indicated in Fig. 9-50 and is called resistor-capacitor-transistor logic (RCTL). The disadvantage of this circuit is that it is susceptible to high-frequency noise (spikes), which passes from the input to the transistor base through a capacitor. There is also cross talk between input channels because of the capacitors. We have already

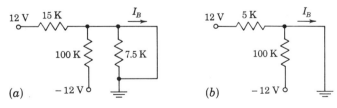

Fig. 9-49 An example of an RTL NOR gate. (*a*) Two inputs are low (0 V) and one input is high (12 V); (*b*) all three inputs are high.

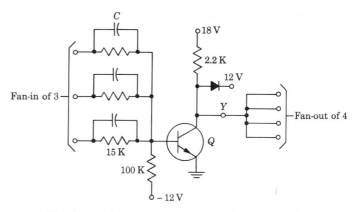

Fig. 9-50 A positive NOR RCTL gate with a clamped output.

noted that in DTL this difficulty is minimized because of the isolation between inputs afforded by the diodes.

The clamping diode indicated in Fig. 9-50 serves several purposes. As already emphasized, it improves the switching speed (Sec. 9-9). It also stabilizes the output voltage level since it makes it independent of the loading (Sec. 8-13). Finally, if this circuit drives similar RTL gates, then the current to saturate the following transistors becomes independent of the fan-out, provided that a diode clamp is used and so long as the loading is not so heavy that the clamp diode stops conducting.

In any logic configuration the number N of inputs is called the *fan-in*, and the number M of outputs is called the *fan-out*. For example, in Fig. 9-50, $N = 3$ and $M = 4$. In the case of resistor-transistor logic the maximum value of N is determined by the condition that, when all the inputs are in the 0 state, the transistor Q must remain OFF even at the highest operating temperature. The maximum value of M is determined by the condition that, when Q is OFF, all M of the following stages must be driven into saturation even at the lowest temperature.

9-17 DIRECT–COUPLED TRANSISTOR LOGIC (DCTL)[2,15,16]

Consider the positive NOR circuit of Fig. 9-51, consisting of the three CE transistors $Q1$, $Q2$, and $Q3$ with collectors tied together. The fan-in is obviously 3 and the fan-out is 2 since the output feeds the two transistors $Q4$ and $Q5$. The input to $Q1$ comes directly from the output Y' of a previous NOR. Since no resistors, capacitors, or diodes are used between stages, such a system is called *direct-coupled transistor logic* (DCTL).

To verify that the circuit implements NOR logic, consider first that all inputs are in the 0 state. Because this low voltage to an input (say to $Q1$)

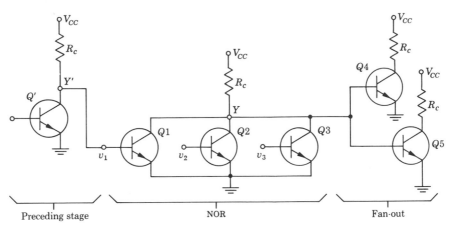

Fig. 9-51 A positive NOR DCTL gate.

comes from a saturated transistor (Q') of a preceding state,

$$v_1 = V_{CE}(\text{sat}) = V(0)$$

Since this voltage is, say, 0.25 V for a heavily saturated silicon transistor, and since the cutin voltage $V_\gamma \approx 0.5$ V, $Q1$ will conduct very little (although theoretically it is not at cutoff, which requires approximately zero bias). Since the current in $Q1$ is almost zero, the output Y tries to rise V_{CC} and $Q4$ and $Q5$ go into saturation. Hence the output Y is clamped at

$$V_{BE}(\text{sat}) = V(1) \approx 0.7 \text{ V}$$

for silicon. Thus, with all inputs in the low state the output is in the high state.

Consider now that at least one input v_1 is in the high state. Since $Q1$ is fed from Q', Q' is cut off and $Q1$ is driven into saturation. Under these circumstances the output Y is $V_{CE}(\text{sat}) = V(0)$. If more than one input is excited, then the output will certainly be low. Hence, we have confirmed that the NOR function is satisfied.

There are a number of difficulties with DCTL. The reverse saturation current for all fan-in transistors adds in the common collector-circuit resistor R_c. At high enough temperatures the total $I_{CBO}R_c$ drop may be large enough so that the output Y is too low to drive the fan-out transistors into saturation. Further, because of the direct connection, the base current is almost equal to the collector current [for $V_{CC} \gg V_{CE}(\text{sat})$ and $V_{CC} \gg V_{BE}(\text{sat})$]. With a transistor so heavily driven into saturation, very large stored base charge will result, with a corresponding detrimental effect on the switching speed. Since the voltage levels are so low—the total output-voltage step is only of the order of 0.5 V for silicon and 0.25 V for germanium—then spurious (noise) spikes can be troublesome. The bases of the fan-out transistors are connected

together.   Since the input characteristics can never be identical let us assume
that $Q4$ has a much lower V_{BE} for a given I_B than does $Q5$.   Under these
circumstances, $Q4$ will "hog" most of the base current, and it is possible that
$Q5$ may not even be driven into saturation.   Hence, transistors suitable for
DCTL must have very close control on uniformity of input characteristics, very
low values of I_{CBO}, as large a differential as possible between $V_{BE}(\text{sat})$ and
$V_{CE}(\text{sat})$, a large h_{FE}, and a small storage time.

The advantages of DCTL are: (1) the need for only one low-voltage supply (operation with 1.5 V is possible), (2) transistors with low breakdown
voltages may be used, (3) the power dissipation is low, and (4) this configuration is advantageous for integrated-circuit manufacture because transistors
are cheaper to fabricate with integrated techniques[15] than are resistors or
capacitors.

We shall conclude this section with a few random observations.   In
DCTL a NOR or NAND circuit is possible in which the fan-in transistors are in
series rather than in parallel as in Fig. 9-51 (Prob. 9-43).   The DCTL FLIP-FLOP
is discussed in Sec. 10-10.   There is no counterpart to DCTL using tubes,
because the grid of one stage cannot be directly connected to the plate of the
preceding triode without drawing excessive grid current.

Finally, the DCTL circuit may be modified in order to circumvent some
of the above disadvantages.   The variability of $V_{BE}(\text{sat})$—the current-hogging
difficulty—may be minimized by including a resistor R_1 in series with the base.
With this addition the base current depends primarily on V_{CC}, R_c, and R_1
as long as $V_{CC} \gg V_{BE}(\text{sat})$.   To reduce storage time, R_1 should be bypassed
with a capacitor.   A further modification consists in returning the base through
a resistor R_2 to a reverse-biasing source.   Such a connection makes the operation of the circuit less sensitive to variations in $V_{CE}(\text{sat})$ and ensures cutoff in
the transistor which is supposed to be reverse-biased.   If these modifications
are made each transistor takes the form of the INVERTER of Fig. 9-15.   Several
inverters, with collectors connected together, can therefore function as a NOR
circuit.   This configuration—it might be called *transistor* or *tube logic* (TL)—
is seldom used because it is more expensive than TRL or DL.

9-18   LOW–LEVEL LOGIC[2,15] (DTL AND TTL)

The DTL circuit of Fig. 9-25 may be modified by replacing the R_1C_1 combination by one or more silicon diodes $D1$, as indicated in Fig. 9-52.   When at least
one input is at $V(0) = V_{CE}(\text{sat}) \approx 0.3$ V, the voltage at P is the sum of $V(0)$
and the diode voltage $V_D \approx 0.6$ V.   At a coincidence the voltage at P rises
to equal the sum of V_{D1} (≈ 0.6 V for one diode, 1.2 V for two series diodes,
etc.) plus $V_{BE}(\text{sat}) \approx 0.7$ V.   Since the swing at point P required to drive the
transistor from cutoff to saturation is of the order of 1 V, this configuration is
called *low-level logic* (LLL).   It is also referred to as *current-switching diode logic*
(CSDL) because the current in a diode D connected to an input in the $V(0)$ state

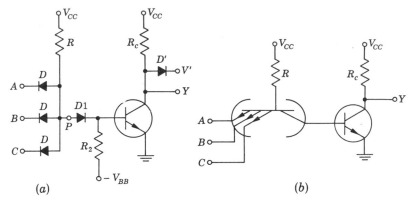

Fig. 9-52 A positive-NAND low-level-logic (LLL) gate. (a) A DTL gate (two or three diodes in series may be used for $D1$; also, V_{BB} may be ground); (b) a TTL gate.

is switched through $D1$ into the base of the transistor whenever all inputs are excited to the $V(1)$ state.

Since, as mentioned above, only a low voltage swing is required, then the output may be clamped at, say, 3 V. As already emphasized, such clamped operation means improved switching speed. In order to make the circuit less susceptible to noise voltages, more than one diode may be connected in series for $D1$. Of course, the required voltage swing is then increased by the drop across these extra diodes. It is also possible to return the base terminal to the emitter through R_2 and thus eliminate the V_{BB} supply. This simplification means that the transistor will not be theoretically cut off, but will operate with $V_{BE} \approx 0$ and hence the collector current will be very small. Even in this cutoff condition the noise immunity is excellent because the series diodes are reverse-biased.

The diodes D and $D1$ of Fig. 9-52a may be replaced by the base-to-emitter diode and the base-to-collector diodes, respectively, of a transistor.[15] This all-transistor configuration is especially simple if integrated-circuit techniques are used because it is then not difficult to fabricate a multiemitter transistor.[17] (See Fig. 17-31 for constructional details.) A one-power-supply version of this *transistor-transistor logic* gate (TTL) is given in Fig. 9-52b. It is capable of high speed at low power levels but may be noisier than DTL.

9-19   CURRENT–MODE LOGIC (CML OR ECTL)[2,15,18]

In order to eliminate minority-carrier storage time, a transistor must not be allowed to go into saturation. In Sec. 7-9 the nonsaturating emitter-coupled clipper circuit of Fig. 7-14b is discussed. It is found that the emitter current

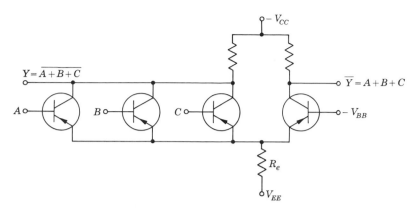

Fig. 9-53 A positive NOR ECTL gate.

remains essentially constant and that this current is switched from one transistor to the other as the input signal varies from below to above the reference voltage $-V_{BB}$. By adding several input transistors in parallel this circuit becomes the *current-mode logic* (CML) or *emitter-coupled transistor logic* (ECTL) of Fig. 9-53. Complementary outputs are available. A positive NOR results when the output is Y and a positive OR is obtained at the $\bar{Y}$ terminal. One of the difficulties with this configuration is that the $V(0)$ and $V(1)$ levels in the output differ from those in the input. Hence, avalanche diodes and emitter followers must be used in the output to provide the proper d-c level shift. The possible higher speed obtainable with ECTL is offset by the increased power dissipation and the increased component count.

9-20   COMPARISON OF LOGIC CIRCUITS[15,19]

In the discussion of each logic configuration some of its advantages and disadvantages have been listed. An exhaustive comparison is extremely difficult because we must take into account all of the following characteristics: (1) speed (propagation time delay), (2) noise immunity, (3) fan-in and fan-out capabilities, (4) power-supply requirements, (5) power density when packaged, (6) suitability for integrated fabrication, (7) reliability, (8) maintainability, and (9) cost. Also to be considered is the personal prejudice of the engineer, who is always strongly influenced by past experience.

It should be clear that there can be no single logic configuration best suited for all applications. Different manufacturers[19] have available in integrated form the types of logic discussed above, there being no decided preference for one type over the other. On the other hand, however, the manufacturers of logic cards using lumped components (with operating repetition rates of up to a few megacycles per second) seem to favor DTL.

REFERENCES

1. Page, C. H.: Digital Computer Switching Circuits, *Electronics*, vol. 21, no. 9, pp. 110–118, September, 1948.

2. Pressman, A. I.: "Design of Transistorized Circuits for Digital Computers," John F. Rider, Publisher, Inc., New York, 1959.

3. Braun, E. L.: "Digital Computer Design," Academic Press Inc., New York, 1963. Ledley, R. S.: "Digital Computer and Control Engineering," McGraw-Hill Book Company, New York, 1960.

4. Richards, R. K.: "Arithmetic Operations in Digital Computers," D. Van Nostrand Company, Inc., Princeton, N.J., 1955.
   Richards, R. K.: "Digital Computer Components and Circuits," D. Van Nostrand Company, Inc., Princeton, N.J., 1957.
   Burroughs Corporation: "Digital Computer Principles," McGraw-Hill Book Company, New York, 1962.

5. American Standard Graphic Symbols for Logic Diagrams, no. 91 (ASA Y32.14), Institute of Electrical and Electronics Engineers, New York, September, 1962.

6. Chen, T. C.: Diode Coincidence and Mixing Circuits in Digital Computers, *Proc. IRE*, vol. 38, pp. 511–514, May, 1950.
   Hussey, L. W.: Semiconductor Diode Gates, *Bell System Tech. J.*, vol. 32, no. 5, pp. 1137–1154, September, 1953.

7. Masher, D. P.: The Design of Diode-Transistor NOR Circuits, *IRE Trans. Electron. Computers*, vol. EC-9, no. 1, pp. 15–24, March, 1960.
   Todd, C. R.: An Annotated Bibliography on NOR and NAND Logic, *IEEE Trans. Electron. Computers*, vol. EC-12, no. 5, pp. 462–464, October, 1963.

8. Computer Control Company, Framingham, Mass.
   Data Technology Corporation, Palo Alto, Calif.
   Digital Equipment Corporation, Maynard, Mass.
   Engineering Electronics Company, Santa Ana, Calif.
   Intercontinental Instrument Incorporated, Farmingdale, N.Y.
   Navigation Computer Corporation, Norristown, Pa.
   Ransom Research, San Pedro, Calif.
   Transitel International, Paramus, N.J.

9. Fairchild, Hughes, Motorola, Philco, Raytheon, Signetics, Siliconix, Sylvania, Texas Instruments, and Westinghouse Companies.

10. Elbourn, R. D., and R. P. Witt: Dynamic Circuit Techniques Used in SEAC and DYSEAC, *Proc. IRE*, vol. 41, pp. 1380–1387, October, 1953.
    Davis, E. M., W. E. Harding, R. S. Schwartz, and J. J. Corning: Solid Logic Technology: Versatile High-performance Microelectronics, *IBM J. Res. Develop.*, vol. 8, no. 2, pp. 102–114, 1964.

11. Gray, H. J.: Logical Description of Some Digital-computer Adders and Counters, *Proc. IRE*, vol. 40, pp. 29–33, January, 1952.

12. Eckert, J. P.: A Survey of Digital Computer Memory Systems, *Proc. IRE*, vol. 41, pp. 1393–1406, October, 1953.

13. "Shift Register 108C and 109A," Navigation Computer Corporation, Norristown, Pa.

14. Brown, D. R., and N. Rochester: Rectifier Networks for Multiposition Switching, *Proc. IRE*, vol. 37, pp. 139–147, February, 1949.

15. Khambata, A. J.: "Integrated Semiconductor Circuits," John Wiley & Sons, Inc., New York, 1963.

16. Beter, R. H., W. E. Bradley, R. B. Brown, and M. Rubinoff: Directly Coupled Transistor Circuits, *Electronics*, vol. 28, no. 6, pp. 132–136, June, 1955.

17. Boulter, B. A.: A New Active Device Suitable for Use in Digital Circuits, *Electron. Eng.*, vol. 35, no. 420, pp. 86–91, February, 1963.

18. "Switching Transistor Handbook," chap. 8, Motorola Inc., Phoenix, Ariz., 1963.

19. Phelps, M. H.: Choice of Logic Forms for Integrated Circuits, *Eng. Design News*, January, 1964, pp. 30–38.
    Holloway, J. A.: On-the-shelf Monolithic Digital Circuits, *Eng. Design News*, February, 1964, pp. 19–20.

10 / BISTABLE MULTIVIBRATORS

A bistable circuit is one which can exist indefinitely in either of two stable states and which can be induced to make an abrupt transition from one state to the other by means of external excitation. In this chapter we consider two-stage regenerative amplifiers which behave as bistable circuits. Such interconnected amplifier pairs are known by a wide variety of names, such as *bistable multivibrator (multi†)*, *Eccles-Jordan circuit* (after the inventors), *trigger circuit, scale-of-2 toggle circuit, flip-flop,†* and *binary*. A bistable multi is used for the performance of many digital operations such as counting and the storing of binary information. The circuit also finds extensive application in the generation and processing of pulse-type waveforms.

10-1   THE STABLE STATES OF A BINARY

The circuit diagram of a flip-flop is shown in Fig. 10-1. The active devices $A1$ and $A2$ are either tubes or transistors, so that the input X is either the grid of a tube or the base of a transistor, the output Y is either the plate of a tube or the collector of a transistor, and Z is either a cathode or an emitter. The indicated supply-voltage polarities are proper for an *n-p-n* transistor or for a tube and must be reversed if a *p-n-p* transistor binary is under consideration. Note that the output of each amplifier is direct-coupled to the input of the other amplifier (inverter). Compare with Fig. 9-39.

Because of the symmetry of the circuit we might expect the

† The jargon terms "multi" and "flip-flop" are firmly entrenched in the literature.

362

Fig. 10-1 A binary circuit. If the active devices are tubes, then $V_{YY} = V_{PP}$, $V_{XX} = V_{GG}$, and $R_y = R_p$. If the devices are transistors, then $V_{YY} = V_{CC}$, $V_{XX} = V_{BB}$, and $R_y = R_c$.

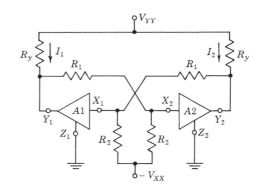

quiescent currents in each amplifier to be the same. Such would indeed be the case if both devices were biased negatively enough to be cut off or if both were biased so positively as to be in clamp (if tubes) or in saturation (if transistors). These extreme cases turn out to be of little practical interest. Let us investigate whether or not it is possible for both devices also to operate normally (within the grid base, if tubes, or within the active region, if transistors) and simultaneously to carry equal currents. In such a circumstance it is possible to find currents $I_1 = I_2$ which are consistent with the device characteristics and with Kirchhoff's laws, and hence such a state of the binary is an equilibrium state. This state, however, is one of *unstable* equilibrium, as may be seen from the following considerations.

Suppose that there should be a minute fluctuation in the current I_1. If I_1 increases, the voltage at the output Y_1 will decrease. This will then decrease the voltage at the input X_2. This change in voltage at X_2 will be amplified and inverted by $A2$ and the output voltage at Y_2 will increase. Hence the voltage at X_1 will become more positive, and as a consequence the current I_1 will increase still further. This cycle of events repeats itself. The current I_1 continues to increase and the current I_2 continues to decrease, the circuit moving progressively further away from its initial condition. This action takes place because of the regenerative feedback incorporated into the circuit and will occur only if the loop gain of the circuit is larger than unity.

From the above discussion it is clear that a *stable* state of a binary is one in which the currents and voltages satisfy Kirchhoff's laws and are consistent with the device characteristics and in which, in addition, the condition is satisfied that the loop gain is less than unity. The condition with respect to the loop gain will certainly be satisfied if either of the two devices is below cutoff or if either device is in clamp or in saturation.† In principle, in order that a flip-flop be in a stable state, it would be sufficient *either* that one of the devices be OFF *or* that one be ON.

For practical reasons, to be discussed now, the arrangement almost invariably employed is one in which one of the devices is OFF *and* the other

---

† A tube in clamp or a transistor in saturation will henceforth be designated as a device which is ON, whereas an OFF device is one biased below cutoff.

is ON.   Consider, for example, that one amplifier is below cutoff and the other is biased in its active region.   As the temperature changes or the devices age and the device parameters vary, the quiescent point changes and the output voltage may change appreciably.   Even more serious is the possibility that a larger drift may cause the device that was assumed to be operating in its active region to be driven below cutoff.   In such a case both devices in the binary would be below cutoff and the circuit would be useless.

In practice, we should like to be able to assemble these flip-flops using components which are held to a tolerance no better than about 10 percent. And we should like to feel confident that the binary will continue to operate as the devices age or are replaced and despite reasonable variations in supply voltages and ambient temperature.   For these reasons the flip-flop is usually adjusted so that in a stable state one device is well below cutoff while the other is well in clamp or in saturation.

10-2   A VACUUM–TUBE BINARY

In the flip-flop of Fig. 10-2 the component values and supply voltages indicated are typical for use with the medium-μ vacuum triode such as the type 5965. The useful signal at the plate, called the *output* or the *plate swing* V_w, is the voltage change corresponding to a transition from one stable state to the other. Since the fraction of the plate swing which is coupled across to the opposite grid depends only on the ratio of the resistances R_1 and R_2, these resistances are usually made large enough compared with the external plate-circuit resistance R_p so as to avoid loading the amplifier output excessively.   The plate swing for a given tube is determined by V_{PP} and R_p.   Since the plate characteristic for $V_G = 0$ can be approximated by a straight line through the origin with a slope given by the plate resistance r_p, then for $R_p = r_p$, $V_w \approx V_{PP}/2$.   If R_p is very much larger than r_p, then the plate swing may approach the full plate supply voltage.   However, too large a value of R_p adversely affects the maximum speed with which the flip-flop may be made to operate (Sec. 10-5).

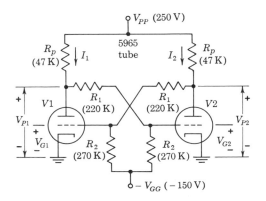

Fig. 10-2   A typical vacuum-tube binary circuit.   The values in parentheses refer to the example on page 365.

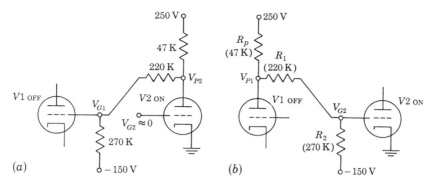

Fig. 10-3 The circuit of Fig. 10-2 redrawn to indicate the connections between (a) the plate of $V2$ and the grid of $V1$ and (b) the plate of $V1$ and the grid of $V2$.

Hence, R_p is usually chosen to be of the order of r_p (tens of kilohms), whereas the magnitudes of R_1 and R_2 are hundreds or thousands of kilohms.

    The procedure for calculating the circuit currents and voltages in a stable state is particularly simple if we take advantage of the fact that R_1 and R_2 are large in comparison with R_p and large also in comparison with the grid-to-cathode resistance r_G. In such a case the tube which is in clamp will be tightly clamped, and we shall not make a serious error if we consider that the grid-to-cathode voltage is zero. Furthermore, in such a case the current in the conducting tube may be considered to be identical to the current through its resistor R_p. A typical calculation is given in the following illustrative example.

EXAMPLE Compute the stable-state currents and voltages for the binary circuit of Fig. 10-2. The triodes are the two sections of a type 5965 vacuum tube (Fig. D-9).

Solution The analysis is clarified if we draw two circuits, Fig. 10-3a—showing the connections between the plate of $V2$ and the grid of $V1$—and Fig. 10-3b—indicating the connections between the plate of $V1$ and the grid of $V2$. Let us assume that $V1$ is cut off and $V2$ is in clamp with a grid-to-cathode voltage equal to zero and then verify that these assumptions are valid.

    To find the plate voltage at $V2$, we neglect the loading of R_1 and R_2 and draw a load line corresponding to 47 K and 250 V on the plate characteristics of the tube. The plate current and voltage for $V_{G2} = 0$ are found to be $V_{P2} = 33$ V and $I_2 = 4.6$ mA.

    We must now check to see whether or not $V1$ is indeed cut off when $V2$ is in clamp. The grid voltage of $V1$ is calculated from the equivalent circuit of Fig. 10-3a. The voltage V_{G1} is calculated by superposition to be

$$V_{G1} = 33\,\frac{270}{220 + 270} + (-150)\,\frac{220}{220 + 270} = -49 \text{ V}$$

The plate voltage V_{P1} of $V1$ is calculated from Fig. 10-3b with $V_{G2} = 0$. We find $V_{P1} = 250 \times 220/(220 + 47) = 206$ V. Since, at a plate-to-cathode voltage of 206 V, cutoff occurs at -8 V (Fig. D-9), $V1$ is well below cutoff. Note that the voltage at the plate of the OFF tube is not equal to the supply V_{PP} ($= 250$ V) because of the loading of the divider R_1.

We must now check to see whether $V2$ is in clamp when $V1$ is cut off. If the grid of $V2$ were not drawing grid current, then, from the equivalent circuit drawn in Fig. 10-3b, the voltage V_{G2} would be

$$V_{G2} = 250 \frac{270}{220 + 270 + 47} + (-150) \frac{220 + 47}{220 + 270 + 47} = 50 \text{ V}$$

This 50 V is applied to the grid of $V2$ from a source whose Thévenin's equivalent impedance is equal to the parallel combination of $R_1 + R_p$ and R_2. This equivalent impedance is 134 K. If we calculate the grid voltage by considering that the grid-to-cathode resistance is $r_G = 250$ Ω (Sec. 6-19), we have

$$V_{G2} = 50 \frac{0.25}{134} = 0.10 \text{ V}$$

This same result can be obtained by multiplying the short-circuit current by the impedance from G_2 to ground. Thus

$$V_{G2} = \left(\frac{250}{267} - \frac{150}{270}\right) 0.25 = 0.10 \text{ V}$$

Hence $V2$ is indeed in clamp, and we have made only a very small error in assuming that the grid-to-cathode voltage is zero. Actually, as we know from the discussion of Sec. 6-19, V_{G2} may well be some tens of millivolts negative. Similarly, if the loading of R_1 and R_2 is taken into account by drawing the Thévenin's equivalent from the plate of $V2$ to ground, the value of V_{P2} is found to be 32 V instead of 33 V. These errors may well be smaller than the error involved in applying the *average* tube characteristics to a *particular* tube. To summarize, in the stable state we have approximately

$$I_1 = 0 \text{ mA} \qquad V_{P1} = 206 \text{ V} \qquad V_{G1} = -49 \text{ V}$$

$$I_2 = 4.6 \text{ mA} \qquad V_{P2} = 33 \text{ V} \qquad V_{G2} = 0 \text{ V}$$

The output swing is $V_w = 206 - 33 = 173$ V.

The binary has *two* stable states. In one state $V1$ is cut off and $V2$ is in clamp. In the second state $V2$ is cut off and $V1$ is in clamp. The principal importance of the flip-flop results from the fact that it is possible, by a variety of means, to transfer the binary from one stable state to the other. Suppose, for example, that the grid of the OFF tube were momentarily shorted to ground. This OFF tube would go ON and in so doing would turn OFF the tube that was initially ON. This condition would again persist permanently even after the short circuit is removed. This means of transferring conduction, however, is not ordinarily useful. More practically useful methods will be considered later.

In most applications the transistor flip-flop is displacing the vacuum-tube binary. Hence, the emphasis in the remainder of this chapter will be on the semiconductor bistable circuit.

10-3   A FIXED–BIAS TRANSISTOR BINARY

A fixed-bias flip-flop using n-p-n transistors is indicated in Fig. 10-4. Nearly the full supply voltage V_{CC} will appear across the transistor that is OFF. The condition that the supply voltage be reasonably smaller than the transistor collector breakdown voltage BV_{CE} (Sec. 6-9) will usually restrict V_{CC} to the order of several tens of volts. This supply voltage is to be compared with the several hundred volts used with vacuum tubes.

Under saturation conditions the collector current I_C is a maximum. Hence, R_c must be chosen so that this value of I_C ($\approx V_{CC}/R_c$) does not exceed the maximum permissible current. The values of R_1, R_2, and V_{BB} must be selected so that in one state the base current is large enough to drive the transistor into saturation whereas in the second state the emitter junction must be below cutoff. The signal at a collector, called the *output swing V_w*, is the change in collector voltage resulting from a transition from one state to the other, or $V_w = V_{C1} - V_{C2}$. If the loading of R_1 can be neglected, then the collector voltage of the OFF transistor is V_{CC}. Since the collector saturation voltage is a few tenths of a volt, then the swing $V_w \approx V_{CC}$ *independently of R_c*.

Manufacturers of switching transistors useful in flip-flop circuits usually specify the cutoff and saturation characteristics. The reverse saturation current I_{CBO} is given as a function of temperature (Sec. 6-8, Fig. 6-14). The d-c CE current transfer ratio h_{FE} is specified as a function of collector current I_C (Sec. 6-12, Fig. 6-23). The saturation voltage V_{CE}(sat) is given as a function of I_C and base current I_B (Sec. 6-12, Fig. 6-22). Also, V_{BE}(sat) versus I_C and I_B is specified (Sec. 6-13, Fig. 6-26). The temperature dependence of the saturation parameters is often indicated (Fig. 6-27). We shall demonstrate by the solution of illustrative problems and by subsequent discussions

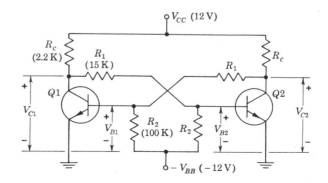

Fig. 10-4   A fixed-bias n-p-n transistor binary. The values in parentheses refer to the example on page 368.

that these parameters [I_{CBO}, h_{FE}, $V_{CE}(\text{sat})$, and $V_{BE}(\text{sat})$] are sufficient to allow an analysis of transistor binary circuits.

EXAMPLE Calculate the stable-state currents and voltages for the flip-flop circuit of Fig. 10-4, consisting of two cross-coupled INVERTER circuits whose parameter values are those given in Fig. 9-16. Assume that the transistors have a minimum h_{FE} value of 20.

Solution The analysis is carried out by drawing two circuits—Fig. 10-5a, showing the connections between the base of $Q1$ and the collector of $Q2$, and Fig. 10-5b, indicating the connection between the collector of $Q1$ and the base of $Q2$. These equivalent circuits are analogous to those in Fig. 10-3 for the vacuum-tube binary. Assume that transistor $Q1$ is OFF and transistor $Q2$ is ON. Since the saturation voltages are small (a few tenths of a volt), let us initially neglect them altogether and assume that $V_{B2} = 0$ and $V_{C2} = 0$. From Fig. 10-5a we can verify that, with $Q2$ in saturation, $Q1$ is OFF. Thus, if we neglect I_{CBO},

$$V_{B1} = -12 \left(\frac{15}{15 + 100} \right) = -1.56 \text{ V}$$

Since a back bias of only about 0.1 V (Ge) or 0 V (Si) is required to cut off an emitter junction (Table 6-1, page 219), $Q1$ is indeed OFF.

To verify that with $Q1$ beyond cutoff, $Q2$ is in saturation, we first calculate I_{C2}. From Fig. 10-5a, and continuing to neglect I_{CBO},

$$I_1 = \frac{12}{2.2} = 5.45 \text{ mA} \qquad I_2 = \frac{12}{15 + 100} = 0.10 \text{ mA}$$

and

$$I_{C2} = I_1 - I_2 = 5.45 - 0.10 = 5.35 \text{ mA}$$

(In this illustration I_2 is negligible compared with I_1, but in many binary designs such is not the case.) If the transistor type had been specified, then the minimum base current I_{B2} required for a collector saturation current of 5.35 mA could have

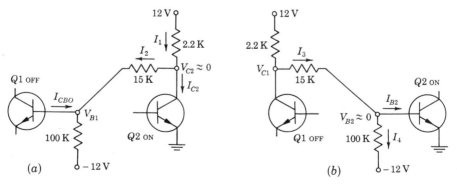

Fig. 10-5 Equivalent circuits for computing the stable states of the binary circuit with the parameters given in Fig. 10-4.

been read from the collector characteristic curves. Since in this example h_{FE} was specified but no curves were supplied, we may use the alternative procedure of finding the minimum I_{B2} for saturation from

$$(I_{B2})_{\min} = \frac{I_{C2}}{h_{FE}} = \frac{5.35}{20} = 0.27 \text{ mA}$$

From Fig. 10-5b we can find the base current of Q2. Thus

$$I_3 = \frac{12}{2.2 + 15} = 0.70 \text{ mA} \qquad I_4 = \frac{12}{100} = 0.12 \text{ mA}$$

and

$$I_{B2} = I_3 - I_4 = 0.70 - 0.12 = 0.58 \text{ mA}$$

Since this value exceeds the minimum base current (0.27 mA) required for saturation, we have verified that Q2 is indeed in saturation.

The collector voltage of Q1 is, from Fig. 10-5b,

$$V_{C1} = 12 - 2.2I_3 = 12 - (2.2)(0.70) = 10.5 \text{ V}$$

In summary, a stable state of the binary is characterized by the following voltages and currents:

$$I_{C1} = 0 \text{ mA} \qquad I_{C2} = 5.35 \text{ mA} \qquad I_{B1} = 0 \text{ mA} \qquad I_{B2} = 0.58 \text{ mA}$$

$$V_{C1} = 10.5 \text{ V} \qquad V_{C2} \approx 0 \text{ V} \qquad V_{B1} = -1.56 \text{ V} \qquad V_{B2} \approx 0 \text{ V}$$

The second stable state is one in which Q2 is OFF and Q1 is ON and the above currents and voltages are interchanged between Q1 and Q2. The output swing is $V_{C1} - V_{C2} = 10.5$ V, which approaches the collector supply voltage of 12 V.

The assumptions (that $V_{B2} = 0$ and $V_{C2} = 0$) made in this example may be removed by referring to the manufacturer's curves. For example, if the transistors are silicon type 2N914, then corresponding to $I_{B2} = 0.58$ mA and $I_{C2} = 5.35$ mA $(I_{C2}/I_{B2} = 9.2)$ we find $V_{CE2}(\text{sat}) = 0.15$ V from Fig. 6-22 and $V_{BE2}(\text{sat}) = 0.7$ V from Fig. 6-26. Using these voltages we can recalculate the stable-state currents and voltages. For example, from Fig. 10-5a with $V_{C2} = 0.15$ V, using the superposition principle,

$$V_{B1} = -12\left(\frac{15}{15 + 100}\right) + 0.15\left(\frac{100}{15 + 100}\right) = -1.43 \text{ V} \tag{10-1}$$

and Q1 is OFF. From Fig. 10-5a we also find

$$I_1 = \frac{12 - 0.15}{2.2} = 5.39 \text{ mA} \qquad I_2 = \frac{0.15 + 12}{15 + 100} = 0.11 \text{ mA} \tag{10-2}$$

$$I_{C2} = I_1 - I_2 = 5.28 \text{ mA} \qquad (I_{B2})_{\min} = \frac{5.28}{20} = 0.26 \text{ mA} \tag{10-3}$$

From Fig. 10-5b, with $V_{B2} = 0.7$ V,

$$I_3 = \frac{12 - 0.7}{2.2 + 15} = 0.66 \text{ mA} \qquad I_4 = \frac{0.7 + 12}{100} = 0.13 \text{ mA} \tag{10-4}$$

and $I_{B2} = I_3 - I_4 = 0.53$ mA.   Since this value of I_{B2} exceeds

$(I_{B2})_{min} = 0.26$ mA

$Q2$ is ON.   Since $V_{C1} = 12 - (0.66)(2.2) = 10.5$ V, the new values of a stable state are

$I_{C1} = 0$ mA	$I_{C2} = 5.28$ mA	$I_{B1} = 0$ mA	$I_{B2} = 0.53$ mA
$V_{C1} = 10.5$ V	$V_{C2} = 0.15$ V	$V_{B1} = -1.43$ V	$V_{B2} = 0.7$ V

If these new values of I_{C2} and I_{B2} are used in Figs. 6-26 and 6-22, we again find $V_{BE2}(\text{sat}) = 0.7$ V and $V_{CE2}(\text{sat}) = 0.15$ V, so that no further calculations are required.

We note by comparing the two sets of values obtained above that a small error results from assuming that a transistor in saturation behaves as an ideal short circuit.   Often this error can be neglected, particularly if the applied voltages are large compared with the junction voltages.

In the calculations given above of the stable states of a flip-flop we neglected initially the voltages across the saturated transistor.   Thereafter, on the basis of the results of this initial calculation, we determined the transistor voltages and made an improved calculation.   We may arrive at our end result more expeditiously if we recognize that over a wide range of operating conditions of low- and medium-power transistors, the saturation voltages are fairly closely approximated by the values in Table 6-1 (page 219).

The peak emitter-junction breakdown voltage BV_{EBO} and the variations of h_{FE} and I_{CBO} with temperature must be taken into account in designing transistor flip-flops.   These limitations are discussed in detail on page 324 in connection with the transistor inverter.

Loading   The binary may be used to drive other circuits, and hence at one or both collectors there are shunting loads which are not indicated in Fig. 10-4. These loads must be taken into account because they reduce the magnitude of the collector voltage V_{C1} of the OFF transistor.   The first effect of the loading is to give a reduced output swing.   More importantly, however, is the fact that a reduced V_{C1} will decrease I_{B2} and it is possible that $Q2$ may not be driven into saturation.   Hence the flip-flop circuit components must be chosen so that under the heaviest load which the binary drives, one transistor remains in saturation while the other is cut off.

Since the resistor R_1 also loads the OFF transistor, we should like to use a value of R_1 which is large compared with R_c.   However, to ensure a loop gain in excess of unity during the transition between states the inequality $R_1 < h_{fe}R_c$ must be satisfied (Prob. 10-6).

For some applications (in computers) the loading varies with the operation (the "logic") being performed.   For such a circuit, the extent to which a transistor is driven into saturation is variable.   A constant output swing

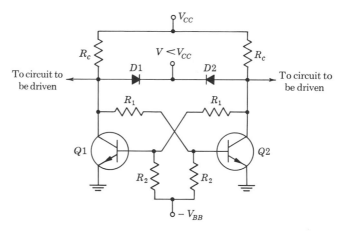

Fig. 10-6 The binary of Fig. 10-4 with collector catching diodes $D1$ and $D2$ added.

$V_w \approx V$ and a constant base saturation current I_{B2} can be obtained by clamping the collectors to an auxiliary voltage $V < V_{CC}$ through the diodes $D1$ and $D2$, as indicated in Fig. 10-6.   As $Q1$ cuts off, its collector voltage rises, and when it reaches V, the "collector catching diode" $D1$ (Sec. 8-13) conducts and clamps the output to V (except for the small drop in the diode). The effect of loading must be taken into account with a vacuum-tube flip-flop as well as with a transistor binary.

10-4   A SELF–BIASED TRANSISTOR BINARY

The need for the negative power supply in Fig. 10-4 may be eliminated by using a common emitter resistor R_e to provide self-bias, as in Fig. 10-7.   The

Fig. 10-7 A self-biased p-n-p transistor binary.   The numerical values refer to the example on page 372.

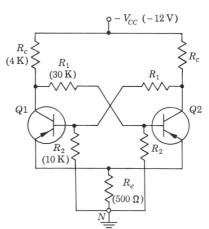

procedure for calculating the stable states is, in principle, the same as is employed for a fixed-bias flip-flop and is given in the following illustrative example.

EXAMPLE Calculate the stable-state currents and voltages for the binary circuit of Fig. 10-7, which uses p-n-p germanium transistors. Find the minimum value of h_{FE} which will keep the ON transistor in saturation.

Solution Assume that $Q1$ is cut off and $Q2$ is in saturation. The connections between the base of $Q1$ and the collector of $Q2$ are indicated in Fig. 10-8a, whereas the connections from the base of $Q2$ to the collector of $Q1$ are given in Fig. 10-8b. In order to proceed further we must find the voltage V_{EN} from the emitter to the ground N. Since $V_{EN} = (I_{B2} + I_{C2})R_e$ we must find the saturation currents from

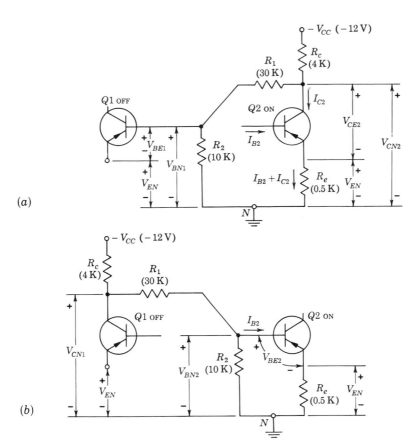

(a)

(b)

Fig. 10-8 The circuit of Fig. 10-7 redrawn to indicate the connections between (a) the base of $Q1$ and the collector of $Q2$ and (b) the base of $Q2$ and the collector of $Q1$.

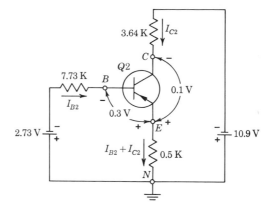

Fig. 10-9 The equivalent circuit when $Q2$ in Fig. 10-7 is in saturation.

the equivalent circuit for $Q2$ given in Fig. 10-9. In this diagram we have replaced the collector circuit of $Q2$ in Fig. 10-8a by its Thévenin's voltage

$$\frac{-V_{CC}(R_1 + R_2)}{R_1 + R_2 + R_c} = \frac{(-12)(30 + 10)}{30 + 10 + 4} = -10.9 \text{ V} \tag{10-5}$$

and its Thévenin's resistance

$$\frac{R_c(R_1 + R_2)}{R_1 + R_2 + R_c} = \frac{(4)(40)}{44} = 3.64 \text{ K} \tag{10-6}$$

Similarly, the Thévenin's equivalent of the base circuit of $Q2$ is obtained from Fig. 10-8b as a voltage

$$\frac{-V_{CC}R_2}{R_1 + R_2 + R_c} = \frac{(-12)(10)}{44} = -2.73 \text{ V} \tag{10-7}$$

in series with a resistance

$$\frac{R_2(R_1 + R_c)}{R_1 + R_2 + R_c} = \frac{(10)(34)}{44} = 7.73 \text{ K} \tag{10-8}$$

Hence the equivalent circuit of $Q2$ is as drawn in Fig. 10-9. Since a germanium transistor is under consideration we shall assume, as in Table 6-1 (page 219), that $V_{BE}(\text{sat}) \approx -0.3$ V and $V_{CE}(\text{sat}) \approx -0.1$ V. The KVL equations are

$$2.73 - 0.3 + I_{B2}(7.73 + 0.5) + I_{C2}(0.5) = 0 \tag{10-9}$$

and

$$10.9 - 0.1 + I_{B2}(0.5) + I_{C2}(3.64 + 0.5) = 0 \tag{10-10}$$

Solving, we find $I_{B2} = -0.138$ mA and $I_{C2} = -2.59$ mA. Hence

$$(h_{FE})_{\min} = \frac{I_{C2}}{I_{B2}} = \frac{-2.59}{-0.138} = 18.8$$

The voltages in the circuit are now found from Fig. 10-8a and b:

$$V_{EN} = (I_{B2} + I_{C2})R_e = (-0.138 - 2.59)(0.5) = -1.36 \text{ V}$$

$$V_{CN2} = V_{CE2} + V_{EN} = -0.1 - 1.36 = -1.46 \text{ V}$$

$$V_{BN2} = V_{BE2} + V_{EN} = -0.3 - 1.36 = -1.66 \text{ V}$$

$$V_{BN1} = V_{CN2} \frac{R_2}{R_1 + R_2} = (-1.46)\left(\frac{10}{40}\right) = -0.37 \text{ V}$$

$$V_{BE1} = V_{BN1} - V_{EN} = -0.37 + 1.36 = +0.99 \text{ V}$$

A positive value of V_{BE} of only about 0.1 V is required to cut off a p-n-p transistor. Hence $Q1$ is certainly OFF.

From Fig. 10-8b,

$$V_{CN1} = \frac{-V_{CC}R_1}{R_c + R_1} + \frac{V_{BN2}R_c}{R_c + R_1} = \frac{(-12)(30)}{34} + \frac{(-1.66)(4)}{34} = -10.8 \text{ V} \quad (10\text{-}11)$$

In summary, the stable state has the following values:

$$I_{C1} = 0 \text{ mA} \quad I_{C2} = -2.59 \text{ mA} \quad I_{B1} = 0 \text{ mA} \quad I_{B2} = -0.14 \text{ mA}$$

$$V_{CN1} = -10.8 \text{ V} \quad V_{CN2} = -1.46 \text{ V} \quad V_{BN1} = -0.37 \text{ V}$$

$$V_{BN2} = -1.66 \text{ V} \quad V_{EN} = -1.36 \text{ V}$$

The output swing is $V_w = V_{CN2} - V_{CN1} = -1.46 + 10.8 = 9.3 \text{ V}$.

In the above example we assumed values for $V_{BE}(\text{sat})$ and $V_{CE}(\text{sat})$ of -0.3 and -0.1 V, respectively, and calculated I_{C2} and I_{B2} of -2.58 and -0.14 mA, respectively. Using these values of current we can find from the manufacturer's saturation curves new values of $V_{CE}(\text{sat})$ and $V_{BE}(\text{sat})$ and can then repeat the above calculations. Since we must work with average transistor characteristics and with resistors known to about 10 percent and since the voltage levels in a flip-flop need seldom be known with precision, then a second approximation is seldom warranted.

The drop across the emitter resistor R_e is nominally the same for the two stable states. However, during the course of a transition, the emitter current I_E will vary by ΔI_E. In order to keep the emitter voltage V_{EN} almost constant during the transition time T_R, a capacitor C_z is used to bypass R_e. The order of magnitude of C_z is given by the condition that the change in voltage across this capacitor is small compared with V_{EN}, or $(\Delta I_E)T_R/C_z \ll V_{EN}$. The stable states are, of course, not affected by the presence of C_z, but the ease of inducing a transition between states and the rapidity with which the flip-flop settles into its new state may be adversely influenced if the capacitor is omitted.

A self-biased multi using vacuum tubes and having a circuit configuration analogous to that of the transistor flip-flop of Fig. 10-7 (but using a positive supply voltage) is also possible (Prob. 10-12). For the ON tube the grid-to-cathode voltage is taken as zero and the plate-to-cathode voltage is found by drawing the load line corresponding to $R_p + R_k$ on the plate characteristics (neglecting the voltage drop across R_k due to grid current and assuming $R_1 + R_2 \gg R_p$).

10-5   COMMUTATING CAPACITORS

A flip-flop will remain in one of its stable states indefinitely until it is induced to make a transition as the result of a "triggering" signal, such as a pulse, applied from some external source.   There are many applications of flip-flops in which it is desired to have a change of state take place as soon after the application of an abrupt triggering signal as possible.   The *transition time* is defined as the interval during which conduction transfers from one transistor (or tube) to the other.   The transition time may be reduced by introducing small capacitances in shunt with the coupling resistors R_1 of the binary.   A flip-flop with such capacitors included is shown in Fig. 10-10.   Because these capacitors assist the binary in making abrupt transitions between states, they are known as *commutating, transpose,* or *speed-up capacitors.*   The usefulness of these capacitors will be seen in the following discussion.   To be specific let us assume that the active devices are tubes or *n-p-n* transistors.

Let us consider that $A2$ is ON and $A1$ is OFF and that, to induce a transition, a negative step is applied at X_2.   The point Y_2 will rise rapidly, and we desire that this rapid rise be transmitted with minimum delay to X_1.   The device $A1$ has an input capacitance C_i, and in the absence of C_1, the circuit configuration consisting of R_1, R_2, and C_i constitutes precisely the uncompensated attenuator discussed in Sec. 2-8.   Even if the voltage at Y_2 rises with negligible rise time, the voltage at X_1 would increase with a time constant RC_i, where R is the parallel combination of R_1 and R_2.   The speed with which X_1 rises may be increased by the addition of the capacitor C_1 in shunt with R_1. If the capacitor C_1 were arbitrarily large, then the waveform at X_1 would rise as rapidly as does the waveform at Y_2 and the full amplitude of the rapid rise

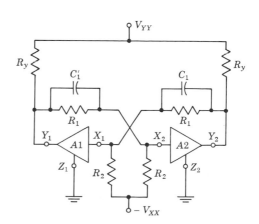

Fig. 10-10   A binary including speed-up capacitors ($C_1' = C_1$).

would be transmitted through the capacitor.   We shall now see, however, that there is also some disadvantage in using a capacitance C_1 of large value.

The voltages across C_1' and C_1 are *not* alike because one stage is ON and the other OFF.   For example, for the circuit of Fig. 10-4, the voltage across C_1' is $V_{C1} - V_{B2} = 9.8$ V and the voltage across C_1 is $V_{C2} - V_{B1} = 1.58$ V with $A1$ OFF and $A2$ ON.   When the circuit is triggered so that $A1$ is ON and $A2$ OFF, then the voltage across C_1' must change to 1.58 V and that across C_1 to 9.8 V.   The flip-flop will not have settled itself completely in its new state until this interchange of capacitor voltages has been completed.   A transition having been induced by a triggering signal, a certain minimum time must elapse before a succeeding trigger will be able reliably to induce the reverse state.   The smallest allowable interval between triggers is called the *resolving time* of the flip-flop, and its reciprocal is the maximum frequency at which the binary will respond.

If the binary has been triggered so as to drive stage $A1$ OFF and $A2$ ON, the equivalent circuit from which to calculate the time constant τ associated with the recharging of C_1 is given in Fig. 10-11a.   (Compare this circuit with that of Fig. 10-5a.)   If the output impedance of $A2$ (including R_y) is R_o, then $\tau = C_1R$, where R is the parallel combination of resistance R_1 and $R_2 + R_o$. For a transistor in saturation, R_o is very small compared with R_2.   For a vacuum tube, it is usually true that $R_2 \gg R_p$, and since R_p is always greater than $R_o = R_p r_p/(R_p + r_p)$, little error is made in taking $R \approx R_1R_2/(R_1 + R_2)$.

Similarly, from Fig. 10-11b we can find the time constant τ' associated with the recharging of C_1'.   The input resistance of A_2 is R_i.   If a vacuum-tube binary is under consideration, then $R_i = r_G$ (the grid resistance) when this electrode is driven positive.   If a flip-flop uses transistors, then $R_i \approx r_{bb'}$ (the base-spreading resistance).   For either device R_i seldom exceeds 1 K and usually $R_i \ll R_2$.   Hence $\tau' \approx C_1'R'$, where R' is the parallel combination of R_1 and $R_y + R_i$.   Since $R_y + R_i$ is usually smaller than either R_1 or R_2, then $\tau > \tau'$ and the larger of the two time constants is

$$\tau = RC_1 \approx \frac{R_1R_2C_1}{R_1 + R_2} \tag{10-12}$$

where $C_1 = C_1'$ is the commutating capacitance.

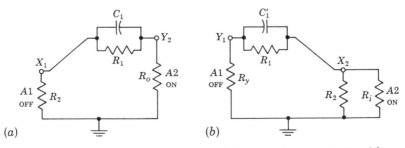

(a)   (b)

Fig. 10-11   Equivalent circuits for computing the time constants with which the commutating capacitors recharge.

As the preceding discussion suggests, the complete transfer of conduction from one device to the other involves two phases. The first of these phases is the *transition* phase, during which conduction transfers from one active device to the other. In order that this transfer take place, the voltage across certain of the capacitances present in the circuit must change. In the binary of Fig. 10-10 there are present capacitances to ground at the points X and Y. These capacitances are the input and output capacitances of the tubes or transistors and also the stray capacitances. None of these capacitances is explicitly indicated in Fig. 10-10. But the voltage across these capacitances must change if the devices are to switch between on and off.

On the other hand, the voltages across the capacitances C_1 and C_1' need not change during the transfer of conduction. If the capacitors C_1 and C_1' were replaced by batteries whose voltages were set equal to the capacitor voltages, then this substitution would in no way restrain the transfer of conduction. Hence these capacitors are permitted to complete their interchange of voltage after the transfer of conduction has taken place. The additional time required for the purpose of completing the recharging of capacitors after the transfer of conduction is called the *settling* time. The resolution time is the sum of the transition time and the settling time. Of course, however, there is no clear-cut separation between the transition phase and the settling phase.

We may now see that if we make the commutating capacitors too small we shall lengthen the transition time and if we make them too large we shall lengthen the settling time. An engineering compromise is called for. In the case of the vacuum-tube binary we find that a reasonable compromise is reached when the capacitors C_1 and C_1' are adjusted so that the R_1R_2 attenuator is nominally compensated for the case where the attenuator is applying a signal to the grid of a tube which is within its grid base. Since under these circumstances the input capacitance is increased as a result of the Miller effect, the attenuator is overcompensated when the grid is at cutoff or in clamp. Such overcompensation is acceptable in the present case since we are not concerned with preserving a waveshape but only with transmitting a signal from a plate to a grid. If the input capacitance taking the Miller effect into account is C_i, then a reasonable choice for C_1 is

$$C_1 = \frac{R_2 C_i}{R_1} \tag{10-13}$$

In practical tube circuits we find that the commutating capacitors rarely exceed 200 pF, are more usually in the range 50 to 75 pF, and may be, if pentodes are used, as low as 10 pF.

In a transistor flip-flop the commutating capacitors serve the same purpose as they do in the tube circuit. They aid the switching by causing the base of one transistor to respond more rapidly to an abrupt change at the collector of the other transistor. The situation is more complicated with transistors than with tubes. This complication arises because, when in its active region, a transistor stores charge in its base and when in the **saturation** region **it** stores even more charge. The transistor cannot be **brought from saturation** to cutoff until all this charge has been removed. The commutating capacitors hasten the removal of this charge. A detailed discussion of the transient switching characteristics of transistors, including an analysis of the delay caused by this base charge, is presented in Chap. 20. For the present, let it suffice to note that the values of commutating capacitors encountered in transistor flip-flops are comparable to those found in tube circuits. They lie in the range from some tens to some hundreds of picofarads (Sec. 20-20).

High-speed transistors are available for which the storage-time delay and the fall time are small compared with the settling time of the circuit. Similarly, for tubes the transition time is generally smaller than the settling time. It seems reasonable to assume that if a time 2τ is allowed between triggers, then all the transients will have died down sufficiently so that the flip-flop can be triggered reliably. In practice, this assumption usually leads to a conservative estimate of the maximum frequency f_{max} of operation, given by

$$f_{max} = \frac{1}{2\tau} = \frac{R_1 + R_2}{2C_1 R_1 R_2} \tag{10-14}$$

where use is made of Eq. (10-12).

10-6   METHODS OF IMPROVING RESOLUTION

We consider first the vacuum-tube binary. A step in the direction of decreasing the resolution time of the binary is clearly to reduce all stray capacitances to a minimum. Beyond this, it is necessary to reduce the values of the resistances R_p, R_1, and R_2. The reduction of R_p will improve the rise time of the waveforms at the tube plates and will also reduce the recharging time of the commutating capacitor connected between the plate of the OFF tube and the clamped grid. The reduction of R_1 and R_2 will reduce the recharging time of the other commutating capacitor.

The price that must be paid for these improvements in resolution time is, first, increased dissipation of power in the circuit since, because of the smaller resistances, the current drain from the supply voltages will increase. Second, unless it is possible to increase the tube current in proportion as the load resistance is reduced, the plate swing will become smaller. Hence, not only will the useful output signal be reduced but the total grid swing will be decreased and it may be difficult to maintain d-c stability in the binary. When the grid swing is reduced, it may become necessary to use 5 percent or perhaps even 1 percent components.

Final measures which may be taken to improve resolution time include the use of pentodes and auxiliary diodes. The pentode reduces the tube input capacitances by suppressing the Miller effect, and the diodes are used as clamps to restrain the total signal excursion at the plates and grids. Such procedures, in spite of their added cost and complexity, will only succeed in reducing the resolution time to the order of 0.1 μsec.[1]

With respect to resolution, the transistor binary has one immediate advantage over the tube circuit. This advantage results from the fact that it is convenient to use in a transistor circuit a much smaller collector-circuit resistance R_c than may be used as a plate-circuit resistance R_p. With a small R_c the collector swing is still very nearly equal to the supply voltage, whereas in a tube circuit the use of a small R_p will result in an important sacrifice of plate swing. Further, plate swing is more at a premium than is collector swing. The plate swing must be several times larger than the grid base, which is of the order of 10 or 20 V, although the collector swing need only be a few times larger than the several tenths of a volt that will carry a transistor from cutoff to saturation. The effect of these considerations is apparent in comparing the values of the components in Fig. 10-2 with those in Fig. 10-4. The fact that R_c is appreciably smaller than R_p allows a reduction also in the size of R_1 and R_2 in the transistor circuit. As we see in Sec. 10-5 and Eq. (10-12), the recovery time is reduced thereby, and smaller values of resistance generally allow all shunting capacitors to charge more rapidly, with a consequent improvement in transition speed.

In a vacuum-tube binary the resolution time is limited by the speed with which interelectrode capacitances and capacitances external to the tube can charge through the resistive circuit components. The situation which normally prevails in a transistor binary is that the speed is limited by mechanisms which are internal to the transistor. As mentioned, and as is discussed in Chap. 20, an inherent and serious limitation on resolution time is the interval required to draw the stored charge out of the base of a transistor which is in saturation. Upon application of a reverse-biasing voltage to the base, there is a delay, called the *storage time* (Sec. 20-14), which elapses before the collector current starts to change. Therefore, where resolution time is at a premium, the transistor must not be allowed to saturate.

A Nonsaturating Binary A method of restraining the transistors from saturating is shown in Fig. 10-12. Here, four diodes have been added to the circuit of Fig. 10-4. These diodes ensure that the collector junctions are reverse-biased and hence that the transistor is always operating in the active region. Both *avalanche diodes* D3 and D4 (also called Zener diodes; Sec. 6-4) are always biased in the breakdown direction and each has a voltage $V_Z < V_{CC}$ across it. The voltage across the diode D1 or D2 is very small in the forward direction. With Q2 ON, its emitter junction is forward-biased and $V_{BE2} \approx 0$. Hence the left-hand side of D2 is at V_Z with respect to ground. The right-hand side is connected to the low voltage of the collector of the ON transistor Q2. Thus D2 is forward-biased and the collector junction of Q2 is reverse-biased by $V_{CB2} \approx V_Z$, therefore preventing saturation. From Fig. 10-12b we see that $V_{BE1} = -V_{BB}R_1/(R_1 + R_2)$, keeping Q1 cut off. Because V_{CE1} is high (approximately equal to $V_{CC} > V_Z$), then D1 is back-biased. The output swing is approximately $V_{CC} - V_Z$. It will be recognized that the configuration used here is the same as that illustrated in Fig. 8-31.

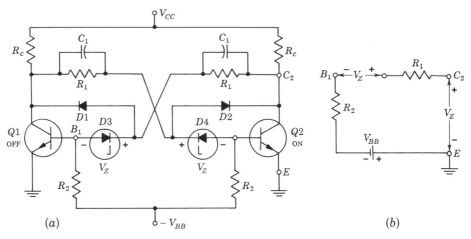

Fig. 10-12   (a) A nonsaturated binary; (b) the equivalent circuit from which to calculate the voltage at the base of the OFF transistor.

If p-n-p transistors are used, then each diode in Fig. 10-12 must be reversed. This same nonsaturating configuration can also be used with the self-biased flip-flop of Fig. 10-7. There exist a number of other circuits for reducing the minority-carrier storage-time delay, one[2] of which is given in Prob. 10-17.

The nonsaturating circuits are more complicated than the saturated flip-flops. Also, the latter dissipate less power, because the ON transistor has low voltages whereas the OFF transistor has low currents. On the other hand, the ON transistor in the nonsaturated configuration operates at a high current and a voltage V_Z which may result in a significant power loss. An additional advantage of the saturated binary is that its voltage swing is more stable with temperature, aging, and component replacement than is the nonsaturating type. For these reasons and because high-frequency transistors with very small storage times are now available, a nonsaturating flip-flop design is justified only in those applications where extreme speed is necessary. Transistor flip-flops have been designed for triggering frequencies in excess of 100 MHz.

10-7   UNSYMMETRICAL TRIGGERING OF THE BINARY

The triggering signal which is usually employed to induce a transition from one state to the other is either a pulse of short duration or a step voltage. This pulse or step may be introduced in such a manner as will produce either *symmetrical* or *unsymmetrical* triggering. In unsymmetrical triggering the triggering signal is effective in inducing a transition in only one direction.

A second triggering signal from a separate source must be introduced in a different manner to achieve the reverse transition. In symmetrical triggering each successive triggering signal induces a transition, regardless of the state in which the binary happens to be. Unsymmetrical triggering, using two triggering sources, is found frequently where the binary is to be used as a generator of a gate whose width equals the interval between triggers. Such triggering also finds extensive application in logic circuitry, as, for example, in the register and coding circuits in Chap. 9. Symmetrical triggering is used in binary-counting circuits (Chap. 18) and in other applications. We shall consider in this present section only the method of unsymmetrical triggering.

It is important to understand that *the sensitivity of the binary to a pulse of such polarity as to turn off the conducting device will appreciably exceed the sensitivity to a pulse of opposite polarity.* To be specific, let us consider the circuit of Fig. 10-10, using either tubes or *n-p-n* transistors. A positive *step* applied through a capacitor to the input of the OFF stage, say $A1$, will give no response until the step amplitude exceeds the voltage by which the device is below cutoff. On the other hand, a negative step applied to the input of the ON stage $A2$ will immediately decrease the current in this device (we are neglecting the minority-carrier storage time in this discussion). Since $A2$ now operates in its active region, the input voltage at X_2 is amplified at Y_2 and a fraction of this voltage appears at X_1. Because of this amplification from X_2 to X_1, a small negative step at X_2 will leave the binary in the condition in which the current in the initially OFF stage is larger than the current in the initially ON triode, and we may expect the flip-flop to complete the transition through regenerative action. A vacuum-tube or *n-p-n* transistor binary will therefore be triggered by a smaller negative than positive voltage step.

Suppose that a positive *pulse* is applied to the input of the OFF stage. The pulse is a combination of a positive step and a delayed negative step. The result to be anticipated is therefore a combination of the response to a positive step applied to the OFF device followed by a negative step applied to the ON stage. If the amplitude of a positive step is large enough to flip the binary at its leading edge, then because of the greater triggering sensitivity to a negative step, the binary will flip back again at its trailing edge.

Next consider a negative *pulse* applied to the ON stage. Since the binary responds to a smaller negative step than positive step, we may adjust the pulse amplitude experimentally to prevent the flip-flop from making a reverse transition on the trailing edge of the pulse.

It is possible to arrange a permanent binary transition in a tube or an *n-p-n* transistor through the use of a positive pulse, provided that the positive pulse is applied to the input of the ON stage through a small capacitor C. At the leading edge of the positive pulse this capacitor will charge through the low input resistance of the amplifier in series with the output impedance R_s of the trigger source. The voltage across C may become as large as the pulse amplitude if R_s is small enough. At the instant of the negative-going trailing edge of the trigger, the input is driven negative by the amount by

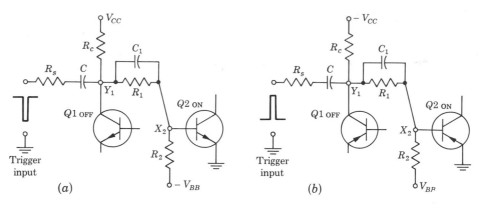

Fig. 10-13 Method of triggering unsymmetrically (*a*) an *n-p-n* or vacuum-tube binary and (*b*) a *p-n-p* flip-flop.

which the voltage across C has changed.   Since the binary is very sensitive to a negative step applied to the ON stage, a transition may result.   It is to be noted, however, that the transition, when it takes place, occurs at the *trailing edge* of the input pulse.   Usually it is desired to avoid this triggering at the trailing edge.   Such suppression may be achieved (if R_s is not large enough) by including a large resistance in series with C.

The triggering signal may be applied at the output of one of the stages of the binary rather than at the input, again preferably through a resistor and capacitor (Fig. 10-13).   Any signal so applied will immediately appear at the input of the other device, being transmitted through the commutating capacitor.   The presence of the series resistor R_s will serve to accentuate even further the relative sensitivity of the tube or *n-p-n* transistor to a negative pulse.   This extra sensitivity results because the positive pulse would have to be introduced at a point where the signal looks directly or through a coupling capacitor at the plate of a tube or the collector of a transistor which is conducting.   For example, in Fig. 10-10, with $A1$ OFF and $A2$ ON a positive step must be applied at Y_2 to bring A_1 out of cutoff.   At Y_2, however, the impedance presented to the triggering signal is low.   A negative signal could be introduced, on the other hand, at the output of a cutoff stage (Y_1 in Fig. 10-13a) where the impedance is high.

If *p-n-p* transistors are used then a positive step must be applied to the base in order to produce cutoff.   Hence, the discussion in this section may be summarized as follows: *An excellent method for triggering a binary unsymmetrically on the leading edge of a pulse is to apply the pulse from a high-impedance source to the output of the nonconducting device.   If the active element is a tube or an n-p-n transistor a negative pulse is required, but for a p-n-p triode the trigger polarity should be positive* (Fig. 10-13b).

10-8   TRIGGERING UNSYMMETRICALLY THROUGH A UNILATERAL DEVICE

A method of triggering a flip-flop which allows the binary to respond to only one polarity of pulse is obtained by adding a diode to the circuit as indicated in Fig. 10-14a. When $Q1$ is conducting, the diode D is back-biased by the drop across R_c and the diode will not transmit a triggering signal (unless it is negative and has an amplitude larger than this voltage drop). When $Q1$ is OFF the drop across D is zero. The diode will still fail to transmit a positive-going trigger but will transmit a negative step or pulse to the input (base) of

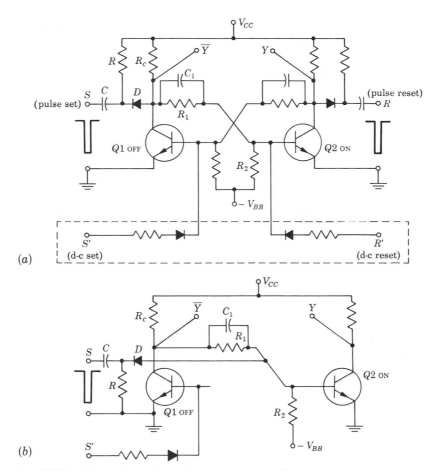

Fig. 10-14 Method of triggering unsymmetrically with a diode. Pulses applied (a) to the collector of the OFF transistor and (b) to the base of the ON triode. Control of the state of the flip-flop is also possible with a binary input at S' (d-c set) and R' (d-c reset).

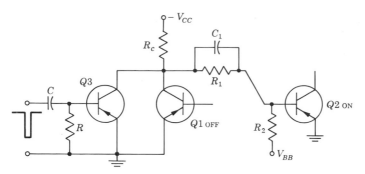

Fig. 10-15 The use of a trigger amplifier in conjunction with a binary.

$Q2$.   Observe that here the binary responds to that signal which effectively applies a negative pulse to the ON device, to which polarity the binary is more sensitive.   The resistance R must be large enough not to load down the trigger source.   On the other hand, R must be small enough so that any charge which accumulates on C during the interval when D conducts will have time to decay during the time between pulses.   If the triggering rate is high, then it may be necessary to replace R with a diode.   The placement of this diode in the circuit is such that it is back-biased during the pulse but it does conduct after the pulse so as quickly to remove the charge accumulated on C.   (See $D3$ in Fig. 10-16.)

The FLIP-FLOP of Fig. 10-14a may be used as a basic logic element (Sec. 9-13).   For a positive logic system the output Y is taken from the collector of the n-p-n transistor, say $Q2$, and the output $\bar{Y}$ is taken from the collector of $Q1$. The triggering signal is a negative pulse and is applied to the *set terminal S*. After the pulse excitation, $Q2$ will be OFF and $Q1$ ON, so that $Y = 1$ and $\bar{Y} = 0$. Similarly, after a triggering signal is applied at the *reset terminal R*, $Q2$ will be ON and $Q1$ OFF, so that $Y = 0$ and $\bar{Y} = 1$.

If a completely d-c positive logic system is under consideration, then the input "signal" to the FLIP-FLOP is the 1 level (the more positive logic voltage). The triggering arrangement is now that indicated in the dashed box in Fig. 10-14a.   With a 1 at the d-c *set terminal S'*, $Q1$ is ON and $Q2$ OFF, so that $Y = 1$ and $\bar{Y} = 0$.   On the other hand, with a 1 at the d-c *reset terminal R'*, $Q1$ is OFF and $Q2$ is ON, so that $Y = 0$ and $\bar{Y} = 1$.

An alternative diode-triggering arrangement is shown in Fig. 10-14b. Here the negative set pulse S is applied through D to the base of the ON stage $Q2$.   Now R is returned to ground rather than to the supply voltage.   The d-c set circuit is identical with that in Fig. 10-14a.   Only one half of the circuit is indicated, the portion not shown being understood to be symmetrical with the configuration which is drawn.

If p-n-p transistors are used, then the diodes in Fig. 10-14a or b must be reversed and the circuits respond only to a positive input signal, which is the proper polarity for turning off the ON transistor.

If the trigger amplitude available is small, it may be necessary to amplify this signal before applying it to the flip-flop. In this case a diode need not be used because the amplifier can provide the unilateral action previously supplied by the diode. Such a configuration is indicated in Fig. 10-15 for a circuit which uses p-n-p transistors. Note that R_c serves as both the collector resistor of the binary transistor $Q1$ and the trigger amplifier $Q3$. Since its emitter-to-base voltage supply is zero, then $Q3$ is virtually at cutoff (Sec. 6-17). Regardless of the state of the binary, a positive signal at the base of $Q3$ will have no effect. However, a negative pulse causes $Q3$ to conduct if $Q1$ is OFF. This signal is amplified and inverted by $Q3$, and the resulting positive waveform at the collector of $Q1$ is coupled to the base of $Q2$. If $Q1$ is cut off, then $Q2$ is ON. Hence, $Q2$ will be driven toward cutoff by the positive pulse at its base, and a transition will occur.

10-9   SYMMETRICAL TRIGGERING

The circuits of Fig. 10-16 show how the binary may be pulsed in a symmetrical manner. They are called complementing FLIP-FLOPS. The input terminal corresponds to the trigger T on the logic symbol of Fig. 9-39. For example, in Fig. 10-16a the diodes $D1$ and $D2$ serve the same function as described in Sec. 10-8 in connection with Fig. 10-14a, and diode $D3$ takes the place of R in Fig. 10-14a. The drop (approximately V_{CC}) across the collector load R_c of the conducting transistor $Q2$ reverse-biases $D2$. Since there is zero voltage across R_c of $Q1$, then $D1$ is at zero bias. Hence, a negative input signal will be transmitted through $D1$ to the collector of $Q1$ and thence to the input

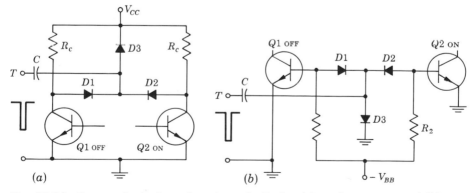

Fig. 10-16 Symmetrical triggering through diodes (a) at the outputs and (b) at the inputs of the amplifiers. These circuits are symmetrical forms of the unsymmetrical circuits of Fig. 10-14, except that R has been replaced by $D3$.

of the ON stage $Q2$ via the R_1C_1 combination connecting the output of $Q1$ to the input of $Q2$. A negative pulse at the base of $Q2$ is appropriate for turning this stage OFF and thus causing a transition. After the transition is completed $D1$ will be reverse-biased and $D2$ will be at 0 V. The next negative pulse will pass through $D2$ instead of $D1$. Hence, these elements are called *steering diodes*. The binary will transfer at *each* successive negative input pulse or step but will not respond to the opposite polarity. If *p-n-p* transistors are used, then the diodes must be reversed and a positive triggering signal is required. If the pulse rate is low, then $D3$ may be replaced by a resistor R.

If the pulse amplitude is too small, a configuration using triggering amplifiers $Q3$ and $Q4$ is used. This circuit is an extension of the unsymmetrical method shown in Fig. 10-15, with the collector of $Q4$ connected to that of $Q2$ and the base of $Q4$ tied to the base of $Q3$. A number of other triggering schemes may be found in the literature (Prob. 10-18).

Triggering of a Bistable Multi Symmetrically without the Use of Auxiliary Diodes or Triodes Symmetrical triggering is possible in this case because of the presence of the commutating capacitors. Such triggering may be successfully achieved only if the commutating capacitances are large enough to predominate over all other capacitances present, so that during the transition the voltages across the commutating capacitors do not change appreciably. We have already seen that large commutating capacitors lengthen the time required to complete the settling from one binary state to the other. Therefore this method of triggering without auxiliary steering diodes or transistors is not employed where the shortest possible resolution time is required.

With the circuit of Fig. 10-17 symmetrical triggering can be achieved by the direct application of a positive step to the common emitters of a self-biased flip-flop. We shall make some assumptions which simplify the analysis but which are not essential to the discussion. We assume that in saturation the collector-to-emitter voltage and the base-to-emitter voltage are zero. We specify only the ratio of the coupling resistances ($R_1 = 3R_2$), and we assume that these resistances are large enough so that they constitute a negligible load on the output of the transistor. Finally, we specify only the ratio of collector circuit resistance to the emitter resistance ($R_c = 4R_e$).

Initially the flip-flop is in a state in which $Q1$ is OFF and $Q2$ is ON. For a 10-V supply and with $R_c = 4R_e$, 8 V appears across R_c and 2 V across R_e. Hence, the transistor voltages are as indicated in Fig. 10-17. (Ignore the values in parentheses for the moment.) Now let there be applied a positive step of amplitude 7 V to the common emitters E through a small capacitance C. Then the voltage of E will rise abruptly to 9 V and both transistors will be turned OFF. The circuit voltages immediately after the input step are given in parentheses. We assume that the voltages across the commutating capacitors have not changed. Observe that the base-to-emitter voltage of $Q2$ is -7 V and the corresponding voltage for $Q1$ is -0.5 V. Hence $Q2$, which was

Fig. 10-17 A symmetrical triggering arrangement which does not employ auxiliary diodes or transistors. The voltages indicated (not in parentheses) are quiescent values. The voltages enclosed in parentheses result immediately after a positive 7-V step is applied at the emitter, assuming that the commutating capacitor voltages do not change.

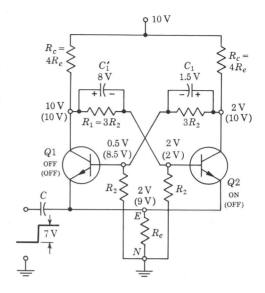

initially ON, is reverse-biased to a larger extent than is $Q1$. Now as the capacitor C charges through R_e, the voltage at E will fall and clearly $Q1$ will enter its active region before $Q2$. Accordingly, when the voltage at E returns to its quiescent level, we shall find $Q1$ ON and $Q2$ OFF. A transition has taken place.

It is necessary that C be selected small enough so that, after the input step, the voltage at E may decay before the commutating-capacitor voltages can change appreciably. Eventually the voltages across C_1 and C_1' will interchange and the flip-flop will settle into its new state.

Essential to the above discussion has been the assumption that the commutating capacitances were large enough to be predominant. If these capacitors were absent or inadequate, the input capacitances to the transistors might well predominate. If these latter capacitors maintained their voltages, then they would serve to keep the ON transistor conducting and the OFF transistor nonconducting and a transfer could not take place.

If a 7-V negative step were applied to E, it would serve only to drive $Q2$ further into saturation. A transition would not take place. If the triggering signal were a square wave, the flip-flop would respond on the positive-going edge but not on the negative-going edge.

Another method of triggering symmetrically without using steering diodes or triodes is indicated in Prob. 10-20.

Before leaving the subject of triggering we shall make one last important observation. Let us call the initial state ($Q1$ OFF, $Q2$ ON) state A and the second state, state B. Immediately after the transition has taken place the predominant commutating capacitors have voltages corresponding to state A rather than to state B. A question which frequently occurs is therefore the following. Since the commutating capacitors have voltages corresponding to state A immediately after the triggering signal, should not the flip-flop be even more susceptible to the next trigger immediately after the first pulse than it would be at a later time when the commutating-capacitor voltages have interchanged? Such is not the case, for, as we see from the numbers in parentheses in Fig. 10-17, immediately after the first trigger $Q2$ is deeply in cutoff. Therefore the next pulse will drive $Q2$ even further into reverse bias and can consequently cause no transition.

10-10   A DIRECT–CONNECTED BINARY CIRCUIT[3]

A DCTL (Sec. 9-17) flip-flop circuit containing very few components (two transistors and two resistors) is indicated in Fig. 10-18. This transistor binary has no vacuum-tube counterpart because, if a direct connection were made from a plate to the opposite grid, the circuit would have only one stable state, in which both tubes would be conducting heavily. We shall now demonstrate that the transistor flip-flop does have two stable states. In one state, transistor $Q2$ is in saturation and $Q1$ is conducting slightly (rather than being beyond cutoff, as in the binaries discussed in the preceding sections), and in the other state $Q1$ is in saturation and $Q2$ is conducting slightly.

Initially, let us assume that $Q1$ is OFF. Then the circuit of Fig. 10-18 consists of transistor $Q2$, whose emitter is grounded and whose collector and base are connected through resistors R_c to the supply voltage V_{CC}. The currents in $Q2$ are given by

$$I_{B2} = \frac{V_{CC} - V_{BE2}}{R_c} \qquad I_{C2} = \frac{V_{CC} - V_{CE2}}{R_c} \qquad (10\text{-}15)$$

If the base current I_{B2} drives $Q2$ into saturation, V_{BE2} and V_{CE2} are small compared with V_{CC}. Consequently, as a first approximation, we have $I_{B2} \approx V_{CC}/R_c \approx I_{C2}$. Since $I_{B2} \gg I_{C2}/h_{FE}$, then we have verified that $Q2$ is indeed driven well into saturation.

From the first-approximation values of I_{B2} and I_{C2} we obtain the values

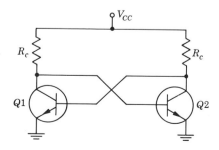

Fig. 10-18   A direct-connected binary.

of V_{BE}(sat) and V_{CE}(sat) from the manufacturer's data. We can use these values in Eqs. (10-15) to find the next approximation for the currents and subsequently for the saturation voltages. This procedure can be repeated again, but it is seldom necessary to do so since the process converges very rapidly.

In a typical case we might obtain for a germanium transistor *under heavy saturation* $V_{CE2} \approx 0.05$ and $V_{BE2} \approx 0.30$ V. Because of the direct connection between the collector of $Q2$ and the base of $Q1$, $V_{BE1} = V_{CE2} = 0.05$ V. Therefore $Q1$ is actually forward-biased by a small amount (50 mV). Although $Q1$ is not cut off, its currents are very small compared with the current in R_c, and the above method of analysis is essentially correct. Since $V_{CE1} = V_{BE2} \approx 0.3$ V the output swing is only $V_w = V_{CE1} - V_{CE2} \approx 0.25$ V.

There are a number of disadvantages to the direct-coupled binary. (1) We have neglected I_{CBO}, but as the temperature increases, this reverse saturation current may increase sufficiently to bring $Q1$ into its active region and may even take $Q2$ out of saturation. (2) Since $Q2$ is driven heavily into saturation, storage-time delay will be large and the switching speed will be low. Hence, a direct-coupled flip-flop can be used only at slow pulse rates and at relatively low temperatures (50°C for germanium and 150°C for silicon transistors). (3) The output voltages are equal to the saturation base and collector voltages, and these parameters may vary appreciably from transistor to transistor. (4) The voltage swing is only a fraction of a volt, and hence the binary is susceptible to spurious (noise) voltages. (5) Since an OFF collector is tied directly to an ON base, it is difficult to trigger the binary by the usual method of applying a pulse to the OFF collector. To supply sufficient current to take the ON transistor out of saturation usually requires an amplifier, and hence the most common triggering method is that indicated in Fig. 10-15.

The advantages of the d-c flip-flop are the following: (1) extreme simplicity; (2) one supply voltage of low value (perhaps only 1.5 V); (3) low power dissipation; (4) transistors with low breakdown voltages may be used; and (5) the binary may easily be constructed as an integrated circuit because of the few elements (resistors and transistors) involved.

10-11   SCHMITT TRIGGER CIRCUIT

A most important bistable circuit is indicated in Fig. 10-19. It differs from the basic Eccles-Jordan configuration of Fig. 10-1 in that the coupling from the output Y_2 of the second stage to the input X_1 of the first stage is missing and that feedback is obtained now through the resistor R_z. If tubes are used for the amplifiers, then Z represents a cathode, and hence the circuit is called a *cathode-coupled binary*. If transistors are used, Z represents an emitter, and the designation *emitter-coupled binary* is appropriate. Quite commonly in the literature either circuit is referred to as a *Schmitt trigger*,[4] after the inventor of the vacuum-tube version.

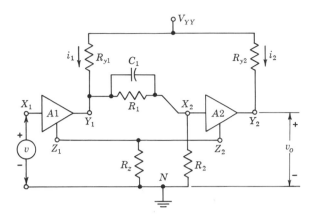

Fig. 10-19 A Schmitt trigger circuit. The supply voltage V_{YY} is of the proper polarity for a tube or an $n\text{-}p\text{-}n$ transistor, but must be reversed for a $p\text{-}n\text{-}p$ transistor.

As in the basic circuit of Fig. 10-1, so here also, the existence of only two stable states results from the fact that positive feedback is incorporated into the circuit and from the further fact that the loop gain of the circuit is greater than unity. We shall obtain additional insight into this circuit if we consider initially that we have adjusted the loop gain to be less than unity. One way, among many other possibilities, to make such an adjustment is simply to reduce the resistance of the resistor R_{y1}. If R_{y1} is small enough, regeneration is not possible. Therefore the circuit will not operate as a binary, but we may use it as an amplifier. Let us then assume that the circuit is an amplifier with input signal v applied as shown in Fig. 10-19 and output v_o.

If device $A2$ is conducting, there will be a voltage drop across R_z which will elevate the emitter or cathode of $A1$. Consequently, if v is small enough in voltage, $A1$ will be cut off. As v rises, the circuit will not respond until $A1$ reaches its cutin point. Until then, the output v_o will be $v_o = V_{YY} - I_2R_{y2}$, where I_2 is the current in R_{y2} for $A1$ cutoff. With $A1$ conducting, the circuit will amplify, and since the gain $\Delta v_o/\Delta v$ is positive, the output will rise in response to the rise of v. As v continues to rise, X_2 continues to fall and Z_2 to rise. Therefore a value of v will be reached at which $A2$ is turned OFF. At this point, $v_o = V_{YY}$ (in the transistor circuit, we neglect the small reverse saturation current), and the output again no longer responds to the input. A plot of v_o against v is shown in Fig. 10-20a, marked "Loop gain <1." The voltage at which $A1$ reaches cutin is marked $v = V_1$.

Let us now increase the loop gain by increasing R_{y1}. Such a change will have negligible effect on the cutin point $v = V_1$. However, in the region of amplification, the amplifier gain $\Delta v_o/\Delta v$ will increase and consequently the slope of the rising portion of the plot in Fig. 10-20a will be steeper. This slope will continue to increase with increasing loop gain until, at a loop gain of unity, where the circuit has just become regenerative, the slope will become infinite. And finally, when the loop gain becomes greater than unity, the slope reverses in sign and the plot of v_o against v assumes the S shape shown in Fig. 10-20b.

This S curve of Fig. 10-20b may be used to describe the behavior of the circuit. As v rises from zero voltage, v_o will remain at the lower of its two levels until v reaches V_1. When v exceeds V_1 a line drawn vertically intersects the plot only at the upper level. Hence when v exceeds V_1 the circuit will make an abrupt transition to this higher level. Similarly, if v is initially greater than V_1, then as v is decreased, the output will remain at its upper level until v attains a definite level V_2, and at this point the circuit will make an abrupt transition to the lower level. We observe that the circuit exhibits *hysteresis*; that is, to effect a transition in one direction we must first pass beyond the voltage at which the reverse transition took place.

A vertical line drawn at $v = V$ which lies between V_2 and V_1 intersects the S curve at three points. The upper and lower points, a and c, are points of stable equilibrium. Point b is a point of unstable equilibrium. The S curve is a plot of values which satisfy Kirchhoff's laws and are consistent with the tube or transistor characteristics. However, a point such as b cannot be attained experimentally. At $v = V$ the circuit will be at either a or c, depending on the direction of approach of v toward V. When $v = V$ in the range between V_2 and V_1 the Schmitt circuit is in one of two possible stable states and hence is a bistable circuit.

Applications   Observe that we may readily describe the circuit of Fig. 10-1 in the manner we have just used to describe the Schmitt circuit. We could develop for that first circuit an S curve such as appears in Fig. 10-20b. As a matter of principle, the Eccles-Jordan circuit of Fig. 10-1 and the Schmitt

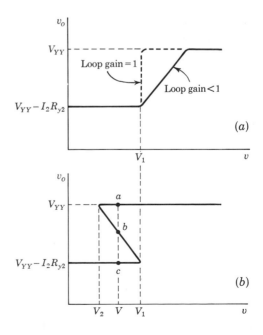

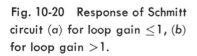

Fig. 10-20 Response of Schmitt circuit (a) for loop gain ≤ 1, (b) for loop gain >1.

circuit of Fig. 10-19 may be used for the same applications. As a matter of practice, however, the Eccles-Jordan circuit, because of its symmetry, is preferred for applications where the circuit is to be triggered back and forth between stable states. The Schmitt circuit has the feature that one device terminal, the base or grid of $A1$, is not involved in the regenerative switching. When the circuit switches between levels, the voltage on this terminal does not change. The Schmitt circuit is therefore preferred for applications in which we desire to take advantage of this free terminal. Observe also that the resistance R_{y2} in the output circuit of $A2$ is not required for the operation of the binary. This resistance may be selected over a wide range to obtain different output-signal amplitudes. Furthermore, capacitive loading at Y_2 will not slow the regenerative action, although such capacitance will increase the rise and fall times of the waveform at Y_2.

A most important application of the Schmitt circuit is its use as an *amplitude comparator* to mark the moment at which an arbitrary waveform attains a particular reference level. Amplitude comparators are considered in Sec. 7-11, where we discuss the use of diodes in amplitude comparison circuits. As a comparator, the Schmitt circuit has some advantages over the diode circuitry. We desire, when the comparison point is reached, that the comparator output make an abrupt and pronounced change. In a diode-resistor comparator, as in Fig. 7-19, the sharpness of the break is limited by the lack of sharpness of the diode break. And after the break point is reached, the rate of change of the output waveform is no faster than the rate of change of the input waveform. As proved in Sec. 7-11, the use of an amplifier after the diode comparator does not improve the sharpness of the break, and in addition the comparison point becomes a function of amplifier gain. When a high-gain amplifier is used *before* the diode comparator the amplifier may have to be d-c-coupled to the comparator, and complications ensue because amplifier drift will change the comparison point.

The situation is otherwise with the Schmitt circuit. As the input v rises to V_1 or falls to V_2 the circuit makes a fast regenerative transfer to its other state. We may increase this speed of response by using all available techniques to improve the rise time of the amplifier stages of which the circuit is composed. Such techniques include minimizing all shunt capacitance, using high-speed transistors or high-figure-of-merit tubes, and even possibly employing high-frequency compensation. In any event, the abruptness of the comparator response is not related to the rate of rise of the input waveform. Also, there is, in the Schmitt comparator, a definite comparison voltage rather than a fairly uncertain comparator region, as in the diode circuit. This improvement in comparator performance results from the regeneration in the Schmitt circuit and is characteristic of all discriminators which employ regeneration.

In a second application, the Schmitt circuit is used as a *squaring circuit*. This application is illustrated in Fig. 10-21. Here the input signal is arbitrary except that it has a large enough excursion to carry the input beyond the limits of the hysteresis range $V_H \equiv V_1 - V_2$. The output is a square wave, as

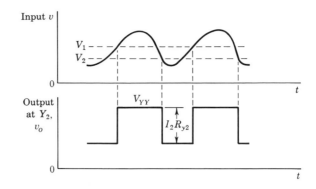

**Fig. 10-21   Response of
the Schmitt binary to an
arbitrary input signal.**

shown, whose amplitude is independent of the amplitude of the input waveform. Observe, further, that the output waveform may have much faster
leading and trailing edges than does the input.

In still another application the circuit is triggered between its two stable
states by alternate positive and negative pulses. Thus if the input is biased
to a voltage V between V_2 and V_1, and if a positive pulse whose amplitude
exceeds $V_1 - V$ is coupled to the input, then $A1$ will conduct and $A2$ will be
driven to cutoff. If there now is applied a negative pulse whose amplitude
is greater than $V - V_2$, the circuit will be triggered back to the state where
$A1$ is OFF and $A2$ is conducting. This behavior is the same as that of a
flip-flop, with alternate positive and negative pulses applied to one input.
However, for the Schmitt binary the possible triggering difficulties discussed
earlier are not encountered because pulses are applied to an input node X_1
which is not connected to any other point in the circuit.

Hysteresis   In many instances, the hysteresis of the Schmitt circuit
is not a matter of concern. Such would be the case if we had a periodic signal
of amplitude large in comparison with the hysteresis range V_H and our
interest lay in using the circuit as a one-way comparator. In Sec. 14-6 an
example is given of a system requiring a large value of V_H. In other applications, a large hysteresis range will not allow the circuit to function properly.
Thus, even when used as a one-way comparator, if the signal were smaller
than V_H, then the comparator, having responded by a transition in one direction, would never reset itself.

As may be seen in Fig. 10-20, hysteresis may be eliminated by adjusting
the loop gain of the circuit to be unity. Such an adjustment may be made
in a variety of ways. The gain will increase or decrease with increase or
decrease in R_{y1}. Another possibility is to add to the circuit a resistor R_{z1} in
series with the emitter or cathode lead of $A1$ or a resistor R_{z2} in series with
the emitter or cathode lead of $A2$. Here the gain will increase or decrease as
either R_{z1} or R_{z2} is decreased or increased. Since R_{y1} and R_{z1} are in series
with $A1$, these resistors will have no effect on the circuit while A_1 is cut off.
Therefore these resistors will not change V_1 but may be used to move V_2
closer to or coincident with V_1. Similarly R_{z2} will affect V_1 but not V_2. A

further method of gain adjustment consists in varying the ratio $R_1/(R_1 + R_2)$. Such an adjustment will change both V_1 and V_2.

Adjusting the gain precisely to unity is not feasible. The device parameters, and hence the gain, are variable over the signal excursion. Hence an adjustment which ensures that the maximum gain is unity would have ranges where the gain is less than unity, with a consequent loss in speed of response of the circuit. Further, the circuit is not stable enough to maintain a gain of precisely unity for a long period of time without frequent readjustment. In practice, therefore, where hysteresis is undesirable we must be content with adjusting the hysteresis to be small and with increasing the signal amplitude, when possible, so that it is large in comparison with the hysteresis range.

A quantitative discussion of the Schmitt circuit is given in the next two sections.

10-12   A CATHODE–COUPLED BINARY

The method of calculating V_1, V_2, and the two stable-state output voltages for a vacuum-tube Schmitt trigger is illustrated in the following example.

EXAMPLE   Analyze the circuit of Fig. 10-22.

Solution   If

$$a \equiv \frac{R_2}{R_1 + R_2} = \frac{200}{200 + 400} = \frac{1}{3}$$

then with $V1$ OFF, the voltage at the grid of $V2$ (assuming zero grid current) is

$$V_{PP}a = \frac{300}{3} = 100 \text{ V}$$

(neglecting the small loading of R_1 and R_2). The equivalent circuit for this state is indicated in Fig. 10-23a. To find the current i_2 we follow the procedure described in Sec. 1-10. We plot the load line corresponding to $R_{p2} + R_k = 30$ K and the

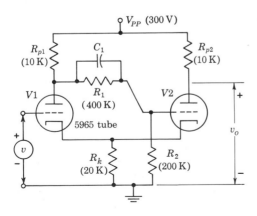

Fig. 10-22   A cathode-coupled binary. The numerical values refer to the illustrative example.

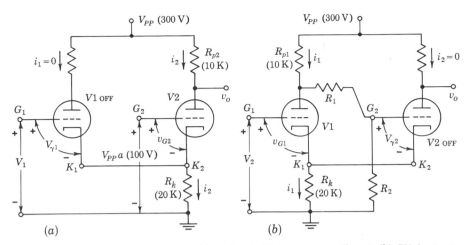

Fig. 10-23 The circuit of Fig. 10-22 with (a) $V1$ just at cutoff and (b) $V2$ just at cutoff.

supply voltage $V_{PP} = 300$ V on the plate characteristics of the 5965 tube (Fig. D-9).   From the grid circuit we see that

$$i_2 = \frac{V_{PP}a - v_{G2}}{R_k} = \frac{100 - v_{G2}}{20} \quad \text{mA} \tag{10-16}$$

This *bias curve*, a plot of i_2 as a function of v_{G2}, is added to the plate characteristics, and the intersection with the load line is found to be

$$i_2 = 5.1 \text{ mA} \equiv I_2 \quad \text{and} \quad v_{G2} = -2.3 \text{ V}$$

$$v_o = V_{PP} - I_2 R_{p2} = 300 - 51 = 249 \text{ V}$$

$$v_{KN} = I_2 R_k = 102 \text{ V}$$

The voltage drop across $V1$ is

$$V_{PP} - I_2 R_k = 300 - 102 = 198 \text{ V}$$

and the cutin grid voltage corresponding to this plate voltage is found from Fig. D-9 to be $V_{\gamma 1} = -8$ V.   Hence

$$V_1 = V_{\gamma 1} + v_{KN} = -8 + 102 = 94 \text{ V}$$

With $V1$ on, the input is now reduced to $v = V_2$ such that $V2$ is just about to come out of cutoff.   The corresponding value of i_1 can be found from the equivalent circuit indicated in Fig. 10-23b.   Kirchhoff's voltage law applied to the grid circuit of $V2$ yields

$$(-V_{PP} + i_1 R_{p1})a + V_{\gamma 2} + i_1 R_k = 0$$

or

$$i_1 = \frac{V_{PP}a - V_{\gamma 2}}{R_{p1}a + R_k} = \frac{100 - V_{\gamma 2}}{23.3} \quad \text{mA} \tag{10-17}$$

As a first approximation take $V_{\gamma 2} \approx V_{\gamma 1} = -8$ V. Then $i_1 = 108/23.3 = 4.63$ mA. The voltage drop across $V2$ is

$$V_{PP} - i_1 R_k \approx 300 - 92.6 = 207 \text{ V}$$

and for this plate voltage we see from Fig. D-9 that a cutin voltage of -8 V is reasonable, so that no correction of the calculated value of i_1 is necessary.

If $i_1 = 4.63$ mA $\equiv I_1$ is located on the load line, a value of $v_{G1} = -2.6$ V is obtained. Hence

$$V_2 = v_{G1} + I_1 R_k = -2.6 + 92.6 = 90 \text{ V} \tag{10-18}$$

For the circuit under consideration

$$V_H = V_1 - V_2 = 94 - 90 = 4 \text{ V}$$

Let us attempt to eliminate the hysteresis by adding a resistor R_{k1} in series with the cathode of $V1$. Such a resistor can have no effect on V_1 because in the equivalent circuit of Fig. 10-23a (from which V_1 is calculated) $V1$ is at cutoff and hence the drop across R_{k1} is zero. Since the current $i_1 = I_1$ in Fig. 10-23b required to bring $V2$ just out of cutoff is determined from the grid circuit of $V2$, it is unaffected by R_{k1}. Hence V_2 will be increased by $I_1 R_{k1}$, and if R_{k1} is chosen properly, then

$$V_2 + I_1 R_{k1} = V_1$$

and hysteresis is eliminated. In the illustrative example a value of

$$R_{k1} = \frac{94 - 90}{4.63} = 0.86 \text{ K} = 860 \text{ }\Omega$$

is needed.

Similarly, if a resistor R_{k2} is added in series with the cathode of $V2$, the value of V_2 is unaffected but V_1 will be decreased. By the proper choice of R_{k2} it is again possible to reduce V_H to zero.

10-13   AN EMITTER–COUPLED BINARY

We shall now analyze the Schmitt transistor binary of Fig. 10-24. First we calculate V_1, defined as the input voltage at which $Q1$ begins to conduct. The circuit for calculating the current in $Q2$ when $Q1$ is just at cutin is shown in Fig. 10-25. We have replaced V_{CC}, R_{c1}, R_1, and R_2 by the Thévenin's equivalent V' in series with R_b between the base of $Q2$ and ground, where

$$V' = \frac{V_{CC} R_2}{R_{c1} + R_1 + R_2} \quad \text{and} \quad R_b = \frac{R_2(R_{c1} + R_1)}{R_{c1} + R_1 + R_2} \tag{10-19}$$

It is possible for $Q2$ to be in its active region or to be in saturation. We shall assume for the present that $Q2$ is in its active region, and hence

$$i_{C2} = h_{FE} i_{B2} \quad \text{and} \quad i_{C2} + i_{B2} = (h_{FE} + 1)i_{B2}$$

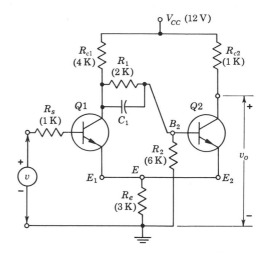

Fig. 10-24 An emitter-coupled binary. The impedance of the input source v is R_s. The numerical values refer to the illustrative example on page 399.

Applying Kirchhoff's voltage law to the base circuit of $Q2$ we find

$$V' - V_{BE2} = [R_b + R_e(h_{FE} + 1)]i_{B2} \qquad (10\text{-}20)$$

Solving for $V_{EN} = (i_{C2} + i_{B2})R_e$, we obtain

$$V_{EN} = V_{EN1} = V_{EN2} = (V' - V_{BE2})\,\frac{R_e(h_{FE} + 1)}{R_b + R_e(h_{FE} + 1)} \qquad (10\text{-}21)$$

and finally

$$V_1 = V_{EN} + V_{\gamma 1} \qquad (10\text{-}22)$$

If $R_e(h_{FE} + 1) \gg R_b$, then the drop across R_b may be neglected and $V_{EN} \approx V' - V_{BE2}$. Then

$$V_1 \approx V' - V_{BE2} + V_{\gamma 1} \qquad (10\text{-}23)$$

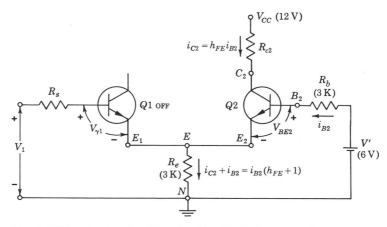

Fig. 10-25 The circuit of Fig. 10-24 with $Q1$ just at cutin.

Since $V_{\gamma 1}$ is the voltage from base to emitter at cutin where the loop gain just exceeds unity, it differs from V_{BE2} in the active region by only about 0.1 V (Table 6-1). If, therefore, we assume that $V_{BE2} = V_{\gamma 1} + 0.1$, we have, for either germanium or silicon,

$$V_1 \approx V' - 0.1 \tag{10-24}$$

This result indicates that V_1 may, in this way, be made almost independent of h_{FE}, of the emitter resistance R_e, of the temperature, and of whether or not a silicon or germanium transistor is used. Hence *the discriminator level V_1 is stable with transistor replacement, aging, or temperature changes, provided that $R_e(h_{FE} + 1) \gg R_b$ and that $V' \gg 0.1$.* Since V' depends on V_{CC}, R_{c1}, R_1, and R_2, where stability is required it is necessary that a stable supply and stable resistors be selected.

It is, of course, possible to adjust the circuit so that, when $Q1$ is cut off, $Q2$ is in saturation. Under these circumstances, regeneration would start not when $Q1$ comes into conduction but rather when $Q2$ comes out of saturation sufficiently so that the loop gain exceeds unity. Because of the presence of the commutating capacitor C_1, the regeneration point will then be a function of the speed of the input waveform, since the response at B_2 to a fast waveform will be greater than for a slow waveform. Since we normally require a fixed comparison voltage point, such operation with $Q2$ in saturation is not desirable. Accordingly we must keep the collector resistor R_{c2} small enough to avoid saturation. Because of this restriction and also because R_e may be comparable to R_{c2}, the swing at the output may be much smaller than V_{CC}.

Calculation of V_2 The voltage V_2, defined as the input voltage at which $Q2$ resumes conducting, is calculated from Fig. 10-26. In order to take into account the loading of R_1 and R_2 at the collector of $Q1$ we have replaced

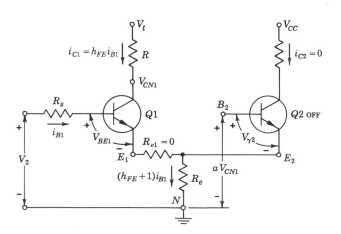

Fig. 10-26 The circuit of Fig. 10-24 with $Q2$ just at cutoff. The resistor R_{e1} will later be set different from zero to eliminate hysteresis.

R_1, R_2, R_{c1}, and V_{CC} by the Thévenin's equivalent voltage V_t and R, where

$$V_t = \frac{V_{CC}(R_1 + R_2)}{R_{c1} + R_1 + R_2} \quad \text{and} \quad R = \frac{R_{c1}(R_1 + R_2)}{R_{c1} + R_1 + R_2} \tag{10-25}$$

The voltage ratio from the first collector to the second base is a, where

$$a = \frac{R_2}{R_1 + R_2} \tag{10-26}$$

In Fig. 10-26 the input signal to $Q1$ is decreasing, and when it reaches V_2 then $Q2$ comes out of cutoff. Kirchhoff's voltage law around the base circuit of $Q2$ is

$$-aV_{CN1} + V_{\gamma 2} + (i_{B1} + i_{C1})R_e = 0 \tag{10-27}$$

where $V_{CN1} = V_t - i_{C1}R$, and $i_{C1} = h_{FE}i_{B1}$ for $Q1$ in the active region. We obtain, using V' from Eq. (10-19),

$$i_{C1} = \frac{aV_t - V_{\gamma 2}}{aR + R'_e} = \frac{V' - V_{\gamma 2}}{aR + R'_e} \tag{10-28}$$

where

$$R'_e \equiv R_e\left(1 + \frac{1}{h_{FE}}\right) \tag{10-29}$$

From Fig. 10-26, the comparator voltage V_2 is given by

$$V_2 = i_{B1}R_s + V_{BE1} + (i_{B1} + i_{C1})R_e = V_{BE1} + i_{C1}\left(R'_e + \frac{R_s}{h_{FE}}\right)$$

$$= V_{BE1} + \frac{R'_e + R_s/h_{FE}}{aR + R'_e}(V' - V_{\gamma 2}) \tag{10-30}$$

Although $V_{BE1} \approx V_{\gamma 2}$, these quantities do not cancel out of Eq. (10-30). Hence, since V_{BE1} is higher for silicon than germanium, V_2 is a few tenths of a volt higher for a Schmitt circuit using silicon transistors than for one using germanium devices.

Since h_{FE} is a large number, $R'_e \approx R_e$, and it may well be that $R_s/h_{FE} \ll R_e$. In this case Eq. (10-30) becomes

$$V_2 = V_{BE1} + \frac{R_e}{aR + R_e}(V' - V_{\gamma 2}) \tag{10-31}$$

From Eq. (10-31) it appears that V_2 is very insensitive to both R_s and h_{FE}. As a matter of fact it is possible for V_2 to be a less sensitive function of h_{FE} than is V_1 (Prob. 10-29).

EXAMPLE   (a) Find V_1 for the circuit of Fig. 10-24. Assume $h_{FE} = 30$ and that silicon transistors are used.   (b) Find V_2.   (c) Find the value of R_{e1} (Fig. 10-26) required to eliminate hysteresis.   (d) Repeat part c for R_{e2} (Fig. 10-27).

Solution   *a.* From Eqs. (10-19),

$$V' = \frac{(12)(6)}{4 + 2 + 6} = 6 \text{ V} \qquad R_b = \frac{(6)(4 + 2)}{4 + 2 + 6} = 3 \text{ K}$$

We have $R_e(h_{FE} + 1) = 3(31) = 93$ K, and $R_b = 3$ K.   Hence $R_e(h_{FE} + 1) \gg R_b$, and as a first approximation $V_1 = V' - 0.1 = 5.9$ V.   For a more accurate calculation we use Eq. (10-21) to calculate V_{EN}.   We have, using $V_{BE2} = 0.6$ V for silicon from Table 6-1,

$$V_{EN} = (6 - 0.6) \frac{3(31)}{3 + 3(31)} = 5.2 \text{ V}$$

and from Eq. (10-22)

$$V_1 = 5.2 + 0.5 = 5.7 \text{ V}$$

which differs from the approximate value by only about 3 percent.   A similar calculation for germanium gives $V_1 = 5.7$ V, just as for silicon.

The above calculation assumes that $Q2$ is operating in its active region.   In order to be sure that $Q2$ is indeed in its active region let us calculate the collector junction voltage.   We find

$$V_{CB2} = V_{CE2} - V_{BE2} = V_{CC} - i_{C2}R_{c2} - V_{EN} - V_{BE2} \tag{10-32}$$

To find i_{C2} note that

$$V_{EN} = i_{B2}(h_{FE} + 1)R_e = i_{C2}R'_e$$

where, from Eq. (10-29),

$$R'_e = (3)\left(1 + \frac{1}{30}\right) = 3.1 \text{ K}$$

$$i_{C2} = \frac{V_{EN}}{R'_e} = \frac{5.2}{3.1} = 1.68 \text{ mA}$$

From Eq. (10-32),

$$V_{CB2} = 12 - (1.68)(1) - 5.2 - 0.6 = 4.5 \text{ V}$$

Since this voltage is positive and an *n-p-n* transistor is under consideration, then the collector is reverse-biased and we have verified that $Q2$ is in its active region. If R_{c2} is increased sufficiently then V_{CB2} becomes negative and $Q2$ is in saturation (Prob. 10-26).

b. From Eqs. (10-26) and (10-25),

$$a = \frac{6}{2 + 6} = 0.75 \qquad R = \frac{(4)(2 + 6)}{4 + 2 + 6} = 2.67 \text{ K}$$

$$aR = (0.75)(2.67) = 2.0 \text{ K}$$

Since $V' = 6$ V and $V_\gamma = 0.5$ V we calculate from Eq. (10-30) that

$$V_2 = 0.6 + \frac{3.1 + 0.03}{2.0 + 3.1}(6 - 0.5) = 4.0 \text{ V}$$

If germanium transistors had been used we would have found

$$V_2 = 0.2 + \frac{3.1 + 0.03}{2.0 + 3.1}(6 - 0.1) = 3.8 \text{ V}$$

a value slightly lower than that found for silicon.

    c. A resistor R_{e1} in series with the emitter of $Q1$ will affect V_2 but not V_1. Hence $V_1 = 5.7$ V. From Fig. 10-26, with R_{e1} in series with E_1, we see that i_{C1}, which was determined by the base circuit of $Q2$ [Eq. (10-27)], is unaffected by R_{e1}. Hence the value of V_{EN2} at which $Q2$ returns to conduction is the same as before. However, in order for the currents to remain unchanged with R_{e1} present, $v = V_2$ must be increased by the amount of the voltage drop across R_{e1}. Hence, this resistance must be chosen so that $(i_{C1} + i_{B1})R_{e1}$ is equal to the value of V_H before the addition of R_{e1}. From Eq. (10-28),

$$i_{C1} = \frac{6 - 0.5}{2.0 + 3.1} = 1.08 \text{ mA} \qquad i_{B1} = \frac{1.08}{30} = 0.04 \text{ mA}$$

so that hysteresis is eliminated when $V_1 = V_2 + (i_{C1} + i_{B1})R_{e1}$, or

$$R_{e1} = \frac{5.7 - 4.0}{1.12} = 1.5 \text{ K}$$

The comparator level is now 5.7 V for either increasing or decreasing voltages.

    d. From Fig. 10-26 we see that if we place a resistor R_{e2} in series with the emitter of $Q2$ it can have no effect on V_2 because $Q2$ is OFF. Hence V_2 remains at 4.0 V. However, from Fig. 10-27 it is clear that R_{e2} will affect V_{EN1} and hence V_1. From the base circuit of $Q2$

$$-6 + 3i_{B2} + 0.6 + (R_{e2} + 3)(i_{B2} + i_{C2}) = 0$$

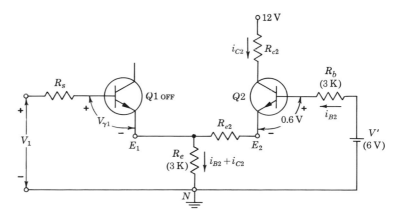

Fig. 10-27 The circuit of Fig. 10-24 with a resistor R_{e2} in the emitter of $Q2$ which may be adjusted to eliminate hysteresis.

Using $i_{B2} = i_{C2}/h_{FE} = i_{C2}/30$, we obtain $i_{C2} = 5.4/(1.03R_{e2} + 3.20)$ and

$$V_{EN1} = (i_{C2} + i_{B2})(3) = 3.10i_{C2} = \frac{16.7}{1.03R_{e2} + 3.20} \quad V$$

$V_1 = V_{EN1} + V_{\gamma 1} = V_2 = 4.0$ for zero hysteresis. Hence

$$1.03R_{e2} + 3.20 = \frac{16.7}{4.0 - 0.5} = 4.77 \quad \text{or} \quad R_{e2} = 1.5 \text{ K}$$

(By coincidence $R_{e1} = R_{e2}$ for this example.)

Hysteresis Considerations   If R_{e1} or R_{e2} is larger than the value required to give zero hysteresis, then the loop gain will be less than unity and the circuit will not change state.   Usually, R_{e1} or R_{e2} is chosen so that a small amount of hysteresis remains, in order to ensure that the loop gain will remain greater than unity even if the circuit drifts somewhat (owing to supply-voltage changes, aging, etc.).   Also, R_{e1} or R_{e2} is usually bypassed with a small capacitor. During the transition interval this capacitor reduces the degeneration caused by these resistors, and hence an output pulse or step with a shorter rise time is delivered.

If a value of R_{e1} (or R_{e2}) is used which is larger than required for zero hysteresis but if the resistor is bypassed so that the a-c loop gain exceeds unity, then it is possible to trigger the circuit from one state to the next. Under these circumstances $|V_2| > |V_1|$, and hence the hysteresis range V_H is negative.   With this condition the circuit produces high-frequency oscillations if the input voltage has a value between V_1 and V_2.   Here, then, is another reason why the resistor R_{e1} (or R_{e2}) is chosen so as to make $|V_1|$ somewhat greater than $|V_2|$.

A few comments about the source resistance R_s are appropriate.   The value of V_1 is independent of R_s.   The value of V_2 does depend upon R_s but only to a small extent as long as $R_s \ll h_{FE}R_e$.   If R_s is too small, then it is possible for large input signals v that the first stage $Q1$ may be driven heavily into saturation.   Under these conditions minority-carrier storage in the base of $Q1$ may limit the maximum speed of operation of this circuit.   On the other hand, too large a value of R_s may make it impossible to trigger the circuit because the loop gain may fall below unity.   Since V_2 increases when R_s is added, whereas V_1 is independent of R_s, then for large enough R_s it is possible for $V_1 = V_2$.   Hysteresis is thus eliminated and the loop gain is unity. If R_s exceeds this critical value, the loop gain falls below unity and the circuit cannot be triggered.   In the example considered above for silicon $V_1 = 5.7$ and V_2 is given by Eq. (10-30).   Hence the maximum value of R_s is found from

$$5.7 = 0.6 + \frac{3.1 + R_s/30}{2 + 3.1}(6 - 0.5)$$

which yields a maximum value of R_s of 49 K.

REFERENCES

1. Millman, J., and H. Taub: "Pulse and Digital Circuits," p. 155, McGraw-Hill Book Company, New York, 1956.

2. Linvill, J. G.: Nonsaturating Pulse Circuits Using Two Junction Transistors, *Proc. IRE*, vol. 43, pp. 826–834, July, 1955.

3. Beter, R. H., W. E. Bradley, R. B. Brown, and M. Rubinoff: Direct-coupled Transistor Circuits, *Electronics*, June, 1955, pp. 132–136.

4. Schmitt, O. H.: A Thermionic Trigger, *J. Sci. Instr.*, vol. 15, no. 1, pp. 24–26, January, 1938.

11 / MONOSTABLE AND ASTABLE MULTIVIBRATORS

The binary circuit, it will be recalled, has two stable states, in either one of which it may remain permanently. The monostable circuit has instead only one permanently stable state and one quasi-stable state. In the monostable configuration, a triggering signal is required to induce a transition from the stable state to the quasi-stable state. The circuit may remain in its quasi-stable state for a time which is very long in comparison with the time of transition between states. Eventually, however, it will return from the quasi-stable state to its stable state, no external signal being required to induce this reverse transition.

Since, when it is triggered, the circuit returns to its original state by itself after a time T, it is known as a *one-shot*, a *single-cycle*, a *single-step circuit*, or a *univibrator*. Since it generates a rectangular waveform and hence can be used to gate other circuits, it is also called a *gating circuit*. Furthermore, since it generates a fast transition at a predetermined time T after the input trigger, it is also referred to as a *delay circuit*.

The astable circuit has two states, both of which are quasi-stable. Without the aid of an external triggering signal the astable configuration will make successive transitions from one quasi-stable state to the other.

Both these circuits find extensive application in pulse circuitry. The basic application of the monostable configuration results from the fact that it may be used to establish a fixed time interval, the beginning and end of which are marked by an abrupt discontinuity in a voltage waveform. The astable circuit is an oscillator and is used as a generator of "square waves" and, since it requires no triggering signal, is itself often a basic source of fast waveforms.

404

Two amplifier stages may be interconnected in such a manner as to possess one stable state or two quasi-stable states. These configurations are called *monostable multivibrators* or *astable multivibrators* (*multis*), respectively. Vacuum-tube and transistor multis will be considered in this chapter. In Chap. 13 other monostable and astable circuits, most of which contain only one active device, will be studied.

11-1   THE MONOSTABLE MULTI

The circuit diagram of a monostable multi is shown in Fig. 11-1. The active devices $A1$ and $A2$ are either tubes or transistors. The supply-voltage polarities indicated are correct for an *n-p-n* transistor or for a tube but must be reversed for a *p-n-p* transistor. Here, as in a binary circuit, the output at Y_2 is coupled to the input at X_1 through a resistive attenuator in which C_1 is a small commutating capacitor. This capacitor serves here the same purpose as is served by the commutating capacitors in the binary. As for a tube-circuit binary, C_1 is given, to a first approximation, by Eq. (10-13). The d-c coupling found in a binary from Y_1 to X_2 is here replaced by capacitive coupling through C. While the resistor R at the input of $A2$ is shown returned to the supply voltage V_{YY}, this feature of the circuit is not essential, and R may be returned to a lower potential. We shall, however, later discuss the advantage of connecting R to the supply voltage.

We shall assume that the circuit parameters have been adjusted properly so that the multi finds itself in its (permanently) stable state with $A1$ OFF and $A2$ ON (in clamp, if a tube; in saturation, if a transistor). The multi may be induced to make a transition out of its stable state by an application of a negative trigger at X_2 or at Y_1 of Fig. 11-1. As with the binary, diode or triode triggering may be used to advantage. It is to be emphasized that the triggering is unsymmetrical, being applied to one device only and not to both simultaneously.

Fig. 11-1   The monostable multi.
If the devices $A1$ and $A2$ are *p-n-p*
transistors, then the supply-voltage
polarities must be reversed.   For
a tube $V_{YY} = V_{PP}$, $V_{XX} = V_{GG}$,
and $R_y = R_p$.   For a transistor
$V_{YY} = V_{CC}$, $V_{XX} = V_{BB}$, and
$R_y = R_c$.

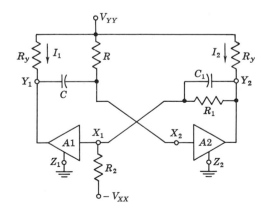

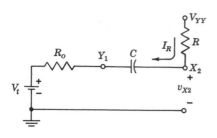

Fig. 11-2 Simplified circuit for computing the voltage v_{X2} at the input to $A2$, during the quasi-stable state. The Thévenin's voltage V_t is the voltage at Y_1 if the capacitor C is disconnected from Y_1. At $t = 0$ the voltage across C is $V_{YY} - V_\sigma$.

Assume that a single trigger is applied to X_2 and that a regenerative action takes place driving $A2$ completely below cutoff. The voltage at Y_2 now rises to approximately V_{YY}, and because of the cross coupling between Y_2 and X_1 the first stage $A1$ comes into conduction. This device may be driven into saturation (or clamp), or it may operate within its active region. In either event a current I_1 now exists in the output-circuit resistor R_y of $A1$, and the voltage at $Y1$ drops abruptly by an amount I_1R_y. The voltage at X_2 drops by the same amount because the voltage across C cannot change instantaneously. The multi is now in its quasi-stable state.

The circuit will remain in this quasi-stable state for only a finite time T because X_2 is connected to V_{YY} through a resistance R. Therefore X_2 will rise in voltage, and when it passes the cutin voltage V_γ of $A2$, a regenerative action will take place, turning $A1$ off and eventually returning the multi to its initial stable state. We look now into the matter of determining the time duration of the quasi-stable state. During this interval $A2$ is OFF, and the voltage changes at X_2 may be calculated from the circuit of Fig. 11-2. In this circuit the stage $A1$ has been replaced by a Thévenin's equivalent generator V_t (the voltage at Y_1 with C disconnected) and a resistance R_o, which represents the amplifier output impedance including the presence of R_y. The voltage waveform at X_2 is indicated in Fig. 11-3. The transition from stable to quasi-stable state occurs at $t = 0$. For $t < 0$, $v_{X2} = V_\sigma$, the saturation base voltage of a transistor (or the grid voltage of a tube in clamp). Since Y_1 and X_2 are coupled by a capacitor, any abrupt change in voltage at Y_1 must result in the same discontinuous change at X_2. At $t = 0+$, the voltage at Y_1 drops by I_1R_y. Hence, at $t = 0+$, $v_{X2} = V_\sigma - I_1R_y$. Thereafter, the voltage v_{X2} will rise exponentially toward V_{YY} with a time constant

$$\tau = (R + R_o)C$$

(Refer to Fig. 11-2.) Since at $t = \infty$, $v_{X2} = V_{YY}$, then the input voltage to the second stage is given by [Eq. (2-3)]

$$v_{X2} = V_{YY} - (V_{YY} - V_\sigma + I_1R_y)\epsilon^{-t/\tau} \tag{11-1}$$

This exponential rise will actually continue, however, only until v_{X2} rises to the cutin voltage V_γ, at which time T a reverse transition will occur. Solving

Eq. (11-1) for $t = T$, when $v_{X2} = V_\gamma$, we obtain

$$T = \tau \ln \frac{V_{YY} + I_1 R_y - V_\sigma}{V_{YY} - V_\gamma} \tag{11-2}$$

In this equation V_σ may be taken as zero for a tube in clamp, whereas for a transistor $V_\sigma = V_{BE}(\text{sat})$ (typically 0.3 V for germanium and 0.7 V for silicon, as on page 219). For a tube V_γ is a negative quantity, the cutin voltage corresponding to a plate voltage V_{PP} ($V_\gamma \approx -V_{PP}/\mu$), whereas for a transistor V_γ is a forward-bias voltage (typically, 0.1 V for germanium and 0.5 V for silicon). The symbol T is referred to as the *delay time* and also as the *gate time, pulse width,* or *duration.*

The delay T may be varied either through the time constant τ or by adjusting I_1. The current I_1, which flows in device $A1$ when this device is ON, is controlled by the base input current or grid input voltage. This input current or voltage is, in turn, dependent on V_{XX}. Therefore T may be varied through variation of V_{XX}. The voltage V_{XX} will affect T up to the point where, when $A1$ comes ON, it finds itself in clamp or in saturation.

The duration T of a monostable multi is ordinarily not particularly stable, depending as it does on the device characteristics through I_1, V_σ, and V_γ. The stability is somewhat better if R is returned to a voltage of large magnitude such as V_{YY} rather than to a low voltage V_1 or to ground. The reason for this feature may be seen in Fig. 11-4. Curve 1 corresponds to returning R to V_{YY}, whereas curve 2 is for R connected to a low voltage V_1. The time constants have been adjusted in the two cases to give the same initial time duration T_o. Suppose that V_γ now changes by ΔV_γ because of, say, a change in ambient temperature, heater voltage, or device replacement. We see from Fig. 11-4 that the change in time $T_o - T_1$ is smaller than $T_o - T_2$.

If the active devices are transistors and if R is returned to ground, then the circuit will not function properly. Under these circumstances $V_{BE2} \approx 0$

Fig. 11-3 Voltage variation at X_2 during the quasi-stable state. The cutin voltage V_γ and the saturation voltage V_σ are positive for an *n-p-n* transistor, but negative for a *p-n-p* device. For a tube V_γ is negative and V_σ is zero.

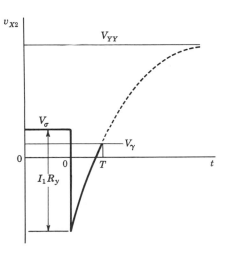

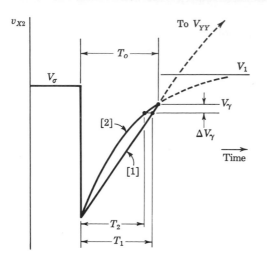

Fig. 11-4 Illustrating the advantage with respect to timing stability of returning the resistor R to V_{YY} rather than to a lower voltage V_1.

in the stable state, and the collector current in $Q2$ is quite small, at most a few times I_{CBO} (Sec. 6-8). If the transistor $Q2$ were reverse-biased then its collector current would change to I_{CBO}. Hence, with R returned to ground, $Q2$ remains virtually cut off at all times. In order for a transition to take place the swing at Y_2 must be large enough to take $Q1$ out of cutoff. But if R goes to ground, then the change in the output of $Q2$ is too small to bring $Q1$ into its active region, and the circuit cannot be triggered.

We may expect that the waveforms generated in the monostable multi will be very similar to those encountered in Sec. 8-8, where we discussed the tube and transistor as a switch. In the multi circuit, tube $V1$ or transistor $Q1$ will be turned on and at a later time turned off. Hence there will appear at the plate or collector a negative gating voltage of the form shown as the input in Fig. 8-18f. This negative gating waveform is applied through a capacitor to the grid or base of a tube or transistor, just as in Fig. 8-18a and b.

11-2   A VACUUM–TUBE MONOSTABLE MULTI

The circuit in Fig. 11-5 is the vacuum-tube version of the circuit in Fig. 11-1. Because of the coupling from the plate of one stage to the grid of the second tube, this configuration is called the *plate-to-grid-coupled monostable multi* or, simply, the *plate-coupled delay multi* or *plate-coupled one-shot*.

We shall now investigate the appearance of the waveforms at both plates and both grids from the time before a trigger is applied to the time the multi has restored itself to its initial stable state. The waveforms are shown in Fig. 11-6. The triggering signal occurs at $t = 0$, and the reverse transition occurs at $t = T$.

The Stable State   For $t < 0$, the current in $V1$ is zero and that in $V2$ is I_2, corresponding to a clamped grid. The plate P_1 is at V_{PP} and the plate P_2

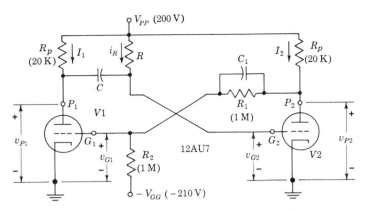

Fig. 11-5 The plate-coupled monostable multi. The component values in parentheses refer to the example on page 411.

(neglecting the loading effect of the attenuator) is at $v_{P2} = V_{PP} - I_2 R_p$. The
grid G_2 is at $V_\sigma = 0$ (approximately), and the grid-to-ground voltage at $V1$ is
calculated by superposition to be

$$v_{G1} = (V_{PP} - I_2 R_p) \frac{R_2}{R_1 + R_2} - \frac{V_{GG} R_1}{R_1 + R_2} \equiv V_F \tag{11-3}$$

The voltage V_F must be negative enough to keep $V1$ below cutoff. These
stable-state voltages are indicated in Fig. 11-6.

The Quasi-stable State As a result of the application of a triggering signal at $t = 0$, $V2$ goes OFF and $V1$ goes ON. The voltages v_{P1} and v_{G2} drop
abruptly by the same amount $I_1 R_p$, where I_1 is the current in R_p of $V1$ when
$V1$ goes ON. At $t = 0+$, a current $i_R = (V_{PP} + I_1 R_p)/R$ flows through R.
Since the grid G_2 is below cutoff, this current flows through C into the plate
terminal of $V1$. This current i_R is not constant but decreases with time as the
capacitor charges. We shall neglect the current i_R in comparison with I_1.
On this basis we may consider that the current in $V1$ is I_1 and we may consider
as well that, so long as the grid G_1 maintains a fixed voltage, so also will the
plate P_1. We shall then be neglecting the voltage drop $i_R R_o$, where R_o is the
output impedance of the amplifier stage $V1$. A method for calculating I_1
taking into account the loading of R at the plate of P_1 is given in Prob. 11-2.
At $t = 0$, also, the voltage v_{P2} rises abruptly by $I_2 R_p$ to V_{PP} and v_{G1}
rises abruptly to

$$v_{G1} = \frac{V_{PP} R_2}{R_1 + R_2} - \frac{V_{GG} R_1}{R_1 + R_2} \equiv V_N \tag{11-4}$$

If this ON voltage V_N of $V1$, as computed from Eq. (11-4), is positive, then $V1$
is in clamp and V_N is to be taken as nominally zero. The voltage at G_2
now starts to rise exponentially with time constant $(R + R_o)C \approx RC$ toward

V_{PP}.   Until v_{G2} reaches the cutin voltage V_γ all voltages at the other electrodes remain unaltered.

Waveforms for $t > T$ Refer to Fig. 11-6a and b.   At $t = 0$, tube $V1$ is driven ON, so that v_{P1} and v_{G2} drop by $I_1 R_p$.   At $t = T$, tube $V1$ will be driven back to cutoff, and if it were not restrained from doing so, the plate P_1 (Fig. 11-5) would rise abruptly by $I_1 R_p$ and thus carry the grid G_2 upward by the same amount.   At $t = T$, however, the grid G_2 is much closer to zero voltage than at $t = 0+$.   Therefore, G_2 is driven positive and appreciable grid current flows.   Hence an overshoot develops in v_{G2} which decays as the capacitor C recharges because of the grid current.   We shall now calculate the magnitude of this overshoot.

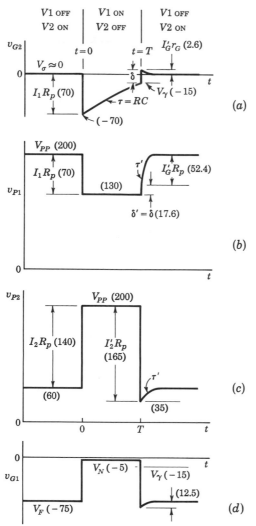

Fig. 11-6   Waveforms of plate-coupled monostable multi.   The exponential portions of the waveforms beginning at $t = T$ all have a time constant $(R_p + r_G)C = \tau'$.   The numerical values (in volts) refer to the circuit of Fig. 11-5.

Fig. 11-7 Circuit for calculating the overshoot in the plate-coupled multi. $V1$ of
Fig. 11-5 is below cutoff and $V2$ is driven
into the positive grid region.

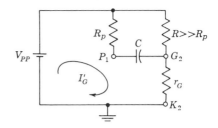

The grid current may be accounted for adequately by using a piecewise
linear and continuous model for the input circuit. We shall assume that the
break point occurs at the origin and that for positive grid voltages the grid
volt-ampere characteristic is linear. The input resistance of $V2$ when its grid
is driven positive is r_G (Sec. 6-19). The grid current immediately after the
reverse transition is designated by I_G', and its path is indicated in Fig. 11-7.
Since R is very much larger than R_p, the current in R has been neglected compared with I_G'. From Fig. 11-7 we see that, at $t = T+$,

$$v_{G2} = I_G' r_G \qquad \text{and} \qquad v_{P1} = V_{PP} - I_G' R_p \tag{11-5}$$

From Fig. 11-6 the jumps in voltage at G_2 and at P_1 are, respectively,

$$\delta = I_G' r_G - V_\gamma \qquad \text{and} \qquad \delta' = I_1 R_p - I_G' R_p \tag{11-6}$$

Since P_1 and G_2 are connected by a capacitor and since the voltage across a
capacitor cannot change instantaneously, then $\delta = \delta'$. Equating these two
voltage changes we obtain from Eq. (11-6)

$$I_G' = \frac{I_1 R_p + V_\gamma}{R_p + r_G} \tag{11-7}$$

The overshoot $I_G' r_G$ decays to zero with a time constant $\tau' = (R_p + r_G)C$,
and as the grid overshoot decays exponentially, the plate P_1 rises exponentially
to V_{PP}. Corresponding to the overshoot at G_2 there is an undershoot at P_2.
The current I_2' in $V2$ at the time of the overshoot may be determined by drawing a load line for R_p on the *positive* grid characteristics of the tube and noting
the current corresponding to a grid voltage $r_G I_G'$. The undershoot at P_2
similarly is reflected in an undershoot at G_1. Of course, all the sharp corners
indicated in Fig. 11-6 are actually slightly rounded by tube and stray shunting
capacitances. Ordinarily, however, this rounding is of a different order of
magnitude from the rounding apparent on the trailing edge of the waveform
at P_1.

The following illustrative example will indicate more specifically how one
may determine the waveforms in a plate-coupled multi.

EXAMPLE Compute the voltage levels for the waveforms of Fig. 11-6 for a plate-
coupled multi whose components and supply voltages are as given in Fig. 11-5.
The tubes employed are the two halves of a type 12AU7.

Solution Drawing a load line for $R_p = 20$ K and $V_{PP} = 200$ V on the negative-grid plate characteristics for the type 12AU7 (Fig. D-5), we find $I_2 R_p = 140$ V and $V_{PP} - I_2 R_p = 60$ V. From Eq. (11-3), we find

$$V_F = 60 \times \tfrac{1}{2} - 210 \times \tfrac{1}{2} = -75 \text{ V}$$

and from Eq. (11-4) we obtain

$$V_N = 200 \times \tfrac{1}{2} - 210 \times \tfrac{1}{2} = -5 \text{ V}$$

Corresponding to this grid voltage we find from Fig. D-5

$$I_1 R_p = 70 \text{ V} \qquad \text{and} \qquad V_{PP} - I_1 R_p = 130 \text{ V}$$

Beginning immediately after $t = 0$, the grid waveform rises from -70 V exponentially toward 200 V with a time constant which is approximately $\tau = RC$. The quasi-stable state persists until v_{G2} reaches the cutoff voltage $V_\gamma = -15$ V, during which time all other voltage levels remain constant.

We must now compute the amplitude of the grid overshoot, for which we may use Eq. (11-7), provided that we are able to decide on a reasonable value of r_G. Examine now the positive-grid tube characteristics for the 12AU7 given in Fig. D-6. Observe that over a broad range of plate voltage the grid current is 10 mA for a grid voltage of 10 V. We therefore tentatively accept for r_G the value $r_G = 1$ K. The grid overshoot is therefore

$$I_G' r_G = \frac{(70 - 15) \times 1}{20 + 1} = \frac{55}{21} = 2.6 \text{ V}$$

The abrupt portion of the rise of the plate voltage v_{P1} has a magnitude

$$\delta = I_G' r_G - V_\gamma = 2.6 + 15 = 17.6 \text{ V}$$

The remainder of the approach to the supply voltage occurs with a time constant $(R_p + r_G)C$, which is also the time constant with which the overshoot decays.

To find $I_2' R_p$, we draw the load line for 20 K and $V_{PP} = 200$ V on the positive grid characteristics, finding approximately that, corresponding to $v_{G2} = +2.6$ V, $I_2' R_p = 165$ V. At the overshoot, then, the voltage v_{P2} drops to 35 V. The amplitude of the undershoot in v_{G1} is

$$(I_2' R_p - I_2 R_p) \frac{R_2}{R_1 + R_2} = (165 - 140) \times \tfrac{1}{2} = 12.5 \text{ V}$$

The delay time is given by Eq. (11-2) with $V_\sigma = 0$, $V_{YY} = V_{PP} = 200$, $I_1 R_y = I_1 R_p = 70$ V, and $V_\gamma = -15$ V, or

$$T = RC \ln \frac{200 + 70}{200 - (-15)} = RC \ln \frac{270}{215} = 0.23 RC$$

It is feasible to select values for R and C which yield values of T from seconds to microseconds.

The least-certain feature of the above calculation has to do with the overshoot amplitude. This difficulty results from the fact that the grid resistance r_G is not constant but is rather a function of the plate voltage, decreasing with decreasing plate voltage. We may, however, note that r_G

does remain fairly constant, provided that the plate voltage remains large in comparison with the grid voltage. This result is borne out by the curves of Fig. D-6, where it appears that for high plate voltage in comparison with grid voltage $r_G = 1,000\ \Omega$. In our present case we seek the value of r_G under the circumstance that $v_{G2} = 2.6$ V and $v_{P2} = 35$ V; hence we may reasonably consider $r_G = 1,000\ \Omega$. This general procedure may be summarized as follows: we select first some reasonable value for r_G and compute, as above, the grid and plate voltage corresponding to the overshoot. Corresponding to this first approximation for grid and plate voltages we note from the tube characteristics a better value for r_G. We may now recalculate the overshoots, leading to a still better value of r_G, etc. Normally, however, the first approximation gives sufficiently good results, and successive calculations are not warranted.

Let us note that there is a minimum allowable value $I_1(\text{min})$ of I_1 which is required in order that there shall be a quasi-stable state. This current $I_1(\text{min})$ is clearly determined by the condition that the drop $I_1(\text{min})R_p$ shall be sufficient to drive G_2 below cutoff. Hence $I_1(\text{min})R_p = |V_\gamma|$, or $I_1(\text{min}) = |V_\gamma|/R_p$. This result is consistent with Eq. (11-2) since this condition makes $T = 0$. Corresponding to $I_1(\text{min})$ there is a maximum value $V_{GG}(\text{max})$ of the bias supply. Similarly, there is a minimum value $V_{GG}(\text{min})$ dictated by the consideration that in the stable state tube $V1$ must be at cutoff. If V_{GG} is adjusted so that $V1$ is not at cutoff, then the multi may have no permanently stable state. In this case, as is described in Sec. 11-11, the multi may become astable and switch back and forth between two quasi-stable states.

11-3   RECOVERY IN A PLATE–COUPLED MONOSTABLE MULTI

After the formation of the gate of duration T, the multi will not have completely returned to its stable state until all overshoots have decayed to zero. The decay time of these overshoots is called the *recovery time*. The recovery time depends on the time constant $(R_p + r_G)C$, whereas T depends on RC. Where a short recovery time is of importance, a fixed required time constant RC is attained by making R as large as possible and C correspondingly small. A practical upper limit for R is of the order of 10 M and is set by the same considerations that limit grid-leak resistors generally. Additionally, if R is large and C small, the effective impedance between G_2 and ground will be large during the interval when G_2 is not in clamp, and the circuit may become excessively sensitive to stray fields. The advantage of a short recovery time may be seen from the following discussion.

Consider what might happen if regularly spaced triggers are to generate gates which are as wide as possible (T to be nearly equal to the interval between triggers). At the end of the time T the capacitance C must recharge through R_p before the next trigger comes along. Suppose that C is not completely recharged and hence that P_1 has not reached V_{PP} before the next trigger is injected. This next pulse will trip the circuit and P_1 will drop

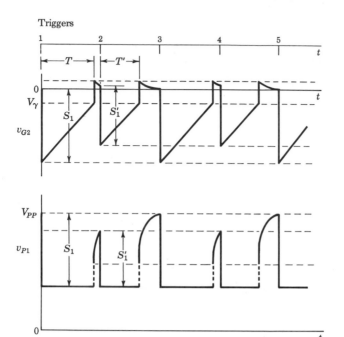

Fig. 11-8 Waveforms of multi when interval between triggers does not allow complete recovery. The dashed line in the v_{P1} waveform represents an abrupt change in voltage. This is followed by an exponential rise to V_{PP}.

to its value for $v_{G1} = V_N$. The change in voltage S_1' at P_1 is less than the full swing $S_1 = I_1 R_p$ because the plate did not reach V_{PP}. This new smaller increment S_1' appears at G_2, as shown in Fig. 11-8. Furthermore, the voltage at G_2 will not have decayed to zero when the second trigger arrives. The result is that, after this second trigger, G_2 starts at a more positive voltage than after the first trigger. Therefore the length of the quasi-stable state T' after the second trigger is less than T after the first trigger. Hence, there is a longer time available for C to charge before the third trigger appears. Thus the drop in G_2 at this third impulse may again be the original S_1. If so, then at the fourth trigger it will be S_1'. This will lead to the peculiar situation pictured in Fig. 11-8 in which not all cycles are alike, but rather alternate cycles have the same character.

There are a number of steps which may be taken if it should become important to reduce the recovery time. The most straightforward method is to reduce the size of the plate load resistance of tube $V1$. It may then become necessary, however, to replace the tube $V1$ by a larger tube since the tube dissipation will increase. Thus, for a given swing at P_1, the current I_1 must increase as R_p is decreased.

A second method is the use of a plate catching diode (Sec. 8-13) to eliminate the slow portion of the rise at the plate of $V1$. Note that in this case the plate resistor of $V1$ must be returned to a higher supply voltage than the plate resistor of $V2$ if the same swing at the plate of $V1$ is desired.

After these first two methods have been employed, some further improvement may result if an additional diode is shunted from grid to cathode of $V2$ in order to reduce r_G. Finally, we may use a cathode follower to couple the plate P_1 to the grid G_2. Since the plate P_1 is coupled directly to the grid of the cathode follower, it is required that the cathode-follower supply voltage be higher than the multi supply voltage. The timing capacitor is connected between the cathode of the cathode follower and the grid G_2. The timing capacitor will now be able to recharge through the low output impedance of the cathode follower. The use of a cathode follower or of a small plate resistance for tube $V1$ will reduce the recovery time as already noted, but the amplitude of the overshoots will be greater than before [Eq. (11-7)].

11-4   GATE WIDTH OF A COLLECTOR–COUPLED MONOSTABLE MULTI

The circuit of Fig. 11-9 is the transistor version of Fig. 11-1 and is called the *collector-to-base-coupled monostable multivibrator* or, more simply, the *collector-coupled monostable multi* or *one-shot*.

The gate width T is given by Eq. (11-2). We shall now prove that T can be made very stable (almost independent of transistor characteristics, supply voltages, and resistance values) if $Q1$ *is driven into saturation*. Under these circumstances $I_1 R_c = V_{CC} - V_{CE}(\text{sat})$. Since $V_\sigma = V_{BE}(\text{sat})$, then from Eq. (11-2),

$$T = \tau \ln \frac{2V_{CC} - V_{CE}(\text{sat}) - V_{BE}(\text{sat})}{V_{CC} - V_\gamma}$$

$$= \tau \ln 2 + \tau \ln \frac{V_{CC} - \dfrac{V_{CE}(\text{sat}) + V_{BE}(\text{sat})}{2}}{V_{CC} - V_\gamma} \tag{11-8}$$

At room temperature we see from page 219 that $V_{CE}(\text{sat}) + V_{BE}(\text{sat}) \approx 2V_\gamma$ for either germanium or silicon. The second term in Eq. (11-8) is zero under these circumstances, and an excellent approximation for T is

$$T = \tau \ln 2 = 0.69(R + R_o)C \approx 0.69RC \tag{11-9}$$

since R_o for a transistor in saturation is small compared with R. The larger V_{CC} is, compared with the junction voltages, the better is this approximation.

The interval T is not particularly stable against temperature variations, as we shall now demonstrate. In Fig. 6-21 we see that the base-to-emitter voltage $V_{BE}(\text{sat})$ (and also V_γ) decrease as temperature increases at the rate of about 2 mV/°C, whereas $V_{CE}(\text{sat})$ has a temperature coefficient which is of opposite sign and substantially smaller. Since the second term in Eq. (11-8) becomes more negative as the temperature increases (Prob. 11-9), the gate width T decreases as the temperature increases. The larger the value of

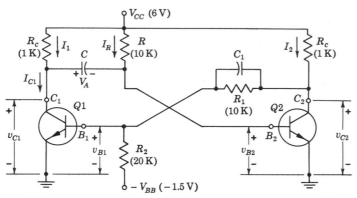

Fig. 11-9 The collector-coupled *n-p-n* transistor monostable multi. The values of the components given in parentheses refer to the illustrative problem on page 420.

V_{CC}, the smaller is this effect. From Eq. (11-8) we find that for $V_{CC} = 6$ V, T decreases about 5 percent as the temperature increases from 25 to 175°C, which is about the highest temperature at which we might expect to use a silicon transistor.

Up to this point we have neglected the reverse saturation current which flows in a transistor when it is cut off. We consider now the effect of the reverse saturation current on the time T of the quasi-stable state. During the interval, when $Q2$ is cut off, a nominally constant current I_{CBO} flows out of the base of the transistor $Q2$ in Fig. 11-9. Assume now that the capacitor C were disconnected from the junction of the resistor R with the base of $Q2$. Then the voltage at the base of $Q2$ with C disconnected would be not V_{CC} but $V_{CC} + I_{CBO}R$. It therefore appears that capacitor C, in effect, charges through R from a source $V_{CC} + I_{CBO}R$. Consequently, since I_{CBO} increases with temperature, the time T will decrease. There is a second way in which the reverse saturation current affects T. The initial voltage v_{C1} in the stable state is not V_{CC} but is $V_{CC} - I_{CBO}R_c$, where now I_{CBO} is the collector current of $Q1$ when it is cut off. Consequently the drop in the voltage v_{C1} when the multi is triggered is smaller by $I_{CBO}R_c$ than it would otherwise be, and consequently a smaller gate duration is obtained. Since, however, $R_c \ll R$, this second effect is negligible in comparison with the first.

To take account here only of the effect of I_{CBO} let us assume that

$$V_{CE}(\text{sat}) = V_{BE}(\text{sat}) = V_\gamma = 0$$

In this case we find that (Prob. 11-8)

$$T = \tau \ln 2 - \tau \ln \frac{1 + \phi}{1 + \phi/2} \tag{11-10}$$

where

$$\phi \equiv \frac{I_{CBO}R}{V_{CC}} \qquad (11\text{-}11)$$

Since I_{CBO} increases with temperature, we conclude from Eq. (11-10), as expected, that the delay T decreases as the temperature increases. This effect is of the same order of magnitude as that discussed above in connection with junction voltages. For example, if I_{CBO} increases from, say, 1 to 100 μA at the extreme ambient temperature and if $R = 10$ K and $V_{CC} = 6$ V, we calculate that T decreases about 12 percent over its room-temperature value.

One method[1] of temperature-compensating a monostable multi for both of the effects discussed above is to connect R not to V_{CC} but to a source V whose value decreases as the temperature increases. The voltage V may be obtained from a voltage divider across V_{CC} in which one resistance varies with temperature. The collector supply remains V_{CC}, which is independent of temperature. The general expression for T taking I_{CBO}, junction voltages, and V into account is given in Prob. 11-10.

11-5　WAVEFORMS OF THE COLLECTOR–COUPLED MONOSTABLE MULTIVIBRATOR

We shall now explain the waveforms (Fig. 11-10) at both collectors and bases. The triggering signal is applied at $t = 0$ and the reverse transition occurs at $t = T$.

The Stable State　The voltages and currents for $t < 0$ are calculated in the same manner as in Sec. 10-3 for the corresponding quantities in a flip-flop. It is possible for $Q2$ to be in its active region, although, just as in a binary, it is usually in saturation for reasons of d-c stability (Sec. 10-1). The base voltage of $Q2$ is $v_{B2} = V_{BE}(\text{sat}) \equiv V_\sigma$. The collector voltage of $Q2$ is $v_{C2} = V_{CE}(\text{sat})$. The collector of $Q1$ is at $v_{C1} = V_{CC}$. The base voltage of $Q1$ is calculated by superposition to be

$$v_{B1} = -\frac{V_{BB}R_1}{R_1 + R_2} + \frac{V_{CE}(\text{sat})R_2}{R_1 + R_2} \equiv V_F \qquad (11\text{-}12)$$

In order that $Q1$ be OFF, we require that $|V_F| \leq 0$ (Si) or $|V_F| \leq 0.1$ V (Ge). These stable-state voltages are indicated in Fig. 11-10.

The Quasi-stable State　As a result of a trigger applied at $t = 0$, $Q2$ goes OFF and $Q1$ conducts. The voltages v_{C1} and v_{B2} drop abruptly by the same amount I_1R_c, where I_1 is the current in R_c of $Q1$. In the preceding section we saw that it is advantageous to drive $Q1$ into saturation. Under these circumstances

$$v_{B1} = V_\sigma \qquad v_{C1} = V_{CE}(\text{sat}) \qquad \text{and} \qquad I_1R_c = V_{CC} - V_{CE}(\text{sat})$$

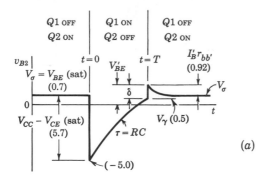

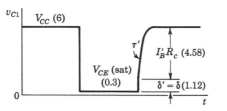

(a)

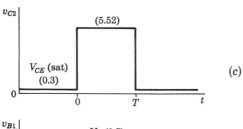

(b)

Fig. 11-10 Waveforms of the collector-coupled monostable multi. The exponential portions beginning at $t = T$ have a time constant $\tau' = (R_c + r_{bb'})C$. The numerical values (in volts) refer to the circuit of Fig. 11-9.

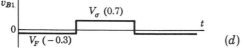

(c)

(d)

The equivalent circuit for $0 < t < T$ from which to calculate v_{C2} is given in Fig. 11-11. Since $Q1$ is ON, $v_{B1} = V_{BE}(\text{sat})$. Taking the loading of R_1 on $Q2$ into account by using the principle of superposition,

$$v_{C2} = \frac{V_{CC}R_1}{R_1 + R_c} + \frac{V_\sigma R_c}{R_1 + R_c} \tag{11-13}$$

The voltage at the base of $Q2$ now starts to rise exponentially toward V_{CC} with a time constant $\tau = RC$. Until v_{B2} reaches the cutin voltage V_γ at $t = T$ all voltages at the other transistor terminals remain unaltered.

Waveforms for $t > T$ At $t = T+$, $Q2$ conducts and $Q1$ is cut off. The collector voltage v_{C2} drops abruptly to $V_{CE}(\text{sat})$ and v_{B1} returns to V_F. The voltage v_{C1} now rises abruptly, since $Q1$ is OFF. This increase in voltage is transmitted to the base of $Q2$ and drives $Q2$ heavily into saturation. Hence, an overshoot develops in v_{B2} at $t = T+$ which decays as the capacitor C

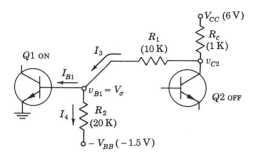

Fig. 11-11 The equivalent monostable multi circuit during the quasi-stable state, from which v_{C2} may be calculated.

recharges because of the base current.   We shall now calculate the magnitude of this overshoot.

The base current may be accounted for adequately by replacing the input circuit of $Q2$ by the base-spreading resistance $r_{bb'}$ in series with the base saturation voltage V_σ, as indicated in Fig. 11-12.   The base current at $t = T+$ is designated by I'_B.   The current in R may usually be neglected compared with I'_B (Prob. 11-20).   The path of I'_B is as shown in Fig. 11-12. From this figure we see that

$$V'_{BE} = I'_B r_{bb'} + V_\sigma \quad \text{and} \quad V_{C1} = V_{CC} - I'_B R_c \qquad (11\text{-}14)$$

From Fig. 11-10 the jumps in voltage at B_2 and C_1 (the collector of $Q1$) are, respectively,

$$\delta = I'_B r_{bb'} + V_\sigma - V_\gamma \quad \text{and} \quad \delta' = V_{CC} - I'_B R_c - V_{CE}(\text{sat}) \qquad (11\text{-}15)$$

Since C_1 and B_2 are connected by a capacitor, these two discontinuous voltage changes must be equal.   From the relationship $\delta = \delta'$ we obtain

$$I'_B = \frac{V_{CC} - V_{CE}(\text{sat}) - V_\sigma + V_\gamma}{R_c + r_{bb'}} \qquad (11\text{-}16)$$

From Fig. 11-12, the time constant with which v_{B2} and v_{C1} decay to their steady-state values is $\tau' = (R_c + r_{bb'})C$.   If I'_B is very much larger than the steady-state base current, then there will be a slight decrease in $V_{CE}(\text{sat})$, which will result in small undershoots in v_{C2} and v_{B1} at $t = T+$.   Since this effect is usually negligible it is not indicated in Fig. 11-10.

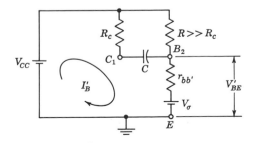

Fig. 11-12 Equivalent circuit for calculating the overshoot at the base B_2 of $Q2$ at $t = T+$ for the collector-coupled multi. $Q1$ is OFF, and $Q2$ is driven heavily into saturation.

If the value of $r_{bb'}$ is not known but if saturation characteristic curves such as those in Figs. 6-22 and 6-26 are given, then an alternative procedure for calculating the overshoots is the following. If V'_{BE} is the value of v_{B2} at the overshoot (Figs. 11-12 and 11-10), then, proceeding as above, we find

$$I'_B = \frac{V_{CC} - V_{CE}(\text{sat}) - V'_{BE} + V_\gamma}{R_c} \tag{11-17}$$

Using as a first approximation for V'_{BE} the value V_σ in Table 6-1, a value of I'_B is found. Corresponding to this I'_B and to the collector current

$$I_C = \frac{V_{CC} - V_{CE}(\text{sat})}{R_c}$$

we find the junction voltages from the characteristic curves. Using these values in Eq. (11-17) we can obtain a better approximation for I'_B. Note that if V_{CC} is large compared with the junction voltages, then $I'_B \approx V_{CC}/R_c$.

EXAMPLE Compute the voltage levels for the waveforms of Fig. 11-10 for a collector-coupled multi whose components and supply voltages are as given in Fig. 11-9. Silicon transistors are used with $r_{bb'} = 200 \ \Omega$ and $h_{FE} = 30$.

Solution We shall assume that $Q2$ is in saturation in the stable state and then justify this assumption. Since $I_{C2} \approx I_2$, and using Table 6-1 (page 219),

$$I_{C2} = \frac{V_{CC} - V_{CE}(\text{sat})}{R_c} = \frac{6 - 0.3}{1} = 5.7 \text{ mA}$$

$$I_{B2} = \frac{V_{CC} - V_{BE}(\text{sat})}{R} = \frac{6 - 0.7}{10} = 0.53 \text{ mA}$$

The minimum base current for saturation is $I_{C2}/h_{FE} = 5.7/30 = 0.19$ mA, and hence a base current of 0.53 mA will indeed keep $Q2$ in saturation. Therefore $v_{B2} = V_\sigma = 0.7$ V and $v_{C2} = V_{CE}(\text{sat}) = 0.3$ V. Since $I_1 = 0$, then the voltage across C is $V_A = V_{CC} - V_\sigma = 5.3$ V. From Eq. (11-12),

$$V_F = -1.5 \frac{10}{10 + 20} + \frac{(0.3)(20)}{10 + 20} = -0.30 \text{ V}$$

which is certainly enough to cut off $Q1$. Hence $v_{C1} = V_{CC} = 6$ V.

If a negative trigger is applied to the collector of $Q1$ or to the base of $Q2$, the multi will make a transition to its quasi-stable state. From the equivalent circuit of Fig. 11-11 we have, assuming $Q1$ is driven into saturation,

$$I_3 = \frac{6 - 0.7}{10 + 1} = 0.48 \text{ mA} \qquad I_4 = \frac{1.5 + 0.7}{20} = 0.11 \text{ mA}$$

$$I_{B1} = I_3 - I_4 = 0.48 - 0.11 = 0.37 \text{ mA}$$

From Fig. 11-9 we now calculate $I_{c1} = I_1 + I_R$. Since

$$I_1 R_c = V_{CC} - V_{CE}(\text{sat}) = 5.7 \text{ V}$$

$I_1 = 5.7$ mA. Since $I_R R = I_1 R_c + V_A = 5.7 + 5.3 = 11.0$ V, $I_R = 1.1$ mA. Hence $I_{C1} = 5.7 + 1.1 = 6.8$ mA **and** $(I_{B1})_{\min} = 6.8/30 = 0.23$ mA. Since

$I_{B1} > (I_{B1})_{min}$, then $Q1$ is indeed in saturation during the quasi-stable state. Hence, $v_{B1} = 0.7$ V, $v_{C1} = 0.3$ V, and at $t = 0+$

$$v_{B2} = V_\sigma - I_1 R_c = V_\sigma - [V_{CC} - V_{CE}(\text{sat})] = 0.7 - 6 + 0.3 = -5.0 \text{ V}$$

From Eq. (11-13)

$$v_{C2} = \frac{(6)(10)}{10 + 1} + \frac{(0.7)(1)}{10 + 1} = 5.52 \text{ V}$$

From Eq. (11-16)

$$I'_B = \frac{6 - 0.3 - 0.7 + 0.5}{1 + 0.2} = 4.58 \text{ mA}$$

Since $I'_B r_{bb'} = (4.58)(0.2) = 0.92$ V and $I'_B R_c = 4.58$ V, then all voltages in Fig. 11-10 are known.

It is interesting to make a comparison of the waveforms of Fig. 11-10 with the waveforms of Fig. 11-6. Observe the extent to which a transistor is more nearly an ideal switch than is a vacuum tube. In a transistor multi the collector voltage swings are very nearly equal to the supply voltage. Note also in the transistor waveforms the absence of a perceptible undershoot. Of course there must be some small undershoot, but this undershoot can be only of the order of tens of millivolts.

The same considerations with respect to recovery time that are discussed in Sec. 11-3 for the plate-coupled delay multi are applicable to the collector-coupled monostable multi. In particular, the anomalous situation depicted in Fig. 11-8 is possible if the second trigger is applied before all the transients arising from the first pulse have died down.

At $t = 0+$, when $Q1$ ($V1$) is driven to saturation and $Q2$ ($V2$) is cut off, the speed-up capacitor C_1 shunts the collector (plate) of $Q2$ ($V2$). Hence, the leading edge of the waveform v_{C2} in Fig. 11-10 (v_{P2} in Fig. 11-6) is not really vertical but rises with a time constant $C_1 R_1 R_c / (R_1 + R_c)$. This time constant is much smaller than τ or τ' and is not indicated in the figures.

The following transistor limitations (page 324) must be taken into consideration: (1) The minimum value of h_{FE} must be large enough to ensure saturation. [For a given $(h_{FE})_{min}$ the value of R cannot be too large (small) or $Q2$ ($Q1$) will not saturate.] (2) The value of I_{CBO} must not be so large as to prevent $Q1$ from being cut off in the stable state. (3) We must be careful that the voltage with which each transistor is driven OFF does not exceed the emitter-to-base breakdown voltage BV_{EBO}. Since at $t = 0+$, $v_{B2} \approx -V_{CC}$ (neglecting saturation voltages, Fig. 11-10a), a transistor with BV_{EBO} equal to the supply voltage must be used. If a transistor with such a large breakdown voltage is not available, then a diode may be inserted in series either with the base or with the emitter of $Q2$. The diode breakdown voltage must exceed $V_{CC} - BV_{EBO}$. The diode will affect some of the voltages in Fig. 11-10, will increase the recovery time τ', and will modify the delay T somewhat (Prob. 11-21).

A mode of operation is possible with R returned to ground, provided that the bias V_{BB} in Fig. 11-9 is so chosen that $Q1$ is ON in the stable state. A trigger is now applied to drive $Q1$ OFF so that its collector voltage rises and carries the base of $Q2$ with it. In this manner $Q2$ is driven ON, and the quasi-stable state begins. The base current of $Q2$ does not remain constant but decreases toward zero with a time constant $C(R_c + r_{bb'})$. When this current falls below that necessary to saturate $Q2$ (when $I_{B2} < V_{CC}/R_c h_{FE}$), then $Q2$ enters its active region. Now the collector of $Q2$ rises and so does the base of $Q1$. When $Q1$ comes out of cutoff, the loop gain exceeds unity, and a reverse transition takes place, thus ending the quasi-stable state. This mode of operation is seldom used because the delay time T depends on parameters which vary widely from transistor to transistor and with temperature (Figs. 6-23 and 6-27).

If p-n-p transistors are used the above analysis remains unchanged, except that all waveforms are inverted. For example, in Fig. 11-10a, v_{B2} starts at -0.7 V for $t < 0$ and jumps positively to $+5.0$ V at $t = 0+$ and then decreases exponentially downward toward $-V_{cc} = -6$ V, with the reverse transition taking place at $v_{B2} = -0.5$ V, etc.

11-6   THE EMITTER–COUPLED MONOSTABLE MULTI

An alternative form of the one-shot, the *emitter-coupled monostable multi*, is shown in Fig. 11-13. Observe that the coupling from the collector C_2 to the base B_1 is lacking and that instead feedback has been provided through a

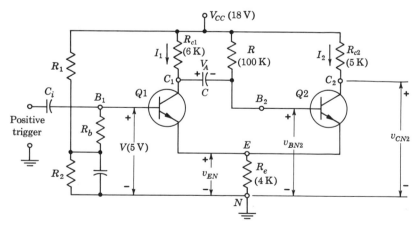

Fig. 11-13 An emitter-coupled monostable multi. Components and supply voltage refer to the illustrative example on page 424. The bias voltage V is obtained from the divider R_1R_2 across the supply V_{CC}.

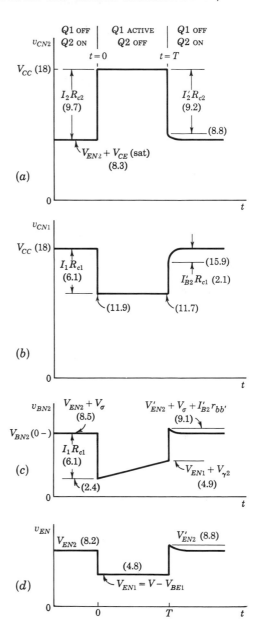

Fig. 11-14 Waveforms of emitter-coupled monostable multivibrator. Numerical values (in volts) in parentheses correspond to the illustrative example.

common emitter resistor R_e. A negative supply is not required. The signal at C_2 is not directly involved in the regenerative loop. Hence, this collector makes an ideal point from which to obtain an output voltage waveform. The base B_1 is a good point at which to inject a triggering signal since this electrode is coupled to no other in the circuit. Hence the trigger source cannot load the circuit.

It was pointed out earlier, in connection with the collector-coupled multi, that the time T of the quasi-stable state can be controlled through I_1, the current in $Q1$ when $Q1$ is conducting. However, in the collector-coupled case, the current I_1 cannot be maintained stable unless I_1 corresponds to saturation. In the present case, when $Q2$ goes OFF and $Q1$ goes ON, $Q1$ operates with a substantial emitter resistance. This emitter resistance will serve to stabilize I_1, and it is therefore feasible to control T through I_1. The current I_1 may be adjusted through the bias voltage V, and it turns out that T varies rather linearly with V. Hence the emitter-coupled configuration makes an excellent gate waveform generator, whose width is easily and linearly controllable by means of an electrical signal voltage. Or, if V is a d-c voltage derived from a linear potentiometer, T will vary linearly with potentiometer rotation.

The waveforms of the emitter-coupled multi are shown in Fig. 11-14. The waveform at base B_1 is now of no interest since it consists only of the triggering pulse and the small voltage drop across R_b due to changes in the base current of $Q1$. Instead, however, we now have a waveform at the emitters. We shall consider only that mode of operation in which $Q1$ is cut off and $Q2$ is in saturation in the stable state. The method of calculating the waveforms (the voltages at all transistor terminals with respect to the ground terminal N) is given in the following illustrative problem.

EXAMPLE Consider the emitter-coupled multi whose components and supply voltages are indicated in Fig. 11-13. If $V = 5.0$ V, calculate the voltage levels of the waveforms in Fig. 11-14. Assume germanium transistors having $h_{FE} = 50$ and $r_{bb'} = 200\ \Omega$.

Solution   Stable-state calculations, $t < 0$ In the stable state $Q1$ is OFF while $Q2$ is in saturation. The equivalent circuit from which to find the currents in $Q2$ is shown in Fig. 11-15. We have again taken advantage of the simplification afforded by using the approximations of Table 6-1. Applying Kirchhoff's voltage law to the two meshes we have

$$104I_{B2} + 4I_2 = 17.7 \tag{11-18a}$$

$$4I_{B2} + 9I_2 = 17.9 \tag{11-18b}$$

Solving, we obtain $I_2 = 1.95$ mA and $I_{B2} = 0.095$ mA. We note that the base current is more than adequate to ensure that $Q2$ is in saturation, since the minimum current required for saturation is $I_2/h_{FE} = 1.95/50 = 0.039$ mA.

We now find that, in the stable state,

$$v_{CN1} = V_{CC} = 18\ \text{V} \tag{11-19}$$

$$v_{EN} \equiv V_{EN2} = (I_2 + I_{B2})R_e = (1.95 + 0.095)(4) = 8.2\ \text{V} \tag{11-20}$$

Since

$$v_{BE1} = v_{BN1} - V_{EN2} = 5.0 - 8.2 = -3.2\ \text{V}$$

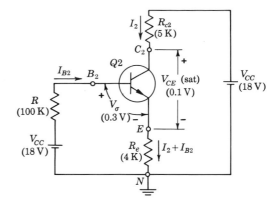

Fig. 11-15 The equivalent circuit of $Q2$ in Fig. 11-13 in the stable state. The saturation voltages are given in Table 6-1.

then $Q1$ is indeed OFF.

$$v_{CN2} = V_{EN2} + V_{CE}(\text{sat}) = 8.2 + 0.1 = 8.3 \text{ V} \tag{11-21}$$

$$v_{BN2} = V_{EN2} + V_\sigma = 8.2 + 0.3 = 8.5 \text{ V} \tag{11-22}$$

We also note from Fig. 11-13 that the capacitor voltage is

$$V_A = v_{CN1} - v_{BN2} = 18 - 8.5 = 9.5 \text{ V} \tag{11-23}$$

Calculations at $t = 0+$ When a triggering signal, applied at $t = 0$, causes a transition from the stable to the quasi-stable state, the current in $Q2$ becomes zero and a current I_{C1} flows in the collector circuit of $Q1$.   This current may be determined by the bias voltage V.   The equivalent circuit for this calculation is shown in Fig. 11-16.   *Assuming that* $Q1$ *is operating in its active region with* $V_{BE1} = 0.2$ V (Table 6-1, page 219),

$$v_{EN} \equiv V_{EN1} = V - V_{BE1} = 5 - 0.2 = 4.8 \text{ V} \tag{11-24}$$

$$I_{C1} + I_{B1} = \frac{V_{EN1}}{R_e} = \frac{4.8}{4} = 1.2 \text{ mA} \tag{11-25}$$

Since

$$I_{C1} + I_{B1} = I_{C1}\left(1 + \frac{1}{h_{FE}}\right) \tag{11-26}$$

then

$$I_{C1} = \frac{h_{FE}}{1 + h_{FE}} \frac{V_{EN1}}{R_e} = \frac{(50)(1.2)}{51} = 1.18 \text{ mA} \tag{11-27}$$

In order to solve for I_1, the current in R_{c1} at $t = 0+$, note that

$$I_1 + I_R = I_{C1} = 1.18 \text{ mA} \tag{11-28}$$

and from Kirchhoff's voltage law applied to the mesh containing R_{c1}, C, and R,

$$R_{c1}I_1 + V_A - RI_R = 0 \tag{11-29}$$

Since the capacitor voltage does not change at the transition, then, from Eq. (11-23), $V_A = 9.5$ V and Eq. (11-29) becomes

$$6I_1 - 100I_R = -9.5 \text{ V} \tag{11-30}$$

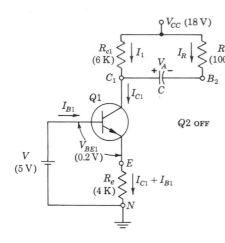

Fig. 11-16   The circuit of Fig. 11-13 during the quasi-stable state when $Q2$ is OFF.

Solving Eqs. (11-28) and (11-30) we obtain

$$I_R = 0.156 \text{ mA} \qquad \text{and} \qquad I_1 = 1.02 \text{ mA} \tag{11-31}$$

Accordingly, at $t = 0+$ the multi voltages are

$$v_{CN2} = V_{CC} = 18 \text{ V} \tag{11-32}$$

$$v_{CN1} = V_{CC} - I_1 R_{c1} = 18 - (1.02)(6) = 11.9 \text{ V} \tag{11-33}$$

$$v_{EN} = V_{EN1} = V - V_{BE1} = 4.8 \text{ V} \tag{11-34}$$

$$v_{BN2} = V_{BN2}(0-) - I_1 R_{c1} = 8.5 - 6.1 = 2.4 \text{ V} \tag{11-35}$$

Equation (11-35) follows from the fact that the abrupt change $I_1 R_{c1}$ in the voltage at C_1 is transmitted unattenuated to B_2 at $t = 0$.   Alternatively, we see from Fig. 11-16 that v_{BN2} may be calculated from

$$v_{BN2} = v_{CN1} - V_A = 11.9 - 9.5 = 2.4 \text{ V} \tag{11-36a}$$

or, also, from

$$v_{BN2} = V_{CC} - I_R R = 18 - 15.6 = 2.4 \text{ V} \tag{11-36b}$$

In the above calculation for v_{CN1} we have explicitly taken into account the effect of the output impedance of the transistor stage $Q1$.   The current I_R flows back into this impedance and increases the voltage v_{CN1}.   Neglecting this effect we would have obtained

$$v_{CN1} = V_{CC} - I_{c1} R_{c1} = 18 - (1.18)(6) = 10.9 \text{ V}$$

which is too low by 1.0 V.   The output impedance is $R_o \approx R_{c1} = 6$ K, since the impedance seen looking into the collector of Q_1 operating in its active region is much larger than 6 K.   Note that $I_R R_o \approx (0.16)(6) \approx 1.0$ V.

The loading of R does not influence the waveforms in the collector-coupled one-shot if $Q1$ is driven into saturation, since R_o is then very small.   The loading only affects the value of $(h_{FE})_{\min}$ required to keep $Q1$ in saturation (page 420).

We must check our assumption that $Q1$ is in its active region.   Since

$$v_{CB1} = v_{CN1} - v_{BN1} = 11.9 - 5.0 = +6.9 \text{ V} \qquad (11\text{-}37)$$

is positive, then the collector junction is reverse-biased (for an n-p-n transistor). Hence, $Q1$ is indeed in its active region.

Calculations at $t = T-$ During the quasi-stable state v_{CN2}, v_{EN1} and i_{C1} remain constant, and v_{BN2} increases exponentially with a time constant $(R + R_o)C$. This voltage starts at 2.4 V at $t = 0+$ and rises to the point at which $Q2$ attains its cutin value at $t = T-$ when

$$v_{BN2} = V_{EN1} + V_{\gamma 2} = 4.8 + 0.1 = 4.9 \text{ V} \qquad (11\text{-}38)$$

The voltage v_{CN1} at $t = T-$ is calculated as follows.   At this time

$$I_R = \frac{V_{CC} - v_{BN2}}{R} = \frac{18 - 4.9}{100} = 0.13 \text{ mA} \qquad (11\text{-}39)$$

$$I_1 = I_{C1} - I_R = 1.18 - 0.13 = 1.05 \text{ mA} \qquad (11\text{-}40)$$

and hence

$$v_{CN1} = V_{CC} - I_1 R_{c1} = 18 - (1.05)(6) = 11.7 \text{ V} \qquad (11\text{-}41)$$

Thus, while at $t = 0+$, $v_{CN1} = 11.9$ V, this voltage falls to 11.7 V at $t = T-$, as indicated in Fig. 11-14b.

The capacitor voltage at $t = T-$ is

$$V_A = v_{CN1} - v_{BN2} = 11.7 - 4.9 = 6.8 \text{ V} \qquad (11\text{-}42)$$

Calculations at $t = T+$ The voltage levels attained at $t = T+$, immediately after the reverse transition, at which time the overshoots occur, may be calculated from the circuit of Fig. 11-17.   Here, I'_{B2} and I'_2 are the base and collector currents at $t = T+$.   To perform the calculation we shall use the fact that the voltage $V_A = 6.8$ V across C is the same immediately before and after the reverse transition.   Since the overshoots are not required with great precision, then for $R \gg R_{c1}$

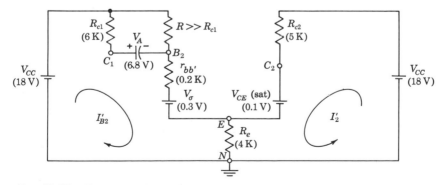

Fig. 11-17   The equivalent circuit at $t = T+$ from which to calculate the overshoots.   The emitter voltage is $V_{EN} = (I'_2 + I'_{B2})R_E$. $Q1$ is OFF, and $Q2$ is driven heavily into saturation.

the current in R may be neglected compared with I'_{B2}. We may write the mesh equations

$$18 - 6.8 - 0.3 = 10.2I'_{B2} + 4I'_2 = 10.9 \tag{11-43a}$$

$$18 - 0.1 = 4I'_{B2} + 9I'_2 = 17.9 \tag{11-43b}$$

Solving, we obtain

$$I'_{B2} = 0.35 \text{ mA} \quad \text{and} \quad I'_2 = 1.84 \text{ mA}$$

Hence at $t = T+$

$$v_{EN} \equiv V'_{EN2} = (I'_2 + I'_{B2})R_e = (1.84 + 0.35)(4) = 8.76 \text{ V} \tag{11-44}$$

$$v_{CN2} = V'_{EN2} + V_{CE}(\text{sat}) = 8.76 + 0.1 = 8.86 \text{ V} \tag{11-45}$$

or, alternatively,

$$v_{CN2} = V_{CC} - I'_2 R_{c2} = 18 - (1.84)(5) = 8.80 \text{ V} \tag{11-46}$$

$$v_{CN1} = V_{CC} - I'_{B2} R_{c1} = 18 - (0.35)(6) = 15.9 \text{ V} \tag{11-47}$$

$$v_{BN2} = V'_{EN2} + V_\sigma + I'_{B2} r_{bb'}$$

$$= 8.76 + 0.3 + (0.35)(0.2) = 9.13 \text{ V} \tag{11-48}$$

From Fig. 11-17 we see that the overshoots decay with a time constant

$$\tau' = C \left(R_{c1} + r_{bb'} + \frac{R_e R_{c2}}{R_e + R_{c2}} \right) \tag{11-49}$$

Extreme Limits of V   In the stable state $Q1$ must be OFF. This condition establishes a maximum allowable value V_{max} for the bias voltage V applied to the base of $Q1$. When $Q1$ goes ON the current I_{C1} must be large enough to drive $Q2$ to cutoff. This condition establishes a minimum allowable bias voltage V_{min}. *For proper operation of the circuit the bias voltage V must lie between these two values V_{max} and V_{min}.* These extreme voltages will now be calculated.

Since in the stable state $Q1$ is OFF, V must not be greater than V_{EN2} by more than the cutin voltage $V_{\gamma1}$. The maximum allowable value of V is therefore

$$V_{max} = V_{EN2} + V_{\gamma1} \tag{11-50}$$

If there is to be a quasi-stable state, then the current I_1 must be large enough to drive $Q2$ OFF. Let us call this minimum current $(I_1)_{min}$ and the corresponding value of emitter voltage $(V_{EN1})_{min}$. Then the minimum input voltage is

$$V_{min} = (V_{EN1})_{min} + V_{BE1} \tag{11-51}$$

where V_{BE1} is the base-to-emitter voltage corresponding to an emitter current V_{EN1}/R_e. To find $(V_{EN1})_{min}$ proceed as follows. The voltage at the base of

$Q2$ at $t = 0+$ must be less than the emitter voltage V_{EN1} by at least the cutin voltage $V_{\gamma 2}$. Using Eq. (11-35),

$$V_{BN2}(0+) = V_{BN2}(0-) - I_1 R_{c1} \leq V_{EN1} + V_{\gamma 2} \tag{11-52}$$

The relationship between I_1 and V_{EN1} is found from Eqs. (11-27) and (11-28):

$$I_{C1} = \frac{h_{FE}}{1 + h_{FE}} \frac{V_{EN1}}{R_e} = I_1 + I_R \tag{11-53}$$

We also have from Fig. 11-16 that

$$R I_R = V_{CC} - V_{BN2}(0+) = V_{CC} - V_{BN2}(0-) + I_1 R_{c1} \tag{11-54}$$

If the equals sign applies in Eq. (11-52), then this equation together with Eqs. (11-53) and (11-54) determines the three minimum values I_1, V_{EN1}, and I_R. If we solve for $(V_{EN1})_{min}$ and substitute this value into Eq. (11-51) we obtain

$$V_{min} = V_{BE1} + \frac{V_{BN2}(0-) - V_{\gamma 2} + (R_{c1}/R)(V_{CC} - V_{\gamma 2})}{1 + (R_{c1}/R) + (R_{c1}/R_e)[h_{FE}/(1 + h_{FE})]} \tag{11-55}$$

For the circuit of Fig. 11-13, Eq. (11-50) gives

$$V_{max} = 8.2 + 0.1 = 8.3 \text{ V}$$

and Eq. (11-55) yields

$$V_{min} = 0.2 + \frac{8.5 - 0.1 + (0.06)(18 - 0.1)}{1 + 0.06 + (6/4)(50/51)} = 3.94 \text{ V}$$

11-7   GATE WIDTH OF AN EMITTER–COUPLED ONE–SHOT

The delay time T of the multi is arrived at by using Eq. (2-3). The voltage v_{BN2} at $t = 0+$ is, from Fig. 11-14c,

$$V_{BN2}(0+) = V_{BN2}(0-) - I_1 R_{c1} \tag{11-56}$$

If $Q2$ did not conduct, then as $t \to \infty$, v_{BN2} would approach V_{CC}. Hence, the instantaneous voltage at the base of $Q2$ is given by

$$v_{BN2} = V_{CC} - [V_{CC} - V_{BN2}(0-) + I_1 R_{c1}]\epsilon^{-t/\tau} \tag{11-57}$$

where $\tau = C(R + R_{c1})$. At $t = T-$, we see from Fig. 11-14c that

$$v_{BN2} = V_{EN1} + V_{\gamma 2}$$

Substituting this value into Eq. (11-57) and solving for T, we obtain

$$T = \tau \ln \frac{V_{CC} - V_{BN2}(0-) + I_1 R_{c1}}{V_{CC} - V_{EN1} - V_{\gamma 2}} \tag{11-58}$$

The delay time is found to vary fairly linearly with the bias voltage V, as will be demonstrated in the following illustrative problem.

EXAMPLE   (a) Find the delay time T as a function of the input voltage V for the one-shot of Fig. 11-13.   (b) For what value of V is $T = 0$?   (c) Expand T in a power series in $V - V_{min}$.

Solution   a. Equation (11-58) gives

$$T = \tau \ln \frac{18 - 8.5 + 6I_1}{18 - V_{EN1} - 0.1} = \tau \ln \frac{9.5 + 6I_1}{17.9 - V_{EN1}} \tag{11-59}$$

From Eq. (11-24)

$$V_{EN1} = V - V_{BE1} = V - 0.2 \tag{11-60}$$

We may obtain I_1 in terms of V_{EN1} and hence V from Eqs. (11-53) and (11-54). For the circuit under consideration these equations are

$$\frac{50}{51} \frac{V_{EN1}}{4} = I_1 + I_R = 0.245 V_{EN1}$$

$$100 I_R = 18 - 8.5 + 6I_1 = 9.5 + 6I_1$$

Eliminating I_R, we obtain

$$I_1 = 0.232 V_{EN1} - 0.090 = 0.232V - 0.136$$

where we have made use of Eq. (11-60).   Substituting for I_1 and V_{EN1} in Eq. (11-59) yields

$$T = \tau \ln \frac{8.68 + 1.39V}{18.1 - V} \tag{11-61}$$

b. If the numerator is equated to the denominator, then $T = 0$:

$$8.68 + 1.39V = 18.1 - V \quad \text{or} \quad V = 3.94 \text{ V}$$

Zero delay occurs at $V = V_{min}$, as might have been anticipated.

c. If we define V_o by

$$V_o \equiv V - V_{min} = V - 3.94$$

then Eq. (11-61) becomes

$$T = \tau \ln \frac{14.2 + 1.39 V_o}{14.2 - V_o} = \tau \ln \frac{1 + 0.098 V_o}{1 - 0.071 V_o} \tag{11-62}$$

From the expansion

$$\ln (1 + x) = x - \frac{x^2}{2} + \frac{x^3}{3} - \cdots \tag{11-63}$$

we obtain, using only the first two terms of the expansion,

$$\frac{T}{\tau} \approx 0.098 V_o - \tfrac{1}{2}(0.098 V_o)^2 + 0.071 V_o + \tfrac{1}{2}(0.071 V_o)^2$$

$$= 0.169 V_o (1 - 0.014 V_o) \tag{11-64}$$

Fig. 11-18 Pertaining to the definition of linearity of delay.

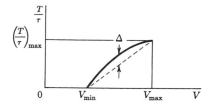

A plot of T/τ versus V is indicated in Fig. 11-18. We have also drawn a straight line between the end points of this graph. If the *maximum* discrepancy between the experimental plot and the line is Δ, then the *linearity* or *displacement error* is defined as

$$e_d = \frac{\Delta}{(T/\tau)_{\max}} \tag{11-65}$$

From Prob. 11-29 we find that if this definition is applied to a general quadratic function $y = Ax(1 - Bx)$, then, if the maximum value of x is x_m,

$$e_d = \tfrac{1}{4}Bx_m \tag{11-66}$$

provided that the deviation from linearity is small. For the circuit under consideration

$$(V_o)_{\max} = V_{\max} - V_{\min} = 8.3 - 3.9 = 4.4 \text{ V}$$

Hence, the maximum deviation from linearity is

$$e_d = (\tfrac{1}{4})(0.014)(4.4) = 0.015$$

or 1.5 percent. The emitter-coupled multi constitutes an excellent *voltage-to-time converter*.

11-8 THE CATHODE–COUPLED MONOSTABLE MULTI

The vacuum-tube form of the emitter-coupled multi of Fig. 11-13 is shown in Fig. 11-19. Since here the resistor R may be made much larger than is ordinarily allowable with transistors, we shall now neglect the resistor current I_R. The calculation of the waveforms in the present case proceeds generally analogously to the calculation for the transistor circuit. A difference is noted in the calculation of the overshoot. In the transistor circuit, the transistor $Q2$ is driven to saturation at the time of the overshoot and hence at that time does not operate as an amplifier. In the tube case, when, at the overshoot, the tube $V2$ is driven to clamp, this tube does continue to operate as an amplifier. Proceeding as outlined in Prob. 11-30, we find that at $t = T+$, the grid current at the overshoot I'_{G2} and the plate current I'_2 of $V2$ are given respec-

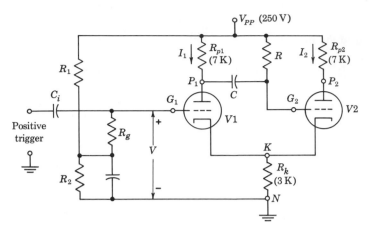

Fig. 11-19   A cathode-coupled monostable multi.   Components and supply voltage specified refer to the illustrative example on page 434.   The bias voltage V is obtained from the divider R_1R_2 across the supply V_{PP}.

tively by

$$I'_{G2} = \frac{(I_1 - I_o)(R_{p1} + R_k)}{R_{p1} + r_G + (1 + \gamma)R_k} \tag{11-67}$$

$$I'_2 = I_2 + \gamma I'_{G2} \tag{11-68}$$

where r_G is the grid-to-cathode resistance, I_1 is the current in $V1$ corresponding to a voltage V between grid and ground (if $V2$ is OFF), and I_2 is the current in $V2$ corresponding to $V_{GK2} = 0$, the clamped stable-state current of $V2$ (with $V1$ OFF),

$$\gamma \equiv \frac{\mu r_G - R_k}{R_k + R_{p2} + r_p} \tag{11-69}$$

and

$$I_o \equiv (I_{p1})_{\min} = (I_1)_{\min} = \frac{I_2 R_k - V_{\gamma 2}}{R_{p1} + R_k} \tag{11-70}$$

is the minimum current in $V1$ required to drive $V2$ to cutoff.   Note that γ may be either positive or negative, although it is usually positive because most frequently μr_G exceeds R_k.   Hence, usually $I'_2 > I_2$, whereas for the transistor circuit, as seen in Fig. 11-14a, $I_2 > I'_2$.

The maximum and minimum allowable values of bias voltage are given by formulas analogous to Eqs. (11-50) and (11-51), namely,

$$V_{\max} = I_2 R_k + V_{\gamma 1} \tag{11-71}$$

and

$$V_{\min} = I_o R_k + V_{GK1} \tag{11-72}$$

where V_{GK1} is the grid-to-cathode voltage corresponding to the current $I_1 = I_o$.

Equation (11-58) for the delay time T now takes the form

$$T = \tau \ln \frac{V_{PP} - I_2 R_k + I_1 R_{p1}}{V_{PP} - I_1 R_k - V_{\gamma 2}} \tag{11-73}$$

Note that, for a tube, $V_\sigma \approx 0$, $V_{BN2}(0-) = V_{EN2} = I_2 R_k$, and $V_{EN1} = I_1 R_k$.

The waveforms for the cathode-coupled circuit are sketched in Fig. 11-20, where the voltage levels are expressed in terms of I_1, I_2, I_2', and I_{G2}'. These

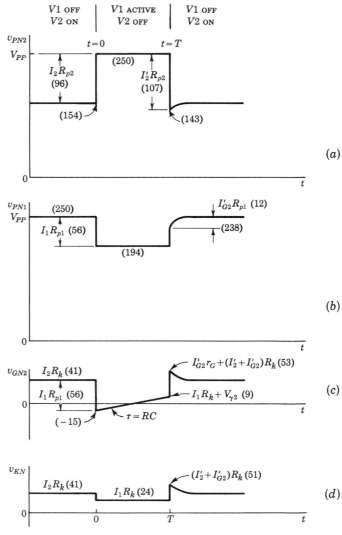

Fig. 11-20 Waveforms of cathode-coupled monostable multi. Numerical values (in volts) in parentheses correspond to the illustrative example with $V = 18$ V.

waveforms have the same shapes as for the emitter-coupled circuit except for the waveform at the output of the second stage (because I_2' exceeds I_2). We shall now illustrate how to calculate these waveshapes.

EXAMPLE Compute the voltages V_{max} and V_{min} for the cathode-coupled multi whose components and supply voltage are as given in Fig. 11-19. For a value of V approximately midway between V_{max} and V_{min}, calculate the voltage levels of the waveforms of Fig. 11-20. The tubes employed are the two half sections of a type 12AU7 tube.

Solution Drawing a load line corresponding to $R_p + R_k = 10$ K and $V_{PP} = 250$ V on the negative-grid plate characteristics for the type 12AU7 (Fig. D-5), we find $I_2 = 13.7$ mA at $V_G = V_{GK} = 0$. Hence

$$I_2R_{p1} = 7 \times 13.7 = 96 \text{ V} \qquad \text{and} \qquad I_2R_k = 3 \times 13.7 = 41 \text{ V}$$

The voltage V_{max} is given by Eq. (11-71), in which $V_{\gamma 1}$ is the cutin voltage corresponding to a plate-to-cathode voltage of $250 - 41 = 209$ V. We find $V_{\gamma 1} = -15$ V, so that

$$V_{max} = I_2R_k + V_\gamma = 41 - 15 = 26 \text{ V}$$

To find V_{min} we must first calculate I_o from Eq. (11-70). Assuming tentatively that $V_{\gamma 2} \approx V_{\gamma 1} = -15$ V, we find

$$I_o = \frac{I_2R_k - V_{\gamma 2}}{R_{p1} + R_k} = \frac{41 + 15}{7 + 3} = 5.6 \text{ mA}$$

It appears, then, that $V_{\gamma 2}$ is actually the cutin voltage corresponding to a plate-to-cathode voltage of $250 - (3 \times 5.6) = 233$ V, so that $V_{\gamma 2}$ is more nearly equal to -17 V. However, the precision with which tube characteristics apply to an individual tube hardly warrants applying this correction. We find now from the tube characteristics that a current $I_o = 5.6$ mA flows when the grid-to-cathode voltage V_{GK1} is -8 V. Hence

$$V_{min} = V_{GK1} + I_oR_k = -8 + 5.6 \times 3 = 8.8 \text{ V}$$

Now let us compute the voltage levels in Fig. 11-20 for $V = 18$ V. Using the method of Sec. 1-10, we find that corresponding to $V = 18$ V, $I_1 = 8.0$ mA, giving $I_1R_p = 56$ V and $I_1R_k = 24$ V. Also, as noted above, $I_2R_p = 96$ V and $I_2R_k = 41$ V. These voltage levels are indicated in parentheses in Fig. 11-20.

At the current corresponding to grid clamping ($I_2 = 13.7$ mA) and at a plate voltage of $250 - 96 - 41 = 113$ V, we have (Fig. D-7) $\mu = 18$ and $r_p = 6$ K. If we assume that $r_G = 1$ K, we find, from Eq. (11-69), that

$$\gamma = \frac{18 - 3}{3 + 7 + 6} = 0.94$$

Equation (11-67) is now

$$I_{G2}' = \frac{(I_1 - I_o)(R_{p1} + R_k)}{R_{p1} + r_G + (1 + \gamma)R_k} = \frac{(8.0 - 5.6)(7 + 3)}{7 + 1 + (1.94 \times 3)} = 1.7 \text{ mA}$$

and Eq. (11-68) is

$$I_2' = I_2 + \gamma I_{G2}' = 13.7 + (0.94)(1.7) = 15.3 \text{ mA}$$

The voltages in Fig. 11-20 at $t = T+$ are

$$I_2' R_{p2} = (15.3)(7) = 107 \text{ V}$$

$$I_{G2}' R_{p2} = (1.7)(7) = 12 \text{ V}$$

$$I_{G2}' r_G + (I_2' + I_{G2}') R_k = 1.7 + (15.3 + 1.7)(3) = 53 \text{ V}$$

$$(I_2' + I_{G2}') R_k = 51 \text{ V}$$

11-9   THE INFLUENCE OF V ON WAVEFORMS

The gate width T of the emitter-coupled or cathode-coupled multi is determined by V and the time constant τ, assuming that all other parameters are held constant. As V is varied from V_{min} to V_{max}, the logarithmic term in Eq. (11-58) varies from zero to a maximum, M. The maximum delay T_{max} is given by $T_{max} = \tau M$. A given delay can be obtained by using either a small V and a large τ or a large V and a small τ. If V is near V_{min} (so that the current in $Q1$ or $V1$ is at its minimum value), then the overshoots in waveform will be small, whereas if V is near V_{max} (large I_1), the overshoots will be emphasized.

If τ is held fixed and V is slowly increased from zero, then the following events will take place. For voltages below V_{min}, the circuit cannot be triggered. For $V > V_{min}$, the waveforms will change from a narrow gate with little overshoot to wider and wider gates with progressively higher and higher overshoots. When V reaches V_{max}, the circuit becomes an astable instead of a monostable multi and it continues to operate even when the triggers are removed (as discussed in Sec. 11-11).

The above assumes that the triggers are widely spaced compared with the maximum multi width. If this is not true, then the above sequence of events takes place until the multi width approaches the time between triggers. At that point the situation pictured in Fig. 11-8 takes place, where alternate cycles have different waveforms. As a matter of fact this anomalous situation may be obtained even at narrow widths if the recovery time constant is comparable to the time between pulses. This emphasizes the importance of keeping C as small as possible, just as with the plate-coupled or collector-coupled monostable multi.

11-10   TRIGGERING OF THE MONOSTABLE MULTI

The cathode-coupled or emitter-coupled circuits may be triggered by applying a pulse of sufficient amplitude to bring the OFF amplifier $A1$ into conduction.

For example, in the emitter-coupled multi, the trigger amplitude V_T when added to $V - V_{EN2}$ must exceed the value $V_{\gamma1}$. Hence, V_T must have a magnitude of at least

$$V_T = V_{EN2} - V + V_{\gamma1}$$

Using Eq. (11-50) we find

$$V_T = V_{\max} - V \qquad\qquad (11\text{-}74)$$

The largest trigger V_{TM} will be required when $V = V_{\min}$ and will have a magnitude given by

$$V_{TM} = V_{\max} - V_{\min} \qquad\qquad (11\text{-}75)$$

If $V_T > V_{TM}$ then the one-shot can be triggered for any V between $V_{\min}$ and $V_{\max}$. If triggers of amplitude less than V_{TM} are employed, then the multi will trigger when V is set for large delays but not when V is set for short delays. Thus, if V is initially near $V_{\max}$ and then progressively decreased, the multi will operate properly with progressively shorter delays until finally a critical value of V will be attained where the multi will fail to respond. At this point, if the trigger amplitude is increased, the multi will once again function properly. These same considerations apply equally to the cathode-coupled multi.

If the trigger amplitude is too large, then appreciable input current will flow. As a result of this current the input capacitor C_i will charge and decrease the average value of the bias voltage to $Q1$. Since effectively the value of V is reduced, then the time of the quasi-stable state is decreased. This shortening of the delay becomes more pronounced as the width of the triggering pulse increases.

The collector-coupled or plate-coupled circuits may also be triggered with pulses of such polarity that the nonconducting device is brought out of

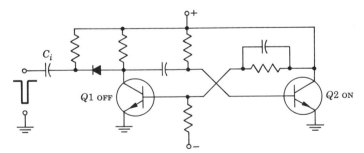

Fig. 11-21   Triggering of a monostable multi through a diode so as to bring the ON triode below cutoff. The indicated polarities are also valid for a plate-coupled multi. If p-n-p transistors are used, the diode and the supply polarities must be reversed and a positive trigger is required.

cutoff, but this method is subject to the difficulties listed in Sec. 10-7 in connection with triggering of a binary in this manner. As emphasized in that section, it is preferable to apply a pulse of the proper polarity to cause the ON device $A2$ to go OFF.

The triggering arrangement shown in Fig. 11-21 for an *n-p-n* transistor multi has a twofold advantage. First, it has the improved sensitivity of the multi to a negative signal applied to the ON device. Second, at the instant of the transition the collector of $Q1$ drops, the diode no longer conducts, and the multi is unresponsive to the triggering signal until the quasi-stable state is completed. This second feature is particularly important in a case where the input signal is not a short trigger but is rather a continuous waveform such as a sine wave.

11-11 THE MONOSTABLE CIRCUIT ADJUSTED FOR FREE-RUNNING OPERATION

The monostable multi circuit has a permanently stable state only if the stage $A1$ is able to remain cut off indefinitely. We inquire now into the matter of what happens when the bias on $A1$ is adjusted so that $A1$ is not able to remain cut off. For this purpose consider the emitter-coupled circuit of Fig. 11-13, with V adjusted so that its value exceeds $V_{\max}$. This configuration has no stable state but instead has two quasi-stable states, between which the multi makes transitions periodically without the aid of external triggers.

We shall now study qualitatively the waveforms which appear at the second base B_2 and the emitter E. Assume that at $t = 0$, because of the past history of the circuit, the transistor $Q1$ is carrying a current I_1 while $Q2$ is below cutoff. Refer to Fig. 11-22. The voltage v_{BN2} is rising exponentially toward V_{CC} while v_{EN} is constant at V_{EN1}. This situation corresponds to the waveforms in Fig. 11-14 just prior to $t = T$. When the voltage at B_2 passes the cutin point of $Q2$, namely, when $v_{BN2} = V_{EN1} + V_{\gamma 2}$, a regenerative transition occurs since now both transistors operate as amplifiers. The voltage at B_2 jumps to a high value, as does also the voltage at E. As long as v_{EN} exceeds $V - V_{\gamma 1} \approx V_{EN1}$, then $Q1$ remains cut off. However, v_{EN} is decreasing toward $V_{EN2} = V_{\max} - V_{\gamma 1}$, from Eq. (11-50). Since we have stipulated that $V > V_{\max}$, then at a time T' after the first transition the voltage at E will decrease to the point where $v_{EN} = V - V_{\gamma 1}$. At this instant a reverse transition will take place, driving $Q2$ below cutoff. We have now returned to our assumed starting conditions, and the cycle of events pictured in Fig. 11-22 repeats itself. Ordinarily the time T' is appreciably smaller than T''. Usually a more symmetrical waveform is of advantage, and the symmetrical configuration of Fig. 11-23 is more commonly employed for a free-running multi.

If R_c or R_e or both are selected to be quite small, it may happen that neither transistor is able even temporarily to keep the other transistor cut off.

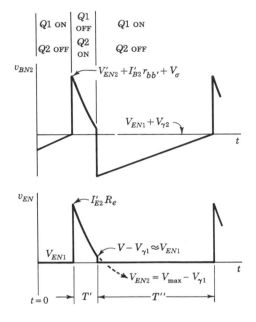

Fig. 11-22   The waveforms at B_2 and E for the emitter-coupled multi with $V > V_{\max}$.

In such a case the circuit will behave approximately as a sinusoidal oscillator. The frequency of oscillation will be determined by the stray capacitances associated with the circuit. These circumstances would correspond to the case in which the loop gain of the circuit is not much in excess of unity.

Another possibility is that V is so much in excess of $V_{\max}$ that it is impossible to cut $Q1$ off. Under this condition both transistors would remain on indefinitely.

The above discussion applies in a general way to the cathode-coupled multi and also to the plate-coupled circuit. In the tube circuit the overshoot at G_2 causes a pronounced undershoot at P_2 and hence at G_1. This undershoot at G_1 keeps $V1$ OFF temporarily. Because there is so little undershoot in the waveform at the collector of $Q2$ and the base of $Q1$ in the collector-coupled multi (Fig. 11-10), it is difficult to adjust this circuit for free-running operation. If the magnitude of V_{BB} is too small, then both transistors will be conducting permanently.

11-12   THE ASTABLE COLLECTOR–COUPLED MULTI

The circuit diagram for a free-running collector-coupled multi using p-n-p transistors is given in Fig. 11-23. Since capacitive coupling is used between stages, neither transistor can remain permanently cut off. Instead, the circuit has two quasi-stable states, and it makes periodic transitions between these states.

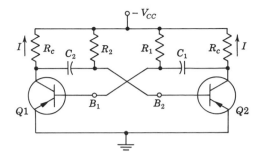

Fig. 11-23 The collector-coupled astable multi.

The waveforms at the bases and collectors for the multi of Fig. 11-23 are shown in Fig. 11-24. We consider that immediately before $t = 0$, transistor $Q2$ is in saturation and carrying the collector current I while $Q1$ is below cutoff. Hence, for $t < 0$, v_{B1} is positive, $v_{C1} = -V_{CC}$, $v_{B2} = V_{BE}(\text{sat})$, and $v_{C2} = V_{CE}(\text{sat})$. The capacitor C_1 charges through R_1, and v_{B1} falls exponentially toward $-V_{CC}$. At $t = 0$, the base B_1 reaches the cutin voltage V_γ and $Q1$ conducts. As $Q1$ goes to saturation, v_{C1} rises by IR_c to $V_{CE}(\text{sat})$, as indicated in Fig. 11-24b. The rise in v_{C1} causes an equal rise of IR_c in v_{B2} since the two are capacitively coupled. The rise in v_{B2} cuts off $Q2$, and its collector falls toward $-V_{CC}$. This fall in v_{C2} is coupled through capacitor C_1 to the base of $Q1$, causing the undershoot δ in v_{B1} of Fig. 11-24a and the

Fig. 11-24 Waveforms of the collector-coupled free-running multi of Fig. 11-23 with p-n-p transistors. At $t = 0-$, the first stage, $Q1$, is OFF and the second stage, $Q2$, is ON.

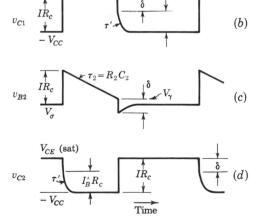

abrupt drop by the same amount δ in v_{C2}.   The equivalent circuit for calculating δ is identical with that in Fig. 11-12.   The analytical expression for δ is the same as that for the collector-coupled monostable multi [Eq. (11-15)], with the base current I'_B at the overshoot given by Eq. (11-16).   The waveforms at the base of $Q1$ and the collector of $Q2$ change exponentially with the time constant $\tau' = (R_c + r_{bb'})C_1$ to the levels V_σ and $-V_{CC}$, respectively.

The voltage v_{B2} is $IR_c + V_\sigma$ at $t = 0+$ and decreases exponentially with time constant $\tau_2 = R_2C_2$ toward $-V_{CC}$.   At $t = T_2$, B_2 reaches the cutin level V_γ and a reverse transition takes place.   The waveshapes in the first stage during the interval T_1 are the same as the waveforms in the second stage during the interval T_2 and are depicted in Fig. 11-24.   If the base time constants are different for the two transistors, the durations of the two portions of a complete cycle are not alike.

For a free-running multi using n-p-n transistors the voltages and currents are the negative of those for a circuit with p-n-p devices.   Hence, all the waveforms in Fig. 11-24 must be inverted if n-p-n transistors are used.

Timing Considerations   The time for each portion of the cycle is found from the equations in Sec. 11-4.   The period T is given approximately by

$$T = T_1 + T_2 = 0.69(R_1C_1 + R_2C_2) \tag{11-76}$$

For a symmetrical circuit with $R_1 = R_2 \equiv R$ and $C_1 = C_2 \equiv C$

$$T = 1.38RC \tag{11-77}$$

The frequency of oscillation may be varied over the range from cycles to megacycles per second by adjusting R or C.   It is also possible to change T electrically by connecting R_1 and R_2 to an auxiliary voltage $-V$ (the collector supply remains $-V_{CC}$).   If V is varied, then T changes in accord with the equation

$$T = 2RC \ln\left(1 + \frac{V_{CC}}{V}\right) \tag{11-78}$$

provided that V is large compared with the junction voltages (Prob. 11-38).   Such a circuit is a *voltage-to-frequency converter*.   If each resistor R is replaced by a transistor which acts as a constant-current source for charging C, then excellent linearity between frequency and voltage may be attained.[3]

As with the monostable collector-coupled circuit (and subject to the same approximations) the period T is independent of the supply voltage, junction voltages, temperature, etc.   However, if the temperature variations of the junction voltages and of the reverse saturation current I_{CBO} are taken into account, then we must make the correction given in Sec. 11-4 which indicates that the period will decrease, or that the frequency will increase with an increase in temperature.

In Fig. 11-24 we note that there is a transient (of time constant τ') associated with the waveforms of a transistor when it is driven heavily into saturation.   Each collector waveform has one rounded edge because of the time required for this transient to die down. This recovery time t_r may be an appreciable fraction of the half period $T/2$ of a symmetrical multi, as we shall now demonstrate. We have

$$t_r = 2.2\tau' = 2.2(R_c + r_{bb'})C \approx 2.2R_cC \qquad \text{and} \qquad \frac{T}{2} = 0.69RC$$

where $R \equiv R_1 = R_2$. Hence

$$\frac{t_r}{T/2} = \frac{2.2R_c}{0.69R} = 3.2\frac{R_c}{R} \tag{11-79}$$

If we neglect the junction saturation voltages, then

$$I_C \approx \frac{V_{CC}}{R_c} \qquad \text{and} \qquad I_B \approx \frac{V_{CC}}{R}$$

To ensure saturation, $I_B \geq I_C/h_{FE}$. Hence

$$R \leq h_{FE}R_c \tag{11-80}$$

and

$$\frac{t_r}{T/2} \geq \frac{3.2}{h_{FE}} \tag{11-81}$$

For example, if $(h_{FE})_{\min} = 16$, then the recovery time will be at least 20 percent of the half period.

Other Astable Multis A collector waveform with vertical edges is obtained by the addition of two diodes and two resistors[2] as indicated in Fig. 11-25. If $Q1$ is driven OFF its collector falls immediately to $-V_{CC}$ so that $D1$ is reverse-biased and $Q2$ goes into saturation. The saturation base current of $Q2$ passes through C_2 and R_3 rather than through R_c. Since I'_B no longer passes through R_c the collector waveform now has vertical edges, as desired.

A disadvantage of the astable multi is the possibility that both transistors will go into saturation simultaneously and remain in this state. This "blocked condition" is likely to take place if the supply voltage is increased slowly from zero to its full value V_{CC}, but not if the voltage is applied suddenly. A circuit which cannot block is given in Prob. 11-39.

A circuit which may be forced to start (or stop) oscillating at definite times is obtained by adding a transistor $Q3$ in series with the emitter of $Q1$

Fig. 11-25 The addition of diodes $D1$ and $D2$ and resistors R_3 to the collector-coupled astable multi of Fig. 11-23 results in a collector waveform with vertical edges.

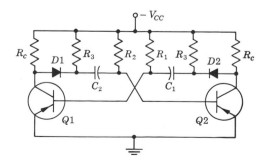

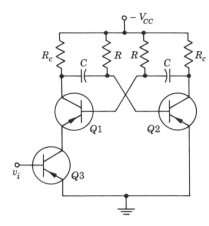

Fig. 11-26 A gated astable multi.

or $Q2$ in Fig. 11-23.   Such a *gated square-wave oscillator* is indicated in Fig. 11-26.   The input voltage v_i to the base of $Q3$ may take on one of two values (as would be the case if v_i were obtained from a flip-flop).   One voltage is such that $Q3$ is OFF.   Then $Q1$ is OFF, $Q2$ is ON, and the circuit is quiescent (not oscillating).   The second binary level is chosen such that $Q3$ is driven into saturation.   Hence, at the instant (say $t = 0$) that this voltage is applied, $Q1$ goes ON and $Q2$ is driven OFF.   The circuit operates as an astable multi with waveforms which are essentially those in Fig. 11-24 starting abruptly at $t = 0$.

11-13   THE ASTABLE PLATE–COUPLED MULTI

If a "square-wave generator" is desired with voltage swings in excess of those which can be tolerated by a transistor (the base-to-emitter reverse voltage is quite low for many transistors), then the vacuum-tube free-running multi of Fig. 11-27 may be used.   The discussion of the latter circuit follows along lines similar to those given in the preceding section for the analogous transistor astable multi of Fig. 11-23.

The waveforms at the plates and grids of the circuit of Fig. 11-27 are shown in Fig. 11-28.   We consider that at the time immediately before $t = 0$,

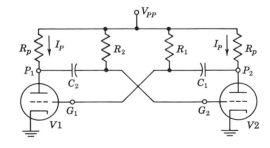

Fig. 11-27   The plate-coupled astable multi.

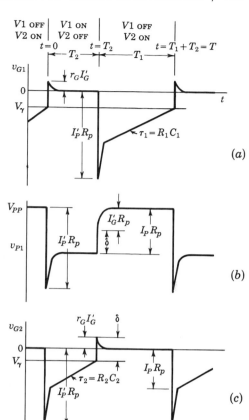

Fig. 11-28 Waveforms of the plate-coupled free-running multi of Fig. 11-26.

tube $V2$ is in clamp and carrying a current I_P while $V1$ is below cutoff. Hence, for $t < 0$, v_{G1} is more negative than V_γ, $v_{P1} = V_{PP}$, $v_{G2} = 0$, and

$$v_{P2} = V_{PP} - I_P R_p$$

The capacitor C_1 charges through R_1, and v_{G1} rises exponentially toward V_{PP}. At $t = 0$ the grid G_1 reaches the cutoff voltage V_γ. Tube $V1$ conducts and its plate voltage falls. Since P_1 is capacitively coupled to G_2, tube V_2 is driven to cutoff, causing P_2 to start to rise toward V_{PP}. The voltage rise at plate P_2 is transferred to G_1, causing the customary grid overshoot at G_1. The amplitude of this overshoot is calculated in precisely the same manner as the grid overshoot which occurs in the plate-coupled monostable multi of

Fig. 11-5. The grid current I'_G is given by Eq. (11-7), with I_1 replaced by the clamped current I_P.

If the tube current at the moment of the overshoot is I'_P, then the corresponding undershoot at the plate P_1 will carry the plate from V_{PP} (at $t = 0-$) to $V_{PP} - I'_P R_p$ (at $t = 0+$), and G_2 will change from zero (clamp) to $-I'_P R_p$. The abrupt portion of the rise at P_2 is the same in amplitude as at the grid G_2 and therefore is of magnitude $r_G I'_G - V_\gamma$. The overshoot at G_1 and the undershoots at G_2 and P_1 decay with a time constant $(R_p + r_G)C_1$, which is the time constant also with which v_{P2} eventually attains the level V_{PP}. After the overshoot has decayed, the grid G_2 is left at the voltage $-I_P R_p$, from which point it rises toward V_{PP} with the time constant $R_2 C_2$. When G_2 reaches the cutoff level, the reverse transition occurs. The second part of the cycle produces the same waveshapes as does the first part described above except for the fact that if the grid time constants are different, the duration of the individual portions of a complete cycle are different.

The essential differences between the vacuum-tube and the transistor astable waveshapes of Figs. 11-28 and 11-24 are the following: (1) The waveshapes of Fig. 11-24 correspond to a device in which the charge carriers are holes, whereas in the tube circuit the carriers are electrons. Hence, the waveforms in Fig. 11-24 should be inverted (as they would be for n-p-n transistors) before they are compared with the waveshapes in Fig. 11-28. (2) When a grid is driven positive a plate current I'_P in excess of the clamped current I_P is drawn. In the transistor circuit there may be an overshoot at the base, but the resultant collector current is essentially the saturation current $I_C \approx V_{CC}/R_c$. In other words, there is no I'_C in Fig. 11-24 corresponding to I'_P in Fig. 11-28. (3) The cutoff voltage V_γ for a tube is of the order of several volts (or tens of volts), whereas the corresponding quantity for a transistor is at most a few tenths of a volt. (4) The voltage swings in the tube circuit are usually at least 10 times those in the transistor multi.

The time durations of the overshoots in Fig. 11-28 are ordinarily very small in comparison with the times required for C_1 to charge through R_1 and C_2 to charge through R_2. Using Eq. (11-2), we find that the time required for a complete cycle is (neglecting the overshoot times, assuming the output impedance R_o to be small compared with R_1 or R_2, and taking $V_\sigma = 0$)

$$T = T_1 + T_2 = (R_1 C_1 + R_2 C_2) \ln \frac{V_{PP} + I_P R_p}{V_{PP} - V_\gamma} \tag{11-82}$$

We may neglect V_γ in comparison with V_{PP}. Then for values of plate swing $I_P R_p$ which vary all the way from one-fourth to three-fourths of V_{PP}, the logarithm in Eq. (11-82) varies only between 0.25 and 0.55. Hence, as a rough but useful general approximation,

$$T \approx \frac{R_1 C_1 + R_2 C_2}{2} = RC \tag{11-83}$$

in the symmetrical case when $R_1 C_1 = R_2 C_2 = RC$.

Since the plate characteristic for $V_G = 0$ can be approximated by a straight line through the origin, $I_P R_p$ is roughly proportional to V_{PP}. As a first approximation $-V_\gamma = V_{PP}/\mu$. Under these circumstances the factor V_{PP} can be canceled in the numerator and the denominator of Eq. (11-82). Thus, the frequency of the multi will vary only on the order of several percent for a supply-voltage variation of the order of 100 V.

If R_1 and R_2 are returned not to V_{PP} but to an auxiliary adjustable supply V, then the period may be varied by changing V. The expression for T is given by Eq. (11-82), with V_{PP} replaced by V but with I_P remaining the clamped current corresponding to V_{PP} (Prob. 11-40). Such a circuit would operate as a *voltage-to-frequency converter*.

11-14   THE ASTABLE EMITTER–COUPLED MULTI[4]

The circuit diagram for a free-running emitter-coupled multi using n-p-n transistors is given in Fig. 11-29, and the waveforms are shown in Fig. 11-30. Three power supplies are indicated for the sake of simplifying the analysis. A more practical circuit using one supply is indicated in Fig. 11-31. We shall assume that the circuit operates in such a manner that $Q1$ switches between cutoff and saturation and $Q2$ switches between cutoff and its active region.

Calculations at $t = t_1 -$ Since $Q1$ is ON and $Q2$ is OFF just before the transition at $t = t_1-$, we have

$$v_{CN2}(t_1-) = V_{CC2} \tag{11-84}$$

$$v_{EN1}(t_1-) = V_{BB} - V_{BE}(\text{sat}) = V_{BB} - V_\sigma \tag{11-85}$$

$$v_{CN1}(t_1-) = v_{BN2} = v_{EN1} + V_{CE}(\text{sat}) = V_{BB} - V_\sigma + V_{CE}(\text{sat}) \tag{11-86}$$

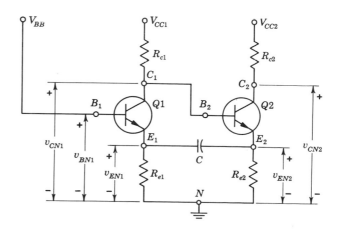

Fig. 11-29 The astable emitter-coupled multivibrator.

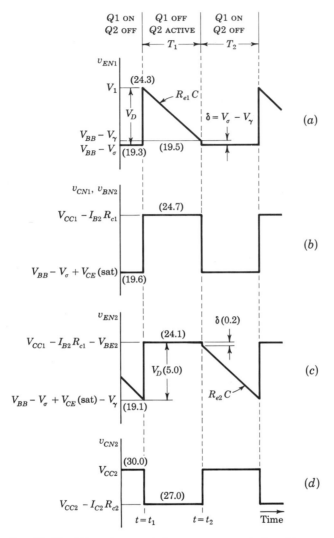

Fig. 11-30 Waveforms of the emitter-coupled multi. The voltage V_1 is given by Eq. (11-92). (The values in volts given in the parentheses refer to the illustrative problem on page 449.)

During the interval preceding $t = t_1$ the capacitor C charges from a fixed voltage $V_{BB} - V_\sigma$ through the resistor R_{e2}. All circuit voltages remain constant except v_{EN2}, which falls asymptotically toward zero. The transistor $Q2$ will begin to conduct when v_{EN2} falls to

$$v_{EN2}(t_1-) = v_{BN2} - V_\gamma = V_{BB} - V_\sigma + V_{CE}(\text{sat}) - V_\gamma \qquad (11\text{-}87)$$

Calculations at $t = t_1+$ When $Q2$ conducts, v_{EN2} and v_{EN1} rise.   As v_{EN1} rises, $Q1$ comes out of saturation and v_{CN1} ($= v_{BN2}$) also increases, causing a further increase in the current in $Q2$.   Because of this regenerative action $Q1$ is driven OFF and $Q2$ is driven into its active region where its base-to-emitter voltage is V_{BE2}, its base current is I_{B2}, and its collector current is I_{C2}.   From Fig. 11-30 we see that after the transition, at $t = t_1+$,

$$v_{CN2}(t_1+) = V_{CC2} - I_{C2}R_{c2} \qquad (11\text{-}88)$$

$$v_{CN1}(t_1+) = v_{BN2}(t_1+) = V_{CC1} - I_{B2}R_{c1} \qquad (11\text{-}89)$$

$$v_{EN2}(t_1+) = v_{BN2}(t_1+) - v_{BE2}(t_1+) = V_{CC1} - I_{B2}R_{c1} - V_{BE2} \qquad (11\text{-}90)$$

At t_1 there is an abrupt change V_D in v_{EN2} (Fig. 11-30c).   Because of the capacitive coupling between emitters there must also be the same discontinuity V_D in v_{EN1} (Fig. 11-30a).   Hence

$$V_D = v_{EN1}(t_1+) - v_{EN1}(t_1-) = v_{EN2}(t_1+) - v_{EN2}(t_1-) \qquad (11\text{-}91)$$

Substituting from Eqs. (11-85), (11-87), and (11-90) into (11-91), we obtain

$$v_{EN1}(t_1+) = V_{CC1} - I_{B2}R_{c1} - V_{BE2} - V_{CE}(\text{sat}) + V_\gamma \equiv V_1 \quad (11\text{-}92)$$

In all the above equations the quantities which appear, aside from the currents I_{B2} and I_{C2}, either are known supply voltages or else may be estimated with adequate accuracy from Table 6-1.   The currents with $Q2$ in the active region are related by $I_{C2} = h_{FE}I_{B2}$.   We shall now estimate I_{C2}.   For this purpose we make the approximations that all transistor-junction voltages are zero and that the voltage $I_{B2}R_{c1}$ is small enough so that it may be neglected in comparison with V_{CC1}.   Subject to these approximations

$$v_{EN1}(t_1+) = v_{EN2}(t_1+) = V_{CC1}$$

At $t = t_1+$, the current in R_{e1} is $v_{EN1}(t_1+)/R_{e1} \approx V_{CC1}/R_{e1}$ and the current in R_{e2} is $v_{EN2}(t_1+)/R_{e2} \approx V_{CC1}/R_{e2}$.   Since at this time $Q1$ is OFF, the sum of the currents in the emitter resistors must be supplied by $Q2$.   Hence

$$I_{C2} \approx \frac{V_{CC1}}{R_e} \qquad I_{B2} \approx \frac{V_{CC1}}{h_{FE}R_e} \qquad (11\text{-}93)$$

where R_e is the parallel combination of R_{e1} and R_{e2}.

The Period   The interval T_1 when $Q2$ conducts and $Q1$ is OFF ends at $t = t_2$.   The transistor $Q1$ will turn ON when the base-to-emitter voltage reaches the cutin value V_γ or when V_{EN1} reaches the voltage

$$v_{EN1}(t_2-) = V_{BB} - V_\gamma \qquad (11\text{-}94)$$

Since the base voltage of $Q1$ is fixed, then to carry the transistor from the cutin point to saturation the emitter must drop.   However, this drop δ in Fig.

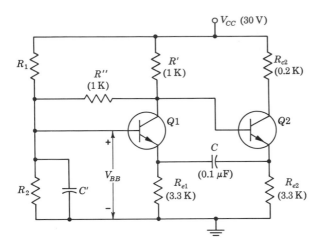

Fig. 11-31 A practical emitter-coupled multi-vibrator. (The values in parentheses refer to the illustrative problem on page 449.)

11-30a is small since, from Table 6-1, $\delta = V_\sigma - V_\gamma \approx 0.2$ V. In Fig. 11-30c there is indicated an identical jump δ in v_{EN2} because the emitters are capacitively coupled. After $t = t_2$, in the interval T_2, conditions are the same as they were for $t < t_1$ at the beginning of the discussion. Therefore the cycle of events described above is repeated and the circuit behaves as an astable multi. In Fig. 11-30 we have indicated all voltage levels in terms of the circuit parameters, the transistor-junction voltages which may be read from Table 6-1, the currents in $Q2$ which may be calculated from Eqs. (11-93), and V_1 as given by Eq. (11-92).

We see from Fig. 11-30a that the voltage v_{EN1} starts at V_1 at $t = t_1+$ and falls to $V_{BB} - V_\gamma$ at $t = t_2-$. Since this decay is exponential with a time constant $R_{e1}C$ and is approaching zero asymptotically, we calculate

$$T_1 = R_{e1}C \ln \frac{V_1}{V_{BB} - V_\gamma} \tag{11-95}$$

Assuming that the supply voltages are large in comparison with the junction voltages and assuming also that $I_{B2}R_{c1} \ll V_{CC1}$ (or equivalently, $h_{FE}R_e \gg R_{c1}$), we find

$$T_1 \approx R_{e1}C \ln \frac{V_{CC1}}{V_{BB}} \tag{11-96}$$

Subject to the same approximations T_2 is also given by Eq. (11-96) with R_{e1} replaced by R_{e2}, and hence the frequency of oscillation is independent of the transistor parameters. If V_{CC1} and V_{BB} are arranged to be proportional to one another, then the frequency is independent of the supply voltages.

When $Q1$ is OFF its collector-to-ground voltage is approximately V_{CC1} and equals the base-to-ground voltage of $Q2$. Since it is desired that $Q2$ be in its active region, then V_{BN2} should be less than V_{CN2} or $V_{CC1} < V_{CC2}$. Since $Q1$ is to be driven into saturation, then its base voltage may be almost as large as its collector supply voltage. However, to avoid driving $Q1$ too deeply into saturation it is better to arrange that $V_{BB} < V_{CC1}$. A circuit which uses a single supply and which satisfies the requirements that V_{BB} be proportional to V_{CC1} and that $V_{BB} < V_{CC1} < V_{CC2}$ is shown in Fig. 11-31. Since C' is a bypass capacitor intended to maintain V_{BB} constant, it is not involved in the operation of the circuit. We assume that R_1 and R_2 are small enough so that the voltage V_{BB} at the junction of R_1 and R_2 remains nominally constant during the entire cycle of operations of the multi. Using Thévenin's theorem we see that the circuit of Fig. 11-31 is of the same form as that of Fig. 11-29 with $V_{CC2} = V_{CC}$ and with

$$R_{c1} = \frac{R'R''}{R' + R''} \tag{11-97}$$

and

$$V_{CC1} = V_{CC} \frac{R''}{R' + R''} + V_{BB} \frac{R'}{R' + R''} \tag{11-98}$$

EXAMPLE In the circuit of Fig. 11-31, $V_{CC} = 30$ V, $R_2 = 2R_1 \ll R''$, $C = 0.1\ \mu$F, $R_{c2} = 0.2$ K, $R' = R'' = 1$ K, and $R_{e1} = R_{e2} = 3.3$ K. Calculate (a) the voltage levels of the waveforms and (b) the frequency of oscillation. Assume silicon transistors with $h_{FE} = 30$.

Solution   a. We shall assume that Q1 saturates and Q2 does not, and we shall justify this assumption later. We have

$$V_{BB} = \frac{R_2}{R_1 + R_2} V_{CC} = \tfrac{2}{3} \times 30 = 20 \text{ V} \qquad V_{CC2} = 30 \text{ V}$$

and from Eqs. (11-97) and (11-98)

$$R_{c1} = \frac{1 \times 1}{1 + 1} = 0.5 \text{ K} \qquad V_{CC1} = \tfrac{1}{2} \times 30 + \tfrac{1}{2} \times 20 = 25 \text{ V}$$

From Eqs. (11-93)

$$I_{C2} = \frac{25}{3.3/2} = 15 \text{ mA} \qquad I_{B2} = \frac{15}{30} = 0.5 \text{ mA}$$

and

$$I_{B2}R_{c1} = (0.5)(0.5) = 0.25 \text{ V} \qquad I_{C2}R_{c2} = (15)(0.2) = 3 \text{ V}$$

From Table 6-1, $V_\gamma = 0.5$ V, $V_{BE2} = 0.6$ V, $V_\sigma = 0.7$ V, and $V_{CE}(\text{sat}) = 0.3$ V. From Eq. (11-92)

$$V_1 = 25 - 0.25 - 0.6 - 0.3 + 0.5 = 24.3 \text{ V}$$

All the voltage levels may now be written down from the expressions given in Fig. 11-30 and are indicated in parentheses on that diagram.

We may now verify that when Q2 is conducting, it is in its active region. At $t = t_1+$ we find from Fig. 11-30 that the collector junction voltage is

$$v_{CB2}(t_1+) = v_{CN2}(t_1+) - v_{BN2}(t_1+) = 27.0 - 24.7 = 2.3 \text{ V}$$

Since we are using n-p-n transistors, then the collector junction of $Q2$ is reverse-biased by 2.3 V and $Q2$ is therefore in its active region.

If R_{c2} is increased (beyond 0.35 K for the circuit under consideration) so that $v_{CN2}(t_1+)$ becomes smaller than $v_{BN2}(t_1+)$, then $v_{CB2}(t_1+)$ reverses sign and $Q2$ goes into saturation. Such a mode of operation is undesirable because in saturation the base current may be large and hence the waveforms and the timing will both depend upon the saturation parameters of the transistor.

Let us verify that $Q1$ is in saturation when conducting. At $t = t_1-$ when $Q1$ is ON and $Q2$ OFF, $v_{EN1} = 19.3$ V and $v_{EN2} = 19.1$ V, and hence the magnitude of the emitter current of $Q1$ is $19.3/3.3 + 19.1/3.3 = 11.6$ mA. If $Q1$ were in its active region we could neglect the base current and then the collector current would be 11.6 mA. Under these circumstances $v_{CN1} = 25 - (11.6)(0.5) = 19.2$ V. Hence $v_{CB1} = v_{CN1} - v_{BN1} = 19.2 - 20 = -0.8$ V, which means that the collector junction is actually forward-biased. Hence, $Q1$ must be in saturation and not in the active region. With $Q1$ in saturation

$$I_{C1} = \frac{V_{CC1} - V_{CN1}}{R_{c1}} = \frac{25 - 19.6}{0.5} = 10.8 \text{ mA}$$

Hence the base of $Q1$ must supply a current of $11.6 - 10.8 = 0.8$ mA.

The lower voltage level of v_{CN1} is quite constant because $Q1$ is in saturation. The lower level of v_{CN2} has a slight upward tilt (not shown in Fig. 11-30d) since I_{C2} decreases slightly during the ON time T_1 of $Q2$. The emitter current of $Q2$ which flows through R_{e1} and C decreases during T_1 from 24.3/3.3 to 19.5/3.3 mA. This 20 percent reduction produces a 10 percent tilt on the lower level of v_{CN2} since only one-half of the current in $Q2$ comes through C.

The sloping portions of the emitter waveforms fall from about 24 to 19 V during an exponential decay which has 0 V as its asymptotic limit. These sloping portions are therefore quite linear in appearance and are so indicated in Fig. 11-30.

b. From Eq. (11-95)

$$T_1 = 3.3 \times 10^3 \times 10^{-7} \ln \frac{24.3}{20 - 0.5} = 7.28 \times 10^{-5} \text{ sec} = T_2 = \frac{T}{2}$$

A value of 7.36 is obtained from the approximate formula (11-96). The frequency is

$$f = \frac{1}{T} = \frac{1}{2 \times 7.28 \times 10^{-5}} \text{ Hz} = 6.87 \text{ kHz}$$

Note that although $Q1$ operates in saturation it does so only because of the additional current delivered to it through C. If the capacitor C were removed (in which case, of course, the circuit would not oscillate) $Q1$ would not be in saturation. Such a situation is fortunate. If it were possible for $Q1$ to be in saturation before the circuit started oscillations, it might never start. For with $Q1$ in saturation the loop gain would not exceed unity and regeneration would not be possible. It is necessary, as a matter of fact, for reliable operation, to arrange d-c operating conditions (i.e., with C removed) so that neither transistor is in saturation.

The advantages of the emitter-coupled astable multi over the collector-coupled circuit of Fig. 11-23 are as follows: (1) It is inherently self-starting. (2) It makes an output available at the collector of $Q2$ which is connected to no other part of the circuit.   Since this collector is not involved in the regenerative action, it may be loaded heavily, even capacitively.   (3) The output is free of recovery transients such as appear in Fig. 11-24.   (4) It has an isolated input at the base of $Q1$ which may be used conveniently for synchronization.   (5) It uses only one timing capacitor.   This feature makes frequency adjustment convenient.   In a multi with two capacitors both capacitors need to be changed if the frequency is to be changed without changing the relative values of T_1 and T_2.   The disadvantages of the circuit of Fig. 11-31 over that of Fig. 11-23 are the following: (1) The former circuit is more difficult to adjust for proper operating conditions.   We require that, in the quiescent state (C removed), neither $Q1$ nor $Q2$ should saturate, but that when the circuit is operating as a multi, $Q1$ should be driven to saturation, while $Q2$ is driven into its active region.   (2) Since there is only a single capacitor and since the emitter resistors cannot be widely different from one another, the present circuit cannot be operated with T_1 and T_2 widely different.   (3) The emitter-coupled multi uses more components than does the collector-coupled circuit.

REFERENCES

1. Bénéteau, P. J., and L. Blaser: A Simple Method of Temperature Stabilizing Monostable Multivibrators, *Appl. Note* 28, Fairchild Semiconductor Corporation, Mountain View, Calif.

2. Gregory, R. O., and J. C. Bowers: Simple Square-wave Generator, *Electronics*, vol. 35, no. 51, p. 47, Dec. 21, 1962.

3. Biddlecomb, R. W.: Latest Multivibrator Improvement—Linear Voltage-to-Frequency Converter, *Electronics*, vol. 36, no. 17, pp. 64–65, Apr. 26, 1963.

4. Bénéteau, P. J., and A. Evangelist: An Improved Emitter-coupled Multivibrator, *Appl. Note* 59, Fairchild Semiconductor Corporation, Mountain View, Calif., February, 1963.

12 / NEGATIVE–RESISTANCE DEVICES

A number of devices which find extensive application in pulse and switching circuitry are most conveniently characterized in terms of a volt-ampere curve which displays, over a limited range, a negative incremental resistance. In this chapter we describe the physical principles which account for this characteristic in the tunnel diode, the unijunction transistor, the p-n-p-n diode, the silicon controlled switch, and the thyristor. In the following chapter, circuits are constructed with these negative-resistance devices and it is demonstrated that bistable, monostable, and astable operation are possible.

12-1   THE TUNNEL DIODE

A p-n junction diode of the type discussed in Sec. 6-1 has an impurity concentration of about 1 part in 10^8. With this amount of doping the width of the depletion layer, which constitutes a potential barrier at the junction, is of the order of 5 microns (5×10^{-4} cm). This potential barrier restrains the flow of carriers from the side of the junction where they constitute majority carriers to the side where they constitute minority carriers. If the concentration of impurity atoms is greatly increased, say to 1 part in 10^3, then the device characteristics are completely changed. This new diode was announced in 1958 by Esaki,[1] who also gave the correct theoretical explanation for its volt-ampere characteristic, which is depicted in Fig. 12-1. The width of the junction barrier varies inversely as the square root of impurity concentration and therefore is reduced from 5 microns to about 100 Å (10^{-6} cm). This thickness is only about one-fiftieth the wavelength of visible light. Classically, a particle must have an energy

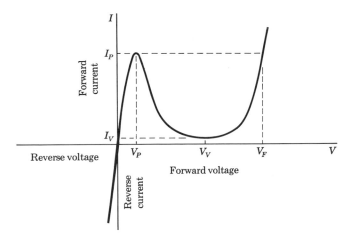

Fig. 12-1   Volt-ampere characteristic of a tunnel diode.

at least equal to the height of a potential barrier if it is to move from one side of the barrier to the other. However, for barriers as thin as those estimated above in the Esaki diode, quantum mechanics dictates that there is a large probability that an electron will penetrate *through* the barrier. The quantum-mechanical behavior is referred to as "tunneling," and hence these high-impurity-density *p-n* junction devices are called "tunnel diodes." This same tunneling effect is responsible for high-field emission of electrons from a cold metal and for radioactive emissions.

As a consequence of the tunneling effect and the band structure of heavily doped semiconductors the volt-ampere characteristic of Fig. 12-1 is obtained.[2,3] The device is an excellent conductor in the reverse direction (*p* side of junction negative with respect to the *n* side). Also, for small forward voltages (up to 50 mV for Ge) the resistance remains small (of the order of 5 Ω). At the *peak current* I_P corresponding to the voltage V_P the slope dI/dV of the characteristic is zero. If V is increased beyond V_P, then the current decreases. As a consequence the dynamic conductance $g = dI/dV$ is negative. The tunnel diode exhibits a *negative-resistance characteristic* between the peak current I_P and the minimum value I_V, called the *valley current*. At the *valley voltage* V_V at which $I = I_V$ the conductance is again zero, and beyond this point the resistance becomes and remains positive. At the so-called *peak forward voltage* V_F the current again reaches the value I_P. For larger voltages the current increases beyond this value. The portion of the characteristic beyond V_V is caused by the injection current in an ordinary *p-n* junction diode. The remainder of the characteristic is a result of the tunneling phenomenon in the highly doped diode.

For currents whose values are between I_V and I_P the curve is triple-valued, because each current can be obtained at three different applied voltages. It is this multivalued feature which makes the tunnel diode useful in pulse and

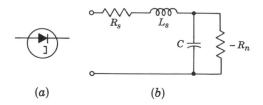

Fig. 12-2 (a) Symbol for a tunnel diode (Ref. 4); (b) small-signal model in the negative-resistance region.

(a) (b)

digital circuitry (Chap. 13). Note that whereas the characteristic in Fig. 12-1 is a multivalued function of current, it is a single-valued function of voltage. Each value of V corresponds to one and only one current. Hence, the tunnel diode is said to be *voltage-controllable*. The vacuum-tube tetrode is another negative-resistance device belonging to the voltage-controllable class. On the other hand, there also exist negative-resistance devices whose character-istics are multivalued functions of voltage but are single-valued with respect to current. These *current-controllable* devices (for example, the unijunction transistor, the p-n-p-n diode, etc.) are discussed later in this chapter.

The standard circuit symbol[4] for a tunnel diode is given in Fig. 12-2a. The small-signal model for operation in the negative-resistance region is indicated in Fig. 12-2b. The negative resistance $-R_n$ has a minimum at the point of inflection between I_P and I_V. The series resistance R_s is ohmic resistance. The series inductance L_s depends upon the lead length and the geometry of the diode package. The junction capacitance C depends upon the bias and is usually measured at the valley point. Typical values for these parameters for a tunnel diode of peak current value $I_P = 10$ mA are $-R_n = -30$ Ω, $R_s = 1$ Ω, $L_s = 5$ nH, and $C = 20$ pF.

Our principal interest in the tunnel diode is its application as a very high speed switch. Since tunneling takes place at the speed of light, the transient response is limited only by total shunt capacitance (junction plus stray wiring capacitance) and peak driving current. Switching times of the order of a nanosecond are reasonable, and times as low as 50 psec have been obtained.

The most common commercially available tunnel diodes are made from germanium or gallium arsenide. It is difficult to manufacture a silicon tunnel diode with a high ratio of peak-to-valley current I_P/I_V. Table 12-1 summa-rizes the important static characteristics of these devices. The voltage values in this table are determined principally by the particular semiconductor used

TABLE 12-1   Typical tunnel-diode parameters

	Ge	GaAs	Si
I_P/I_V	8	15	3.5
V_P, V	0.055	0.15	0.065
V_V, V	0.35	0.50	0.42
V_F, V	0.50	1.10	0.70

and are almost independent of the current rating. Note that gallium arsenide has the highest ratio I_P/I_V and the largest voltage swing $V_F - V_P \approx 1.0$ V as against 0.45 V for germanium.

The peak current I_P is determined by the impurity concentration (the resistivity) and the junction area. A spread of 20 percent in the value of I_P for a given tunnel-diode type is normal, but tighter-tolerance diodes are also available. For computer applications, devices with I_P in the range of 1 to 100 mA are most common. However, it is possible to obtain diodes whose I_P is as small as 100 μA or as large as 100 A.

The peak point (V_P, I_P), which is in the tunneling region, is not a very sensitive function of temperature. Commercial diodes are available[3] for which I_P and V_P vary by only about 10 percent over the range -50 to $+150°$C. The temperature coefficient of I_P may be positive or negative, depending upon the impurity concentration and the operating temperature, but the temperature coefficient of V_P is always negative. The valley point V_V, which is affected by injection current, is quite temperature-sensitive. The value of I_V increases rapidly with temperature and at 150°C may be two or three times its value at $-50°$C. The voltages V_V and V_F have negative temperature coefficients of about 1.0 mV/°C, a value only about half that found for the shift in voltage with temperature of a p-n junction diode or transistor. These values apply equally well to Ge or GaAs diodes. Gallium arsenide devices show a marked reduction of the peak current if operated at high current levels in the forward injection region. However, it is found empirically[3] that negligible degradation results if, at room temperature, the average operating current I is kept small enough to satisfy the condition $I/C \leq 0.5$ mA/pF, where C is the junction capacitance. Tunnel diodes are found to be several orders of magnitude less sensitive to nuclear radiation than are transistors.

The advantages of the tunnel diode are low cost, low noise, simplicity, high speed, environmental immunity, and low power. The disadvantages of the diode are its low output-voltage swing and the fact that it is a two-terminal device. Because of the latter feature, there is no isolation between input and output, and this leads to serious circuit-design difficulties. Hence, a transistor (an essentially unilateral device) is usually preferred for frequencies below about 1 GHz (a kilomegacycle per second) or for switching times longer than several nanoseconds. The tunnel diode and transistor may be combined advantageously, as indicated in Sec. 13-11.

12-2   THE BACKWARD DIODE

A tunnel diode designed to have a small peak current (I_P of the order of I_V) may be used to advantage, in the reverse direction, for purposes for which the conventional diode is employed in the forward direction. The volt-ampere characteristic of such a "tunnel rectifier" is shown in Fig. 12-3. Because this device is a better conductor in the reverse than in the forward direction it is also called a "backward diode" or simply a "back diode." In the neighborhood of zero voltage, in response to either a forward-biasing or reverse-biasing voltage, the tunnel diode responds with a current which is large in comparison to the corresponding current in a conventional diode. These large currents are a result of the tunneling effect. In the back diode, the current due to tunneling is large only in the reverse direction. For this reason the back diode is also called the "unitunnel diode."

The high-conduction portion of the volt-ampere characteristic of Fig. 12-3 is in the third quadrant. Since this portion of the characteristic corresponds to the region of forward conduction in a conventional diode, it is customary to plot the back diode with the voltage and current scales both reversed. In the back diode, the "forward direction" of applied voltage is actually the direction where the p side of the diode is negative with respect to the n side. The appearance of the characteristics as normally supplied by manufacturers may be seen by turning the page upside down.

The merits of the back diode are made clear in Fig. 12-4, where the "forward" characteristics at various temperatures of a typical silicon back diode are compared with the forward characteristics of a conventional diode. We note that the temperature sensitivity of the back diode is appreciably less than the sensitivity of the conventional diode. The back diode has a sensitivity of about -0.1 mV/°C (both silicon and germanium), compared, as we have seen, with about -2 mV/°C for the conventional diode. We observe further that while the conventional silicon diode has a break point, at room temperature, between 0.6 and 0.7 V, the back diode has a break point at 0 V. The back diode is therefore very useful when the rectifying action of a diode is required in connection with small-amplitude waveforms. Suppose, by way

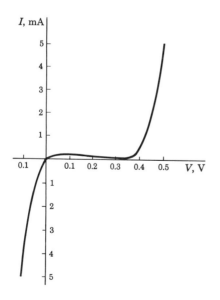

Fig. 12-3 A typical germanium backward-diode characteristic.

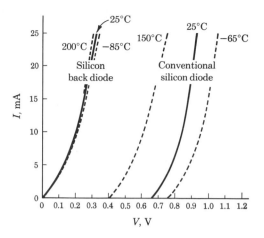

Fig. 12-4 The characteristics of a typical silicon back diode at various temperatures compared with the corresponding characteristics of a conventional silicon diode. (Courtesy of Hoffman Semiconductor.)

of example, that a sinusoidal signal is applied to a rectifying circuit which consists of a diode and resistor in series. If the signal has an amplitude of, say, 200 mV, and the diode is a conventional device (silicon or germanium), the diode will hardly conduct at any point in the cycle and the efficiency of rectification will be very poor. With a back diode the efficiency will be greatly improved.

12-3   THE UNIJUNCTION TRANSISTOR[5]

The construction of this device is indicated in Fig. 12-5a. A bar of high-resistivity n-type silicon of typical dimensions $8 \times 10 \times 35$ mils, called the *base B*, has attached to it at opposite ends two ohmic contacts $B1$ and $B2$. A 3-mil aluminum wire, called the *emitter E*, is alloyed to the base to form a p-n rectifying junction. This device was originally described in the literature as the *double-base diode* but is now commercially available under the designation

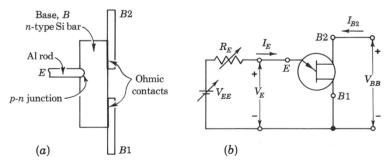

Fig. 12-5 Unijunction transistor. (a) Constructional details; (b) circuit symbol (Ref. 4).

unijunction transistor (UJT). The standard symbol for this device is shown in Fig. 12-5b. Note that the emitter arrow is inclined and points toward $B1$, whereas the ohmic contacts $B1$ and $B2$ are brought out at right angles to the line which represents the base.

As usually employed, a fixed interbase potential V_{BB} is applied between $B1$ and $B2$. The most important characteristic of the UJT is that of the input diode between E and $B1$. If $B2$ is open-circuited so that $I_{B2} = 0$, then the input volt-ampere relationship is that of the usual p-n junction diode as given in Eq. (6-1). In Fig. 12-6 the input current-voltage characteristics are plotted for $I_{B2} = 0$ and also for a fixed value of interbase voltage V_{BB}. The latter curve is seen to have the current-controlled negative-resistance characteristic which is single-valued in current but may be multivalued in voltage. A qualitative explanation of the physical origin of the negative resistance will now be given.

If $I_E = 0$ then the silicon bar may be considered as an ohmic resistance R_{BB} between base leads. Usually R_{BB} lies in the range between 5 and 10 K. Between $B1$ (or $B2$) and the n side of the emitter junction the resistance is R_{B1} (or R_{B2}, respectively), so that $R_{BB} = R_{B1} + R_{B2}$. Under this condition of zero or very small emitter current the voltage on the n side of the emitter junction is ηV_{BB}, where $\eta \equiv R_{B1}/R_{BB}$ is called the *intrinsic stand-off ratio*. This parameter is specified by the manufacturer and usually lies between 0.5 and 0.75. If V_E is less than ηV_{BB}, then the p-n junction is reverse-biased and the input current I_E is negative. As indicated in Fig. 12-6, the maximum value of this negative current is the reverse saturation current I_{EO}, which is of the order of only 10 μA even at 150°C. If V_{EE} is increased beyond ηV_{BB} the input diode becomes forward-biased and I_E goes positive. However, as already noted in connection with Fig. 12-6, the current remains quite small until the forward bias equals the cutin voltage V_γ (≈ 0.6 V), and then increases very rapidly with small increases in voltage. We must now take account of the conductivity modulation of the base region due to I_E.

The emitter current increases the charge concentration between E and $B1$ because holes are injected into the n-type silicon. Since conductivity

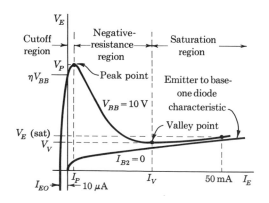

Fig. 12-6 Idealized input characteristic of a unijunction transistor. (Courtesy of General Electric Company.)

is proportional to charge density, the resistance R_{B1} decreases with increasing emitter current. Hence, for voltages above the threshold value V_γ, as I_E is increased (by either increasing V_{EE} or decreasing R_E in Fig. 12-5b) the voltage V_E between E and $B1$ decreases because of the decrease in the value of the resistance R_{B1}. Since the current is increasing while the voltage is decreasing, then this device possesses a negative resistance.

After the emitter current has become very large compared with I_{B2}, then we may neglect I_{B2}. Hence, for very large I_E the input characteristic asymptotically approaches the curve for $I_{B2} = 0$. As indicated in Fig. 12-6, this behavior results in a minimum or valley point where the resistance changes from negative to positive. For currents in excess of the valley current I_V the resistance remains positive. This portion of the curve is called the *saturation region*. The voltage at $I_E = 50$ mA is arbitrarily called the *saturation voltage* $V_E(\text{sat})$ and is of the order of 3 V.

At the maximum voltage or peak point V_P the current is very small ($I_P \approx 25$ μA), and hence the region to the left of the peak point is called the *cutoff region*. For many applications the most important parameter is the peak voltage V_P, which, as explained above, is given by

$$V_P = \eta V_{BB} + V_\gamma \tag{12-1}$$

In Sec. 6-1 we noted that $V_\gamma \approx 0.6$ and decreases about 2 mV/°C. (Both of these facts are approximated by replacing V_γ by $200/T$, where T is the junction temperature in degrees Kelvin.) Since the temperature coefficient of R_{B1} is the same as that of R_{B2}, then $\eta = R_{B1}/(R_{B1} + R_{B2})$ should be independent of temperature. Experimentally it is found that the temperature coefficient of η is less than 0.01 percent/°C and may be either positive or negative. To illustrate that V_P is quite insensitive to temperature, assume $\eta = 0.5$, $V_{BB} = 20$ V, and a temperature change from 25 to 125°C. At 25°C, $V_P = 10.6$ V. At 125°C the change due to V_γ is 0.2 V and due to η is a maximum of 0.2 V. Hence V_P will decrease no more than 4 percent over the 100°C increase in temperature. The peak voltage can be made even much less sensitive to temperature by adding a small resistance R_2 in series with the $B2$ lead (Prob. 12-12).

The peak current I_P varies inversely with the interbase voltage V_{BB} and decreases with increasing temperature. Typically, a peak current of 10 μA at 25°C will reduce to about 6 μA at 125°C and increase to 12 μA at -55°C.

A family of input characteristics for commercially available UJTs is indicated in Fig. 12-7a. Note that the peak voltage increases linearly with V_{BB}, and observe also that the valley is very broad. Hence it is difficult to give the exact value of the valley current I_V, but the valley voltage V_V can be determined fairly accurately. The valley current has about the same temperature coefficient as the peak current.

A family of output characteristics are given in Fig. 12-7b. The straight line for $I_E = 0$ indicates that with the emitter open-circuited the n-type silicon bar is essentially ohmic. The reciprocal slope of this line gives R_{BB}.

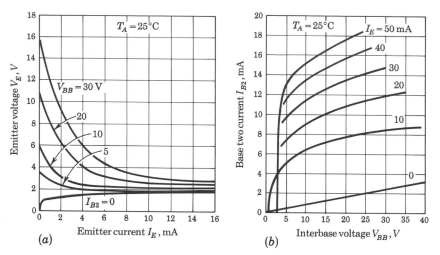

Fig. 12-7 Unijunction characteristics for types 2N489 through 2N494. (a) Input; (b) output. (Courtesy of General Electric Company.)

For $I_E = 50$ mA the drop across R_{B1} is 3 V even if $I_{B2} = 0$. As I_{B2} is increased, R_{B1} decreases, and the decreased drop across R_{B1} offsets the increase in voltage across R_{B2}. Hence the voltage V_{BB} remains almost constant for small values of I_{B2} (up to about 10 mA) and then rises with $B2$ current.

We shall find when we discuss applications of the unijunction transistor that the input characteristics in Fig. 12-7a are much more important than the output curves of Fig. 12-7b. The most useful features of the UJT are its stable firing voltage V_P which depends linearly on V_{BB}, the low (microampere) firing current I_P, the stable negative-resistance characteristic, and the high pulse-current capability.

12-4   THE FOUR–LAYER DIODE[6]

Another device which exhibits a negative resistance and which finds extensive applications in switching circuits is represented in Fig. 12-8, together with its circuit symbol. The device consists of four layers of silicon doped alternately

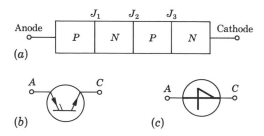

Fig. 12-8 (a) A four-layer p-n-p-n diode; (b) standard circuit symbol (Ref. 4); (c) alternative symbol (not recommended).

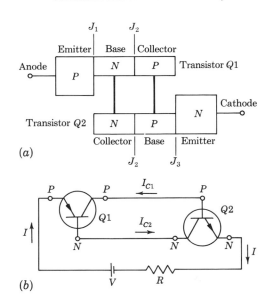

Fig. 12-9 (a) The p-n-p-n diode is redrawn to make it appear as two interconnected "transistors." (b) The two interconnected transistors are supplied current from a source through a resistor.

with p- and n-type impurities. Because of this structure it is called a p-n-p-n (often pronounced "pinpin") diode or switch. The terminal P region is the anode, or p emitter, and the terminal N region is the cathode, or n emitter. When an external voltage is applied to make the anode positive with respect to the cathode, junctions J_1 and J_3 are forward-biased and the center junction J_2 is reverse-biased. The externally impressed voltage appears principally across the reverse-biased junction, and the current which flows through the device is small. As the impressed voltage is increased, the current increases slowly until a voltage called the *firing* or *breakover* voltage V_{BO} is reached where the current increases abruptly and the voltage across the device decreases sharply. At this breakover point the p-n-p-n diode switches from its OFF (also called *blocking*) state to its ON state.

In Fig. 12-9a, the p-n-p-n switch has been split into two parts which have been displaced mechanically from one another but left electrically connected. This splitting is intended to illustrate that the device may be viewed as two transistors back to back. One transistor is a p-n-p type, whereas the second is an n-p-n type. The N region that is the base of one transistor is the collector of the other, and similarly for the adjoining P region. The junction J_2 is a common collector junction for both transistors. In Fig. 12-9b the arrangement in Fig. 12-9a has been redrawn using transistor-circuit symbols, and a voltage source has been impressed through a resistor across the switch, giving rise to a current I. Collector currents I_{C1} and I_{C2} for transistors $Q1$ and $Q2$ are indicated. In the active region the collector current is given by Eq. (6-13),

$$I_C = -\alpha I_E + I_{CO} \tag{12-2}$$

with I_E the emitter current, I_{CO} the reverse saturation current, and α the short-circuit common-base forward current gain. We may apply Eq. (12-2), in turn, to $Q1$ and $Q2$. Since $I_{E1} = +I$ and $I_{E2} = -I$, we obtain

$$I_{C1} = -\alpha_1 I + I_{CO1} \tag{12-3}$$

$$I_{C2} = \alpha_2 I + I_{CO2} \tag{12-4}$$

For the p-n-p transistor I_{CO1} is negative, whereas for the n-p-n device I_{CO2} is positive. Hence, we write $I_{CO2} = -I_{CO1} \equiv I_{CO}/2$. Setting equal to zero the sum of the currents into transistor $Q1$ we have

$$I + I_{C1} - I_{C2} = 0 \tag{12-5}$$

Combining Eqs. (12-3) through (12-5) we find

$$I = \frac{I_{CO2} - I_{CO1}}{1 - \alpha_1 - \alpha_2} = \frac{I_{CO}}{1 - \alpha_1 - \alpha_2} \tag{12-6}$$

We observe that as the sum $\alpha_1 + \alpha_2$ approaches unity Eq. (12-6) indicates that the current I increases without limit; that is, the device breaks over. Such a development is not unexpected in view of the regenerative manner in which the two transistors are interconnected. The collector current of $Q1$ is furnished as the base current of $Q2$, and vice versa. When the p-n-p-n switch is operating in such a manner that the sum $\alpha_1 + \alpha_2$ is less than unity, the switch is in its OFF state and the current I is small. When the condition $\alpha_1 + \alpha_2 = 1$ is attained, the switch transfers to its ON state. The voltage across the switch drops to a low value and the current becomes large, being limited by the external resistance in series with the switch.

The reason why the device can exist in either of two states is that at very low currents α_1 and α_2 may be small enough so that $\alpha_1 + \alpha_2 < 1$, whereas at larger currents the α's increase, thereby making it possible to attain the condition $\alpha_1 + \alpha_2 = 1$. Thus, as the voltage across the switch is increased from zero, the current starts at a very small value and then increases because of avalanche multiplication (not avalanche breakdown) at the reverse-biased junction. This increase in current increases α_1 and α_2. *When the sum of the small-signal avalanche-enhanced alphas equals unity, $\alpha_1 + \alpha_2 = 1$, breakover occurs.* At this point, the current is large, and α_1 and α_2 might be expected individually to attain values in the neighborhood of unity. If such were the case, then Eq. (12-6) indicates that the current might be expected to reverse. What provides stability to the ON state of the switch is that in the ON state the center junction becomes forward-biased. Now all the transistors are in saturation and the current gain α is again small. Thus, stability is attained by virtue of the fact that the transistors enter saturation to the extent necessary to maintain the condition $\alpha_1 + \alpha_2 = 1$.

In the ON state all junctions are forward-biased, and so the total voltage across the device is equal very nearly to the algebraic sum of these three saturation junction voltages. The voltage drop across the center junction

J_2 is in a direction opposite to the voltages across the junctions J_1 and J_3. This feature serves additionally to keep quite low (of the order of 0.7 V) the total voltage drop across the switch in the ON state.

The operation of the p-n-p-n switch depends, as we have seen, on the fact that at low currents, the current gain α may be less than one-half, a condition which is necessary if the sum of two α's is to be less than unity. This characteristic of α is not encountered in germanium but is distinctive of silicon, where it results from the fact that at low currents an appreciable part of the current which crosses the emitter junction is caused by recombination of holes and electrons in the transition region rather than the injection of minority carriers across the junction from emitter to base.[7]   In germanium it is not practicable to establish $\alpha_1 + \alpha_2 < 1$.   Therefore germanium structures incline to settle immediately in the ON state and have no stable OFF state.   Accordingly, germanium p-n-p-n switches are not available.   We shall see in the discussion below that the p-n-p-n structure and mechanism are basic to a large number of other switching devices and in all but one (Sec. 12-9) the material employed is silicon.   Thus we shall encounter silicon controlled rectifiers (SCR) but no germanium controlled rectifiers, etc.

12-5 p-n-p-n CHARACTERISTICS

The volt-ampere characteristic of a p-n-p-n diode, not drawn to scale, is shown in Fig. 12-10.   When the voltage is applied in the reverse direction, the two outer junctions of the switch are reverse-biased.   At an adequately large voltage, breakdown will occur at these junctions, as indicated, at the "reverse avalanche" voltage V_{RA}.   However, no special interest is attached to operation in this reverse direction.

When a forward voltage is applied, only a small forward current will flow until the voltage attains the breakover voltage V_{BO}.   The corresponding current is I_{BO}.   If the voltage V which is applied through a resistor as in Fig. 12-9b is increased beyond V_{BO}, the diode will switch from its OFF (blocked) state to its ON (saturation) state and will operate in the saturation region. The device is then said to *latch*.   If the voltage is now reduced, the switch will remain ON until the current has decreased to I_H.   This current and the corresponding voltage V_H are called the *holding* or *latching* current and voltage, respectively.   The current I_H is the minimum current required to hold the switch in its ON state.

There are available p-n-p-n switches with voltages V_{BO} in the range from tens of volts to some hundreds of volts.   The current I_{BO} is of the order, at most, of some hundreds of microamperes.   In this OFF range up to breakover, the resistance of the switch is the range from some megohms to several hundred megohms.

The holding current varies, depending on the type, in the range from several milliamperes to several hundred milliamperes.   The holding voltage is found to range from about 0.5 V to about 20 V.   The incremental resistance R_H in the saturation state is rarely in excess of 10 Ω and decreases with increasing current.   At currents of the order of amperes (which can be sustained briefly under pulsed operation), the incremental resistance may drop to as low as some tenths of an ohm.

The switching parameters of the four-layer diode are somewhat temperature-dependent.   A decrease in temperature from room temperature to $-60°C$ has negligible effect on V_{BO}, but a temperature increase to $+100°C$ will decrease V_{BO} by about 10 percent. I_H decreases substantially with increase in temperature and increases to a lesser extent with decrease in temperature.

Rate Effect   We can see that the breakover voltage of a p-n-p-n switch depends on the rate[8] at which the applied voltage rises.   In Fig. 12-11 we have represented the switch in the OFF state as a series combination of three diodes, two forward-biased and the center one reverse-biased.   Across this latter diode we have placed a capacitance which represents the transition capacitance across this reverse-biased junction.   When the applied voltage v increases slowly enough so that the current through C may be neglected, we must wait until the avalanche-increased current through $D2$ (which is also the current through $D1$ and $D3$) increases to the point where the current gains satisfy the condition $\alpha_1 + \alpha_2 = 1$.   When, however, v changes rapidly, so that the capacitor voltage changes at the rate dv_C/dt, a current $C\,dv_C/dt$ passes

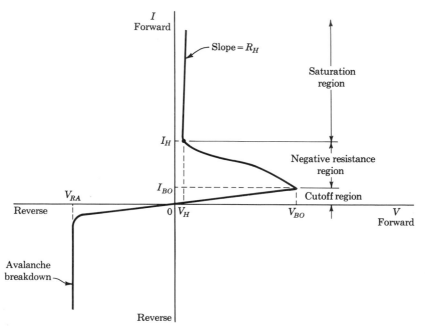

Fig. 12-10   Volt-ampere characteristic of the p-n-p-n diode.

Fig. 12-11 *p-n-p-n* diode in OFF
state to illustrate the origin of the
rate effect.

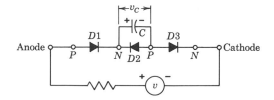

through C and adds to the current in $D1$ and $D3$. The current through $D2$
need not be as large as before to attain breakover, and switching takes place
at a lower voltage. The capacitance at the reverse-biased junction may lie
in the range of some tens of picofarads to over 100 pF, and the reduction in
switching voltage may well make itself felt for voltage rates of change dv_C/dt
of the order of tens of volts per microsecond.

The discussion above suggests that V_{BO} would continue to become smaller
as dv_C/dt increases. Actually, as the rate dv_C/dt increases and becomes very
rapid the reduction of switching voltage may become much less pronounced.
The reason for this apparently anomalous situation is that, before switching
takes place, not only must the switch current reach a certain level but also a
time interval is required for the redistribution of base charge in the two
"transistors" to allow the end junctions to function as emitting junctions of a
transistor. If the applied voltage rises rapidly enough it may well have
reached the d-c breakover voltage before this redistribution of base charges
has been completed. This matter of stored base charge is considered in detail
in Chap. 20, where there is also a discussion of matters relating to the speed
with which a transistor can be turned ON and OFF. For the present we simply
note that for typical *p-n-p-n* switches the time required to complete the transition from OFF to ON is about 0.1 μsec, and the time to complete the reverse
transition is about 0.2 μsec.

12-6   THE SILICON CONTROLLED SWITCH[9,10]

The structure of the silicon controlled switch (SCS) consists of four alternate
p- and *n*-type layers, as in the four-layer diode. In the SCS (also called a
p-n-p-n transistor or *n-p-n-p switch*) connections are made available to the inner
layers, which are not accessible in the diode. The circuit symbols for the
SCS are shown in Fig. 12-12. The terminal connected to the P region nearest
the cathode is called the *cathode gate*, or *p base*, and the terminal connected
to the N region nearest the anode is called the *anode gate*, or *n base*. In very
many switch types both gates are not brought out. Where only one gate
terminal is available it is ordinarily the cathode gate. These three-terminal
devices are available under a variety of commercial names (Sec. 12-8).

The usefulness of the gate terminals rests on the fact that currents introduced into one or both gate terminals may be used to control the anode-to-
cathode breakover voltage. Such behavior is to be expected on the basis

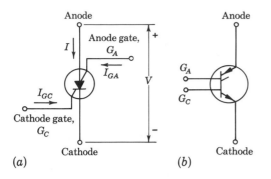

(a)   (b)

Fig. 12-12 (a) Circuit symbol used by most manufacturers for the SCS; (b) alternative symbol (Ref. 4).

of the earlier discussion of the condition $\alpha_1 + \alpha_2 = 1$ which establishes the firing point. If the current through one or both outer junctions is increased as a result of currents introduced at the gate terminals, then α increases and the breakover voltage will be decreased. In Fig. 12-13 the volt-ampere characteristic of an SCS is shown for various cathode-gate currents. We observe that the firing voltage is a function of the gate current, decreasing with increasing gate current and increasing when the gate current is negative and consequently in a direction to reverse-bias the cathode junction. The current after breakdown may well be larger by a factor of a thousand than the current before breakdown. When the gate current is very large, breakover may occur at so low a voltage that the characteristic has the appearance of a simple p-n diode.

The breakover voltage may be increased by applying reverse voltage to the cathode junction or equivalently by injecting a reverse current into the gate terminal. So long as firing is determined by the condition $\alpha_1 + \alpha_2 = 1$

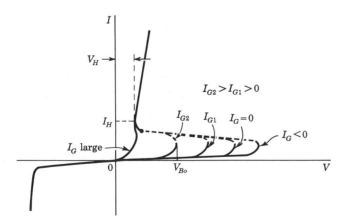

Fig. 12-13 Volt-ampere characteristics of a three-terminal SCS illustrating that the forward breakover voltage is a function of the cathode-gate current. (Not drawn to scale.)

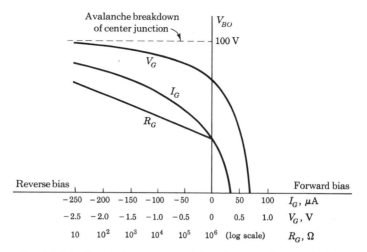

Fig. 12-14 Typical variation of breakover voltage V_{BO} in an SCS as a function of cathode-gate bias V_G, gate current I_G, and gate resistance R_G.

the breakover voltage V_{BO} will be increased. Eventually, however, a point is reached where firing is the result of an avalanche breakdown of the center junction. In this case there is no additional increase in V_{BO} even if the cathode junction is further reverse-biased. The form of the variation of breakover voltage for a typical SCS is sketched in Fig. 12-14 for three separate circumstances of bias provided by a voltage source V_G, a current source I_G, or a resistor R_G connected from the cathode gate to the cathode. Note that very large values of R_G are equivalent to $I_G = 0$, and very small values of R_G to $V_G = 0$. These curves are temperature-dependent, the breakover voltage at fixed I_G decreasing with an increase in temperature.

From Fig. 12-14 we see that in the forward-bias region V_{BO} changes very rapidly with V_G or I_G. Hence we may expect great variability in the firing voltage from unit to unit. For this reason the manufacturer does not supply curves of the sort indicated in Fig. 12-14. Instead he provides information concerning the forward voltage or current (as a function of temperature), which, even at an anode-to-cathode voltage of only a few volts, is "guaranteed to cause triggering in all units." For the SCS depicted in Fig. 12-14 this maximum gate firing signal is 30 μA or 0.6 V d-c or a pulse of this magnitude (with a minimum pulse width, as discussed in the following section).

Three commonly employed methods of providing bias to a silicon controlled switch are shown in Fig. 12-15. In Fig. 12-15a a resistor R_G is connected from gate to cathode. In Fig. 12-15b the resistor is returned to a negative supply voltage to raise the firing voltage. The diode is employed to limit the possible back-biasing voltage across the cathode junction. This precaution is necessary since the maximum allowable reverse voltage at the

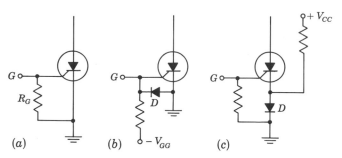

Fig. 12-15 Biasing methods for the SCS. The switch is fired by applying a signal (either d-c or pulse) to gate input terminal G.

n-emitter junction is often not more than a few volts. In Fig. 12-15c the voltage drop across the forward-biased diode provides negative bias for the switch. The positive supply serves to provide current to keep the diode forward-biased.

Suppose that a supply voltage is applied through a load resistor between anode and cathode of a silicon controlled switch. Consider that the bias is such that the applied voltage is less than breakover voltage. Then the switch will remain OFF and may be turned ON by the application to the gate of a triggering current or voltage adequate to lower the breakover voltage to less than the applied voltage. The switch having been turned on, it latches, and it is found to be impractical to stop the conduction by reverse-biasing the gate. For example, it may well be that the reverse gate current for turn-off is nearly equal to the anode current. Ordinarily the most effective and commonly employed method for turn-off is, temporarily at least, to reduce the anode voltage below the holding voltage V_H or equivalently to reduce the anode current below the holding current I_H. The gate will then again assume control of the breakover voltage of the switch.

12-7   SCS CHARACTERISTICS

Silicon controlled switches are ordinarily available in types that allow continuous anode currents up to about 1.0 A and maximum breakover voltages in the range from about 30 to 200 V.

The anode current which flows when reverse voltage is applied between anode and cathode is small, being of the order of several thousandths of a microampere in a unit with low leakage to about 1 μA in other units. The magnitude of this current is quite comparable to the current which flows when the anode-cathode voltage is in the forward direction but the switch is in an OFF state. Both these currents, forward and reverse, increase with temperature in a manner similar to the reverse saturation current in a transistor.

The ratio of the continuous allowable anode current to the forward gate current required to turn the switch ON even at low anode-to-cathode voltage is rarely less than several hundred, and in specially sensitive switches may attain values approaching 50,000. Thus, triggering currents of tens of microamperes may be enough to turn ON a switch which will then carry continuously hundreds of milliamperes. The impedance seen looking into the cathode-gate terminals is that of a forward-biased silicon diode. As we observe from Fig. 6-2, no appreciable diode current flows in a silicon diode except for voltages in excess of about 0.6 V. Accordingly, we find that required triggering voltages are usually of the order of 0.6 V. The firing current and voltage required decrease with increasing temperature.

The holding voltage at room temperature is approximately 1.0 V and the holding current lies in the range from less than a milliampere to several tens of milliamperes depending on the size of the unit. Both the holding voltage and holding current decrease with temperature. The incremental resistance between anode and cathode on the ON state is usually less than 1 Ω and may be as low as several tenths of an ohm. The holding current is affected by the gate bias. Increasing this bias increases the holding current because the more negative bias diverts out of the gate terminal some of the internal feedback current that the switch requires to maintain itself in the ON state.

Rate Effect   Silicon controlled switches suffer from the same *rate effect* as do four-layer diodes (page 464). The inclination of a switch to fire prematurely because of the rate effect may be suppressed by operating the switch with a larger reverse bias on the gate and by reducing the resistance between gate and cathode. Both these measures, of course, reduce the sensitivity of the switch to an externally impressed triggering signal. An additional effective method is to bypass the gate to the cathode through a small capacitance. This component will shunt current past the cathode-gate junction in the presence of a rapidly varying applied voltage, but will have no effect on the d-c operation of the switch.

When both gate terminals G_A and G_C are externally available, the circuit[11] of Fig. 12-16 may be employed to suppress the rate effect entirely. In this circuit, the anode gate has been returned to the supply voltage through a resistor R_{GA}. The switch S is not essential to the circuit and has been included only to facilitate the discussion to follow.

Assume initially that switch S is open and that the SCS is OFF. The resistance R_{GA} is large enough not to affect materially the voltages on the various layers of the silicon switch. The capacitor C represents the capacitance across the center junction of the switch when this junction is reverse-biased. When a triggering signal is applied at the cathode gate G_C, the voltage at the anode A drops, as does also the voltage at the anode gate G_A. The SCS may now be reset to the OFF state by closing switch S, since, with S closed, the anode voltage and current are reduced below the holding values. As long as switch S is closed the anode must remain at ground voltage even though the

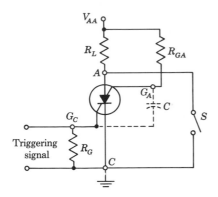

Fig. 12-16 A silicon-controlled-switch circuit which eliminates the rate effect.

SCS is OFF. The anode gate, however, is not so restrained and will begin to rise toward V_{AA} with a time constant nominally equal to $R_{GA}C$ (for $R_G \ll R_{GA}$). The rising anode-gate voltage will charge the capacitor C. It will also reverse-bias the anode-to-anode-gate junction because A is at ground potential and G_A is rising toward V_{AA}.

If the switch S is now opened, the anode voltage will rise abruptly, being limited in its speed dv/dt only by whatever incidental capacitance may be present at the anode. If it were not for the fact that the anode gate G_A is already at the supply voltage, this rapid rise in anode voltage might cause a premature firing. As noted in Sec. 12-5, premature firing results from the current which passes through the two forward-biased junctions and the capacitance C of the reverse-biased junction. In the present case, however, this rapid rise in voltage is transmitted through a reverse-biased rather than a forward-biased anode junction. Consequently a smaller cathode junction current flows. Furthermore, so long as the anode junction is reverse-biased the negative-resistance p-n-p-n characteristic which is responsible for the switching cannot develop since forward-biased anode and cathode junctions are required. The net result is that, provided enough time is allowed after the closing of switch S for the anode-gate voltage to rise, the switch is entirely free of rate effect at the anode. In a typical case $R_{GA} \approx 100$ K and $C \approx 5$ pF, so that the time constant is $R_{GA}C = 0.5$ μsec.

The connection of R_{GA} to the supply V_{AA}, as in Fig. 12-16, will suppress the rate effect against a rapid rise in anode voltage such as would result from the opening of switch S. However, with switch S open the rate effect might again make itself felt if a positive transient appears at the anode supply voltage. In this case the capacitor C would restrain the anode gate from rising as rapidly as the anode, and the anode junction would not be reverse-biased. This situation may be corrected by simply returning the resistor R_{GA} to a supply voltage higher than the anode supply.

Gate ON and OFF Times The process by which the SCS changes state occupies a finite time interval. When a triggering signal is applied to a gate to turn a switch ON, a time interval, the *turn-on* time, elapses before the transition is completed. This turn-on time decreases with increasing amplitude of trigger signal, increases with temperature, and increases also with increasing anode current. If the triggering signal is a pulse, then, to be effective, not only must the pulse amplitude be adequate but the pulse duration must be at least as long as a critical value called *gate time to hold*. Otherwise, at the termination of the gating pulse the SCS will fall back to its original state. A similar situation applies in driving the switch OFF by dropping the anode voltage. At a minimum, the anode voltage must drop below the holding voltage. If, however, the anode voltage is driven in the reverse direction the *turn-off* time may thereby be reduced. The turn-off time increases with temperature and with increasing magnitude of anode current. Further, the anode voltage must be kept below the maintaining voltage for an interval at least as long as a critical value called the *gate recovery time* if the transition is to persist after the anode voltage rises.

In fast units, all of the time intervals are in the range of tenths of microseconds, whereas in slower units these times may be as long as several microseconds. In general, the time required to turn a switch OFF is longer than the time required to turn it ON. The time intervals involved are required to allow for the establishment and dissipation of stored charge in the base regions, very much as in the case of the transistor. The matter of charge storage in transistors is discussed in detail in Chap. 20.

12-8   ADDITIONAL FOUR–LAYER DEVICES[12]

The *Binistor, Trigistor,* and *Transwitch* (commercial names) are *p-n-p-n* silicon devices which are in almost every respect identical to the SCS described above. They differ only in that, by design, they are more readily turned off by a negative triggering signal at the gate terminal. Their sensitivity to a turn-off trigger is an order of magnitude or so smaller than to a turn-on signal. Voltage ratings, temperature dependence, and switching-speed characteristics are not unlike those described for the SCS. Bistable operation of these switches results when the gate is driven by a train of alternate positive and negative pulses. When the speed attainable with these devices is adequate, they offer the great convenience of providing a bistable circuit which involves only a single active device, a minimum of components, and the feasibility of triggering to either state at one triggering input terminal.

The silicon controlled rectifier (SCR)[10] is a three-terminal silicon controlled switch with the principal difference that the rectifier is mechanically larger and designed to operate at higher currents and voltages. Currents in excess of 100 A and operating voltages approaching 1,000 V are possible. The rectifiers are used to control large amounts of power, whereas the switches are intended for low-level logic and switching applications. The switches are characterized not only by smaller voltage and current ratings but also by comparatively low leakage and holding currents. In addition, the switches are designed to require smaller triggering signals and have triggering characteristics which are more uniform from sample to sample of a given type.

12-9　　　THE THYRISTOR[13]

This germanium three-layer device is similar in construction to a mesa diffused-junction p-n-p transistor.　The essential difference between a transistor and a thyristor is in the nature of the collector contact.　In the thyristor the contact consists of a nickel tab soldered at high temperature to the germanium of the collector with an alloy of lead, tin, and indium.　At low currents this combination behaves as an ohmic contact which serves simply to collect the holes that have diffused across the base and entered the collector region.　The special nature of this contact is that at high currents it acts to inject electrons into the collector.　Thus, at high currents the collector tab behaves like an n-type layer, making the transistor, in effect, a four-layer device.　Typical collector characteristics for a thyristor are shown in Fig. 12-17.　The dashed portions represent regions of negative resistance.　Observe that the curves for low currents are those of a conventional p-n-p transistor, whereas for higher currents the characteristics are generally similar to the silicon controlled switch.

It turns out to be feasible to turn the thyristor OFF as well as ON by the application of a triggering signal at the base.　The thyristor is like the SCS in that a larger triggering signal (by a factor of about 10) is required to turn it OFF than to turn it ON.　The thyristor has the advantage of being faster.　Turn-on times of 25 nsec and turn-off times of 100 nsec are not uncommon.　Since the device material is germanium instead of silicon, the holding collector-to-emitter voltage (≈ 0.3 V) is smaller than in an SCS.

The name *thyristor* is intended to indicate a transistor replacement for the thyratron (the gas-filled hot-cathode triode).　All the switches discussed above, however, are analogous to the thyratron and are replacing it in most applications.　These semiconductor devices have the advantages of smaller triggering requirements, microsecond switching, no heater power,

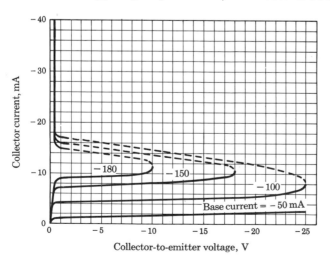

Fig. 12-17　Typical volt-ampere characteristics of a 2N1214 thyristor for various values of base current.　(Courtesy of Radio Corporation of America.)

no vacuum seal, and smaller size. In addition, unlike the thyratron, some of the switches may be turned off with a gate signal. In terms of the concepts introduced in this chapter a thyratron may be described as a device possessing a current-controlled negative-resistance characteristic which latches upon the application of a positive pulse of voltage on its grid, but which can be turned off only by reducing its anode current below the value required to maintain the ionization.

12-10 AVALANCHE–MODE OPERATION OF TRANSISTORS[14]

We observe in Figs. 6-18 and 6-20 that when the collector-to-emitter voltage of a transistor is raised to the level where avalanche breakdown occurs, the volt-ampere characteristic of the transistor exhibits a current-controlled negative-resistance characteristic. The similarity between Fig. 6-18 and Fig. 12-13 for the silicon controlled switch is readily apparent. Thus a transistor may be induced to make an abrupt switching transition from an OFF state to an ON state much in the manner of any of the other current-controlled devices we have considered in this chapter. There are two important differences between an avalanche-mode-operated transistor and the other switches. The first difference concerns the switching time. The OFF to ON switching time of the other switches is in the range of about 0.1 μsec (100 nsec), whereas the switching time for an avalanche transistor may be of the order of a few nanoseconds. The only competitors for speed with an avalanche transistor are the tunnel diode and snap-off diode (Sec. 20-7). The ON to OFF switching time of an avalanche transistor is quite long in comparison with the OFF to ON time, and such turn-off switching is not used when speed is at a premium. The normal procedure for turning OFF an avalanche transistor is to drop the collector voltage and, correspondingly, the collector current to a point where the current is no longer able to sustain the avalanche discharge, that is, to a point below the holding current.

The second important difference concerns the holding voltage of the avalanche transistor. In the four-layer switch the maintaining voltage is of the order of 1 V down to several tenths of a volt. Thus when the switch is conducting, the power dissipation may be quite small. In the avalanche transistor the holding voltage may be comparable to the breakdown voltage. Thus in a transistor biased to avalanche at, say, 100 V, the latching voltage may be 50 V. As a consequence, when the transistor is ON the dissipation may be comparatively large and thus limit the ON time of the transistor. For example, if an avalanche transistor is used to generate a train of pulses, and even if the pulse duration is short, being of the order of tens of nanoseconds, the allowable dissipation may severely limit the pulse repetition frequency.

Many transistors originally intended for other purposes will "avalanche," that is, respond with speed in avalanche-mode operation. Generally, those transistors do best which were designed for fast switching applications. In addition, there are available a small number of transistor types such as the Texas Instruments 2N3033 which have been specifically designed for high-current, high-speed, avalanche-mode operation.

REFERENCES

1. Esaki, L.: New Phenomenon in Narrow Ge *p-n* Junctions, *Phys. Rev.*, vol. 109, p. 603, 1958.

2. Millman, J., and C. C. Halkias: "Electronic Devices and Circuits," McGraw-Hill Book Company, New York, 1967.

3. "Tunnel Diode Manual, TD-30," Radio Corporation of America, Semiconductor and Materials Division, Somerville, N.J., 1963.
   "Tunnel Diode Manual," 1st ed., General Electric Company, Semiconductor Products Dept., Liverpool, N.Y., 1961.

4. MIL-STD-15-1A, May, 1963.

5. Sylvan, T. P.: *Appl. Note* 90.10, General Electric Company, Syracuse, N.Y., May, 1961.

6. Harding, M., and R. Windecker: Small Signal Planar PNPN Switch, Texas Instruments Company, Inc., August, 1963.
   Shockley, W., and J. F. Gibbons: Introduction to the 4-layer Diode, *Semicond. Prod.*, 1958.
   Shockley, W.: The Four Layer Diode, *Electron. Ind.*, vol. 16, no. 8, pp. 58–60, August, 1957.
   Gibbons, J. F.: A Critique of the Theory of *p-n-p-n* Devices, *IEEE Trans. Electron Devices*, vol. ED-11, no. 9, pp. 406–413, September, 1964.

7. Moll, J. L., M. Tannenbaum, J. M. Goldey, and N. Holonyak: *p-n-p-n* Switches, *Proc. IRE*, vol. 44, pp. 1174–1182, 1956.
   Sah, C. T., R. N. Noyce, and W. Shockley: Carrier Generation and Recombination in *p-n* Junctions and *p-n* Junction Characteristics, *Proc. IRE*, vol. 45, pp. 1228–1243, September, 1957.

8. Rate Effect, the Voltage-Current Characteristics of Four-layer Diodes at High Frequencies, *Appl. Data Note*, Shockley Transistor Corporation, May, 1959.

9. "Transistor Manual," 7th ed., chap. 16, General Electric Company, Syracuse, N.Y., 1964.
   Stasior, R. A.: Silicon Controlled Switches, *Appl. Note* 90.16, General Electric Company, Syracuse, N.Y., September, 1964.

10. "Silicon Controlled Rectifier Manual," 3d ed., General Electric Company, Auburn, N.Y., 1964.

11. Stasior, R. A.: How to Suppress Rate Effect in PNPN Devices, *Electronics*, vol. 37, no. 2, pp. 30–33, Jan. 10, 1964.

12. A Survey of Some Basic Trigistor Circuits, *Bull.* D410-02, Solid State Products, Inc., Salem, Mass., March, 1960.
    The Transwitch, Circuit Design Information and Application Notes, *Bull.* AN-1357A, March, 1960, and The Binistor, Circuit Design Information and Application Notes, *Bull.* AN-1360A, August, 1960, Transitron Electronic Corp., Wakefield, Mass.

13. RCA Thyristors, *Bull.* 1CE-208, Radio Corporation of America, Somerville, N.J., March, 1960.
    Mueller, C. W., and J. Hilibrand: The "Thyristor"—A New High-speed Switching Transistor, *IRE Trans. PGED*, vol. ED-5, pp. 2–5, January, 1958.

14. Henebry, W. M.: Avalanche Transistor Circuits, *Rev. Sci. Instr.*, vol. 32, no. 11, pp. 1198–1203, November, 1961.
    Miller, S. L., and J. J. Ebers: Alloyed Junction Avalanche Transistors, *Bell System Tech. J.*, vol. 34, pp. 883–902, September, 1955.

13 / NEGATIVE–RESISTANCE SWITCHING CIRCUITS

We consider, in this chapter, circuit applications of the negative-resistance devices introduced in Chap. 12. Certain generalized circuit properties of negative-resistance devices are described first. We then establish the principles which allow us to determine the mode of operation of the circuit: bistable, monostable, or astable. Finally, these general principles are applied to account for the particular behavior of circuits employing the negative-resistance devices described in the preceding chapter.

13-1   THE NEGATIVE–RESISTANCE CHARACTERISTIC[1]

The devices of interest to us display, between a selected set of terminals, a volt-ampere characteristic such as is represented, in somewhat generalized form, in either Fig. 13-1a or b. In Fig. 13-1a, between 0 and A and to the right of the point B the device has a positive incremental resistance, whereas between A and B the incremental resistance is negative since, as may be noted, an *increase* in voltage causes a *decrease* in current. Similarly, in Fig. 13-1b, the portion of the characteristic between A and B displays a negative incremental resistance. For the sake of simplicity, and with no loss in generality of principle, we have made the characteristics piecewise linear and have arranged that they pass through the origin. This feature is in no way essential to the present discussion, although actually we see in Chap. 12 that all characteristics except that of the unijunction transistor do indeed pass through the origin. In addition, the characteristics have the general form shown in Fig. 13-1 without being piecewise linear.

476

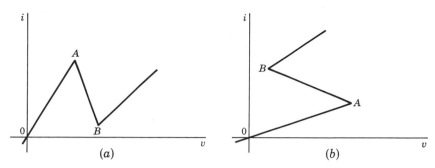

Fig. 13-1 (a) A voltage-controllable negative-resistance characteristic; (b) a current-controllable negative-resistance characteristic.

We observe in Fig. 13-1a that associated with each voltage there is a unique current, but the plot does not everywhere associate a unique voltage with each current. In the plot of Fig. 13-1b the inverse applies. To distinguish the one from the other we call the characteristic in Fig. 13-1a *voltage-controllable* and in Fig. 13-1b *current-controllable*. The tunnel diode falls into the voltage-controllable class, whereas all of the other devices discussed in the preceding chapter (the UJT, *p-n-p-n* diode, SCS, thyristor, and avalanche transistor) have a current-controllable characteristic.

13-2 BASIC CIRCUIT PRINCIPLES

We shall show that a device with a region of negative incremental resistance may be used to construct a switching circuit. The two classes of negative-resistance (NR) devices must be considered separately since the basic circuit of one type is essentially the dual of the other. We shall consider the current-controlled NR device first. The basic circuit is indicated in Fig. 13-2a, where a source of voltage V and a resistor R are shown connected to an NR device with the volt-ampere characteristic of Fig. 13-2b. Shunted across the NR device is a capacitor C, which may represent stray capacitance or capacitance deliberately introduced.

The voltage v across the device is the supply voltage V less the drop across R. If i_R is the current through R, then

$$v = V - i_R R \tag{13-1}$$

The device current is i and the capacitor current is i_C. We are interested in the steady state corresponding to a particular value of supply voltage, say $V = V_1$. When the currents and voltages in the circuit stop changing, $i_C = 0$ and $i_R = i$. Hence, on the same set of coordinate axes on which is plotted the device characteristic we have drawn the load line corresponding to Eq. (13-1) with $i_R = i$. This line is plotted in Fig. 13-2b as the solid line

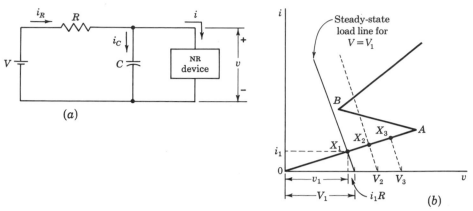

Fig. 13-2   (a) A circuit involving a current-controllable negative-resistance (NR) device; (b) the load line corresponding to supply voltage $V = V_1$ and resistance R is superimposed on the NR device characteristic in a manner to yield one stable equilibrium point at X_1.

passing through the point $v = V_1$, $i = 0$, and having a slope $-1/R$.   Under steady-state conditions, the point of operation of the circuit must lie on the device characteristic and simultaneously on the load line.   Hence, corresponding to the particular supply V_1 the steady-state current and voltage are i_1 and v_1, respectively, corresponding to point of intersection X_1 of the load line and the device characteristic.

Now let there be added to the supply a step of voltage which makes the new supply voltage $V = V_2$.   The new load line is shown dashed, and the new steady-state operating point will be at X_2.   A time will elapse before this new steady-state condition is reached, since the capacitor must change its voltage.   The capacitor charges through an equivalent resistance which is equal to the parallel combination of R and the (positive) resistance of the NR device over the region OA.   We shall now prove that in response to the change in supply voltage from V_1 to V_2 the operating point of the device moves from its original position at X_1 in the direction toward X_2.   This result seems obvious enough at the present time, but in establishing the proof, we shall arrive at a result which will be quite useful in the following discussion. We have from Fig. 13-2a that

$$i_R = i_C + i = C\frac{dv}{dt} + i \tag{13-2}$$

Combining Eq. (13-2) with Eq. (13-1) we find that

$$RC\frac{dv}{dt} = V - (iR + v) \tag{13-3}$$

Now suppose that the NR device is operating at a particular point X_E where $i = i_E$ and $v = v_E$. Then $V_E = i_E R + v_E$ is the supply voltage which would make operation at i_E and v_E an equilibrium point. Suppose, however, that the supply voltage is not V_E but is instead V_S. Then Eq. (13-3) may be written

$$RC \frac{dv}{dt}\bigg|_{X_E} = V_S - V_E \tag{13-4}$$

Expressing Eq. (13-4) in words, we have the following theorem: *If the device is operating at X_E, which would be an equilibrium point if the supply voltage were V_E, but if in reality $V = V_S$, then dv/dt is positive if $V_S > V_E$. Alternatively, if $V_S < V_E$ then dv/dt is negative.* Applying this result to Fig. 13-2b we note that when the supply voltage is abruptly increased to $V_2 (= V_S)$, then since $V_2 > V_1 (= V_E)$, dv/dt is positive and the operating point moves to the right along the device characteristic in the direction of increasing v.

We may use Eq. (13-4) to prove that a point such as X_2 in Fig. 13-2b, corresponding to the intersection of the device characteristic with the load line passing through X_2, is a *stable* operating point. That is, if the circuit is perturbed in some manner so that the operating point is caused to depart from X_2, it will return to X_2. Such a perturbation might be caused by removing the capacitor, changing the voltage across it, and then replacing it in the circuit. Suppose that as the result of such a maneuver the operating point is established momentarily at $X = X_1$ in Fig. 13-2. Then Eq. (13-4) with $V_S = V_2$ and $V_E = V_1$ indicates $dv/dt\,|_{X_1} > 0$. Hence there will be an increase in v which will carry the operating point back to X_2. Similarly, if X were located to the right of X_2, say at $X = X_3$, then Eq. (13-4) would become, with X_E replaced by X_3, $V_E = V_3$, and $V_S = V_2$,

$$RC \frac{dv}{dt}\bigg|_{X_3} = V_2 - V_3 \tag{13-5}$$

Since $V_2 < V_3$, then $dv/dt < 0$ and v decreases so as to carry the operating point from X_3 back to X_2. We have thus verified that X_2 is a stable point regardless of whether a disturbance momentarily increases or decreases the voltage across the device.

We shall now apply the above principles to show that the basic circuit of Fig. 13-2a may be made to function in a monostable, bistable, or astable mode, depending upon the biasing method.

13-3   MONOSTABLE OPERATION

Consider the situation depicted in Fig. 13-3. The supply voltage is initially V_1, and the steady-state operating point is at X_1. A voltage step added to V increases the supply voltage to V_2, carrying the new load line beyond the critical point A, so that a new steady state will be established at X_2. The

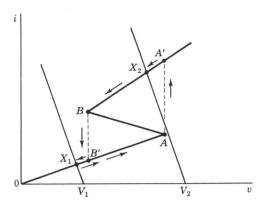

Fig. 13-3 Illustrating the abrupt transitions which occur when the load line moves beyond a critical point such as A or B.

operating point of the device must, of course, remain on the device characteristic as it moves toward X_2. Accordingly, the operating point initially moves to the point A. We may now apply the principle of Eq. (13-4) to establish that dv/dt is positive at A. Hence, the operating point cannot continue its approach toward X_2 by moving along AB, because along this path dv/dt would be negative. The only alternative is to have the operating point jump abruptly vertically from A to A'. By applying Eq. (13-4) at A' we find that dv/dt is negative. The operating point may therefore continue toward X_2 by moving toward the left along $A'B$. The device has, by taking this operating path, avoided behaving inconsistently with Eq. (13-4), but there now exists an anomalous situation in that the path has departed from the device characteristic. We may, however, take a philosophical attitude toward this matter and judge that if the transition from A to A' is completed in *zero time*, then actually the operating point has not really left its characteristic. Returning to the circuit of Fig. 13-2a we find no inconsistency in making the abrupt jump from A to A'. Specifically, the voltage has not changed instantaneously, as indeed it may not, because of the capacitor. The current has changed abruptly. There is nothing in this idealized circuit to restrain it from so doing.

With a physical NR device, we expect and observe, in a corresponding situation, a rapid but hardly instantaneous transition. First, realistic circuits and devices have all sorts of stray inductances and capacitances not contemplated in the discussion above. Second, a physical device has an NR region which does not really exist initially at all but is rather generated when a certain critical voltage or current is reached. To complete creating this region will ordinarily require a finite time interval. For example, in an avalanche transistor a negative resistance results from the cumulative generation of current carriers resulting from the disruption of atomic bonds. A finite time is required for this cumulative avalanche process to build up and for certain other redistributions required of the stored charge in the transistor. Thus when stray inductance or capacitance is reduced to a minimum the speed of the transition depends very greatly on the mechanism internal to the device by which the NR region is generated.

A discussion analogous to that given above leads to the conclusion that if, after X_2 has been attained, the supply voltage is returned to V_1, then the operating point will return to X_1. From X_2 the operating point will move relatively slowly to B, make an abrupt transition to B', and finally return to X_1. The complete path of the operating point is indicated by the arrows.

In order to induce this transition from the original stable low-current operating point X_1 to a temporary high-current state and back again, it is only necessary that the step of voltage have an amplitude adequate to carry the load line beyond A and last long enough to allow the operating point to reach the point A'. Thus the total transition to a quasi-stable state and back again may be induced by a pulse. Accordingly, we have just described the operation of a *monostable* generator of fast waveforms.

One-shot operation is also possible by establishing a quiescent voltage more positive than that at point A and then applying a negative pulse of large enough magnitude and long enough duration to carry the operating point to the left of B.

We have achieved triggering by superimposing a pulse on the supply voltage to carry the load line beyond a critical point A. In many physical devices (those with additional terminals other than the ones between which the negative-resistance characteristic appears) a triggering pulse applied to one of these extra terminals will achieve triggering by modifying the device characteristic. Thus instead of moving the load line to the right of point A, the triggering pulse may achieve triggering by moving the critical point to the left of X_1 while the load line remains stationary.

13-4   BISTABLE OPERATION

A bistable circuit is generated by selecting a supply voltage and load resistor that give rise to a load line which intersects the device characteristic in three places, X, Y, and Z, as in Fig. 13-4. If the circuit is initially at X, then a positive pulse added to the supply voltage will carry the operating point

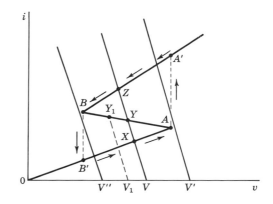

Fig. 13-4 The load line passing through V and intersecting the characteristic in three places allows bistable operation.

in the direction of the arrows to the stable operating point Z, where the circuit will remain permanently. Again the pulse must be adequate in amplitude to carry the load line beyond point A, as, for example, is the case with the load line passing through V'. Also, the pulse duration must be long enough to allow the operating point to reach A'. After the circuit is established at Z a negative pulse that carries the load line to the left, as is the case with the load line passing through V'', will return the circuit to X. Thus the circuit has two stable states and may be triggered from one state to the other by applying alternate positive and negative pulses.

The intersection point at Y on the negative-resistance portion of the characteristic is also a stable point, as may readily be verified. Consider, for example, that a momentary disturbance has shifted the operation to point Y_1. Then, since $V_1 < V$, Eq. (13-4) indicates that dv/dt is positive. Hence v increases and the operating point moves from Y_1 toward Y, thus confirming that Y is a stable point. Operation at Y, however, will ordinarily not be attained because in the triggering scheme we have described it is simply not accessible.

13-5   ASTABLE OPERATION

An astable waveform is generated by selecting a supply voltage and load resistor such that the load line intersects the characteristic at a single point X on the NR portion of the characteristic, as in Fig. 13-5. The important distinction between the present case and the situation which gave rise to the bistable circuit is that in the present case the equilibrium point at X is unstable. If a perturbation causes an initial departure of the operating point from X, the subsequent response of the circuit is to carry the operating point still farther from the equilibrium point. That such is the case in the present instance may be seen by an application of Eq. (13-4). Thus, suppose that with the equilibrium point at X a perturbation moves the operating point to X_1. Since now $V_1 > V$, we see, from Eq. (13-4), that dv/dt is correspondingly negative, and the operating point which momentarily was displaced to X_1

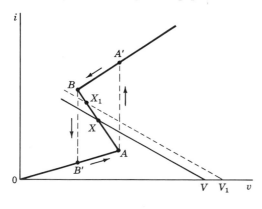

Fig. 13-5   Adjustment of the load line for astable operation.

Fig. 13-6 Circuit equivalent to the network of Fig. 13-2a for small voltage excursions in the negative-resistance region.

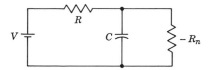

moves still farther from X toward B. Similarly, if X_1 were below X, the operating point would move toward A.

We now take up a second way of establishing the instability of the equilibrium point at X. For this purpose consider the equivalent circuit of Fig. 13-6, where, since the NR device is operating in its NR region, we have replaced the device by a negative resistance of magnitude R_n. This configuration, with $V = 0$, is an equivalent circuit for the purpose of computing incremental changes from an initial equilibrium state. The circuit has a single time constant given by $R_{\parallel}C$, where $R_{\parallel}$ is the parallel combination of R and $-R_n$, that is,

$$R_{\parallel} = \frac{-R_n R}{R - R_n} \tag{13-6}$$

Now $R_{\parallel}$ is positive or negative as R_n is larger or smaller than R. If an equilibrium state is perturbed, the subsequent response will be of the form $\epsilon^{-t/R_{\parallel}C}$. If $R_n > R$, then $R_{\parallel} > 0$, and the circuit responds in accordance with a negative exponential back to its equilibrium state. The equilibrium state is then stable. If $R_n < R$, then $R_{\parallel} < 0$, and the response has the form of an exponential with a positive exponent. The operating point therefore moves farther away from its initial equilibrium state, picking up speed as it goes. We observe that in Fig. 13-5, $R_n < R$, which is consistent with our previous conclusion that the equilibrium state is unstable. We may also note that again, as expected, on the basis of our present manner of establishing stability, the state at Y in Fig. 13-4 is stable because $R_n > R$.

Returning now to Fig. 13-5 we observe that, starting from an initial state at X, a small disturbance will start the circuit toward either B or A, depending on the direction of the disturbance. Thereafter, the system will trace out the path $BB'AA'B$ in the direction of the arrows, with no need for continued external triggering. The circuit is astable and generates two abrupt current transitions per cycle.

We noted above that a current-controlled NR device would be stable even in its negative-resistance region provided that $R_n > R$. If we desire stability no matter what the value of R_n may be, we require that $R = 0$. For this reason a current-controlled NR device is often characterized as being "short-circuit stable." Similarly, we shall see later that a voltage-controlled device is "open-circuit stable."

Both of the arguments given above to prove that X is an unstable point depended upon the existence of a capacitance across the NR device. If, ideally, C could be reduced to zero, then stable operation at point X would be possible. In practice it turns out that if C is smaller than a certain "critical" value, then a stable state at X is indeed obtained. This critical capacitance for most NR devices is of the order of some tens of picofarads, although its exact value is difficult to calculate. In order to do so it is necessary to include in Fig. 13-6 the inductance of the device along with wiring inductances and then solve for the minimum value of the capacitance that gives a positive real part to the roots of the equation determining the transient response (or, equivalently, to the poles of the network function). In summary, if the shunt capacitance across a current-controlled NR device is below some tens of picofarads, then a point on the NR portion of the characteristic is stable. However, for capacitances larger than critical it becomes unstable.

13-6   VOLTAGE–CONTROLLED NEGATIVE–RESISTANCE SWITCHING CIRCUITS

We consider now the voltage-controlled NR device whose characteristic appears again in Fig. 13-7b. The appropriate circuit for switching operation is now shown in Fig. 13-7a. Comparing Fig. 13-7a with Fig. 13-2a we note that the shunt capacitor has been replaced by a series inductor. Analogously to the current-controlled case, load lines passing through supply voltages V, V', and V'' are appropriate to astable, bistable, and monostable operation, respectively. Note that the monostable line intersects the device characteristic at one point along a positive-resistance portion. The bistable line has three points of intersection, two along positive-resistance branches and the third on the negative-resistance section. The astable line intersects the characteristic at one point along the negative-resistance segment.

The stability of an operating point may be investigated in one of two ways. First, from Fig. 13-7a we have

$$L \frac{di}{dt} = V - iR - v \tag{13-7}$$

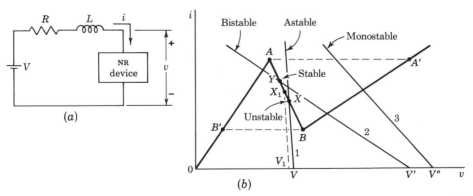

Fig. 13-7 (a) A circuit involving a voltage-controlled negative-resistance device; (b) characteristic and load lines for bistable, astable, and monostable operation.

If the operating point is momentarily at $X_E(i_E, v_E)$, where $i_E R + v_E = V_E$, and if the supply voltage is V_S, then this equation may be written

$$L \frac{di}{dt}\bigg|_{X_E} = V_S - V_E \qquad (13\text{-}8)$$

analogously to Eq. (13-4). Expressing Eq. (13-8) in words, we have the following theorem: *If the device is operating at X_E, which would be an equilibrium point if the supply voltage were V_E, but if in reality $V = V_S$, then di/dt is positive if $V_S > V_E$. Alternatively, if $V_S < V_E$ then di/dt is negative.* Let us apply this principle to Fig. 13-7b. Consider, for example, load line 1, which intersects the NR characteristic at X. Is this a point of stable or unstable equilibrium? If a disturbance takes the circuit to point X_1, then since

$$V_S = V > V_1 = V_E$$

di/dt is positive. Hence, the operating point X_1 must move in the direction of increasing current, away from X and toward A. This argument establishes that X is an unstable point. Hence, if the circuit is adjusted to have a load line such as 1 in Fig. 13-7b, it will operate in an astable mode. Applying the above principle to the conditions depicted by load line 2, we can verify that Y is a stable point.

Second, we recognize that Fig. 13-8 gives the equivalent circuit from which to calculate the response to small perturbations. Again we have a one-time-constant circuit which will yield a response of exponential form $\epsilon^{-(R-R_n)t/L}$. The exponent will be positive (unstable case) if $R_n > R$ (load line 1) and negative (stable case) if $R_n < R$ (load line 2). The reverse condition applies for a current-controlled device since in Sec. 13-5 we find that a stable point corresponds to $R_n > R$.

We may note that to ensure stability of equilibrium point on the NR portion of the characteristic, no matter how large R_n may be, we require that R be arbitrarily large. Hence the present case is "open-circuit stable," in comparison with the previous circuit which was "short-circuit stable."

A basic difference between the present voltage-controlled device and the previously discussed current-controlled case is seen in the discontinuous jumps B to B' and A to A'. In the present case the jumps are changes in voltage, whereas in the previous circuit current discontinuities were encountered. It is the reactive element which was added to the circuit (the inductor in the voltage- and the capacitor in the current-controllable-device circuit) which determines the type of abrupt transition. As we shall see, the time intervals established in the monostable and astable circuits are determined by this reactive element. In the bistable circuit, where no timing interval is to be

Fig. 13-8 Equivalent circuit for calculating response of circuit of Fig. 13-7a for small excursions in the negative-resistance region.

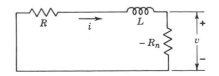

established, the addition of a reactive element is theoretically unnecessary. However, it is found in Sec. 13-10 that an improvement in the transition time results if a small amount of reactance is added in some circuits: for example, a small inductance in series with a tunnel diode.

The remainder of this chapter makes use of the principles enunciated above to explain the operation of switching circuits using the NR devices discussed in the preceding chapter. Calculations of the timing intervals established in monostable and astable waveforms and of the transition time in a bistable circuit are also given.

13-7   TUNNEL–DIODE MONOSTABLE CIRCUIT[2,3]

A monostable tunnel-diode circuit is shown in Fig. 13-9. Bias is provided by the source V in Fig. 13-9a, which is adjusted so that the load line intersects the characteristic at one point on the positive-resistance portion. The operating point is initially at point O, where $v = V_o$ and the diode current is $i = I_o$. A positive voltage pulse v_s is applied to raise the load line so that it clears the peak at A. This trigger must have a time duration t_p adequate to allow the current in the inductor to change from I_o to I_P. The operating point having been raised to A, the circuit, of its own accord, follows the path indicated by the arrows, returning eventually to the starting point at O.

Waveforms of output voltage and diode current, somewhat idealized, are shown in Fig. 13-10. The application of the voltage pulse carries the diode from O to A, increasing the voltage from V_o to V_P and the current from I_o to I_P. If the pulse amplitude were large enough so that the asymptotic limit to which the current were headed was much larger than I_P and if the diode resistance in the range from O to A were constant, then the rise of voltage from O to A would be linear as shown.

At A the voltage jumps abruptly to B, where the voltage is V_F, while the current remains constant at I_P. As the operating point now moves from B to C, the voltage drops from V_F to V_V and the current from I_P to I_V. If the

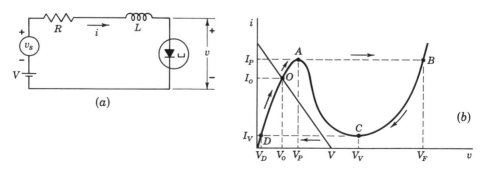

Fig. 13-9   Monostable operation of a tunnel-diode (a) circuit, (b) operating path.

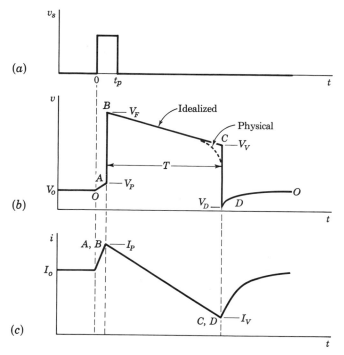

Fig. 13-10 Waveforms of monostable tunnel-diode circuit. (a) Triggering pulse; (b) output voltage; (c) tunnel-diode current.

pulse persisted indefinitely, the load line would establish a stable equilibrium point on the high-voltage positive-resistance portion of the characteristic somewhat below B. We have assumed, however, that long before this equilibrium point is approached the pulse has terminated. Hence the operating point continues from B to C. There is then an abrupt transition in voltage from V_V to V_D at point D, and finally the circuit settles down to initial point O in an asymptotic manner.

If the diode characteristic in the region from B to C were approximated by a constant resistance, the waveform between these points would be exponential and could be readily calculated (see below). We have represented, as an approximation, that the current falls linearly. Since the resistance increases as the operating point moves from B to C, then the time constant L/R decreases as C is approached. Hence, in a physical circuit, the principal respect in which the voltage waveform differs from that of Fig. 13-10b is that the negative slope increases in magnitude as point C is approached, as is shown by the dashed waveform. Also, of course, the voltage rise from V_P to V_F is not instantaneous but is determined by the time required to charge any shunt capacitance present.

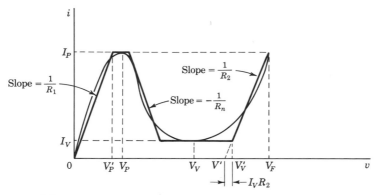

Fig. 13-11 A piecewise linear approximation to a tunnel-diode characteristic (Ref. 2).

A calculation of the duration T of the quasi-stable state, which agrees with experiment[2] to within about 10 percent, can be easily carried out if the tunnel-diode characteristic is represented by the piecewise linear approximation of Fig. 13-11. A reasonable fit with the tunnel-diode curve is obtained if we choose

$$V'_P = 0.75 V_P \quad \text{and} \quad V'_V = \frac{V_F + V_V}{2} \tag{13-9}$$

If the diode resistance of the portion passing through the origin is called R_1 and if the second positive-resistance region is designated by R_2, then

$$R_1 = \frac{V'_P}{I_P} \quad \text{and} \quad R_2 = \frac{V_F - V'_V}{I_P - I_V} \tag{13-10}$$

The line with the negative slope is drawn so as to form a reasonable approximation to the NR segment. Fortunately, neither its location nor its slope affects the delay time T.

To calculate T, the time duration from I_P to I_V, we replace the device by a resistor R_2 in series with a battery $V' \equiv V'_V - I_V R_2$ (as dictated by the piecewise linear approximation of Fig. 13-11). The equivalent circuit is now indicated in Fig. 13-12a and b, which are equivalent since

$$R_T \equiv R + R_2 \qquad V_Y \equiv V' - V = V'_V - V - I_V R_2 \tag{13-11}$$

Let us shift the time origin so that $t = 0$ when $i = I_P$. If the circuit of Fig. 13-12b were valid indefinitely, then at $t = \infty$, $i = -V_Y/R_T$. Since this is a single-time-constant circuit,

$$i = -\frac{V_Y}{R_T} + \left(I_P + \frac{V_Y}{R_T}\right)\epsilon^{-R_T t/L} \tag{13-12}$$

Since i decreases to I_V at $t = T$ we can solve this equation to obtain

$$T = \frac{L}{R_T} \ln \frac{V_Y + I_P R_T}{V_Y + I_V R_T} \tag{13-13}$$

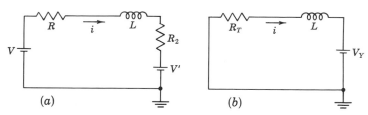

Fig. 13-12   (a) In the region from B to C where the current decreases from I_P to I_V the tunnel diode is replaced by a battery V' in series with a resistor R_2; (b) simplified circuit.

It is clear that the apparently linear decay in Fig. 13-10 is really exponential. Note that the delay time T is proportional to L.   A rough idea of the inductance needed for a desired value of T can be obtained by using the following reasonable order of magnitudes: $V_Y = 0.25$ V, $I_P = 5$ mA, $I_V = 0.5$ mA, and $R_T = 100$ Ω.   We find $T \approx 0.01L$, so that for $L = 100$ μH, $T = 1.0$ μsec.

If it should be true that $I_P R_T \ll V_Y$, then T is independent of the resistance R_T.   To verify this statement we write Eq. (13-13) in the form

$$T = \frac{L}{R_T} \ln \left(1 + \frac{I_P R_T}{V_Y} \right) - \frac{L}{R_T} \ln \left(1 + \frac{I_V R_T}{V_Y} \right) \approx \frac{L}{R_T} \left(\frac{I_P R_T}{V_Y} - \frac{I_V R_T}{V_Y} \right)$$

(13-14)

where use has been made of the expansion $\ln (1 + x) \approx x$.   Hence

$$T \approx \frac{L(I_P - I_V)}{V_Y}$$

(13-15)

The recovery time constant at the end of the waveform is $L/(R_1 + R)$.

A tunnel-diode one-shot which allows one side of the trigger source to be grounded is shown in Fig. 13-13.   Additionally, the voltage division through

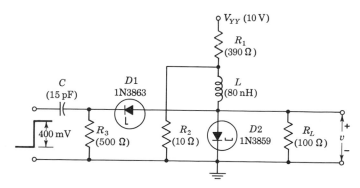

Fig. 13-13   A tunnel-diode monostable multivibrator which allows the trigger source to be grounded.   Typical values are shown for components, supply voltage, and triggering-signal amplitude.   (Courtesy Radio Corporation of America.)

the resistors R_1 and R_2 provides the small supply voltage required for proper bias of the diode even from a relatively large 10-V supply. The quiescent voltage across the diode is only about 250 mV and the input trigger only 400 mV in amplitude. For this reason the input trigger is introduced through a backward diode $D1$ (Sec. 12-2).

13-8   TUNNEL–DIODE ASTABLE CIRCUIT

In Fig. 13-14a the supply voltage and load line have been selected to yield an equilibrium point at O. This point is unstable, and the operating point having moved, say, to point A, it will thereafter follow the circuital path indicated by arrows. Waveforms of diode voltage and diode current are shown in Fig. 13-14b and c. Again these waveforms have been idealized somewhat. If the diode characteristic were piecewise linear the rises and falls of voltage and current in the waveforms would be exponential. In a physical circuit the voltage waveform departs from the idealized waveform much in the manner shown in Fig. 13-10.

The tunnel-diode astable-circuit waveform is not necessarily exactly symmetrical ($T_1 \neq T_2$) because the portions DA and BC of the device characteristic are not identical. On the other hand, the two portions of the cycle are often not markedly different.

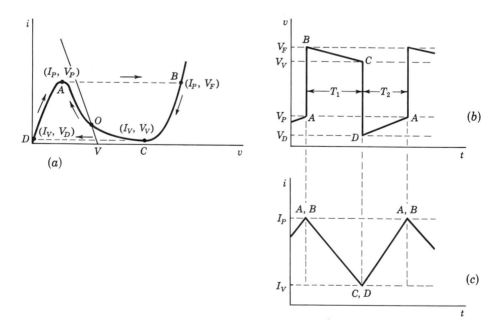

Fig. 13-14 Tunnel-diode astable multivibrator. (a) Adjustment of load line; (b) voltage waveform; (c) current waveform.

Using the piecewise linear approximation of Fig. 13-11, the time T_1 is given by Eq. (13-13). The expression for T_2 will now be found. In the range from D to A, where the current increases from I_V to I_P, the diode may be replaced simply by a resistor R_1. Hence the equivalent circuit of Fig. 13-12b is applicable if $V_Y = -V$ and R_T is replaced by

$$R_T' \equiv R + R_1 \tag{13-16}$$

where R is the load resistance. Proceeding as we did in deriving Eq. (13-13), the duration T_2 of the second portion of the waveform is found to be given by

$$T_2 = \frac{L}{R_T'} \ln \frac{V - I_V R_T'}{V - I_P R_T'} \tag{13-17}$$

The total period is $T_1 + T_2$ and the free-running frequency is the reciprocal of this time. For the parameters given in the preceding section and for a symmetrical waveform the frequency of oscillation is $f = 0.5$ MHz for $L = 0.1$ mH, and f varies inversely with L.

If a square wave of the form shown in Fig. 13-14b is required with the exception that the tilt is not acceptable, such a waveform may be generated by using the astable tunnel-diode circuit to drive a bistable tunnel-diode circuit. Thus, if the waveform of Fig. 13-14b is differentiated with a small RC time constant and if the resulting spikes (alternatively positive and negative) are applied to the flip-flop circuit of Sec. 13-10, an excellent square wave is obtained.

13-9   TUNNEL–DIODE COMPARATOR[4]

A comparator or discriminator (Sec. 7-11) is a circuit which yields an abrupt response when a critical reference voltage or current is attained by a waveform. A tunnel-diode comparator circuit is shown in Fig. 13-15. Bias is provided by the voltage source V in Fig. 13-15a or the current source I in Fig. 13-15b, corresponding to which is the load line 1 in Fig. 13-15c joining I to V with slope $-1/R$. Consider that the initial equilibrium point is at X. When the signal v_s or i_s raises the load line to position 2, where it just clears the peak of the tunnel-diode characteristic, an abrupt transition to V_F will take place and then the diode will settle to the equilibrium point X'. The abrupt transition in output voltage from V_P to V_F constitutes the sharp comparator response. The signal current i_s in Fig. 13-15b at which the comparator responds is $I_P + V_P/R - I$, and the corresponding signal voltage v_s in Fig. 13-15a is $V_P + RI_P - V$.

The tunnel-diode comparator is fundamentally different from the diode comparator discussed in Sec. 7-13. In the rectifying diode comparator the output-voltage change is, at most, equal to the input-voltage change, and comparator action is obtained by suppressing the input signal until the diode

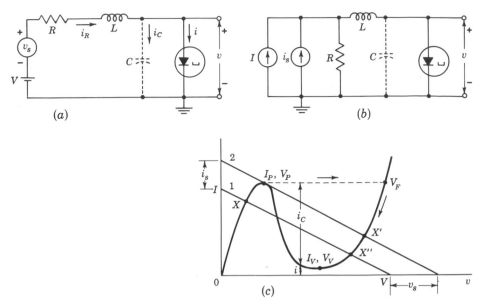

Fig. 13-15 The tunnel diode used as a comparator. In (a) and (b) the sources are represented by Thévenin's and Norton's equivalents, respectively. (c) Illustrating the abrupt response when the load line just clears the peak of the characteristic.

passes through its break point. In the tunnel-diode comparator the output response bears, in principle at least, no relationship to the change of input signal required to activate the comparator. Ideally, the most insignificant change in input voltage produces a change $V_F - V_P$ in the output. Even with physical tunnel diodes the energy required to carry the comparator from one side to the other is remarkably small, being of the order of 10^{-15} J (a pulse 5 mV in amplitude of 2 nsec duration in a 50-Ω load dissipates 10^{-15} J).

The speed of transition of the comparator is of interest. The speed is limited by the inherent junction capacitance of the tunnel diode, which appears dashed in Fig. 13-15a and b. Let us assume that the effect of the inductance predominates in the circuit, so that we may reasonably make the approximation that the inductor current remains constant ($i_R = I_P$) in spite of the presence of the capacitance C. From Fig. 13-15c we see that, over most of the transition, the diode current i is much smaller than the capacitive current i_C. If we neglect i altogether compared with i_C, then the capacitor charges at a constant current I_P, and since the capacitor voltage changes by $V_F - V_P$ the time of transition is $t_r = C(V_F - V_P)/I_P$. The value of $V_F - V_P$ is approximately 0.5 V for germanium and about 1 V for gallium arsenide. We have then

$$t_r = \frac{C}{2I_P} \quad \text{for Ge} \qquad t_r = \frac{C}{I_P} \quad \text{for GaAs} \qquad (13\text{-}18)$$

Manufacturers normally specify I_P/C as a figure of merit for tunnel diodes in connection with switching applications. Typical transition times are in the range of a few nanoseconds down to possibly 0.1 nsec in an extreme case.

13-10   TUNNEL–DIODE BISTABLE CIRCUIT

With the load line selected as in Fig. 13-15c the circuit has two stable states at the points of intersection of the load line with the positive-resistance portions of the device characteristic at X and X''. If the circuit is at X and the signal source furnishes a *positive* pulse adequate to raise the load line to position 2, a transition will occur, and when the pulse has passed, the circuit will find itself at X''. Similarly, a *negative* pulse adequate to drop the load line to the point where it clears the bottom of the characteristic will reset the circuit to its original stable point at X. Thus the circuit has two permanent stable states and may be used to store binary information or for any of the other purposes for which binary devices are employed.

Suppose that we desired to trigger the tunnel-diode flip-flop with a train of pulses (of alternating polarity) at the maximum possible rate. The minimum time between pulses must be the sum of the transition time t_r and the settling time t_s. The former, which is limited by the shunt capacitance, is the time required to change the voltage from V_P to V_F and was estimated to be of the order of 1 nsec in Sec. 13-9. The time t_s, which is determined by the series inductance, is the time it takes the diode to settle from the voltage V_F to the voltage at X'' in Fig. 13-15c. If the current at X'' is $I_{X''}$, then i decreases exponentially from I_P to $I_{X''}$ with a time constant $\tau = L/R_T$. We shall assume, as we did in Sec. 10-5 in connection with the transistor binary, that $t_s \approx 2\tau = 2L/R_T$. For $R_T = 200\ \Omega$, $t_s = 0.01L$. If the settling time is to be no longer than the transition time of 1 nsec, then L must not exceed 100 nH.

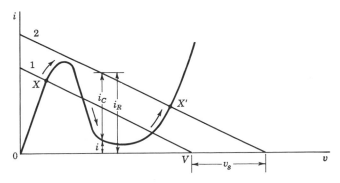

Fig. 13-16   Tunnel-diode bistable operating path if no series inductance is present.

Since t_s is proportional to L and we want t_s to be as small as possible, why not omit L altogether? We shall now show that if no inductance is added to the circuit it will still possess bistable properties but that t_r will be increased. Consider the circuit of Fig. 13-15 but with $L = 0$. The operation is indicated in Fig. 13-16. With $v_s = 0$ load line 1 is applicable and the quiescent point is X. Then an input step is applied and the load line moves to position 2. The steady-state condition is at X', but since the operating point must not leave the tunnel-diode characteristic for any finite interval then it must move from X to X' along the curve, as indicated by the arrows. The time to charge the shunt capacitance is

$$t_r = \int_{V_X}^{V_{X'}} \frac{C\,dv}{i_C} = \int_{V_X}^{V_{X'}} \frac{C\,dv}{i_R - i} \tag{13-19}$$

where i_C, i_R, and i are indicated in Fig. 13-16. In general the integration must be carried out graphically. If a piecewise linear approximation is used for the diode characteristic, then t_r can be evaluated analytically (Prob. 13-22). In Fig. 13-15c, i_C equals the difference between I_P and the diode current i. In Fig. 13-16, i_C equals the difference between the current on the load line corresponding to v and the diode current. We see that i_C is smaller in Fig. 13-16. Hence, t_r will be larger (perhaps by a factor of 2) if no inductor is used in the circuit. Therefore, a small amount of inductance should be added to the circuit so that the rise time may be improved without appreciably increasing the settling time. The required value of inductance is best found experimentally, but the above arguments indicate that a reasonable value to start with is of the order of 100 nH. Of course, the larger the triggering signal v_s, the larger will be i_C and the smaller t_r, and hence the necessity for adding inductance decreases as v_s increases.

13-11   TUNNEL–DIODE—TRANSISTOR HYBRID CIRCUITS[2,3]

A tunnel diode has two qualitites of great merit in switching applications It switches extremely rapidly (≈ 1 nsec) and responds to a pulse of very small energy. We noted earlier that a pulse of 10^{-15} J could be counted on to move the tunnel diode past a critical point and initiate a transition. In comparison, a transistor may require a pulse of 10 times larger energy to switch states. Moreover, a single tunnel diode may be used to construct a circuit with two stable states, whereas two transistors are required for this purpose.

On the other hand, a transistor has the advantage over the tunnel diode that the transistor operates at appreciably higher voltages. The voltages and voltage changes encountered in tunnel diodes are of the order of a few tenths of a volt, but in transistor circuits these voltages are at least tens of volts. Additionally, the tunnel diode has the very disconcerting feature that, having only two terminals, the input and output ports are not isolated from one another.   As a consequence, in circuits that involve cascades of tunnel diodes, it is difficult to ensure that the signal will proceed in one direction only.   It is not surprising, then, that advantages accrue from the employment of combinations of tunnel diodes and transistors in circuits that exploit the merits of each device.

A basic *hybrid* circuit is shown in Fig. 13-17a.   In Fig. 13-17b are plotted the tunnel-diode characteristic (i_N against v), the input characteristic of the transistor (i_B against v), and the composite characteristic ($i = i_N + i_B$ against v).   We observe that the composite input characteristic of the hybrid tunnel-diode–transistor combination retains the general shape of a tunnel-diode characteristic.   That this feature persists depends on the fact that a germanium transistor is being employed.   For in a germanium transistor the base current becomes appreciable in comparison with the diode current at forward voltages of the order of 0.2 V.   This voltage, as indicated in Fig. 13-17b, falls normally between the peak and valley of a tunnel diode (Table 12-1).   If a silicon transistor had been used, the transistor would not yet be turned ON even when the tunnel diode were operating well up on the high-voltage positive-resistance portion of its characteristic.   It is, however, possible to use hybrid circuits with silicon transistors by returning the cathode of a tunnel diode to a positive voltage of some tenths of a volt.   In this way the tunnel-diode characteristic will be shifted in the direction of positive voltage to the point where the rising base current crosses the tunnel-diode characteristic between its peak and valley.   Either a Ge or GaAs tunnel diode may be used.

Since the input characteristics of the hybrid circuit and the tunnel diode have the same form, we may use this input characteristic for all the same types of switching functions.   Thus the load line in Fig. 13-17b is appropriate for flip-flop operation, and the hybrid circuit of tunnel diode and transistor together operates as a bistable device.   To drive a transistor from cutoff to saturation requires that the input current increase from nominally zero to some hundreds of microamperes.   Observe, then, that as required, the base current corresponding to point X, the lower voltage of the two stable points, is

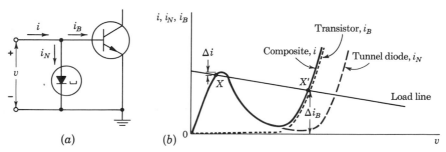

Fig. 13-17   (a) A basic tunnel-diode–transistor hybrid circuit; (b) composite characteristic formed by combining individual characteristics of tunnel diode and transistor.

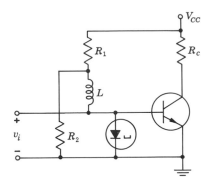

Fig. 13-18 A hybrid circuit which may be adjusted to function as a comparator or to operate in an astable, monostable, or bistable mode.

negligible, and at X', the other stable point, the base current may be comparable to the peak current I_P. Observe, further, that the tunnel diode provides *current gain* at wide bandwith between the circuit input and the base. A change in current Δi at the input triggers the circuit to point X', and the corresponding change in i_B is Δi_B as shown. In principle, by setting point X arbitrarily close to the peak, we may increase the current gain without limit. Of course, as a matter of practicality, requirements for stability against aging, temperature changes, noise, etc., will establish some minimum separation between X and the peak.

Now a second transistor substituted for the tunnel diode would have provided the current gain just as well. The use of a second transistor, which involves additional components and a more complicated coupling arrangement, would have lost for us the delightful simplicity of the hybrid circuit. But, putting this relatively unimportant matter aside, let us compare the responses to an input switching signal of a cascade of transistor stages and a cascade of hybrid circuits. Suppose that in a cascade of transistors a step of current or voltage of zero rise time is applied to the first transistor. The output of the first transistor will have a finite rise time. This finite-rise-time signal applied to the second transistor will cause the output of the second transistor to have an even longer rise time, and so on. However, the speed of transition of a tunnel diode is nominally not influenced by the speed of the applied signal, because the signal does not drive it from one state to another but only acts as a trigger. Since, also, the speed of tunnel diodes is comparable to or faster than even the highest-speed transistors, the interposition of tunnel diodes between transistors serves in effect to keep the speed and amplitude of the drive signal at each transistor base adequate to cause the transistor to respond at its fastest speed.

A basic circuit which can be adjusted to function as a comparator or to operate in a bistable, monostable, or astable mode is shown in Fig. 13-18. The load line at the input passes through an equivalent supply voltage $R_2 V_{CC}/(R_1 + R_2)$, and the equivalent load resistance is the parallel combination of R_1 and R_2. Thus we may adjust the load line to intersect the hybrid input characteristic in whatever manner is appropriate to attain the

type of operation desired.   The terminals at the left are for a triggering
signal when required.   The circuit of Fig. 13-18 is that of Fig. 13-13 with the
load R_L across the tunnel diode replaced by the base circuit of the transistor.

The hybrid-circuit multivibrators have an advantage over the two-
transistor multivibrators with respect to responding to a narrow triggering
pulse.   When, in a two-transistor multivibrator, a pulse is applied, say at
the base of one transistor, to induce a transition, the pulse must persist
long enough to allow the collector current to change substantially so that a
transition takes place.   Otherwise, at the termination of the pulse the multivibrator will simply settle back into its original state.   In the hybrid circuit
the pulse need only last long enough to get the load line to clear the peak or
valley of the characteristic.   If the pulse lasts long enough to get the tunnel
diode irreversibly started to its other state the transition is ensured.   Even
if the collector current has, in this time, not changed appreciably, it does not
matter.   The collector current will catch up later.

13-12   CIRCUIT APPLICATIONS OF p-n-p-n DIODES[5]

A p-n-p-n diode (Secs. 12-4 and 12-5) may be operated in an astable, monostable, or bistable mode, depending, as we have seen, on the relationship of the
load line to the device characteristic.   In some cases, for the sake of the waveforms thereby generated or for convenience in triggering, two p-n-p-n diodes
are used rather than one.

Sawtooth Generator   The circuit of Fig. 13-19a, when adjusted for
astable operation, generates the waveform in Fig. 13-19b.   This "sawtooth"-
type waveform (which may be generated by any current-controlled device)
finds extensive use in timing applications and will be discussed in detail in Chap.
14.   To consider the operation of Fig. 13-19 we disregard, for the moment, the

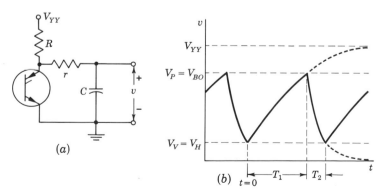

Fig. 13-19 (a) A p-n-p-n diode sawtooth generator; (b) output
waveform.

small resistance r and assume that the capacitor is initially uncharged. Since the voltage across the diode is low, the device is in its nonconducting state, so that the current supplied by V_{YY} flows through R into the capacitor. The capacitor charges toward V_{YY} until the output voltage reaches V_{BO}, the break-over voltage of the diode. At this peak $V_P = V_{BO}$, the device switches abruptly to its high-current–low-voltage state and the capacitor discharges through the diode. The diode remains in this state until the capacitor has discharged to the voltage V_H, where it can no longer supply the holding current. From the minimum or valley voltage $V_V = V_H$ the diode then returns to its OFF state and the cycle repeats. The small resistor r has been included to limit the peak current through the diode to within its rating during the interval of capacitor discharge.

The capacitor charges from an initial voltage V_V exponentially with a time constant $\tau = (R + r)C$ toward V_{YY}. At $t = T_1$ the voltage reaches V_P. Solving for T_1 we find

$$T_1 = \tau \ln \frac{V_{YY} - V_V}{V_{YY} - V_P} \tag{13-20}$$

The fall time T_2 is difficult to calculate since it depends upon the holding current I_H, which is often not specified and which may be quite variable from unit to unit. The capacitor voltage decreases exponentially from V_P toward zero, and the diode becomes nonconducting when its current falls below I_H. Let us assume that a reasonable interval for the current to fall to I_H is three time constants, or $T_2 = 3rC$, a quantity which may be as small as some tens of nanoseconds.

The frequency of the oscillation is $f = 1/(T_1 + T_2)$. This frequency may be increased by decreasing R. However, R was originally selected to intersect the characteristic in the manner indicated in Fig. 13-5. If R is decreased, a point will eventually be reached where the intersection occurs on the high-current positive-resistance portion of the characteristic. The circuit then latches, with a stable point in the high-current region. This manner of operation is to be avoided on the grounds that the continuous ON current will ordinarily exceed the allowed average diode current. After R has been decreased just short of moving the equilibrium point out of the negative-resistance region, the frequency may be increased further by reducing C. With a physical diode we find that when the capacitance C is progressively decreased to the point where the time constant becomes comparable to the time intervals required for the diode to make transitions between states, the sawtooth waveform looks rather more sinusoidal. Also, the oscillation amplitude decreases with increasing frequency.

Pulse Generator The circuit of Fig. 13-19 may be used as a pulse generator since, each time the capacitor discharges abruptly, a pulse of voltage appears across r. When intended as a pulse generator the load resistance is larger than the small resistance r used to limit the current and is located, as

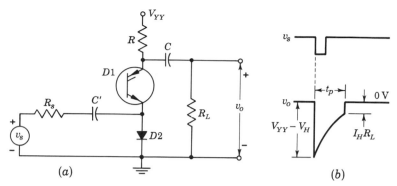

Fig. 13-20   (a) A p-n-p-n diode pulse generator; (b) input and output waveforms with $R \gg R_L$.

shown in Fig. 13-20, so that one side of the pulse output may be at ground. The circuit in Fig. 13-20 allows for the possibility of monostable operation. In such operation the supply voltage is adjusted to be somewhat less than the breakover voltage. A negative pulse introduced as indicated increases the voltage across the p-n-p-n diode $D1$ to the point where switching occurs. The capacitor discharges, a voltage pulse appears across R_L, and the circuit restores itself to its initial state. In Fig. 13-20b the output-pulse width is $t_p \approx 3R_LC$.

When the source v_s applies a negative pulse through C', the p-n diode $D2$ is momentarily back-biased while the voltage across $D1$ is raised to the point where it fires. The capacitor C then discharges, the path of the discharge being through $D1$, C', R_s, and R_L. Almost immediately, however, the voltage across R_s and C' rises to the point where $D2$ becomes forward-biased, and thereafter the discharge proceeds through $D2$ rather than the source. The usefulness of $D2$ (rather than a resistor) is that it limits the loading on the pulse source.

The output-pulse waveform starts at ground level. When the diode $D1$ fires, the anode voltage drops from V_{YY} to the holding voltage V_H. This abrupt voltage change is transmitted through C to the output. Since $R \gg R_L$ the diode current and the current through R_L are very nearly the same. This current decreases as the capacitor discharges, until the holding current I_H is reached. At this point diode $D1$ turns OFF abruptly and consequently there is an abrupt voltage change $I_H R_L$ at the output. Although it is true that, in monostable operation, an input pulse is required to produce an output pulse, the output pulse may be larger in amplitude, may be wider, and is available at a much lower impedance level than the triggering pulse.

Bistable Multivibrator   In the two previous p-n-p-n diode circuits, the circuit operated in an astable and in a monostable mode, respectively. We can also construct a bistable circuit with a single diode, but for the sake of

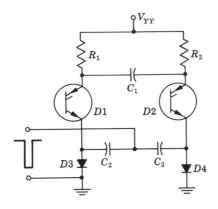

Fig. 13-21 A multivibrator using two p-n-p-n diodes. Nominally $R_1 = R_2$ and $C_2 = C_3$.

convenience of triggering with a pulse of one polarity at one point in the circuit, we use two p-n-p-n diodes.

In the circuit of Fig. 13-21, the supply voltage is less than the breakover voltage and the resistors $R_1 = R_2$ are small enough so that a stable state exists in both the high-current and in the low-current regions. Assume, then, that $D1$ and $D3$ are ON while $D2$ and $D4$ are OFF. The negative pulse, applied to the cathode of $D2$ through C_3, turns $D2$ ON. Consequently, the anode of $D2$ drops abruptly from the supply voltage V_{YY} to the maintaining voltage. This negative step of voltage is applied through C_1 to the anode of $D1$, and $D1$ is thereby turned OFF. The current which flows through C_1 in response to the voltage change at the anode of $D2$ must be supplied through R_1. This resistor R_1 is selected to be large enough so that the supply voltage cannot furnish through it both the capacitor C_1 current and the holding current for $D1$. Successive negative triggering signals will switch conduction back and forth between $D1$ and $D2$.

Other p-n-p-n diode multivibrators using two diodes are given in Probs. 13-27 and 13-28.

13-13   APPLICATIONS OF THE UNIJUNCTION TRANSISTOR[6,7]

A unijunction transistor may be operated in an astable, monostable, or bistable mode depending on the relationship of the load line to the device characteristic.

Sawtooth Generator A relaxation-oscillator circuit, similar in operation to the p-n-p-n diode circuit of Fig. 13-19, is shown in Fig. 13-22. The circuit must be biased for astable operation; that is, the load line determined by V_{YY} and R must intersect the input characteristic in the negative-resistance region. The resistors R_{b1} and R_{b2} are not essential to the circuit but are included because the voltages developed across these resistors may prove useful (see Sec. 14-4). The capacitor C charges to the peak voltage V_P, the device turns ON, and the capacitor discharges to the valley voltage V_V,

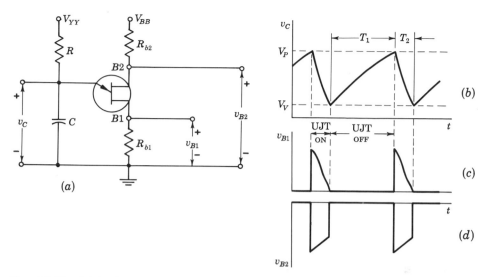

Fig. 13-22 (a) Unijunction transistor sawtooth-waveform generator; (b) capacitor voltage waveform; (c) voltage waveform at $B1$; (d) voltage waveform at $B2$.

whereupon the cycle repeats.   The capacitor voltage appears as in Fig. 13-22b. The charging time T_1 is given by Eq. (13-20).

Pulses develop as shown in Fig. 13-22c and d across R_{b1} and R_{b2} during the interval when the device is ON and the capacitor is discharging.   The pulse at $B1$ has an abrupt leading edge, but the anticipated abrupt drop at the trailing edge may not be easily apparent if the holding current is small in comparison with the current at the moment of breakdown.   The $B2$ current is smaller, being the $B1$ current minus the emitter current, but is more nearly constant during the capacitor discharge.   Improved linearity of the triangular waveform may be obtained by using feedback methods, as explained in Chap. 14.   An improved pulse waveform is obtained by connecting the base of a CE n-p-n transistor amplifier to $B1$ of Fig. 13-22a and taking the output from the collector of the transistor.

The resistor R_{b1} is in series with the device terminals between which appears the negative resistance.   In order that the effect of the negative resistance not be deemphasized it is necessary to limit the magnitude of the positive resistance R_{b1} to some tens of ohms.   The resistor R_{b2} may be larger, ranging up to some hundreds of ohms.

Astable Circuit with Controllable ON and OFF Times   The total time of a cycle of the circuit of Fig. 13-22 includes the interval during which C discharges. This interval is difficult to control, depending as it does on the characteristics of the device when it is conducting heavily.   Such characteristics are ordinarily neither specified nor controlled by the manufacturer.   However, for the case where $R_{b1} = 0$ the following empirical relationship has been found[7]

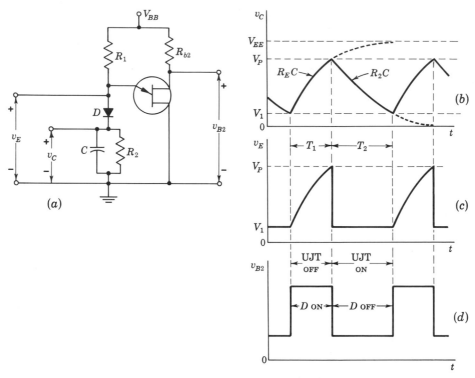

Fig. 13-23 (a) Unijunction transistor astable multivibrator yielding two comparable controllable timing intervals; (b) waveform across C; (c) emitter voltage waveform; (d) waveform at $B2$.

for the time T_2 for the emitter voltage to fall from V_P to V_V:

$$T_2 \approx (2 + 5C)V_E(\text{sat}) \tag{13-21}$$

where T_2 is in microseconds, C is in microfarads, and $V_E(\text{sat})$ is in volts. For example, with $C = 0.01\ \mu\text{F}$ and $V_E(\text{sat}) = 2\ \text{V}$, $T_2 \approx 4\ \mu\text{sec}$. It is also found that T_2 increases with temperature.

A circuit modification which yields a waveform consisting of two comparable controllable intervals is shown in Fig. 13-23. During the interval T_1, when the UJT is OFF, the capacitor C charges through diode D and an effective resistance R_E toward the voltage V_{EE}, where

$$R_E \equiv \frac{R_1 R_2}{R_1 + R_2} \qquad V_{EE} \equiv \frac{R_2 V_{BB}}{R_1 + R_2} \tag{13-22}$$

if we neglect the forward voltage across the diode. This asymptotic voltage is selected, as indicated in Fig. 13-24, to be larger than the peak voltage, $V_{EE} > V_P$, so that the UJT goes ON when $v_E = v_C = V_P$. At this point the emitter voltage drops abruptly, diode D becomes reverse-biased, and the capacitor begins to discharge through R_2. During this discharge the voltage v_C is approaching zero asymptotically.

During the interval T_2, when diode D is reverse-biased, the supply V_{BB} furnishes emitter current to the UJT through R_1. The supply V_{BB} and resistor R_1 have been selected, as in Fig. 13-24, so that the point of intersection of the corresponding load line with the device characteristic is in the negative-resistance region at point P_2. In order that this point should be so located it is necessary that

$$R_1 > \frac{V_{BB}}{I_V} \qquad\qquad (13\text{-}23)$$

where I_V is the valley current. The equilibrium voltage is V_1, which is somewhat larger than the valley voltage V_V. Thus, as shown in Fig. 13-23c, when the UJT goes ON, v_E falls to V_1. It should be recalled from Sec. 13-5 that even though P_2 is on the NR portion of the characteristic it is a stable equilibrium point, provided that the capacitance C' (not C) shunting the emitter with the diode open is below the critical capacitance.

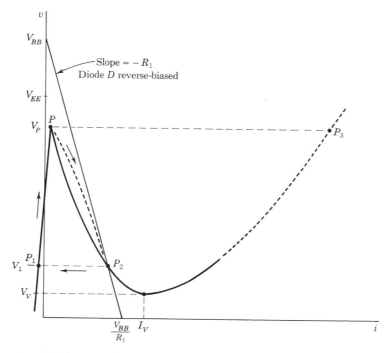

Fig. 13-24 Illustrating the path of operation of the astable unijunction transistor circuit of Fig. 13-23. Note that current is plotted along the abscissa, as in Fig. 12-6, rather than along the ordinate, as in Fig. 13-1b.

From Fig. 13-23 we observe that when v_C drops to V_1, diode D conducts, the capacitor C appears again across the negative-resistance terminals, and the point P_2 now becomes unstable.   At the moment diode D conducts, the emitter current is diverted through the diode and the UJT goes OFF.

It is of interest to trace the path of operation of the circuit on the volt-ampere characteristic of the device.   The characteristic and load lines are shown in Fig. 13-24.   Starting at the point P_1, where $v = V_1$, the operating point moves up the characteristic as the capacitor charges until the break point P is reached.   In an ideal case (where there is no series inductance L), at this point there would be an abrupt transition at constant voltage to P_3, which is the intersection of the peak current I_P and the characteristic.   Thereafter, the operating point would move along the characteristic to the equilibrium point at P_2.   Actually, because of the presence of L and the small capacitance C' which shunts the emitter when D is reverse-biased, the transition is not instantaneous.   The capacitance C' discharges during the transition and the equilibrium point P_2 is approached by some shorter path, as indicated. In any event, the circuit comes to rest at P_2 and remains at this point until it is rendered unstable by the reconnection of the capacitance C as the diode D becomes forward-biased.   At this time a fast transition takes place from P_2 to the initial point P_1, the current through R_1 transferring from the emitter to the diode while the voltage across C remains at V_1.

The charging interval T_1 is given by Eq. (13-20), which now takes the form

$$T_1 = \frac{R_1 R_2 C}{R_1 + R_2} \ln \frac{V_{EE} - V_1}{V_{EE} - V_P} \tag{13-24}$$

The discharge interval T_2 is found from Fig. 13-23b to be

$$T_2 = R_2 C \ln \frac{V_P}{V_1} \tag{13-25}$$

The voltage levels of the waveform v_{B2} may be determined by drawing a load line corresponding to V_{BB} and R_{B2} on the characteristics of Fig. 12-7b. The OFF level corresponds to the current $I_E = 0$, and the ON level to the current $I_E = (V_{BB} - V_1)/R_1$.   Since the emitter current is constant during each of the two intervals, the waveform of Fig. 13-23d displays no tilt.

Bistable Circuit[6]   A bistable UJT circuit is shown in Fig. 13-25a.   The use of the two resistors R_1 and R_2 allows some flexibility of adjustment since they make possible separate adjustments of emitter-supply voltage V_{EE} and resistance R_E.   That such is the case is seen in Fig. 13-25b, where the values of V_{EE} and R_E are given by Eqs. (13-22).   The load line passing through the points P_1, P_2, and P_3 corresponds to bistable operation since, of the three stable equilibrium states, only two, P_1 and P_3, are accessible.   The load line passing through P_1', P_2', and P_3' may also yield bistable operation.   For again, in the absence of capacitance, the point P_3' on the NR portion may

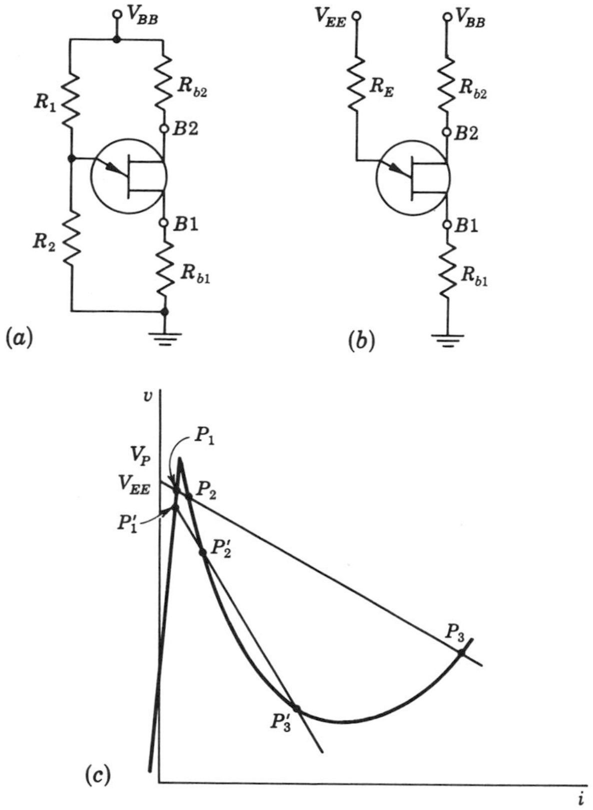

Fig. 13-25 (*a*) A UJT bistable multivibrator; (*b*) circuit equivalent to (*a*); (*c*) two possible load lines which will yield bistable operation.

actually be stable. Note that unlike the astable condition ($V_{EE} > V_P$), it is necessary for bistable operation that $V_{EE} < V_P$.

If the circuit is at point P_1 or P_1', a transition may be induced by raising the voltage at the emitter above V_P. In Fig. 13-25c the difference in voltage level between V_{EE}, P_1, and P_1' has been exaggerated. Actually the three voltages are so close that if a triggering voltage in excess of $V_P - V_{EE}$ is introduced either in series with V_{EE} or applied directly to the emitter (say through a small capacitance), then switching will have been accomplished. The circuit may also be triggered by the application of a negative pulse at $B2$. From Eq. (12-1) we see that if the voltage at $B2$ is lowered by ΔV_{B2}, then V_P will be lowered by $\eta \Delta V_{B2}$. To cause triggering we need to lower V_P by the amount $V_P - V_{EE}$. Thus triggering to the ON condition may be achieved with a negative pulse applied at $B2$ and of magnitude $(V_P - V_{EE})/\eta$, which is larger than that required at the emitter because $\eta < 1$. In the same way a negative pulse at $B1$ of magnitude $(V_P - V_{EE})/(1 - \eta)$ will cause triggering. Observe that at both $B1$ and $B2$ a negative pulse is required.

When the circuit is at P_3 or P_3' a transition may be induced by lowering the load line to the extent that only a single equilibrium point remains, and this in the low-current region. A negative pulse applied at the emitter will lower the effective V_{EE} and so accomplish switching. Alternatively, a positive pulse at $B1$ will raise the device characteristic and so achieve the same end. In either case the advantage is apparent of operating at P_3' rather than at P_3, since in the former case a smaller trigger will be required. Once the UJT is ON the voltage at $B2$ does not have a marked effect on the forward-biased emitter junction, so that triggering to the OFF state at $B2$ is normally not very effective.

A monostable UJT configuration is given in Fig. 14-8.

13-14   SILICON–CONTROLLED–SWITCH CIRCUITS[8]

All of the circuits which can be designed around p-n-p-n diodes and UJTs can be constructed as well with silicon controlled switches. The SCS offers the advantage of a gating terminal. The p-n-p-n diode has no separate gating terminal, and any gating voltage to be applied must be introduced as a superposition on the voltage across the two diode terminals. Thus there is no isolation between the triggering source and that part of the circuit which consists of supply voltage, load resistor, and the terminals between which there appears the negative resistance. The situation is not much better with the UJT. Here it is possible to trigger at $B2$ while the negative resistance appears between emitter and $B1$. But we have seen that the triggering pulse required at $B2$ is multiplied by the factor $1/\eta$ (≈ 2) over the trigger required at the emitter. On the other hand, as described in Sec. 12-7, an SCS may be extremely sensitive to a trigger signal on the gate.

A number of applications of the SCS are indicated in Fig. 13-26. In Fig. 13-26a an SCS is employed as a comparator. When the signal v_i attains the triggering level, the switch fires and a large and abrupt response occurs in v_o at the output. The firing voltage of the switch has a negative temperature coefficient of about 3 mV/°C. For application over a wide temperature range some form of temperature compensation will be required, such as the use of temperature-sensitive resistors (thermistors) at the input. It should be remembered in this and other applications that the SCS can handle considerable power. Thus the low-level output from a magnetic core or a sensitive transducer applied at the gate in Fig. 13-25a can excite power loads placed directly in the anode circuit, for example, a card-punch solenoid, a magnetic clutch, an indicator lamp, etc.

A pulse generator is shown in Fig. 13-26b. This circuit is so similar in operation to the circuit of Fig. 13-20 that we need only call attention to the convenience of triggering.

In the sawtooth-generating circuit in Fig. 13-26c the capacitor charges through R, and with time both the anode and gate voltage rise to the point where the switch fires. The capacitor then discharges and the cycle begins again.

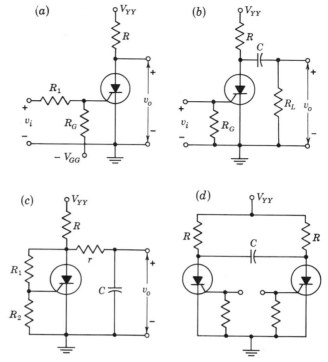

Fig. 13-26 Circuit applications of the silicon controlled switch. (a) Comparator; (b) pulse generator; (c) sawtooth-waveform generator; (d) multivibrator or flip-flop.

In Fig. 13-26d we have a circuit which may be astable or bistable. It will be astable if the supply voltage is larger than the firing voltage and bistable if the supply voltage is smaller. In the latter case triggering signals will have to be supplied to the two bases through small capacitors, as in the corresponding p-n-p-n diode circuit of Fig. 13-21.

13-15   TRANSWITCH, TRIGISTOR, AND THYRISTOR TRIGGERING

We had noted that the principal feature of these devices is the feasibility of turning them OFF with a triggering signal. Two methods of triggering are shown in Fig. 13-27. In Fig. 13-27a a positive pulse at the base turns the device ON and a negative pulse OFF. The device is ON for the time interval between the leading edges of the two pulses. In Fig. 13-27b both input pulses are negative; one is applied to the base and one to the cathode. As in Fig. 13-20 the diode D allows the device current to flow when it is ON and presents a high impedance to the turn-on pulse when the device is OFF. Applications of these devices in counting circuits are given in Chap. 18.

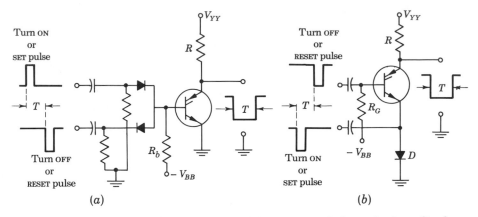

Fig. 13-27   Two methods of triggering a trigistor, transwitch, or thyristor flip-flop.

13-16   AVALANCHE–MODE TRANSISTOR CIRCUITS[9,10]

Most mesa and MADT switching transistors may be used in an avalanche mode as pulse generators. The transistor volt-ampere characteristic of Fig. 6-18 or 6-20 is repeated in Fig. 13-28a, and a basic circuit is indicated in Fig. 13-28b. The load line is selected to yield a single stable point in the low-current region. The supply voltage charges the capacitor through R_c to a voltage V_{CC} slightly less than the breakdown voltage BV_{CER}. A pulse shown in Fig. 13-28c (or some other positive-going signal) applied to the base lowers the breakdown voltage, and the capacitor discharges rapidly through the transistor and the small resistance R_L. The voltage across R_L shown in Fig. 13-28d drops very rapidly, as does the collector voltage. The speed with which these voltages change is determined by how quickly the transistor makes the transition from its low-current state to the state in which an avalanche discharge is established. It is in this low transition time (a few nanoseconds) that the transistor excels.

Having reached a peak, the output voltage now decays to zero as the capacitor discharges. The collector voltage, as in Fig. 13-28e, starts close to V_{CC} and drops at the same high speed to the latching voltage LV_{CER}, which, as the symbol indicates, is a function of the base resistance. Even after the capacitor has discharged, the collector voltage remains for a time at the latching voltage since an interval is required to allow the transistor to recover and return to its initial state. During this interval a small transistor current flowing now through R_c maintains the collector at the lower voltage. Finally, as the transistor approaches complete recovery, the collector voltage returns again to V_{CC} but at a much slower rate than it fell. We may note, in passing, that if our interest is in a rapid voltage step rather than a pulse we may set $R_L = 0$ and use the collector voltage itself.

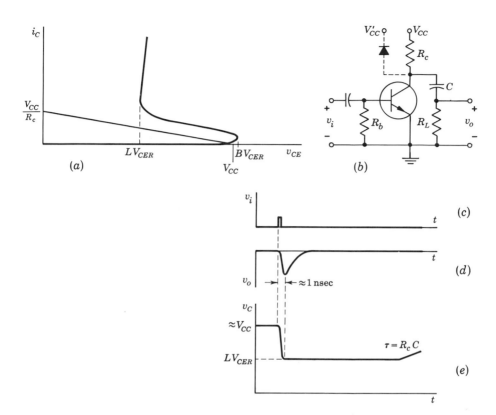

Fig. 13-28 (a) Volt-ampere characteristic of avalanche-mode transistor; (b) circuit of a pulse generator $(R_L \ll R_c)$; (c) triggering pulse; (d) output waveform; (e) waveform at collector.

The pulse width depends upon R_L and C and increases as C increases. The rate at which the circuit may be driven is determined by R_c and C. Typically $R_c \approx 10$ K, whereas $R_L \approx 50$ Ω, so that nanosecond pulses with microsecond spacing between pulses may be obtained. Some precautions must be taken when adjusting the circuit for a high repetition rate. If C has been fixed by pulse-width requirements, the repetition rate may be increased by decreasing R_c. But R_c must not be made so small that the transistor will latch and remain permanently in avalanche. Under these circumstances the transistor will exceed its dissipation rating. As noted in Sec. 6-10, latching voltages are in the range of tens of volts and currents in the avalanche region may range up to many hundreds of milliamperes.

A procedure for increasing the repetition rate which avoids the danger of latching is indicated by the diode (shown dashed) in Fig. 13-28b. This collector catching diode is returned to a supply voltage V'_{cc} which is less than the breakdown voltage. Thus the supply V_{CC} may be increased well

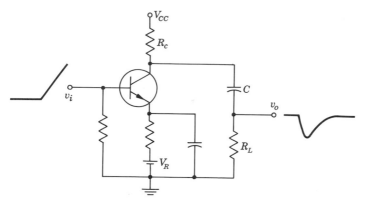

Fig. 13-29   An avalanche-transistor comparator circuit.

beyond the breakdown voltage, thereby increasing the speed of charging of the capacitor and hence the repetition frequency.

A second precaution to be observed also has to do with the transistor dissipation. The maximum frequency of operation is limited by the dissipation rating of the transistor. In physical circuits a 10-MHz repetition frequency is nonetheless possible. The pulse may be made progressively narrower by reducing the size of the capacitor until such time that the pulse width becomes comparable to the rise time. Beyond this point the pulse amplitude will become progressively smaller.

The comparator circuit of Fig. 13-29 constructed with an avalanche transistor is useful for very fast waveforms. A (nanosecond) output pulse is produced when the input exceeds a critical level, which depends upon the reference voltage V_R.

13-17   AVALANCHE TRANSISTOR DELAY–LINE PULSE GENERATOR[10]

An avalanche pulse generator whose pulse amplitude and width are separately controllable and which provides a more rectangular pulse than is available from the circuit of Fig. 13-28b is shown in Fig. 13-30a. Here the capacitor has been replaced by an open-circuited delay line of characteristic impedance R_o. An equivalent circuit for the purpose of calculating the waveforms is shown in Fig. 13-30b, where avalanche breakdown is represented by the closing of the switch S. We now describe the operation of this circuit.

When, in response to a triggering signal, the transistor breaks down (S closes), a negative voltage step $\Delta v = -V$ is applied to the line through the resistor $R_L \ll R_c$. This step has an amplitude $V = V_{CC} - LV_{CER}$. Since the initial line current is zero, then a current step

$$i(0+) = \frac{-V}{R_o + R_L} \equiv I \tag{13-26}$$

starts down the line at $t = 0+$. When this current wave reaches the end of the line at $t = t_d$ it is reflected as a current step $-I$ (the reflection coefficient ρ for current at an open circuit is -1). At $t = 2t_d$ this step $-I$ reaches the beginning of the line and, as indicated in the reflection chart of Fig. 13-30c, is again reflected as $(-I)\rho'$, where ρ' for a current wave is the negative of the reflection coefficient for a voltage wave given in Eq. (3-42). Hence, the total line current at $t = 2t_d+$ is

$$i(2t_d+) = I - I - I\rho' = -I\rho' = \frac{-V}{R_o + R_L}\frac{R_L/R_o - 1}{R_L/R_o + 1} \qquad (13\text{-}27)$$

Neglecting the small current through R_c, we see from Fig. 13-30b that the transistor collector current i_C is the negative of the line current i. Hence, the collector current will be positive at $t = 2t_d+$ as long as $R_L > R_o$. Under these circumstances the transistor remains ON and further reflections take place. This mode of operation is undesirable because the waveform across R_L will consist of a series of different voltage levels separated by intervals $2t_d$, rather than a pulse output.

We note from Eq. (13-27) that if $R_L = R_o$ (the line is matched at the input end), then at $t = 2t_d$ the transistor current i_C is brought to zero and consequently the transistor goes OFF and the switch S is opened. However, if $R_L < R_o$, then Eq. (13-27) indicates a reversal of collector current. Actually, in this case the transistor will again simply go OFF. In either case, then, $R_L = R_o$ or $R_L < R_o$, S opens at $t = 2t_d$, and any current through R_L must now flow as well through the resistor R_c, which is very large in comparison with R_L. Therefore the voltage v_o across R_L will drop nominally to zero and remain there. Altogether, we have the result that for $R_L \leq R_o$ there will develop across R_L a pulse whose duration $2t_d$ is controllable through adjustment of the transmission-line delay time and whose amplitude is adjustable through R_L and V in accordance with the relationship

$$v_o = i(0+)R_L = \frac{-VR_L}{R_o + R_L} \qquad (13\text{-}28)$$

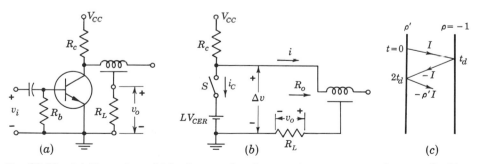

Fig. 13-30 (a) The pulse width of an avalanche-transistor generator is controlled by an open-circuited delay line; (b) an equivalent circuit after the transistor is triggered; (c) the reflection chart for the line current.

The output voltage step is negative and has a maximum magnitude of $V/2$ because $R_L \leq R_o$.

The pulse which appears across R_L in Fig. 13-28b results from the differentiation of the step generated by the avalanche transistor by the $R_L C$ differentiating circuit. On this basis we may see that, in going from the circuit of Fig. 13-28b to the circuit of Fig. 13-30, we have replaced the $R_L C$ differentiation by delay-line differentiation, as described in Sec. 3-16.

We have assumed that the circuits of Figs. 13-28b and 13-30a were externally triggered. Either circuit may be rendered astable by raising V_{CC} or returning R_b to a bias voltage such that avalanche breakdown occurs without the need for an external trigger.

REFERENCES

1. Strauss, L.: "Wave Generation and Shaping," chap. 11, McGraw-Hill Book Company, New York, 1960.
   Zimmerman, H. J., and S. J. Mason: "Electronic Circuit Theory," chap. 9, John Wiley & Sons, Inc., New York, 1959.

2. "Tunnel Diode Manual," General Electric Company, Syracuse, N.Y., 1961.
   "Transistor Manual," 7th ed., chap. 14, General Electric Company, Syracuse, N.Y., 1964.

3. "Tunnel Diode Manual," Radio Corporation of America, Somerville, N.J., 1963.

4. Cleverley, D. S.: How to Design a Tunnel-diode Threshold Detector, *Electron. Design*, vol. 11, pp. 58–62, Sept. 13, 1963.

5. Multivibrator Circuits, *Appl. Note* AD-6, Shockley Transistor Corp., October, 1959.

6. Crawford, B., and R. T. Dean: The Unijunction Transistor in Relaxation Circuits, *Electro-Technol.*, pt. 1, vol. 73, no. 2, pp. 34–38, February, 1964; pt. 2, vol. 73, no. 3, pp. 40–45, March, 1964.

7. Sylvan, T. P.: Notes on the Application of the Silicon Unijunction Transistor, *Appl. Note* 90.10, General Electric Company, Syracuse, N.Y., May, 1961.

8. A Survey of Some Circuit Applications of the Silicon Controlled Switch and Silicon Controlled Rectifiers, *Bull.* D420-02-9-60, Solid State Products, Inc.
   Silicon Controlled Rectifiers and Switches, *Bull.* AN-1356C, Transitron Electronic Corp., August, 1961.
   "Transistor Manual," 7th ed., chap. 16, General Electric Company, Syracuse, N.Y., 1964.
   Stasior, R. A.: Silicon Controlled Switches, *Appl. Note* 90.16, General Electric Company, Syracuse, N.Y., June, 1964.

9. Bell, B. H.: Avalanche Circuits Are More Versatile Than You Think, *Electron. Design*, vol. 12, pp. 56–63, June 8, 1964.

Bramson, M.: Application of Avalanche Transistors to 10-megacycle Pulse Generator, *IRE Trans. PGNS*, vol. NS-9, no. 4, pp. 35–37, August, 1962.

Haas, I.: Millimicrosecond Avalanche Circuits Utilizing Double-diffused Silicon Transistors, *Appl. Note* APP-8/2, Fairchild Semiconductor Corporation, December, 1961.

10. "High-speed Switching Transistor Handbook," chap. 9, Motorola Inc., Semiconductor Products Div., Phoenix, Ariz., 1963.

14 / VOLTAGE TIME–BASE GENERATORS

A linear time-base generator is one that provides an output waveform, a portion of which exhibits a linear variation of voltage or current with time. An application of first importance of such a waveform is in connection with a cathode-ray oscilloscope. The display on the screen of a scope (cathode-ray oscilloscope) of the variation with respect to time of an arbitrary waveform requires that there be applied to one set of deflecting plates a voltage which varies linearly with time. Since this waveform is used to *sweep* the electron beam horizontally across the screen, it is called a *sweep voltage*. There are, in addition, many other important applications for time-base circuits, such as in radar and television indicators, in precise time measurements, and in time modulation.

14-1   GENERAL FEATURES OF A TIME–BASE SIGNAL

The typical form of a time-base voltage is shown in Fig. 14-1. Here it appears that the voltage, starting from some initial value, increases linearly with time to a maximum value, after which it returns again to its initial value. The time required for the return to the initial value is called the *restoration time*, the *return time*, or the *flyback time*. Very frequently the shape of the waveform during the restoration time and the restoration time itself are matters of no special consequence. In some cases, however, a restoration time is desired which is very short in comparison with the time occupied by the linear portion of the waveform. If it should happen that the restoration time is extremely short and that a new linear voltage is initiated at the instant the previous one is terminated, then the waveform will appear as in Fig. 14-2.

514

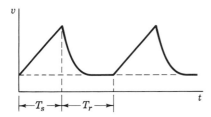

Fig. 14-1   A general sweep
voltage.   The sweep time is T_s
and the return time is T_r.

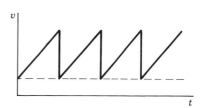

Fig. 14-2   A sawtooth voltage
waveform.

This figure suggests the designation *sawtooth generator* or *ramp generator*.   It is customary to refer to waveforms of the type indicated in Figs. 14-1 and 14-2 as *sweep* waveforms even in applications not involving the deflection of an electron beam.

We shall see that generators of time-base signals do not ordinarily provide sweep voltages that are precisely linear.   Moreover, a nominally linear sweep may be distorted in the course of transmission through a coupling network (Secs. 2-2 and 2-6).   The three most useful ways of expressing the deviation from linearity, and the correlations among them, are given below.

The Slope or Sweep-speed Error e_s In the case of a general-purpose cathode-ray oscillograph, an important requirement of the sweep is that the sweep speed (i.e., the rate of change of sweep voltage with time) be constant. A reasonable definition of the deviation from linearity is

$$e_s \equiv \frac{\text{difference in slope at beginning and end of sweep}}{\text{initial value of slope}} \tag{14-1}$$

The Displacement Error e_d In connection with other timing applications, a more important criterion of linearity is the maximum difference between the actual sweep voltage and linear sweep which passes through the beginning and end points of the actual sweep, as in Fig. 14-3.   Here we may define

$$e_d \equiv \frac{(v_s - v_s')_{\text{max}}}{V_s} \tag{14-2}$$

The Transmission Error e_t If a ramp voltage is transmitted through a high-pass RC network, the output falls away from the input, as indicated in Fig. 2-10a and in Fig. 14-4.   The *transmission error* is defined as the difference between the input and output divided by the input.   Thus, with reference to Fig. 14-4, we have (at time $t = T_s$)

$$e_t = \frac{V_s' - V_s}{V_s'} \tag{14-3}$$

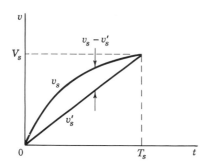

Fig. 14-3 Relating to the definition of displacement error.

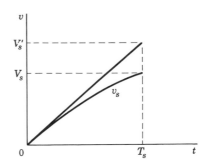

Fig. 14-4 Relating to the definition of transmission error.

If the deviation from linearity is small, so that the sweep voltage may be approximated by the sum of a linear and a quadratic term in t, then

$$e_d = \tfrac{1}{8}e_s = \tfrac{1}{4}e_t \tag{14-4}$$

14-2   METHODS OF GENERATING A TIME–BASE WAVEFORM

We shall discuss time-base circuits in which sweep linearity is achieved by one of the following seven methods:

1. *Exponential charging.* A capacitor is charged through a resistor to a voltage which is small in comparison with the supply voltage.
2. *Constant-current charging.* A capacitor is charged linearly from a constant-current source.
3. *The Miller circuit.* An operational integrator is used to convert a step into a ramp waveform.
4. *The phantastron circuit.* This configuration is a version of the Miller circuit which requires only a pulse input and not an external step or gating waveform.
5. *The bootstrap circuit.* A constant current is approximated by maintaining nearly constant voltage across a fixed resistor in series with a capacitor.
6. *Compensating networks.* A compensating circuit is introduced to improve the linearity of the bootstrap and the Miller time-base generators.
7. *An inductor circuit.* An *RLC* series circuit is used to yield more linear capacitor charging than is possible without the use of the inductor.

Vacuum tubes, gas tubes, transistors, or other semiconductor devices are used with each of the above circuits, depending upon the application, convenience, speed, and other practical considerations.

14-3   EXPONENTIAL SWEEP CIRCUIT

The exponential sweep circuit is illustrated in Fig. 14-5a. At $t = 0$ the switch S is opened and the sweep voltage v_s is given by

$$v_s = V(1 - \epsilon^{-t/RC}) \tag{14-5}$$

For the present discussion the physical form of the switch S (tube, transistor, etc.) is unimportant. Suffice it to say that after an interval T_s, when the sweep amplitude has attained the value V_s, the switch again closes. The resultant sweep waveform is indicated in Fig. 14-5b (assuming zero switch resistance).

If the definition of the sweep-speed error of Eq. (14-1) is applied to Eq. (14-5) we find that e_s is given exactly by

$$e_s = \frac{V_s}{V} \tag{14-6}$$

The linearity improves as the ratio V_s/V decreases. Hence, the simple circuit of Fig. 14-5 is useful only in applications requiring sweep voltages of the order of volts or tens of volts. For example, a 20-V sweep can be obtained with a sweep-speed error of less than 10 percent by using a supply voltage of at least 200 V. Time-base voltages of hundreds of volts require power supplies of thousands of volts, which are inconveniently large.

If $t/RC \ll 1$, it is convenient to expand the exponential in Eq. (14-5), so that

$$v_s = \frac{Vt}{RC}\left(1 - \frac{t}{2RC} + \frac{t^2}{6R^2C^2} - \cdots\right) \tag{14-7}$$

Since $v_s = V_s$ when $t = T_s$, we have to a first approximation that

$$\frac{V_s}{V} = \frac{T_s}{RC} \tag{14-8}$$

It follows from Eqs. (14-6) and (14-8) that if a voltage increases exponentially for a time T_s with a time constant $\tau = RC$, the sweep-speed error is approximately

$$e_s = \frac{V_s}{V} = \frac{T_s}{\tau} \tag{14-9}$$

Fig. 14-5 (a) Charging a capacitor through a resistor from a fixed voltage; (b) the resultant exponential waveform.

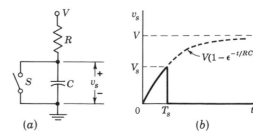

(a)   (b)

We shall apply this result many times to the various circuits to be discussed in this chapter. If a sweep is to be reasonably linear, the time constant τ must be large compared with the sweep duration T_s.

From the linear term in Eq. (14-7) we find that V_s', defined in Fig. 14-4, is given by $V_s' = VT_s/RC$. If the first two terms in Eq. (14-7) are retained, we find

$$V_s = \frac{VT_s}{RC}\left(1 - \frac{T_s}{2RC}\right) \tag{14-10}$$

Applying the definition [Eq. (14-3)] of transmission error to Eq. (14-10) gives

$$e_t = \frac{T_s}{2RC} = \frac{1}{2}\frac{T_s}{\tau} \tag{14-11}$$

Note from Eqs. (14-9) and (14-11) that $e_s = 2e_t$, which is consistent with Eq. (14-4). From Eq. (14-4) the displacement error is given by

$$e_d = \frac{1}{8}\frac{T_s}{\tau} \tag{14-12}$$

If a capacitor C is charged by a constant current I, then the voltage across C is It/C. Hence, the rate of change of voltage with time is given by

$$Sweep\ speed = \frac{I}{C} \tag{14-13}$$

This relationship will be used many times throughout this chapter.

14-4   NEGATIVE–RESISTANCE SWITCHES

Many devices are available to serve as the switch S in Fig. 14-5. In this section we consider the use of a negative-resistance device for S. In the

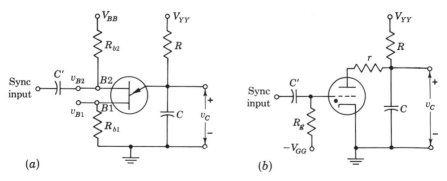

(a)   (b)

Fig. 14-6 Sweep circuits which use for the switch S of Fig. 14-5 (a) a UJT and (b) a thyratron (the dot inside the tube symbol indicates that the tube contains gas).

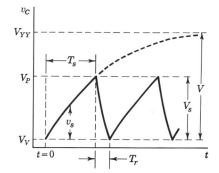

Fig. 14-7 The capacitor waveform of the circuits of Fig. 14-6.

following section we consider circuits in which the vacuum tube and the transistor are employed as the switch.

Any current-controlled negative-resistance device (Sec. 13-1) may be used to discharge the sweep capacitor. In Fig. 14-6a a unijunction transistor (Sec. 12-3) is indicated for this purpose, and in Fig. 14-6b the thyratron[1] plays the role of the switch. The waveform of the capacitor voltage for either switch is shown in Fig. 14-7. The UJT base-current waveforms are plotted in Fig. 13-22. This astable operation of the UJT is explained in Sec. 13-13, where V_P and V_V are identified, respectively, as the peak and valley voltages of the emitter characteristic of Fig. 12-6. The supply voltage V_{YY} and the charging resistor R must be chosen so that the load line intersects the volt-ampere characteristic in the negative-resistance region. If the intersection is at some voltage V' on one of the positive-resistance regions, then the circuit will "latch" or "block"; i.e., the oscillations will cease and the voltage across C will remain constant at V'.

The Thyratron Switch The voltage V_P now represents the plate break-down potential corresponding to the grid voltage V_{GG}. For the type 884 thyratron the critical grid breakdown characteristic is given by $V_P \approx 8V_{GG}$. The minimum value V_V of the sweep in Fig. 14-6b is the maintaining volt-age of the gas tube. It will be recalled that once the arc has formed in a thyratron, the grid loses its ability to control the tube current. Variations in grid voltage accomplish nothing but a variation in the thickness of the positive-ion sheath surrounding the grid. The arc will persist just as long as the current through the tube is large enough to maintain an adequate supply of positive ions to replace those ions which are lost through the process of recombination. If the current is large enough to maintain the ion supply and is less than the saturation current of the tube, the tube drop will remain essentially constant at the maintaining voltage V_V independently of the cur-rent. Over this range of currents, the tube current is determined by the circuit external to the tube. The arc may be extinguished only by reducing the tube current below the minimum required to maintain ionization. When the arc has been extinguished, the grid once again regains control and determines the plate voltage that must be applied to cause breakdown. The maintaining and extinction voltages are approximately equal.

The thyratron circuit generates the waveform of Fig. 14-7 in the following manner. The capacitor C charges through R, approaching asymptotically the supply voltage V_{YY} as shown. When the plate attains the voltage V_P, the thyratron ignites. At this point the thyratron may be considered to be replaced by a battery whose terminal voltage is equal to the maintaining voltage of the tube. The capacitor C will discharge through the tube and series resistor r until the capacitor voltage drops to V_V. The arc will extinguish itself at the instant the tube current is less than the minimum required to maintain the arc, and the charging of the capacitor through R from the supply voltage will begin again. The resistor r is made small enough to permit a rapid discharge (small T_r) of the capacitor but not so small that it permits a larger discharge current through the tube than the tube can safely handle. The maintaining voltage of the 884 thyratron is about 16 V, the minimum current required to maintain the arc is of the order of magnitude of 1.0 mA, and the maximum peak current which should be permitted to flow at the discharge of the capacitor is about 0.5 A. To limit the positive-ion grid current to a safe value, a resistor R_g (≈ 10 K) is inserted in the grid circuit.

The Free-running Mode The sweep-voltage waveform, $v_s = v_C - V_V$ in Fig. 14-7, is given by Eq. (14-5) with $V = V_{YY} - V_V$. The sweep amplitude is $V_s = V_P - V_V$ and is given by Eq. (14-8). The linearity errors are given by Eqs. (14-9), (14-11), and (14-12). For good linearity V_s must be much smaller than V, or, since usually $V_P \gg V_V$ and $V_{YY} \gg V_V$, we require that $V_P \ll V_{YY}$. If the unijunction transistor is used as the switch, then there is, in addition, a restriction on the magnitude of V_{BB}. From Eq. (12-1) $V_P = \eta V_{BB} + V_\gamma$, where the intrinsic stand-off ratio $\eta \approx 0.5$ and $V_\gamma \approx 0.6$ V. Therefore for a UJT it is necessary that $V_{YY} \gg V_{BB}$. If, for example, one power supply is used, so that $V_{BB} = V_{YY}$, then a very poor sweep is obtained (Prob. 14-5).

In the sweep circuit presently being considered the waveform is repetitive; one sweep is initiated immediately at the termination of the previous sweep and the circuit does not wait for some external signal to initiate the sweep. This mode of operation is termed *free-running*, *astable*, or *recurrent*, and it is customary to calibrate the control dials of the scope in frequency rather than sweep time. The frequency is a function of V_{YY}, V_{BB} or V_{GG} (since V_P is a function of V_{BB} or V_{GG}), R, and C. In practice, the voltages are kept constant and the frequency is varied through R and C. In this way the amplitude and linearity of the sweep are kept constant. Continuous variation of frequency is accomplished through varying R, and the ranges are changed by switching the value of C.

If a periodic signal of frequency f_v is applied to the vertical axis of a scope while a sweep of frequency f_s is applied to the horizontal axis, a stationary pattern of n cycles will appear if $f_v = nf_s$. A portion of the last cycle occurs

during the return time.   If the return time T_r is very small in comparison
with the sweep time, the beam will move much more rapidly during the return
than during the sweep, and the return trace on the CRT (cathode-ray tube)
screen will be very faint or even invisible.   This disappearance of the return
trace is ordinarily desirable for the sake of clarity of presentation of the pattern
on the scope screen.   In both the thyratron and the UJT sweep circuits, keeping the capacitor C small will shorten the retrace time.   In Prob. 14-8 there
is presented a circuit employing an additional transistor in a manner to hasten
the capacitor discharge in the UJT circuit.

Where a sweep operates with T_r not very small in comparison with T_s,
the return trace may be made invisible by turning off, or *blanking*, the CRT
beam during retrace.   In the thyratron sweep this blanking may be achieved
by applying to the grid of the CRT the negative spike which results when the
sweep waveform is passed through a differentiating circuit.   For the UJT
sweep the voltage developed at $B2$, v_{B2} in Fig. 13-22, applied to the CRT will
accomplish the required blanking.

To maintain the condition $f_v = nf_s$ exactly for long periods of time,
it is necessary to *synchronize* the sweep generator to the signal.   If f_v is
only very slightly different from nf_s, the waveform will drift slowly across
the screen.   Synchronization is accomplished by applying to the base of
the UJT or the grid of the thyratron the vertical-deflecting signal, increased
or reduced in amplitude as may be required.   The process of synchronization
is explained in detail in Chap. 19.

The Triggered Mode   A waveform may not be periodic but may occur
rather at irregular intervals.   In such a case it is desirable that the sweep
circuit, instead of running continuously, should remain quiescent and wait to be
initiated by the waveform itself.   Even if it should happen that the waveform
does recur regularly, it may be that the interesting part of the waveform is
short in time duration in comparison with the period of the waveform.   For
example, the waveform might consist of 1-msec pulses with a time interval of
100 msec between pulses.   In this case the fastest recurrent sweep which will
provide a synchronized pattern will have a period of 100 msec.   If, typically,
the time base is spread out over 4 in. (on a 5-in. CRT), the pulse will occupy
0.04 in. and none of the detail of form of the pulse will be apparent.   If, on the
other hand, a sweep of period 1 msec or somewhat larger could be used, the
pulse would be spread across the entire screen.   Therefore, what is required
here is a sweep set for, say, a 1.5-msec interval which remains quiescent until
it is initiated by the pulse.   Such a monostable circuit is known as a *driven*
sweep or a *triggered* sweep.

The circuit for a driven sweep is shown in Fig. 14-8a.   The block S
represents either the UJT or the thyratron switch of Fig. 14-6.   The only
modification required in the circuit of Fig. 14-6 to convert it from an astable
to a monostable time-base generator is the addition of the clamping diode D.
The mechanical switch S' is open (position B), and the clamping reference

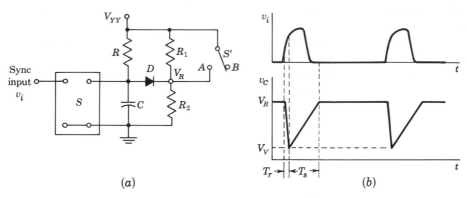

(a)   (b)

Fig. 14-8 (a) A driven sweep, where the switch S is a negative-resistance device and S' is open; (b) a pulse-type sync signal v_i (positive for a thyratron and negative for a UJT) and the triggered time-base voltage v_C across the capacitor.

voltage V_R is determined by the resistors R_1 and R_2. The voltage V_R is adjusted to be less than V_P. Accordingly, as the capacitor voltage increases, a point is reached before the thyratron or the UJT fires where the diode conducts and prevents further rise of the capacitor voltage. Thus D is a catching diode (Sec. 8-13) which clamps the sweep to V_R. Now let a signal be applied to the sync (synchronizing) input (a negative signal to the UJT base $B2$ or a positive signal to the thyratron grid) which even instantaneously lowers the firing voltage V_P below V_R. Then the capacitor will discharge quickly (in the retrace time T_r) and charge again exponentially toward V_{YY}, but it will be clamped when it reaches V_R.

Figure 14-8b shows the operation of a triggered sweep for a case in which the waveform to be observed on the scope consists of a train of pulses. This signal v_i is applied to the vertical-deflection amplifier of the scope and is used also to trigger the sweep circuit. Observe that at the occurrence of the leading edge of a pulse the circuit capacitor first discharges, after which a linear sweep occurs. Note also that the sweep speed has been adjusted so that the pulse will be spread out over a large portion of the sweep trace. The signal will have to make some departure from its quiescent value before the sweep circuit responds. Thereafter a portion of the signal occurs during the retrace. Therefore this type of triggered sweep is suitable for use with slow waveforms only where we may tolerate losing a small interval of the first portion of the waveform.

For the thyratron switch the minimum retrace time T_r is determined by the deionization time of the gas and is of the order of 10 μsec. For the UJT switch, T_r is given by Eq. (13-21), which yields a minimum value for small C of a few microseconds. Because of the relatively large retrace times obtained with a UJT or thyratron switch, most modern high-frequency scopes with fast

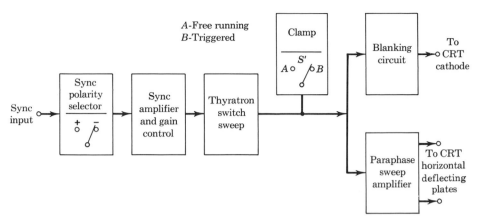

Fig. 14-9 A block diagram of a general-purpose-scope sweep system. (Switch S' is detailed in Fig. 14-8a.)

sweeps use vacuum-tube or transistor switches (described in the following section).

The triggered sweep of Fig. 14-8 may readily be converted to free-running operation by using a mechanical switch S' (in position A) to short out R_1. Under these circumstances $V_R = V_{YY}$ and D never conducts. The circuit is then equivalent to the astable time-base generator of Fig. 14-6.

A block diagram of a slow-speed general-purpose scope[2] using a thyratron switch is indicated in Fig. 14-9. The sync polarity selection is made by taking the output across either a plate or a cathode resistor. The sync amplifier is conventional and has an input gain control. The thyratron sweep is essentially that of Fig. 14-6b, and the clamp is the diode network of Fig. 14-8a. The blanking circuit consists of a high-pass RC differentiating network followed by an amplifier. The blanking pulse is applied to the CRT cathode so that the CRT grid may be available at a front-panel terminal for external intensity modulation.

14-5   SWEEP CIRCUIT USING A VACUUM–TUBE
OR TRANSISTOR SWITCH

The basic circuit of a vacuum-tube or transistor-switch sweep circuit is indicated in Fig. 14-10a, where A represents either a vacuum-tube switch, as in Fig. 8-26, or a transistor switch, as in Fig. 8-27. This capacitor-loaded switch is discussed in detail in Sec. 8-12. However, there is an essential difference in emphasis now, since the sweep capacitor C may be quite large instead of simply representing the small inevitable output shunt capacitance C_s of Sec. 8-12. The input "gating" waveform v_i may be derived from a single-shot circuit, in which case a monostable or triggered sweep results. On the

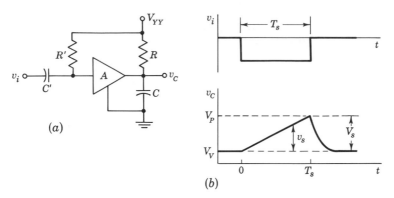

Fig. 14-10 (a) A transistor or vacuum tube A used as a switch for a triggered sweep; (b) the input gate voltage v_i and the voltage waveform v_C across the capacitor.

other hand, if v_i is the output of an astable multivibrator, then a free-running sweep is generated. Without sacrificing generality we need consider only the driven time-base waveform.

In the quiescent state, $t < 0$, the input (grid or base) is clamped near ground and the output is at its lowest magnitude V_V, as indicated in Fig. 14-10b. This valley voltage V_V equals the collector-to-emitter saturation voltage $V_{CE}(\text{sat})$ for a transistor or the plate voltage for zero grid voltage if a tube is under consideration. At $t = 0+$ and for a gating time T_s, the switch is cut off and the capacitor charges through R toward V_{YY} to a peak value V_P. From Eqs. (14-13) and (14-8) the initial sweep speed and the (approximate) sweep amplitude are

$$\frac{I}{C} = \frac{V}{RC} \quad \text{and} \quad V_s = \frac{VT_s}{RC} \tag{14-14}$$

where

$$V = V_{YY} - V_V \qquad V_s = V_P - V_V \tag{14-15}$$

The slope error is, from Eq. (14-9),

$$e_s = \frac{V_s}{V} = \frac{T_s}{RC} \tag{14-16}$$

At the end of sweep time T_s, the capacitor discharges and its final value is V_V, as indicated in Fig. 14-10b. The recovery waveform is quite different with a tube switch from that obtained with a transistor switch. These two situations are discussed in detail in Sec. 8-12, and the waveforms are pictured in Figs. 8-26d and 8-27d (in the neighborhood of $t = T_2+$).

If a negative-going sweep is desired, then a p-n-p transistor may be used, with V_{YY} replaced by $-V_{CC}$. With a tube switch it is not possible to obtain a negative-going sweep directly, but an (operational) amplifier may be used to invert the output.

A Time-base System The essential blocks of many commercial[3] high-frequency scopes (built with either tubes or transistors) are shown in Fig. 14-11. The circuit is more complicated but more versatile than one using a negative-resistance switch and is capable of the much higher sweep speeds which are required in a high-frequency oscilloscope. The sweep circuit is intended to be used as a triggered (driven) sweep at all times. Typically, the vertical amplifier response extends from zero to some tens of megacycles per second (rise time equals a few tens of nanoseconds), and the sweep time is continuously adjustable from 1 sec/cm to 0.1 μsec/cm.

Ignoring for the moment the refinement introduced by the dashed block marked "Hold-off circuit" (explained in the next section), we shall describe the remainder of the system. The first two blocks in Fig. 14-11 serve the same purpose as the corresponding circuits in Fig. 14-9. The sync amplifier need not operate linearly, since all that is required is that the output be large and fast enough to be able to trigger the gate generator. In some scopes a Schmitt discriminator (Sec. 10-11) is used (not indicated in Fig. 14-11) to obtain a pulse at some definite voltage on either the rising or falling portion of the signal, as desired. Since this trigger is used to start the sweep, a selected portion of the input signal appears on the scope face.

The third block in Fig. 14-11 may be a monostable multivibrator (Chap. 11) whose gate width is determined by a resistor R_1 and a capacitor C_1. A negative gating waveform (v_i of Fig. 14-10b) obtained from output Y_1 of the multi is applied to the sweep generator of Fig. 14-10a, the sweep speed of which depends upon R and C. It may happen that the sweep circuit is adjusted in such a manner that the interval between triggering pulses is only slightly longer than the sweep time T_s. In this case, there will not be adequate time for the sweep capacitor to discharge to V_V between sweeps. If the sweep

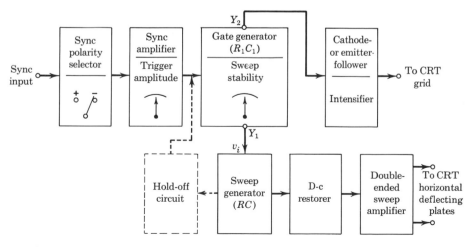

Fig. 14-11 A block diagram of a time-base system using transistors or vacuum tubes.

generator is d-c-coupled through the amplifier to the CRT, the starting point of the sweep on the CRT screen will depend on the extent to which the sweep capacitor has been able to discharge. To avoid this shift in starting point the sweep generator is coupled to the amplifier through a d-c restorer.

In a case in which the sweep time is short in comparison with the time between sweeps the CRT beam will remain in one place most of the time. If the intensity is reduced to prevent screen burns, the fast trace will be very faint. To intensify the trace during the sweep, a positive gate which is derived from the output Y_2 of the multi is applied to the CRT grid. As a matter of fact, in the presence of this "unblanking" or "intensifier signal" the beam brightness may be adjusted so that the spot is initially invisible but the trace will become visible as soon as the sweep starts.

The gating circuit has a bias gain control labeled "Sweep stability," which may be set so that the monostable multi functions in an astable mode (Sec. 11-11). In operation the "trigger-amplitude" control is initially set so that no sync signal reaches the multi. The sweep-stability control is adjusted so that the multi is astable and a trace appears on the CRT screen. The sweep-stability control is then adjusted until the trace just disappears. The multi is now biased for monostable operation. Under these circumstances, the trigger amplitude is increased just to the point where the sweep reappears, and the sweep-time controls are adjusted until a stationary pattern is observed.

We have already noted that the sweep speed is determined by RC in the sweep generator, whereas the gate width is determined by R_1C_1 in the gate generator. If the sweep amplitude is to remain nominally constant, the gate controls R_1 and C_1 must be adjusted whenever the sweep speed controls R and C are varied. Capacitors C_1 and C are switched simultaneously to change the range of sweep speed, and resistor R, which is used for continuous variation of sweep speed, is ganged to R_1. No attempt is made to maintain constant amplitude with any precision. The sweep amplitude is deliberately made so large that the end of the sweep occurs at a point well off the CRT screen, so that variations of amplitude are not observed.

14-6   A FIXED–AMPLITUDE SWEEP

An excellent method[4] of obtaining a fixed-amplitude time base is to replace the monostable gate with a bistable circuit and to add a feedback loop from the output back to the input of the binary via the "hold-off circuit," as indicated by the dashed block of Fig. 14-11. This part of the system is detailed in Fig. 14-12 for a transistorized sweep, although the same technique has been implemented with tubes. The description of the circuit operation now follows.

The gate generator is an emitter-coupled binary (a Schmitt circuit) with a large hysteresis voltage V_H (Sec. 10-11). Initially transistor $Q5$ is cut off and diode $D2$ is conducting, so that the stability-control P may be used to

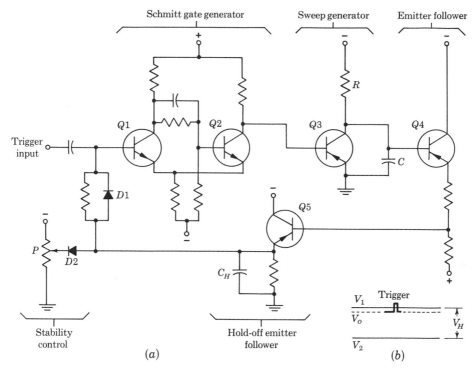

Fig. 14-12 (a) A technique for generating a time-base waveform of fixed ampli-tude. (b) The Schmitt critical levels are V_1 and V_2. The base of $Q1$ is initially at V_o. An input trigger causes a transition, and then the voltage fed back from the sweep drives the base of $Q1$ toward V_2, where the reverse transition takes place.

adjust the bias of $Q1$ to be V_o just below the upper triggering which is level V_1, as indicated in Fig. 14-12b. Hence, $Q1$ is OFF, $Q2$ is ON, and $Q3$ is in saturation, so that the voltage across the sweep capacitor C is clamped to $V_{CE}(\text{sat})$. The trigger input waveform consists of positive and negative sync pulses (obtained by shaping the input signal). A negative pulse is bypassed by $D1$ but a positive signal larger than $V_1 - V_o$ causes a transition in the binary. The resulting positive step at $Q2$ reverse-biases $Q3$ and initiates a negative-going sweep voltage across C. The emitter follower $Q4$ is used for the sake of its high input impedance. This emitter follower transmits the sweep to the hold-off emitter follower $Q5$, which, as noted, is initially cut off. After some time elapses, the negative-going sweep turns on $Q5$. The falling voltage at the emitter of $Q5$ eventually causes $D2$ to become reverse-biased. Thereafter, the base of $Q1$ is free to follow the negative-going output voltage of $Q5$. Because of the hysteresis of the binary circuit, this reduction in voltage at the input to $Q1$ will not immediately cause a reversal of state of the binary. But eventually, at a definite sweep amplitude the input to $Q1$

is brought to the lower triggering level V_2 of Fig. 14-12b.   The emitter-coupled binary makes a reverse transition at this time T_s, so that $Q3$ is again driven into saturation and capacitor C starts to discharge.

As the capacitor C discharges, the transistor $Q5$ returns to cutoff.   Eventually the circuit will settle itself again in its initial state, with diode $D2$ conducting and the base of $Q1$ at the level V_o just slightly below V_1.   However, before this initial state may be attained the hold-off capacitor C_H must discharge through the emitter resistor of $Q5$ to allow the anode side of $D2$ to return to its initial level.   The presence of C_H, therefore, delays the return to the initial state and allows time for capacitor C to discharge completely before the gate generator is able to respond to the next sync pulse.   Since the time required to allow C to discharge depends on its capacitance, the hold-off capacitor C_H is switched when C is switched to change sweep speed.

It is possible to operate the time base in a free-running fashion by rendering inoperative the clamping action of the diode $D2$.   In this case, at the termination of one sweep and without the need for a triggering signal, the base of $Q1$ will eventually rise to the point where a second sweep is initiated.   Ordinarily, however, driven operation is employed.   In this case, in the absence of a triggering signal, the sweep-stability control is adjusted just slightly below the voltage V_1, where free-running operation is restrained.   Thereafter, as explained above, a triggering signal initiates the formation of the time base.

The technique just described not only establishes a fixed sweep amplitude but also ensures that the excessively large supply voltage used to minimize sweep-speed error [Eq. (14-6)] never appears across the transistor, where it could cause breakdown.   Some care must be exercised in using this circuit with vacuum tubes, for in that case, the voltage limitations of the switch tube and sweep capacitor may well be exceeded before the comparator tube or tubes warm up.

The remainder of this chapter is concerned with the generation of time bases possessing a high degree of linearity.   The technique just described for obtaining a fixed-amplitude sweep is applicable to all of these if the time base under consideration is substituted for the simple sweep generator of Fig. 14-12.

14-7   A TRANSISTOR CONSTANT–CURRENT SWEEP

Except for very small values of collector-to-base voltage, the collector current of a transistor *in the common-base configuration* is very nearly constant when the emitter current is held fixed (Fig. 6-12).   We may use this characteristic to generate a quite linear sweep by causing a constant current to flow into a capacitor.   In the circuit of Fig. 14-13a, if V_{EB} is the emitter-to-base voltage, the emitter current is

$$I_E = \frac{V_{EE} - V_{EB}}{R_e} \tag{14-17}$$

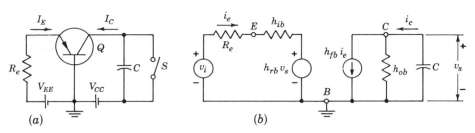

Fig. 14-13 (a) The constant collector current of a CB transistor is used to charge a capacitor linearly; (b) the small-signal model from which to calculate the slope error ($v_i = V_{EE} - V_\gamma \equiv V_i$).

Assuming, for the moment, that V_{EB} remains constant with time after the switch S is opened, then the collector current will be a constant whose nominal value is $I_C = h_{FB}I_E \approx -\alpha I_E$, and the capacitor will charge linearly with time.

To inquire into the sources of nonlinearity we shall replace the transistor by an equivalent circuit using CB hybrid parameters (Sec. 1-3). This model is valid only for changes from the quiescent values. Hence, let us define the initial condition as one where the transistor is just brought to the point of conduction by an emitter threshold bias of V_γ (Sec. 6-17) and a collector voltage V_{CC}. The equivalent circuit from which to determine the sweep voltage v_s is indicated in Fig. 14-13b, where the effective input signal v_i (the change from the quiescent value) is $v_i = V_{EE} - V_\gamma \equiv V_i$. Since only variations from the initial condition are under consideration, then the supply V_{CC} is replaced by its internal impedance (assumed to be negligible).

Kirchhoff's voltage law applied to the input mesh and KCL applied to the output node of Fig. 14-13b yield, respectively,

$$v_i = i_e(R_e + h_{ib}) + h_{rb}v_s = V_i \tag{14-18}$$

$$i_c = i_e h_{fb} + h_{ob}v_s = -C\frac{dv_s}{dt} \tag{14-19}$$

Subject to the initial condition that $v_s = 0$ at $t = 0$, the solution to these equations is

$$v_s = \frac{\alpha \tau V_i}{C(R_e + h_{ib})}(1 - \epsilon^{-t/\tau}) \tag{14-20}$$

where $\alpha = -h_{fb}$, $V_i = V_{EE} - V_\gamma$, and

$$\frac{1}{\tau} = \frac{1}{C}\left(h_{ob} + \frac{\alpha h_{rb}}{R_e + h_{ib}}\right) \tag{14-21}$$

Expanding the exponential into a power series in t/τ and retaining only the first term,

$$v_s = \frac{\alpha V_i t}{C(R_e + h_{ib})} \tag{14-22}$$

This result is consistent with Fig. 14-13b, which gives an emitter current of $i_e = V_i/(R_e + h_{ib})$ at $t = 0+$ and a short-circuit collector current α times as large. If the capacitor current were to remain constant at this value, then $v_s = \alpha i_e t/C$, in agreement with Eq. (14-22).

The sweep amplitude V_s is obtained from Eq. (14-22) with $t = T_s$. The slope error, given by Eq. (14-9), is

$$e_s = \frac{T_s}{\tau} = \frac{V_s}{V_i}\left[h_{rb} + \frac{h_{ob}}{\alpha}(R_e + h_{ib})\right] \tag{14-23}$$

where use was made of Eqs. (14-21) and (14-22). The generator $h_{rb}v_s$, representing the reaction of the collector voltage on the input circuit, causes the emitter current to change as the sweep forms. The first term in Eq. (14-23) results from this change in emitter current. The fact that the collector current is not precisely constant (even for constant emitter current) because of the shunting effect of h_{ob} gives rise to a slope error whose value is given by the second term in Eq. (14-23).

This circuit cannot be loaded appreciably without seriously deteriorating the linearity. If a load R_L is placed across C, then this resistance appears in parallel with $1/h_{ob}$ in Fig. 14-13b. Hence, in Eq. (14-23), h_{ob} must be replaced by $h_{ob} + 1/R_L$. Since $1/h_{ob} \approx 2$ M, then even if $R_L = 2$ M (an unreasonably large value) the second term in Eq. (14-23) is doubled. It is therefore clear that the sweep voltage must be applied to the load by means of an emitter follower and that the input impedance of the follower must be taken into account since it shunts C.

A source of concern in the circuit of Fig. 14-13, as in transistor circuits generally, is the temperature dependence of the device parameters. This variation does not cause sweep nonlinearity, but does make the sweep rate a function of temperature. In Sec. 6-2 it is pointed out that the emitter-junction voltage V_{EB}, for a fixed current, decreases about 2 mV/°C. Hence, from Eq. (14-17), the sweep speed $|I_C|/C = \alpha I_E/C$ increases with temperature.

The circuit of Fig. 14-13 may be modified[5] as in Fig. 14-14 to employ a single supply V_{YY} and to achieve considerable temperature compensation. The emitter supply voltage V_{EE} of Fig. 14-13a is now equal to the voltage V_Z across the Zener diode D_Z plus the drop V_D across D. The collector supply voltage V_{CC} is now equal to the drop across R, so that $V_{CC} = V_{YY} - V_{EE}$. The diode D is of the same material as the transistor (both silicon or both germanium). Hence, D serves to compensate for the temperature-dependent

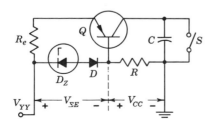

Fig. 14-14 A constant-current sweep circuit involving temperature compensation and using a single supply V_{YY}.

emitter-to-base voltage V_{EB}. If the drops across D and across the emitter junction are always the same, then the voltage across R_e remains equal to V_Z. The emitter current is then V_Z/R_e, which is constant provided that a temperature-compensated avalanche diode (Sec. 6-4) is employed. Under these circumstances the sweep speed is essentially independent of temperature.

EXAMPLE   The circuit of Fig. 14-14 has the following parameter values: $R_e = 1$ K, $R = 6.8$ K, $C = 0.05\,\mu\text{F}$, $V_{YY} = 24$ V, and $V_Z = 8.0$ V. The transistor CB h parameters are $\alpha = 0.98$, $h_{ib} = 20\,\Omega$, $h_{rb} = 3 \times 10^{-4}$, and $h_{ob} = 0.5\,\mu\text{mho}$. The temperature coefficient of the avalanche diode is 0.002 percent/°C. All the devices are silicon.

Calculate (a) the sweep speed, (b) the maximum sweep voltage and the corresponding sweep duration, (c) the slope error at no load and with a 100-K load, (d) the percentage change in sweep speed over a 100°C variation in temperature. (e) Repeat part d if a fixed supply voltage of 8.6 V were used for V_{EE}.

Solution   *a.* The sweep speed is

$$\frac{|I_C|}{C} \approx \frac{\alpha I_E}{C} = \frac{\alpha V_Z}{R_e C} = \frac{(0.98)(8.0)}{(10^3)(0.05 \times 10^{-6})} = 1.57 \times 10^5 \text{ V/sec}$$

b. The capacitor C charges at a constant rate only as long as the collector-to-base voltage is negative (for a p-n-p transistor). Hence the sweep must be terminated when the capacitor voltage approaches V_{CC}, where

$$V_{CC} = V_{YY} - V_{EE} = 24 - 8.0 - 0.6 = 15.4 \text{ V}$$

and where the voltage drop across D is taken as 0.6 V. To allow additionally for the drop from collector to base to keep the transistor in the active region let us take $V_s = 14$ V. Then

$$T_s = \frac{V_s C}{|I_C|} = \frac{14}{1.57 \times 10^5} \text{ sec} = 89\,\mu\text{sec}$$

c. Since the drop V_D across the diode is about 0.6 V and the cutin base-to-emitter voltage V_γ is about 0.5 V (Table 6-1), then

$$V_i = V_{EE} - V_\gamma = V_Z + V_D - V_\gamma \approx 8.0 + 0.6 - 0.5 = 8.1 \text{ V}$$

From Eq. (14-23)

$$e_s = \frac{14.0}{8.1}\left[3 \times 10^{-4} + \frac{0.5 \times 10^{-6}}{0.98}(1,020)\right] \times 100 = 0.05 + 0.09$$

$$= 0.14 \text{ percent}$$

This excellent linearity worsens if an appreciable load resistance R_L is placed across C. For example, if $R_L = 100$ K, we must replace $h_{ob} = 0.5 \times 10^{-6}$ with $h_{ob} + 1/R_L = 0.5 \times 10^{-6} + 10^{-5} = 10.5 \times 10^{-6}$, so that the second term in e_s is multiplied by $10.5/0.5 = 21$. The slope error would then become

$$e_s = 0.05 + (21)(0.09) = 1.9 \text{ percent}$$

d. Assuming perfect temperature tracking of V_D and V_γ, then the slope error is due to the variation in V_Z. Over 100°C this is $0.002 \times 100 = 0.2$ percent.

e. If $V_{EE} = 8.6$ V = constant, and since V_{EB} changes by 200 mV = 0.2 V in 100°C (Eq. 6-7), then the emitter current [Eq. (14-17)], and hence also the sweep speed, changes by

$$\frac{0.2}{8.1} \times 100 = 2.5 \text{ percent}$$

We note that this is worse by a factor of 12.5 than the corresponding variation in part *d*, where D and D_Z were used to simulate the supply V_{EE}.

14-8   A VACUUM–TUBE CONSTANT–CURRENT SWEEP

We note from Eqs. (1-44) that the resistance seen between the points P and N in a tube circuit with a cathode resistor R_k is $R = r_p + (\mu + 1)R_k$. This resistance will be very large if μ and R_k are both large. If a constant voltage is placed in series with this very large resistance, the resultant circuit will constitute a source of constant current. Such a constant-current source is used in Fig. 14-15 to charge the capacitor C linearly with time. With switch S closed, the diode D conducts until the voltage across C_g is established at the reference voltage V_R determined by R_1, R_2, R_g, and V_{PP}. We assume that V_R is much greater than the diode drop or the grid-to-cathode voltage V_G. Then V_R appears across R_k, and the plate current is

$$I \approx \frac{V_R}{R_k} = \frac{V_{PP}}{R_k} \frac{R_2'}{R_1 + R_2'} \tag{14-24}$$

where $R_2' = R_2 \| R_g \approx R_2$ if $R_g \gg R_2$. If V_G is not negligible compared with V_R, then I is determined by the graphical method outlined in Sec. 1-10. When S opens, and assuming $R_g C_g \gg T_s$, the voltage across C_g remains constant at V_R. As C charges, the voltage from grid to ground increases

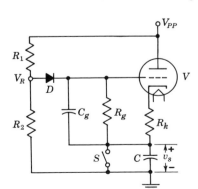

Fig. 14-15 A time-base generator in which a constant current is generated through the use of a vacuum tube operating with a large cathode resistance.

above V_R and the diode becomes reverse-biased. Now the tube current I flows into C. Hence, C charges at a sweep speed of I/C.

The charging time constant is $\tau = RC$, and the slope error is, from Eq. (14-9),

$$e_s = \frac{T_s}{\tau} = \frac{T_s}{[r_p + (\mu + 1)R_k]C} \tag{14-25}$$

If the finite grid-leak time constant is taken into account, then it is found that to the above expression for e_s we must add the term T_s/R_gC_g (Prob. 14-18). At a time T_s the switch S closes, D conducts, and the charge lost by C_g is replaced through D.

The sweep amplitude is

$$V_s = \frac{IT_s}{C} \approx \frac{V_R T_s}{R_k C} \tag{14-26}$$

If $\mu \gg 1$ and if $\mu R_k \gg r_p$, then, from Eqs. (14-25) and (14-26),

$$e_s \approx \frac{T_s}{\mu R_k C} = \frac{V_s}{\mu V_R} \tag{14-27}$$

Comparing this result with Eq. (14-6) we see that the circuit behaves as if the capacitor were being charged from a supply whose value is μV_R through a resistance μR_k, so that the charging current is $\mu V_R/\mu R_k = V_R/R_k$, as indicated in Eq. (14-24).

One difficulty with the circuit is the possibility of breakdown between the cathode and the grounded heater if V_R is too large. A second disadvantage is that there may be power-line hum across R_k due to heater-to-cathode leakage, and this hum will modulate the sweep current. A third difficulty is that a very long recovery time must be allowed if the sweep speed is not to depend on the sweep repetition rate. During the formation of the sweep, capacitor C_g loses some charge through R_g. When switch S closes, C_g recharges through D with a time constant $R'C_g$, where R' is the parallel combination of R_1 and R_2 in Fig. 14-15. Since, during recharging, the capacitor voltage is completing the last portion of its asymptotic approach to the limiting voltage V_R, a recharging time must be allowed which is very long in comparison with $R'C_g$. The recharging cycle may be hastened by returning diode D to a low-impedance supply.

EXAMPLE Design the constant-current time-base generator of Fig. 14-15 to furnish a 100-V sweep in 100 μsec. Find the sweep-speed error.

Solution Assume that a 300-V supply is available for V_{PP}. From Eq. (14-27) we see the advantage of choosing μ and V_R as large as possible. We shall select a type 12AX7 tube, for which $\mu = 100$ and $r_p = 100$ K. At the end of the sweep the voltage across the tube is $V_{PP} - V_R - V_s$, and in order for the tube to behave linearly let us not allow the tube voltage to fall below 50 V. Then the maximum value for V_R is

$$V_R = 300 - 100 - 50 = 150 \text{ V}$$

To obtain this value of V_R from a 300-V supply requires that $R_1 = R_2$, assuming that $R_g \gg R_1$. For the sake of shortening the recovery time it is advantageous to make R_1 and R_2 small. But to limit the power dissipated in the divider we select $R_1 = R_2 = 20$ K.

The value of R_k is determined by the tube current. With 50 V across the tube, we see from the 12AX7 characteristics of Fig. D-8 that the tube is operating linearly at a plate current of 0.5 mA and a grid voltage of 0.5 V. Therefore, a reasonable choice for R_k is

$$R_k = \frac{V_R}{I} = \frac{150}{0.5} = 300 \text{ K}$$

From Eq. (14-26)

$$C = \frac{V_R T_s}{V_s R_k} = \frac{(150)(100)}{(100)(300,000)} \mu\text{F} = 500 \text{ pF}$$

Since $\mu R_k = 3 \times 10^7 \gg r_p = 10^5$, we may use Eq. (14-27) instead of Eq. (14-25) for the slope error,

$$e_s = \frac{V_s}{\mu V_R} = \frac{100}{100 \times 150} = 0.0067$$

or the sweep-speed error is 0.67 percent if we neglect $T_s/R_g C_g$ compared with 0.0067. Hence, let us choose $R_g C_g$ so as to make $T_s/R_g C_g = 0.0001$. Using $R_g = 1$ M we have

$$C_g = \frac{T_s}{10^{-4} R_g} = \frac{100}{10^{-4} \times 10^6} = 1 \ \mu\text{F}$$

We may check the sweep-speed error directly from the tube characteristics rather than use the small-signal parameter μ. At the end of the sweep we choose $V_G = -0.5$ V, so that $I = 150.5/300$ mA. At the beginning of the sweep $v_s = 0$ and the effective supply voltage is 300 V. Drawing a load line corresponding to this voltage and to a 300-K resistance and using the graphical construction of Sec. 1-10, we find that $V_G \approx -1.5$ V. Hence I at the beginning of the sweep is $I = 151.5/300$ mA. Therefore the sweep-speed error is

$$e_s = \frac{\Delta I}{I} = \frac{1}{151} = 0.0066$$

in excellent agreement with the value found above. In summary, we find, using a 12AX7 and any low-current diode that can withstand a reverse voltage of 100 V, that $R_1 = R_2 = 20$ K, $R_g = 1$ M, $R_k = 300$ K, $C_g = 1 \ \mu\text{F}$, and $C = 500$ pF.

The plate current in a pentode, for fixed screen-to-cathode voltage, is largely independent of plate voltage except for quite low voltages. Therefore a capacitor charging from a supply voltage through a pentode will charge at approximately constant current. For example, in a 6AU6, if the supply voltage is 300 V, the grid bias -2 V, and the screen voltage 150 V, the initial charging current is 5.8 mA (Fig. D-1). When the capacitor voltage is 200 V, the tube voltage is 100 V, and the charging current is 5.7 mA. The percentage slope error is therefore $100 (5.8 - 5.7)/5.8 \approx 2$ percent. Without the pentode the slope error would be $100 \times 200/300 = 67$ percent. An arrangement in which a capacitor appears in series with a pentode and in shunt with a switch tube is given in Prob. 14-20.

14-9   MILLER AND BOOTSTRAP TIME–BASE GENERATORS—GENERAL CONSIDERATIONS

We now consider circuits involving feedback for improving the linearity of time-base waveforms. The basic sweep circuit is shown in Fig. 14-16a, in which S opens to form the sweep. If, as in Fig. 14-16b, an auxiliary variable generator v is introduced and if v is always kept equal to the voltage drop across C, the charging current will be kept constant at $i = V/R$ and perfect linearity will have been achieved. Methods of simulating the fictitious generator v with an amplifier will now be given.

It is common practice to build electronic circuits on a metallic chassis (Sec. 3-13) and to have one point in the circuit electrically connected to the chassis. The chassis is then referred to as *ground*, and customarily the voltage at any point in the circuit is given with respect to ground. Differences in the point selected to be grounded yield different descriptions of the mechanism whereby the linearity is improved. Suppose that the point Z of Fig. 14-16b is grounded, as in Fig. 14-17a. A linear sweep will appear between Y and ground and will increase in the negative direction. Let us now replace the fictitious generator by an amplifier with output terminals YZ and input terminals XZ, as in Fig. 14-17b. Since we have assumed that the magnitude of the generator voltage v equals the voltage v_C across the capacitor at every instant of time, then the input v_i to the amplifier is zero. In other words, point X behaves as a virtual ground, and in order to obtain a finite

Fig. 14-16 In (a) the current varies exponentially with time, whereas in (b) it remains constant, provided that v is the instantaneous voltage v_C across C.

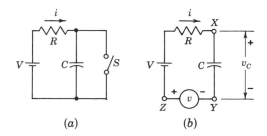

(a)   (b)

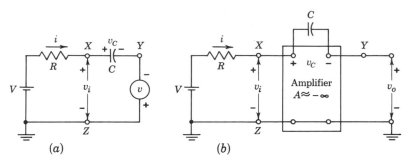

Fig. 14-17 (a) Figure 14-16b with point Z grounded; (b) the same circuit simulated with an operational (Miller) integrator with $A \approx -\infty$.

output, the amplifier gain A should ideally be infinite. Figure 14-17b should be recognized as the operational integrating amplifier of Sec. 1-8 and is customarily referred to as a *Miller integrator* or *Miller sweep*.

Suppose that point Y of Fig. 14-16b is grounded. A linear sweep will appear between Z and ground and will increase in the positive direction. Let us now replace the fictitious generator by an amplifier with output terminals ZY and input terminals XY, as in Fig. 14-18. Since we have assumed that $v = v_C$, then $v_o = v_i$ and the amplifier voltage gain A must equal unity. The circuit of Fig. 14-18 is referred to as a *bootstrap sweep*, since the voltage V is lifted, as it were, by its own bootstraps.

The Miller Sweep The circuit of Fig. 14-17b has been redrawn in Fig. 14-19a. We have included a switch, at the closing of which the time-base waveform is to start, and also a resistor R_i, which represents the input impedance of the amplifier. The base amplifier has been replaced, with respect to its output terminals, by its Thévenin's equivalent. Here R_o is the output resistance and A is the *open-circuit voltage gain* of the base amplifier. In Fig. 14-19b we have replaced the input circuit, consisting of V, R, and R_i, by its Thévenin's equivalent, in which

$$V' = \frac{V R_i}{R_i + R} = \frac{V}{1 + R/R_i} \quad \text{and} \quad R' = \frac{R_i R}{R_i + R} \qquad (14\text{-}28)$$

The circuits of Fig. 14-19 are to be used only to calculate departures from an initial quiescent state.

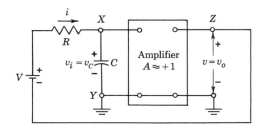

Fig. 14-18 The circuit of Fig. 14-16b with point Y grounded implemented with a noninverting unity-gain amplifier (a bootstrap circuit).

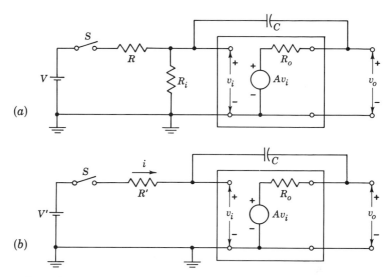

Fig. 14-19 (a) A Miller integrator showing starting switch, amplifier input resistance, and a Thévenin's circuit for the base amplifier; (b) the input circuit is replaced by a Thévenin's equivalent.

Let us, for the moment, neglect R_o. We consider that the capacitor voltage is zero and that the switch S is closed at $t = 0$. Since the capacitor voltage was zero at $t = 0-$, it must be zero at $t = 0+$. The voltage across the capacitor at $t = 0+$ is $v_i - Av_i = (1 - A)v_i = 0$. Hence

$$v_i = Av_i = v_o = 0$$

at $t = 0+$. Thus, immediately after the closing of the switch the output is unchanged.

At $t = \infty$, when the capacitor is completely charged and no current flows through it, we may remove the capacitor for the purpose of finding the output voltage. Then we find at $t = \infty$ that $v_i = V'$ and therefore $v_o = AV'$.

The output waveform v_o is a simple exponential since the circuit involves just the single capacitor. Thus, altogether, we have the result that the output is exponential with total excursion AV', starting at zero. The sweep is nega-tive-going since A is a negative number.

We have already noted that for an exponential waveform the slope error is given exactly by $e_s = V_s/V$ [Eq. (14-6)], where V_s is the sweep ampli-tude and V is the total peak-to-peak excursion of the exponential. Therefore the time-base waveform v_o has the slope error, using Eq. (14-28),

$$e_s(\text{Miller}) = \frac{V_s}{|A|V'} = \frac{V_s}{V}\frac{1 + R/R_i}{|A|} \tag{14-29}$$

In summary we find that the deviation from linearity is $(1/|A|)(1 + R/R_i)$ times that of a resistance-capacitance circuit charging directly from a source V.

Now let us take R_o into account. The final value attained by v_o remains, as before, $AV' = -|A|V'$. The initial value, however, is slightly different. To find v_o at $t = 0+$ we write KVL for the mesh in Fig. 14-19b, including V', R', C, R_o, and the generator Av_i. Again assuming zero voltage across the capacitor we have

$$V' - R'i - R_o i - Av_i = 0 \tag{14-30}$$

and

$$v_i = V' - R'i \tag{14-31}$$

From Eqs. (14-30), (14-31), and (14-28) we find

$$v_i(t = 0+) = \Delta V_i = v_o(t = 0+) = \Delta V_o = \frac{(R_o/R')V'}{1 - A + R_o/R'} \tag{14-32}$$

$$v_i(t = 0+) \approx \frac{R_o V'}{R'|A|} \tag{14-33}$$

since very often $R_o \ll R'$ and $|A| \gg 1$. We thus find that, taking R_o into account, the output continues as a simple exponential with the same asymptotic limit but that it starts at $v_o = \Delta V_o$ rather than at zero. Thus the negative-going ramp is preceded by a positive jump. Because of this jump the total excursion of the exponential is slightly larger than before, and consequently the linearity is improved thereby. However, this jump is ordinarily too small in comparison with the excursion AV' to be of any consequence with respect to linearity.

We may calculate the sweep speed of the Miller circuit by recalling, as discussed in Sec. 1-8, that a virtual short circuit exists at the input terminals to the base amplifier. Then, in Fig. 14-19b the current $i = V'/R'$ very nearly. This current continues through C, so that the sweep speed from Eq. (14-13) is

$$\text{Sweep speed} = \frac{i}{C} \approx \frac{V'}{R'C} = \frac{V}{RC} \tag{14-34}$$

Thus the sweep speed is the same as in the case where the capacitor charges through R directly from the source V.

The Bootstrap Sweep   The bootstrap circuit has been redrawn in Fig. 14-20. Again R_i is the amplifier input resistance, A the *open-circuit gain of the amplifier*, and R_o its output impedance. Proceeding as in the case of the Miller circuit we find, if S opens at $t = 0$, that at $t = 0+$, $v_o = -R_o V/(R_o + R)$. The output v_o has this same value at $t = 0-$, and hence there is no jump in the output voltage at $t = 0$. At $t = \infty$ we find

$$v_o(t = \infty) = \frac{V(R_i A - R_o)}{R_i(1 - A) + R_o + R} \approx \frac{V}{(1 - A) + R/R_i} \tag{14-35}$$

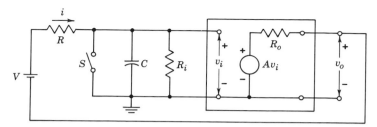

Fig. 14-20 The bootstrap sweep showing starting switch, amplifier input resistance, and a Thévenin's equivalent of the base amplifier.

The approximations in Eq. (14-35) are reasonable because the amplifier in Fig. 14-20 will be either an emitter follower or a cathode follower for which $A \approx 1$, $R_i \gg R_o$, and $R \gg R_o$. Further, since $R_o \ll R$, we may neglect v_o at $t = 0+$ in comparison with v_o at $t = \infty$. Then the sweep amplitude is given by Eq. (14-35), and the slope error is

$$e_s(\text{bootstrap}) = \frac{V_s}{V}\left(1 - A + \frac{R}{R_i}\right) \qquad (14\text{-}36)$$

In summary, we find that the slope error is $(1 - A + R/R_i)$ times the slope error that would result if the capacitor charged directly from V through a resistor. It appears in comparing Eq. (14-36) with Eq. (14-29) that it is more important to keep R/R_i small in the bootstrap circuit than in the Miller circuit. Thus, even if we assume $A = 1$, if $R = R_i$ the bootstrap will provide no improvement in linearity. Therefore the Miller integrator enjoys some advantage over the bootstrap circuit in that in the Miller circuit a high amplifier input impedance is less important.

In Fig. 14-20, at $t = 0+$, $v_i = 0$ and $i = V/(R + R_o) \approx V/R$, since $R \gg R_o$. This current flows into the capacitor C, so that the sweep speed at the input to the amplifier is i/C, and at the output

$$\text{Sweep speed} = \frac{Ai}{C} \approx \frac{AV}{RC} \qquad (14\text{-}37)$$

We have neglected here the very small rate of change of voltage across R_o. Since $A \approx 1$, the sweep speed is the same as would result if the capacitor C had been charged directly from V through R. The sweep speed does not depend on the input resistance R_i.

When vacuum tubes are used in the amplifiers of the Miller and bootstrap circuits the input resistance R_i is quite independent of the amplifier load and the output impedance is independent of the source impedance. In the Miller circuit, when transistors are used, the amplifier may well consist of a number

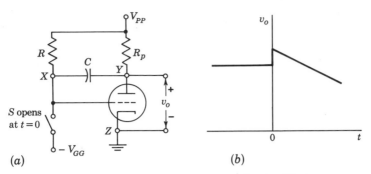

Fig. 14-21 (a) A Miller run-down sweep circuit; (b) the output waveform.

of stages beginning and ending with an emitter follower. Hence, again, input and output are well isolated. A bootstrap amplifier normally involves just a single emitter follower, and, in this case, we shall have to take account of the effect of the loading on the input impedance.

14-10   THE VACUUM–TUBE MILLER TIME–BASE GENERATOR[6]

A simple tube sweep circuit is shown in Fig. 14-21a. The negative bias is chosen so that the tube is operating within its grid base. Observe that V_{PP} is used both to charge C and to supply tube current. When S opens, a negative-going sweep (a so-called Miller *run-down* waveform) will appear at the plate. As indicated in Fig. 14-21b, the sweep is preceded by a positive jump whose magnitude is given by Eq. (14-32). The jump may be eliminated by the addition of a resistor $r = 1/g_m$ in series with the capacitor C, as we shall now demonstrate.

If, when S opens at $t = 0$, there is a jump ΔV_i at the input (grid), then the Thévenin's open-circuit voltage at the output (plate) will be $A \Delta V_i$, where A is the voltage gain. The equivalent circuit at $t = 0+$ is indicated in Fig. 14-22. If the jump in voltage at the output is to be zero, then the current I in a short circuit from output to ground must also be zero [Eqs. (1-1)]. Hence,

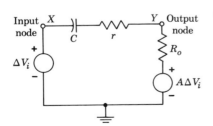

Fig. 14-22 Relating to the elimination of the initial jump in voltage in a Miller integrator (with either tubes or transistors).

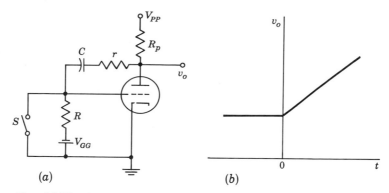

Fig. 14-23 (a) A Miller run-up sweep circuit; (b) the output waveform.

since C acts as a short circuit at $t = 0+$,

$$I = \frac{\Delta V_i}{r} + \frac{A \, \Delta V_i}{R_o} = 0 \qquad (14\text{-}38)$$

or

$$r = -\frac{R_o}{A} \qquad (14\text{-}39)$$

This derivation is equally valid for a transistor stage, where the input node is the base and the output node is the collector. For a single-stage tube amplifier $A = -g_m R_o$, and hence $r = R_o/g_m R_o = 1/g_m$, as anticipated above.

A circuit with a positive-going sweep (a so-called Miller *run-up* waveform) and including the resistor r to eliminate the initial step is indicated in Fig. 14-23. Note that now two power supplies are required, one for the charging current and the other to supply the tube current.

A Miller Sweep with Symmetrical Outputs In the circuit of Fig. 14-24 triode $V1$ with its grid clamped to ground acts as the closed switch S of Fig. 14-21. A negative gate applied to the grid cuts $V1$ off and allows C to charge from V_{PP} through R and the Miller tube $V2$. The output v_{o1} from the plate of $V2$ is a negative-going sweep. This voltage is applied to the grid of $V3$ through the resistor R_2, and an equal resistor R_2 is used for feedback from plate to grid of $V3$. Hence, $V3$ acts as an operational phase inverter (Sec. 1-8), and the output v_{o2} is a positive-going sweep. The symmetrical voltages v_{o1} and v_{o2} may be used to drive CRT horizontal-deflecting plates.

Note that the negative bias of Fig. 14-21 has been replaced by the voltage across the cathode resistor R_k. This resistor does *not* introduce degeneration because the current through it remains constant. Thus, as the current in $V2$ increases, the symmetrical current in $V3$ decreases by the same amount, leaving the current in R_k unchanged. The grid leak for $V3$ is R_3, and the blocking

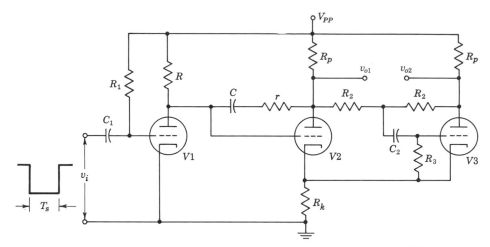

Fig. 14-24 A symmetrical Miller sweep. The switch tube is $V1$, the Miller integrator is $V2$, and the operational inverter is $V3$.

capacitor C_2 keeps the high plate voltage of $V2$ and $V3$ from reaching the grid of $V3$. The time constant R_3C_2 must be large enough to introduce negligible transmission error (Sec. 14-1).

From Eq. (14-29) with $R_i = \infty$, the slope error is

$$e_s = \frac{V_s}{V}\frac{100}{|A|} \text{ percent} \tag{14-40}$$

where $V = V_{PP} - V_V$, and V_V is the clamped plate voltage of $V1$. If we choose the swing V_s at each plate to equal $V/2$, a push-pull sweep of amplitude V is obtained. It is not difficult to realize a voltage gain of $|A| = 50$ (say with a 12AX7, for which $\mu = 100$), and Eq. (14-40) gives $e_s = 1$ percent. Here, then, is a relatively simple circuit that gives a push-pull sweep which approaches the supply voltage in magnitude (for $V_V \ll V_{PP}$) and has a sweep-speed error of only 1 percent. Note that the displacement error is only 0.125 percent [Eq. (14-4)].

Recovery Considerations At the end of the sweep, C discharges and returns to its quiescent voltage. The discharge path is through the output impedance of the amplifier and through a switching tube such as $V1$ of Fig. 14-24. If the switch resistance is $R_s \ll R$ and since $r \ll R_o$, the discharge path is indicated in Fig. 14-25a. This circuit is in the form of Fig. 1-7 with Z' consisting of C in series with R_o. Applying the Miller theorem, the equivalent circuit from which to calculate the recovery time constant τ_r is given in Fig. 14-25b. Hence

$$\tau_r = C(1 - A)\left(R_s + \frac{R_o}{1 - A}\right) = C[R_o + (1 - A)R_s] \tag{14-41}$$

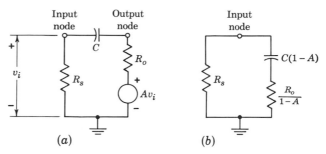

Fig. 14-25 (a) The circuit through which the timing capacitor C discharges at the end of the sweep; (b) the retrace time-constant network. (These equivalent circuits are equally valid for tube or transistor configurations.)

Note that, unfortunately, the switch resistance is effectively increased by the gain of the stage and hence that the retrace time may be quite long unless a high-current switching tube with extremely low plate resistance is used. If low recovery time is important, then C should also be kept as small as possible and R chosen sufficiently large to give the desired sweep speed. For practical reasons it is advisable not to permit R to exceed several megohms. Additionally, the use of a cathode follower interposed between the capacitor C and the amplifier output will reduce the recovery time still further. The recovery may also be hastened by selecting the time constant R_1C_1 (Fig. 14-24) to be comparable to the width of the gating signal, since under these circum-

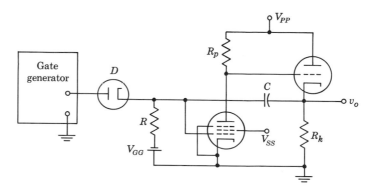

Fig. 14-26 A Miller run-up sweep using a pentode for high gain and a cathode follower to reduce recovery time. The diode D clamps the pentode within its grid base in the quiescent state and disconnects the gate generator from the pentode during **the sweep.**

stances there will be a pronounced overshoot at the grid at the termination of the gate.

A run-up Miller sweep using a cathode follower to speed recovery and using a pentode amplifier for higher gain is shown in Fig. 14-26. Note that the low output impedance of the cathode follower makes the resistor r in series with C (Fig. 14-23) unnecessary [Eq. (14-39)]. The switch S of Fig. 14-23 is now simulated with a diode D connected to the gate generator, say a Schmitt circuit. In one state of the binary the output level is slightly negative, and the pentode is clamped within its grid base by D. When a trigger causes a transition, the output of the Schmitt goes negative enough to reverse-bias the diode. Hence D is disconnected from the circuit and the sweep begins. The switch resistance R_s, of interest during the recovery period, is the diode forward resistance plus the output impedance of the gate circuit; R_s may be as low as 1 or 2 K. A disconnect-diode switch could also have been used in place of $V1$ in the rundown Miller sweep of Fig. 14-24.

The block diagram of a time-base system for a CRO using a Miller sweep is given in Fig. 14-11. In a commercial CRO, the box marked "Sweep generator" may well be the circuit of Fig. 14-26 and the "Hold-off circuit" a tube version of Fig. 14-12. The sweep speed may be changed by adjusting R, C, or the supply V_{GG}.

14-11   THE TRANSISTOR MILLER TIME–BASE GENERATOR

From the linearity considerations given in Sec. 14-9 it is essential that a high-input-impedance amplifier be used for the transistor Miller integrator circuit. Hence, the first stage $Q1$ of Fig. 14-27 is an emitter follower. The transistor $Q2$ supplies the voltage amplification. The output stage $Q3$ is chosen as an emitter follower for a number of reasons. First, because of its low output impedance R_o it can drive a load such as the horizontal amplifier, the hold-off circuit of Fig. 14-11, etc. Second, because of its high input impedance it does not load the collector circuit of $Q2$ appreciably, and the gain provided by this stage can therefore be quite high (in the illustrative problem below it is ~1,000). Finally, from Eq. (14-39) we see that if the gain $|A|$ is large and R_o is small enough, no resistance r need be added in series with C and yet the output will start with virtually no step before the sweep. The timing capacitor C is placed between the base of $Q1$ (the input of the three-stage amplifier) and the emitter of $Q3$ (the output of the amplifier). The sweep speed is changed from range to range by switching R and C and may be varied continuously by varying V_{BB} (using a voltage divider across a supply voltage).

It is possible to employ an arrangement in which the switch appears between amplifier input and ground, as in Fig. 14-24 or 14-26. For gains as high as 1,000, Eq. (14-41) indicates that the retrace time may be prohibitive. Hence, in order to facilitate the rapid discharge of the capacitor the switch transistor $Q4$ and disconnect diode D are placed directly across C. This man-

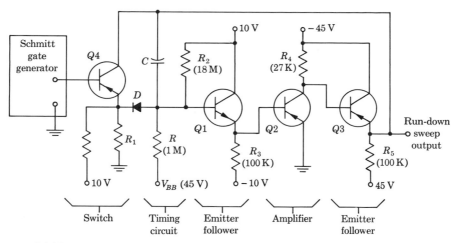

Fig. 14-27 A transistorized Miller time-base generator. The parameter values in parentheses refer to the illustrative example on page 546. (Courtesy of Tektronix, Inc.—Type 321A CRO.)

ner of locating the switch is not so easily realizable with vacuum tubes. In a tube circuit there is always some small leakage of the filament heating power, either through capacitive or inductive coupling from the filament wiring directly to other parts of the circuit or through capacitive coupling or leakage from the heater to the tube cathode. Thus in tube circuits it is necessary to ensure that this leakage is always to a low-impedance point in the network. With transistors, not only is there no heating power, but also because of their small size, transistors may be floated off ground without introducing into the circuit undesirable capacitances to ground.

In the quiescent condition $Q4$ is held in conduction as a result of the state of the Schmitt gate. The emitter voltage of $Q4$ is negative because of the emitter current through R_1, and therefore D is also conducting. The current through R is bypassed around the capacitor C and flows instead through D and $Q4$. Hence C is prevented from charging. A triggering signal initiates the sweep by changing the state of the Schmitt-binary gate generator. The corresponding rise in voltage at the base of $Q4$ turns off this transistor. The positive voltage across R_1 now causes D also to become reverse-biased. At this time the upper terminal of C is connected to the collector of $Q4$, which is in cutoff. The small reverse saturation current which must be supplied to the collector of $Q4$ is easily furnished from the low output impedance of the emitter follower $Q3$. The junction of R and C is connected to a reverse-biased diode. The leakage current through this diode must be supplied through R. Consequently D is selected to be a diode with extremely low leakage. If fast sweeps are to be generated, then D is also required to have a very small storage recovery time (Sec. 20-2).

At the end of the sweep, capacitor C discharges rapidly through $Q4$ and D (Sec. 8-12). The sweep output might go to a hold-off circuit such as that shown in Fig. 14-12 in order to ensure a fixed time-base amplitude. Such a hold-off circuit would also allow adequate time for C to discharge completely between sweeps, so that the time-base calibration would be independent of the duty cycle.

EXAMPLE The Miller integrator of Fig. 14-27 has the following parameters: $R = 1$ M, $V_{BB} = 45$ V, $R_2 = 18$ M, $R_3 = 100$ K, $R_4 = 27$ K, and $R_5 = 100$ K. The transistor h parameters have the values given in Table 1-2. The output sweep amplitude is 25 V. The longest sweep is 5 sec and the shortest is 5 μsec. (a) For $T_s = 5$ sec, find C. (b) Find the slope error e_s. (c) Repeat (a) and (b) for $T_s = 5$ μsec. (d) Repeat (c) if R is changed to 20 K.

Solution *a.* The quiescent base-to-ground voltage of $Q1$ is $(V_{BE})_{Q1} + (V_{BE})_{Q2}$ for both transistors in the active region. Since $Q1$ is an *n-p-n* transistor and $Q2$ is a *p-n-p* type, then V_{BE} is of opposite sign for the two transistors and hence the base voltage of $Q1$ must be almost at ground potential. Hence, the capacitor charging current is $I \approx V_{BB}/R$ and

$$C = \frac{IT_s}{V_s} = \frac{V_{BB}T_s}{RV_s} = \frac{(45)(5)}{(10^6)(25)} \text{ F} = 9 \; \mu\text{F}$$

b. In order to find e_s we must first calculate the voltage gain A_V and input impedance R_i for each stage and then, from these quantities, we can evaluate A_V and R_i for the three-stage amplifier. For the output emitter follower $Q3$ we find, from Eqs. (1-6) to (1-8),

$$A_{I3} = \frac{h_{fe} + 1}{1 + h_{oe}Z_{e3}} = \frac{51}{1 + 100/40} = 14.5$$

$$Z_{i3} = h_{ie} + A_{I3}Z_{e3} = 1.1 + (14.5)(100) = 1,450 \text{ K} = 1.45 \text{ M}$$

$$1 - A_{V3} = \frac{h_{ie}}{Z_{i3}} = \frac{1.1}{1,450} = 0.00076 \quad \text{or} \quad A_{V3} = 0.9992 \approx 1$$

The load Z_{L2} of $Q2$ is $R_4 = 27$ K in parallel with $Z_{i3} = 1.45$ M, or $Z_{L2} = 27$ K to within 2 percent. From Eqs. (1-2) to (1-4) we find for the second stage

$$A_{I2} = \frac{-h_{fe}}{1 + h_{oe}Z_{L2}} = \frac{-50}{1 + 27/40} = -29.8$$

$$Z_{i2} = h_{ie} + h_{re}A_{I2}Z_{L2} = 1.1 + (2.5 \times 10^{-4})(-29.8)(27) = 0.90 \text{ K}$$

$$A_{V2} = \frac{A_{I2}Z_{L2}}{Z_{i2}} = \frac{(-29.8)(27)}{0.90} = -894$$

The effective emitter impedance of $Q1$ is $R_3 = 100$ K in parallel with $Z_{i2} = 0.90$ K, or $Z_{e1} = 0.90$ K to within 1 percent. For the input emitter follower $Q1$

we find, from Eqs. (1-6) to (1-8),

$$A_{I1} = \frac{51}{1 + 0.90/40} = 49.9$$

$$Z_{i1} = 1.1 + (49.9)(0.90) = 46.0 \text{ K}$$

$$1 - A_{V1} = \frac{1.1}{46.0} = 0.024 \quad \text{or} \quad A_{V1} = 0.976$$

The voltage gain of the three stages in cascade is

$$A_V = A_{V1}A_{V2}A_{V3} = (0.976)(-894)(1.00) = -871$$

The input resistance R_i of the amplifier is Z_{i1} in parallel with $R_2 = 18$ M, or $R_i \approx Z_{i1} = 46.0$ K.

Finally, then, from Eq. (14-29),

$$e_s = \frac{V_s}{V_{BB}} \frac{1}{|A|} \left(1 + \frac{R}{R_i}\right) = \left(\frac{25}{45}\right)\left(\frac{1}{871}\right)\left(1 + \frac{1,000}{46.0}\right) = 0.0145$$

The sweep-speed error is 1.45 percent.

It is interesting to note that if transistors with $h_{fe} = 100$ instead of 50 were used, A_V and R_i each would be approximately doubled and e_s would be about one-quarter the value found above, giving a slope error of only 0.36 percent.

If two emitter followers are used in cascade (this connection is discussed in Sec. 14-15), then an input impedance $R_i = 1$ M may be obtained. With this configuration, since $R = 1$ M,

$$e_s = \left(\frac{25}{45}\right)\left(\frac{1}{871}\right)(1 + 1) = 0.00128$$

or only 0.13 percent. Even better linearity can be obtained if a vacuum tube or a field-effect transistor (FET, Sec. 17-19) is used for the input stage. Since the input impedance of the cathode follower or the FET is virtually infinite, then the term $R/R_i = 0$ and e_s is half the above value, or 0.06 percent.

c. For $T_s = 5$ μsec, then

$$C = \frac{V_{BB}T_s}{RV_s} = \frac{(45)(5 \times 10^{-6})}{(10^6)(25)} \text{ F} = 9 \text{ pF}$$

Since the slope error does not depend on the value of the capacitance, the slope error is again $e_s = 1.45$ percent.

d. The value $C = 9$ pF obtained in part *c* is impractically small since it is of the same order of magnitude as the collector-junction capacitance and stray wiring capacitance. A much larger value of C may be used if R is reduced. If R is changed from 1 M to 20 K (a factor of 50), then C is multiplied by this same factor, or $C = (50)(9) = 450$ pF.

This change in R also improves the linearity because R_i will have been increased relative to R. Thus

$$e_s = \frac{V_s}{V_{BB}} \frac{1}{|A|} \left(1 + \frac{R}{R_i}\right) = \left(\frac{25}{45}\right) \left(\frac{1}{871}\right) \left(1 + \frac{20}{46.0}\right) = 0.0009$$

or $e_s = 0.09$ percent, a decided improvement over the value 1.45 percent obtained with $R = 1$ M.

14-12   THE PENTODE MILLER SWEEP WITH SUPPRESSOR GATING

If a pentode is used as the amplifier tube of a Miller sweep, then the gating voltage may be applied to the suppressor grid instead of to the control grid. A suppressor-gated Miller integrator is indicated in Fig. 14-28. A tube with a sharp cutoff suppressor characteristic is used. These include the types 6AS6, 5725, 5915, 6BE6, 6CS6, 5636, and 6BH6. The 6SA7 converter tube[7] has also been used in this application. Initially the suppressor grid is biased to plate-current cutoff, and the control grid is clamped to the cathode. The cathode current flows to the screen, and therefore the screen voltage is low. The waveforms at all the electrodes are given in Fig. 14-29. A positive gate applied to the suppressor drives this electrode either slightly positive or to clamp. Clamping may occur either because the impedance of the driving source is large in comparison with the suppressor-cathode resistance or because a diode is added to the circuit from suppressor to ground. This increased suppressor voltage permits plate current to flow, and the plate voltage drops. Since the voltage across the Miller capacitor C cannot change instantaneously, the grid voltage must drop by the same amount V_1 that the plate falls. The grid voltage is now $-V_1$, the tube finds itself operating above cutoff, and a negative-going sweep forms at the plate. The load resistor R_p is large, so that bottoming will take place (Sec. 7-5). The load line is drawn on the plate characteristics in Fig. 14-30. Since $-V_1$ is very close to the cutoff bias, we have assumed that the tube characteristic corresponding to the grid voltage $-V_1$ is coincident with the abscissa. For a type 6AS6 tube, the order of magnitude of V_1 is 5 V, and bottoming begins when the grid has increased by

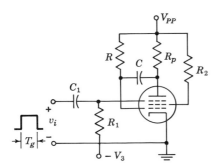

Fig. 14-28   A suppressor-gated Miller time-base generator.

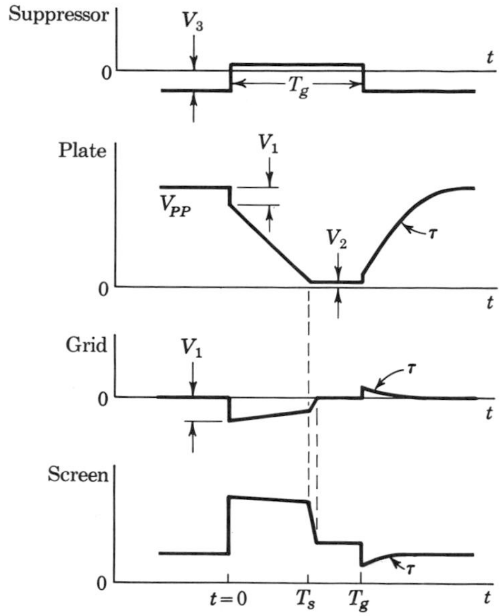

Fig. 14-29 The waveforms for a suppressor-gated Miller sweep generator.

only 1 or 2 V.   For example, if $V_{PP} = 300$ and the amplifier gain is 150, then the grid will increase by $300/150 = 2$ V for complete "run-down."

When the grid voltage drops from zero to $-V_1$, the cathode current falls, the screen current drops, and hence the screen voltage rises, as indicated in Fig. 14-29.   During the formation of the sweep the grid rises slightly, as noted above, and the increased screen current results in a slight decrease in screen voltage.   When the plate voltage bottoms, the grid voltage increases to zero with a time constant RC, the space current and hence the screen current increase, and the screen voltage drops, as indicated in Fig. 14-29.   The screen voltage does not quite fall to its value for $t < 0$ because some of the cathode current is now being collected by the plate, whereas for $t < 0$ all the space current goes to the screen.

At the end of the gate the suppressor again cuts off the plate current. The capacitor C whose voltage has fallen almost to zero recharges toward V_{PP} through R_p and the grid-cathode resistance r_G with a time constant $\tau = (R_p + r_G)C \approx R_pC$.   The grid voltage will be driven positive by approximately $r_G V_{PP}/R_p$.   This positive grid voltage will increase the cathode current above its value for $t < 0$, and hence there will be a dip in screen

Fig. 14-30 Illustrating bottoming in a pentode and the fact that the grid voltage changes by only a few volts during the entire sweep voltage.

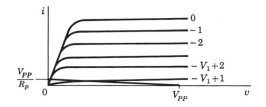

voltage below its value for $t < 0$. The overshoot in grid voltage and under-shoot in screen voltage are indicated in Fig. 14-29.

At $t = 0+$, the voltage across R is $V_{PP} + V_1$, and since the current through R passes through C, the initial sweep speed is $(V_{PP} + V_1)/RC$. As long as the amplifier gain remains high, the sweep speed remains essentially constant. Hence, a linear ramp results for almost the entire plate-voltage run-down except near the very bottom.

If the gate width T_g is less than the time T_s for the capacitor to dis-charge completely, then there will be no bottoming, and the flat portions in Fig. 14-29 between T_s and T_g are missing. The screen voltage is itself a gating voltage and, if the sweep is being used in connection with a scope display, can be used as an intensifier to brighten the CRT trace during the sweep time and to cut off the CRT beam during the retrace time. The recovery time may be made quite small by driving the capacitor C from a cathode follower, as in Fig. 14-26. Under these circumstances the recovery time constant is $\tau = C(R_o + r_G)$, where R_o is the output impedance of the cathode follower and r_G is the grid-cathode resistance.

The step in the plate voltage at $t = 0+$ *cannot* be eliminated by adding a resistor r in series with C in Fig. 14-28, as was done for the grid-gated Miller integrator. The use of the resistor r is effective because the amplifier of the grid-gated circuit is initially biased within its grid base. The suppressor-gated circuit, however, is held beyond cutoff in the quiescent condition. When the gate is applied, the tube must draw some plate current, and hence the plate voltage must drop somewhat.

14-13   PHANTASTRON CIRCUITS[7, 8]

The screen waveform of Fig. 14-29 is a positive step for the interval of the linear run-down. Hence, it is possible to start the sweep by means of a narrow pulse or trigger and to couple the output from the screen to the suppressor so that the positive gate needed at the latter grid is supplied internally. A circuit in which a Miller time base is initiated with a trigger and the circuit then supplies its own gate is called a *phantastron.*†

The screen-coupled phantastron is drawn in Fig. 14-31. This circuit differs from the suppressor-gated Miller sweep (Fig. 14-28) only in that the screen and suppressor voltages are obtained from a divider arrangement R_1-R_2-R_3. These resistors are so chosen that in the quiescent state the suppressor grid is sufficiently negative (say $-V_3$) so that no plate current flows, all the space current going to the screen. Assume now that a positive trigger is applied at $t = 0$ to the suppressor, so as to allow the plate to draw current. The plate voltage drops, the grid voltage drops the same amount

---

† The British considered the operation of this circuit as *fantastic* and dignified it with the name *phantastron*.

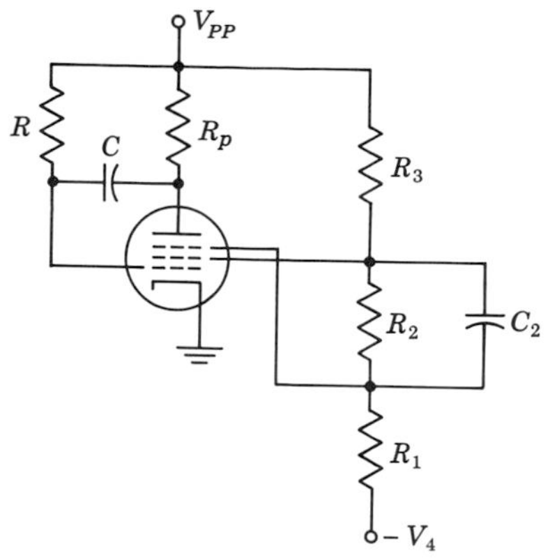

Fig. 14-31   The screen-coupled phantastron. A trigger is applied at the suppressor.

(say V_1), the cathode current falls, and the screen voltage rises.   This causes the suppressor voltage to rise because it is obtained from the same voltage divider as the screen..   The capacitor C_2 is a speed-up capacitor (Sec. 10-5). The action described above is regenerative, the tube is rapidly driven from cutoff to its normally ON condition, and the Miller sweep is initiated.

Waveforms   The waveforms generated are given in Fig. 14-32 and are identical with those in Fig. 14-29 except that the flat portions at the end of the run-down are missing.   As soon as the tube bottoms and the plate can fall no further, the plate side of C in Fig. 14-31 remains at a fixed potential. Thereafter, the grid side of C rises with respect to ground (with a time constant RC) because current continues to flow from V_{PP} through R through C into the tube plate.   As the grid rises, the cathode current increases, the screen current increases, and the screen voltage drops.   Because of the action of the divider arrangement the suppressor voltage also falls and the plate current decreases.   A reduction of plate current means an increase in plate voltage, which in turn causes the grid to rise still further.   This action is regenerative, and the suppressor voltage is rapidly driven negative.   The waveshapes for $t > T_s$ in Fig. 14-32 are explained in exactly the same manner as the waveshapes for $t > T_g$ in Fig. 14-29.   There is one difference between these two figures in the region under consideration.   Since the suppressor voltage comes from the same divider as the screen voltage, there will be a tilt and an undershoot at the suppressor, as indicated in Fig. 14-32.

Triggering may also be done with negative pulses applied to the plate and hence fed to the grid through the capacitor C.   A negative trigger so applied reduces the cathode current and consequently raises the screen voltage. The rise of screen voltage is transmitted to the suppressor to bring the suppressor above the point of plate-current cutoff.   The trigger size should be large enough to start the regenerative action but not so large as to cut off the tube current.   If the tube should be driven below cutoff, the grid voltage will

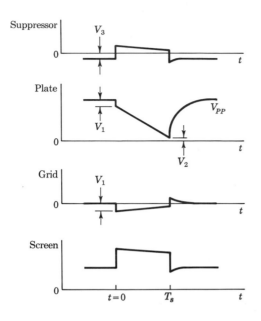

Fig. 14-32 Waveforms in the screen-coupled phantastron.

rise initially with a time constant RC into the conducting region, and there will be a delay between the application of the trigger and the start of the sweep.

The phantastron circuit has a recovery time constant $\tau = R_p C$. If a shorter retrace time is desired, then a cathode follower can be interposed between the plate of the amplifier and the capacitor C, as was done for the externally control-grid-gated Miller sweep of Fig. 14-26. Since the sweep voltage at the plate of the pentode starts near V_{PP} and since this voltage is applied directly to the grid of the cathode follower, a separate supply V'_{PP} must be used for the cathode follower, with $V'_{PP} > V_{PP}$. The magnitude of the overshoots and undershoots is greatly increased (perhaps by a factor of 10) if a cathode follower is used. The reason for this feature is that the plate of the amplifier is no longer loaded down by the low grid-cathode resistance r_G, which allowed the grid to overshoot to only $r_G V_{PP}/R_p$. With a cathode follower in the circuit, when the plate rises at the end of the sweep it carries the grid of the cathode follower up with it, and hence the grid of the pentode can be driven several volts positive.

Sweep Speed and Linearity The sweep speed is, as with the simple suppressor-gated Miller integrator, $(V_{PP} + V_1)/RC$ and hence can be adjusted by changing V_{PP}, R, or C. If the run-down proceeds to within V_2 of ground (Fig. 14-32), then the amplitude of the sweep is $V_{PP} - V_1 - V_2$. The sweep time T_s is the amplitude divided by the speed, so that

$$\frac{T_s}{RC} = \frac{V_{PP} - V_1 - V_2}{V_{PP} + V_1} \tag{14-42}$$

If $V_{PP} \gg V_1 + V_2$, then $T_s \approx RC$, a result which is independent of variations in V_{PP}. The next approximation is obtained by dividing the numerator in Eq. (14-42) by the denominator, with the result

$$x \equiv \frac{T_s}{RC} = 1 - \frac{2V_1 + V_2}{V_{PP} + V_1} \approx 1 - \frac{2V_1 + V_2}{V_{PP}} \tag{14-43}$$

Taking the derivative, we find

$$\frac{dx}{x} = \frac{2V_1 + V_2}{V_{PP}} \frac{dV_{PP}}{V_{PP}} \tag{14-44}$$

For example, if $V_1 = V_2 = 5$ V and $V_{PP} = 150$ V, then a 10 percent change in supply voltage $(dV_{PP}/V_{PP} = 0.1)$ gives

$$\frac{dx}{x} = \frac{15}{150} \times 0.1 = 0.01$$

or a 1.0 percent change in sweep time.

A diode ($D2$ in Fig. 14-33) may be used to clamp the suppressor to ground during the time when the time base is being formed. Hence the negative supply $-V_4$ plays no part in determining conditions during the interval T_s. The voltage $-V_4$ is needed only to ensure plate-current cutoff before the circuit is triggered. Variations in the negative supply have negligible effect on the sweep time T_s.

Variations in the heater voltage should affect V_1 and V_2 to some extent. Experimentally it is found that a 10 percent change in heater voltage results in only a few tenths of a percent change in T_s and in a direction opposite to the change caused by a plate-supply variation. If tubes are changed, then T_s may change by a few percent, because $2V_1 + V_2$ varies from tube to tube.

From Eq. (14-29)

$$e_s = \frac{V_s}{V} \frac{1}{|A|} \approx \frac{1}{|A|} \tag{14-45}$$

If the amplifier gain is 100, then $e_s \approx 1$ percent. (Incidentally, if linearities under 1 percent are to be realized, then the capacitance C must be independent of voltage to this precision. A mica capacitor is usually satisfactory, whereas a paper capacitor may not be.) Here, then, is a circuit possessing many fine characteristics: excellent linearity of sweep and a time-base duration whose value is not very sensitive to positive, negative, or heater supply voltages and whose sweep speed is readily adjusted. With a trigger input a gate-type output is obtained at the screen, in addition to the linear output at the plate, and hence the circuit is analogous to the plate-coupled monostable multi discussed in Chap. 11.

Circuit Variations   One of the principal applications of the phantastron is that of a delay unit. If the output at the screen is differentiated (peaked), then a negative output pulse is obtained, delayed T_s sec from the triggering pip.

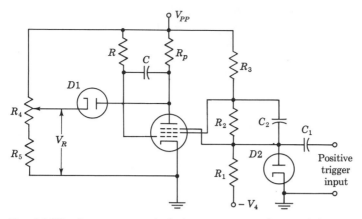

Fig. 14-33 A screen-coupled phantastron used as a delay unit.

The delay is adjusted by controlling the voltage V_R from which the run-down begins. A plate catching diode $D1$ is ideal for this purpose, and the complete circuit is shown in Fig. 14-33. The waveforms are given in Fig. 14-32, except that the plate voltage starts at V_R rather than V_{PP}. The overshoot at the grid is approximately $V_{PP}r_G/R_p$, where r_G is the static grid-cathode resistance and is independent of V_R and therefore T_s. This characteristic is different from the corresponding one for the plate-coupled multivibrator, where the overshoot increases with delay. The delay T_s is a linear function of V_R except for small delays, where curvature due to bottoming becomes important. Incidentally, $D1$ also serves the useful purpose of reducing the recovery time because it catches the plate which is rising toward V_{PP} when it reaches V_R.

Analogous to the cathode-coupled monostable multi a cathode-coupled monostable phantastron can be constructed (Prob. 14-32). The cathode-coupled phantastron has the following advantages over the screen-coupled circuit: no negative supply is needed; the screen is a free (unloaded) electrode, from which a positive gate is obtained; and a negative gate is available at the cathode. The principal disadvantages are that there is a larger initial step in plate voltage and that the gain of the amplifier is smaller because of the cathode degeneration introduced by R_k. This decreased gain means that the linearity is somewhat poorer.

The phantastron has the advantage over the cathode-coupled multi that it is much less sensitive to tube characteristics and to supply-voltage variations than the latter. For example, if the V_{PP} supply changes by 10 percent in a multi circuit, we may expect the delay to change by perhaps 5 percent, which is five or 10 times what can be expected in the phantastron circuit. Also, the phantastron delay can be made more linear than that of the multi if sufficient gain is used. An astable phantastron circuit is suggested in Prob. 14-30.

The phantastron circuit is limited to the generation of linear sweeps of duration of the order of 10 μsec or longer because of the effect of the stray capacitance to ground at the various tube electrodes.   Williams and Moody[7] describe circuits of the Miller type, called *sanatrons* or *sanaphants*, which are capable of giving precise delays as short as 1 μsec.   These circuits are essentially phantastrons in which a separate tube is used to generate the necessary gate from the input trigger.

14-14   BOOTSTRAP TIME–BASE GENERATORS—BASIC PRINCIPLES

The basic bootstrap configuration of Fig. 14-18 is repeated for convenience in Fig. 14-34a.   Since the amplifier gain must be close to unity, a vacuum-tube amplifier may consist of a cathode follower and a semiconductor amplifier may consist of an emitter follower.   The latter configuration is indicated in Fig. 14-34b, but the general considerations given below apply equally well to the tube circuit.

The voltage V in Fig. 14-34a is equal in Fig. 14-34b to the sum of V' and the small initial voltage V_{oi} across R_e when S is closed.   The practical disadvantage in Fig. 14-34b is that neither side of the supply V' is grounded.   This difficulty may be remedied by replacing V' with a charged capacitor C_1. This capacitor is charged through a resistor R_1, as indicated in Fig. 14-35.   It is necessary that C_1 be large enough so that the voltage across it shall not change appreciably during the sweep time.   If the voltage across C_1 were precisely constant and if the emitter follower had precisely unity voltage amplification, then point P_2 in Fig. 14-35 would exactly follow point P_1. Because of this bootstrapping, the voltage difference between P_2 and P_1 would be invariant and the current i_R through R would be constant.   If the input impedance of the emitter follower were infinite, then the constant current through R would pass into C and cause the voltage across C (and hence also v_o) to increase linearly with time.

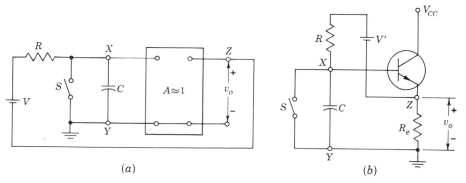

Fig. 14-34   (a) The basic bootstrap circuit; (b) the amplifier in (a) is an emitter follower.

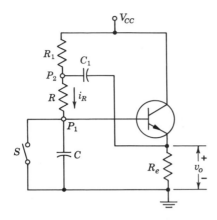

Fig. 14-35 A practical form of a bootstrap sweep circuit.

The presence of the resistor R_1 makes the circuit of Fig. 14-35 look different from the circuits in Fig. 14-34. However, C_1 is to be so large that the voltage across it remains constant, and the supply-voltage node V_{CC} is ground so far as signal is concerned. Therefore we may, when calculating changes from a quiescent operating point, move R_1 from its original position to a position directly in parallel with R_e. The circuit of Fig. 14-35 will then have the form of Fig. 14-34b. However, we must keep in mind, in calculating the voltage gain A of the emitter follower, that the emitter load R_e' consists of the parallel combination of R_e and R_1.

Sweep Speed and Linearity The sweep speed is given by Eq. (14-37), in which

$$V = V_{CC} \frac{R}{R + R_1} \tag{14-46}$$

so that

$$\text{Sweep speed} = \frac{A V_{CC}}{(R + R_1)C} \tag{14-47}$$

Alternatively, we may calculate the sweep speed in the following manner. With S closed, the steady-state current through R is $i_R = V_{CC}/(R_1 + R)$. When S opens, this current flows into C since initially the voltage across C is zero, so that the current flowing into the input resistance R_i is zero. Hence the sweep speed at the transistor input is i_R/C and at the output $A i_R/C$, giving the same result as directly above.

The ramp output will not be exactly linear because of the finite values of R_i and C_1 and because the emitter-follower voltage gain A is not exactly unity. The dependence of sweep-speed error on R_i and A is given in Eq. (14-36). The effect of the finite value of C_1 is considered in Prob. 14-36, where it is found that a term C/C_1 must be included in the parentheses. The expression for e_s is

(with R_1 replaced by a diode as in Fig. 14-36)

$$e_s = \frac{V_s}{V}\left(1 - A + \frac{R}{R_i} + \frac{C}{C_1}\right) \tag{14-48}$$

or, using Eq. (14-46),

$$e_s = \frac{V_s}{V_{CC}}\left(1 + \frac{R_1}{R}\right)\left(1 - A + \frac{R}{R_i} + \frac{C}{C_1}\right) \tag{14-49}$$

The resistor R_1 appears in Eq. (14-49) in three places, once explicitly in the factor $(1 + R_1/R)$ and twice implicitly in the gain A and in the input resistance R_i. It appears in the gain A since, as already noted, R_1 is in shunt with R_e. The resistance R_1 appears in R_i since the input impedance of an emitter follower is approximately $h_{fe}R_e'$. While R_1 should be selected so that $R_1 \ll R$, it is even more important that R_1 be selected large enough in comparison with the emitter-follower output impedance so that it does not appreciably reduce the gain. Since the gain is very nearly unity, and is subtracted from unity in Eq. (14-49), a very small percentage decrease in A will cause a large percentage increase in e_s. Such a selection of R_1 will also serve to keep the input resistance high. It also follows from Eq. (14-49) that, if R_i and C_1 are not to have too deleterious an effect on e_s, we must satisfy the inequalities

$$R_i > \frac{R}{1 - A} \quad \text{and} \quad C_1 > \frac{C}{1 - A} \tag{14-50}$$

In calculating the input resistance we should, in principle, take into account that the emitter follower is loaded not only by R_e and R_1 but also by R. The extent to which R loads the emitter follower is calculated as follows. The emitter (top) end of R is at a voltage A times as large as is the base end of R. From Fig. 1-7b, illustrating Miller's theorem, the effective resistance seen looking into R from the emitter is not R but, exaggerated by the Miller effect, is

$$R_M = \frac{AR}{A - 1} \tag{14-51}$$

Since A is positive and slightly less than unity, then R_M is a (negative) resistance of large magnitude. Since R_M is paralleled with the appreciably smaller resistors R_1 and R_e, the effect of R will usually be quite negligible.

Recovery Considerations At the termination of the sweep, C_1 must recharge through the emitter-follower output impedance in series with the parallel combination of R and R_1. In a repetitive sweep, if the restoration time is not long enough to permit the capacitor C_1 to recharge fully, the sweep speed will be a function of the recovery time and hence of the repetition rate. The difficulties with respect to nonlinearity and restoration time caused by R_1 may be minimized by replacing R_1 by a diode, as in Fig. 14-36. For the sake of variety this bootstrap circuit is shown using vacuum tubes instead of semiconductors.

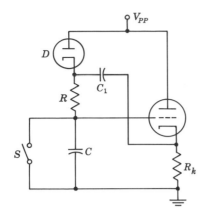

Fig. 14-36   A tube version of the circuit of Fig. 14-35 with R_1 replaced by a diode.

Referring to Fig. 14-36 we see that initially (with S closed) the diode conducts and the charging current is $I \approx V_{PP}/R$, since the diode forward resistance $R_f = R_1$ is much smaller than R. Hence in Eq. (14-49) the term R_1/R may be neglected. With S open, the cathode voltage rises and D is cut off. When this happens, R_1 represents the diode reverse resistance and $R_1 \gg R_o$, the cathode-follower output resistance. Hence, the maximum possible gain is obtained from the cathode follower. Since the cathode voltage of the diode must rise a few tenths of a volt before its resistance becomes very large compared with the output impedance of the cathode follower, then the beginning (~ 0.2 V) of the sweep may not be linear. This effect may be eliminated by adding a small resistance R' in series with C. When the sweep starts, the charging current through R' produces a voltage drop which reverse-biases the diode. Of course, the sweep waveform is now trapezoidal (an initial step followed by the ramp).

At the termination of the time base, S closes and discharges C through the switch resistance, whereas the charge lost by C_1 during the sweep is replaced through the low resistance R_f of the conducting diode in series with the output impedance of the cathode follower. Thus the diode not only improves the linearity but also shortens the recovery time. For minimum retrace time, C should be kept as small as possible. For a given value of C the desired sweep speed is obtained by the proper choice of R. For a tube circuit, the maximum value of R is limited to a few megohms by practical considerations. For a transistor circuit, R is limited to a few tens of kilohms because of the restriction of Eq. (14-50). For example, if with the aid of a Darlington circuit (Sec. 14-15) an input impedance of 1 M is obtained and if $A = 0.99$, then $R < 10$ K to satisfy Eq. (14-50).

14-15   THE TRANSISTOR BOOTSTRAP TIME–BASE GENERATOR

We shall now consider in detail the bootstrap sweep generator[9] of Fig. 14-37, which is the semiconductor version of the circuit of Fig. 14-36 in which switch S has been replaced by transistor $Q1$. The input to $Q1$ is the gating waveform

from a monostable multi (although it could equally well be a repetitive waveform, say a square wave).

Quiescent Conditions   Before the application of the gating waveform at
$t = 0$, transistor $Q1$ is in saturation, as indicated in Fig. 14-37b.   The voltage
across the capacitor C and at the base of $Q2$ is $V_{CE}(\text{sat})$.   The quiescent output
voltage at the emitter of $Q2$ is lower than the base voltage by V_{BE2} and hence
with respect to ground is at $V_{CE}(\text{sat}) - V_{BE2}$, as indicated in Fig. 14-37d.
If we neglect this small voltage (a few tenths of a volt negative) as well as the
small diode drop, then the voltage across C_1 and also across R is V_{CC}.   Hence
the current i_R is V_{CC}/R.   Since the quiescent output voltage is close to zero,
then the emitter current for $Q2$ for $t < 0$ is V_{EE}/R_e and the base current of $Q2$
is $i_{B2} \approx V_{EE}/R_e h_{FE}$.   Since i_{B2} is usually small in comparison with i_{C1},

$$i_{C1} \approx i_R = \frac{V_{CC}}{R} \tag{14-52}$$

In order that $Q1$ should indeed be in saturation for $t < 0$ it is necessary
that its base current ($\approx V_{CC}/R_b$) be at least equal to i_{C1}/h_{FE}, so that

$$\frac{V_{CC}}{R_b} > \frac{V_{CC}}{h_{FE}R} \qquad \text{or} \qquad R_b < h_{FE}R \tag{14-53}$$

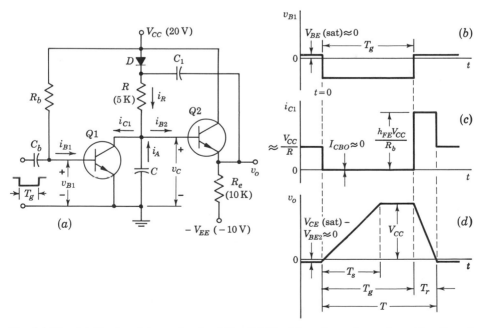

Fig. 14-37   (a) The bootstrap circuit of Fig. 14-35 with resistor R_1 replaced by D and
switch S by $Q1$; (b) the base voltage of $Q1$; (c) the collector current of $Q1$; (d) the
output-voltage waveform.   Parameter values in parentheses refer to the illustrative
problem on page 563.

Formation of Sweep With the application of the gating waveform at $t = 0$, $Q1$ is driven OFF. The current i_{C1} now flows into C, and, assuming unity gain in the emitter follower, the output increases linearly with time in accordance with

$$v_o \approx \frac{V_{CC}t}{RC} \tag{14-54}$$

When the sweep starts, the diode is reverse-biased, as already explained above, and the current through R is supplied by the capacitor C_1, which is charged to V_{CC}. Equation (14-54) is valid only if the gate duration T_g is small enough so that the calculated value of v_o does not exceed V_{CC}. From Fig. 14-37 it is seen that as v_o approaches V_{CC}, the collector-to-emitter voltage of $Q2$ approaches zero, and it enters the saturation region, where it no longer acts as an emitter follower. Then, as drawn in Fig. 14-37d, v_o (and also v_C) remain constant at V_{CC}. The current V_{CC}/R through C_1 and R now flows from base to emitter of $Q2$. If the output v_o reaches the voltage V_{CC} in a time $T_s < T_g$, then, from Eq. (14-54),

$$T_s = RC \tag{14-55}$$

whereas if the sweep amplitude V_s is less than V_{CC}, then the maximum ramp voltage is given by

$$V_s = \frac{V_{CC}T_g}{RC} \tag{14-56}$$

Retrace Interval At the termination of the gate, the current V_{CC}/R_b flows into the base of $Q1$. Until the collector of $Q1$ drops nominally to zero, at which point the transistor will be in saturation, the collector current of $Q1$ remains reasonably constant at the value $i_{C1} = h_{FE}V_{CC}/R_b$, as indicated in Fig. 14-37c. (Compare with the collector-current waveform in Fig. 8-27d.) Neglecting i_{B2}, part of this current is supplied through R, and the remainder i_A is supplied by C. While this capacitor is discharging, the current i_R through R remains at V_{CC}/R. Such is the case because, as the voltage across C falls, so does the voltage v_o (since $Q2$ functions as an emitter follower), and the voltage across R remains fixed at the value V_{CC} across C_1 (neglecting the small voltage change across C_1 during the sweep). Hence

$$i_A \approx i_{C1} - i_R = \frac{h_{FE}V_{CC}}{R_b} - \frac{V_{CC}}{R} \tag{14-57}$$

Since this current which discharges C is constant, then C falls linearly with time. By emitter-follower action the output v_o also returns to its initial value in a linear manner, as shown in Fig. 14-37d. If the retrace time is T_r, then $i_A T_r/C = V_s$, or, using Eq. (14-57),

$$T_r = \frac{CV_s/V_{CC}}{(h_{FE}/R_b) - (1/R)} \tag{14-58}$$

After C is discharged, the collector current is now supplied completely through R and becomes established at the value V_{CC}/R.

The Recovery Process During the course of the time $T = T_g + T_r$ (Fig. 14-37d) the capacitor C_1 has been discharging at a constant rate because the current V_{CC}/R has passed through C_1 during this entire interval. Therefore the circuit will not have recovered completely to its initial state until C_1 has regained the charge it lost. At the time T, when the voltage across C and at the base of $Q2$ returns to its value for $t < 0$, the voltage across C_1 is smaller than it was initially at the beginning of the sweep. The diode D starts conducting at $t = T$ (actually, slightly sooner), and the terminal of C_1 which is connected to D returns to its initial voltage ($\approx V_{CC}$). Therefore, the other terminal of C_1, which is connected to the emitter of $Q2$, is at a more positive potential than initially and $Q2$ is cut off. The capacitor consequently must now charge through the resistor R_e, which for the sake of good linearity (high input impedance) must be kept large. It is for this reason that R_e is returned to the source $-V_{EE}$, as we shall now demonstrate.

We may calculate the minimum recovery time T_1 for C_1 as follows. Since the discharge current of C_1 is V_{CC}/R for the total duration T, then the lost charge is $V_{CC}T/R$. The charging current is V_{EE}/R_e, and hence the recovered charge is $V_{EE}T_1/R_e$. Equating charge lost to charge gained we find

$$T_1 = \frac{V_{CC}}{V_{EE}} \frac{R_e}{R} T \tag{14-59}$$

We note that T_1 is independent of C_1 and varies inversely with V_{EE}.

If the sweep is operated recurrently and if the recovery interval allowed is not adequate, a voltage deficiency will develop across C_1, with two undesirable results. First, the sweep slope will change and be smaller than it was when a full recovery was allowed. Second, since transistor $Q2$ is OFF until recovery is complete, the appearance of a sweep voltage at the output v_o will be delayed. For, when transistor $Q1$ is cut off and C begins to charge, a signal v_o will not appear until the voltage at the base of $Q2$ has risen to the point where $Q2$ again begins to conduct.

The delay in the onset of conduction of $Q2$ may be prevented, as already noted, by including a small resistor in series with C. In this way, when the switching transistor $Q1$ opens, there will appear at the base of $Q2$ an initial jump which will turn this transistor on immediately. The output sweep v_o will similarly also be preceded by an initial jump. This method does not, however, affect the change in sweep speed which results from incomplete recovery. Accordingly, if the economy of the situation allows it, there is much to be gained from using an additional transistor, as in Fig. 14-38, to assist in the rapid recharging of C_1. In this circuit, transistor $Q3$ and $Q1$ are driven OFF simultaneously by an input gating waveform so that the sweep voltage may be generated. When, at the end of the sweep, $Q1$ turns ON to discharge C, so also does $Q3$, and C_1 may recharge through $Q3$. Observe, in

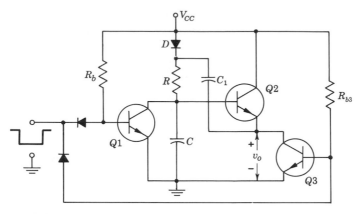

Fig. 14-38 A transistor $Q3$ is used in the bootstrap time-base circuit of Fig. 14-37 to allow rapid recharge of C_1.

this circuit, that the emitter resistor R_e of Fig. 14-37 has been eliminated entirely. The transistor $Q2$ does not need an emitter resistor to operate properly, since, as has already been described, a transistor current V_{CC}/R will flow through C_1 into R. The emitter resistor in Fig. 14-37 served principally to allow rapid recharging of C_1. But since $Q3$ is now available for this purpose, R_e may be dispensed with entirely.

The Emitter Follower   From Eq. (1-8) we see that $1 - A$ for an emitter follower varies inversely as the input resistance R_i. Because the slope error [Eq. (14-49)] contains the term $1 - A + R/R_i$, it is clear that for best linearity R_i must be as large as possible. A large input impedance requires a large emitter resistance [Eq. (1-7)]. An arrangement which, in effect, permits an emitter follower to operate into a much larger emitter resistance than could be conveniently directly employed is shown in Fig. 14-39. Here the input impedance R_{i4} of the emitter follower $Q4$ is used as the emitter load R_{e2} of the emitter follower $Q2$. These two transistors jointly are called a *composite-* or *compound-transistor* or *Darlington configuration*. Since $R_{e2} = R_{i4} \approx h_{fe}R_e$ is large, then the input impedance R_{i2} to the composite-transistor configuration may be very high (~ 1 M). (Incidentally, the largest possible value of input impedance is obtained for infinite R_e, or an open-circuited emitter. The impedance from base to ground is then the impedance from base to collector, because the collector is grounded. But the impedance from base to collector with the emitter open is by definition $1/h_{ob} \approx 2$ M, which therefore is the maximum R_i attainable.)

When the composite-transistor configuration is used in the bootstrap sweep, the output at the emitter of $Q2$ provides the feedback, as indicated in Fig. 14-39, and v_o is the sweep signal. Taking the output from the emitter of $Q2$ would shunt the high impedance seen looking into $Q4$ by the input imped-

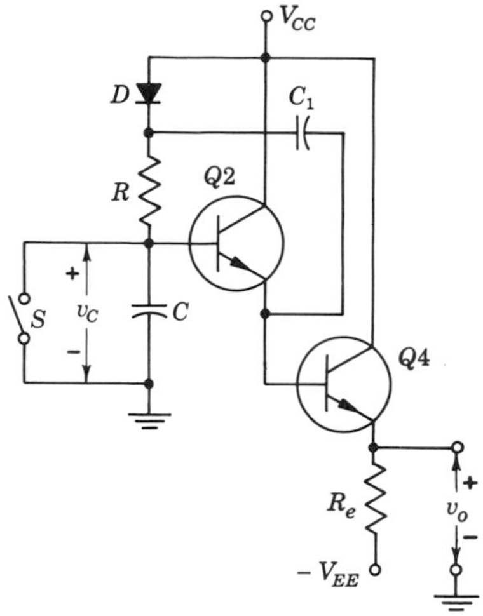

Fig. 14-39 A bootstrap circuit using a composite-transistor emitter follower (Darlington circuit).

ance of the sweep amplifier and undo the advantage of the high impedance R_{i4}. When the ultimate in performance is not required from the Darlington configuration, the output v_o may be used to supply the feedback voltage. However, there are two reasons why it is not advantageous to do so. First, the gain of the composite amplifier is smaller than the gain of either stage, since the gain of each stage is less than unity. Second, when no feedback signal is taken from $Q2$, the emitter current of $Q2$ can be no larger than the small base current of $Q4$. This restriction on the current of $Q2$ will adversely affect its properties as an emitter follower because h_{fe} decreases and h_{ie} increases with decreasing emitter current. Since such variations decrease the gain and R_i, it is clearly advantageous to increase the emitter current of $Q2$ above the small base current of $Q4$. The use of $Q2$ to provide the feedback signal serves to allow this extra current because, as noted above, at the moment the sweep begins, the feedback source is called upon to furnish the current V_{CC}/R which flows through R.

When the composite-transistor emitter follower is used, the feedback signal is taken from $Q2$ in Fig. 14-39 from a high-impedance point. It is reasonable to ask whether this feedback connection itself does not constitute a heavy load. The effective resistive load on $Q2$ is R_M, given by Eq. (14-51), and is seen to be a very large (negative) resistance because the gain is very close to unity. Hence, the loading of R is usually negligible.

The Darlington connection aggravates the recovery problem. Capacitor C_1 must recharge through the base current of $Q4$. In order to obtain reasonable recovery times the additional transistor $Q3$ of Fig. 14-38 should be used.

EXAMPLE The circuit of Fig. 14-37 has the following parameters: $V_{CC} = 20$ V, $V_{EE} = 10$ V, $R_e = 10$ K, $R = 5$ K, and $T_g = 700$ μsec. The transistor h parameters are those given in Table 1-2. A 20-V sweep in 500 μsec is desired. (a)

Find a reasonable value for R_b.   (b) Calculate C.   (c) Calculate e_s, assuming C_1 arbitrarily large.   (d) Choose C_1 so that e_s is increased by no more than 10 percent over its value in part c.   (e) Calculate the retrace time T_r for C to discharge completely at the end of the gating waveform.   (f) Calculate the recovery time T_1 for C_1 to recharge completely.   (g) What would be the slope error if the single-stage emitter follower were replaced by the Darlington composite configuration?

Solution   a. From Eq. (14-53), and assuming $h_{FE} \approx h_{fe}$,

$$R_b < h_{FE}R = (50)(5) = 250 \text{ K}$$

Hence, a reasonable value for R_b is 100 K.

b. Since $T_s = 500 < T_g = 700$ μsec, then Eq. (14-55) is valid, and

$$C = \frac{T_s}{R} = \frac{500}{5,000} = 0.1 \text{ μF}$$

c. To find e_s we must first calculate R_i and $A \equiv A_V$. From Eqs. (1-6) to (1-8) we have

$$A_I = \frac{h_{fe} + 1}{1 + h_{oe}R_e} = \frac{51}{1 + 10/40} = 40.8$$

$$R_i = h_{ie} + A_I R_e = 1.1 + (40.8)(10) = 409 \text{ K}$$

$$1 - A_V = \frac{h_{ie}}{R_i} = \frac{1.1}{409} = 0.0027 \quad \text{or} \quad A_V = 0.9973$$

The effective load R_M on the emitter follower is, from Eq. (14-51),

$$R_M = \frac{A_V R}{A_V - 1} \approx \frac{5}{-0.0027} = -1,850 \text{ K}$$

which we may neglect since it is in parallel with $R_e = 10$ K.   From Eq. (14-49), with the diode resistance $R_1 = 0$, $V_s = V_{CC}$, and assuming C_1 arbitrarily large,

$$e_s = 1 - A + \frac{R}{R_i} = 0.0027 + \frac{5}{409} = 0.0149 \quad \text{or} \quad 1.5 \text{ percent}$$

d. If e_s increases by no more than 10 percent, then, from Eq. (14-49),

$$\frac{C}{C_1} \leq 0.0015 \quad \text{or} \quad C_1 \geq \frac{0.1}{0.0015} = 67 \text{ μF}$$

We choose $C_1 = 100$ μF, which is quite a large capacitance.

e. From Eq. (14-58)

$$T_r = \frac{CV_s/V_{CC}}{\dfrac{h_{FE}}{R_b} - \dfrac{1}{R}} = \frac{0.1}{\dfrac{50}{10^5} - \dfrac{1}{5 \times 10^3}} = 333 \text{ μsec}$$

The retrace time can be reduced by choosing a value of R_b smaller than 100 K. However, if R_b is reduced too greatly, then the collector current $i_C = h_{FE}V_{CC}/R_b$ may increase to the point where the transistor dissipation becomes excessive.

f. From Eq. (14-59)

$$T_1 = \frac{V_{CC}}{V_{EE}} \frac{R_e}{R} T = \frac{20}{10} \frac{10}{5} (700 + 333) = 4,130 \ \mu\text{sec}$$

We see that the recovery time T_1 is far greater than the retrace time T_r. Without affecting the other quantities calculated above we can reduce T_1 by increasing V_{EE}. However, this modification will increase the quiescent current in $Q2$ and hence its dissipation.

g. From part *c* the effective emitter resistance of $Q2$ in Fig. 14-39 is $R_{e2} = R_{i4} = 409$ K. Hence, from Eqs. (1-6) to (1-8)

$$A_{I2} = \frac{51}{1 + 409/40} = 4.55$$

$$R_{i2} = 1.1 + (4.55)(409) = 1,860 \text{ K} = 1.86 \text{ M}$$

$$1 - A_{V2} = \frac{1.1}{1,870} = 0.00059 \quad \text{or} \quad A_{V2} = 0.99941$$

The effective load R_M at the emitter of $Q2$ is given by (14-51):

$$R_M = \frac{5}{-0.00059} = -8,470 \text{ K}$$

which we may neglect since it is in parallel with 409 K. From Eq. (14-47)

$$e_s = 0.00059 + \frac{5}{1,860} + \frac{0.1}{100}$$

$$= 0.00059 + 0.0027 + 0.0010 = 0.0043$$

or $e_s = 0.43$ percent, which is an improvement over the 1.5 percent obtained with a single emitter follower.

A monostable bootstrap (requiring only a pulse input instead of a gate) is given in Prob. 14-45. An astable bootstrap time-base generator is obtained if a negative-resistance device such as the unijunction transistor is used as the switch across C, as in Prob. 14-44.

Short-recovery-time Bootstrap Circuits For long sweep durations (seconds), C and hence C_1 must be very large and C_1 may be of an impractical size. Also, the recovery time may become prohibitively long. A method of avoiding the use of such a large C_1 and yet ensuring that the sweep speed be independent of the repetition rate is to replace C_1 by an avalanche diode,

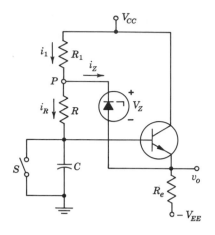

Fig. 14-40 A bootstrap sweep circuit with short recovery time.

as indicated in Fig. 14-40. Neglecting the base-to-emitter voltage and the base current of the transistor, the quiescent conditions with S closed are as follows:

$$v_o = 0 \qquad i_R = \frac{V_Z}{R} \qquad i_1 = \frac{V_{CC} - V_Z}{R_1} \tag{14-60}$$

$$i_Z = i_1 - i_R = \frac{V_{CC} - V_Z}{R_1} - \frac{V_Z}{R} \tag{14-61}$$

When S opens and the output rises to V_s, point P rises to $V_Z + V_s$ and the current i_R remains at the above value, provided that V_Z is independent of current. At the end of the sweep

$$i_Z = i_1 - i_R = \frac{V_{CC} - V_Z - V_s}{R_1} - \frac{V_Z}{R} \tag{14-62}$$

Hence, the Zener current decreases as the sweep progresses. Perfect linearity requires that V_Z be strictly constant over the range of currents from that given in Eq. (14-61) down to that in Eq. (14-62). If V_Z changes by, say, 1 percent in this range, then this causes a 1 percent change in sweep speed. Also, if V_Z drifts with temperature, then e_s will drift by the same percentage.

Even if the breakdown diode voltage is truly constant with current it is difficult to obtain very low values of e_s for the following reason. The effective emitter resistance R_e' is R_e in parallel with R_1. However, R_1 cannot be too large because, if it were, i_Z might fall below the knee of the characteristic (Fig. 6-6a), where V_Z varies with current. Since the effective emitter resistance R_e' is low, then $R_i \approx h_{fe}R_e'$ may not be large relative to R and e_s will suffer [Eq. (14-49)]. A Darlington connection will not be too effective since the effective emitter resistance of the first stage will be R_1. It should also be clear that R_1 cannot be replaced by a diode because, if it were, then in the quiescent state approximately the full supply voltage would appear across the avalanche diode.

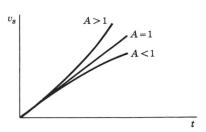

Fig. 14-41 If the amplifier gain in a boot-strap circuit is different from unity, the sweep voltage curves away from the ideal linear sweep.

A small neon lamp ($\frac{1}{25}$ W) may be used[10] in place of the avalanche diode. This lamp has a voltage drop of about 65 V and hence is used with the vacuum-tube version of the bootstrap circuit. The lamp voltage is essentially constant over the range from 300 to 30 μA. Very slow sweeps (of the order of 40 sec) have been obtained, with retrace times as short as 10 μsec. The restoration time is now determined by how fast C can be discharged through the resistance of the switch.

Instead of a follower circuit we could use a noninverting amplifier and adjust the gain to be exactly unity. The linearity would then be perfect. However, as the gain varied in the neighborhood of unity because of drift in parameters of the active device, temperature effects, line-supply variation, etc., the output would vary, as shown in Fig. 14-41. To minimize the drift in gain, we might use a two-stage negative-feedback amplifier. This circuit would have to be stabilized so that it would not oscillate. Also, the amplifier band-width would have to be large enough so as to pass the sweep undistorted. As we have mentioned before, this latter requirement is not an easy one to fulfill.

14-16   A COMPENSATING NETWORK [11]

Methods of linearity improvement exist which are used principally to supple-ment linearization by one of the feedback circuits discussed above. The circuit of Fig. 14-42 is the bootstrap configuration modified by splitting the sweep capacitor into two and connecting a resistor R_2 from the emitter-follower output to the junction of the two capacitors. We shall now explain qualitatively how this scheme serves to improve linearity, and we shall deduce an approximate value for R_2.

Neglecting the small supplementary feedback current i_f (compared with i_R), we find by referring to Eqs. (14-7) and (14-9) that the sweep capacitor voltage from base to ground is approximately

$$v_{BN} = \frac{V_{cc}t}{RC}\left(1 - \frac{t}{2T_s}\,e_s\right) = \frac{V_{cc}t}{RC} - \frac{V_{cc}t^2 e_s}{2RCT_s} \tag{14-63}$$

where $C = C_2C_3/(C_2 + C_3)$ and e_s is given by Eq. (14-49). This voltage appears as the sum of a term linear in t and a quadratic term in t which is

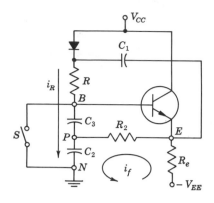

Fig. 14-42 A compensation method is used to improve the linearity of a bootstrap time base.

negative in sign. We may, by proper adjustment of R_2, cause the appearance across C_2 of a voltage quadratic in t exactly equal to the second term in Eq. (14-63), with the exception that the sign is reversed. The two quadratic terms will then cancel, and the nonlinearity will thereby be greatly reduced.

Neglecting the voltage from the base to the emitter, the voltage drop across R_2 is

$$v_{EP} = v_{BP} = \frac{C_2}{C_2 + C_3} v_{BN} \approx \frac{C_2}{C_2 + C_3} \frac{V_{CC}t}{RC} = \frac{V_{CC}t}{RC_3} \tag{14-64}$$

The voltage Δv_2 developed across C_2 as a result of the current $i_f = v_{EP}/R_2$ is

$$\Delta v_2 = \frac{1}{C_2} \int i_f\, dt = \frac{V_{CC}t^2}{2C_2 R_2 RC_3} \tag{14-65}$$

Equating the quadratic term in Eq. (14-63) with Eq. (14-65), we have

$$R_2 = \frac{T_s}{(C_2 + C_3)e_s} \tag{14-66}$$

Using the parameter values in the illustrative example in the preceding section for the single emitter follower with $e_s = 0.016$, $T_s = 500$ μsec, and $C_2 = C_3 = 2C = 0.2$ μF,

$$R_2 = \frac{500}{(0.4)(0.016)} \ \Omega = 78 \text{ K}$$

Because of the approximations made in this derivation and because the transistor or tube parameters vary over the swing of the sweep, the final value of R_2 should be determined experimentally, starting with the resistance given by Eq. (14-66) as a nominal value.

A diode may be placed across R_2 (with the anode at point P, so that it is reverse-biased during the generation of the sweep), which will allow rapid discharge of the capacitors when S closes at the end of the sweep period.

14-17   AN INDUCTOR CIRCUIT [12]

An inductor L may be used to improve the linearity of a simple RC sweep, as indicated in Fig. 14-43. The inductor also allows a sweep to be obtained whose amplitude is larger than the supply voltage because of the oscillatory nature of the circuit. The improvement in linearity results from the fact that there is no quadratic term in the power-series expansion of the capacitor voltage versus time, as we shall now prove.

From Fig. 14-43, with S closed, $v_s = 0$ and $i = V/R$. Assume S opens at $t = 0$. Then, since neither the capacitor voltage nor the inductor current can change instantaneously, it must be true that $v_s = 0$ and $i = V/R$ at $t = 0+$. Since

$$V = L \frac{di}{dt} + iR + v_s \qquad (14\text{-}67)$$

then at $t = 0+$,

$$V = L \frac{di}{dt} + \frac{V}{R} R + 0 \qquad \text{or} \qquad \frac{di}{dt} = 0$$

Since $i = C \, dv_s/dt$, then $di/dt = 0$ means $d^2v_s/dt^2 = 0$ at $t = 0+$. If the second derivative of a function is zero at $t = 0$, then its power-series expansion can contain no quadratic term. It is found that the first two terms in the expansion are

$$v_s = \frac{Vt}{RC}\left(1 - \frac{t^2}{6LC}\right) \qquad (14\text{-}68)$$

The inductor current is smaller at the end of the sweep period than at the beginning. The initial inductor current depends on the restoration period allowed between sweeps, and therefore the sweep speed will be a function of the repetition rate of the sweep. The time constant associated with the restoration period is L/R, but, on the other hand, the percentage difference in the inductor current between beginning and end of the sweep is small. Therefore the restoration period need not be large in comparison with L/R to make the initial inductor current approximately independent of repetition rate.

Fig. 14-43 A large inductance will improve the sweep linearity because an inductor serves to keep constant the current through it.

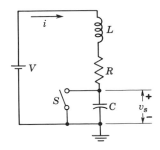

REFERENCES

1. Millman J.: "Vacuum-tube and Semiconductor Electronics," pp. 314–319, McGraw-Hill Book Company, New York, 1958.

2. For example, the Du Mont (Fairchild) type 304 scope.

3. For example, the Tektronix type 511, 514, 515A, 321, or Hewlett-Packard Company type 175A scope.

4. Day, J. E.: Recent Developments in the Cathode-ray Oscilloscope, "Advances in Electronics and Electron Physics," vol. 10, pp. 239–299, Academic Press Inc., New York, 1958.

5. Friend, C. K., and S. Udalov: Stabilized Delay Circuit Provides High Accuracy, *Electronics*, vol. 34, no. 15, pp. 78–80, Apr. 14, 1961.

6. Chance, B., et al.: "Waveforms," pp. 278–285, Massachusetts Institute of Technology Radiation Laboratory Series, vol. 19, McGraw-Hill Book Company, New York, 1949.

7. Williams, F. C., and N. F. Moody: Ranging Circuits, Linear Time-base Generators and Associated Circuits, *J. IEE*, pt. IIIA, vol. 93, no. 7, pp. 1188–1198, 1946.

8. Ref. 6, pp. 195–204 and 285–288.

9. Nambiar, K. P. P., and A. R. Boothroyd: Junction-transistor Bootstrap Linear-sweep Circuits, *Proc. IEE (London)*, pt. B, vol. 104, no. 15, pp. 293–306, 1957.

10. Miller, R. Z., and W. R. Kincheloe, Jr.: Neon Lamps as Bootstrap Circuit Elements, *Tech. Rept.* 10, Electronics Research Laboratory, Stanford University, Stanford, Calif., June 25, 1952.

11. Ref. 6, pp. 274–278.

12. Ref. 6, pp. 261–264.

15 / CURRENT TIME–BASE GENERATORS

In Chap. 14 we considered methods and circuits by means of which we may generate a voltage that varies linearly with time. Such a voltage may, of course, be applied to a resistor, thereby generating a current that increases linearly with time. In this chapter we shall discuss methods of causing a linearly time-varying current to flow instead through an *inductor*. The inductor of interest is most frequently a coil used to produce a magnetic field which serves to cause deflection of the electron beam[1] in a cathode-ray tube. The coil or set of coils, called a *yoke*, is mounted external to the tube near the gun structure and furnishes a magnetic field which is perpendicular to the direction of the electron beam. Such magnetic deflection finds its principal application in connection with radar and television displays on large-screen tubes, where electrostatic deflection would require deflecting voltages which are inconveniently large. For these magnetic-deflection applications, yoke currents are required which vary nominally linearly with time, and such currents yield deflections which vary fairly linearly with time.

15-1   A SIMPLE CURRENT SWEEP

If, at time $t = 0$, a voltage V is applied to a coil of inductance L in which the current is initially zero, then the inductor current i_L will increase linearly with time according to $i_L = (V/L)t$. [This result has as its dual the statement that the voltage v_C across a capacitor C charged from a constant-current source I will vary in accordance with $v_C = (I/C)t$.] A time-base circuit using this elemental principle is shown in Fig. 15-1*a*. Here an inductor in series with a transistor, which is used as a switch, is bridged across the supply voltage. The

571

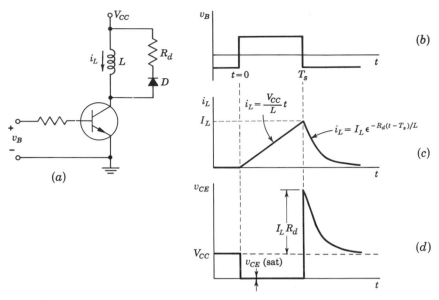

Fig. 15-1 (a) A current sweep circuit; (b) the gating waveform; (c) the inductor current waveform; (d) the waveform of the collector voltage.

gating waveform at the base operates between two levels. The lower level keeps the transistor in cutoff while the upper level drives the transistor into saturation. When the transistor switch is turned ON, then, neglecting the effect of the small saturation resistance of the transistor, the current i_L increases linearly with time. During this sweep interval the diode D does not conduct, since it is reverse-biased. The sweep terminates at $t = T_s$, when the gating signal drives the transistor to cutoff. The inductor current then continues to flow through the diode D and resistor R_d until it decays to zero. This decay is exponential with a time constant $\tau = L/R_d$, where R_d is understood to represent the sum of the damping resistance and the diode forward resistance.

The inductor-current waveform is shown in Fig. 15-1c, where it is indicated that the gate duration T_s is such that the current attains a maximum value I_L. Of interest also is the waveform of the voltage v_{CE} across the transistor. Before the transistor is turned ON, and some time after it has been turned OFF, $v_{CE} = V_{CC}$. While the transistor is ON, the voltage v_{CE} is quite low, being only some tens of millivolts for a germanium unit driven well into saturation. At the moment the transistor is turned OFF a spike of amplitude $I_L R_d$ appears across the inductor L. This peak voltage must be limited so that the transistor is below its collector-to-base breakdown voltage. Since I_L may well be determined by deflection requirements, there is clearly an upper limit to the size of R_d. The spike decays with the same time constant as does the inductor current. Note, however, that while the spike duration depends on the inductance of L, the spike amplitude does not (for a fixed peak current, see also Sec. 8-10).

We have not considered the fact that a physical yoke may well have associated with it a resistance R_L which is too large to neglect. If R_{CS} is the collector saturation resistance (Sec. 6-12) of the transistor, we find that the current increases in accordance with the equation

$$i_L = \frac{V_{CC}}{R_L + R_{CS}} (1 - \epsilon^{-(R_L + R_{CS})t/L}) \approx \frac{V_{CC}}{L} t \left[1 - \frac{1}{2} \frac{(R_L + R_{CS})t}{L}\right]$$

(15-1)

This equation is entirely analogous to Eq. (14-5) for the voltage sweep. The current then departs from a linear increase with time just as does the voltage sweep, as shown in Fig. 14-5. We may use the same measures of nonlinearity as in Chap. 14. Thus, if the current increases to a maximum value I_L, the slope error, in correspondence with Eq. (14-6), is

$$e_s = \frac{I_L}{V_{CC}/(R_L + R_{CS})} = \frac{(R_L + R_{CS})I_L}{V_{CC}}$$

(15-2)

To maintain linearity the voltage $(R_L + R_{CS})I_L$ across the total circuit resistance must be kept small in comparison with the supply voltage V_{CC}. This requirement corresponds, in connection with voltage ramps, to the necessity for keeping the capacitor voltage small in comparison with the resistor voltage.

A method of linearity correction against the effect of the resistance $R_L + R_{CS}$ which has been used effectively in the sweep generator of one radar display system[2] involves the use of a "linearity compensation coil." This coil, which is not used for beam deflection, is placed in series with the deflection yoke in the circuit of Fig. 15-1a. The compensation coil consists of a winding mounted on a magnetic core. The core is designed and the core air gap adjusted so that the coil saturates somewhat as the current increases. As a result of this saturation the inductance of the compensation coil decreases. A decrease in total inductance increases the rate of change of current and is therefore in a direction to compensate for the decrease in rate of change of current due to the resistor. In practice, best linearity is obtained experimentally by an adjustment of the air gap of the compensation coil.

15-2   LINEARITY CORRECTION THROUGH ADJUSTMENT OF DRIVING WAVEFORM

The nonlinearity encountered in the circuit of Fig. 15-1 results from the fact that as the yoke current increases, so also does the current in the series resistance. Consequently the voltage across the yoke decreases and the rate of change of current decreases as well. We may compensate for the voltage developed across the resistor in the manner indicated in Fig. 15-2. The driving voltage source has a Thévenin's resistance R_s, and the total circuit

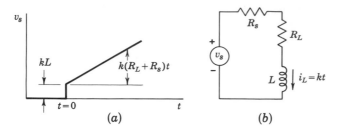

Fig. 15-2 The trapezoidal voltage waveform shown in (a) produces a linear current ramp in the circuit of (b).

resistance is $R_s + R_L$. If the current is to be $i_L = kt$, then the voltage-source waveform must be

$$v_s = L\frac{di}{dt} + (R_s + R_L)i = Lk + (R_s + R_L)kt \tag{15-3}$$

This applied waveform consists of a step followed by a ramp. Such a waveform is called *trapezoidal*.

We may find it more convenient to use a Norton's representation for the driving source, as shown in Fig. 15-3. In this case the current source must furnish a current

$$i_s = \frac{v_s}{R_s} = \frac{k}{R_s}L + kt\left(1 + \frac{R_L}{R_s}\right) \tag{15-4}$$

The waveform of the current source is also trapezoidal, being a step followed by a ramp.

At the end of the sweep the current will return to zero exponentially with a time constant $\tau = L/(R_s + R_L)$. Often we find that $R_s \gg R_L$, so that $\tau \approx L/R_s$. If R_s is small, the current will decay slowly and a correspondingly long period will elapse before another sweep will be possible. But, as a compensation, the peak voltage developed across the current source (which may well be a transistor) will be small. Alternatively, if R_s is large, the current will decay rapidly but a large peak voltage will appear across the source.

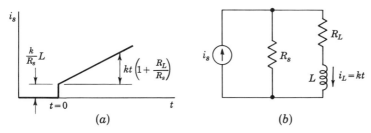

Fig. 15-3 The trapezoidal current waveform in (a) produces a linear current ramp in the inductor of the circuit of (b).

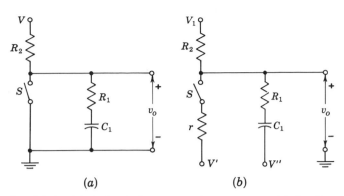

(a)　　　　　　(b)

Fig. 15-4 (a) A circuit for generating a trapezoidal voltage waveform; (b) a more general circuit.

Generally some happy medium will be sought and will often lead to the necessity of bridging a damping resistor R_d across the yoke, as in Fig. 15-1, to limit the peak voltage. Let R be the parallel combination of R_s and R_d. Then the retrace time constant is $\tau = L/R$.

The trapezoidal waveform required is generated by a voltage sweep circuit modified, as in Fig. 15-4a, by the addition of a resistor R_1 in series with the sweep capacitor C_1. If the switch S is opened at $t = 0$, the output v_o is given by

$$v_o = V - \frac{R_2 V}{R_1 + R_2} \epsilon^{-t/(R_1+R_2)C_1} \tag{15-5}$$

Since usually it turns out that $R_2 \gg R_1$, we find on expanding the exponential in Eq. (15-5) that

$$v_o \approx \frac{R_1 V}{R_2} + \frac{Vt}{R_2 C_1}\left(1 - \frac{t}{2R_2 C_1}\right) \tag{15-6}$$

As long as $t/2R_2 C_1 \ll 1$ the waveform v_o is trapezoidal, consisting of a step of amplitude $R_1 V/R_2$ on which is superimposed a ramp of slope $V/R_2 C_1$.

A more general form of the trapezoidal waveform generator is shown in Fig. 15-4b. Here a switch resistance r has been included, and the switch and capacitor C_1 have been returned to arbitrary voltages V' and V'', respectively. Equations (15-5) and (15-6) continue to apply, however, provided that V is taken to be the quiescent voltage across R_2 and that v_o is interpreted as the departure of the output voltage from its quiescent value.

If the inductor were placed directly across the output terminals of Fig. 15-4, the signal v_o would no longer be given by Eq. (15-5). Therefore the signal generated by the circuit of Fig. 15-4 is not to be applied directly to the yoke but rather through an active device—tube or transistor. If the input impedance of the device is R_i, then a Thévenin's equivalent can be made for V_1, R_2, and R_i. It is then found that Eq. (15-6) remains valid, except that the second term in the parentheses is changed to

$$\frac{t}{2R_2C_1}\left(1 + \frac{R_2}{R_i}\right)$$

Hence, R_i must be much larger than R_2 if the linearity of the trapezoid is not to suffer appreciably.

15-3   A TRANSISTOR CURRENT TIME–BASE GENERATOR

The current sweep circuit of Fig. 15-5 is illustrative of the principles set forth in the preceding section.   Transistor $Q1$ is a switch which serves the function of S in Fig. 15-4.   The switch is maintained in saturation as a result of the connection of its base to V_{CC1} through R_b.   The sweep is formed when the input gating signal turns OFF $Q1$ and a trapezoidal voltage waveform appears at the base of $Q2$.   Transistors $Q2$ and $Q3$ are connected as a Darlington composite transistor in order that the input impedance at the base of $Q2$ be high enough not to load the trapezoidal-waveform source.   Such loading would cause nonlinearity in the ramp part of the trapezoid.

The emitter resistor R_e introduces negative current feedback into the output stage and thereby improves the linearity with which the collector current responds to the base voltage (Sec. 1-9).   For best linearity it is necessary to make this emitter resistor as large as possible consistent with the available supply voltage.

Characteristically in a transistor, a plot of collector current against base-to-emitter voltage shows a marked nonlinearity at small collector currents. To avoid this region of operation it is advantageous to establish a quiescent current in the output transistor which is comparable to (say 25 percent of) the peak current to be driven through the yoke.   It is for this reason that the resistor r has been included in the circuit.   In the quiescent condition, the voltage across r will provide some forward bias for $Q2$ and hence for $Q3$.

Transistors $Q2$ and $Q3$ will have different current ratings.   Transistor $Q3$ has to be selected with a rating adequate to supply the required yoke current.   Transistor $Q2$ need only supply the base current of $Q3$, and this base current will equal the yoke current divided by the current gain h_{fe} of transistor $Q3$.   Even quite independently of the matter of the economy involved, it is advantageous to select for $Q2$ a transistor with a current rating at least an order of magnitude smaller than the current rating of $Q3$.   Otherwise, the operating current of $Q2$ may be so small in comparison with its rated value that we shall encounter in $Q2$ the characteristic nonlinearity at low currents.

The impedance R_i seen looking into the base of a transistor operating with an emitter resistor is $R_i \approx h_{fe}R_e$, provided that $h_{fe}R_e \gg h_{ie}$ and $h_{oe}R_e \ll 1$ [Eqs. (1-50) and (1-56)].   In Fig. 15-5 we view the input resistance looking into the base of $Q3$ as the emitter resistance of $Q2$ and find that the input

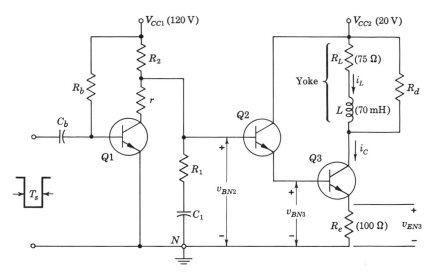

Fig. 15-5   A transistor current sweep circuit.   Supply voltages and component values refer to the illustrative problem on page 578.

resistance at the base of $Q2$ is $\approx h_{fe2}h_{fe3}R_e$, where h_{fe2} and h_{fe3} are the current gains, respectively, of $Q2$ and $Q3$. In a circuit where the supply voltage V_{CC2} is only of the order of tens of volts and the yoke current ranges into amperes, R_e would be limited to a value of only several ohms. In this case the input impedance at the base of $Q2$ may not be large enough to avoid loading the trapezoidal voltage generator, and a Darlington cascade of three transistors may be required. In such a case the current ratings of the transistors would progressively decrease from output transistor to input transistor of the Darlington cascade.

The emitter resistor R_e is selected so that the voltage developed across it will be comparable to the supply voltage V_{CC2}. Since the voltage gain from the base of $Q2$ to the emitter of $Q3$ is of the order of unity, the voltage v_{BN2} will similarly be required to attain a maximum comparable to V_{CC2}. Therefore, in order that the linearity of the ramp portion of the waveform v_{BN2} shall not be adversely affected, it is necessary that V_{CC1} be appreciably larger than V_{CC2}. The maximum voltage attained at the collector of $Q1$ will ordinarily be limited to a safe value by the duration of the gate. As a precaution we may bridge across the transistor a Zener diode which will break down before the collector voltage exceeds its safe value.

The damping resistor R_d is limited by the allowable peak voltage which may be permitted to develop across the yoke current driver $Q3$. It may be necessary to make R_d so small that it carries, during the sweep interval, a current comparable to the yoke current. In such a case it may be advisable to introduce a diode in series with R_d so that R_d carries current only during the retrace. Alternatively R_d may be dispensed with entirely and the transistor protected by bridging an avalanche diode across it. The Zener diode must be selected so that it is adequate to carry, at least for a short time, the peak yoke current. Suppose, then, that there is no damping resistor R_d or that no current flows through R_d during the sweep interval. Then if the output impedance R_s (Fig. 15-3) of the driving transistor is high enough so that it takes negligible current in comparison with the yoke, the step in the input trapezoidal waveform is not needed. In this case the resistor R_1 may be omitted.

EXAMPLE In the current sweep of Fig. 15-5 consider that $V_{cc1} = 120$ V, $V_{cc2} = 20$ V, and $R_e = 100$ Ω. The yoke has $L = 70$ mH and $R_L = 75$ Ω. It is required that the yoke current change by 50 mA in a time 10^{-3} sec. Assume that germanium transistors are used throughout and let $h_{fe} = 50$ for all transistors. Let the output transistor operate with a quiescent current of 20 mA. (a) Select the damping resistor to limit the peak collector voltage to 20 V higher than the quiescent collector voltage, and calculate the damping time constant. (b) Calculate the waveform of the collector current of $Q3$ and the waveform of the voltage across R_e. (c) Calculate the value to which the voltage v_{CE} across the transistor $Q3$ drops just before the termination of the sweep. (d) Calculate the voltage gain $v_{en3}/v_{bn2} \equiv A_{VD}$ of the Darlington circuit. (e) Calculate the waveform v_{BN2}. (f) Select reasonable values for R_2 and r. (g) Calculate R_1 and C_1. (h) Select reasonable values for R_b and C_b.

Solution   a. We assume that the impedance seen looking into the collector of $Q3$ is large enough so that we may view this transistor as an ideal current source. At the end of the sweep period, the yoke current will be larger by 50 mA than it is at the beginning of the sweep. When, at the end of the sweep, the transistor returns to its initial current, the 50-mA current must continue flowing through R_d. This extra current is superimposed on whatever current flows in R_d in the quiescent condition. If the extra voltage across R_d is to be limited to a maximum of 20 V, we must select R_d so that $R_d = 20/0.05 = 400$ Ω. This calculation is based on the assumption that the return of the current in $Q3$ to its initial value is abrupt. Actually, since C_1 must discharge through R_1 and r (as well as through $Q1$), the return will not be abrupt and the peak collector overshoot will be less than 20 V.

The quiescent voltage across the transistor $Q3$ is

$$V_{CE}(\text{quiescent}) = 20 - 20 \times 10^{-3}\left(100 + \frac{400 \times 75}{475}\right)$$

$$= 20 - 3.3 = 16.7 \text{ V} \tag{15-7}$$

The maximum voltage across the transistor is $16.7 + 20 = 36.7$ V. The transistor $Q3$ will have to be selected with a breakdown voltage in excess of this figure.

The damping time constant is

$$\tau_d = \frac{L}{R_L + R_d} = \frac{70}{475} = 0.15 \text{ msec} \tag{15-8}$$

b. We use Eq. (15-4), with $R_s = R_d$, $i_s \equiv \Delta i_C = i_c$, and $k = 50 \times 10^{-3}/10^{-3} = 50$ A/sec. We have

$$i_c = \frac{50}{400} \times 70 \times 10^{-3} + 50t\left(1 + \frac{75}{400}\right) \text{ A}$$

$$= 8.8 + 59.4 \times 10^3 t \text{ mA} \tag{15-9}$$

This current i_c is superimposed on a quiescent current of 20 mA, or $i_C = i_c + 20$ mA.

To calculate v_{en3} we neglect the small difference between the collector and emitter currents. The approximation involved here is equivalent to neglecting unity in comparison with h_{fe}. We shall make this approximation consistently in what follows. The voltage trapezoid across R_e ($= 100 \ \Omega = 0.1$ K) is

$$v_{en3} = 0.1 i_c = 0.88 + 5.94 \times 10^3 t \text{ V} \tag{15-10}$$

This voltage is superimposed on a quiescent voltage of $(20)(0.1) = 2$ V.

c. In the quiescent condition, the yoke current is

$$\frac{400}{400 + 75} \times 20 = 16.8 \text{ mA}$$

Therefore at the end of the sweep the yoke current is $50 + 16.8 = 66.8$ mA. At this time the collector-to-ground voltage is

$$v_{CN3} = V_{CC2} - (kL + R_L i_L)$$

$$= 20 - 50 \times 0.07 - 75 \times 66.8 \times 10^{-3} = 11.5 \text{ V} \tag{15-11}$$

and the emitter voltage is, using Eq. (15-10) with $t = 10^{-3}$ sec,

$$v_{EN3} = 2.00 + 0.88 + 5.94 = 8.82 \text{ V}$$

leaving a voltage $v_{CE3} = 11.5 - 8.82 = 2.68$ V.

The point of this calculation is to show that the voltage across the transistor drops to about the lowest possible value considering that $Q3$ must stay in the active region and that some margin of safety is, of course, desirable. If v_{CE3} had turned out negative it would be necessary to modify the design by decreasing R_e or increasing V_{CC2}. In general, R_e is made as large as possible so as to increase the current feedback and thereby improve the linearity of the driver stage.

d. The voltage gain, base to emitter, of a transistor stage with an emitter resistor R_e is, from Eqs. (1-8) and (1-12),

$$A_V = 1 - \frac{h_{ie}}{R_i} = 1 - \frac{r_{bb'} + r_{b'e}}{R_i} \tag{15-12}$$

The impedance seen looking into the base of $Q3$ is $R_{i3} \approx h_{fe}R_e = 50 \times 0.1 = 5$ K. The impedance seen looking into the base of $Q2$ is $R_{i2} \approx h_{fe}R_{i3} = 50 \times 5 = 250$ K. From Eqs. (1-10) and (1-11) we find that, with I_E in milliamperes,

$$r_{b'e} = \frac{26 h_{fe}}{I_E} \tag{15-13}$$

Since transistor $Q3$ operates with a minimum emitter current of 20 mA, the maximum value of $r_{b'e} = (26)(50)/20 = 65\ \Omega$. The base-spreading resistance may be of the order of about $100\ \Omega$. On this basis we may conservatively estimate h_{ie} to be about $165\ \Omega$. We now find for the voltage gain of $Q3$

$$A_{V3} = 1 - \frac{165}{5,000} = 0.967 \tag{15-14}$$

Transistor $Q2$, operating at a minimum emitter current of about $20/50 = 0.4$ mA, will have

$$h_{ie} \approx 100 + \frac{(26)(50)}{0.4} = 3,350\ \Omega = 3.35\ \text{K}$$

This transistor, however, operates with an effective input resistance of $R_{i2} = 250$ K. Therefore

$$A_{V2} = 1 - \frac{3.35}{250} = 0.987$$

The overall voltage gain of the Darlington cascade $Q2$ and $Q3$ is

$$A_{VD} = A_{V2}A_{V3} = (0.987)(0.967) = 0.954$$

This value corresponds to minimum current (where $r_{b'e}$ is a maximum). As the current increases, A_{VD} will increase slightly and hence introduce some nonlinearity into the sweep. Since A_{VD} cannot exceed unity, then the change in voltage gain over the sweep is only a few percent. The discussion demonstrates that the current feedback through the emitter resistor R_e improves the linearity of response of the driver stage.

e. Since the voltage gain from the base of $Q2$ to the emitter of $Q3$ is $A_{VD} = 0.954$, we have, from Eq. (15-10),

$$v_{bn2} = \frac{1}{A_{VD}}\, v_{en3} = \frac{1}{0.954}\,(0.88 + 5.94 \times 10^3 t)$$

$$= 0.92 + 6.2 \times 10^3 t \tag{15-15}$$

This trapezoidal voltage must be superimposed on a quiescent voltage appropriate to establish the voltage v_{EN3} at a quiescent level of 2.0 V. From the entries in Table 6-1 we judge that for germanium transistors $Q2$ and $Q3$, the base-to-emitter voltage will be about 0.2 V for each transistor. Therefore the waveform of Eq. (15-15) must be superimposed on a quiescent level of 2.4 V.

f. We have seen that the input impedance looking into $Q2$ is 250 K. To minimize the nonlinearity of the trapezoidal-voltage-waveform generator, we make $R_2 = \frac{1}{10}R_{i2} = 25$ K. We may now set r so that V_{BN2}(quiescent) = 2.4 V. We have, assuming that the drop across $Q1$ in saturation is 0.1 V,

$$120\,\frac{r}{r + R_2} = 2.4 - 0.1 = 2.3\ \text{V}$$

$$r = 490\ \Omega$$

g. Comparing now Eqs. (15-6) and (15-15) we may set

$$\frac{R_1 V}{R_2} = 0.92 \qquad \frac{V}{R_2 C_1} = 6.2 \times 10^3 \tag{15-16}$$

in which V, the quiescent voltage across R_2, is $120 - 2.4 = 117.6$ V. We find $R_1 = 196\ \Omega$ and $C_1 = 0.76\ \mu\text{F}$.

h. The transistor $Q1$ in saturation need only carry a collector current of $117.6/25 \approx 5$ mA. If we assume also that $h_{FE} = 50$ for $Q1$, then at a minimum the base current of $Q1$ should be $\frac{5}{50} = 0.1$ mA. If R_b is selected to be 400 K, the base current will be about 0.3 mA, which is more than adequate to drive $Q1$ into saturation. The capacitor C_b should be selected so that the time constant $R_b C_b$ is long in comparison with the sweep duration, which is 10^{-3} sec. Thus we may set $R_b C_b = 0.01$ and select $C_b = 0.01/(4 \times 10^5)$ F $\approx 0.03\ \mu\text{F}$.

15-4   A VACUUM–TUBE CURRENT SWEEP

We consider now an example of a vacuum-tube current sweep, as shown in Fig. 15-6. The circuit is similar to the transistor configuration of Fig. 15-5. The tube circuit is simpler since we need have no concern about raising the impedance seen looking into the output tube. Also, we have placed the yoke in the cathode rather than in the plate circuit. The output impedance of the cathode follower is low enough so that no additional damping resistor is needed.

EXAMPLE In Fig. 15-6 it is required that the sweep duration be 10^{-3} sec, during which time the coil current is to change by 50 mA, so that $k = 50$ A/sec. Choose reasonable values for V_{GG}, R_2, R_1, C_1, R_g, and C_g.

Solution Initially the grid of $V1$ is clamped to the cathode, and the sweep starts when $V1$ is driven beyond cutoff by the application to the grid of a negative gate, as shown. We arbitrarily select R_2 to be 1 M, so that with $V1$ in clamp

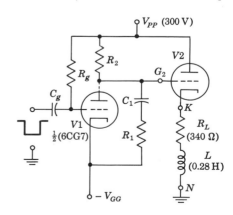

Fig. 15-6 A vacuum-tube current sweep. Supply voltages and component values refer to the illustrative problem.

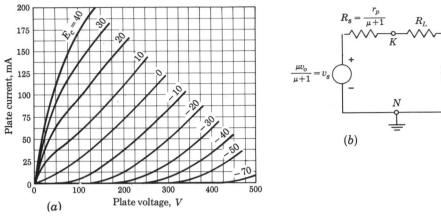

(a)

(b)

Fig. 15-7 (a) Plate characteristics of the power triode $V2$, used as the coil driver in Fig. 15-6; (b) the equivalent circuit, for changes from the quiescent value, of the cathode follower $V2$.

the plate voltage of $V1$ is close to the cathode voltage. From the characteristics of $V2$ given in Fig. 15-7a we find that $V2$ will be cut off when the grid voltage is about -50 V. We bias $V2$ at -25 V *with respect to ground*. We then find that, in the presence of the 340-Ω cathode resistance, the tube current is about 20 mA. In this way the tube nonlinearity at low current is avoided. The drop through R_2 must then be 325 V, so that the current in $V1$ is about 0.3 mA. At zero grid-to-cathode voltage $V1$ will conduct 0.3 mA when the plate-to-cathode voltage is about 5 V (Fig. D-2). Accordingly we set $-V_{GG} = -30$ V. The quiescent voltage across C_1 is 5 V.

From Eq. (1-48) with $R_p = 0$ we see that the equivalent circuit for the cathode follower $V2$ is given in Fig. 15-7b, in which v_o is the voltage applied to the grid of $V2$. From the plate characteristics in Fig. 15-7a we estimate that $r_p = 2,300 \ \Omega$ and $\mu = 7.5$. $R_s = r_p/(\mu + 1) = 270 \ \Omega$, and $\mu/(\mu + 1) = 0.88$. Hence, from Eq. (15-3),

$$v_s = \frac{\mu v_o}{\mu + 1} = kL + k(R_s + R_L)t \tag{15-17}$$

Substituting $k = 50$ A/sec, $L = 0.28$ H, $R_s + R_L = 270 + 340 = 610 \ \Omega$, and $\mu/(\mu + 1) = 0.88$, we find

$$v_o = 15.9 + 34.7 \times 10^3 t \tag{15-18}$$

Comparing this equation with Eq. (15-6) we find

$$\frac{R_1}{R_2} V = 15.9 \quad \text{and} \quad \frac{V}{R_2 C_1} = 34.7 \times 10^3 \tag{15-19}$$

Since the quiescent voltage across R_2 is 325 V, then $V = 325$ V. For this value of V and for $R_2 = 1$ M, we find $R_1 = 49$ K and $C_1 = 0.0094 \ \mu$F.

Excellent clamping is obtained with $R_g = 1$ M. The $R_g C_g$ time constant should be long compared with the sweep time of 1 msec. Hence, we may choose $C_g = 0.01/R_g = 0.01$ μF. This value is far from critical; a smaller value will give an overshoot at the grid which will help discharge C_1 more quickly (Fig. 8-26d). At the termination of the sweep interval the inductor current will decay from 70 mA to its original quiescent value of 20 mA with a time constant $L/(R_L + R_s) = 0.28/610$ sec $= 0.46$ msec.

15-5   COIL CAPACITANCE

The inevitable stray capacitance which shunts any physical coil introduces into current sweeps a minor complication which has no counterpart in voltage sweeps. Up to the present we have neglected the effect of this capacitance, but now we shall see how the driving waveform must be modified to take account of this capacitance. In Fig. 15-8 a current source is shown driving a yoke. A shunt capacitance C is indicated, and R, as usual, represents the parallel combination of the source impedance and any additional damping resistance. If a linear time base is desired, $i_L = kt$ for $t \geq 0$ and hence the voltage v is

$$v = L \frac{di_L}{dt} + R_L i_L = kL + kR_L t \tag{15-20}$$

Hence, the current taken by the damping resistor is

$$i_R = \frac{v}{R} = \frac{kL + kR_L t}{R}$$

The sweep is to begin at $t = 0$. The capacitor is uncharged at $t = 0-$ and, as indicated by Eq. (15-20), the capacitor voltage must jump to kL at $t = 0+$. Hence, a current i_C must be furnished by the generator that will charge the capacitor to the voltage kL in zero time. This current must be infinite in magnitude but must last for only an infinitesimal time duration and must have the property that $(1/C)\int i_C \, dt = kL$, or

$$\int i_C \, dt = kLC$$

Also, an additional capacitor current must flow which is equal to $kR_L C$ so that the voltage across the capacitor will continue to rise at the same rate at which the voltage v is rising.

Fig. 15-8 A current generator i_s feeds a magnetic deflection coil. The yoke capacitance C is shown.

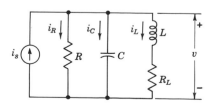

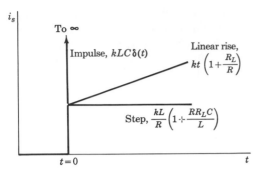

Fig. 15-9 In order to obtain a linear sweep in the presence of yoke capacitance, a current generator must supply three components of current: an impulse, a step, and a linear rise.

It is convenient to introduce the unit impulse or delta function $\delta(t)$, defined by $\delta(t) = \infty$ when $t = 0$, $\delta(t) = 0$ when $t \neq 0$, and

$$\int_{-\infty}^{+\infty} \delta(t)\, dt = 1$$

We may now write for the current source that

$$i_s = i_L + i_R + i_C = kt + (kL + kR_Lt)\frac{1}{R} + kR_LC + kLC\delta(t)$$

or

$$i_s = kLC\delta(t) + \frac{kL}{R}\left(1 + \frac{RR_LC}{L}\right) + kt\left(1 + \frac{R_L}{R}\right) \tag{15-21}$$

Altogether the current generator must furnish a current that consists of an impulse, a step, and a linear rise, as indicated in Fig. 15-9.

At the termination of the sweep the energy stored in the inductor must be dissipated. It is usually desired that the decay of inductor current shall not be accompanied by an oscillation in the circuit. The resistor R serves to provide damping for the circuit and is usually adjusted so that the circuit is either critically damped or overdamped. If we neglect the effect on the damping of the small resistance R_L (Prob. 15-12), the value of R for critical damping is $R = \frac{1}{2}\sqrt{L/C} \equiv R_{cr}$. The parameters for some typical deflection coils are given in Table 15-1.[3]

The coil represented nominally by the second entry in this table, neglecting the capacitance, was used in the illustrative example of Sec. 15-3, and the coil represented by the first entry was used in the illustrative example of Sec. 15-4.

TABLE 15-1   Deflection-coil parameters

Core	L, mH, at 1,000 Hz	R_L, Ω	C, pF	$R_{cr} = \frac{1}{2}\sqrt{\dfrac{L}{C}}$, Ω	$\sqrt{LC}$, μsec	$\dfrac{R_{cr}R_LC \times 10^3}{L}$
Iron.........	280	340	200	19,000	7	5
Iron.........	70	73	250	8,000	4	2
Air.........	97	408	25	31,000	1.5	3

**15-6   EFFECT OF THE OMISSION OF THE
           IMPULSE COMPONENT OF CURRENT**

It is physically impossible to generate exactly the impulse term of Eq. (15-21).
Let us investigate the effect of omitting it completely. If i_s is given by Eq.
(15-21) except that the term $kLC\delta(t)$ is missing, the differential equation for
i_L for the circuit of Fig. 15-8 is found from

$$R(i_L + i_C + i_R) = R\left(i_L + C\frac{dv}{dt} + \frac{v}{R}\right) = Ri_s$$

to be

$$RLC\frac{d^2i_L}{dt^2} + (RR_LC + L)\frac{di_L}{dt} + (R + R_L)i_L$$

$$= kL + RR_LCk + (R + R_L)kt \quad (15\text{-}22)$$

The solution of the inhomogeneous equation is $i_L = kt$, as is to be anticipated.
The transient part of the solution is obtained by setting the right-hand member
of Eq. (15-22) equal to zero.

Critical Damping   Assume first that R has been selected for critical
damping, $R = R_{cr}$.   Then $RR_LC \ll L$ and also $R_L \ll R$ (Table 15-1), so that

$$RLC\frac{d^2i_L}{dt^2} + L\frac{di_L}{dt} + Ri_L = 0 \quad (15\text{-}23)$$

The single root of Eq. (15-23) for critical damping is $s = -1/\sqrt{LC}$, and the
form of the complete solution is

$$i_L = (A + Bt)\epsilon^{-t/\sqrt{LC}} + kt \quad (15\text{-}24)$$

At $t = 0$, the coil current i_L is zero and the capacitor voltage v is zero. Since
$v = L\,di_L/dt + R_Li_L = 0$, then the initial conditions are $i_L = 0$ and $di_L/dt = 0$.
Subject to these conditions we find that

$$i_L = kt(1 - \epsilon^{-t/\sqrt{LC}}) \quad (15\text{-}25)$$

which is plotted as a solid line in Fig. 15-10a. The maximum deviation
between the actual sweep and the ideal sweep is $0.37k\sqrt{LC}$ and occurs at
a time $t = \sqrt{LC}$. The sweep is temporarily delayed for an interval which is
several times $\sqrt{LC}$. Values of $\sqrt{LC}$ are tabulated in Table 15-1.

Overdamped   Consider the case of very heavy damping. Since R is
now even smaller than R_{cr}, then certainly $RR_LC \ll L$, but it may not be that
$R_L \ll R$. Replacing the term Ri_L in Eq. (15-23) by $(R + R_L)i_L$, we find for
the roots

$$s = -\frac{1}{2RC} \pm \frac{1}{2RC}\sqrt{1 - \frac{4(R + R_L)RC}{L}} \quad (15\text{-}26)$$

If R is smaller than $\frac{1}{10}R_{cr}$, then, for the typical coil parameters in Table 15-1, the second term under the square-root sign is less than 0.01. Neglecting this term compared with unity, we have for the two roots

$$s_1 = -\frac{1}{RC} \quad \text{and} \quad s_2 = 0$$

The form of the complete solution is

$$i_L = A + B\epsilon^{-t/RC} + kt$$

Subject to the initial condition $i_L = di_L/dt = 0$ at $t = 0$, we find for the complete solution

$$i_L = kt + RCk(\epsilon^{-t/RC} - 1) \tag{15-27}$$

The current i_L is plotted (solid line) in Fig. 15-10b, where it is seen that the sweep is permanently delayed by a time RC. The time delay will be smaller than the delay in the case of critical damping. For example, say R is reduced to $R_{cr}/10 = 1{,}900\ \Omega$ for the first entry in Table 15-1. Then the delay is of the order of 0.38 μsec rather than several times 7 μsec. This result is to be expected since C may now charge from a lower-impedance source.

The effective capacitance across a deflection coil should, of course, always be kept as low as possible by using a type of winding which gives the minimum distributed capacitance and by keeping stray circuit capacitances at a minimum. One very effective procedure is to reduce the number of turns on the coil, but in this case the deflection produced per unit coil current is correspondingly low. In applications where fast sweeps are required, it is not uncommon to use high-power transistors or small transmitting tubes to provide the necessary current.

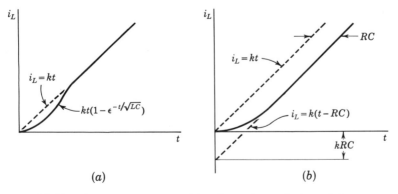

Fig. 15-10 The coil current resulting from the omission of the impulse-current term in Eq. (15-21). In (a) R equals the critically damped value R_{cr} and in (b) $R \ll R_{cr}$.

Fig. 15-11 A voltage generator v_s used to drive a deflection coil.

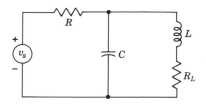

Alternatively, we may use the circuit of Fig. 15-11, in which the Norton's current generator of Fig. 15-8 is replaced by a Thévenin's voltage generator v_s. It is required that $v_s = i_s R$, where i_s is given by Eq. (15-21). Thus

$$v_s = kRLC\delta(t) + k(L + RR_LC) + (R + R_L)kt \qquad (15\text{-}28)$$

15-7 METHODS OF LINEARITY IMPROVEMENT[3]

The sweep circuits discussed in Secs. 15-3 and 15-4 will not provide precisely linear sweeps for the following four essential reasons:

1. The impulse term required by Eq. (15-21) is lacking.
2. The driver transistor or tube which provides the inductor current does not operate with sufficient linearity, especially over the large current ranges required.
3. The nominally linear portion of the trapezoidal waveform provided by the circuit of Fig. 15-4 is actually exponential in form.
4. The inductance of an iron-core coil varies with current. This non-linearity of the iron is avoided by using an air-core coil. The first three nonlinearities mentioned above will now be discussed.

Circuits for Generating an Impulse In Fig. 15-12 is shown a transistor driver with the deflection coil in the collector circuit. A resistor R_e bypassed with a small capacitance C_e is placed in the emitter circuit. The input voltage is a trapezoidal voltage v_s as shown. The degenerative effect of the emitter resistor R_e will not make itself felt until C_e has charged. The output current i of the transistor will appear as in Fig. 15-13. During the relatively slow rise of the linear part of the trapezoidal voltage the presence of C_e will have little influence on the output current and the effect of C_e may be neglected. The time constant R_eC_e is taken of the order of magnitude of the sweep delay resulting from the omission of the impulse. Since the spike of Fig. 15-13 is only a crude approximation of an impulse, the final value of C_e is obtained experimentally for optimum linearity.

Alternatively, instead of adjusting the current driver so that it provides a current spike, we may instead produce a spike in the voltage applied to the driver tube. For example, we may invert the gating square wave in Fig. 15-5, differentiate it with a small RC circuit (Sec. 2-3), and apply the resultant positive pip at the beginning of the sweep across R_1. The voltage at the base

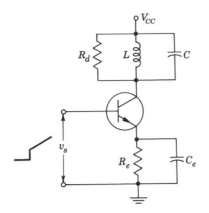

Fig. 15-12   An emitter R_eC_e combination is used to yield an impulse-type
component of collector current.

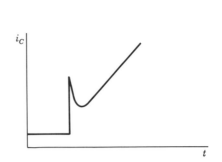

Fig. 15-13   The collector current in Fig.
15-12 contains a *spike* because of the
small R_eC_e time constant.

of $Q2$ will then have the waveshape depicted in Fig. 15-13.   Another circuit
which produces the same result is given in Prob. 15-21.

Improvement of Linearity of Current Driver for Yoke   We see in Sec. 1-9
that negative current feedback may be used to suppress the effect of the nonlinearity of the active devices in an amplifier.   A method of using current
feedback in a yoke driver is shown in Fig. 1-15, where Z_L represents the
deflection yoke.   On replacing the phasors I_L and V_s in Eq. (1-36) by their
instantaneous values i and v_s, this equation indicates that the output current
will be given by

$$i = \frac{v_s}{R_f} \tag{15-29}$$

and is independent of the amplifier characteristics.   The negative current
feedback has transformed the amplifier into a device which acts as a current
generator whose output current is proportional to the applied input signal v_s.
The coil current will vary linearly with time, provided that the input voltage
is given by $v_s = iR_f$, where i is indicated in Eq. (15-21).

Negative current feedback is used in the circuit of Fig. 15-5, where the
emitter resistor R_e of the driver stage $Q3$ is the feedback resistor R_f.   In
Sec. 15-3 we verify that the signal voltage $v_s = v_{bn2}$ at the input to the
base of $Q2$ is almost identical with the voltage drop v_{en3} across R_e [Eq. (15-15)].
Hence, $v_s = iR_e$ is in agreement with Eq. (15-29).

Negative voltage feedback may also be used to suppress the nonlinearity
of the active devices in an amplifier.

Linearization of Trapezoidal Voltage   Referring to Fig. 15-4a, it is clear
that a perfect trapezoid will be developed if the current through R_2 is kept

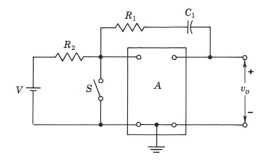

Fig. 15-14 An operational amplifier used to generate a trapezoidal waveform.

constant. The current will be constant at V/R_2 if the top of the resistor R_2 is bootstrapped to the output voltage v_o. This bootstrapping is conveniently done in the case of an amplifier with current feedback. From Eq. (15-29), the voltage developed across the feedback resistor is $R_f i = v_s$, and since the output voltage in Fig. 15-4 is $v_o = v_s$, it is only necessary to bootstrap the top of R_2 to the voltage drop across the feedback resistor.

A second method for improving the linearity of the trapezoid is through the use of an operational amplifier or Miller integrator, as shown in Fig. 15-14. When the switch S opens at $t = 0$, the output is

$$v_o = -\frac{R_1}{R_2} V - \frac{V}{R_2 C_1} t \tag{15-30}$$

just as required, provided only that the gain A of the amplifier is very large.

A third method for obtaining a linear voltage sweep is to charge the capacitor from a constant-current source such as a common-base transistor. Each of these linearizing techniques is explained in detail in Chap. 14.

15-8   ILLUSTRATIVE CURRENT SWEEP CIRCUITS

Because of the feedback, the circuit of Fig. 15-14 not only provides an output of the correct form but, in the case of large A, has an output which is independent of device characteristics, etc. The deflection coil may therefore be placed directly across the output terminals. Since the output impedance is nominally zero, the required output voltage [Eq. (15-3)] is

$$v_o = kL + R_L k t \tag{15-31}$$

Comparing Eqs. (15-30) and (15-31) and neglecting the arbitrary minus signs in Eq. (15-30), we may compute the required values of R_1 and C_1 as

$$R_1 = \frac{kLR_2}{V} \qquad C_1 = \frac{V}{kR_L R_2} \tag{15-32}$$

where V is the quiescent voltage across R_2.

A practical circuit which is patterned after the ideal circuit of Fig. 15-14 is obtained from Fig. 15-5 by changing one connection: the terminal of C_1

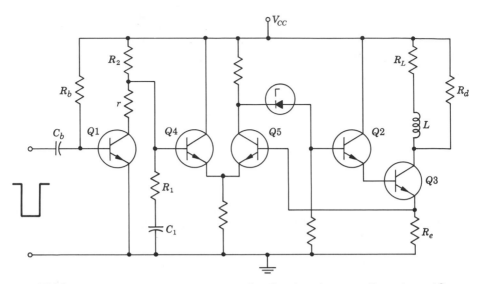

Fig. 15-15 A current sweep using current feedback to improve linearity. (Courtesy of CBS Laboratories, Stamford, Conn.)

which is grounded is now tied to the collector of $Q3$. This alteration places R_1 and C_1 in series between the input and the output of the amplifier consisting of $Q2$ and $Q3$ in cascade, so that the circuit now functions as an operational amplifier.

A more elaborate current sweep involving negative current feedback is shown in Fig. 15-15. This circuit is the configuration of Fig. 15-5 modified to incorporate a difference-amplifier stage $Q4$ and $Q5$ in the lead connected to the base of $Q2$. Figure 15-15 falls into the block-diagram form of Fig. 1-15, with input 1 corresponding to the base of $Q4$, input 1' to the base of $Q5$, and $R_f = R_e$. Because of the additional gain provided by the difference amplifier there is a further improvement in the yoke-driver linearity over that attained with the simpler circuit of Fig. 15-5. In a high-current deflection system where R_e is very small, additional gain as provided by Fig. 15-15 is necessary in order to satisfy the restriction $A'R_e \gg R_o$ upon which Eq. (15-29) is based. The difference amplifier is used to combine the feedback voltage applied at the base of $Q5$ with the trapezoid applied at $Q4$. The avalanche diode is a coupling element between the collector of $Q5$ and the base of $Q2$, which operate at different quiescent voltages.

Since, before application to the base of $Q2$, the trapezoidal voltage developed across R_1C_1 is amplified by the difference amplifier, the trapezoid may have an appreciably smaller amplitude. For this reason, it is now not necessary to supply the transistor $Q1$ from a supply voltage appreciably higher than is used for the remainder of the circuit.

An improvement in linearity of the trapezoid waveform is possible if C_1 is charged not through R_2 but rather by placing C_1 in the collector of a transistor operating in a common-base configuration (Sec. 14-7). Furthermore, the linearity of the current sweep may be improved considerably if a push-pull configuration is employed to drive the yoke.

15-9   A TRANSISTOR TELEVISION SWEEP CIRCUIT[4]

The frequency of a television horizontal sweep circuit is 15,750 Hz, corresponding to a total time of 63.5 μsec for the combined sweep and retrace times. Since the flyback should be a small fraction of the complete cycle, one of the special problems associated with this television sweep is that of obtaining a fast retrace time (of the order of a few microseconds). Another important problem is that of conserving power. The peak energy stored in the inductance L of a deflecting coil is $\frac{1}{2}LI_m^2$, where I_m is the peak current. If this energy were dissipated in each cycle, the power lost in the yoke would be $P = \frac{1}{2}LI_m^2 f_h$, where f_h is the horizontal scanning frequency. Typical values are $L = 30$ mH, $I_m = 300$ mA, and $f_h = 15,750$ Hz, so that $P = 21$ W. This value represents a large portion of the total power taken by an entire television set with a large picture tube.

The basic mechanism by means of which we may generate a current

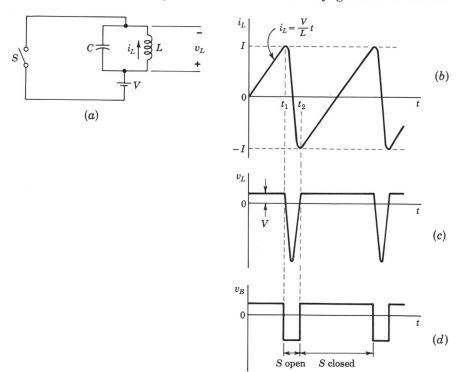

Fig. 15-16 (a) A basic fast-retrace lossless current sweep; (b) the inductor current waveform; (c) the waveform of the inductor voltage; (d) the gating waveform required at the base of the transistor circuit of Fig. 15-17.

sweep with a fast retrace and with a minimum loss of power may be understood from a consideration of the circuit of Fig. 15-16a. Here, a parallel combination of an inductor L and a capacitor C is connected to a voltage source V through a switch S. Assume that initially there is no energy stored in C or L and that switch S is closed at time $t = 0$. Then, as shown in Fig. 15-16b, the current i_L in the inductor will start at zero and increase in accordance with $i_L = (V/L)t$. Of course, as shown in Fig. 15-16c, so long as S is closed, the voltage across L (and C) is V.

At time $t = t_1$, when the inductor current has reached a value $i_L = I$, the switch S is opened. The LC circuit now begins to oscillate; that is, it "rings" as described in Sec. 2-11. The waveform of the current is cosinusoidal with an angular frequency $\omega = 1/\sqrt{LC}$. At the end of one half cycle of oscillation the inductor current will have reversed, being $i_L = -I$. At this moment (time $t = t_2$), the switch S is closed again and the current i_L increases linearly from $-I$ to I. Then S is again opened, and the cycle beginning at t_1 repeats.

The inductor current during the retrace is

$$i_L = I \cos \omega(t - t_1) \tag{15-33}$$

and the peak voltage across the switch has the magnitude

$$L \left(\frac{di_L}{dt} \right)_{\text{max}} = \omega L I \tag{15-34}$$

The retrace time is $1/2f$. Since the resonant frequency of a coil used in a commercial television set is at least 70 kHz, the flyback time is less than $1/(2 \times 7 \times 10^4)$ sec ≈ 7 μsec. This calculation demonstrates that the circuit does indeed have the desired short recovery time.

In Fig. 15-17 the switch of Fig. 15-16 is replaced by a transistor. The input signal makes transitions between two levels such that the transistor

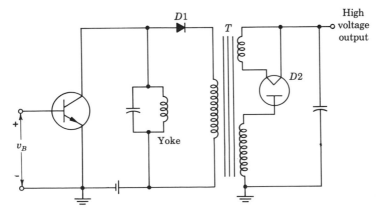

Fig. 15-17 A television horizontal-deflection circuit and flyback power supply. The gating waveform v_B is indicated in Fig. 15-16d.

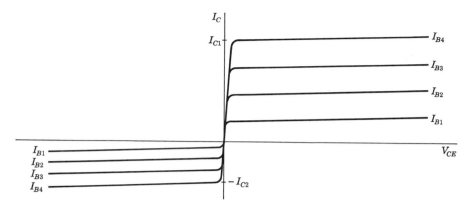

Fig. 15-18 Common-emitter collector characteristics illustrating the bi-directional property of a transistor switch.

is driven from cutoff to saturation. The timing of the base waveform is shown in Fig. 15-16d. The train of large amplitude pulses (Fig. 15-16c) which appears across the yoke is coupled through $D1$ to the primary of the transformer T. The transformer steps up the voltage, and the pulse waveform is rectified by the thermionic diode $D2$. This arrangement is called a *flyback* (or *kickback*) power supply. The high d-c voltage developed at the output of the rectifier is used to accelerate the beam in the television picture tube. Power is supplied from the sweep circuit to provide filament heating power as well. The pulse waveform across the yoke appears across the transistor as well as across the transformer and may, for this reason, have to be limited to a peak voltage of the order of 100 V or so. However, T may have a step-up ratio as high as 100 or more, so that a rectified output voltage of the order of 10 kV is not unreasonable. Diode $D2$ must have a correspondingly high peak inverse voltage. This diode is thermionic, because suitable single semiconductor rectifiers with so large a peak inverse voltage are not presently available.

In using a transistor as a switch in the circuit of Fig. 15-17 we find a requirement imposed on the switch which we have not previously encountered. We observe in Fig. 15-16 that during the interval of the linear sweep the current through the inductor, and hence through the transistor, flows first in one direction and then in the other. The transistor must operate as a bidirectional switch.[5] That a transistor does indeed have such bidirectional properties may be seen in Fig. 15-18. Here, typical collector characteristics have been plotted for collector-to-emitter voltages of both polarities. The normal region of operation is in the first quadrant. In the third quadrant, the collector-to-emitter voltage and the transistor current are both reversed but the base current is not reversed. In this third-quadrant region the roles of the emitter and collector junctions are interchanged, the emitter acting as a collector and vice versa. This inverted mode is discussed in Sec. 6-16. The closer spacing of the characteristics indicates a reduced current gain for the inverted connection. It appears from Fig. 15-18 that if the base current is set at I_{B4}, the transistor will remain in saturation over the range of currents $-I_{C2}$ to $+I_{C1}$. If the transistor is reasonably symmetrical, as is the case with alloy-junction transistors, I_{C2} and I_{C1} will be comparable to one another. Special transistors that have been symmetrically constructed for use as bidirectional switches are commercially available.

15-10   A VACUUM–TUBE TELEVISION SWEEP CIRCUIT[6]

This sweep operates on the same basic principle as does the transistor sweep. Relative to the transistor case, however, the vacuum-tube circuit is beset by two complications which make the latter circuit slightly less simple. First, a vacuum tube will not carry current bidirectionally. Therefore, to allow current flow in two directions as is required, two tubes must be used. Secondly, the vacuum tube is not adequate as a switch to allow the generation of a linear sweep by the simple mechanism of connecting the yoke to a supply voltage through the tube switch. The tube resistance is too high. Therefore the yoke must be placed in series with the tube, and a sweep generated by the application to the grid of the tube of an appropriate trapezoidal voltage.

The basic circuit structure of the vacuum-tube sweep is shown in Fig. 15-19 together with its waveforms. The diode replaces the damping resistor and is called the *damper diode*. This diode provides the path through which the yoke current may reverse. The operation of the circuit is the following.

Assume that the input to the grid of the amplifier is a single-stroke trapezoidal sweep lasting for a time T_s, as indicated in Fig. 15-19b. The

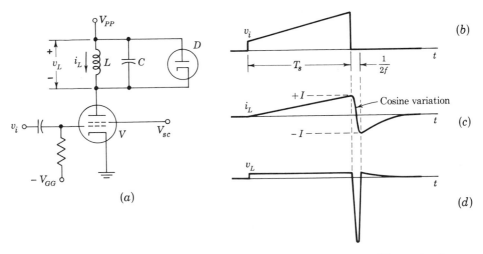

Fig. 15-19   (a) Basic vacuum-tube television deflection amplifier; (b) required trapezoidal input driving waveform; (c) inductor current for single-sweep operation; (d) inductor voltage.

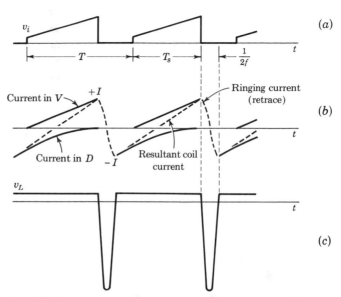

Fig. 15-20   A recurrent sweep.   (a) The input waveform; (b) the coil current is made up of the superposition of two currents; (c) the coil voltage.

current i_L in the coil is approximately linear.   If the coil resistance is neglected, the voltage v_L across the coil is positive and of constant value for the interval T_s.   The diode conducts a constant current during this interval.   At the end of the sweep, the switch V is cut off.   The plate current immediately falls to zero, and the coil current now flows through the capacitor C.   The circuit rings for a half cycle, during which time $v_L = L \ di/dt$ is negative and the diode is cut off.   The coil current changes in a cosinusoidal manner from a positive peak I to a negative value $-I$ in a time $1/2f$, where f is the resonant frequency corresponding to L and C.   At the end of this half cycle the voltage v_L reverses sign (Fig. 15-19d), and D conducts again.   The current i_L and the voltage v_L now decrease to zero with a time constant L/R_f, where R_f is the forward resistance of the diode.

For the sake of simplicity, we assumed a single-stroke sweep in the above discussion.   For a recurrent sweep, the situation is as pictured in Fig. 15-20. The sweep time T_s is chosen so that the period T of the input voltage is only slightly longer than $T_s + 1/2f$.   Under these circumstances tube V begins to conduct again before D stops conducting.   Hence, the total coil current (shown dashed in Fig. 15-20) is now made up of two components: the positive one is that part of the coil current which flows in V and the negative one is that portion of the coil current which flows in D.

By adding a step-up transformer and a high-voltage diode to Fig. 15-19, in the manner indicated in Fig. 15-17, the flyback power-supply voltage to accelerate the picture-tube beam is obtained.

We also see from Fig. 15-20b that the total sweep corresponds to a current of $2I$ and yet the peak energy stored in the inductor is $\frac{1}{2}LI^2$ and *not* $\frac{1}{2}L(2I)^2$. Hence the use of the damper diode to replace a damping resistor has reduced (to one-fourth its previous value) the magnetic coil energy which is dissipated in each cycle. The ideal transistor television sweep described in Sec. 15-9 dissipates no energy. Of course, taking account of the yoke resistance, the transistor leakage, etc., we should find energy lost in the transistor sweep as well. However, these losses may be of the order of magnitude of only one-tenth the losses in the tubes of a vacuum-tube sweep. Hence transistor sweeps are generally more efficient.

REFERENCES

1. Millman, J., and C. C. Halkias: "Electronic Devices and Circuits," McGraw-Hill Book Company, New York, 1967.

2. Schwartz, S. A.: Transistor Range Marker and Sweep Generator for Radar Display, *Fairchild Appl. Data*, APP-77, September, 1963.

3. Soller, T., M. A. Starr, and G. E. Valley, Jr.: "Cathode-ray Tube Displays," Massachusetts Institute of Technology Radiation Laboratory Series, vol. 22, McGraw-Hill Book Company, New York, 1948.

4. Goodrich, H. C.: A Transistorized Horizontal Deflection System, *RCA Rev.*, vol. 18, no. 3, pp. 293–307, September, 1957.
   Sziklai, G. C., R. D. Lohman, and G. B. Herzog: A Study of Transistor Circuits for Television, *Proc. IRE*, vol. 41, no. 6, pp. 708–717, June, 1953.

5. Sziklai, G. C.: Symmetrical Properties of Transistors and Their Applications, *Proc. IRE*, vol. 41, no. 6, pp. 717–724, June, 1953.

6. Anner, G. E.: "Elements of Television Systems," chap. 4, Prentice-Hall, Inc., Englewood Cliffs, N.J., 1951.

16 / BLOCKING–OSCILLATOR CIRCUITS

The output of an active device may be coupled back to the input through a pulse transformer. If the relative winding polarities are properly chosen, the feedback will be regenerative and the circuit can be made to generate a single pulse (monostable operation) or a pulse train (astable mode). A transformer-coupled configuration of considerable practical importance, called the *blocking oscillator*, is described in this chapter. Also considered is a blocking-oscillator type of circuit, called a *multiar*, which functions as a regenerative comparator.

16-1 A TRIGGERED TRANSISTOR BLOCKING OSCILLATOR (BASE TIMING)[1,2]

A monostable blocking-oscillator circuit which may be triggered by a slowly varying input voltage is shown in Fig. 16-1. Depending upon the parameters of the pulse transformer and the circuit, the pulse width may lie in the range from nanoseconds to microseconds. The transformer has n times as many turns in the base circuit as in the collector circuit and is connected into the circuit so as to provide polarity inversion as indicated by the dots on the windings. A resistor R is included in series with the base of the transistor, and it will be seen that this resistor controls the timing, that is, the pulse duration.

In the quiescent state the transistor is OFF. Since the cutin base-to-emitter voltage at room temperature is approximately 0.1 V positive for a germanium and 0.5 V positive for a silicon (n-p-n) transistor, then V_{BB} may be reduced to zero. However, in order to avoid triggering by noise pulses and to prevent free-running operation

597

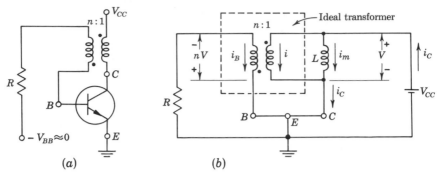

Fig. 16-1   (a) A triggered transistor blocking oscillator with base timing;
(b) the equivalent circuit during the pulse formation.   (In blocking oscillators, iron-cored pulse transformers are invariably employed.   However,
to simplify the drawings the symbol for these iron cores has not been
indicated.)

at elevated temperatures, V_{BB} is selected to be of the order of a few tenths
of a volt.   Since $V_{BB} \ll V_{CC}$ and does not basically affect the operation of
the circuit we shall neglect V_{BB} in the discussion below.

Suppose that a triggering signal is momentarily applied to the collector
to lower its voltage.   By transformer action and with the winding polarities
indicated in Fig. 16-1a the base will rise in potential.   When the base-
to-emitter voltage exceeds the cutin voltage, the transistor starts to draw
current.   The increase in collector current lowers the collector voltage, which
in turn raises the base voltage.   Hence, still more collector current flows,
resulting in a further drop in collector potential.   If the a-c loop gain exceeds
unity, regeneration takes place and the transistor is quickly driven into
saturation.

Let us ignore the small time required for the transistor to enter saturation.
The pulse amplitude, width, and waveshape are obtained from the equivalent
circuit of Fig. 16-1b, where the pulse transformer is replaced by an ideal transformer in shunt with the magnetizing inductance L of the collector winding.
The leakage inductance σ and shunt capacitance C of the transformer model
of Fig. 3-5 affect the rise time but not the relatively slowly varying voltages
during the pulse.   Hence σ, C, and the small ohmic winding resistance are
omitted from Fig. 16-1b.   The transistor saturation junction voltages $V_{CE}(\text{sat})$
and $V_{BE}(\text{sat})$ could easily be included in the circuit, but they are small compared with V_{CC} and for the sake of simplicity we shall neglect them.

For an ideal transformer the sum of the ampere-turns is a constant and
the induced voltages are proportional to the turns.   Since in the quiescent state
all currents are zero, then the ampere-turns within the dashed box of Fig.
16-1b must add algebraically to zero.   If i is the current in the ideal trans-

former collector winding, then, taking winding polarities into account,

$$i - ni_B = 0 \tag{16-1}$$

If V is the amplitude of the step across the collector winding, then nV is the corresponding voltage across the base winding. Observe in Fig. 16-1b that the polarities assigned to the winding voltages V and nV are consistent with the polarity inversion indicated by the dots. From the collector circuit of Fig. 16-1b we see that

$$V = V_{CC} \tag{16-2}$$

From the base circuit

$$i_B = \frac{nV}{R} = \frac{nV_{CC}}{R} \tag{16-3}$$

and from Eq. (16-1)

$$i = \frac{n^2 V_{CC}}{R} \tag{16-4}$$

If i_m is the magnetizing current, and since V is a constant, then

$$L \frac{di_m}{dt} = V \qquad \text{or} \qquad i_m = \frac{Vt}{L} \tag{16-5}$$

Since $i_C = i + i_m$, then, from Eqs. (16-4), (16-5), and (16-2),

$$i_C = \frac{n^2 V_{CC}}{R} + \frac{V_{CC}t}{L} \tag{16-6}$$

Note that the collector current is trapezoidal, whereas the base current is constant, as indicated in Fig. 16-2.

At $t = 0+$ the operating point on the collector characteristics of Fig. 16-3 is at point P, where $i_C = n^2 V_{CC}/R$ and $i_B = nV_{CC}/R$. The transistor is in

Fig. 16-2 (a) The collector current and (b) the base current in the circuit of Fig. 16-1. (The dashed curve is due to minority carriers stored in the base.)

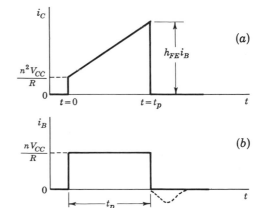

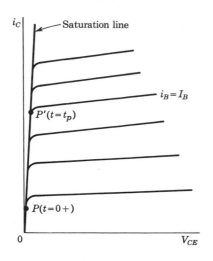

Fig. 16-3 The path of the collector current is along the saturation line from P to P'. The pulse ends at P', where the transistor comes out of saturation.

saturation, provided that

$$h_{FE}i_B > i_C \quad \text{or} \quad h_{FE} > n \tag{16-7}$$

a condition which is very easy to satisfy since the turns ratio n is usually close to unity. For $t > 0$ the collector current increases and the operating point moves up the curve in Fig. 16-3, corresponding to the constant base current $i_B = nV_{CC}/R \equiv I_B$. When point P' is reached, V_{CE} increases rapidly, and this decreases the transformer voltage and hence the base current. At this point P' the transistor comes out of saturation. Because the loop gain exceeds unity in the active region, the transistor is quickly driven to cutoff by regenerative action and the pulse ends. Since the regeneration which terminates the pulse starts when the transistor comes out of saturation, then the pulse width t_p is determined by the condition

$$i_C = h_{FE}i_B \tag{16-8}$$

or, from Eqs. (16-6) and (16-3),

$$\frac{n^2 V_{CC}}{R} + \frac{V_{CC}t_p}{L} = h_{FE}\frac{nV_{CC}}{R} \tag{16-9}$$

Solving yields

$$t_p = \frac{nL}{R}(h_{FE} - n) \approx \frac{nLh_{FE}}{R} \tag{16-10}$$

This result indicates that the pulse width is a linear function of h_{FE}, a parameter which varies with temperature (Fig. 6-27c) and which may change by a large factor (Fig. 6-23) with transistor replacement. Hence, the circuit of Fig. 16-1 with the timing resistor R in the base circuit is impractical if predictable and stable pulse widths are desired. The circuit is presented here because it illustrates very simply the basic principles involved in determining the pulse shape in a transistor blocking oscillator.   In the next section we show that by moving the timing resistor R from the base to the emitter leg, the pulse width of the blocking oscillator can be made quite insensitive to the parameter h_{FE}.

In Fig. 16-2b it appears that at $t = t_p$ the current i_B returns to zero and remains at this level.   Actually, because of minority-carrier storage (Chap. 20) there may be an undershoot, as indicated by the dashed waveform.

16-2   A TRIGGERED TRANSISTOR BLOCKING OSCILLATOR (EMITTER TIMING)[2,3]

The blocking oscillator of Fig. 16-4 consists of a transistor with an emitter resistor and a three-winding pulse transformer.   One winding is in the collector circuit; the second winding, with n times as many turns, is in the base circuit; the third winding, with n_1 times as many turns as the collector winding, feeds a resistor R_L which may be the load or may be required for damping, as we shall see below.   The base and collector turns must be connected for regenerative feedback, but the relative winding direction of the third leg of the transformer is arbitrary.   It may be chosen to obtain either a positive or negative output pulse across the load.   Of course, such a third winding could also have been included in the blocking-oscillator circuit of Fig. 16-1.

Pulse Amplitude, Width, and Waveforms   Subject to the same approximations made in the preceding section, the equivalent circuit from which to calculate the current and voltage waveforms is that indicated in Fig. 16-4b.

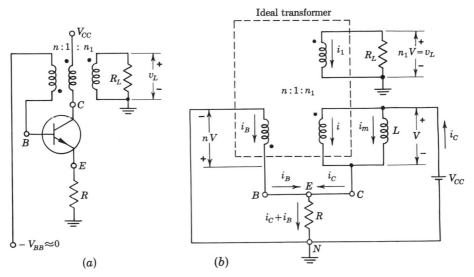

Fig. 16-4 (a) A monostable transistor blocking oscillator with emitter timing; (b) the equivalent circuit during the pulse formation.

Applying KVL to the outside loop encompassing both the collector and base meshes gives

$$-V_{CC} + V + nV = 0$$

or the drop V across the collector winding during the pulse is

$$V = \frac{V_{CC}}{n + 1} \tag{16-11}$$

From the base circuit we see that the drop across R is

$$V_{EN} = nV = (i_C + i_B)R \tag{16-12}$$

or

$$-i_E = i_C + i_B = \frac{nV}{R} = \frac{n}{n+1}\frac{V_{CC}}{R} \tag{16-13}$$

Note that the emitter current is a constant. In order to find the collector and base currents individually, we seek one more relationship between i_C and i_B. Since the sum of the ampere-turns in the ideal transformer is zero,

$$i - ni_B + n_1 i_1 = 0 \tag{16-14}$$

From the load circuit,

$$i_1 = -\frac{n_1 V}{R_L} \tag{16-15}$$

From KCL at the collector node and Eq. (16-5),

$$i = i_C - i_m = i_C - \frac{Vt}{L} \tag{16-16}$$

Substituting from Eqs. (16-15) and (16-16) into (16-14),

$$i_C - \frac{Vt}{L} - ni_B - \frac{n_1^2 V}{R_L} = 0 \tag{16-17}$$

This is the desired second relationship between i_C and i_B. Solving Eqs. (16-13) and (16-17) and using Eq. (16-11) we obtain

$$i_C = \frac{V_{CC}}{(n+1)^2}\left(\frac{n^2}{R} + \frac{n_1^2}{R_L} + \frac{t}{L}\right) \tag{16-18}$$

and

$$i_B = \frac{V_{CC}}{(n+1)^2}\left(\frac{n}{R} - \frac{n_1^2}{R_L} - \frac{t}{L}\right) \tag{16-19}$$

Note that the collector-current waveform is trapezoidal with a positive slope, the base current is also trapezoidal but with a negative slope, and the emitter current is constant during the pulse. These waveforms and also the base, collector, and load voltages as a function of time are pictured in Fig. 16-5.

At $t = 0+$, $i_C < h_{FE}i_B$ and the transistor is in saturation. As time passes, i_C increases and the operating point moves up the saturation line

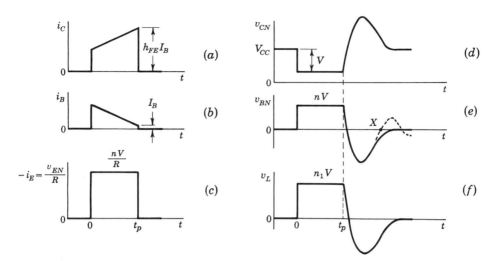

Fig. 16-5 The current and voltage waveforms in a transistor blocking oscillator with an emitter resistor $[V = V_{CC}/(n + 1)]$.

in Fig. 16-3. While i_C grows with time, the base current is decreasing and eventually point P' is reached at $t = t_p$, where $i_B = I_B$ and $i_C = h_{FE}I_B$. Equation (16-8) is satisfied and, as explained in the preceding section, the transistor then comes out of saturation and enters its active region. By regenerative action the transistor quickly returns to the OFF state and the pulse ends. Applying Eq. (16-8) to Eqs. (16-18) and (16-19) with $t = t_p$ we obtain

$$t_p = \frac{nL}{R}\frac{h_{FE}-n}{h_{FE}+1} - \frac{n_1{}^2 L}{R_L} \tag{16-20}$$

Since usually $\frac{1}{5} \leq n \leq 1$, then $h_{FE} \gg n$ and

$$t_p \approx \frac{nL}{R} - \frac{n_1{}^2 L}{R_L} \tag{16-21}$$

Subject to the reasonable approximations made above, the pulse width t_p is independent of h_{FE} and depends only upon passive elements, n, L, R, etc. Here, then, is a simple circuit which yields a pulse of very stable duration.

If the second term in Eq. (16-21) exceeds the first term, then t_p is negative, which situation is obviously impossible. It turns out (Prob. 16-2) that in order for the loop gain to exceed unity, which is the necessary condition for regeneration to take place and a pulse to form, the following inequality must be valid:

$$R_L > \frac{n_1{}^2 R}{n}\frac{h_{FE}+1}{h_{FE}-n}$$

If this inequality is indeed satisfied, then t_p in Eq. (16-20) cannot be negative.

The effect of the transistor saturation voltages on the pulse width (Prob. 16-3) is to multiply the first term in Eq. (16-20) by the factor

$$F \equiv \frac{V_{CC} - V_{CE}(\text{sat}) - [V_{BE}(\text{sat})/n]}{V_{CC} - V_{CE}(\text{sat}) + V_{BE}(\text{sat})} \tag{16-22}$$

Hence, as long as $V_{CC} \gg V_{CE}(\text{sat}) + [V_{BE}(\text{sat})/n]$, then the pulse duration is almost independent of supply voltage and is given by Eq. (16-21). We should also mention that the effect of the base-spreading resistance $r_{bb'}$ has been neglected in the derivation of Eq. (16-21). It turns out that we must add to R the small resistance $r_{bb'}/(h_{fe} + 1)$. However, if R exceeds 100 Ω, very little error is made in neglecting the effect of $r_{bb'}$.

Recovery and Loading Considerations   We shall now examine the waveform for $t > t_p$.   At the termination of the pulse there exists a current i_m in the magnetizing inductance of the transformer.   Since the current through an inductor cannot change instantaneously, the current must continue to flow even after $t = t_p$, when the transistor currents have dropped to zero.   The path for the magnetizing current is through the capacitance of the transformer. Since this capacitance (not shown in the figure) is small, then i_m decays rapidly and hence induces large voltage overshoots at the collector, base, and at the load, as indicated in Fig. 16-5.   These overshoots must not be so large as to exceed the breakdown voltage BV_{CE} or BV_{EB}.

It is important to note that adequate damping of the backswing which occurs at the termination of the pulse is absolutely essential to the operation of the blocking oscillator.   In Fig. 16-5e the solid curve represents a typical waveform when the damping is sufficient to cause the backswing to die down in half a cycle.   If the damping is inadequate, the backswing may oscillate as indicated by the dashed curve.   In such a case, regeneration would start again at the point marked X, where the base voltage is slightly positive and the transistor reenters the active region.   The blocking oscillator would then be free-running and would generate a continuous oscillation whose shape would resemble a very distorted sinusoid.   If the core losses of the transformer are low, as they are in a ferrite core, and if the load R_L does not supply adequate

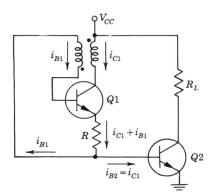

Fig. 16-6   The load R_L is switched ON and OFF through $Q2$, which is driven by the blocking-oscillator collector current.

damping, then *an external resistor must be shunted across the transformer in order to obtain monostable operation.*

In order to suppress the transformer oscillations without loading the blocking oscillator during the pulse interval, a damping resistor R' in series with a diode may be shunted across the transformer. The diode is introduced with such polarity that it does not conduct during the pulse interval but does conduct during the overshoot. With an *n-p-n* transistor, as in Fig. 16-4, the anode of the diode is at the collector side and the cathode at the supply-voltage side. The resistor R' is selected to be smaller than the critical damping resistance. At the end of the pulse the diode conducts and the energy stored in the magnetizing inductance is dissipated in R' with a time constant L/R'. Allowing four time constants for the quiescent condition to be established, the maximum frequency at which the circuit can be triggered is $f_{max} = R'/4L$. A method which allows a higher triggering rate is given in Prob. 16-15.

The load to be driven by the blocking oscillator may be supplied from a tertiary winding, as indicated in Fig. 16-4, or placed across R from emitter to ground. In either case we see from Eq. (16-21) that the pulse width depends upon the value of the load resistance. There are applications where the blocking oscillator must drive a variable load: for example, a ferrite-core memory where the load depends upon the number of cores to be excited. A method[2] of obtaining a pulse duration which is independent of the loading is to place R_L in the collector leg of a second transistor $Q2$ whose base current is the collector current of the blocking-oscillator transistor $Q1$, as indicated in Fig. 16-6. The effective supply voltage of $Q1$ is $V_{CC} - V_{BE2}(\text{sat}) \approx V_{CC}$. The load current is $[V_{CC} - V_{CE2}(\text{sat})]/R_L \approx V_{CC}/R_L$ for a time $t_p \approx nL/R$ and zero outside this interval.

Triggering An excellent way to trigger a blocking oscillator is indicated in Fig. 16-7. Transistor $Q2$ provides amplification for the applied triggering signal. In this circuit there is little reaction of the blocking oscillator back on the triggering source. The applied trigger must have a sufficiently steep leading edge so that the induced transformer voltage brings the blocking-oscillator base out of cutoff. The resistor R'' is very large compared with R

Fig. 16-7 A positive triggering signal is introduced through transistor $Q2$ to the collector of $Q1$.

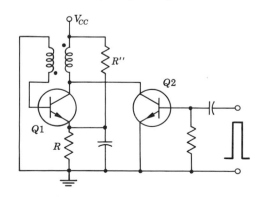

and is included so that under quiescent conditions the voltage across R is of the order of 0.5 V and serves the same function as did V_{BB} in Fig. 16-4. The small capacitor across R improves the rise time of the pulse (Sec. 16-4).

Alternative methods of triggering the blocking oscillator include applying a positive pulse through a diode to the base or a negative pulse through a diode to the collector. While the pulse is formed the diode becomes reverse-biased and hence prevents the blocking oscillator from reacting back on the trigger source or the trigger source from influencing the blocking oscillator during the formation of the pulse.

If the input signal is a pulse, it is advantageous to have this triggering pulse wider than the blocking-oscillator pulse. Otherwise, if the trailing edge of the triggering waveform has a short fall time, it may prematurely terminate the blocking-oscillator pulse.

16-3   ADDITIONAL METHODS FOR CONTROLLING THE PULSE DURATION OF A BLOCKING OSCILLATOR

In the common-emitter circuits of the preceding sections the transformer coupling has been between the collector and the base. It is also possible to place one winding of the transformer in the collector and the second winding in the emitter, resulting in a common-base configuration. A third possibility employs transformer coupling between the base and the emitter, a common-collector configuration. In each case it is found that if the timing resistor R is in the base, t_p varies drastically with h_{FE}, whereas if R is in the emitter, t_p is almost independent of h_{FE}. Clearly the timing resistor should be placed in the emitter, but there is no distinct advantage in one coupling method over the others. However, the CE configuration generates output pulses of opposite polarity at the two windings, whereas the CB and CC do not. These various circuits are considered in Probs. 16-7 to 16-9.

In deriving Eq. (16-21) for t_p we implicitly assume that the magnetizing inductance L is a constant. Suppose now that we use a core which saturates

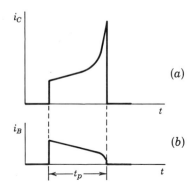

(a)

(b)

Fig. 16-8 The collector and base currents in the blocking oscillator of Fig. 16-4 if the transformer core saturates.

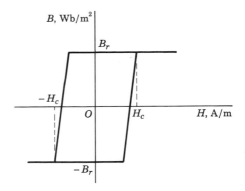

Fig. 16-9 A rectangular hysteresis loop for an iron core.

when the flux density B reaches a maximum value B_m. The inductance L now decreases with current and $L \rightarrow 0$ as $B \rightarrow B_m$. From Eqs. (16-18) and (16-19) we may anticipate that the collector and base currents will no longer be trapezoidal, as in Fig. 16-5a and b, but rather will have waveshapes more like those indicated in Fig. 16-8. As $B \rightarrow B_m$, i_C increases very rapidly, whereas i_B decreases quickly toward zero. Hence, as this maximum flux density is approached, the condition $i_C = h_{FE}i_B$ is satisfied and the pulse ends. Let us assume that the pulse terminates when the core is completely saturated. Let N equal the number of turns in the collector winding, A equal the cross-sectional area of the core, and ϕ the magnetic flux. Equation (16-11) continues to apply to the present case where the core saturates. Hence the voltage across the collector winding is a constant V, and

$$V = N \frac{d\phi}{dt} = NA \frac{dB}{dt} = \frac{V_{CC}}{n+1}$$

where we have used Eq. (16-11). Integrating between the limits 0 and t_p for t, corresponding to 0 and B_m for B, gives

$$t_p = \frac{(n+1)NAB_m}{V_{CC}} \tag{16-23}$$

Note that now the pulse duration depends upon the supply voltage and the properties of the core but not upon the transistor parameters (under the assumption made above that saturation takes place abruptly at $B = B_m$).

Now consider a core which, in addition to saturation, also exhibits hysteresis, as indicated in the B-H loop of Fig. 16-9. Assume that at $t = 0$, with no current in the windings, $B = -B_r$, and that when the blocking oscillator is triggered, H increases beyond H_c, so that B reaches B_r. Then t_p is given by Eq. (16-23) with B_m replaced by $2B_r$. At the end of the pulse B retains the residual flux density $+B_r$. An attempt to trigger the blocking oscillator again is now doomed to failure because $dB/dt = 0$, no voltages are induced in the transformer, and regeneration is impossible. In order for the circuit to operate properly the core must be reset so that its flux is $-B_r$ after each pulse. A convenient method of obtaining this reset flux is to use

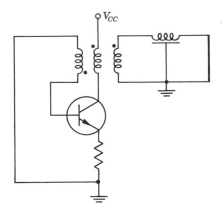

Fig. 16-10 A shorted delay line is used to determine the pulse width of a blocking oscillator.

an auxiliary winding in series with a resistor R_r across the supply voltage. When the transistor is OFF, the current V_{CC}/R_r in this reset winding supplies the necessary ampere-turns so that H becomes more negative than $-H_C$ and hence $B = -B_r$ (Prob. 16-10).

The last method to be discussed for controlling the pulse width uses a shorted delay line in the circuit of Fig. 16-10. When the blocking oscillator is triggered, a positive step is generated at the input to the line. This discontinuity upon reaching the shorted end of the line at $t = t_d$ is reversed in polarity. When this reflected wave reaches the input to the line, the positive step at the collector of the transistor starts the regenerative action which terminates the blocking-oscillator pulse. The width of the pulse is $2t_d$, provided that t_p, as determined by Eq. (16-21), is greater than $2t_d$.

16-4   THE BLOCKING–OSCILLATOR RISE TIME

The rise time of the blocking oscillator may be calculated in principle by replacing the transistor and the transformer by their high-frequency small-signal models. The characteristic equation of the resulting equivalent circuit is solved for its roots. One of these roots, say s_1, must be positive, and the term $\epsilon^{s_1 t}$ in the expression for the waveforms will predominate. This exponentially increasing function of time represents the regenerative action taking place during the formation of the pulse. The time t_r it takes the pulse to rise from 0.1 to 0.9 of its peak value is then given by $t_r = 2.2/s_1$.

In practice the above indicated analysis is very difficult to carry out because of the complexity of the resulting equivalent circuit. A digital computer has been used[4] to obtain numerical solutions for a wide range of circuit-parameter values. Common-emitter operation is found to have the fastest rise time. The analysis also shows that there is an optimum turns ratio, but that this minimum in t_r is quite broad and for the CE configuration occurs for n of the order of $\frac{1}{2}$ to $\frac{1}{3}$. Clearly, the better the high-frequency response of the transistor and the transformer, the smaller will be the rise time. Because of the mathematical difficulties mentioned above and also because it is a gross approximation to assume that the active element behaves linearly for the large swings involved during the pulse formation, the value of rise time should be determined experimentally. A rise time of about one-tenth the pulse width is a reasonable order of magnitude.

Experience shows that if the emitter resistor R in Figs. 16-4 and 16-7 is bypassed by a small capacitance (of the order of 100 pF), then a rise-time improvement by a factor of about 2 can be obtained. Such a rise-time improvement is to be expected on the basis that during the rise time the capacitor eliminates the degenerative feedback resulting from the resistor. The loop gain is thereby increased and, correspondingly, the rise time decreased. Too large a capacitance introduces undesirable ringing at the top of the pulse.

16-5   AN ASTABLE TRANSISTOR BLOCKING OSCILLATOR (DIODE–CONTROLLED)

One form[5] of a free-running blocking oscillator is indicated in Fig. 16-11a. The symbol D is used to represent a diode network which may consist of one or more p-n junction germanium or silicon diodes in series or a p-n diode $D1$ in series with either a battery or an avalanche diode $D2$, as in Fig. 16-11b. The piecewise linear approximation for the diode combination is indicated in Fig. 16-11c, where V_γ is the offset voltage of the p-n diodes or the Zener voltage or the battery voltage and R_f is the diode forward resistance. The divider supplies a small bias V_{BB} (≈ 1 V) which is in the direction to forward-bias the

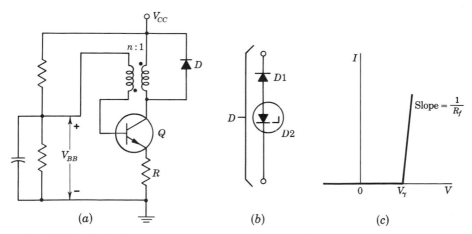

Fig. 16-11   (a) A free-running blocking oscillator; (b) the diode D in (a) may consist of a p-n diode $D1$ in series with a breakdown diode $D2$; (c) the piecewise linear model for diode D.

emitter junction. As soon as base current starts to flow, the regeneration discussed in Sec. 16-2 takes place and the waveforms indicated in Fig. 16-5 for $0 \leq t \leq t_p$ are generated. During this interval the diode is OFF, and hence the amplitude and duration of the pulse have the values previously calculated for the monostable circuit (provided that $V_{BB} \ll V_{CC}$).

At the end of the pulse the magnetizing current i_m, obtained from Eqs. (16-5) and (16-11), is given by

$$i_m = \frac{V_{CC}}{n+1} \frac{t_p}{L} \equiv I_o \tag{16-24}$$

Using Eq. (16-21) with $R_L = \infty$ we obtain

$$I_o = \frac{n}{n+1} \frac{V_{CC}}{R} \tag{16-25}$$

which indicates that the peak magnetizing current is independent of the inductance. At $t = t_p$ the transistor is cut off and the inductor current must flow through the diode and through the transformer capacitance, as shown in Fig. 16-12a. We shall neglect the small diode forward resistance R_f, so that V_γ appears directly across L and the shunt capacitance C. Hence, the collector voltage rises above V_{CC} by V_γ, as indicated in Fig. 16-13a. Since, for $t > t_p$, $L \, di_m/dt = -V_\gamma$, then, shifting the time origin to the end of the pulse, we obtain

$$i_m = \frac{-V_\gamma t}{L} + I_o \tag{16-26}$$

The current decreases linearly with time after the pulse ends, as indicated in Fig. 16-13c. The diode current falls to zero at a time t_f given by Eqs. (16-26) and (16-25):

$$t_f = \frac{L I_o}{V_\gamma} = \frac{n}{n+1} \frac{L}{R} \frac{V_{CC}}{V_\gamma} \tag{16-27}$$

After i_m is reduced to zero the diode becomes an open circuit, and the underdamped ringing circuit of Fig. 16-12b results. Since the shunt capacitor

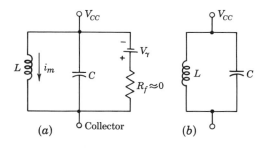

(a)   Collector   (b)

Fig. 16-12   (a) At the end of the pulse the magnetizing current i_m passes through the diode; (b) when i_m is reduced to zero the diode becomes an open circuit and the magnetizing inductance rings with the shunt capacitance.

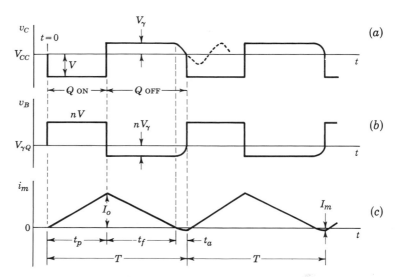

Fig. 16-13   Waveforms for the astable blocking oscillator of Fig. 16-11. (a) Collector voltage; (b) base voltage; (c) magnetizing current.   (The symbol $V_{\gamma Q}$ is used for the cutin voltage of the transistor to distinguish it from V_γ, which represents the break-point voltage of the diode network of Fig. 11-6.)

C is initially charged to V_γ, a sinusoidal oscillation of amplitude V_γ and period $2\pi\sqrt{LC}$ begins, as indicated by the dashed curve in Fig. 16-13a.   After one-quarter of a cycle, in a time

$$t_a = 1.57\sqrt{LC} \tag{16-28}$$

v_C falls below V_{CC}.   This negative swing at the collector is reflected in a positive swing at the base.   When v_{BE} exceeds the cutin value $V_{\gamma Q}$ of the transistor, there is initiated the regenerative action needed to turn the transistor ON again.   Hence, without the aid of an external trigger a new cycle starts.   Clearly, the diode has caused the circuit to function in an astable mode with period $T = t_p + t_f + t_a$.   The collector and base voltages (Fig. 16-13a and b) are nearly rectangular without the overshoots shown in Fig. 16-5d and e.   The transistor current waveforms are those indicated in Fig. 16-5a to c over the interval t_p and zero for the next $t_f + t_a$, and then the cycle repeats.

EXAMPLE   The circuit of Fig. 16-11a has the following parameters: $L = 5.2$ mH, $C = 90$ pF, $V_{CC} = 10$ V, $R = 500\ \Omega$, $V_\gamma = 6$ V, $n = 1$, and $V_{BB} = 0.5$ V.   Calculate (a) the period and the duty cycle of the free oscillations, (b) the peak voltages and currents, and (c) the current in the magnetizing inductance at the end of one cycle.   Neglect saturation junction voltages.

Solution   Since $V_{BB} \ll V_{CC}$, Eq. (16-21) for t_p is valid:

$$t_p = \frac{nL}{R} = \frac{5.2 \times 10^{-3}}{500} \text{ sec} = 10.4 \ \mu\text{sec}$$

From Eq. (16-27)

$$t_f = \frac{n}{n+1} \frac{L}{R} \frac{V_{CC}}{V_\gamma} = \frac{1}{2} \left(\frac{5.2 \times 10^{-3}}{500} \right) \left(\frac{10}{6} \right) \text{ sec} = 8.7 \ \mu\text{sec}$$

From Eq. (16-28)

$$t_a = 1.57(5.2 \times 10^{-3} \times 90 \times 10^{-12})^{\frac{1}{2}} \text{ sec} = 1.1 \ \mu\text{sec}$$

The period T is given by

$$T = t_p + t_f + t_a = 10.4 + 8.7 + 1.1 = 20.2 \ \mu\text{sec}$$

The duty cycle is defined by the ratio of the ON time t_p to the period T, or

$$\frac{t_p}{T} = \frac{10.4}{20.2} = 0.51$$

Since the ON to OFF time, the so-called "mark-space ratio," is approximately unity, the output is almost a symmetrical square wave. Despite this high duty cycle the power dissipation of the transistor is not excessive, because when the current is high the transistor is in saturation and the saturation voltages are low, whereas when the output voltage is high the transistor currents are zero.

   b. From Eq. (16-11)

$$V = \frac{V_{CC}}{n+1} = \frac{10}{2} = 5 \text{ V}$$

The collector voltage in Fig. 16-13*a* extends from $V_{CC} - V = 5$ to $V_{CC} + V_\gamma = 16$ V. The base voltage in Fig. 16-13*b* extends from $+5$ to -6 V. The emitter current is constant, and from Eq. (16-13)

$$-i_E = \frac{nV}{R} = \frac{5}{0.5} = 10 \text{ mA}$$

The base current is a maximum at $t = 0$, and from Eq. (16-19)

$$i_B \ (t = 0) = \frac{nV_{CC}}{(n+1)^2R} = \frac{10}{(4)(0.5)} = 5 \text{ mA}$$

The collector current is a maximum at $t = t_p = nL/R$, and from Eq. (16-18)

$$i_C \ (t = t_p) = \frac{V_{CC}}{(n+1)^2} \left(\frac{n^2}{R} + \frac{t_p}{L} \right) = \frac{V_{CC}}{(n+1)^2} \left(\frac{n^2}{R} + \frac{n}{R} \right) = \frac{nV_{CC}}{(n+1)R}$$

$$= \frac{10}{(2)(0.5)} = 10 \text{ mA}$$

The peak magnetizing current is, from Eq. (16-25),

$$I_o = \frac{nV_{CC}}{(n+1)R} = 10 \text{ mA}$$

These calculations are consistent with the fact that the peak magnetizing current equals the collector current at the end of the pulse. Also, since $h_{FE} \gg 1$ the base current at the end of the pulse is close to zero and hence the emitter current must be the negative of the collector current. Also note that the peak base current is of the same order of magnitude as the collector current.

    c. At the time $t_p + t_f$ the capacitor is charged to a voltage V_γ and the magnetizing current is zero. The *LC* circuit then rings and in one-quarter of a cycle (in time t_a) the capacitive energy is transformed to magnetic energy. Since

$$\tfrac{1}{2}LI_m{}^2 = \tfrac{1}{2}CV_\gamma{}^2$$

then

$$I_m = V_\gamma \sqrt{\frac{C}{L}} = (6)\sqrt{\frac{90 \times 10^{-12}}{5.2 \times 10^{-3}}} \text{ A} = 0.79 \text{ mA}$$

The current i_m goes below zero (reverses) by the amount I_m, as indicated in Fig. 16-13c. Hence, at the initiation of the next pulse the magnetizing current is not zero and the values for the times calculated above must be modified. Since I_m is only about 8 percent of I_o this is a minor complication. Experimentally obtained values agree with those calculated above to within about 5 percent.

    If we are interested in a lower duty cycle (a pulse rather than a square-wave output), then we must increase t_f relative to t_p. Since from Eqs. (16-21) and (16-27)

$$\frac{t_p}{t_f} = (n+1)\frac{V_\gamma}{V_{CC}} \tag{16-29}$$

then the duty cycle will be a minimum if we use for D a single *p-n* germanium diode, for which $V_\gamma \approx 0.1$ V. Then with $V_{CC} = 20$ V and $n = 1$ a duty cycle of about $(2)(0.1)/20 = \frac{1}{100}$ is possible. However, operation with such a low mark-space ratio is none too stable because V_γ varies with diode replacement and also changes with temperature (about -2 mV/°C). On the other hand, high-duty-cycle operation obtained with a temperature-compensated Zener diode is very stable. It is also possible to obtain a high-duty-cycle blocking oscillator without using an avalanche diode, by placing a germanium diode in series with a tertiary winding across the supply voltage (Prob. 16-14).

16-6   AN ASTABLE TRANSISTOR BLOCKING OSCILLATOR (*RC*–CONTROLLED)

Another form of astable blocking oscillator may be obtained by adding an R_1C_1 network either in the emitter circuit of a monostable blocking oscillator, as indicated in Fig. 16-14a, or in the base circuit, as in Fig. 16-14b. These circuits differ from that of Fig. 16-4 not only in the addition of R_1C_1 but also in the reversal of the polarity of V_{BB}. Qualitatively the operation of the circuit of Fig. 16-14a is as follows. Assume initially, as in Fig. 16-15a, that there is a voltage v_1 on C_1 larger than $V_{BB} - V_\gamma$, where V_γ is the cutin base-to-emitter voltage. Hence, the transistor is OFF and C_1 discharges exponentially toward ground with a time constant R_1C_1. When v_1 reaches $V_{BB} - V_\gamma$ the base starts to draw current, as does the collector, and the regenerative action discussed in connection with the monostable circuit begins. The collector (Fig. 16-15b) and base waveforms generated are similar to those in Fig. 16-5. During the pulse duration t_p, the capacitor C_1 is recharged and attains a voltage V_1, which is larger than the value it had at the beginning of the pulse (Fig. 16-15a). The transistor now remains OFF for a time t_f, during which C_1 discharges to the voltage at which Q again enters its active region. At this point the cycle repeats itself. From Fig. 16-15a we find

$$t_f = R_1C_1 \ln \frac{V_1}{V_{BB} - V_\gamma} \tag{16-30}$$

The period of the free-running oscillations is $T = t_p + t_f$.

The value of V_1 and t_p may be found from the equivalent circuit of Fig. 16-4b, to which are added V_{BB}, R_1, and C_1. After considerable algebraic manipulation, but without involving any new principles, we obtain. (Prob. 16-16), if $h_{FE} \gg 1$, $R_1 \gg R$, and neglecting V_{BE}(sat) and V_{CE}(sat),

$$\frac{t_p}{L} - \frac{n}{R} \epsilon^{-t_p/RC_1} = -\frac{n_1^2}{R_L} \tag{16-31}$$

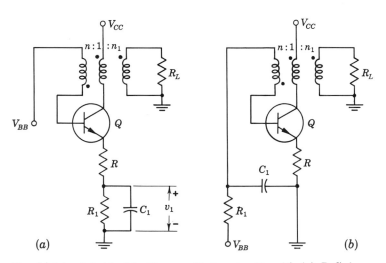

(a)   (b)

Fig. 16-14 Astable-blocking-oscillator circuits with (a) R_1C_1 in the emitter and (b) R_1C_1 in the base. In (b) the capacitor C_1 may be connected directly across R_1.

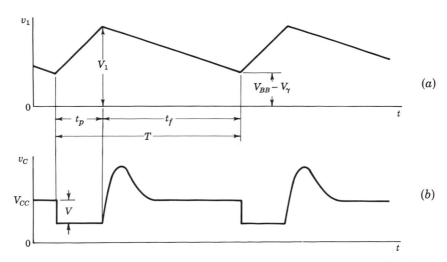

Fig. 16-15 (a) The waveform across the R_1C_1 combination of Fig. 16-14a;
(b) the collector waveform [$V = (V_{CC} - V_{BB})/(n + 1)$].

Unfortunately this is a transcendental equation in t_p. If all parameter values
are given we can find t_p graphically. If a desired value of t_p is specified,
however, we can solve directly for the required inductance L to give this t_p.
Very often $t_p/RC_1 \ll 1$, and then by retaining only the first two terms in the
power-series expansion of the exponential, Eq. (16-31) can be solved explicitly
for t_p and yields

$$\frac{t_p}{L}\left(1 + \frac{nL}{R^2C_1}\right) \approx \frac{n}{R} - \frac{n_1^2}{R_L} \qquad (16\text{-}32)$$

Note that the pulse duration depends upon C_1 in addition to all the other
parameters which determined t_p in the monostable case. For very large C_1,
Eq. (16-32) reduces to Eq. (16-21).

The peak capacitor voltage V_1 is obtained from the same analysis which
gives t_p and is found to be given by

$$(n + 1)V_1 = nV_{CC} + V_{BB} - n(V_{CC} - V_{BB})\epsilon^{-t_p/RC_1} \qquad (16\text{-}33)$$

Using this value of V_1 in Eq. (16-30) the value of t_f may be calculated.

The interval t_f is terminated when the voltage v_1 falls to the point where
the transistor comes out of cutoff. At this time a succeeding pulse is generated.
When the rate at which v_1 falls is very slow, the blocking-oscillator pulse is
preceded by a number of cycles of oscillation of amplitude small in comparison
with the pulse, as is shown in Fig. 16-16. A qualitative explanation of this
phenomenon follows. As the transistor comes gradually out of cutoff the
loop gain of the circuit increases from zero. This gain variation is to be
expected in view of Eq. (1-10), which indicates that at zero emitter current
the transconductance g_m is zero and that g_m increases with increasing current.

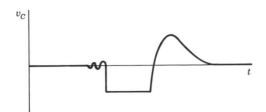

Fig. 16-16 Illustrating the oscillations which precede the collector pulse when the transistor comes out of cutoff very gradually.

When the loop gain becomes larger than unity the circuit will begin to oscillate, the amplitude of the oscillations being limited by the nonlinearity of the transistor in the neighborhood of cutoff. We may note, say, in connection with Fig. 16-1a, that if the transformer capacitance is included, the circuit of the blocking oscillator is precisely the same in form as a tuned-plate oscillator. As a matter of fact, the only essential difference between the tuned oscillator and the blocking oscillator is in the tightness of coupling between the transformer windings.

As the transistor moves further into the active region, the loop gain increases and the amplitude of the oscillation increases correspondingly. Finally, the decrease of v_1 as well as the increasing oscillation itself will, during some particular cycle, carry the transistor to an operating point where the loop gain becomes very large. During this cycle the amplitude will grow so large that it will be limited only when the transistor is driven to saturation. This distorted cycle constitutes the blocking-oscillator pulse. If the voltage v_1 falls rapidly, the transistor will be carried rapidly through the low-loop-gain region and no time will be available for the generation of small-amplitude oscillations before the main pulse.

The oscillations preceding the pulse ordinarily appear when we try to adjust the astable-blocking-oscillator circuit to operate with a very low duty cycle, say with t_f/t_p greater than about 200. For it is precisely in this case that we must adjust the R_1C_1 time constant (Fig. 16-14) to be long, and consequently v_1 will fall very slowly. This low-duty-cycle circuit will also suffer from instability in that there may be erratic fluctuation in the interval t_f. This instability results from the fact that from cycle to cycle there may be small variations in the precise voltage v_1 at which the loop gain becomes large enough to generate the pulse. Since the approach to this critical point is very slow, a small change in v_1 will produce a large change in t_f.

The circuit of Fig. 16-14b behaves similarly to that of Fig. 16-14a. However, the analytical expressions for t_p and V_1 for Fig. 16-14b are even more formidable than those given for Fig. 16-14a.

A comparison of the diode-controlled blocking oscillator with the resistance-capacitance-controlled circuit follows. The advantages of the diode circuit are: (1) Voltage waveforms are better, since the overshoots at the end of the pulse are missing. Compare Figs. 16-13a and 16-15b. (2) The design equations are simpler, and hence it is easier to synthesize a circuit to meet definite specifications. (3) There is much less possibility of the behavior described above (Fig. 16-16) for the circuits of Fig. 16-14. The advantages of the R_1C_1 circuit are: (1) Low-duty-cycle operation, $t_p/t_f \approx \frac{1}{200}$, is possible. For the diode circuit, stable operation below $t_p/t_f \approx \frac{1}{20}$ cannot be achieved. [See the discussion given below Eq. (16-29).] (2) The frequency of oscillation can be varied simply and continuously by varying R_1 or C_1. For the diode circuit, for a given pulse width t_p, we can change t_f only by changing the diode V_γ or the supply voltage V_{CC}. (3) Since the timing does not depend upon a diode it is quite stable with temperature changes, provided that $V_{BB} \gg V_\gamma$ (of the transistor) and the passive elements R_1, C_1, etc., are independent of temperature.

16-7   VACUUM–TUBE BLOCKING OSCILLATORS[6]

The circuit configurations for a vacuum-tube blocking oscillator are identical with those using a transistor. For example, Fig. 16-7 for the monostable mode and Figs. 16-11 and 16-14 for astable operation may be constructed with a tube replacing the transistor. The qualitative operation of these tube circuits is very similar to that of their transistor counterparts. For example, the plate (grid) waveform has the shape of the collector (base) waveform of Fig. 16-11 for a diode-controlled blocking oscillator or of Fig. 16-15 for an R_1C_1-controlled blocking oscillator.

Quantitatively, the calculations for a tube circuit are much more involved than for a transistor circuit. The transistor is a much better switch than is the tube. When the former goes into saturation the collector and base voltages are so low that very little error is made if they are neglected. However, when a pulse is generated in a tube circuit, the plate and grid voltages are not negligible but rather turn out in a typical case to be about half the supply voltage in magnitude. Because of the regeneration, the tube is driven into the highly nonlinear positive-grid region of the voltage-ampere characteristics, where analysis is extremely difficult. We shall now indicate how to estimate the pulse amplitude and width for the circuit of Fig. 16-17a, which corresponds to Fig. 16-14b with $R = 0$ and $R_L = \infty$.

Assume that initially the voltage v_1 across C_1 is negative enough so that the tube is at cutoff. Then v_1 increases toward V_{GG}, and when v_1 reaches the cutin grid-to-cathode voltage V_γ the tube starts conducting. Because of the regenerative connection the currents rise rapidly. Because of the short rise time the voltage across C_1 cannot change, and v_1 remains equal to V_γ. If V_P and V_G are the plate and grid voltages, respectively, when the pulse has reached its full amplitude, then $V_{PP} - V_P \equiv V$ is the change in plate voltage and $V_G - V_\gamma$ is the change in grid voltage. If there are n times as many turns in the grid winding as in the plate winding, then, from Fig. 16-17b,

$$V_G - V_\gamma = nV = n(V_{PP} - V_P) \tag{16-34}$$

For each value of V_G given on the plate characteristics of the tube, the value of V_P is calculated from Eq. (16-34) and the locus of corresponding values of V_G and V_P is plotted on the plate characteristics.

Let I_P and I_G be the peak plate and grid currents, respectively. Equating ampere-turns in the primary to those in the secondary gives (since the transformer currents are zero at the beginning of the pulse)

$$I_P = nI_G \qquad (16\text{-}35)$$

For each value of V_G given on the plate characteristics, the value of V_P may be found (by trial and error) such that Eq. (16-35) is satisfied. The locus of the corresponding values of I_P and V_P is plotted. Then the intersection of the two curves which have been constructed in accordance with Eqs. (16-34) and (16-35) gives the value of V_P, V_G, and I_P corresponding to the peak of the pulse. The peak grid current is $I_G = I_P/n$.

The above-outlined construction has been carried out in Fig. 16-18 for one section of a 6CG7 tube with $V_{PP} = 140$ V and a 1:1 transformer. Since $n = 1$, Eq. (16-35) is satisfied at that value of V_P where the plate-current curve for a given value of V_G intersects the grid-current curve for the same V_G. The result is

$$V_P = 57 \text{ V} \qquad V_G = 75 \text{ V} \qquad I_P = I_G = 0.25 \text{ A}$$

and hence the cathode current is 0.5 A. Oxide-coated cathodes such as are found in receiving-type tubes are capable of furnishing pulsed currents even up to some amperes. Of course, however, the average tube current must be kept within the rating, which for a 6CG7 is about 5 or 10 mA. Experimentally the following values were measured:

$$V_P = 50 \text{ V} \qquad V_G = 75 \text{ V} \qquad I_P = I_G = 0.25 \text{ A}$$

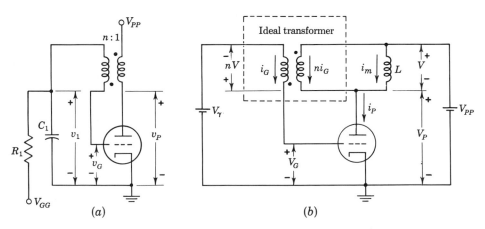

Fig. 16-17 (a) A free-running tube blocking oscillator; (b) the equivalent circuit during the pulse for C_1 extremely large. At the beginning of the pulse $v_G = V_G$, $v_P = V_P$, $v_1 = V_\gamma$, $i_m = 0$, $i_G = I_G$, and $i_P = I_P$.

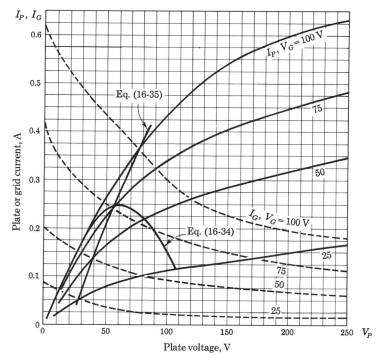

Fig. 16-18   Positive-grid characteristics for a 6CG7 tube.   Illustrating the construction for obtaining the currents and voltages at the peak of the pulse.

At the beginning of the pulse the loop gain is greater than unity and regeneration takes place, driving the grid highly positive.   This regenerative action continues until limited by the tube nonlinearities, as depicted in Fig. 16-18, so that the loop gain drops to unity, and finally at the top of the pulse the loop gain is less than unity.   The plate and grid voltages cannot remain constant at the values calculated above because of the finite size of C_1 and L. During the pulse, the grid current charges C_1 and makes the grid voltage less positive and causes the plate voltage to increase.   Also, we saw in Sec. 3-7 that a tilt appears at the output of a pulse transformer because of the finite value of L when the transformer input is constant.   Since the loop gain is less than unity, then the changes in plate and grid voltages due to C_1 and L occur relatively slowly until the circuit drifts back to the point where the loop gain once again equals unity.   Then a regenerative action occurs in the direction to turn the tube off, and the pulse terminates.

From Fig. 16-17b

$$V = L \frac{di_m}{dt} = V_{PP} - V_P$$

If we assume V_P is approximately constant over the pulse duration, we can integrate the above equation and obtain

$$t_p = \frac{LI_m}{V_{PP} - V_P} \tag{16-36}$$

where I_m is the magnetizing current at the end of the pulse. From Fig. 16-17b $i_m = i_P - ni_G$, and hence we must find the plate and grid currents at the end of the pulse—a difficult task. Let us assume that C_1 is so large that the voltage across it remains constant at V_γ during the pulse. Then the operating point moves along the curve marked "Eq. (16-34)" in Fig. 16-18 in the direction of increasing loop gain A. When a point is reached for which $A = 1$ the pulse ends. However, such a point can only be estimated crudely from the very nonlinear curves of Fig. 16-18. Whereas in principle Eq. (16-36) gives t_p, in practice it is obtained experimentally much more readily, particularly since positive-grid characteristics such as those of Fig. 16-18 are seldom supplied by the manufacturer. Recall how much simpler the calculation of t_p is for a transistor blocking oscillator, where the termination of the pulse is given by the simple condition $i_C = h_{FE}i_B$.

A resistor R may be placed in the cathode of the tube in Fig. 16-17 corresponding to the emitter resistor R in the transistor blocking oscillator. The following effects of the addition of R are found experimentally. Increasing R reduces the operating currents and the pulse amplitude but leaves the waveshapes essentially unchanged. The rise time is increased somewhat but may be brought back to the value obtained with $R = 0$ if R is bypassed with a capacitor. The pulse width is very insensitive to the value of R, but the time between pulses decreases with increasing R. The addition of R stabilizes the width t_p against filament-voltage changes, tube changes, etc. An excellent pulse waveform without ringing is obtained across R.

16-8   APPLICATIONS OF BLOCKING OSCILLATORS

Among the most important applications of the blocking oscillator are the following:

1. The astable circuit is used as a master oscillator to supply triggers for synchronizing a system of pulse-type waveforms—square waves, sweep voltages, etc.

2. The monostable circuit is used to obtain abrupt pulses from a slowly varying input triggering voltage.

3. Either form of blocking oscillator is capable of generating a pulse of large peak power. For example, it is possible to obtain 0.5 A at 100 V or 50 W from a receiving-type tube. Of course, the average power is small since the duty cycle is low.

4. Using a tertiary winding, output pulses of either polarity may be obtained depending upon which end of the winding is grounded. Also the output winding may be isolated from ground where required.

5. The use of the blocking oscillator as a frequency divider or counter is discussed in Chap. 19.

6. The blocking oscillator as a low-impedance switch used to discharge a capacitor quickly is considered in Sec. 18-14.

7. The blocking-oscillator output may be used as a gating waveform with a very small mark-space ratio. For example, in some television receivers the voltage across C_1 is used as the gating waveform for the vertical sweep-voltage generator.

16-9   THE MULTIAR[7]

If a diode is used as a series switching element in the feedback loop of a regenerative circuit, the resulting circuit makes an excellent comparator. A circuit of this type, known as a *multiar*, is indicated in Fig. 16-19. The feedback loop consists of a diode-resistor network (D and R), a cathode follower V, and a pulse transformer in cascade. The difference between the input waveform v_i and the reference voltage V_R controls the gain of the network D-R. The main function of the tube V is to provide cathode-follower transmission from the output of the D-R network to the transformer. The plate-load resistor R_p is not essential but is convenient if an output gate-type waveform is required. The pulse-transformer windings must be connected into the circuit with the polarities as shown in order to have regeneration similar to that in a blocking oscillator. Since the combined maximum gain of the D-R network and the cathode follower is always less than unity, there must be a

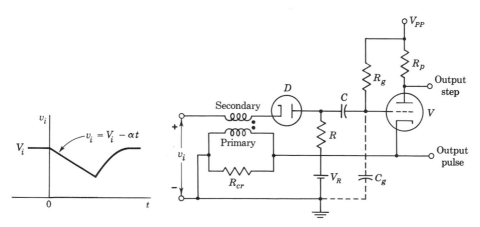

Fig. 16-19 A multiar circuit for negative-going input signals. The total shunting capacitance at the grid is C_g.

voltage step-up in the transformer in completing the loop from the cathode of V back to the diode. Hence, the number of turns in the secondary winding must exceed the number in the primary winding—a ratio of 1.5 or 2 is usually sufficient.

We shall assume that the input is a negative-going ramp voltage. Initially, the grid of V is clamped to the cathode and the tube is conducting heavily. The diode D is nonconducting and hence the quiescent voltage across C is V_R. The network D-R has zero transmission until the input waveform drops sufficiently to bring the diode D into its break region. At some point of this region the gain of the circuit consisting of the source impedance, the diode, and the resistance R becomes sufficiently high to cause an overall loop gain of unity. Regeneration will take place and the grid of V will drop sharply. The waveforms in the circuit are indicated in Fig. 16-20, where we see that a negative pulse is generated at the cathode (or in a tertiary transformer winding, not indicated) and a gate waveform at the plate.

The Waveforms   We shall now give a detailed explanation of the wave-shapes. Initially the rapid fall of cathode and grid voltages results from regenerative action. When tube V reaches cutoff, regeneration ceases. As a result, however, of the abrupt turn-off of tube current, the circuit in the cathode of V rings (Sec. 2-11). If the primary inductor is critically damped by the resistor R_{cr}, then the cathode voltage of V executes a single oscillation without a back-

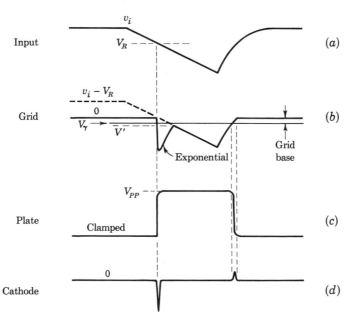

Fig. 16-20   Waveforms in the circuit of Fig. 16-19 at (a) input, (b) grid, (c) plate, and (d) cathode of V.

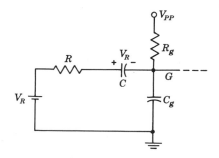

Fig. 16-21 Equivalent circuit at the grid of
the triode V shortly after regeneration
begins.

swing.   In this manner the pulse waveform indicated in Fig. 16-20d is formed.
During the rapid drop of cathode voltage the polarity of voltage induced in
the secondary winding is in a direction to cause diode D to conduct.   Hence
the charging current for the capacitor C_g is supplied by the secondary winding
of the transformer, and the grid voltage drops abruptly.   After the trailing
edge of the cathode pulse in Fig. 16-20, the diode becomes back-biased.   The
grid of V is therefore isolated from the transformer secondary winding, and
hence the grid voltage must change relatively slowly.

After the diode becomes back-biased, the equivalent circuit for computing
the rise of the grid voltage of V is indicated in Fig. 16-21.   If $C \gg C_g$, we may
assume that the voltage across C remains constant at V_R during the interval
while C_g is charging.   Hence, the voltage at G rises toward $V_{PP}R/(R + R_g)$
with a time constant $C_g R R_g/(R + R_g)$.   When the voltage at G rises to a
value V' (Fig. 16-20b), at which the rising grid voltage meets the falling input
voltage $v_i - V_R$, then the diode conducts again.   Once the diode begins to
conduct, we may neglect the presence of the small capacitance C_g.   We shall
assume tentatively that V' is below the cutoff voltage V_γ of the tube.   Hence
the grid continues to follow the incoming negative-going voltage to the end
of the ramp, as indicated in Fig. 16-20b.

The requirement that the grid voltage remain below cutoff to the end
of the ramp places a restriction on the time constant $R_g C$.   Thus, if we neglect
the cutoff voltage V_γ compared with V_{PP}, the current in R_g must be at least
V_{PP}/R_g.   If the magnitude of the current through C is smaller than the current
in R_g, then there must be some grid current.   Hence, to keep the grid below
cutoff, we require that $C|dv_i/dt| > V_{PP}/R_g$, or

$$CR_g > \frac{V_{PP}}{\alpha} \tag{16-37}$$

where α is the absolute value of the slope of the sweep voltage.   For slow
sweeps, it may be desirable to return R_g to a voltage lower than the plate-
supply voltage V_{PP}.

When V comes out of cutoff at some time during the rise toward its initial
level, the cathode of V rises with the grid voltage.   Since the transformer
step-up ratio is larger than unity, the voltage at the cathode of the diode rises
more rapidly than the grid voltage.   As a result the diode is driven into the

back-biased condition. Therefore, the feedback loop is opened and there is no regeneration as the tube is brought into conduction. The plate voltage of V falls to its clamped value relatively slowly. The duration of this fall equals the time it takes the grid voltage to pass through the grid base of the amplifier, as indicated in Fig. 16-20c. A positive pulse of this same duration is formed at the cathode because the transformer voltage is proportional to the rate of change of current. Since there is no regeneration at the end of the sweep, and since the tube V passes slowly through its grid base, the positive pulse has a smaller amplitude and a longer duration than the negative pulse.

The negative pulse (formed during the falling portion of the input voltage) is timed with a stability which depends primarily on the D-R combination. The stability of this network is affected principally by changes in the filament supply if D is a thermionic diode or by changes in temperature if D is a semiconductor diode. Changes in the gain of the cathode follower V affect the stability of the negative pulse only slightly if the break region of the D-R network is sharp.

The voltage at the plate of D must return to its steady-state value before each comparison. This condition places a restriction upon the product RC, because after the grid of V has returned to zero the plate of D will approach V_R with this time constant RC. Hence, the value of RC must be small compared with the interval between the instant the cathode follower cuts in and the comparison point in the next cycle.

Multiple Oscillations We have made the assumption that after the first negative grid pulse, the grid voltage rises to meet the input voltage v_i at a value V' which is more negative than the cutoff value V_γ (Fig. 16-20b). If this restriction is not satisfied, then the circuit may execute several oscillations before the tube is finally driven below cutoff. This situation is pictured in Fig. 16-22. For times smaller than t_1, the waveforms are identical with those in Fig. 16-20. Between times t_1 and t_2 the grid of V is above cutoff and the amplifier current increases from zero. Hence, the plate voltage falls, as in Fig. 16-22c, and the cathode voltage increases with the rate of change of current, as shown in Fig. 16-22d. At time t_2, the input voltage equals the grid voltage V', the diode D conducts, and regeneration again takes place. Since at this instant of time the current in V is less than the clamped current, the resultant drops in the plate, cathode, and grid voltages are less than the corresponding values at t_o. The grid voltage now rises again, reaching cutin at t_3 and becoming equal to the input voltage V'' at t_4. Between t_3 and t_4 there is a dip in plate voltage and another oscillation takes place at the cathode, as shown in Fig. 16-22d. It is possible for a large number of oscillations to be generated in this manner (with successively smaller amplitudes) until finally the grid voltage rises to the input voltage V''', which is below the cutoff voltage V_γ in Fig. 16-22b.

In order to minimize the possibility of generating several oscillations of the type discussed above, one or more of the following conditions should be

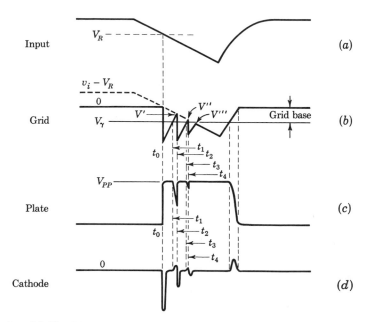

Fig. 16-22 The waveforms in the multiar circuit under the conditions where oscillations take place.

satisfied: a tube V with a small grid base should be used; the top of R_g should be tied to a voltage smaller than V_{PP}; also, the value of C_g should be increased by adding external capacitance between the grid of V and ground. However, this additional capacitance will result in a smaller output pulse and a slower rise time of plate waveform.

In the above discussion we have assumed that the transformer was critically damped. If the resistor R_{cr} which shunts the transformer in Fig. 16-19 is removed so that ringing occurs, then multiple oscillations may also result. Instead of putting a shunting resistor across the transformer, a damper diode may be used to prevent the undesired oscillations.

General Considerations   A solid-state version of the multiar is obtained by using a junction diode for D and an n-p-n transistor for V. In a transistor circuit the transistor should be biased in its active region. Otherwise, if it were initially in saturation, there would be a delay in the response of the multiar due to minority-carrier storage (Chap. 20). The tube multiar is not useful for positive-going signals because multiple oscillations are obtained. However, a multiar constructed with a p-n-p transistor can be used if the input is a positive-going signal.

The temperature-compensation techniques of Sec. 7-10 may be used to ensure that the comparator voltage will be highly stable with changes in temperature.

If it is desired that the input signal waveform be capacitively coupled to the multiar, then a d-c restorer may be used at the input terminals. The restorer-diode uncertainties must now be taken into consideration in a precision circuit.

If a gate voltage is not desired, the transformer primary can be moved to the plate (collector) circuit. The circuit then closely resembles a blocking oscillator, except that the grid (base) cannot be driven highly positive because of the diode.

We may use the same input ramp voltage for several multiar comparator circuits since there is small loading of the input by this comparator if R is large. In radar ranging the multiar is used to obtain many ranges from a single master ramp circuit; hence the name *multi-r* (*r* for range).

REFERENCES

1. Hamilton, D. J.: A Transistor Pulse Generator for Digital Systems, *IRE Trans. Electron. Computers*, vol. EC-7, no. 3, pp. 244–249, September, 1958.

2. Rolfe, J. J.: Designing a Common-emitter Blocking Oscillator, *Electron. Ind.*, vol. 20, no. 7, pp. 110–114, July, 1961.

3. Norman, P., and E. J. E. Smith: The Design of Transistor Blocking Oscillators, *Proc. IEE (London)*, pt. B, vol. 106, no. 18, pp. 1251–1259, May, 1959.

4. Martin, J. D.: Rise Time of the Transistor Blocking Oscillator, *Electron. Eng.*, vol. 35, no. 422, pp. 250–253, April, 1963.

5. Chen, W. I. H., G. Golan, and J. Millman: Astable Blocking Oscillators—They Can Be Practical, Part 1, *Electron. Design*, vol. 13, pp. 22–24, Mar. 1, 1965; Part 2, *ibid.*, pp. 42–44, Mar. 15, 1965. See also Ref. 3.

6. Benjamin, R.: Blocking Oscillators, *J. IEE*, pt. IIIA, vol. 93, no. 7, pp. 1159–1175, 1946.
   Chance, B., et al.: "Waveforms," chap. 6, Massachusetts Institute of Technology Radiation Laboratory Series, vol. 19, McGraw-Hill Book Company, New York, 1949.
   Millman, J., and H. Taub: "Pulse and Digital Circuits," chap. 9, McGraw-Hill Book Company, New York, 1956.

7. Williams, F. C., and N. F. Moody: Ranging Circuits, Linear Time Base Generators and Associated Circuits, *J. IEE*, pt. IIIA, vol. 93, no. 7, pp. 1188–1198, 1946.

17 / SAMPLING GATES

An ideal *sampling gate* is a transmission circuit in which the output is an exact reproduction of an input waveform during a selected time interval and is zero otherwise. The time interval for transmission is selected by an externally impressed signal which is called the *gating signal* and is usually rectangular in waveshape. These gates are also referred to as *transmission gates* or as *time-selection circuits*. In many applications, a less than ideal gate is entirely acceptable. It may be, for example, that the input signal consists essentially of a unidirectional pulse. In such a case a gate will be required to respond to an input signal of only one polarity. In other applications, on the other hand, a gate is required which will handle signal-input excursions of both polarities. In the present chapter we discuss both unidirectional and bidirectional gates.

Sampling gates are distinct from the logic gates discussed in Chap. 9. A logic gate has to provide at the output a pulse or no pulse (or, alternatively, one voltage level or another), depending on the pulses (or voltages) present at the many gate inputs. A sampling gate ordinarily (but not always) has one signal input, and during the selected time interval, the output must reproduce faithfully the input waveform, be it a pulse, a sinusoid, or any other waveform. Hence, a sampling gate is also referred to as a *linear gate*.

17-1 BASIC OPERATING PRINCIPLE OF GATES

The basic principle of a linear gate is illustrated in Fig. 17-1a and b. In Fig. 17-1a, switch S is normally open and is closed during the desired transmission interval. In Fig. 17-1b, switch S is normally closed and

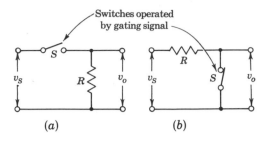

Fig. 17-1   Illustrating the principle of a linear gate (a) using a series switch, (b) using a shunt switch.

is opened only during the desired transmission interval. In practice, the switches will be replaced by diodes or triodes (semiconductor or thermionic) or by multigrid tubes which will be biased in the conducting or nonconducting direction as required. Ideally, the switches should have zero resistance when closed and infinite resistance when open, but, of course, in practice such will not be the case. When thermionic devices are used as the switches, the circuit of Fig. 17-1a is usually favored over the circuit of Fig. 17-1b. The reason for this preference is that in the nonconducting direction a thermionic device may be counted on to have a nominally infinite resistance. Hence, in Fig. 17-1a when S is open, the output will be zero as required. In Fig. 17-1b the output should be zero when S is closed. Since, however, the conducting or forward resistance R_{sw} of the switch will range from several hundred to several thousand ohms, it will be necessary that R be quite large; that is, it is required that $R \gg R_{sw}$ in order for the shunting effect of the switch to be effective. Even in this latter case a small residual output will continue to persist. Also, during the transmission interval, the input and output will be separated by the large resistance R. If there then is some stray capacitance shunting the output, it will not be possible to transmit fast waveforms without deterioration of the waveform. The disadvantages of the circuit of Fig. 17-1a over the circuit of Fig. 17-1b are the following: (1) The inevitable stray capacitance across the switch will permit some signal transmission even when the switch is open; and (2) since the signal is transmitted through S there will be some attenuation and distortion introduced by the nonlinearity of the device used for the switch.

Semiconductor devices do not have infinite back resistances, and their forward resistance may lie in the range of only several ohms. When such devices are used there is no generally apparent advantage in either the series or shunt switch position, and the decision with respect to the circuit of choice must depend on the particular application.

17-2   UNIDIRECTIONAL DIODE GATE

The gate of Fig. 17-2 is suitable for a positive-going input signal. The gate signal (also called a *control* pulse, a *selector* pulse, or an *enabling* pulse) is a

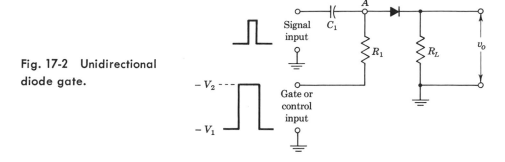

Fig. 17-2   Unidirectional diode gate.

rectangular waveform which makes abrupt transitions between the negative levels $-V_1$ and $-V_2$. When the gate voltage is $-V_1$, the diode is heavily back-biased and there will be no response at the output to an input signal unless the peak amplitude of the input signal is larger than the magnitude of the back-biasing voltage. (Actually, because of the capacitive coupling, the signal-input voltage will appear at point A with an average value $-V_1$. Hence the peak positive excursion of the signal at A, with respect to $-V_1$, will be smaller than the peak-to-peak voltage of the waveform at the input. For simplicity, we shall neglect this feature and consider simply that the input signal consists, say, of a very-low-duty-cycle pulse train, in which case this effect would be negligible. We assume also, in the present discussion, that the diode is ideal, with $V_\gamma = 0$.) When the gate rises to its higher level $-V_2$, a time-coincident signal-input pulse may be transmitted to the output. The effect on the output of the level $(-V_2)$ attained by the gate waveform is illustrated in Fig. 17-3. In Fig. 17-3a, $-V_2 = -5$ V, and for a 10-V input pulse a 5-V output pulse appears. Operation in this manner is often advantageous when the base line of the input signal has some noise signal superimposed. The level $-V_2$ may be adjusted so that only that part of the signal above the noise threshold appears at the output. When used in this manner, the circuit is referred to as a *threshold gate*. In Fig. 17-3b, $-V_2 = 0$ and the entire input pulse is transmitted to the output, whereas in Fig. 17-3c, $-V_2$ is positive and the signal appears superimposed on a 5-V *pedestal*.

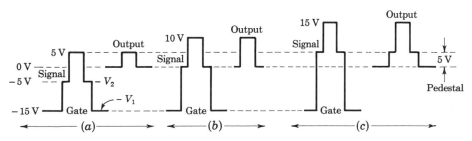

Fig. 17-3   Illustrating the effect of control voltage ($-V_2$) on gate output.

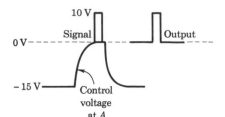

Fig. 17-4 Illustrating the distortion of the effective control waveform at A in Fig. 17-2.

The waveforms of Fig. 17-3 are unrealistic in that they have neglected the fact that the R_1C_1 network constitutes an integrating network for the gate waveform. Hence, the gate voltage will not appear abruptly at A as required, but rather will rise exponentially with a time constant R_1C_1 and fall at a similar rate. Hence, this type of gate is not particularly suitable for selecting a portion of a continuous waveform. If, however, the signal is a pulse whose duration is reasonably smaller than the gate width, the result may be entirely satisfactory, as shown in Fig. 17-4.

The advantages of this gate are the following: (1) It is extremely simple. (2) There is very little time delay through the gate since the input is coupled directly to the output through C_1 and the diode. (3) The gate draws no current in its quiescent condition (i.e., no "stand-by" current). This feature becomes very important in a system requiring many gates. In this respect, also, the present gate should be compared with the AND gate of Sec. 9-4, which will accomplish the same general result as the present gate but which does draw stand-by current. (4) As is shown in Sec. 17-3, this gate is easily extended into a multi-input OR circuit with an INHIBITOR or NOT terminal.

The disadvantages of the gate are the following: (1) There will be inter-action between the signal source and control-voltage source. (2) The gate is of limited use because of the slow rise of the control voltage at the diode. The rise time of the control voltage at A may be improved by reducing R_1 but only at the expense of increasing the coupling between signal and gate. If an attempt is made to improve the time constant by reducing C_1, other complications ensue. For example, suppose a gate time constant of 0.2 μsec

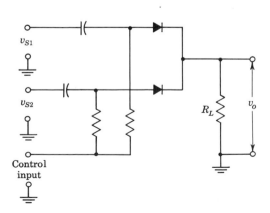

Fig. 17-5 The unidirectional diode gate adapted for more than one signal input.

is required and that values of R_1 less than 10 K permit too much of the control signal to couple back to the signal input. Then for $R_1 = 10$ K, $C_1 = 20$ pF. Now suppose that the output capacitance C_o shunted across the output terminals is $C_o = 10$ pF. Then the gate will attenuate fast signals, allowing only two-thirds of the signal to pass through.

The rise-time difficulties of the control pulse may be eliminated by replacing the capacitor C_1 by a resistor R_2. A first disadvantage of such an arrangement is that any signal will now be attenuated. For example, if $R_2 = R_1$, the amplification will be $\frac{1}{2}$ (assuming that $R_L \gg R_1$). Additionally, in such a case, d-c coupling of the signal will be required, and such coupling may not be feasible.

17-3   OTHER FORMS OF THE UNIDIRECTIONAL DIODE GATE

The unidirectional diode gate may be adapted to accept more than one signal input, as in Fig. 17-5. Here two signal inputs v_{S1} and v_{S2} are indicated, but, of course, more than two may be used. When the control signal is at its higher level (say zero voltage), the circuit is recognized as a capacitively coupled OR gate (Sec. 9-3). When the control voltage is at its lower level, the gate transmits no inputs. Hence, the negative level of the control pulse may be considered an inhibitor signal. Here, then, is a multi-input OR circuit with an INHIBITOR terminal.

A difficulty associated with the arrangement of Fig. 17-5 is that the loading on the control input becomes increasingly heavy as the number of inputs increases. This difficulty may be corrected through the use of one additional diode, as indicated in Fig. 17-6. In this latter circuit the control-input voltage does not feed into the signal sources. In neither case, however, is there any stand-by current.

It is possible to arrange the diode gate so that it will transmit a signal only as the result of the simultaneous occurrence of a number of gate voltages.

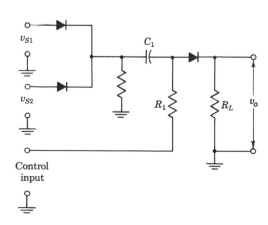

Fig. 17-6 A two-input gate in which the signal sources do not load the control signal.

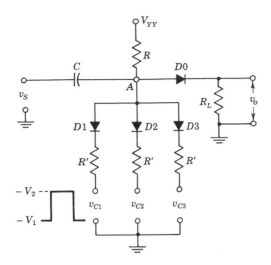

Fig. 17-7 A unidirectional gate which delivers an output pulse only at a coincidence of a number of control voltages.

Such a circuit is shown in Fig. 17-7. Suppose that the control-signal levels are $-V_1$ and $-V_2$. Then when *any one* of the control signals v_C is at $-V_1$, point A is negative with respect to ground by an amount, say, V', and no part of the waveform is transmitted unless the input signal is larger than V'. When all control voltages are at $-V_2$, the back bias on diode $D0$ is removed, and the signal is sampled. It will be recognized that this circuit, except for the signal-input connection and capacitor, is an AND circuit (Sec. 9-4). This circuit differs from the gates previously described in that it draws a quiescent current from the V_{YY} supply, although this current may be kept low.

A threshold gate which may be sampled by any one of a number of control signals (an OR circuit) is given in Prob. 17-5.

All the forms of the unidirectional diode gate considered up to the present have a common feature which may, on some occasions, constitute a disadvantage. In Fig. 17-3 it appears that unless the upper level of the gating waveform is exactly zero, then either a portion of the input signal will not get through the gate or else the transmitted signal will be superimposed on a pedestal. This situation may be corrected as in Fig. 17-8 by the use of an additional diode. Initially, in the absence of an enabling signal at the control terminals, diode $D1$ conducts and the consequent voltage drop across R keeps

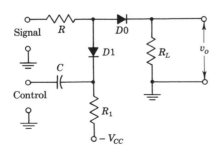

Fig. 17-8 A gate whose response is not sensitive to the upper level of the control voltage.

diode $D0$ back-biased. A positive-going control gate causes $D1$ to cease conduction, and the signal is sampled for the duration of the control pulse. The signal must be d-c-coupled, but the gate may be either a-c-coupled, as indicated, or d-c-coupled. If a-c coupling is employed, then, of course, the time constant R_1C must be large in comparison with the gate duration.

17-4   BIDIRECTIONAL GATES USING TRANSISTORS OR TUBES

All the above gates have the limitation that they pass only unidirectional signals. We shall defer consideration of diode bidirectional gates until we have examined first some simpler bidirectional gates which employ multi-element devices.

Several examples of linear bidirectional sampling gates are shown in Fig. 17-9. In Fig. 17-9a the signal voltage v_S and the control voltage v_C are applied through summing resistors R_1 and R_2 to the base of a transistor or

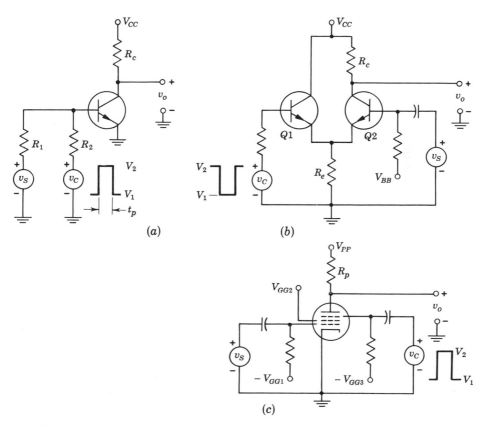

Fig. 17-9 Examples of bidirectional sampling gates.

the grid of a triode (or pentode). The gating voltage is again a pulse waveform between the levels V_1 and V_2, with a duration t_p equal to the required transmission-interval duration. When the control voltage is at its lower level V_1, the transistor or tube is biased well below cutoff. However, when the selector voltage is at its upper level V_2, this bias brings the base or grid up out of cutoff and into the active region or grid base. So long as the gate persists, the transistor (or tube) will sample the signal voltage, which will then appear amplified at the output.

In the gating circuit of Fig. 17-9b separate bases are available for the signal and gating voltages. When the control voltage v_C is at its upper level, the current through R_e is large enough to raise the emitter voltage to the point where Q2 is cut off. When the gating signal is at its lower level, Q1 is cut off and Q2 is free to operate as an amplifier stage. During the sampling interval, when Q1 is cut off, the gating and signal sources are uncoupled from one another. During this interval, as well, the loading on the signal source is reduced, since the presence of an emitter resistor R_e increases the impedance seen looking into the base of Q2 [Eq. (1-50)]. Further, the signal may be applied directly to the base rather than through a resistor, and the loss of high-frequency response due to the capacitive component of the input impedance at the base is thereby avoided. An analogous vacuum-tube circuit operates in a similar manner.

In Fig. 17-9c is illustrated the use of a multigrid vacuum tube as a sampling gate. The signal is applied to the control grid. The gating pulse is applied to the suppressor. The levels V_2 and V_1, respectively, are adequate to allow plate current during the sampling interval and to turn off the plate current otherwise. Pentodes suitable for this type of service have suppressor grids with rather more tightly spaced wires than are customarily found in ordinary pentodes. This suppressor construction allows the plate current to be turned on or off with a small gating signal. We have already seen an application of such a gating tube in connection with the suppressor-gate Miller sweep of Sec. 14-12. The circuit of Fig. 17-9c has no analog in a single-transistor circuit.

17-5   REDUCTION OF PEDESTAL IN A GATE CIRCUIT

The gate circuits of Fig. 17-9 share a common feature which may on occasion constitute a disadvantage. Initially, the voltage level at the output is V_{CC} or V_{PP}. When the gating signal is applied, the transistor or tube draws current and the output therefore establishes itself at a new, lower quiescent level. When, now, the signal is applied, the output signal is superimposed on this new quiescent level. The appearance, typically, of the output during a gating interval is as shown in Fig. 17-10, where we see that the sampled portion of the signal is superimposed on a pedestal.

The pedestal can be largely suppressed by the symmetrical arrangement

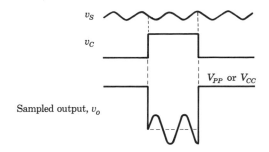

Fig. 17-10 Illustrating the pedestal associated with the sampling gates of Fig. 17-9.

shown in Fig. 17-11 or in an analogous tube circuit. Here the gating circuit employed is essentially that indicated in Fig. 17-9a, except that, for simplicity, the gating and signal voltages have been placed directly in series. The gating transistor is $Q1$, and $Q2$ is used to minimize the pedestal. Gating voltages of opposite polarity are applied to the transistor bases. During the nongating time, $Q2$ draws current and $Q1$ does not. The bias voltages $-V_{BB1}$ and $-V_{BB2}$ and the gate-signal amplitude have been adjusted so that the two transistor currents are identical, and as a result the quiescent output voltage level will remain constant.

If, as is usually the case, the gate waveform has a nonzero rise time, then the arrangement of Fig. 17-11 does not solve completely the problem of the pedestal, as may be seen from the following considerations. Let us assume, as in Fig. 17-12a, that the gate pulse is large in comparison with the active-region base-voltage range, so that each transistor, when it is not conducting, is biased far below cutoff. Then when the gate voltage appears, $Q2$ will be driven to cutoff before $Q1$ starts to conduct, whereas at the end of the gate, $Q1$ will be cut off before $Q2$ starts to conduct. Hence, as a result of the gate signals themselves, the output will appear as in Fig. 17-12b. The gated signal voltage will appear superimposed on this waveform. If the gate-waveform rise time is small in comparison with the gate duration, these spikes may not be very objectionable.

Fig. 17-11 A linear-gate circuit with provision to cancel the pedestal.

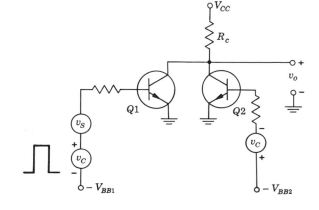

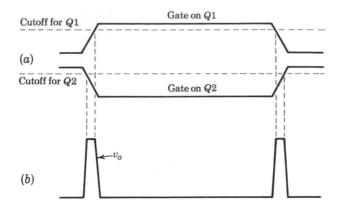

(a)

(b)

For positive unidirectional signals, the circuit of Fig. 17-9b may be used without generating a pedestal if transistor $Q2$ is operated at cutoff. However, in order to be certain not to lose small signals, if may be necessary to adjust the bias so that a slight quiescent current exists. Such a bias will result in a small pedestal, which may be acceptable. Even though the transistor is operated near cutoff, the circuit will be quite linear because of the degeneration introduced by the large emitter resistor. Gain is sacrificed for linearity. To adapt this circuit for use with negative unidirectional signals we need but to replace the n-p-n transistors by p-n-p types.

17-6   A BIDIRECTIONAL DIODE GATE[1]

A bidirectional gate may be constructed with the use of diodes instead of transistors or multielement tubes. Such diode gates have advantages of linearity of operation and ease of adjustment to ensure zero pedestal. A bidirectional gate is shown in Fig. 17-13. We may observe that the gate consists essentially of two gates of the type shown in Fig. 17-2 with the modification that C_1 is replaced by a resistor. The circuit is redrawn in the form of a bridge in Fig. 17-14. Two symmetrical gating voltages $+v_C$ and $-v_C$

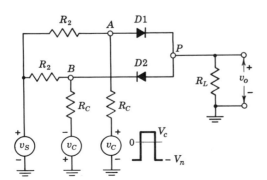

Fig. 17-13 A bidirectional diode sampling gate.

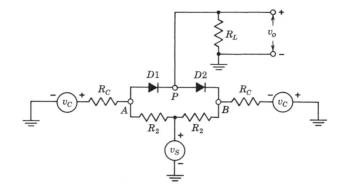

Fig. 17-14   The circuit
of Fig. 17-13 redrawn in
the form of a bridge
network.

are now required.   When the gating or control voltages are at the levels V_n
and $-V_n$, respectively, the diodes are nonconducting and there is no transmission of the signal v_S.   When the control signals are at the levels V_c and
$-V_c$, then a sample of v_S appears at the output.   Provided that the diodes are
identical in characteristics (the diodes need not be perfect), it follows from
the complete symmetry of the circuit that no pedestal can appear at the output in response to the gating voltages.   In the following discussion we shall
compute some of the properties of this gating circuit.   For simplicity, we
shall neglect the impedance of the signal source.   The equations given below
are modified in Ref. 1 to take into account this signal-source impedance R_s.

Gain   The *gain* of the gate is defined as the ratio v_o/v_S during the transmission interval.   The gain is easily calculated from the equivalent circuit
of Fig. 17-15.   This circuit is derived from that of Fig. 17-13 through the application of Thévenin's theorem, where R_1, R_3, and α are defined as follows:

$$R_1 \equiv \frac{R_2 R_C}{R_2 + R_C} \qquad R_3 \equiv R_1 + R_f \qquad \alpha \equiv \frac{R_C}{R_2 + R_C} \qquad (17\text{-}1)$$

Each diode has been replaced by its piecewise linear model of Fig. 6-10, where

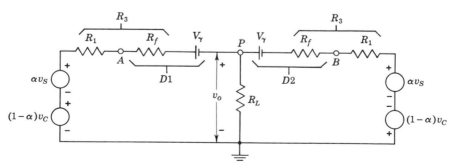

Fig. 17-15   An equivalent circuit for the bidirectional gate of Fig. 17-14.

R_f is the forward resistance and V_γ is the offset voltage. Note that the control voltages as well as the offset voltages give no resultant current through R_L. The open-circuit voltage from P to ground is αv_S and the Thévenin's resistance is $R_3/2$. Hence, the gain is given by

$$A = \alpha \frac{R_L}{R_L + R_3/2} = \frac{R_C}{R_C + R_2} \frac{R_L}{R_L + R_3/2} \tag{17-2}$$

The Control Voltage V_c Suppose that the signal voltage attains a maximum voltage V_s. Then there is a minimum control voltage V_c that is required to ensure that *both* diodes will continue to conduct. That there is such a minimum required value for V_c may be seen from the following considerations. Initially, in the presence only of the gate voltages, the diodes $D1$ and $D2$ conduct equal currents. The load current is zero and the pedestal is zero. Now assume, say, that v_S is a positive-going signal. Then the current in $D1$ increases and the current in $D2$ decreases, the difference current flowing through R_L. As v_S continues to increase, eventually the current in $D2$ will become zero.

To compute the required minimum V_c, assume that diode $D2$ has just stopped conducting. Then the voltage across R_3 associated with $D2$ is zero. For the sake of simplifying the ensuing equations we shall make the reasonable assumption that $V_\gamma \ll V_s$ and shall set $V_\gamma = 0$. Hence, the output voltage must equal the voltage $\alpha V_s - (1 - \alpha)V_c$, or

$$\frac{R_L}{R_3 + R_L}[\alpha V_s + (1 - \alpha)V_c] = \alpha V_s - (1 - \alpha)V_c \tag{17-3}$$

from which we find

$$V_c = \frac{R_C}{R_2} \frac{R_3}{R_3 + 2R_L} V_s \equiv (V_c)_{\min} \tag{17-4}$$

We note from Eq. (17-4) that $(V_c)_{\min}$ decreases with increasing R_L. There is, however, an important consideration which limits the size of R_L. Suppose we assume that a total stray capacitance C shunts the output terminals in Fig. 17-13 or 17-15. Suppose also that at the end of the gating interval the output is at some finite voltage level because of the signal v_S. Then, when the diodes cut off at the end of the gating interval, the output voltage must decay to zero with a time constant R_LC. This difficulty is of the same type described in Sec. 9-3 in connection with logic gates.

The Control Voltage V_n Just as there is a minimum control voltage $(V_c)_{\min}$ required to keep both diodes conducting over the full range of the input signal, so there is also a minimum control voltage $(V_n)_{\min}$ required to ensure that *both* diodes are back-biased when no sampling is to take place. If the diodes are reverse-biased, then point P in Fig. 17-15 is at ground potential. Hence, the voltage across $D1$ is $\alpha V_s - (1 - \alpha)V_n$, where V_n is the magnitude of the voltage for the lower level of v_C. Since the voltage across $D1$

must be negative or, at a minimum, zero,

$$(V_n)_{\min} = \frac{\alpha V_s}{1 - \alpha} = \frac{R_C}{R_2} V_s \tag{17-5}$$

where use was made of Eq. (17-1) for α.

In practice, the control voltages V_c and V_n should be somewhat larger, say 25 percent larger, than the minimum values given in Eqs. (17-4) and (17-5). In the case of V_n this additional voltage is essentially a safety factor. In the case of V_c the additional voltage not only is a safety factor but also serves to improve the gate linearity. The larger V_c, the greater is the ratio of control current to signal current in the diodes. Hence, the more nearly constant will be the diode forward resistance R_f over the range of the signal current.

17-7   BALANCE CONDITIONS IN A BIDIRECTIONAL DIODE GATE

The gate circuit of Fig. 17-13 has been redrawn in Fig. 17-16a to make the symmetry of the circuit more evident. Here the signal-source impedance R_s (not previously shown) has been included in the circuit. We may note now that, as a result of this symmetry, not only is the output free of pedestal but also no current flows through the signal-source impedance R_s in response to the control voltage. The converse is, of course, not true. The signal source in series with R_s (not indicated in Fig. 17-16) does cause a current to flow through the control-signal generators.

If the two control voltages are equal in amplitude, the points A and B in Fig. 17-16a are at voltages which are equal but opposite in polarity. Hence, the junction point P of the diodes and load and the junction point of the resistors R_2 are both at zero potential, and balance is ensured at both output and input. If it is not convenient to establish precisely equal control-signal amplitudes or if the circuit components are not precisely matched, balance may be restored by either of the voltage dividers r_f or r_2 indicated in Fig. 17-16b.

An initial adjustment to make the control voltages equal or a slight adjustment of the gating to correct for small unbalance is usually not difficult. We have considered that the gating waveform is rectangular in shape. However, unless it is important that the gate open and close rapidly, there is no need for the gating waveform to make an abrupt transition between levels. Similarly, if the control-voltage levels are able to keep the diodes either definitely conducting or definitely nonconducting, there is no need for the control voltage to remain absolutely constant between transitions. What is important, with respect to balance, is that whatever the shape of the control voltages, these voltages must be *identical in waveshape*, except, of course, for polarity. If, for example, the two control voltages have different rise times, spikes will appear at the output at the times of opening and closing of the gate.

When the gate-signal duration is long in comparison with the gate-signal

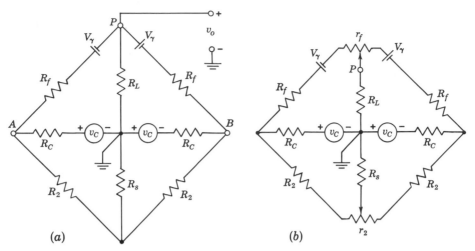

Fig. 17-16 (a) The bidirectional gate drawn to exhibit symmetry; (b) voltage dividers r_f and r_2 are used for balancing of the gate.

rise time, the spikes which appear may not be objectionable. However, when gates of the order of a microsecond or less duration are required, it may well be difficult to reduce the gate rise times sufficiently. In such cases it may be necessary to resort to extreme measures to ensure identical waveforms. One method which often proves effective is to derive the control signals from identical and bifilar-wound windings of a pulse transformer. In order that the diode-bridge sampling gate be balanced it is necessary for the diode characteristics to be identical, just as it is for the control signals to be identical. Consequently, with just such applications in mind as the diode sampling gate, various manufacturers make available *matched* diode assemblies. These diodes, consisting usually of a pair or group of four diodes (quads), are matched with respect to forward current, reverse current, temperature dependence, and sometimes also frequency response. The quads find application in four-diode bridge circuits, such as the four-diode gate discussed in Secs. 17-10 and 17-11. The assemblies may consist of individual diodes or of diodes in a common encapsulation or diodes formed by integrated techniques for the purpose of ensuring a common temperature dependence.

17-8   SIGNAL–INPUT RESISTANCE AND CONNECTIONS

The current drawn from the input-signal source is not a function of the control voltage except in so far as it depends on whether or not the diodes are conducting. This result follows from the fact that, as we have seen, no current flows through R_s in response to the control voltages. Assuming that the diode forward and reverse resistances R_f and R_r are zero and infinity, respectively,

Fig. 17-17   An equivalent circuit of the bidirectional gate which gives correctly the
gain and input impedance.

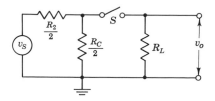

we may write the input resistances by inspection from Fig. 17-13 or 17-14 as

$$R_i = \frac{R_C + R_2}{2} \qquad \text{diodes open} \tag{17-6}$$

and

$$R_i = \frac{R_L R_C}{2R_L + R_C} + \frac{R_2}{2} \qquad \text{diodes conducting} \tag{17-7}$$

An equivalent circuit which gives the correct gain and input impedance and which takes into account the absence of pedestal and reaction of control signals on input is shown in Fig. 17-17. Switch S is closed or open, depending on whether or not the diodes are conducting.

So far we have considered only d-c coupling of the signal input. We may, however, also use a-c signal-input coupling. In the case of the unidirectional gate of Fig. 17-2, the use of the input coupling capacitor C_1 had the markedly disadvantageous effect that the capacitor prevented the gate voltage at the diode from rising abruptly. In the present case, however, no such difficulty exists and the gate will operate basically in the same manner, regardless of whether or not an input capacitor is employed. This feature results from the fact that no current flows through the input capacitor because of the control signals.

17-9   EFFECT OF CIRCUIT CAPACITANCES.   EXAMPLE

The capacitances which have a principal effect on the operation of the gate circuit are the following: (1) the capacitance C_o across the gate output terminals, (2) the capacitance C_d across each diode, and (3) the stray capacitances C_s to ground from each of the junctions of the resistors R_2 and R_C. The capacitance C_o has an adverse effect on the ability of the gate to transmit fast waveforms. The capacitance C_s not only accentuates this inability to transmit fast waveforms but also limits the speed with which the gate can be opened and closed. Finally, the diode capacitances C_d provide a transmission path through the gate even when the diodes are not conducting.

When the diodes are forward-biased, the shunt capacitances affect the high-frequency response. From the equivalent circuit of Fig. 17-18 (derived from Fig. 17-17 through the use of Thévenin's theorem), we see that the gate behaves as a low-pass resistance-capacitance circuit whose time constant is $(C_o + 2C_s)$ times a resistance equal to the parallel combination of R_L and $R_1/2$.

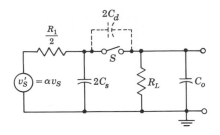

Fig. 17-18 Illustration of the capacitances which influence transmission in a bidirectional diode gate. [The symbols R_1 and α are defined in Eq. (17-1).]

The residual transmission which results, from the capacitance which shunts the diodes, may be calculated from Fig. 17-18 by opening switch S and shunting across the switch a capacitance equal to the sum of the diode capacitances.

EXAMPLE In the circuit of Fig. 17-14 assume that $R_L = R_C = 100$ K, $R_2 = 50$ K, and that the signal has a peak value of 20 V. Find A, $(V_c)_{\min}$, $(V_n)_{\min}$, R_i, and the 3-dB frequency of the gate.

Solution In view of the order of magnitude of the resistance involved, we may assume that the diodes are perfect (that is, $V_\gamma = 0$, $R_f = 0$, and $R_r = \infty$). From Eq. (17-1),

$$R_1 = \frac{R_2 R_C}{R_2 + R_C} = 33.3 \text{ K} = R_3$$

From Eqs. (17-2), (17-4), and (17-5), we find

$$A = 0.57 \qquad (V_c)_{\min} = 5.7 \text{ V} \qquad (V_n)_{\min} = 40 \text{ V} \tag{17-8}$$

The resistances seen by the signal source are given by Eqs. (17-6) and (17-7):

$$R_i = 75 \text{ K (diodes open)} \qquad R_i = 58 \text{ K (diodes conducting)} \tag{17-9}$$

Now $R_1/2$ ($= 16.7$ K) in parallel with R_L ($= 100$ K) is equivalent to 14.3 K. If we assume a total shunting capacitance $C_o + 2C_s$ of 20 pF, then the time constant is $\tau = (14.3 \times 10^3)(20 \times 10^{-12})$ sec $= 0.29$ μsec and the upper 3-dB frequency f_2 is given by $f_2 = 1/2\pi\tau = 0.55$ MHz.

17-10   FOUR–DIODE GATE[1]

Among the disadvantages of the two-diode gate of Fig. 17-13 are its low gain, its sensitivity to control-voltage unbalance, the possibility that the voltage $(V_n)_{\min}$ may be excessive, and the fact that there may be appreciable leakage through the diode capacitances. These features may be improved by the use of two additional diodes, as in Fig. 17-19. Two balanced voltages $+V$ and $-V$ are required, but since these are fixed d-c voltages they need occasion no difficulty.

Qualitatively, the operation is as follows. When the control voltages are V_c and $-V_c$, respectively, the diodes $D3$ and $D4$ are back-biased. The diodes $D1$ and $D2$ are conducting because of the voltages $+V$ and $-V$, and the signal source is coupled to the load through the resistors R_2 and the conducting diodes. Since, under these circumstances, the control voltages are disconnected from the gate by the back-biased diodes, an imbalance in control signals cannot result in a pedestal at the output.

When the control voltages are V_n and $-V_n$, respectively, the points P_1 and P_2 are clamped to these voltages and the diodes $D1$ and $D2$ are back-biased. Under these circumstances, the output is zero.

The gain of the gate of Fig. 17-19 is the same as that of the gate of Fig. 17-13 and is given by Eq. (17-2). In connection with the circuit of Fig. 17-13 we found earlier that a gain close to unity is not conveniently attained. The reason for this feature is that to achieve such a gain it is necessary to make $R_C \gg R_2$ and $R_L \gg R_3$. However, as is apparent from Eq. (17-5), under such circumstances the required value of $(V_n)_{min}$ becomes prohibitively high. In the gate of Fig. 17-19 the principal requirement on R_2 is that it be large in comparison with the diode-conducting resistance, so that the diodes may be effective as clamps. The minimum value of R_2 is also limited by the fact that when the diodes $D3$ and $D4$ conduct, the control source must furnish a current V_n/R_2, and the signal source must furnish a current $2V_s/R_2$, approximately. Hence, R_2 must not be so low as to draw excessive control or signal current.

We now compute the required minimum values of V, V_c, and V_n. When diodes $D3$ and $D4$ are not conducting, the gates of Figs. 17-14 and 17-19 are identical with the voltages $+V$ and $-V$, replacing the voltages V_c and $-V_c$. Hence, V_{min} is given by Eq. (17-4).

The voltage $(V_c)_{min}$ is computed as follows. We shall assume that $R_f \ll R_L$. Then, for a positive-going signal of amplitude V_s, the voltage at point P_1 becomes AV_s, where A is the circuit gain. If the diode $D3$ is to con-

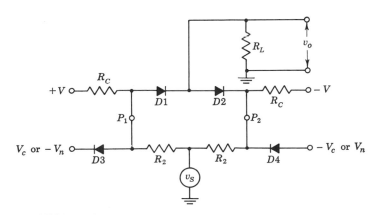

Fig. 17-19   A four-diode gate.

tinue to be back-biased, the voltage V_c must be at least

$$(V_c)_{min} = A V_s \tag{17-10}$$

The voltage V_n must be selected not only to keep the transmission diodes reverse-biased but also to keep the clamp diodes conducting in the presence of a signal V_s. The voltage at P_2 for a positive signal V_s, and hence the minimum value of V_n, is

$$(V_n)_{min} = V_s \frac{R_C}{R_C + R_2} - V \frac{R_2}{R_2 + R_C} \tag{17-11}$$

EXAMPLE In the circuit of Fig. 17-19 consider that $R_L = R_C = 100$ K and that $R_2 = 1$ K. (Assume semiconductor diodes with $R_f = 25 \ \Omega$, in which case the value selected for R_2 should give adequate clamping.) For $V_s = 20$ V. compute A, V_{min}, and $(V_c)_{min}$. Compute $(V_n)_{min}$ for $V = V_{min}$.

Solution From Eq. (17-1) we obtain $R_1 = 990 \ \Omega$ and $R_3 = 1{,}015 \ \Omega$. From Eqs. (17-2), (17-4), (17-10), and (17-11), we have

$$A = 0.985 \qquad V_{min} = 10.1 \text{ V} \qquad (V_c)_{min} = 19.7 \text{ V} \qquad (V_n)_{min} = 19.7 \text{ V}$$

The two auxiliary diodes not only improve the gate circuit in the manner described previously, but, as is seen from the example above, an improvement results also in the gain and the value of $(V_n)_{min}$.

17-11   FOUR–DIODE GATE (ALTERNATIVE FORM)

An alternative form of the four-diode gate is shown in Fig. 17-20. When the control voltages are at the levels V_c and $- V_c$, all four diodes are conducting

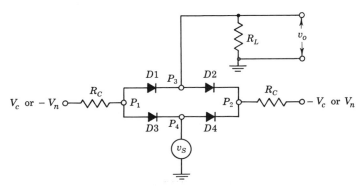

Fig. 17-20 An alternative form of four-diode gate.

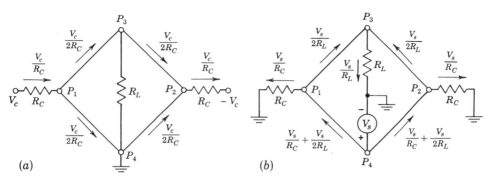

Fig. 17-21   The diodes in Fig. 17-20 are replaced by short circuits.   (a) The currents due to V_c; (b) the currents due to V_s.

and the source is connected to the load through two parallel paths, each consisting of two diodes in series.   In this gate the supply voltages V and $-V$ are not required, but, to avoid a pedestal at the output, it is necessary that the control voltages V_c and $-V_c$ be balanced.

The required voltages V_c and $-V_c$ depend on the amplitude V_s of the signal and are determined by the condition that the current be in the forward direction in each of the diodes $D1$, $D2$, $D3$, and $D4$.   The current in each diode consists of two components, one due to V_c (as indicated in Fig. 17-21a) and the other due to V_s (as indicated in Fig. 17-21b).   The current due to V_c is $V_C/2R_C$ and is in the forward direction in each diode, but the current due to V_s is in the reverse direction in $D3$ (between P_1 and P_4) and in $D2$ (between P_3 and P_2).   The larger reverse current is in $D3$ and equals $V_s/R_C + V_s/2R_L$, and hence this quantity must be less than $V_c/2R_C$.   The minimum value of V_c is therefore given by

$$(V_c)_{min} = V_s \left(2 + \frac{R_C}{R_L} \right) \tag{17-12}$$

The above derivation assumes that R_f is much less than R_C or R_L.   A balancing voltage divider may be inserted between $D3$ and $D4$ so as to give zero output for zero input.   If the divider is assumed to be set at its midpoint, if its total resistance is R, and if R and R_f are both much less than R_C or R_L, then, proceeding as in Fig. 17-21, we find (Prob. 17-14) that

$$(V_c)_{min} = V_s \left(2 + \frac{R_C}{R_L} \right) \left(1 + \frac{R}{4R_f} \right) \tag{17-13}$$

The voltage $(V_c)_{min}$ may become excessive if the resistance R is too large relative to R_f.   Under these circumstances the value of $(V_c)_{min}$ in Eq. (17-13) should be used rather than the approximate equation (17-12), which may easily be incorrect by quite a large factor.

If R_C and R_L are large compared with R_f and R, we see, from Fig. 17-21b, that the gain will be very close to unity:

$$A \approx 1 \tag{17-14}$$

The exact expression for A is given in Prob. 17-15.

The diodes in Fig. 17-20 are to remain reverse-biased when the control voltages are V_n and $-V_n$. If the signal has a peak amplitude V_s, then it is required that, at a minimum,

$$(V_n)_{\min} = V_s \tag{17-15}$$

As an example, assume that $V_s = 20$ V, $R_f = 25$ Ω, $R_L = R_C = 100$ K, and $R = 100$ Ω. Then we find from Eqs. (17-13) to (17-15) that

$$(V_c)_{\min} = 120 \text{ V} \qquad A \approx 1 \qquad (V_n)_{\min} = 20 \text{ V}$$

The deviation from linearity of such a gate may be less than 0.1 percent. Note that a large control voltage is required.

17-12   SIX–DIODE GATE

A six-diode gate is shown in Fig. 17-22. This gate combines the features of the four-diode gates of Figs. 17-19 and 17-20. When the gate transmits no signal, diodes $D5$ and $D6$ are conducting and acting as clamps while all other diodes are open. During transmission, diodes $D1$ through $D4$ conduct while $D5$ and $D6$ are reverse-biased.

During transmission, when diodes $D5$ and $D6$ are back-biased, the six-diode gate becomes equivalent to the four-diode gate of Fig. 17-20, with the exception that the fixed voltages V and $-V$ replace the control voltages V_c and $-V_c$. Therefore Eqs. (17-12) and (17-13) for $(V_c)_{\min}$ apply in the six-diode

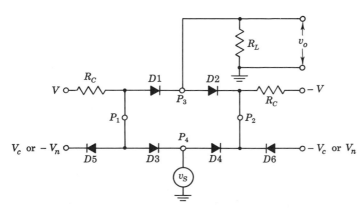

Fig. 17-22 A six-diode gate.

case to $V_{\min}$. We note in the typical case cited above that $(V_c)_{\min}$ for the four-diode gate or $V_{\min}$ for the six-diode gate may be as large as 120 V for a 20-V signal. In such a case the advantage of the six-diode gate is apparent in the fact that this large voltage need appear only as a fixed voltage and not as a control signal.

As in the case of the four-diode gate of Fig. 17-19, so in the present case the control signal need not be balanced. From Fig. 17-22 we see that if the clamping diodes $D5$ and $D6$ are to remain reverse-biased for a signal amplitude V_s, then V_c must be at least equal to $(V_c)_{\min} = V_s$. On the other hand, it is also apparent that if the points P_1 and P_2 are clamped at a voltage V_n, then none of the transmission diodes will conduct until V_s exceeds V_n. Hence, the minimum required value of V_n is $(V_n)_{\min} = V_s$. To summarize, we have

$$(V_c)_{\min} = (V_n)_{\min} = V_s \qquad (17\text{-}16)$$

Again, if R_C and R_L are large compared with R_f and R, we see, from Fig. 17-22, that the gain will be very close to unity:

$$A \approx 1 \qquad (17\text{-}17)$$

The exact expression for A is given in Prob. 17-15.

17-13   CHOPPER AMPLIFIERS[2]

The circuits of Fig. 17-1a and b have a particularly interesting application which we shall now discuss. Let us assume that we are required to amplify a small signal (say of the order of millivolts) and that the signal $v(t)$ is one in which dv/dt is very small. By way of example, if the signal is periodic, it may be that the period is many minutes or even hours in duration. An a-c amplifier with the customary capacitive coupling between stages would not be feasible, since these coupling capacitances would be impractically large. Instead it would be necessary to use direct coupling between stages. With such a d-c amplifier we would not be able to distinguish between a change in output voltage as the result of a change in input voltage or as the result of a drift in some active device or component. If the amplifier has high gain, even small changes in the operating point of the first stage, amplified by the remaining stages, might cause a large change in output.

A method for circumventing this difficulty is shown in Fig. 17-23a. The low-frequency input signal v is shown in Fig. 17-23b. Assume that switch S_1 is being driven so that it is alternately open and closed. Then the signal v_i at the amplifier input will appear as in Fig. 17-23c. When S_1 is open $v_i = v$, and when S_1 is closed $v_i = 0$. Observe that the waveform v_i is a "chopped" version of the waveform v. It is for this reason that the circuit consisting of R and S_1, in the present application, is called a *chopper*.

In Fig. 17-23c we note that when the switch is open the signal v_i reproduces the input signal v. As we have drawn the figure, a perceptible voltage change

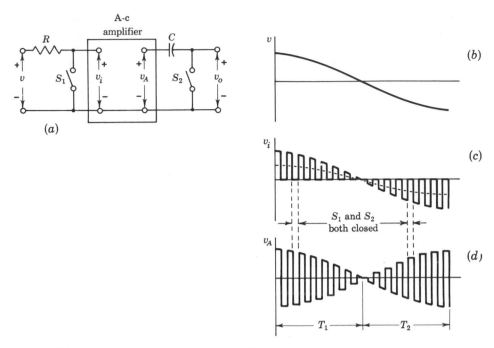

Fig. 17-23 (a) A chopper-stabilized amplifier; (b) input signal; (c) chopped signal; (d) signal-modulated square wave.

takes place in v during any interval when S_1 is open. Thus, when v is positive, the positive extremities of the waveform v_i are not at a constant voltage and similarly for the negative extremities when v is negative. But this feature is in no way essential to the operation. More customarily, the frequency of operation of the switch is very large (typically 100 times) in comparison with the frequency of the signal v. Therefore no appreciable change takes place in v during the interval when S_1 is open. Accordingly, it is proper to describe the waveform v_i as a square wave of amplitude proportional to v and having an average value (shown dashed) that is also proportional to the signal v. Alternatively stated, the waveform v_i is a square wave at the switching frequency, amplitude-modulated by the input signal and superimposed on a signal which is proportional to the input signal v itself.

The low-frequency cutoff of the a-c amplifier is such that the relatively high frequency square wave passes with small distortion while the signal frequency is well below the cutoff point. Consequently, at the output of the amplifier the waveform v_A is as shown in Fig. 17-23d. Here we are left with only the modulated waveform itself. Because of this process of modulation which has been accomplished, the chopper (or the chopper together with a filter to eliminate the signal itself) is often called a *modulator*.

The signal is recovered through the mechanism of the capacitor C and the switch S_2. The switch S_2 closes and opens in synchronism with S_1. Thus, during the interval T_1, the negative extremity of v_A is restored to zero, whereas during T_2 the positive extremity is restored to zero. As a result, except for an increase in amplitude, the signal v_o across S_2 assumes again the form of the signal v_i. If now this signal v_o is passed through a low-pass filter which rejects the square wave and transmits the signal frequency, at the filter output we shall have an amplified replica of the original signal. If S_2 operates anti-synchronously with S_1 (S_2 closed while S_1 is open and vice versa), then at the output the signal will appear with reversed polarity. In either case the combination of the capacitor C, the switch S_2, and the filter constitutes a *synchronous demodulator*. The amplifier of Fig. 17-23a is called a *chopper-stabilized* amplifier. Note that the amplifier is not stabilized by the choppers but rather that the choppers eliminate the necessity for a direct-coupled stabilized amplifier.

17-14   THE TRANSISTOR AS A CHOPPER SWITCH[3]—ON STATE

Many devices and circuits are feasible for use as chopper switches. For example, the four-diode gate of Fig. 17-20 has found wide use in this application. Electrically driven mechanical relays have similarly been effectively employed in this service. Our present interest, however, is in the use of transistors. The collector and emitter terminals of the transistor are to be used as the switch terminals. The switch is closed by supplying a base current adequate to drive the transistor into saturation, and the switch is opened by applying a base voltage adequate to drive the transistor beyond cutoff.

A transistor chopper is shown in Fig. 17-24. In Fig. 17-24a and c the transistor switch is located in series with the signal lead, as in Fig. 17-1a. In Fig. 17-24b and d the transistor is in shunt across the output, as in Fig. 17-1b. In Fig. 17-24a and b the transistor is operated in the "normal" manner, with the gating square wave v_G applied between base and emitter. In Fig. 17-24c and d the transistor is operated in the "inverted" manner, with the gating signal applied between base and collector. The relative merits of normal and inverted operation will appear in the discussion to follow. When the transistor is ON we should like the transistor to behave as a short circuit. Of course, the transistor will only approach this ideal condition, and we shall now examine in what respects the short-circuit approximation fails.

In Chaps. 6 and 8 we consider the transistor switch in saturation and in Sec. 15-9 we point out that the transistor switch is bidirectional. We now study the matter anew, however, because in our present application the currents and voltages encountered are entirely different in magnitude. In the preceding chapters the voltage v is a supply voltage V_{CC} of the order of tens of volts and the collector or emitter current is of the order of milliamperes. In the present case v is of the order of millivolts and the collector or emitter current may be in the range of microamperes. In the previous case the base

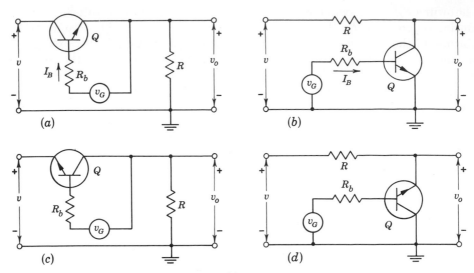

Fig. 17-24 Various forms of chopper circuits excited by a square-wave gate v_G. (a) Series type, normal transistor connection; (b) shunt type, normal connection; (c) series type, inverted connection; (d) shunt type, inverted connection.

current I_B was somewhat larger than I_C/h_{FE}, whereas in the present case I_B is very much larger than I_C or I_E.

We consider first the circuits in which the transistor is used in the normal manner. Figure 17-25a gives plots of I_C against V_{CE} for various base currents I_B, showing again the bidirectional nature of the transistor switch. In Fig. 17-25b we have drawn, greatly enlarged, a small portion of one of these plots in the neighborhood of the origin. We note that the plot does not pass through the origin. This feature is not apparent in Fig. 17-25a on account of the coarseness of the scale. We can calculate the value V_{CE} corresponding to the

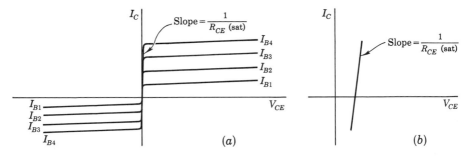

Fig. 17-25 (a) Collector characteristics showing the bidirectional nature of the transistor; (b) enlargement in the neighborhood of the origin of a characteristic for constant I_B.

point of intersection of the plot with the abscissa by starting with the Ebers and Moll equations and proceeding as suggested in Prob. 17-18. The result for the magnitude of the collector-to-emitter voltage is

$$(V_{CE})_N \approx \frac{\eta V_T}{\beta_I} \tag{17-18}$$

where $\eta = 1$ for germanium, $\eta \approx 2$ for silicon, V_T is defined in Eq. (6-2) and equals 26 mV at room temperature, and β_I is the current gain h_{FE} corresponding to inverted operation.

We now consider the transistor in the inverted mode, where not only are the collector and emitter interchanged but also the base current is returned to the collector, as indicated in Fig. 17-24c and d. The emitter-to-collector voltage magnitude is now found to be

$$(V_{EC})_I \approx \frac{\eta V_T}{\beta_N} \tag{17-19}$$

where $\beta_N = h_{FE}$. We observe that V_{CE} for inverted operation is smaller than for the normal mode because $\beta_N > \beta_I$. By way of an example, if $\beta_N = 50$, $\eta = 1$ (Ge), and $V_T = 26$ mV at 25°C, we find $(V_{EC})_I = 0.5$ mV. Typically, we find that $(V_{EC})_I$ lies in the range from about 0.25 to 2.0 mV. In the normal connection we find typically that $(V_{CE})_N$ is in the range 5 to 25 mV.

The need for a low value of V_{CE} will appear in the present discussion. Consider the chopper of Fig. 17-24b and assume for simplicity that the only defect of the transistor switch Q is that $V_{CE} \equiv V' \neq 0$. Let the input signal $v = 0$. Then, when Q is open, $v_o = 0$, as it should be. But when Q closes, v_o will equal V'. Thus, even with $v = 0$, a square wave of amplitude V' will appear at the output. The smallest signal that may be effectively chopped without serious error is therefore limited by V'. Now, of course, we may introduce in series with the transistor a fixed voltage which balances out V', but since V' will vary with time, temperature, etc., this compensation will have limited value. Further, since we may expect that the percentage variability of $(V_{CE})_N$ and $(V_{EC})_I$ may be about the same, it is advantageous, even in the presence of compensation, to use the inverted connection, since a given percentage change will produce a smaller absolute change.

Considering now the inverted mode, we find that there is a second source of voltage between emitter and collector, even in the absence of an emitter current. This voltage results from the passage of the base current I_B through the collector ohmic resistance r_C. This voltage drop, which is in the same direction as the drop V_{EC}, is not negligible because of the relatively large base current. Thus, suppose r_C is only 1 Ω but that $I_B = 500$ μA. Then the drop $r_C I_B = 0.5$ mV. Altogether the voltage V_Q across the transistor at zero emitter current is

$$V_Q = V_{EC} + I_B r_C = \frac{\eta V_T}{\beta_N} + I_B r_C \tag{17-20}$$

This voltage V_Q is called the *offset voltage*. The temperature variation of this

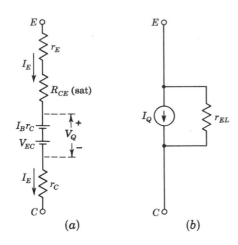

Fig. 17-26 Equivalent circuit of a transistor operating in the inverted mode. (a) In the ON state, with the base current much larger than the emitter current; (b) in the OFF state.

offset voltage V_Q may be positive or negative, depending on the temperature and on I_B. However, it is rarely in excess of 2 or 3 μV/°C and in many situations may be appreciably smaller.

If the collector current is not zero, then, as seen from Fig. 17-25b, V_{CE} is increased by $I_C R_{CE}(\text{sat})$, where $R_{CE}(\text{sat})$ is the collector saturation resistance. This parameter is discussed in Sec. 6-12 and in Prob. 17-19. Typically $R_{CE}(\text{sat})$ is of the order of a few ohms. The preceding discussion may be summarized by the equivalent circuit of Fig. 17-26a, which represents all sources and resistors between the emitter and collector of a transistor chopper switch in the ON condition. The offset voltage V_Q consists of two parts. The resistors r_C and r_E are the spreading or body resistances of the collector and emitter, respectively.

17-15   TRANSISTOR CHOPPER—OFF STATE

We consider now the current which flows in the load circuit when the transistor should be OFF. Since the transistor is in the inverse connection, our interest is in the emitter current when a reverse-biasing voltage is applied at the base, since the emitter current is the load current in Fig. 17-24c and d. From Eqs. (6-38) we have that when both junctions are reverse-biased

$$I_E = \frac{I_{EO}(1 - \alpha_N)}{1 - \alpha_N \alpha_I} \equiv I_Q \tag{17-21}$$

This current I_Q is called the *offset current*. For an n-p-n transistor I_{EO} and I_Q are positive numbers, so that the direction of flow of the offset current is into the emitter. The corresponding offset current for the normal connection is given by I_C in Eqs. (6-38). Since $|I_{CO}| > |I_{EO}|$ and $\alpha_N > \alpha_I$, then the offset current I_{QN} in the normal connection exceeds the corresponding current I_{QI}

in the inverted mode. Since $|I_{QN}| > |I_{QI}|$, we see a second advantage of inverted over normal operation.

The offset current causes an error of the same type as is caused by the offset voltage. Thus, assume that when the transistor is ON it is a perfect short circuit, but that when it is OFF an offset current I_Q flows. This current will flow through R (Fig. 17-24) and result in an output signal even when $v = 0$. The offset current, being a reverse saturation current, has the temperature sensitivity already discussed in Sec. 6-8. Of course, where a minimum offset current is required, silicon rather than germanium transistors should be used.

A second source of emitter current, in addition to the reverse saturation current, is the leakage current across the emitter junction. Accordingly, a more realistic representation of the transistor in the OFF condition is as shown in Fig. 17-26b. Here r_{EL} represents the leakage resistance across the emitter junction.

It may well turn out, in particular cases, that the leakage current exceeds by far the reverse saturation current. Since the leakage current is proportional to the junction voltage, it increases with increasing reverse-biasing voltage applied during the OFF interval. These considerations prompt us to inquire whether it is feasible to operate the chopper transistor in the OFF condition without the application of an externally applied reverse-biasing voltage. Thus, instead of applying a reverse bias we may return the base to the collector through a large resistance. If the resistance is large enough, we shall make only a small error if we compute the transistor OFF current on the assumption that the base is open-circuited.

In Fig. 17-27 we have plotted the emitter-to-collector current I_E under the circumstance that the base connection is open-circuited. Depending on the polarity of V_{EC}, one or the other of the transistor diodes will be back-biased, and the current will thereby be limited. At $V_{EC} = 0$ the current I_E is, of course, zero. At a value of V_{EC} which substantially reverse-biases a junction, the transistor current will be $I_{EO}/(1 - \alpha_I)$ for the inverted connection or $I_{CO}/(1 - \alpha_N)$ for the normal mode [Eq. (6-40)]. These currents are larger

Fig. 17-27 A plot of current I_E (I_C) for a transistor operating with an open base in the inverted (normal) connection as a function of the voltage between emitter and collector. The current would be I_{QI} (I_{QN}) if the transistor were reverse-biased in the inverted (normal) connection.

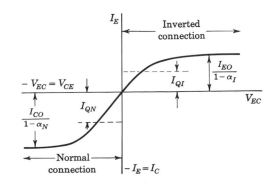

than the corresponding offset currents I_{QI} and I_{QN}, given in Eq. (17-21) and indicated in Fig. 17-27. However, in the neighborhood of $V_{EC} = 0$, the plot of Fig. 17-27 is a straight line whose reciprocal slope represents a resistance. The value of this resistance r_o presented by the transistor between emitter and collector in the open-circuited-base condition is given by the following expression (Prob. 17-20):

$$r_o = \eta V_T \left| \frac{1 - \alpha_I}{I_{EO}} + \frac{1 - \alpha_N}{I_{CO}} \right| \qquad (17\text{-}22)$$

If this resistance r_o is very large in comparison with the load resistance, then the transistor will be as effectively OFF as if it had been reverse-biased. At 25°C, for example, for germanium, with $\eta = 1$, $I_{EO} = 1$ μA, $\alpha_I = 0.7$, and $\alpha_N = 0.9$, $r_o = 10$ K. For silicon, with $\eta = 2$, $I_{CO} \approx I_{EO} = 2.0$ nA, and $\alpha_I \approx \alpha_N \approx 0$, $r_o = 52$ M. Since I_{EO} increases with temperature, r_o decreases with temperature. Thus, whereas operation of germanium transistors without reverse bias may be limited to moderate temperatures and in low-impedance circuits, silicon transistors will be serviceable even at elevated temperatures.

17-16   BALANCED CHOPPERS

The offset errors may be reduced materially through the use of two transistors in a chopper circuit in such a manner that the offset voltages and currents compensate for one another. Such a combination of transistors, shown in Fig. 17-28a, may be substituted directly for the single transistor in any of the circuits of Fig. 17-24. An equivalent circuit for calculating the combined offset voltages is shown in Fig. 17-28b. This equivalent circuit is constructed as the series combination of two circuits, as in Fig. 17-26. Observe that the offset voltages cancel. The voltages across the resistors $R' \equiv r_E + r_C + R_{CE}(\text{sat})$ caused by the signal current do not cancel but rather add. Since R' is invaria-

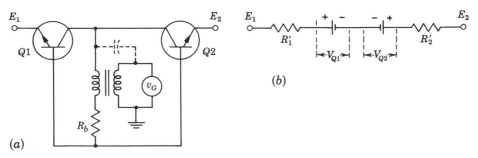

Fig. 17-28 (a) A balanced-transistor chopper pair; (b) equivalent circuit in the ON state where R' (R_1' or R_2') is given by $R' = r_E + r_C + R_{CE}(\text{sat})$.

bly insignificant in comparison with the resistance R in the circuit, its presence is of no consequence.

As a matter of practice, the cancellation of the offset voltage will not be perfect. It is found, however, that with matched transistors, and when care is taken to maintain both at equal temperatures, the combined offset voltage may readily be maintained at less than 50 μV over extended periods.

Note the use of the transformer to introduce the gating signal. Without the transformer, it is found that in none of the circuits of Fig. 17-24 is it possible simultaneously to ground one side of the signal source v and one side of the gating source v_G. The transformer may well have a significant capacitance between windings, as shown by the dashed capacitor in Fig. 17-28a. The gating signal v_G may be 10 V or more in peak-to-peak amplitude. A portion of this gating signal will be transmitted through the capacitor to the entire structure of the two transistors and will therefore appear at the chopper output. This undesirable gating waveform at the output may be comparable to the small (millivolt) amplitude of the signal being transmitted. For this reason it is necessary to take pains to minimize the transformer capacitance. A grounded copper sheet or array of wires arranged so as not to constitute a closed circuit is often interposed as an electrostatic shield between windings.

The effect on the offset current of a balanced-chopper arrangement may be seen in the simplified circuit of Fig. 17-29. Here, a circuit has been completed through the two chopper transistors, which are reverse-biased. Since the currents are very small, we have omitted all resistors as not being essential to the discussion. The back-biasing source appears directly across the collector junction of $Q1$ and $Q2$, and the corresponding currents I_1 and I_2 are shown. We may now see that no current may flow through the emitter junctions, provided that I_1 and I_2 are equal. It is apparent from symmetry that the voltages across the emitter junctions are equal and that if any current were to flow across these junctions it would have to be in the same direction across each junction. However, if current were to flow in the outer circuit in Fig. 17-29, it would cross the emitter junctions in opposite directions. Thus the current in the outer circuit must be zero. Of course, again, with physical transistors the compensation will not be perfect. Since the reverse saturation current of a particular transistor type is a rather variable parameter and since such a current is very temperature-sensitive, the improvement in offset current due to balancing is ordinarily not as good as the improvement in offset voltage. However, with silicon transistors particularly, the offset-current error is not important.

The circuit of Fig. 17-30 is intended to take advantage of the fact that

Fig. 17-29 Simplified circuit for determining the effect of a balanced transistor pair on offset current.

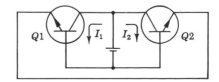

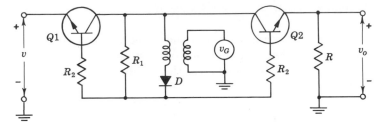

Fig. 17-30 A circuit which does not reverse-bias the transistors during the OFF period but instead effectively reduces the base current to zero.

a transistor will offer a high resistance to the flow of current with the base returned to the collector through a resistor and without reverse bias. This method of operation is especially useful in transistors with large leakage currents. In the circuit of Fig. 17-30, diode D conducts when the gating signal is in the direction to turn the transistors ON. When the gating signal is in the direction to reverse-bias the transistors, diode D opens and the bases are returned to the collectors through R_1 and R_2. Another chopper circuit is presented in Prob. 17-22.

17-17   THE DOUBLE–EMITTER TRANSISTOR[4]

We note in Fig. 17-28a that the two collectors are joined, as are also the two bases. There are commercially available for chopper service *double-emitter* transistors, which consist of a single base, a single collector, and two symmetrically placed emitters. Such a double-emitter chopper is shown in Fig. 17-31. Since the individual "transistors" are fabricated by integrated techniques at the same time and under identical conditions a better match is possible than with two individual transistors. The emitter junctions are

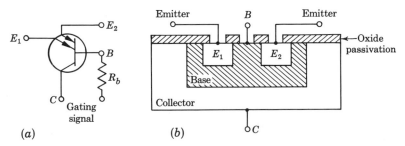

Fig. 17-31 Double-emitter transistor. (a) Circuit representation; (b) physical construction.

located within a few thousandths of an inch of one another. This proximity gives extremely good thermal tracking. The single base-to-collector junction avoids the external collector-to-collector connections, temperature differences between collector junctions, and voltage drops produced by currents flowing in contacts. We may also note in Fig. 17-31b that the signal current proceeds from emitter to emitter while the gating current in the ON condition flows between base and collector. These currents, then, are generally perpendicular to one another, and we avoid thereby that part of the offset voltage $I_B r_C$ which appears in single transistors. Offset voltages less than 20 μV are possible with such a double-emitter integrated chopper.

17-18   CHOPPER TRANSIENTS

In Fig. 17-32a a single-transistor chopper circuit is indicated. The signal has been set equal to zero, and the base-to-emitter capacitance C_{ib} has been indicated. The reason for our interest in this capacitance is that it is principally responsible for spikes which appear at the chopper output each time the gating signal makes a transition from one level to another. When the gating signal goes positive, the signal is transmitted through C_{ib} and a positive spike appears at the output v_o. The amplitude of the output spike is determined by C_{ib} and the capacitance C_s which shunts the output. If, say, the input gating waveform has an amplitude of 1 V, $C_{ib} = 5$ pF, and $C_s = 50$ pF, then the spike amplitude is $\frac{5}{55}$ V = 91 mV. Since simultaneously the transistor has been turned ON, a low resistance is presented from emitter to collector and the spike decays very rapidly. When the transistor is turned OFF a negative spike appears, but now, since the transistor is OFF, the spike decays with a relatively long time constant $\sim R(C_{ib} + C_s)$. For $R = 1$ K the decay time constant is

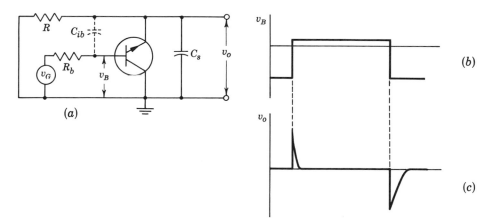

Fig. 17-32   (a) A chopper circuit showing the emitter-to-base capacitance C_{ib}; (b) gating waveform; (c) output spikes.

55 nsec.   We have assumed here that the intrinsic transistor has a switching time which is small in comparison with the time constants of the spikes.   If such is not the case, then the analysis of the switching transients[5] must take account of the base-charge storage, which then becomes principally responsible for the transients (Chap. 20).   In any event it is apparent that these switching transients limit the maximum speed with which a transistor sampling circuit may operate.   It is presently possible, with choppers using individual transistors, to operate up to about 100 kHz and with integrated double-emitter choppers to about 1 MHz.

17-19   THE FIELD–EFFECT TRANSISTOR[6]

The structure of an n-channel field-effect transistor (FET) is shown in Fig. 17-33.   Ohmic contacts are made to the two ends of a bar of n-type silicon. (If p-type silicon is used, the device is referred to as a p-channel FET.)   Current is caused to flow across the length of the bar because of the voltage source connected between the ends.   This current is due to majority carriers which in the present case are electrons.   The terminal through which the majority carriers enter the bar is called the *source S*, and the terminal through which they leave is called the *drain D*.

On both sides of the n-type bar of Fig. 17-33, regions of p-type impurities have been formed by alloying, by diffusion, or by any other procedure available for creating p-n junctions.   These regions of p-type impurity are called the *gate G*.   Between the gate and source a voltage is applied in the direction to reverse-bias the p-n junction.

To understand the operation of the FET it is necessary to recall that on

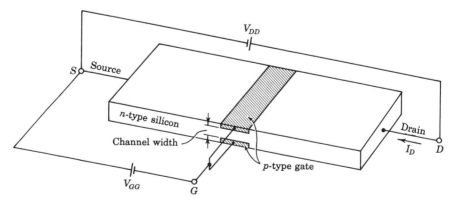

Fig. 17-33 The basic structure of an n-channel field-effect transistor.   The normal directions of drain-to-source and gate-to-source voltages are shown.   In a p-channel transistor the voltages would be reversed.

the two sides of the transition region of a reverse-biased p-n junction there are space-charge regions. The current carriers have diffused across the junction, leaving only uncovered positive ions on the n side and negative ions on the p side. The electric lines of field intensity which now originate on the positive ions and terminate on the negative ions are precisely the source of the voltage drop across the junction. As the reverse bias across the junction increases, so also does the thickness of the region of immobile uncovered charges. The conductivity of this region is nominally zero because of the unavailability of current carriers. Hence we see that the effective width of the channel in Fig. 17-33 will become progressively decreased with increasing reverse bias. Accordingly, for a fixed drain-to-source voltage, the drain current will be a function of the reverse-biasing voltage across the gate junction. The term *field effect* is used to describe this device because the mechanism of current control is the *effect* of the extension, with increasing reverse bias, of the *field* associated with the region of uncovered charges.

FET Characteristics The circuit symbol and polarity conventions for an FET are indicated in Fig. 17-34. The common-source drain characteristics for a typical p-channel FET shown in Fig. 17-35 give I_D against V_{DS} with V_{GS} as a parameter. To see qualitatively why the characteristics have the form shown consider, say, the case for which $V_{GS} = 0$. For $I_D = 0$ the channel between the gate junctions is entirely open. In response to a small applied voltage V_{DS} the p-type silicon bar acts as a simple semiconductor resistor and the current I_D increases linearly with V_{DS}. With increasing current the ohmic voltage drop between the source and the channel region reverse-biases the junction and the conducting portion of the channel begins to constrict. Because of the ohmic drop along the length of the channel itself, the constriction is not uniform but is more pronounced at distances farther from the source, as indicated in Fig. 17-36. Eventually a voltage V_{DS} is reached at which the channel is "pinched off." This is the voltage, not too sharply defined in Fig. 17-35, where the current I_D begins to level off and approach a constant value. It is, of course, in principle not possible for the channel to close completely and thereby reduce the current I_D to zero. For if such indeed could be the case, the ohmic drop required to provide the necessary

Fig. 17-34 Circuit symbol for a p-channel FET. (For an n-channel FET the arrow at the gate junction points in the opposite direction.) For a p-channel FET, I_D and V_{DS} are negative and V_{GS} is positive. (For an n-channel FET, I_D and V_{DS} are positive and V_{GS} is negative.)

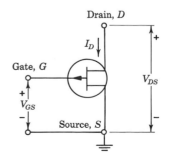

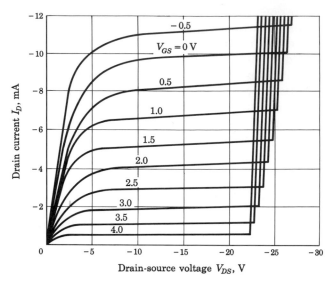

Fig. 17-35 Common-source drain characteristics of the p-channel type 2N2499 field-effect transistor at 25°C. (Courtesy Texas Instruments, Inc.)

back bias would itself be lacking. Instead the overall characteristic curve has an ohmic region where I_D is proportional to V_{DS} followed by a constant-current region in which I_D responds very slightly to V_{DS}. The two regions are separated by the reverse bias corresponding to "pinch-off."

If now a gate voltage V_{GS} is applied for smaller values of $|V_{DS}|$ in the direction to provide additional reverse bias, pinch-off will occur and the maximum drain current will be smaller. This feature is brought out in Fig. 17-35. Note that a plot is given even for $V_{GS} = -0.5$ V, which is in the direction of forward bias. We note from Table 6-1 (page 219) that actually the gate current will be very small because at this gate voltage the junction is barely at the cutin voltage V_γ. The similarity between the FET characteristics and those of a vacuum-tube pentode need hardly be belabored.

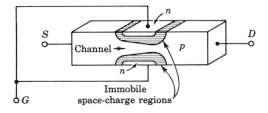

Fig. 17-36 Schematic representation of the space-charge region (shaded) at the gate p-n junction of an FET. The larger the magnitude of the reverse bias, the wider the region of uncovered immobile charge and hence the more constricted the channel.

We observe also in Fig. 17-35 that when the junction avalanche break-down voltage is exceeded the current I_D rises precipitously. Breakdown occurs at a fixed junction voltage, and when the gate voltage supplies part of this voltage, breakdown occurs at a lower value of $|V_{DS}|$. For example, we see in Fig. 17-35 that at breakdown $V_{DS} = -26$ V for $V_{GS} = 0$, whereas $V_{DS} = -22$ V for $V_{GS} = 4$ V.

A number of distinctions between FETs and conventional transistors are worthy of being singled out. First, as already noted, current is carried by majority rather than minority carriers and the carriers do not cross a junction in going from one end of the transistor to the other. Since only one type of carrier is involved, the FET is referred to as a *unipolar* device in distinction to the conventional (bipolar) transistor in which current is carried both by holes and by electrons. Finally let us identify the gate and source terminals of the FET as the input terminals, just as most frequently we so identify the base and emitter terminals of a transistor or the grid and cathode terminals of a vacuum tube. Then the FET has a very high input impedance which is many orders of magnitude larger than that of the transistor and easily comparable to that of the vacuum tube operating within its grid base. This feature results from the fact that the input gate current is the leakage current across a reverse-biased silicon semiconductor junction.

At the present time (1965) the field-effect transistor does not compete favorably with the transistor as a switching element in lumped-circuit applications principally because of the greater speed with which a transistor operates. However, in slower switching applications, in chopper circuits, and in certain types of integrated circuits the FET has appreciable merit.

17-20   THE FIELD–EFFECT TRANSISTOR AS A CHOPPER[7]

We shall now discuss the use of the FET in chopper circuit applications. When a large reverse-biasing voltage is applied to the FET gate, the device will be pinched off and only a small leakage current will flow between source and drain. The leakage current may be of the order of 1 nA, and the "pinch-off voltage" required to reduce the current to this level may range from several volts to several tens of volts. When the externally impressed voltage across the gate junction is in the neighborhood of zero there appears between source and drain terminals an equivalent ohmic resistor. In an FET intended for chopper service this resistance may be as low as several hundred ohms.

A typical FET chopper circuit in which the FET appears as a shunt element is shown in Fig. 17-37 and should be compared with Fig. 17-24b or d. (We shall neglect for the moment the dashed part of the circuit involving the diode D.) Of course, it is equally feasible to use the FET as a series element analogous to the chopper transistor in Fig. 17-24a or c. With the FET there is no advantage to using the device in inverted fashion, as is the case with the transistor. A gating signal is applied between gate and source to turn the

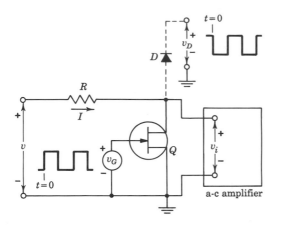

Fig. 17-37 An FET chopper circuit, including (dashed) provision for leakage-current compensation.

transistor ON and OFF. The slowly varying input signal v appears at the chopper output as a chopped signal v_i, as in Fig. 17-23.

Like the transistor chopper, so also the FET chopper suffers from the leakage current which flows when the device is OFF. The leakage currents of the FET and the silicon transistor are comparable in magnitude. The effect of leakage current in transistors is minimized by the balanced arrangement described in Sec. 17-16. A method for compensating for the leakage current of an FET is shown by the dashed portion of Fig. 17-37. A compensating square waveform is applied through the diode D. The levels of this waveform are selected to be such that when the FET is ON the voltage across D is small enough so that negligible diode current flows. On the other hand, when the FET is OFF the diode D is reverse-biased and the reverse-biased diode current supplies the leakage current of the FET. A rather exact compensation may be achieved either by a careful selection of the silicon diode or through an adjustment of the square-wave amplitude or through both. In any event the leakage current need not be supplied through R and a leakage-current offset error is thereby avoided. Note in Fig. 17-37 the reversed relative polarity of the waveform v_G and v_D. Observe also the polarity of diode D. If the FET had been a p-channel FET rather than an n-channel FET as indicated, diode D would have to be reversed. The compensation scheme of Fig. 17-37 has the merit that it will hold over a large temperature range. For, since both FET and diode are of silicon, the leakage current of both increases nominally at the same rate with temperature.

In the matter of leakage current the FET has no advantage over a silicon transistor except for the relatively minor consideration that in the former a diode may be used for leakage compensation whereas in the latter a second transistor is required. As a matter of fact, in silicon choppers generally leakage currents are not a major problem since they give rise to errors of the order of magnitude of microvolts. The really important advantage of the FET in comparison with the transistor is that the FET *has no offset voltage*. When the

FET is ON it is a simple ohmic resistance and not a resistance in series with an offset voltage. In transistor choppers, offset voltages give rise to the largest error by far. In a typical case with a single transistor an offset error may be of the order of 1,000 μV. Even in a balanced compensated transistor chopper, offset drifts as large as 50 μV may occur, particularly during the first several hours of warm-up. To attain the ultimate in stability with transistor choppers it is normally necessary to enclose the chopper transistors in an oven in which the temperature is maintained to within a few degrees. On the other hand, the warm-up drift of an FET chopper rarely exceeds 3 or 4 μV and is completed in 10 to 20 min. The long-term drift of a transistor chopper is rather unpredictable, whereas the long-term drift of an FET chopper may be as low as 0.5 μV/month.

When the ultimate in performance of an FET chopper is not required we may simplify the leakage compensation scheme of Fig. 17-37. Thus we may return the diode not to a square-wave source but to a fixed voltage so that the diode is always reverse-biased. The voltage would be selected for proper compensation when the FET is OFF. Then when the FET is ON there will appear across the FET a voltage due to the diode leakage current. This "offset voltage," however, may be quite small. For, assuming a leakage current of 10 nA and an ON resistance of 500 Ω, the "offset voltage" will be only 5 μV.

The FET chopper, like the transistor chopper, suffers from the common difficulty that because of the transition capacitance between gate and channel a spike appears at each transition of the gating waveform. In this respect the FET may even be at some disadvantage because its capacitances are larger. Therefore FET choppers, like transistor choppers, are limited to low chopping frequencies, say of the order of some hundreds of cycles per second.

17-21   TRANSISTOR SAMPLING CIRCUITS

We have considered the bipolar transistor as a chopper in connection with very small signals. Clearly the very same circuits may be used as gates to sample larger signals, say for the purpose of multiplexing. When larger signals are applied, certain voltage limits must be observed. When a single-transistor circuit is used, the signal voltage V_s must be limited to be not more than the gating voltage used to back-bias the transistor. Otherwise, for one or the other polarity of input signal it will not be possible to maintain the transistor in the OFF state.

If, however, a pair of transistors is used, the maximum allowable input signal V_s is limited only by the breakdown voltage of the transistors. For even if the signal should become large enough to cause one transistor to go ON, the signal will drive the other transistor further toward reverse bias. Similar precautions must be taken with the FET to limit the applied voltage to a value smaller than the breakdown voltage.

At best, transistor sampling circuits, particularly in the inverted connection, are not especially fast (because of the considerations noted in Sec. 17-18), and when speed of operation is at a premium, diode gates are invariably better.

17-22   SAMPLING SCOPE

As an example of an important application of a sampling gate we consider now the basic principle of the *sampling scope*.[8]   In this scope the display consists of a sequence of samples of the input waveform, each sample taken at a time progressively delayed with respect to some reference point in the waveform. We shall explain the principle with the aid of the waveforms of Fig. 17-38 and the block diagram of Fig. 17-39.

We assume that the waveform to be displayed is the pulse in the train of pulses in Fig. 17-38b.   We assume further that we have available, as in Fig. 17-38a, a train of triggers whose time of occurrence precedes somewhat the pulses in Fig. 17-38b.   Such a train of triggering signals may be produced in either of the following ways.   It may be that the pulses in Fig. 17-38b are the signals generated by some monostable circuit which is excited by the triggers in Fig. 17-38a.   If such is the case, and if the trigger source is under our control, then we may apply the trigger signal to the CRO sweep circuit directly, but use a delay line to delay the triggers before application to the monostable circuit.   If there is no trigger source (the pulse generator is astable) or if the trigger source is not under our control, we may generate triggers by applying the signal itself to a comparator, the output of which provides a trigger each time the pulse waveform attains some reference level.   To attain the relative timing indicated in Fig. 17-38a and b we may now pass the signal (rather than the triggering waveform) through a delay line.

In any event, the triggers being available, we may use them to trigger a sweep signal and also a stairstep signal.   The latter is a waveform which remains at a constant level between triggers and which, at each trigger, jumps

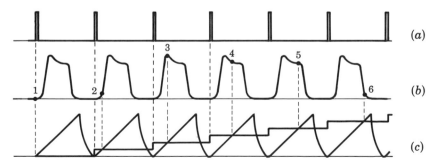

Fig. 17-38   The sampling-scope principle.   (a) The triggering signal;
(b) the signal to be observed; (c) the ramp and stairstep signals.

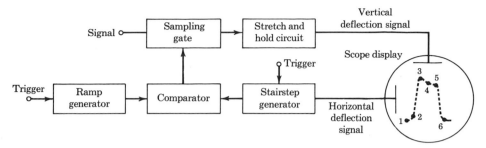

Fig. 17-39 Block diagram of the essential elements required for a sampling-scope display.

from one level to another. Such a stairstep signal (with jumps of constant magnitude) is shown in Fig. 17-38c together with the sweep signal. Circuits for generating stairstep signals are described in Sec. 18-14.

As shown in Fig. 17-39, the stairstep and ramp signals are applied to a comparator. The stairstep serves as the reference voltage, and, in each cycle, whenever the ramp attains the stairstep level, the comparator furnishes a pulse output which is used as the control signal of the sampling gate. At each such control signal, the gate furnishes at its output a sample of the signal, the sample having a duration equal to the width of the control pulse. The control signal has so short a duration that during its interval no sensible change takes place in the input signal. Thus the gate output, at each control signal, is a voltage equal to the signal voltage at the time of sampling. The points at which samples are taken have been marked by dots 1, 2, . . . on the waveform of Fig. 17-38b. We observe that the samples are taken at a time which is progressively delayed by equal increments.

The sample consists, then, of a pulse whose duration is equal to the duration of the sampling-gate control signal and whose amplitude is determined by the magnitude of the input signal at the sampling time. We need to "hold" this voltage for a time comparable to but smaller than the interval between samples. This "holding" operation may be performed by using the gate output to charge a capacitor through a diode to the peak value of the sample in such a manner that when the sample is completed the capacitor holds its charge. As already noted, however, the sample is so short in duration that it turns out not to be feasible to charge a capacitor in this small interval. Therefore, before the sample is applied to the "hold" diode-capacitor combination, it is first passed through an amplifier stage whose output time constant is rather long. The sample pulse is thereby widened, i.e., "stretched," and now will have a much broader peak. The stretching and holding operations are performed in Fig. 17-39 by the block so labeled before the sample is applied as a vertical-deflection signal to the scope. Shortly before each new sample is to be taken, the hold capacitor must be discharged. Provision for this discharge operation is not indicated in the simplified block diagram of Fig. 17-39.

The stairstep generator furnishes the horizontal-deflection signal for the scope. Thus the CRT spot moves horizontally across the screen in jumps, and at each new position the spot is deflected vertically by an amount proportional to the sample height. For the sake of clarity of presentation the CRT beam is normally blanked and is unblanked only at the time of the display of the sample. Consequently the presentation of the CRT screen consists of a series of dots which trace out the form of the original signal, as shown in Fig. 17-39. The dashed signal waveform itself, which is shown in the CRT display of Fig. 17-39, does not actually appear. If the dots (samples) are spaced sufficiently closely, the presentation will indicate quite precisely the original signal waveform. It is to be noted that the real time required for the beam to cross the CRT is very much longer than the effective time as judged from the time interval of the displayed waveform. For example, suppose that in Fig. 17-38b the pulse width is 10 nsec, the pulse repetition frequency is 50 kHz, and one pulse is to be displayed with 100 samples. Then the effective sweep time is 10 nsec, but the time required for the sweep to cross the screen will be 100/50,000 sec = 2 msec.

The sampling principle finds application in a scope used to display very fast *periodic* waveforms, that is, waveforms with rise times in the nanosecond range. Its usefulness lies in the fact that it makes unnecessary the use of high-gain wideband amplifiers. By way of example, presently available sampling scopes have equivalent rise times of about 0.35 nsec, corresponding [Eq. (2-33)] to an amplifier bandwidth of 1,000 MHz. On the other hand, of course, the circuits represented by the blocks in Fig. 17-39 must operate with high speed. But the problems in constructing such high-speed circuits are appreciably simpler than those of constructing a linear high-gain 1,000-MHz amplifier. The fast ramp generator is simply a small capacitance charging through a small resistance. Linearity is attained by using a supply voltage very much larger than the ramp amplitude, which need only be several volts. The very short control pulse is generated either in an avalanche transistor comparator (Sec. 13-16) circuit or through the use of a snap-off diode whose fast step is shaped into a pulse through delay-line differentiation (Sec. 20-8). We may note, in passing, that the operation of the sampling scope is not unlike the operation of a stroboscope. For this reason the gate-control pulse is frequently referred to as the "strobe" pulse. The diode threshold gate of Fig. 17-2 (except with C_1 replaced by the signal-generator source impedance) and the four-diode gate of Fig. 17-20 have both been used effectively for the sampling gate.

REFERENCES

1. Millman, J., and T. H. Puckett: Accurate Linear Bidirectional Diode Gates, *Proc. IRE*, vol. 43, pp. 29–37, January, 1955.

2. Chaplin, G. B. B.: sec. 13 in L. P. Hunter (ed.), "Handbook of Semiconductor Electronics," 2d ed., McGraw-Hill Book Company, New York, 1962.
Bright, R. L., and A. P. Kruper: Transistor Chopper for Stable D.C. Amplifiers, *Electronics*, vol. 28, no. 4, pp. 135–137, April, 1955.

3. Holcomb, S. W., F. Opp, and J. A. Walston: Choppers: Their Characteristics and Applications, *Appl. Rept.*, Texas Instruments, Inc., July 15, 1963.
Simpson, J. H.: Temperature Effects in Low-level Transistor Choppers, *Solid State Design*, vol. 5, pp. 22–28, March, 1964.

4. Boulter, B. A.: A New Active Device Suitable for Use in Digital Circuits, *Electron. Eng.*, vol. 35, no. 420, pp. 86–91, February, 1963.
Mitchell, B., and B. Brian: The Inch, Discussion and Applications, *Gen. Eng. Memo* 7, National Semiconductor Corp., June 1, 1962.

5. Ekiss, J. A., and J. W. Halligan: The Application, Characterization and Performance of the SPAT as a Chopper, *Philco Appl. Lab. Rept.* 768, May 7, 1962.

6. Dacey, G. C., and I. M. Ross: The Field Effect Transistor, *Bell System Tech J.*, vol. 34, pp. 1149–1189, November, 1955.

7. Barton, K.: The Field-effect Transistor Used as a Low-level Chopper, *Electron. Eng.*, vol. 37, no. 444, pp. 80–83, February, 1965.

8. Sugarman, R.: Sampling Oscilloscope for Statistically Varying Pulses, *Rev. Sci. Instr.*, vol. 28, no. 11, pp. 933–938, November, 1957.
Bushor, W. E.: Sample Method Displays Millimicrosecond Pulses, *Electronics*, vol. 32, no. 31, pp. 69–71, July 31, 1959.

18 / COUNTING AND TIMING

In this chapter we discuss circuits which are used for the purpose of counting events. Applications of the counting process are also given. In particular, the use of counters in conjunction with a source of precisely determined frequency to measure the time interval which elapses between two events is described.

18-1   A CASCADE OF BINARIES AS A COUNTER

Let us consider the cascade of binaries shown in Fig. 18-1. In Fig. 18-1a the binaries, or flip-flops, are represented by their logic symbols. The output Y of one binary in the chain is coupled to the trigger input T of the next binary. Some details of the circuitry employed are shown in Fig. 18-1b. For simplicity, in this latter circuit, the cross coupling in each binary from the collector to the base has been omitted. The outputs Y and $\bar{Y}$ are taken from the collectors (or plates if tubes are used). The trigger input terminal is a point at which the binary may be triggered symmetrically, as described in Sec. 10-9, so that each successive triggering signal will reverse the state of the binary. For binaries constructed of n-p-n transistors, as in Fig. 18-1b, a negative-going voltage change or negative pulse is required for triggering.

Let us arbitrarily agree that a binary is to be considered in state 0 when the n-p-n transistor (tube) conducts whose collector (plate) is connected to the Y output. The binary is in state 1 when this device is not conducting. When the binary is in state 1 the voltage at Y will be high, and when the binary is in state 0 the voltage at Y will be low. Accordingly, our arbitrary decision in the present case about the definition of the binary state is equivalent to having decided to use *positive* logic to describe our system.

668

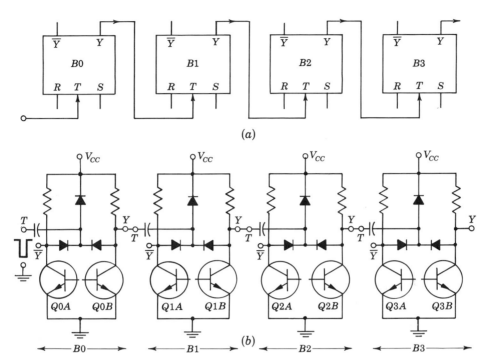

(a)

(b)

Fig. 18-1   (a) A chain of flip-flop circuits, each represented by its logic symbol; (b) illustrating how the circuit in (a) would be realized with transistor flip-flops.   (The symbol B for *binary* is used instead of *FF* for *flip-flop* throughout this chapter.)

Let us start from a reference configuration in which all binaries are in the state 0, so that all the transistors $Q0B, \ldots , Q3B$ are conducting.   The waveforms which appear at the outputs Y of the individual flip-flops as the result of 16 successive input negative triggering pulses are shown in Fig. 18-2. The first external pulse applied to $B0$ causes this binary to make a transition

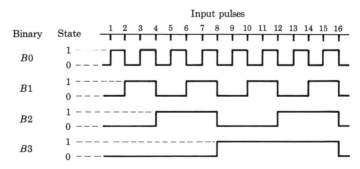

Fig. 18-2   Waveform chart for the binary counting system of Fig. 18-1.

from state 0 to state 1.   Referring to Figs. 18-1 and 18-2, we see that as a result of this transition a positive voltage step is applied at the T input of $B1$. Because of the arrangement of diodes in $B1$, this positive step will not induce a transition in $B1$.   The overall result is that $B0$ has changed its state to 1, and all other flip-flops remain in state 0, as indicated in Fig. 18-2.

The second externally applied pulse causes flip-flop $B0$ to return from state 1 to state 0.   Flip-flop $B1$ now receives a *negative* step voltage, to which the flip-flop is sensitive, and responds by making a transition from state 0 to state 1.   The flip-flop $B2$ does not respond to the transition in flip-flop $B1$ because flip-flop $B2$ receives a positive step.   The overall result of the application of two input pulses is that flip-flop $B1$ is in state 1 while all other flip-flops are in state 0.   We may similarly verify that the remainder of the *waveform chart* of Fig. 18-2 is correct by applying the following principles:

1. Flip-flop $B0$ makes a transition at each externally applied pulse.
2. Each of the other flip-flops makes a transition when and only when the preceding flip-flop makes a transition from state 1 to state 0.

If the chain were constructed of binaries using *p-n-p* transistors, then the voltage level at Y corresponding to the 1 state would be lower than the voltage level corresponding to the 0 state.   It would then be appropriate to describe the operation in terms of negative logic.   The rules given directly above would continue to apply without modification.   However, the waveforms in Fig. 18-2 would have to be inverted and the input waveform would have to be a positive pulse train.

Table 18-1 lists the states of all the binaries of the chain as a function of the number of externally applied pulses.   This table may be verified directly by comparison with the waveform chart of Fig. 18-2.   Note that in Table 18-1 the binaries have been ordered in the reverse direction from their order in Fig. 18-1.   We observe that the ordered array of states 0 and 1 in any row in Table 18-1 is precisely the binary representation of the number of input pulses as given in Table 9-2.   Thus the chain of flip-flops *counts* in the binary system. This feature indicates the appropriateness of the designation *binary*, which we use interchangeably with the term *flip-flop*, to designate a bistable circuit.

A chain of n binaries will count up to the number 2^n before it resets itself into its original state.   Such a chain is referred to as a counter *modulo* 2^n. It is clear that the counting range may be extended indefinitely by adding more binaries.   A single flip-flop is a counter *modulo* 2.

To read the count of a binary chain, it is necessary to determine the state of each individual binary in the chain.   A very rudimentary indicator may be used for this purpose since we need only to know which transistor (or tube) is conducting.   When tubes are employed and the voltages encountered in the circuit are in the range of some hundreds of volts, a small neon bulb ($\frac{1}{25}$ W) in series with a resistor may be used as an indicator.   This indicator is connected between the output plate Y and ground.   When the plate voltage is

TABLE 18-1   States of the binaries

No. of input pulses	State of binary			
	$B3$	$B2$	$B1$	$B0$
0	0	0	0	0
1	0	0	0	1
2	0	0	1	0
3	0	0	1	1
4	0	1	0	0
5	0	1	0	1
6	0	1	1	0
7	0	1	1	1
8	1	0	0	0
9	1	0	0	1
10	1	0	1	0
11	1	0	1	1
12	1	1	0	0
13	1	1	0	1
14	1	1	1	0
15	1	1	1	1
16	0	0	0	0
17	0	0	0	1

low, the voltage across the neon tube is not enough to make it glow, but the neon tube will light up when the plate voltage is high. The neon bulbs connected to binaries $B0$, $B1$, $B2$, and $B3$ are assigned values 1, 2, 4, and 8, respectively. To determine the count it is only necessary to add the numbers assigned to those neon bulbs that are lit. An alternative arrangement, suitable for transistor circuits, uses high-voltage transistors to drive neon bulbs.

18-2   A REVERSIBLE BINARY COUNTER[1]

A counter which can be made to count in either the forward or reverse direction is called a *reversible counter* or a *forward-backward* counter. Forward counting is accomplished, as we have seen, when the trigger input of a succeeding binary is coupled to the Y output of a preceding binary. The count will proceed in the reverse direction if the coupling is made instead to the $\bar{Y}$ output, as we shall now verify.

If a binary makes a transition from state 0 to 1, then the output $\bar{Y}$ will make a transition from state 1 to 0. This negative-going transition in $\bar{Y}$ will induce a change in state in the succeeding binary.   Hence, for the reversing connection the following rules apply:

1. Binary $B0$ makes a transition at each externally applied pulse.

2. Each of the other binaries makes a transition when and only when the preceding binary goes from state 0 to state 1.

If these rules are applied to any of the numbers in Table 18-1, the next smaller number in the table results.   For example, consider the number 12, which is 1100 in binary form.   At the next pulse, the rightmost 0 (corresponding to $B0$) becomes 1.   This change of state from 0 to 1 causes $B1$ to change state from 0 to 1, which in turn causes $B2$ to change state from 1 to 0.   This last transition is in the direction not to affect the following binary, and hence $B3$ remains in state 1.   The net result is that the counter reads 1011, which represents the number eleven.   Since we started with twelve and ended with eleven, a reverse count has taken place.

18-3   APPLICATION OF COUNTERS

We describe here briefly some fundamental applications of counters.

Direct Counting[2]   Direct counting finds application in many industrial processes.   Counters will operate with reliability where human counters fail because of fatigue or limitations of speed.   It is required, of course, that the event which is to be counted first be converted into an electrical signal, but this requirement usually imposes no important limitation.   For example, objects may be counted by passing them single file on a conveyor belt between a photoelectric cell and a light source.

A *preset* feature may be incorporated into a counter to allow control of industrial processes.   A *preset counter* is one which has been modified (at the expense of additional active devices and components) so that it will deliver an output pulse when the count reaches a predetermined number.   Such a counter may be used, for example, to count the number of pills dropped into a bottle.   When the preset count is attained, the output pulse is used to divert the pills to the next container and at the same time to reset the counter for counting the next batch.

Scaling   If we should differentiate each of the waveforms of Fig. 18-2, a positive pulse would appear at each transition from 0 to 1 and a negative pulse at each transition from 1 to 0.   If now we count only the negative pulses (the positive pulses may be eliminated, say, by using a diode), then it appears that each binary divides by 2 the number of negative pulses applied to it.   The four binaries together accomplish a division by a factor $2^4 = 16$.   A single negative pulse will appear at the output for each 16 pulses applied at the input. A chain of n binaries used for this purpose of dividing or scaling down the number of pulses is referred to as a *scaler*. Thus a chain of four binaries constitutes a *scale-of*-16 scaler, etc.

One of the earlier applications for scalers arose in connection with research into the properties of the nucleus of an atom. In order to determine the radioactivity of a source, it is necessary to *count individually* the particles emitted in a given time interval. The emission of a radioactive particle may be converted into an electrical pulse through the use of a device known as a *geiger tube*. With a weak radioactive source, a mechanical register may, in many instances, act with sufficient speed, but more usually an electrical counter will have to be interposed between source and mechanical counter. For example, a scale-of-64 scaler followed by a mechanical register will be able to respond to geiger pulses which occur at a rate 64 times greater than the maximum rate at which the mechanical register will respond. The net count in such a case will be 64 times the reading of the mechanical register plus the count left in the scale-of-64 counter.

Measurement of Frequency The basic principle by which counters are used for the precise determination of frequency is illustrated in Fig. 18-3. The input signal whose frequency is to be measured is converted into pulses and applied through a gate to an electronic counter. To determine the frequency, it is now only required to keep the gate open for transmission for a known time interval. If, say, the gating time is 1 sec, then the counter will yield the frequency directly in cycles per second. The *clock* for timing the gate interval is an accurate crystal oscillator whose frequency is, say, 1 MHz. The crystal oscillator drives a chain of divider circuits (Chap. 19) which divides the crystal frequency by a factor of 1 million. The divider output consists of a 1-Hz signal whose period is as accurately maintained as the crystal frequency. This divider output signal controls the gating time. The system is susceptible to only slight errors. One source of error results from the fact that a variation of ± 1 count may be obtained, depending on the instant when the first and last pulses occur in relation to the sampling time. A second source of error arises from *jitter* in the divider, which may produce a small uncertainty in the sampling period. Beyond these, of course, the accuracy depends on the accuracy of the crystal oscillator.

Measurement of Time The time interval between two pulses may also be measured with the circuit of Fig. 18-3. The first input pulse is used to open

Fig. 18-3 A counter used in a frequency meter.

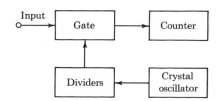

the gate for transmission and the second pulse to close the gate.   The crystal-oscillator signal (or some lower frequency from the divider chain) is converted into pulses, and these are passed through the gate into the counter.   The number of counts recorded is proportional to the length of time the gate is open and hence gives the desired time interval.

Measurement of Speed   A speed determination may be converted into a time measurement.   For example, if two photocell–light-source combinations are set a fixed distance apart, the average speed of an object passing between these points is proportional to the time interval between the generated pulses. Projectile velocities have been measured in this manner.

Digital Computer   In a digital computer a problem is solved by subjecting data to a sequence of operations in accordance with the program of instructions introduced into the computer.   Counters may be used to count the operations as they are performed and to call forth the next operation from the memory when the preceding one has been completed.

Waveform Generation   The waveforms which occur at the collectors or bases of binary counters may be combined either directly or in connection with other circuits (Prob. 18-1) to generate complex pulse-type waveforms.

Conversion between Analog and Digital Information[3]   In Fig. 18-4 the solid blocks indicate a method, using a counter, for generating a voltage proportional to the number of pulses in a pulse train.   Such a system is one example of a *digital-to-analog converter*.   By adding the dashed blocks, the inverse operation of *analog-to-digital conversion* may be performed.   In other words, the modified system establishes in a counter a registration proportional to a voltage.

Consider initially only that part of the block diagram which is drawn in solid lines.   Suppose that, say, a train of 23 clock pulses is applied to the input of the counting chain.   In binary notation this number is written 10111. Accordingly, $B4$ is in state 1, $B3$ in state 0, $B2$ in state 1, etc.   Assume that when a binary is in state 1 its Y output is higher by ΔV than when it is in state 0.   Since at the input to the operational amplifier we see a virtual ground (Sec. 1-8), then the change in amplifier input current ΔI over that corresponding to the situation when all binaries are in state 0 is

$$\Delta I = \frac{\Delta V}{R/16} + \frac{\Delta V}{R/4} + \frac{\Delta V}{R/2} + \frac{\Delta V}{R/1}$$

$$= (16 + 4 + 2 + 1)\frac{\Delta V}{R} = 23\frac{\Delta V}{R} \tag{18-1}$$

Hence we see that the input current change to the amplifier and its output voltage v_o are proportional to the count registered in the counter.   As the

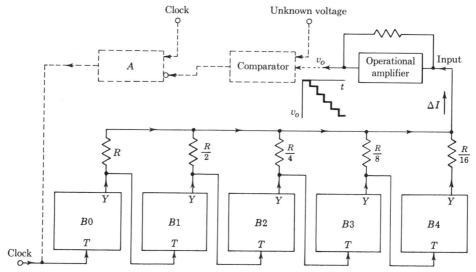

Fig. 18-4 A digital-to-analog converter—a voltage is generated which is proportional to the number of pulses in a train (solid lines). An analog-to-digital converter—the addition of the dashed components provides a method of registering a count proportional to an unknown voltage.

input pulses are applied, the amplifier output is the stairstep voltage indicated. The output will jump a fixed increment each time the count increases by unity.

The inverse process, by which a count is registered proportional to the unknown voltage, is achieved by adding the components in Fig. 18-4 that are dashed. Now the clock pulses are *not* applied directly to $B0$ but are introduced at the AND circuit with the inhibitor terminal. Initially the negation input is not excited and the AND gate is open for transmission. As the clock pulses pass through the gate and are registered in the counter, the comparator-input voltage increases in magnitude with each count. When the comparator input reaches the unknown voltage which acts as the comparator reference, the circuit responds. The comparator output then excites the inhibit terminal of the AND gate, and the clock pulses are no longer transmitted to binary $B0$. Therefore the counter stops at a count proportional to the unknown voltage.

18-4   COUNTING TO A BASE[4] NOT A POWER OF 2

It may be desired to count to some base R which is not a power of 2. We may prefer, for example, to count to the base 10, since the decimal system is the one with which we are most familiar. To construct such a counter we start with a cascade of n binary counters such that n is the smallest number for which $2^n > R$. Then feedback is introduced from later binary stages to

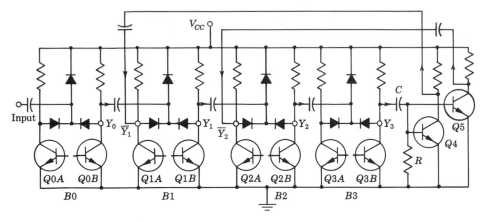

Fig. 18-5 A scale-of-16 binary chain modified by feedback into a scale-of-10.

earlier stages so that $2^n - R$ of the states are bypassed by advancing the counter by $2^n - R$ at some time during the counting of the first R pulses. The advance may be made at one time or in several stages.

The schematic of Fig. 18-5 shows in some detail one of many possible techniques employing feedback to convert a scale-of-16 counter into a scale-of-10. In this schematic the cross-coupling connections from collectors to bases have been omitted. The signal at Y_3 of the last binary is differentiated by R and C. The negative pulse which results when this flip-flop goes from state 1 to 0 is ineffective because the coupling transistors $Q4$ and $Q5$ are virtually at cutoff. However, the positive pulses which result when $B3$ goes from 0 to 1 are inverted and fed back to $\bar{Y}_1$ and $\bar{Y}_2$.

To see the effect of the feedback on the count, we now examine the waveform chart of Fig. 18-6. We must remember that a binary does not respond instantaneously to an input pulse and that there is some additional delay before the flip-flop output reaches full amplitude. This feature has been taken into account in Fig. 18-6 by drawing the binary transitions with a finite

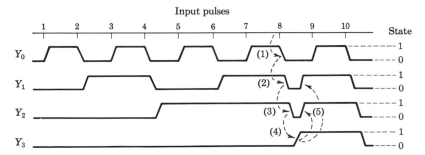

Fig. 18-6 Waveform chart of the feedback counter of Fig. 18-5.

slope and by starting the transition in a succeeding binary only at the completion of the transition in a preceding flip-flop. The counting proceeds in the customary way through the seventh input pulse. At the eighth pulse, binary $B3$ responds and negative pulses are fed back to $\bar{Y}_1$ and $\bar{Y}_2$. Since $B1$ and $B2$ are in state 0 after the eighth pulse, the right-hand transistors of these flip-flops are conducting. Hence, negative pulses at $\bar{Y}_1$ and $\bar{Y}_2$ (which are coupled to the opposite bases) will cause transitions in $B1$ and $B2$, respectively, and these flip-flops are forced back to state 1 by this fedback pulse. At the ninth pulse $B0$ changes to state 1 and after the tenth pulse all binaries are again in state 0, so that the count is complete.

Observe that after the eighth pulse and before the feedback has had a chance to be effective, the chain of flip-flops is in the state $1000 = 8$. The feedback to binary $B1$ reverses this flip-flop and advances the count by $2^1 = 2$. The feedback to $B2$ advances the count by $2^2 = 4$. The two feedback paths advance the count by $2 + 4 = 6$. We can confirm from the waveforms in Fig. 18-6 that the counter now reads $1110 = 14$. Two additional pulses are required to reset the counter to 0000. Hence the counter recycles at pulse 10 instead of 16.

In a binary chain without feedback the delay between the transition of one binary and the next following is of no importance. The resolution of the binary chain depends only on the resolution of the first binary and not on the total number of binaries in the chain. Observe, however, that in Fig. 18-6 there are five separate consecutive delays between the time of occurrence of the eighth pulse and the time when binaries $B1$ and $B2$ have been completely reset. These five delays are indicated by the dashed lines, which are numbered consecutively from 1 to 5. If the time interval between input pulses is not sufficient to allow binaries $B1$ and $B2$ to be reset before the occurrence of the tenth pulse, the counter will not operate correctly. (The ninth pulse has only to reset binary $B0$, so that the eighth and ninth pulses have only to be separated by the resolution time of a binary.)

Feedback may be used in a wide variety of ways to change the scale of a counter. By way of an additional example[5] we consider the system shown in Fig. 18-7, which is very commonly employed in decade counters. The feedback connections are shown by the heavy lines in Fig. 18-7a, and the waveform chart is shown in Fig. 18-7b. In this chart, for simplicity, we have neglected the time delay of each binary. It should be kept in mind, however, that this delay is essential to the operation of a feedback counter. The count proceeds in normal fashion through the third pulse. At the fourth pulse Y_1 goes from 1 to 0 and Y_2 goes from 0 to 1. The negative transition at $\bar{Y}_2$ is applied to $B1$ through the *set* terminal (Fig. 10-14) and immediately causes a reverse transition in $B1$, returning Y_1 to state 1. The heavy vertical line in the waveform Y_1 at the time of the fourth pulse represents a transition from 1 to 0 followed immediately by a reverse transition from 0 to 1. After the fourth pulse the chain of binaries reads $0110 = 6$ rather than 4, so that the count has been advanced two steps by the feedback. In a similar manner the feedback from

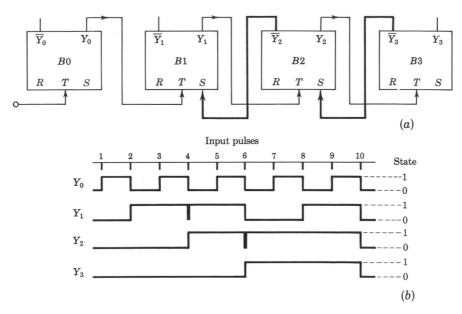

Fig. 18-7 (a) An alternative technique for converting a scale-of-16 counter into a decade counter; (b) waveform chart.

$B3$ to $B2$ at the time of the sixth pulse causes a reverse transition in $B2$ and advances the count by four. We can confirm from the waveforms in Fig. 18-7 that the counter now reads $1100 = 12$. Four additional pulses are required to bring the counter to 0000. Hence the entire unit resets to its original stage at the tenth pulse.

18-5   DECADE COUNTER WITHOUT FEEDBACK[6]

The use of feedback to change the scale of a binary counting chain may severely limit the maximum counting rate (Sec. 18-4). We consider now a method of changing the scale which avoids the use of feedback. As an example, the modification of a four-binary chain into a decade counter is represented in principle in Fig. 18-8a by means of logic circuits. In this circuit the flip-flops are induced to make transitions in some instances as the result of a triggering signal at the *complementing* input T and in other instances by signals at the S (*set*) or R (*reset*) terminals. These latter signals are derived from the output of AND gates. Such a signal will appear or not depending on the state of the outputs of several of the flip-flops.

Assume that the flip-flops are constructed of n-p-n transistors, in which case positive logic applies. Now, however, the triggering signal at S, T, or R should be a negative pulse. Therefore the AND gates must operate with negative logic.   That is to say, the negative input pulse must be transmitted through the gate when and only when all the other inputs are at their lower level.   Hence, each block marked A in Fig. 18-8a could be a sampling gate controlled by an AND circuit, as illustrated in Fig. 17-7 (but for negative logic).

The waveform chart of this counter is shown in Fig. 18-8b and is identical with that in Fig. 18-7b but with Y_2 and Y_3 interchanged.   Since $B0$ changes state at each input pulse P, then P must be applied to T_0.   Since $B2$ makes a transition each time B_1 changes from state 1 to state 0, then T_2 must be

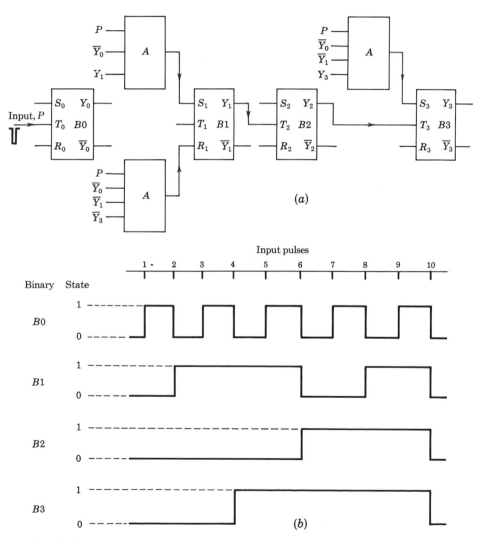

Fig. 18-8   (a) A decade counter that does not use feedback (similarly labeled points are physically connected); (b) waveform chart.

connected directly to Y_1. The transitions of flip-flops $B1$ and $B3$ are more involved. As an example, consider the transition in $B3$ in response to pulse 4. Observe that immediately before this pulse $Y_0 = 1$, $Y_1 = 1$ and $Y_3 = 0$, whereas at no other time does this combination of bits occur. Hence, a negative AND can be used preceding S_3 with inputs P, $\bar{Y}_0$, $\bar{Y}_1$, and Y_3, and it will deliver a trigger which will set $B3$ at the fourth input pulse. At the tenth pulse and only at this pulse does $B2$ make a transition from 1 to 0. Hence $B3$ may be forced to change from the 1 to the 0 state at the tenth pulse by connecting Y_2 to either R_3 or T_3. In a similar manner the required inputs to the AND circuits preceding $B1$ can be ascertained in order to obtain the waveform Y_1. A decade counter[6] capable of counting at a 50-MHz rate, similar in principle to the one presently under discussion, is commercially available.

18-6   CARRY TIME IN A BINARY CHAIN[7]

The *carry time* in a binary chain is the time required for the chain to complete its response to an input triggering pulse. The carry time is longest when each binary is in the state 1. For, in this situation, the next input must cause all binaries to change state. Any particular binary will not respond until the previous binary has nominally completed its transition. Hence, the carry time will be of the order of the sum of the transition times of all the binaries. If the chain is long, the carry time may well be longer than the interval between the input pulses. In such a case, it will not be possible to read the counter between pulses.

The carry time may be reduced appreciably by employing AND gates between flip-flops, as shown in Fig. 18-9. Now the triggering signal for each flip-flop is the input pulse itself or the input signal transmitted through one or more gates. The state of a flip-flop serves to determine whether the gate which it excites allows or restrains transmission of a pulse. Again, if positive logic

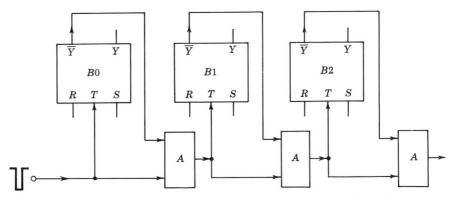

Fig. 18-9 To reduce carry time, AND gates are used between binaries.

applies to the flip-flops and negative triggering pulses are used, then the gate logic must be negative. As is seen from Fig. 18-9, a binary receives a pulse only if immediately prior to the trigger all preceding binaries are in the state 1. The trigger pulse will arrive at a binary and its immediately following gate at the same time. It is, of course, necessary that the pulse pass through the gate before the binary has changed state. In this respect a long binary transition time may be an advantage. If the transition time is too short it will be necessary artificially to lengthen it or even to introduce some small delay in the signal connection from the binary output to the gate input. In the binary chain without gates, each binary generates a triggering signal. In the circuit of Fig. 18-9 the trigger passes only through the passive AND gates and may eventually become attenuated to the point where it is no longer effective. In this case it will be necessary to introduce amplifiers in the gate chain to raise the trigger-signal level.

When the AND gates are used, the maximum carry time will be of the order of the sum of the AND-gate propagation times (plus amplifier propagation times if amplifiers are used) rather than the sum of the longer binary transition times. Still further improvement is possible through the circuit of Fig. 18-10, in which the delay of the triggering pulse is the delay resulting from transmission through a *single* AND gate. Here the trigger is delivered directly to each AND gate. If all preceding binaries are in state 1, then the gate will allow the transmission of the triggering pulse to the next binary. A complication of this circuit arrangement is the large fan-in of the gates in long chains. Thus, in a chain of N binaries, the last gate must have N input terminals. Further, the binaries at the beginning of the chain must supply signals to many gates. In a chain of N binaries, the first binary must have a fan-out of $N - 1$.

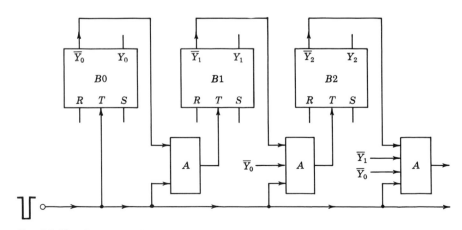

Fig. 18-10   A circuit which uses AND gates arranged so that the triggering signal of any binary is delayed only by the propagation time of one gate.

18-7   IMPROVEMENT OF RESOLUTION[8]

The *resolution* of a counter is the minimum time separation between two successive triggering signals to which the counter can respond separately. The resolution may be specified by stating this time interval or by stating the maximum allowable input frequency for which the counter will be able to distinguish each cycle. In all of the counting circuits discussed so far the first binary was required to change state at each input trigger. Hence the resolution of the counter is the resolution of this first counter. Methods of improving the resolution of a binary are discussed in Chap. 10, and it is found that the resolution time is made up of a *transition time* and a *settling time*. We shall now discuss a counting scheme in which the resolution of the counter is limited not by the settling but only by the transition time of a binary. This transition time may be appreciably smaller than the total resolution time especially since, within limits, it is possible to shorten the transition time at the expense of the settling time, about which we no longer are concerned.

To illustrate the technique we shall use the decade counter of Fig. 18-11, which consists of five flip-flops. To simplify the diagram and description we shall assume that in this case all circuits operate with positive logic. Thus we assume that the input triggers are positive pulses and that a positive pulse applied to the set (S) terminal of a binary sets $Y = 1$ and a positive pulse at R resets $Y = 0$. Similarly, a positive pulse on one input to an AND gate will be transmitted when the other input to the gate is in the state 1.

Observe that the trigger input terminal T of each flip-flop is not used and instead the external trigger appears at the R or S terminal of each flip-flop, depending on which AND gate allows transmission. Each binary output Y is connected through a gate to S and $\bar{Y}$ to R, except that in the connection from the output of $B5$ to the input of $B1$ the connections are crossed. In other words, Y_5 is connected to R_1 and $\bar{Y}_5$ to S_1 through AND gates.

Let us consider that initially each binary has $Y = 0$ and $\bar{Y} = 1$. Then the gate connected to the input S_1 will allow pulse transmission, as will also the gates at R_2, R_3, R_4, and R_5. The first trigger will cause $B1$ to make a transition to state 1. The pulses which are transmitted through the other

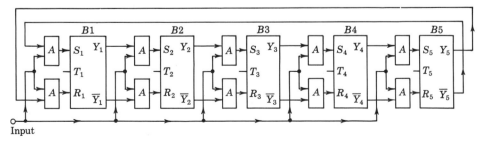

Fig. 18-11   A high-resolution counting circuit.

gates appear at a reset input R and are therefore ineffective. Now since $Y_1 = 1$, the second pulse will pass through S_2 and set $B2$ in state 1. This same second pulse, in going through S_1, does not affect $B1$, which already is in state 1. The overall result is that the first pulse has set $B1$ in state 1, the second pulse has set $B2$ in state 1, and none of the other binaries has been affected. Continuing the analysis we may see that the third, fourth, and fifth pulses will transfer in turn $B3$, $B4$, and $B5$ to state 1, so that at the end of five pulses all binaries are in state 1. The sixth pulse passes through the AND gate preceding R_1 and hence resets $B1$ to state 0. The seventh pulse resets $B2$ to state 0, and so on until, at the tenth pulse, all binaries have been reset to state 0 and the counting cycle is complete.

The improvement in resolution afforded by the circuit of Fig. 18-11 may be seen to be the following. Suppose that a first pulse has caused a transition in binary $B1$. The second pulse is to serve to change the state of $B2$. To allow this second pulse to be effective it is necessary that the gate at S_2 be enabled. This gate will be enabled as soon as the output Y_1 reaches the voltage level corresponding to state 1. Therefore the resolution time of the counter is the transition time of one binary. Observe further that during the entire counting cycle of the counter each binary makes only two transitions and that the frequency with which each binary changes state is only one-fifth of the frequency of the input triggers.

A precaution must be taken in connection with the width of the pulse applied to trigger the circuit of Fig. 18-11. For suppose that the width of the input pulse is longer than the transition time of a binary and that the pulse has just changed the state of B_1 from state 0 to state 1. Then the gate at S_2 will be enabled while the input line is still excited, and binary B_2 will respond. Hence the pulse duration must be smaller than the binary transition time.

When any of the binaries is in state 0 and a pulse appears at the reset input terminal, the binary does not change state. However, this reset trigger may well induce transients, and until the transients have decayed, the sensitivity of the binary to a trigger at the set terminal may be reduced. When the ultimate in speed is being sought, additional circuits are introduced into the system to avoid altogether the appearance of a pulse at the R terminal except when needed to reset the binary.

18-8   VERNIER COUNTING[9]

We have already referred to the method by which a counter may be used to measure the time interval between two events. This method is indicated in Fig. 18-12. The beginning and end of the interval are assumed to be marked, respectively, by the occurrence of pulses, a *start pulse* and a *stop pulse*. The start pulse allows transmission of the oscillator signal through the gate, and the stop pulse closes the gate against such transmission. The number of cycles or pulses generated by the oscillator during this interval is recorded on

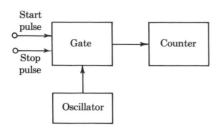

Fig. 18-12   A method of determining the time interval between two events.

the counter.   The elapsed time is the product of the counter reading and the oscillator period.

This system has a maximum uncertainty of two counts since the gate might open just immediately after the occurrence of a pulse and might close just before the occurrence of a pulse.   The uncertainty in time may be minimized by increasing the oscillator frequency but only at the expense of requiring that the counter have a correspondingly smaller resolution time.   We shall now describe a *vernier-count timing system* which reduces the timing error (at the expense of circuit complexity) but avoids the need for high-resolution counters.   At the present time such systems are capable of yielding a timing accuracy corresponding to the use in the system of Fig. 18-12 of an oscillator whose frequency is well in excess of 1,000 MHz.

For the sake of simplicity we shall assume an internally synchronized system—one which generates its own start pulse.   The system is to be used to measure the time interval between this start pulse and a stop pulse provided from some external source.   For the sake of being specific we shall assume that we have available counters adequate to only about 1 MHz but that we require a time measurement accurate to 0.01 μsec = 10 nsec.   The method is illustrated in block-diagram form in Fig. 18-13.   As in the simple system of

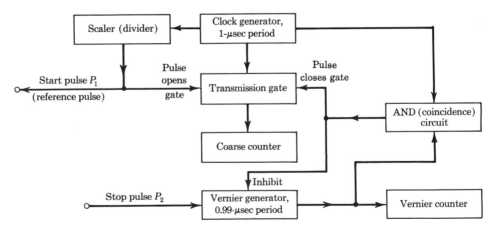

Fig. 18-13   A vernier system for very accurate time measurements.

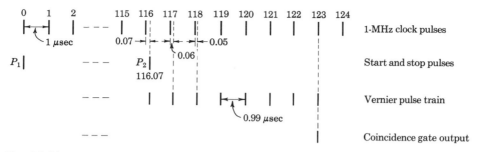

Fig. 18-14 The timing waveforms in the vernier system of Fig. 18-13.

Fig. 18-12, pulses from an oscillator, called a *clock generator*, are passed through a transmission gate and counted by the so-called *coarse counter*. The clock-generator pulses are also applied to a scaling or dividing chain (Chap. 19). The start pulse appears and the timing starts when the scaler yields an output pulse. For example, with a scale-of-10^6 divider, the timing operation will be repeated once per second.

The stop pulse is applied to the *vernier generator*. This generator is an oscillator which is initially quiescent and begins to oscillate in response to the stop-pulse signal. The vernier generator has a period 0.99 μsec, that is, 0.01 μsec shorter than the clock period. The *vernier counter* counts the pulses furnished by the vernier generator. The outputs from both generators are applied to a *coincidence*, or AND, circuit which delivers a pulse when a clock and a vernier pulse coincide in time. The output from the coincidence circuit inhibits the vernier-count generator so that it delivers no further pulses, and it also closes the transmission gate so that no further clock pulses enter the coarse counter. If the coarse-count reading is C and the vernier-count reading is V, then the time interval T between P_1 and P_2 is given by

$$T = (C - V) \times 1 \text{ μsec} + V \times 0.01 \text{ μsec} \tag{18-2}$$

as will now be demonstrated.

The operation of the circuit may be understood by referring to the timing sequences in Fig. 18-14. For purposes of illustration, a timing interval of 116.07 μsec has been assumed. The pulse P_2 occurs 0.07 μsec after the 116th clock pulse. The first pulse delivered by the vernier generator occurs 0.06 μsec after the 117th clock pulse, the second vernier pulse occurs 0.05 μsec after the 118th pulse, etc. In other words, because of the 0.01-μsec difference in the periods of the two generators, each succeeding vernier pulse slides back 0.01 μsec relative to the corresponding clock pulse. Therefore, after seven vernier pulses are generated, there will be a coincidence between a clock pulse and a vernier pulse. This condition obtains at clock pulse 123. Hence, the number of integral multiples of the clock period in the interval between P_1

and P_2 is

$$C - V = 123 - 7 = 116$$

The remainder of the interval is given by $0.01V = 0.07$ μsec. This explanation justifies Eq. (18-2). A mechanical vernier ruler operates on precisely the above principle, and this is the reason this method is called a *vernier system*.

It might appear that the system could be simplified by using the stop pulse directly to close the transmission gate and hence stop the coarse counter. However, if the pulse P_2 were used to stop the coarse count, it is possible that the counter may read too high by one count. This error will occur if a clock pulse follows immediately after P_2 so that the transmission gate does not close quickly enough to eliminate this unwanted clock pulse.

In this system the pulse generators must be extremely stable. Consider, for example, the vernier-pulse generator which may have to deliver as many as 99 pulses. If the total error is to be less than 0.01 μsec, then the delay of each pulse must be in error by less than $0.01/99 \approx 10^{-4}$ μsec over an interval $99 \times 0.99 = 98$ μsec. By increasing the frequency of the clock generator the accuracy requirement may be made less stringent and also the total time required for the vernier count may be decreased. For example, if a 2-MHz clock is used, the time between pulses is 0.5 μsec, the vernier interval is 0.49 μsec, and the maximum number of vernier counts is 49. Hence, the maximum time required for the vernier count is $49 \times 0.49 = 24$ μsec instead of 98 μsec for a 1-MHz clock. Also, now the delay of each pulse must be in error by less than $0.01/49 \approx 2 \times 10^{-4}$ μsec over an interval of 24 μsec. Some engineering compromise must be made in the selection of a clock frequency since, as already noted, an increase in clock frequency makes proportionally greater demands on counter resolution.

A critical element in the system is the coincidence gate, which must be able to establish the moment of coincidence to within 10 nsec, or even less, in faster systems. The coincidence circuit may be an AND gate followed by a comparator and amplifier, which will then furnish the power to operate the transmission gate and vernier generator. Alternatively, the coincidence circuit may simply add linearly the clock and vernier pulses and apply the sum to a comparator. The comparator reference level is set so that neither pulse individually will operate the comparator but the two together will have a large enough amplitude to cause the comparator to respond.

In Fig. 18-15a is shown a clear coincidence between a clock and vernier pulse, which gives a single pulse large enough to operate the comparator. In Fig. 18-15b is shown a situation where there is no precise coincidence but where, because of the width of the pulse, there is an overlapping between two neighboring clock and two neighboring vernier pulses. In this case, as shown, it is possible to get two pulses large enough to operate the comparator. As a result an error is possible, which in the worst case is equal to the difference between the clock and vernier periods.

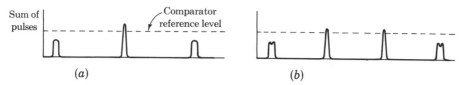

Fig. 18-15 Two possible coincidence patterns of clock and vernier pulses. (a) A true coincidence; (b) two adjacent clock pulses overlap two vernier pulses, so that the sum contains two large output pulses.

It is possible, with added complexity, to use the vernier method in an externally synchronized system. Since the start pulse in such a system is randomly timed with respect to the clock pulse, an additional vernier measurement must be made. This measurement gives the interval between the start pulse and the preceding clock pulse.

18-9   TUNNEL–DIODE COUNTERS[10]

We shall first demonstrate that it is possible to construct a complementing flip-flop using a pair of tunnel diodes. Thereafter we shall show how counters are constructed using such circuits.

Flip-Flop Operation The tunnel-diode characteristic of Fig. 18-16 shows two operating ranges in which the device displays a positive resistance and where operation is consequently always stable. One range extends to the left of the peak current I_P and the other extends to the right of the valley voltage V_V. Suppose that two tunnel diodes are placed in series and that across the pair is applied a voltage larger than $2V_P$ but less than $2V_V$. Then it will not be possible to have an equilibrium condition in which both diodes operate to the left of V_P or one in which both diodes operate to the right of V_V. It will be

Fig. 18-16 A tunnel-diode volt-ampere characteristic.

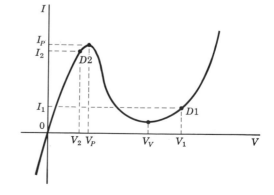

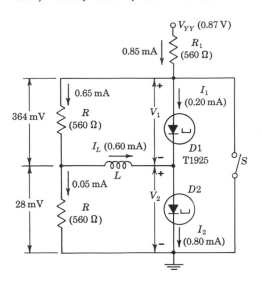

Fig. 18-17 A tunnel-diode trigger-able binary (Ref. 10).

possible in general, however, to establish an equilibrium condition in which one diode operates in the low-voltage region and the other in the high-voltage range.   Such a two-diode circuit will obviously have two stable states since the operating conditions of the diodes may be interchanged.

The simple series circuit requires that the diode currents be the same, and thereby our freedom to adjust separately the operating points is limited.   The restriction of equal currents may be removed by shunting resistors R across the diodes, as indicated in the circuit of Fig. 18-17 due to Chow.[10]   The volt-ages, currents, and components specified in Fig. 18-17 are included to convey some idea of the order of magnitude of these parameters in a typical circuit. The components have been selected so that $D1$ operates at a low current I_1 and high voltage V_1, whereas $D2$ operates at high current I_2 and low voltage V_2. These operating points are indicated in Fig. 18-16.   The inductor L has no effect on the equilibrium state, but as we shall see, it is required to ensure reliable triggering.   The resistor R_1 is included to allow some latitude in the selection of the supply voltage and also because the triggering pulse is to be applied at the anode of $D1$.   In the absence of R_1 this triggering signal would be shorted to ground.

We shall now demonstrate that the momentary closing of the switch in Fig. 18-17 induces an interchange of the operating points of the tunnel diodes. Immediately after the switch closes and before the inductor current has been able to change appreciably the situation is as shown in Fig. 18-18a and b. The sum $V_1' + V_2'$ of the voltages across the diodes must add to zero.   There-fore one diode must be forward-biased and the other must be reverse-biased and consequently carrying current in the reverse direction.   Since the inductor current cannot change instantaneously, then $I_L = I_2 - I_1 = I_2' - I_1'$. Since $I_2 > I_1$ in Fig. 18-16, then the constancy of I_L requires that I_2' be more positive than I_1'.   Hence, as indicated in Fig. 18-18b, it is diode $D1$ rather than

$D2$ which becomes reverse-biased, and I_1' is a negative current. We observe that whereas the operating point of $D2$ moves somewhat down the characteristic, the operating point of $D1$ changes very substantially since it goes from an operating point of high forward voltage V_1 to a point of reverse voltage V_1'.

When the switch is opened, the situation is as indicated in Fig. 18-18c. Added to the diode currents I_1' and I_2' is the current I' supplied by the source V_{YY} through R_1. Initially this current flows into the stray capacitances and the diode capacitances present. But, as these capacitors charge, the current is diverted into the diodes and the operating points of the diodes move in the direction of increasing current, as shown by the arrows in Fig. 18-18d. It seems intuitively obvious that diode $D2$ will reach the peak first. At this point a rapid transition will occur, in which the diodes establish themselves at the equilibrium points indicated in Fig. 18-16. However, the operating points of the diodes will be interchanged, with $D2$ now operating at high voltage and

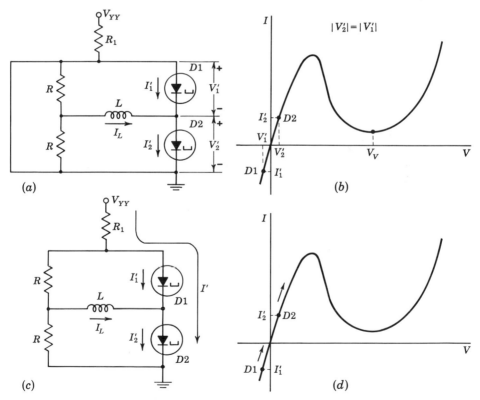

Fig. 18-18 Immediately after closing switch S (Fig. 18-17) the currents are as shown in (a) and the diode operating points as in (b). Immediately after reopening the switch, the currents are as shown in (c) and the directions in which the operating points are moving are as shown in (d).

low current. The momentary closing of S has resulted in a change of state of the flip-flop.

Switching Times The switch S must remain closed for a time long enough to allow the diodes to change from their initial operating points to those shown in Fig. 18-18b. The speed with which this transfer takes place is limited by the rapidity with which the diode (and stray) capacitances may change their voltages. When the switch S in Fig. 18-17 is closed, the diodes together with their total shunting capacitances are thrown in parallel. Initially, impulsive capacitor currents flow in order abruptly to equalize the capacitor voltages. Thereafter the speed of approach to the final operating points indicated in Fig. 18-18b is determined by the rate at which the capacitors may charge with current supplied through the diodes. (As long as the inductor current remains constant, this inductor current will affect only the location of the final operating points and not the speed with which they are approached.) In practice the circuit is triggered not by the closing of a switch but rather by the application of a step voltage. We have therefore also to allow for the rise time of this triggering step. Altogether it turns out in a typical case, for a tunnel diode with $I_P \approx 1$ mA, that the time T required for the transfer is of the order of 20 nsec. This transition time T may be reduced by using diodes with larger I_P and smaller capacitances.

The switch S must not remain closed for too long an interval because after the diodes have established themselves at the operating points indicated in Fig. 18-18b both operating points will move toward the origin as the inductor current decays to zero. If the operating points are allowed to approach one another quite closely, the new state of the circuit will not be predictable at the opening of the switch S. Therefore the switch ought to be kept closed for a time T of the order indicated above and the inductor L should be large enough to maintain approximately constant current for that period. During the interval when $D1$ is making a rapid transition from an operating voltage V_1 to an operating voltage V_1', diode $D2$ moves only slightly along that portion of the diode characteristic which is nominally a straight line passing through the origin. On this part of the characteristic the diode appears as a resistance which is much smaller than R. Let us then assume that, during this interval, $D2$ is a short circuit and the time constant of the circuit is of the order of $2L/R$. Let us require that this time constant be 5 times the time $T \approx 20$ nsec. Then with $R = 560\ \Omega$, $L = 28\ \mu$H.

Like the bistable multivibrator using transistors or tubes, so also the tunnel-diode binary has a settling time, because after the transition has been made, an interval is required for all transients to decay. This settling time is proportional to the size of the inductance L. If I_L flows in L before a transition has occurred, then after the transition when the circuit has settled, a current of the same magnitude $|I_L|$ will be flowing in L but in the opposite direction. Immediately after the transition both diodes operate in the low-voltage positive-resistance region and hence each diode may be considered to be nearly a short circuit. Hence, the time constant τ with which the settling takes place is L/R', where R' is the parallel combination of R and R, or $R/2$. If, say, $L = 28\ \mu\text{H}$ and $R = 560\ \Omega$, so that $R' = 280\ \Omega$, then

$$\tau = \frac{L}{R'} = \frac{28}{280} \approx 0.10\ \mu\text{sec} = 100\ \text{nsec} \tag{18-3}$$

In Sec. 10-5 we make the reasonable assumption that if a time 2τ is allowed between pulses, then all transients will die down sufficiently so that the flip-flop can be triggered reliably. Using this same criterion and neglecting the transition time in comparison with the settling time, we estimate that the maximum frequency at which the tunnel-diode binary may be driven is $1/0.2 \approx 5$ MHz. We deduce in general from this discussion that while a certain minimum inductance is required to allow triggering, the inductance should not greatly exceed this minimum requirement. The analogy to be drawn between the role of the inductor in the tunnel-diode circuit and the commutating capacitors in the bistable-multi circuits is apparent.

Triggering We found that closing the switch S in Fig. 18-17 for a short time will induce a transition in the tunnel-diode binary. This brief closing of the switch is equivalent to the application of a negative pulse at the anode of $D1$. This pulse need only be large enough to bring the anode of $D1$ below the valley of the tunnel-diode characteristic. As is generally the case, so here too it is advantageous to use a pulse no larger than is necessary. For in this way transients associated with the triggering signal are avoided and the resolution of the binary is thereby improved. If a pulse is larger than necessary and its amplitude is not adjustable it is advantageous to apply the triggering pulse through a resistor comparable, say, to R.

A discussion analogous to the one above establishes that a tunnel-diode flip-flop will respond as well to a *positive* triggering pulse. In this case the pulse transfers both diodes to operating points on the high-voltage portion of the characteristic. Since, before the excitation, $I_2 > I_1$ and $I_L = I_2 - I_1$ is constant during the transition, then after the transition $D2$ will be operating at a higher current than $D1$. As both diodes begin to move down toward the valley, the diode $D1$ initially in the high-voltage state will reach the valley first and a transition will ensue which leaves all operating points of the diodes interchanged, with $D2$ now in the high-voltage state.

It is, of course, possible to arrange that the tunnel-diode binary respond only to pulses of one polarity by the addition of rectifying diodes to transmit to the binary pulses of only one polarity. Alternatively, such discrimination can be built into the flip-flop itself. Suppose, for example, we adjust a binary so that in the stable state diode $D2$ (Fig. 18-16) is very close to the peak while $D1$ is rather appreciably higher in voltage than the valley voltage. Then we would expect that such a binary would respond to a smaller positive pulse than a negative pulse. Alternatively, if $D1$ is near the valley and $D2$ well below the peak, the binary would respond to a smaller negative pulse.

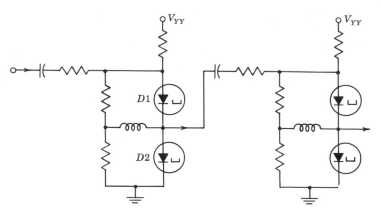

Fig. 18-19 The first two stages of a tunnel-diode counting chain.

Circuits using tunnel diodes with peak currents in the range of tens of milliamperes, resistances R of the order of 50 Ω, and inductances L of the order of hundreds of nanohenrys are able to respond at frequencies in the range of hundreds of megacycles per second.

A Scaler   A tunnel-diode counter is shown in Fig. 18-19.   Depending on the design of the binaries, a succeeding binary will respond when it receives either a positive or negative trigger from the preceding binary.   So long, however, as the binary responds to a pulse of only one polarity the counter will operate precisely in accordance with the principles established in Sec. 18-1.   The circuit of Fig. 18-19 is somewhat sensitive to supply-voltage changes.   Improved performance may be obtained by using a transistor for interstage coupling.

18-10   COUNTERS USING CURRENT–CONTROLLED NEGATIVE–RESISTANCE DEVICES[11]

In Secs. 12-8 and 12-9 current-controlled negative-resistance devices are considered which can be turned off by a negative triggering signal at the gate terminal.   These include the transwitch, trigistor, and thyristor.   We shall now indicate how these devices may be incorporated into counters.   Consider the circuit of Fig. 18-20.

The output levels are 10 V (device OFF) and 1 V (device ON).   We intend this circuit to be driven by an identical circuit and have therefore indicated in Fig. 18-20b an input signal v_i which consists of transitions between these same two levels.   Let us start with the input at the 10-V level and the output at this same level.   Then the voltages at points A, C, and D are equal:

$$V_A = V_C = V_D = 10 \text{ V}$$

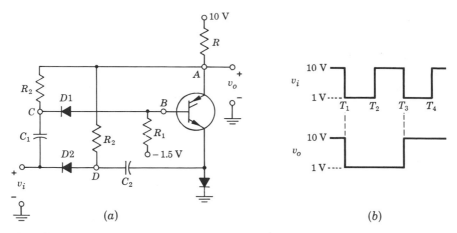

Fig. 18-20 (a) A binary circuit using a current-controlled negative-resistance device; (b) input and output waveforms.

The voltage $V_B = -1.5$ V. Diode $D1$ is back-biased to the extent of 11.5 V while diode $D2$ has zero voltage across it. When, therefore, the input signal makes a 9-V negative-going transition at $t = T_1$, this change is transmitted through $D2$ and C_2 and the device is turned ON. Now the voltages are

$$V_A = V_C = V_D = 1 \text{ V}$$

The voltage V_B is somewhat uncertain since gate current is flowing. We may reasonably estimate that $V_B \approx 0$. Now $D1$ is slightly reverse-biased and again $D2$ has zero bias. When, at $t = T_2$, a positive-going transition occurs at the input, this transition is in the direction to reverse-bias the diodes. The signal is therefore not transmitted and the device does not respond. With the input again at 10 V, diode $D2$ is back-biased to the extent of 9 V while diode $D1$ is reverse-biased to the extent of only about 1 V. When, then, the next negative input transition occurs at $t = T_3$, the change is transmitted through $D1$ and not through $D2$. As a result the device returns to its OFF state. The overall result is that the circuit does not respond to positive-going transitions but responds only to negative-going transitions by alternately changing states from ON and OFF. The circuit therefore satisfies the conditions required; so it may be used as the binary device in a counting chain.

18-11   RING COUNTERS[12]

A chain of binaries in which the first is coupled to the second, the second to the third, and so on, with the last coupled back to the first, is called a *ring counter*. In such a counter each binary device receives its triggering signal directly from the external triggering source. The coupling between flip-flops is not for the purpose of triggering but rather for the purpose of favoring or "priming" only one binary so that it alone will respond to the triggering signal. One stage of a ring counter is in the state 1 and all other stages are in the state 0. With each successive trigger the 1 state moves to the following flip-flop.

Consider a ring counter with N stages. If the interval between triggers is T, then the output from any binary stage is a pulse train of period NT with each pulse of duration T. The output pulse of one stage is delayed by a time T from a pulse in the preceding stage. These pulses may be used where a set of sequential gating waveforms is required. Thus a ring counter is analogous to a stepping switch, where each triggering pulse causes an advance of the switch by one step. We consider now some examples of ring counters.

Ring Counters Using Transistors Flip-flops using transistors (or tubes) may be used as elements in ring counters. The configuration required is a slight modification of that of a shift register. Consider, for example, that in the shift register of Fig. 9-41 the Y output of the last flip-flop $FF0$ in the chain is coupled back to the S input of the first flip-flop $FF3$ through a small delay Δ. Assume also that all flip-flops are in state 0 except $FF3$, which is in state 1. Then a pulse on the shift line will reset $FF3$ to state 0 and the output of $FF3$ will set $FF2$ to state 1. Succeeding pulses will transfer the state 1 progressively around the ring.

Tunnel-diode Ring Counter[13] A tunnel-diode ring counter is shown in Fig. 18-21. Assume that binary A is in a state where diode $D1A$ is at high voltage and low current while $D2A$ is at low voltage and high current. Assume also that the remaining binaries are in the other state. Let us consider initially only the coupling of the input positive pulse to the junction of the two diodes. That is, assume that the leads between binaries have been opened at the points marked X.

The positive pulse will reverse the state of binary A, for this pulse is of polarity such that it raises the voltage across and the current through $D2A$ and may therefore carry this diode above the peak of the characteristic. Similarly it lowers the voltage across and the current through $D1A$ and may therefore carry this diode below the valley. Both diodes, then, are being forced in a direction to induce a transition. The applied positive pulse can induce a transition in only one direction. Hence, the pulses which appear at the other binaries are not effective since $D2B$, $D2C$, etc., are already in the high-voltage state.

Now let us reestablish the connections at the points X. The positive pulse transmitted to each binary through this connection may, if sufficiently large, induce a transition no matter what the state of the binary. Suppose, however, that we adjust the pulse amplitude and the resistors through which this signal is transmitted so that the pulse is not adequate to cause a transition. Then the situation is the following. The pulse applied to circuit A at the junction of the two diodes induces a transition, as a result of which a positive

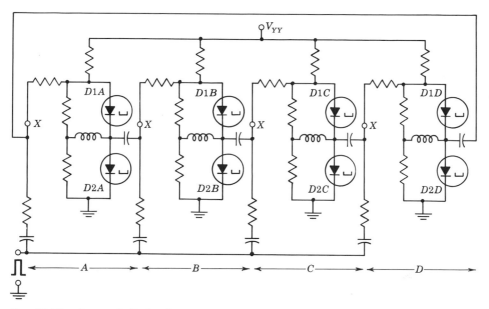

Fig. 18-21   A tunnel-diode ring counter.

step appears at this diode junction. The combination of this positive step together with the externally applied pulse induces a transition in B. Binaries C and D, not having the advantage of this combined effect of external pulse and signal from the preceding binary, do not change state. The overall result is that the state of binary A has been transferred to B, and A is in the same state as C and D.

Unijunction-transistor Ring Counter[14] A *ring-of-*4 UJT counter is shown in Fig. 18-22. The base-to-base supply voltage is V_{BB}, and the emitter supply voltage V_{EE} is provided through $Q0$, which is initially in saturation so that the voltage across it is negligible. Let us assume that initially $Q1$ is ON and all other UJTs are OFF. The voltage at the emitter of $Q1$ will be in the neighborhood of the valley voltage V_V (Fig. 12-6), and all other emitters are at V_{EE}. For the sake of being specific let us use UJTs whose characteristics are as given in Fig. 12-7, so that for $V_{BB} = 25$ V we obtain $V_V \approx 3$ V. Let us select $V_{EE} = 10$ V, which is comfortably below the emitter breakdown voltage corresponding to $V_{BB} = 25$ V. Neglecting the small voltage across $D1$ and R_{b1} we see that C_1 and C_4 are charged to $10 - 3 = 7$ V but with opposite polarities as indicated. The other capacitors have zero voltage across them.

Now let a positive pulse applied through C_0 turn $Q0$ OFF. Then the emitter supply voltage has been removed and all UJTs go OFF. Capacitor C_4 will discharge through R_{E4}, R_{E1}, and $D1$. Assume that the triggering pulse has a duration adequate to allow C_4 to discharge. The capacitor C_1 cannot

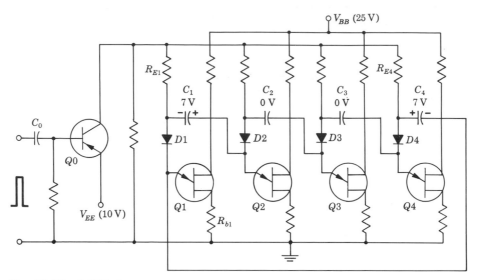

Fig. 18-22   A UJT ring counter.

discharge because of the polarity of $D2$. Accordingly, when the trigger pulse terminates, UJTs $Q1$, $Q3$, and $Q4$ will have the supply voltage $V_{EE} = 10$ V appear at their emitters. However, the voltage at the emitter of $Q2$ will be $10 + 7 = 17$ V. We may see from Fig. 12-7 that 17 V is more than enough to cause $Q2$ to fire. The capacitor C_1 may now discharge through the emitter of $Q2$ and thereafter recharge to 7 V in the opposite direction. Each succeeding trigger transfers the condition to the next UJT.

The counter has as many states as it has UJTs. To set the counter in operation properly it is necessary to apply the base supply voltage before the emitter supply voltage. Otherwise, more than one UJT may go into conduction initially. On the other hand, having observed this precaution with respect to the order of applying voltage, we shall find that no UJT has gone ON. A momentary-contact switch in series with a resistor bridged between V_{BB} and the emitter of one UJT will start the counter.

Ring Counters Using Silicon Controlled Switches[15]   Ring counters using silicon controlled switches and other similar devices may be constructed in a variety of ways. Two representative circuits are shown in Fig. 18-23. In Fig. 18-23a it is assumed that the device (trigistor, thyristor, etc.) responds to both a turn-on and a turn-off pulse applied at the gate. When the device cannot be relied upon to respond to a turn-off pulse at the gate the circuit in Fig. 18-23b is used.

In the ring-of-4 counter in Fig. 18-23a assume that all the devices are OFF except one, say $Q1$. Then all anode voltages are at 10 V except the anodes of $Q1$ and $D1$, which are at about 1 V. The reference level of the input negative

triggering pulse is $+10$ V and the pulse makes excursions to $+5$ V. The negative pulse will be transmitted through all the diodes except $D1$. The pulse will
have no effect on $Q3$ or on $Q4$, which are already OFF. It will, however, turn
OFF $Q1$. The anode of $Q1$ will rise abruptly toward the positive supply voltage,
as will also the anode of $D1$. Neglecting the drop in $D1$, its anode rises from
1 to 5 V. Hence, an abrupt change of $+4$ V will have been applied through
the capacitor to the base of $Q2$, which consequently will turn ON. The overall

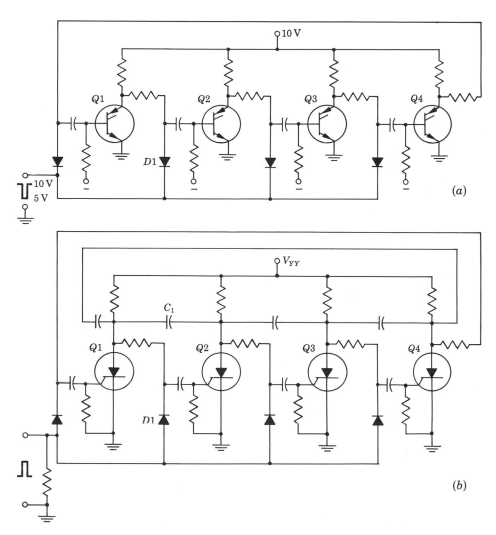

(a)

(b)

**Fig. 18-23 Ring counters using silicon controlled switches. (a) Circuit which
depends on the ability of a gate trigger to turn the device** OFF; **(b) circuit in which
the gate is not used to turn the device** OFF.

result is that the ON state has been shifted from $Q1$ to $Q2$. Successive pulses will continue this sequential shift around the ring.

In the ring counter of Fig. 18-23b, the diodes are reversed from those in Fig. 18-23a, the triggering pulse is positive, and each anode is coupled through capacitors to the anode of each neighboring SCS. Assume $Q1$ is ON and all other devices are OFF. Then all diodes are reverse-biased to the extent of the supply voltage V_{YY} except $D1$, which is only slightly reverse-biased. A positive trigger pulse of amplitude smaller than V_{YY} will be transmitted only through $D1$ to the gate of $Q2$ and will turn it ON. The abrupt drop in voltage at the anode of $Q2$ will be transmitted through C_1 to the anode of $Q1$, which will drop below its maintaining voltage, and $Q1$ will go OFF. This method of turn-off is discussed in Sec. 13-15.

18-12   A SPECIAL GAS–FILLED COUNTER TUBE[16]

The basic decade counter, as we have seen, requires a minimum of four double triodes. The wide range of applicability of counters has naturally prompted investigations into the design of special tubes which would permit the construction of counters with a greater economy. In the present section we describe the *Dekatron*, a gas-filled counter tube which has many advantages, if counting speed is not an important consideration.

The mechanical structure of the gas-filled counter tube is shown in Fig. 18-24, and the circuit and typical component values are indicated in Fig. 18-25.

The central anode in Fig. 18-24 is surrounded by 30 similar electrodes. Of these, 10 are cathodes, nine of which, $K_1, K_2, \ldots, K_9$, are brought out to a common tube terminal, while the tenth, K_0, is brought out separately. Ten electrodes are referred to as *guide No.* 1 electrodes and are labeled $1G_1$, $1G_2$, etc. The remaining ten electrodes are *guide No.* 2 electrodes and are labeled $2G_1$, $2G_2$, etc. The $1G$ electrodes are brought out to a single tube pin and the $2G$ electrodes are connected to another terminal.

The tube operates as a cold-cathode glow-discharge tube. Let us assume initially that a glow discharge is taking place between the anode and the

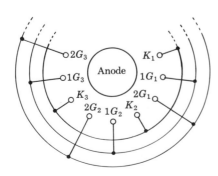

Fig. 18-24 Mechanical structure of a gas-filled counter tube.

cathode K_1.   Since the maintaining voltage of a glow discharge is smaller than the breakdown voltage, there is no likelihood that the discharge will of its own accord transfer to some other one of the cathodes.   Since, furthermore, the guide electrodes are biased positively, there is similarly no possibility of a transfer of the glow to a guide electrode.   Now suppose that a negative pulse is applied to the $1G$ electrodes of sufficient amplitude so that the voltage difference between anode and the $1G$ electrodes exceeds the breakdown voltage. Of the ten $1G$ electrodes, the $1G_1$ electrode closest to K_1 will have the lowest breakdown voltage since it is closest to the region of the discharge where the gas is most heavily ionized.   The discharge will therefore transfer preferentially to the $1G_1$ electrode.   The increased current which now flows through the anode-circuit resistor will lower the voltage between anode and K_1 to a value below the maintaining voltage, so that the discharge to K_1 will be extinguished. The net result is that the glow has been transferred from K_1 to $1G_1$.

Now suppose that before the pulse applied to $1G$ has decayed, a negative pulse is applied to $2G$ which persists after the termination of the first pulse. Then the discharge will transfer to electrode $2G_1$.   When finally this second pulse decays, the discharge will transfer to the nearest cathode, which in the present case is K_2.   In a similar way the discharge may be transferred to K_3, K_4, etc.

The single triggering pulse which is to be used to drive the counter must first be transformed before application to the counter tube into two pulses, which bear the time relationship indicated in Fig. 18-25.   Actually, except where maximum counting speed is required, the shapes of the driving pulses for the tube are not critical and a very simple circuit will suffice.   One such elementary circuit is shown in Fig. 18-26.   The pulse to be counted is used to trigger some type of pulse generator such as a monostable multi.   The output

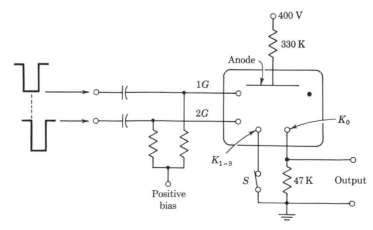

Fig. 18-25   The method of driving and obtaining output from the Dekatron tube.   The switch S is a normally closed reset switch.

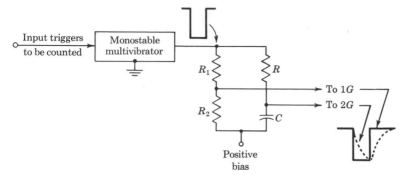

Fig. 18-26 One method of obtaining the guide waveforms for the Dekatron tube.

from the multi is passed through an RC integrating circuit to obtain the required delayed-pulse-type waveform. The delayed pulse is applied to $2G$, and the undelayed pulse is applied to $1G$. The attenuator (R_1R_2) for the undelayed pulse is used to make the two pulses comparable in amplitude. The duration of the multivibrator output pulse must be long enough to allow adequate time for deionization of the gas in the counter tube along the previous discharge path. Since there are three operations involved in a transfer, i.e., transfer from K_1 to $1G_1$, $1G_1$ to $2G_1$, and $2G_1$ to K_2, the multivibrator pulse duration should be roughly one-third the time interval between input triggering pulses at the maximum counting speed. The deionization time is the principal factor which limits counting speed. The fastest multicathode counter tube operates with a counting speed of 20 kHz.

Each trigger input pulse advances the glow from one cathode to the next so that every 10th pulse results in an output signal across the cathode load resistor in Fig. 18-25. This output signal may be used, if desired, to drive a succeeding counter. The current drain of the tube is comparatively small, a tube drawing typically only 0.6 mA from the supply voltage. The tube drop is of the order of 160 V, and the pulses required at $1G$ and $2G$ are of the order of magnitude of 100 V. The count of the tube is easily read since the glow at the cathode is visible through the top of the tube. A ring bearing numbers is mounted external to the tube, and the position of the glow indicates the count. Special-purpose tubes are also available in which all 10 cathodes are brought out individually. Such a tube makes available across 10 cathode resistors 10 sequential gating-type waveforms. Under these circumstances the tube behaves as a stepping switch or ring counter. Tubes are also available with 12 separate cathodes. Such a tube could be used, depending on the external cathode connections, to divide by 2, 3, 4, 6, or 12. Lastly, it is worth noting that if the connections to $1G$ and $2G$ are reversed, the tube will count backward, that is, it will subtract instead of add.

18-13   A VACUUM–TYPE COUNTER TUBE[17]

We now discuss a vacuum-type counter tube which is designated generally as a *trochotron, beam switching tube,* or *Beam-X switch.* It will be necessary, first, to make a short digression to review the nature of the motion of an electron in perpendicular electric and magnetic fields.[18]

Electron Trajectories   The path of an electron in uniform perpendicular electric and magnetic fields, illustrated in Fig. 18-27, is termed a *trochoid.* The solid-line path *a* of Fig. 18-27 corresponds to the circumstances in which the electron is introduced into the fields with zero initial velocity. In this case the path is a common cycloid and has sharp cusps. In the more general case of nonzero initial velocity the path may have subsidiary loops, as in curve *b,* or blunted cusps, as indicated in curve *c.*

It is to be noted particularly that the motion of the electron in the direction of the electric field (Fig. 18-27) is oscillatory and of *restricted amplitude* which is proportional to the electric field intensity. The general motion of the electron (upon which the oscillatory motion is superimposed) is in the direction perpendicular to the electric field, that is, parallel to the electric equipotential surfaces. Assume that a cathode at O in Fig. 18-27 serves as a source of electrons, the electrons being emitted with a distribution of initial velocities. Then individual electrons will follow trochoidal paths, but the electrons generally will form a broad beam which follows an equipotential.

An electrode placed in a vacuum tube which has a trochoidal electron beam has two important properties. The first of these is its ability to guide the direction of the beam; the second is referred to as the *self-locking* feature. We examine first the guiding property. Consider the simple configuration indicated in Fig. 18-28. Here two parallel plates maintained at $+V'$ and $-V'$, respectively, provide an electric field, while a magnetic field exists perpendicular to the figure. A cathode K maintained at 0 V, and situated midway between the plates, emits electrons. These electrons leave the cathode and individually follow a trochoidal path. The beam, however, is guided by the 0-V equipotential. Suppose now that the electrode S is also maintained at

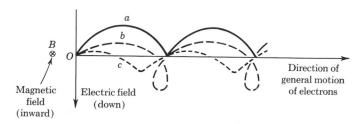

Fig. 18-27 The trochoidal paths of electrons in perpendicular electric and magnetic fields.

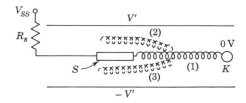

Fig. 18-28   The guiding effect on an electrode S near a trochoidal beam.

0 V.   Then the 0-V equipotential intersects the electrode S and the beam follows path 1 and will be collected.   If, however, the electrode S is maintained at less than 0 V, the 0-V equipotential passes above the electrode and the beam follows path 2.   Similarly, if the electrode is maintained at a voltage higher than 0 V, the beam follows path 3.

The *self-locking* feature may be appreciated by considering the volt-ampere characteristic of the electrode S in the presence of the trochoidal beam. A plot of current, I versus V, for the electrode S will clearly have a maximum at $V = 0$ V, since at this voltage the beam goes directly toward the electrode. As V is varied in either direction, the current decreases as the beam moves to one or the other side of the electrode.   The general appearance of the volt-ampere characteristic is shown in Fig. 18-29.   If the electrode voltage V results from the application of a voltage V_{SS} through a resistor R_s, the resultant electrode current may be determined by superimposing the load line on the volt-ampere characteristic.   There are three equilibrium points, given by A, B, and C in Fig. 18-29.   Of these, point B is unstable, since here a change in electrode current in either direction changes the electrode voltage in such a direction as to cause an additional current change in the same direction.   In other words, to the right of the maximum the volt-ampere characteristic exhibits a negative resistance.   We are left, then, with two stable points A and C.   Suppose that the initial operating point is at A.   If the electrode

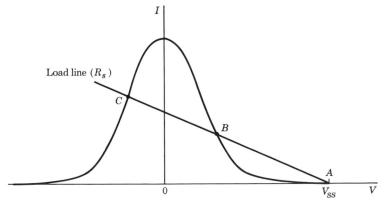

Fig. 18-29   The volt-ampere characteristic of an electrode near a trochoidal beam.

voltage is lowered temporarily, the beam may then be directed toward the electrode. The current collected by this electrode may now maintain the electrode at reduced voltage and the operating point will be shifted permanently, i.e., locked at point C. The current at operating point C is less than the total beam current and may, as a matter of fact, be made a very small part of the total current if R_s is large. The beam then barely skims along the surface of the electrode S and the rest of the beam current may be collected, if desired, by an auxiliary collector plate, not indicated in Fig. 18-28. Observe that the beam is in a stable position when it skims along one side of the electrode S but not when it skims along the other side.

The Beam-X Switch The structure of the tube is shown in Fig. 18-30. It consists of ten identical groups of four electrodes each: a *spade*, a *target*, a *shield*, and a *switching grid*. The spade serves the function of the electrode S in Fig. 18-28. In operation, the beam skims along one side of the spade. The spade collects a small part of the beam current, and the remainder of the current is collected by the target, which provides the output signal. The target electrodes are bar magnets and therefore serve also to provide the required magnetic field parallel to the tube axis. The shields serve to isolate each target from the fields of the neighboring targets and to collect any current not collected by the target when the tube operation carries the target to a very low voltage. The switching grids are used to direct the beam from one electrode group to the next with each successive input pulse.

To understand qualitatively the operation of the tube let us assume that all the electrodes are initially maintained at some common positive voltage with respect to the cathode. Then, at least near the cathode, the equipotential surfaces will be cylinders about an axis passing through the cathode. The electron path follows the equipotentials, and therefore the electrons will encircle the cathode and at the same time oscillate back and forth radially.

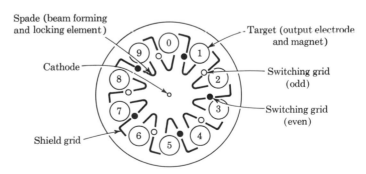

Fig. 18-30 Cross-sectional structure of the Beam-X switch. (Courtesy of Burroughs Corporation, Plainfield, N.J. The designation Beam-X is a registered trade name.)

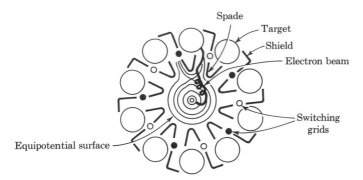

Fig. 18-31 Illustrating beam formation and equipotential surfaces in the presence of the beam.

The amplitude of these trochoids increases with increasing voltage between cathode and electrodes.   We assume that this voltage is not adequate to cause the radial amplitude to extend to the electrodes, so that no current is collected.

Now suppose that the voltage of one spade is temporarily lowered.   Then the equipotential surfaces must be modified to the form shown in Fig. 18-31. In going from the cathode to the spade of lowered voltage we must cross a "saddle point" (a point where two equipotential surfaces of common potential touch).   For in going from cathode to spade we move initially in the direction of higher electrical potential and then, at the saddle point, the potential begins to decrease.   Consider, then, electrons which were initially encircling the cathode in the neighborhood of the equipotential which passes through the saddle point.   As they travel along this path and reach the saddle point, they need not continue around the cathode but may instead follow the equipotential which carries them along the side of the spade.   Because of the oscillatory motion of the electrons in the direction perpendicular to the equipotentials, some current will be captured by the spade.   The spade is connected to a power supply through a resistor.   Hence, when some external agency lowers the spade voltage to the point where the spade intercepts some beam current, the locking action described above will keep the beam in place.   The current not intercepted by the spade will be collected by the target.

Five alternate switching grids are connected together and constitute the *odd grids*.   The remaining alternate switching grids are also connected together and are called the *even grids*.   Suppose that the beam, in approaching the spade and target which are collecting the electrons, passes near an odd grid.   Then if this grid is driven negative, it will repel the beam and drive the electrons to the next spade and target, whose corresponding even grid is not negative.   The application of a negative pulse to the odd grids has advanced the beam from one target to an adjacent target.   Since the beam is now locked in the neighborhood of an even grid, the subsequent application of a negative pulse to the even grids will repel the electrons and force the beam to lock on the adjacent spade. Clearly, successive switching is achieved by applying negative pulses alternately to the odd and even grid groups.

A Beam-X Counter A counting circuit is shown in Fig. 18-32. The supply V_{SS} is of the order of 100 V. The spade load resistance R_s is of the order of 150 K, and the target load resistance R_t is of the order of 3.3 K. The switching-grid bias voltage is of the order of 50 V, and negative pulses of the order of 55 V are used to achieve the switching. The triggered flip-flop is coupled to the grids through differentiating circuits that include diodes to suppress the positive pulse which would otherwise result when a flip-flop output made a positive-going excursion. A sequential series of gating voltages comparable in amplitude to the supply voltage is available at the targets. The maximum operating speed is about 2 MHz.

When the supply voltage is first applied, the tube may be at cutoff, with the beam circling the cathode and with no electrode collecting current. A negative reset pulse applied to one spade, say No. 0, will lock the beam to that spade. After this condition is achieved it is necessary that the state of the binary be such that the first trigger applied to the flip-flop will advance the beam to the next target. Hence the reset pulse is applied to the binary as well as to spade No. 0.

The count in the Beam-X switch may be indicated visually by employing a Nixie tube (trade name of Burroughs Corporation). The Nixie indicator is

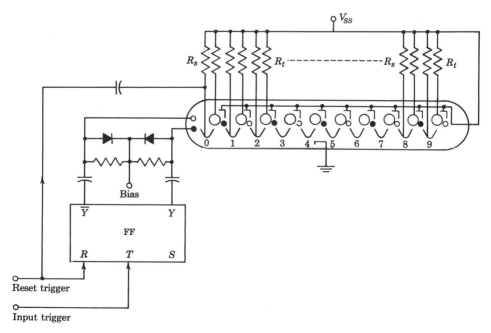

Fig. 18-32 A Beam-X counting tube circuit. (Courtesy of Burroughs Corporation.)

a cold-cathode gas-discharge tube with a single anode and 10 cathodes, which are wires shaped in the form of the numerals "0" through "9." These cathodes are connected to targets "0" through "9," respectively, of the Beam-X switch and the anode is connected to a fixed supply voltage. When, for example, the seventh pulse has caused target "7" to collect current, then the cathode glow in the Nixie tube causes the numeral "7" to be visible.

18-14   STORAGE COUNTERS[19]

We consider now a circuit, known as a *storage counter*, which operates on a principle that is basically different from the counters previously described. A two-diode storage-counter circuit is shown in Fig. 18-33. Let us neglect, for the present, the voltage-operated discharge switch and consider the output waveform v_o which results from the application to the circuit of a train of negative pulses of amplitude V. To simplify the discussion we assume that the pulse starts at a base level of 0 V. Since the input is applied through a capacitor, this assumption has no essential bearing on the discussion.

Principle of Operation Assume initial zero charge on the capacitor C_1 or C_2. The first input pulse will cause the capacitor C_1 to charge through the diode $D1$. The time constant with which C_1 charges is the product of C_1 times the sum of the diode and generator resistances. If this time constant is very small in comparison with the duration of the pulse, then C_1 will charge fully to the value $v_1 = V$, with the polarity indicated. During the charging time of C_1, the diode $D2$ does not conduct and the voltage across C_2 remains at zero. At the termination of the input pulse, the capacitor C_1 is left with the voltage $v_1 = V$, which now appears across $D1$ and across the series combination of $D2$ and C_2. The polarity of this voltage is such that $D1$ will not conduct. The capacitor C_1 will, however, discharge through $D2$ into C_2 until the voltages across the two capacitors are equal. The time constant with which this transfer of charge takes place must be quite small in comparison with the interval between pulses in order to allow equilibrium to be established between the capacitor voltages. The capacitor C_2 is ordinarily quite large in comparison with C_1. As a consequence the voltage change across C_2 is small in comparison with the voltage $v_1 = V$ across C_1. The next input pulse restores the voltage on C_1 to V, and at the termination of the pulse C_1 discharges again into C_2. Since now, however, C_2 has some initial charge, the amount of charge trans-

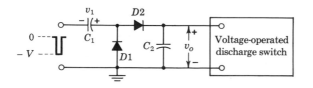

Fig. 18-33 A two-diode storage counter.

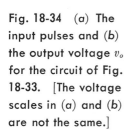

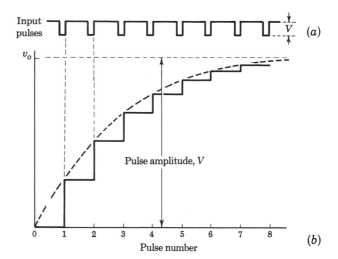

Fig. 18-34 (a) The input pulses and (b) the output voltage v_o for the circuit of Fig. 18-33. [The voltage scales in (a) and (b) are not the same.]

ferred from C_1 to C_2 will be smaller than before.   Hence the second increment of voltage across C_2 will be smaller than the first.   Each successive input pulse will cause a progressively smaller step in voltage at the output, the output approaching asymptotically the voltage $v_o = V$, as in Fig. 18-34.   In this figure the voltage is plotted as a function of pulse number, and not time.   If, however, the pulses are regularly spaced in time, the plot of Fig. 18-34 also gives the output waveform.   Note, too, as indicated, that the voltage steps occur at the trailing edge rather than the leading edge of the pulse.

The counter is completed by the addition of a circuit which operates as a switch shunted across C_2.   This switch is normally open, but it closes when v_o attains some preestablished reference value.   One such switch circuit is shown in Fig. 18-35.   The emitter of the blocking-oscillator transistor is maintained at some positive voltage so that the transistor is initially below cutoff. The blocking oscillator responds when the voltage across C_2 becomes high enough to bring the transistor out of cutoff.   At this point the capacitor C_2

Fig. 18-35   A blocking oscillator is used as a voltage-controlled discharge switch in a storage counter.

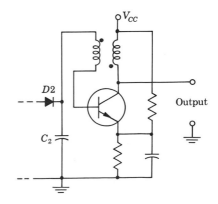

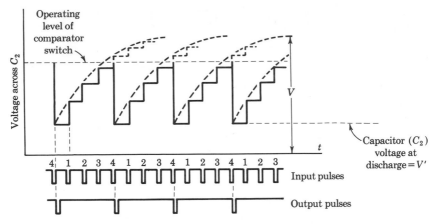

Fig. 18-36 The voltage across C_2 in a 4:1 counter, showing the time relationships to the input and output pulses.

discharges rapidly because of the large base current during the pulse formation, and at the same time an output pulse is furnished. In the present application the blocking oscillator serves as a comparator. Another device which may be used for this purpose is the unijunction transistor (Fig. 13-22). Quite generally, then, when some type of comparator-operated switch is included in the circuit, a counter results and the waveform across C_2 has the typical appearance shown in Fig. 18-36. Here the waveform has been drawn taking into account the possibility that the discharge switch may not discharge the capacitor C_2 to zero. The countdown ratio shown is 4:1, since the comparator response voltage has been set between the voltage levels corresponding to the third and fourth pulses.

Step Size The counting ratio may be increased by raising the voltage at which the comparator responds. As the ratio is increased, however, the counter operation becomes progressively less reliable since the size of the steps decreases. This lack of reliability results from uncertainties in the level to which the voltage across C_2 drops at discharge and uncertainties in the operating voltage of the comparator. An expression showing how the step size decreases with pulse number may be derived as follows. Suppose that after the nth pulse has produced the nth step the net voltage across C_2 is v_n. Now let the $(n + 1)$st pulse be applied. After the trailing edge of this pulse, diode $D1$ is open, $D2$ is closed, and the voltage across C_1 is V. An equivalent circuit for computing the jump in voltage across C_2 when $D2$ conducts is shown in Fig. 18-37, in which S represents the diode $D2$. From elementary electrostatics we find that the voltage v_{n+1} across C_2 after the $(n + 1)$st pulse is

$$v_{n+1} = v_n + (V - v_n) \frac{C_1}{C_1 + C_2} \qquad (18\text{-}4)$$

Fig. 18-37 Equivalent circuit for computing the step size at C_2.

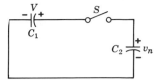

If we introduce y_n, defined by

$$y_n \equiv v_n - V \tag{18-5}$$

then Eq. (18-4) reduces to the homogeneous equation

$$y_{n+1} = y_n - \frac{C_1}{C_1 + C_2} y_n \tag{18-6}$$

or

$$y_{n+1} = x y_n \tag{18-7}$$

where

$$x \equiv \frac{C_2}{C_1 + C_2} \tag{18-8}$$

Since the change in y_n is proportional to y_n, we may anticipate that y_n will vary exponentially with n. Hence, we shall seek a solution of the form

$$y_n = A \epsilon^{\alpha n} \tag{18-9}$$

where A and α are to be determined so as to satisfy Eq. (18-7) and the proper initial condition. Substituting Eq. (18-9) into Eq. (18-7), we find

$$A \epsilon^{\alpha(n+1)} = x A \epsilon^{\alpha n} \qquad \text{or} \qquad \epsilon^\alpha = x \tag{18-10}$$

Hence $y_n = A x^n$ and, from Eq. (18-5),

$$v_n = V + A x^n \tag{18-11}$$

The initial condition is that the capacitor voltage v_n equals V' for $n = 0$, and hence $A = V' - V$. Finally, the complete solution is

$$v_n = V - (V - V')x^n \tag{18-12}$$

The difference between the $(n + 1)$st voltage level and the nth level is

$$v_{n+1} - v_n = (V - V')(1 - x)x^n \tag{18-13}$$

If, for example, $V - V' = 100$ V and $C_2 = 9C_1$, then $x = \frac{9}{10}$ and the first step has an amplitude of $(100)(1 - \frac{9}{10}) = 10$ V. The 11th step, on the other hand, is $100(1 - \frac{9}{10})(\frac{9}{10})^{10} = 3.5$ V.

Practical Considerations Because step size decreases with n and because of the instability in the value of V', the comparator response voltage, and amplitude of input pulse, storage counters are normally not used for counting ratios larger than about 10. Storage counters, of course, may be cascaded to secure large counting ratios in the same manner that binary counters are cascaded. An important limitation of storage counters is that they may be used ordinarily only to count pulses which occur fairly regularly. This feature results from the fact that the count is determined by the charge on a capacitor, and this charge will leak off slowly because of capacitor leakage or through the resistance of any device which is used to read the count. The leakage problem is complicated by the fact that the capacitors must be able to charge rapidly and must therefore be quite small. Still, where storage counters are applicable, they often effect a worthwhile economy of components over comparable binary counters. In comparing the binary and storage counters for economy, one must keep in mind that the storage counter must be driven from a pulse source of large and constant amplitude and of low impedance. The pulses must be of reasonably rectangular shape.

The counter of Fig. 18-33 may be operated with a positive input pulse. In this case C_1 charges first through $D2$ and thereafter discharges through $D1$. The step at C_2 then occurs at the leading edge of the input pulse. Similarly, with either polarity of input pulse, the counter may be operated with both diodes reversed. In these latter cases the steps in voltage across C_2 are in the negative direction.

18-15   LINEARIZATION OF STORAGE COUNTERS

It is possible to extend somewhat the counting ratio of a storage counter by linearizing the envelope of the step waveform of Fig. 18-34 to yield steps of more nearly equal amplitude. The methods used to achieve this end are essentially the same as those used to linearize time-base waveforms, that is, bootstrapping and Miller integration (Chap. 14).

The Bootstrap Configuration Consider the circuit of Fig. 18-33 modified so that the anode of $D1$ is returned to a positive voltage v_R rather than to ground. Under these circumstances, immediately after the input pulse the voltage across C_1 is $V + v_R$ rather than V. Hence from Eq. (18-4)

$$\Delta v_n \equiv v_{n+1} - v_n = (V + v_R - v_n)\frac{C_1}{C_1 + C_2} \tag{18-14}$$

where Δv_n gives again the size of the nth step. If, then, we arrange to make v_R equal to v_n, the step size will be independent of n. A cathode follower with nominally unity gain, as in Fig. 18-38, will achieve the desired end. It is not necessary that v_R be instantaneously equal to v_n, since it is sufficient that v_R follow the general rise of the output voltage v_n. A capacitor C may therefore be shunted across the output of the cathode follower to assist the rapid charging of C_1.

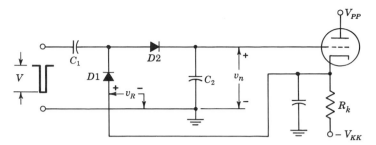

Fig. 18-38   Use of the bootstrap principle to make counter steps equal in amplitude.   (See text for modifications necessary to convert this circuit into a practical form.)

Although the circuit of Fig. 18-38 illustrates the bootstrapping principle, it is not a practical circuit.   Let us consider the quiescent condition when no pulses are present and the supply voltages are first applied to the cathode follower.   Since the grid-to-ground voltage v_n is initially zero, the cathode-to-ground voltage v_R of the triode is a few volts positive.   This positive voltage causes both $D1$ and $D2$ to conduct, and the voltage v_n increases.   This rise in grid voltage causes v_R to increase, which, in turn, raises v_n still further.   This action continues until v_n equals v_R, so that the grid-to-cathode voltage is zero.   If a pulse were now applied, this would tend to drive the grid positive and the circuit would not operate properly.

The above difficulty can be avoided by returning the anode of $D1$ to a tap on the cathode resistor R_k such that the voltage v_R is less than zero when the voltage $v_n = 0$.   This connection will reduce the gain of the cathode follower and cause the output to be somewhat less linear with the number of input pulses.   However, if V_{KK} is large, the tap will be close to the cathode and the reduction in linearity will not be great.

A second method for avoiding the d-c instability of the circuit of Fig. 18-38 mentioned above is to reverse both diodes $D1$ and $D2$.   With no input pulses and the voltage across C_2 equal to zero, the voltage v_R will be a few volts positive, say V_o.   Hence, diodes $D1$ and $D2$ cannot conduct, and the action described above does not take place.   The amplitude of the input pulse must exceed V_o or $D1$ will never conduct.   It should be mentioned that in the two circuits just discussed the voltage across C_2 will become negative in the quiescent condition because some small grid current flows even for negative grid voltages.

An emitter follower [or if higher input resistance is required, a Darlington cascade (Fig. 14-39)] may be used in place of the cathode follower in Fig. 18-38. Since the emitter of an n-p-n transistor operating in the active region is a few tenths of a volt *negative* with respect to the base, then diodes $D1$ and $D2$ will be reverse-biased in the quiescent state.   Hence this circuit, unlike the tube version, possesses d-c stability.

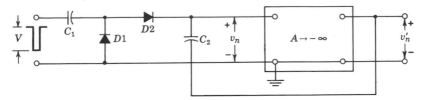

Fig. 18-39 Use of the Miller integrating principle with storage counters.

The Miller Circuit A Miller integrator used to improve linearity is illustrated in Fig. 18-39. Here the voltage v_n between the cathode of $D2$ and ground remains nominally zero. If the voltage across C_2 increases, then the bottom side of C_2 falls by this same amount in order to keep v_n constant. The virtual ground at the input terminals of the operational amplifier takes no current (Sec. 1-8). Hence, all the charge $C_1 V$ which leaves capacitor C_1 must transfer to capacitor C_2. The increase in voltage across C_2 is, therefore,

$$\Delta v_{C_2} = -\Delta v_n' = \frac{C_1 V}{C_2} \tag{18-15}$$

This equation verifies the fact that the output decreases by a constant amount for each input pulse.

18-16   APPLICATIONS OF STORAGE COUNTERS

Among the most important uses of storage counters are the following:

A Divider For every n pulses into the counter, one appears at the output. The waveform in such an application is illustrated in Fig. 18-36 for $n = 4$. The pulses must be fairly regularly spaced, although an exactly constant interval between pulses is not required. A storage counter is more economical as a divider than is a binary counter. Frequency division is considered in more detail in Chap. 19.

Staircase Voltage Generator The staircase waveshape of Fig. 18-36 is frequently useful to vary some voltage in a step fashion. The staircase generator of the sampling oscilloscope (Fig. 17-39) is such an application. The storage-counter waveform may also be used to trace out a family of transistor (or tube) volt-ampere characteristics on a CRO. In this application each step of the staircase corresponds to a particular constant value of base current (or grid voltage).

Frequency or Counting-rate Meter If the capacitor C_2 in Fig. 18-33 is shunted by a resistor R, then the counter may be used as a frequency meter.

Let V be the pulse amplitude and f be the pulse-repetition frequency. If $C_1 Rf \ll 1$ and $C_2 \gg C_1$, then (Prob. 18-23) the average output voltage is

$$V_{\text{d-c}} = VC_1 Rf \tag{18-16}$$

Thus, for given values of C_1, R, and V a high-impedance voltmeter placed across the output may be calibrated in *frequency* or in pulses per unit time. Such a meter has found extensive application in nuclear radiation measurements[20] and in frequency-modulation radar systems.[21]

Capacitance Meter If a known frequency is applied to a storage counter which has been modified so that Eq. (18-16) is valid, then the output voltage is proportional to the capacitance of C_1. Hence, this device may be used as a capacitance meter.

REFERENCES

1. Barney, K. H.: The Binary Quantizer, *Elec. Eng.*, vol. 68, pp. 962–967, November, 1949.

2. Thomason, T. H.: A Preset Counter for Time and Quantity Measurements, *Tele-Tech*, vol. 12, p. 82, August, 1953.
Wild, J. J.: Predetermined Counters, *Electronics*, vol. 20, pp. 120–123, March, 1947.
Blume, R. J.: Predetermined Counter for Process Control, *Electronics*, vol. 21, pp. 88–93, February, 1948.

3. "Introduction to Digital Techniques and Applications," Navigation Computer Corporation, Norristown, Pa.
"Instruction Manual, Model 580A Digital-Analog Converter," Hewlett-Packard Company, Palo Alto, Calif.

4. "Instruction Manual, Model AC-4 Decade Counter," Hewlett-Packard Company, Palo Alto, Calif.

5. "Instruction Manual, Types 700A, 705A, and 707A Decimal Counting Units," Berkeley Division, Beckman Instrument Company, Richmond, Calif.

6. "Instruction Manual, Model 5245L Electronic Counter," Hewlett-Packard Company, Palo Alto, Calif., 1962.

7. Rhoades, W. T.: Ultra High Speed Counting Techniques, *Electron. Design*, vol. 12, pt. I, pp. 58–63, Sept. 14, 1964; pt. II, pp. 48–52, Sept. 28, 1964; pt. III, pp. 62–67, Oct. 12, 1964.

8. Tarczy-Hornoch, Zoltan: Five-Binary Counting Technique Makes Faster Decimal-counting Units, *Electron. Design*, vol. 9, pp. 34–37, Jan. 18, 1961.

9. Baron, R. G.: The Vernier Time-measuring Technique, *Proc. IRE*, vol. 45, pp. 21–30, January, 1957.

10. Chow, W. F.: Tunnel-diode Digital Circuitry, *IRE Trans. Electron. Computers*, vol. EC-9, no. 3, pp. 295–301, September, 1960.
Bush, E. G.: A Tunnel-diode Counter for Satellite Applications, *NASA Tech. Note* D-1337, June, 1962.

11. Solid State Products Inc., Salem, Mass.: A Survey of Some Basic Trigistor Circuits, *Bull.* D410-02, March, 1960.

12. Sharpless, T. K.: High-speed *N*-scale Counters, *Electronics*, vol. 21, pp. 122–125, March, 1948.
Beckwith, H.: Flip-Flop Counter Has Expanded Range, *Electronics*, vol. 28, pp. 149–151, January, 1955.
Seren, L.: Decade Ring Scaling Circuit, *Rev. Sci. Instr.*, vol. 18, pp. 654–659, September, 1947.
Regener, V. H.: Decade Counting Circuits, *Rev. Sci. Instr.*, vol. 17, pp. 185–189, May, 1946.
Burks, A. W.: Electronic Computing Circuits of the ENIAC, *Proc. IRE*, vol. 35, pp. 756–767, August, 1947.

13. Spiegel, P.: Basic Digital Applications of the T1925 and T1975 Tunnel Diodes, *Appl. Lab. Rept.* 681, Philco Corporation, Lansdale, Pa.

14. Crawford, B., and R. T. Dean: The Unijunction Transistor in Relaxation Circuits, *Electro-Technol.*, pt. 1, vol. 73, pp. 34–38, February, 1964; pt. 2, vol. 73, pp. 40–45, March, 1964.
Sylvan, T. P.: Notes on the Application of the Silicon Unijunction Transistor, *Appl. Note* 90.10, General Electric Company, Syracuse, N.Y., May, 1961.

15. Stasior, R. A.: Silicon Controlled Switches, *Appl. Note* 90.16, General Electric Company, Syracuse, N.Y., June, 1964.

16. Bacon, R. C., and J. R. Pollard: The Dekatron—A New Cold Cathode Counting Tube, *Electron. Eng.*, vol. 22, pp. 173–177, May, 1950.
McAsulan, J. H. L., and K. J. Brimley: Polycathode Counter Tube Applications, *Electronics*, vol. 26, pp. 138–141, November, 1953.
"Dekatrons Containing Neon," Baird-Atomic Company, Cambridge, Mass., January, 1963.

17. Kuchinsky, S.: Multi-output Beam-switching Tubes for Computers and General Purpose Use, *IRE Natl. Conv. Record*, pt. 6, *Electron Devices*, 1953, pp. 43–45.
Fitzpatrick, A. G.: PCM Coding System Uses Special Tubes, *Electronics*, vol. 26, pp. 173–175, November, 1953.
Romanus, H., and H. Alfren: Trochotrons—A New Family of Switching Tubes, *Tele-Tech*, vol. 13, p. 94, June, 1954.
Jörkman, J. B., and L. Lindberg: Development of Trochotrons, *Trans. Roy. Inst. Technol.*, *Stockholm*, no. 80, 1954.
Kandiah, K.: Decimal Counting Tubes, *Electron. Eng.*, vol. 26, pp. 56–63, February, 1954.
Fan, S. P.: The Magnetron Beam Switching Tube, *Brit. Inst. Radio Engrs.*, vol. 15, no. 7, pp. 335–354, July, 1955.
Beam-X, *Tech. Brochure* BX535B, Burroughs Corporation, Plainfield, N.J.

18. Millman, J.: "Vacuum-tube and Semiconductor Electronics," chap. 1, McGraw-Hill Book Company, New York, 1958.

19. Easton, A., and P. H. Odessey: Design of Counter Circuits for Television, *Electronics*, vol. 21, pp. 120–123, May, 1948.
Wintle, M. F.: Precision Calibrator for Low-frequency Phase-meters, *Wireless Eng.*, vol. 28, pp. 197–208, July, 1951.
Bedford, A. V., and J. P. Smith: Precision Television Synchronizing Signal Generator, *RCA Rev.*, vol. 5, pp. 51–68, July, 1951.

20. Elmore, W. C., and M. Sands: "Electronics—Experimental Techniques," pp. 249–256, McGraw-Hill Book Company, New York, 1949.

21. Luck, D. G. C.: "Frequency Modulation Radar," chap. 4, McGraw-Hill Book Company, New York, 1949.

19 / SYNCHRONIZATION AND FREQUENCY DIVISION

A pulse or digital system may involve several different basic waveform generators. Such a system may require that these generators run synchronously, that is, in step with one another, so that each generator arrives at some reference point in its cycle at the same time. The frequency stability of waveform generators is never adequate to ensure synchronism. Even a very small frequency difference between generators will eventually cause the accumulation of a large error. In many pulse systems it is required that the individual generators be synchronized but be permitted to operate at different frequencies. We may require, say, that one generator complete exactly some integral number of cycles while a second generator executes only one cycle. Such a situation is very common in pulse and digital systems and is described as synchronization with frequency division.

This chapter discusses the mechanism of synchronization on a one-to-one basis and also synchronization with frequency division. The two processes are basically so nearly alike that no clean-cut distinction will be drawn between them.

The counting circuits of Chap. 18 may, of course, be used for frequency division. These counting circuits, with the exception of the storage counters, do not depend for their operation on the regularity of recurrence of the input waveform. If, say, the input signal consists of pulses, the counters will divide correctly, independently of whether the pulses occur regularly or in a random fashion. In the present chapter, however, we contemplate only the case of an input waveform of a nominally fixed recurrence rate. This feature, as we shall see, permits considerable economy in the circuits which may be used to achieve division.

716

19-1   PULSE SYNCHRONIZATION OF RELAXATION DEVICES

The term "relaxation circuit" is applied to any circuit in which a timing interval is established through the gradual charging of a capacitor, the timing interval being terminated by a relatively abrupt discharge (relaxation) of the capacitor. The relaxation circuits which have been described in earlier chapters include the sweep generator, the blocking oscillator, the multivibrator, and the phantastron circuit. Each of these circuits has in common a timing period and a relaxation (or recovery) period and each exists in an astable or monostable form.† The mechanism of synchronization and frequency division is basically the same for all of these relaxation devices.

In the monostable form, the matter of synchronization (1:1 division) to a pulse-type waveform is a trivial one. The circuit normally remains in a quiescent condition and awaits the arrival of a triggering pulse to initiate a single cycle of operation. It is only necessary that the interval between triggers be larger than the timing interval and recovery period combined.

The important features of pulse synchronization of an astable relaxation device may be exposed by examining the mechanism in connection with any of the circuits mentioned above. Let us select for consideration the sweep generator of Fig. 19-1a since this circuit is slightly simpler than the others. The block S represents a current-controlled negative-resistance switch such as a unijunction transistor, a silicon controlled switch, a thyristor, a gas thyratron, etc. (Fig. 14-6). In the absence of an external signal the capacitor stops charging when the voltage v_C reaches the peak or breakdown voltage V_P of the negative-resistance device. Thereafter the capacitor discharges abruptly through the negative-resistance device (Fig. 14-7). In order to simplify the waveform drawings to follow we shall neglect the recovery time T_r and assume that the capacitor discharges in zero time. When the capacitor voltage v_C falls to the valley voltage V_V, the negative-resistance device goes OFF and the capacitor begins to recharge.

Synchronization to an external signal is possible because this signal may be introduced at the sync terminal in Fig. 19-1a in such a manner as to change the peak voltage V_P. Thus, in the UJT, a negative pulse applied at $B2$ (as in Fig. 14-6) will lower V_P, whereas in the SCS, the thyristor, and the thyratron a positive pulse applied at the gate, the base, or the grid will serve the same purpose.

The situation which results when synchronizing pulses are applied is illustrated in Fig. 19-1b. The effect of the sync pulse is to lower, for the duration of the pulse, the peak or breakdown voltage as indicated. A pulse train of regularly spaced pulses is shown starting at an arbitrary time $t = 0$.

---

† The so-called "monostable multi" can be rendered astable, as described in Sec. 11-11. Similarly the so-called "astable multi" may be rendered monostable by applying an appropriate fixed negative bias to one triode. The essential distinction between these circuits is that the monostable multi has a single timing period and a single recovery interval, whereas the astable circuit has two timing periods and two recovery intervals per complete cycle.

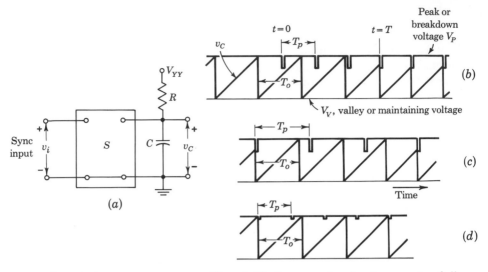

Fig. 19-1 (a) A sweep generator; (b) an initially unsynchronized generator falls into synchronization shortly after the application of synchronizing pulses; (c) illustrating that, for synchronization to result, T_p must be less than T_o; (d) illustrating failure of synchronization due to inadequate amplitude of sync pulses.

The first several pulses have no influence on the sweep generator, which continues to run unsynchronized. Eventually, however, the exact moment at which the negative-resistance device goes ON is determined by the instant of occurrence of a pulse (at time T in Fig. 19-1b), as is also each succeeding beginning of the ON interval. From this point on, the sweep generator runs synchronously with the pulses.

In order that synchronization may result, it is necessary that each pulse shall occur at a time when it may serve to terminate the cycle *prematurely*. This requirement means that the interval between pulses, T_p, must be *less* than the natural period, T_o, of the sweep generator. In Fig. 19-1c the case is shown in which $T_p > T_o$. Here synchronization of each cycle does not occur. The pulses do serve to establish that four sweep cycles shall occur during the course of three pulse periods, but synchronization of this type is normally of no value. Even if the requirement $T_p < T_o$ is met, synchronization cannot result unless the pulse amplitude is at least large enough to bridge the gap between the quiescent breakdown voltage and the sweep voltage v_C. In Fig. 19-1d we have the case where T_p is less than T_o as required, but the pulse amplitude is too small and again synchronization does not result.

19-2   FREQUENCY DIVISION IN THE SWEEP CIRCUIT

In Fig. 19-2 we have a case in which $T_p < T_o$ but in which the pulse amplitude is too small to permit each pulse to terminate a cycle. The sweep cycles are therefore terminated only by the alternate pulses marked "2" in the figure. The pulses marked "1" would be required to have an amplitude at least equal to V_1 if they were to be effective. The pulses marked "2" are effective because they occur closer to the time when the cycle would terminate of its own accord. The sweep generator now acts as a divider, the division factor being 2, since exactly one sweep cycle occurs for each two synchronizing pulses. If T_s is the sweep generator period after synchronization, we have $T_s/T_p = 2$. Note that the amplitude V_s' of the sweep after synchronization is less than the unsynchronized amplitude V_s.

Suppose, referring again to Fig. 19-2, that T_p is progressively decreased. Eventually a point would be reached where even the alternate pulses would be too small in amplitude to fire the switch device. At this point it might be that $T_o > 3T_p$, in which case division by a factor of 3 would result. If the condition $T_o > 3T_p$ were not met, then again we would have no synchronization. If, on the other hand, we make the pulse amplitude large enough, we may make sure that every $(n + 1)$st pulse is in a position to ensure synchronization before the nth pulse loses control.

The general behavior of the circuit for regularly spaced pulses of varying pulse period and pulse amplitude is illustrated in Fig. 19-3a. This diagram may be verified by making sketches such as that in Fig. 19-2. The amplitude scale extends from 0 to V_s, where V_s is the total sweep amplitude. (The amplitude and polarity of the pulse at the sync input of Fig. 19-1a corresponding to the pulse amplitude in Fig. 19-2 will depend on which switching device is being used.) The shaded areas of the diagram represent the regions of synchronization, unshaded areas regions of lack of synchronization. The diagram is to be interpreted in the following way. Suppose that the pulse amplitude is fixed at V_1. Then as the ratio T_p/T_o decreases from 1, there will be a range during which 1 : 1 synchronization will hold, followed by a range of no synchronization, followed by a range of synchronization in which 2 : 1 division will result, followed again by a range of no synchronization, and so on. If, on the other hand, the pulse amplitude is large, say V_2, then with decreasing T_p/T_o, synchronization will always hold, the division changing abruptly from 1 : 1 to 2 : 1 to 3 : 1, and so on.

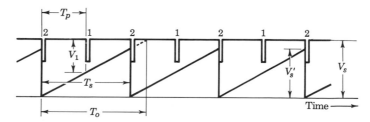

Fig. 19-2 Frequency division by a factor of 2 in a sweep generator.

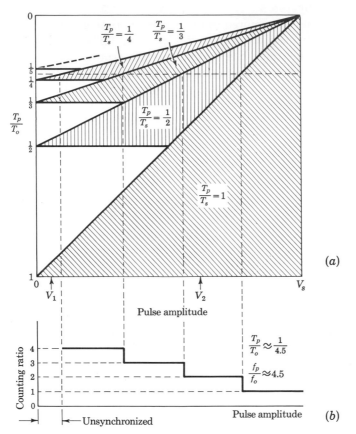

Fig. 19-3 (*a*) The ranges of synchronization for a sweep generator as a function of pulse amplitude or frequency. The sweep waveform is assumed to be linear. (*b*) The counting ratio as a function of pulse amplitude for $f_p/f_o \approx 4.5$.

Next suppose that we maintain a fixed value of T_p/T_o, say $\frac{1}{5} < T_p/T_o < \frac{1}{4}$. Then as the pulse amplitude is increased from zero, we shall initially have no synchronization, followed by ranges of synchronization where the counting ratio is first 4:1, then 3:1, then 2:1, and finally 1:1. The transitions in counting ratio are abrupt. This last-described characteristic is of much interest and is often represented in the form shown in Fig. 19-3*b*. Here the counting ratio is plotted as a function of pulse amplitude for the case where $T_p/T_o \approx 1/4.5$, the ratio of pulse frequency to natural frequency being $f_p/f_o \approx 4.5$. We emphasize that as the pulse amplitude increases, the counting ratio decreases and the sweep amplitude also decreases.

19-3   OTHER ASTABLE RELAXATION CIRCUITS

The synchronization and use for counting purposes of other types of relaxation oscillators differ only in detail and not in basic principle from the synchronization of the time-base generator.

Blocking Oscillator   The use of a blocking oscillator to accomplish pulse-recurrence-frequency (PRF) division by a factor of 4 is illustrated in Fig. 19-4. Positive sync triggers are introduced through a separate transistor $Q2$, as in Fig. 16-7. These positive triggers appear at the common collector as negative triggers and, inverted by the transformer, appear as positive triggers at the base of $Q1$. The waveform v_1 across the R_1C_1 combination (Fig. 16-15) is shown in Fig. 19-4b. During the interval t_p of the pulse generated by the blocking oscillator the capacitor charges and at the end of the pulse, $v_1 = V_1$. Then the voltage v_1 decreases, and in the absence of synchronizing pulses, a new blocking-oscillator pulse would form when v_1 falls to the level $V_{BB} - V_\gamma$. The injected triggers are shown superimposed on the level $V_{BB} - V_\gamma$. The important point is that a pulse (No. 4) occurs at a time and has a sufficient amplitude to cause a premature firing of the oscillator. The oscillator therefore fires at a moment dictated by the occurrence of a trigger and is not permitted to terminate its cycle naturally. The synchronizing characteristics given in Fig. 19-3 for the sweep generator may be applied directly to the blocking oscillator, provided only that the sweep amplitude V_s is replaced by the corresponding amplitude $V_1 - V_{BB} + V_\gamma$ for the blocking oscillator and provided that $t_p \ll T_s$.

Astable Multi[1]   The astable multi in Fig. 19-5a may be synchronized or used as a divider by applying positive or negative triggering pulses to either

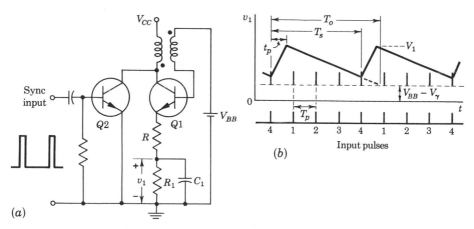

Fig. 19-4   (a) Sync signal is applied to blocking-oscillator transistor $Q1$ through $Q2$; (b) the waveform v_1 across R_1C_1 of $Q1$ showing PRF division by 4.

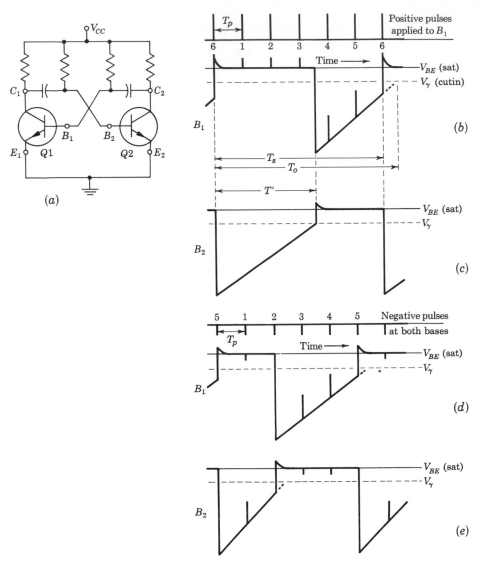

Fig. 19-5 (a) An astable multi; (b, c) base waveforms for division by 6 through the application of positive pulses to one base; (d, e) base waveforms for division by 5 through the application of negative pulses to both bases.

transistor or to both transistors simultaneously. These pulses may be applied to the collector, base, or emitter. If, for example, positive pulses are applied to B_1 or C_2 or negative pulses to E_1, these triggers may produce synchronization by establishing the exact instant at which $Q1$ comes out of cutoff. If negative pulses are applied to B_2 or to C_1, or positive pulses are applied to E_2, then when $Q2$ conducts, *these pulses will be amplified and inverted and appear as positive pulses at* B_1. Hence again the pulses may establish the instant when $Q1$ comes out of cutoff. It must, however, be borne in mind that negative pulses will not be effective unless they succeed in moving the transistor at least slightly into the active region. Therefore such negative pulses must be large enough in amplitude and be supplied from a low enough impedance source to divert enough current from the base to draw the transistor out of saturation. The polarities referred to above are appropriate for the *n-p-n* transistors indicated and must be reversed if *p-n-p* devices are under consideration.

In Fig. 19-5*b* and *c* are shown waveforms for the case where positive pulses are applied to one base, say B_1. The division ratio is 6. The cycle would normally have terminated at $t = T_o$ when the base voltage reached the cutin level V_γ, as shown by the dashed extension of the base waveform in Fig. 19-5*b*. The cycle is prematurely terminated at the sixth pulse since the amplitude of the sixth pulse added to the base waveform B_1, at the time of the sixth pulse, raises the base voltage above V_γ. Observe that while the complete multi period has been synchronized, the individual portions have not been synchronized. Thus T' in Fig. 19-5*c* is the same as it would be without synchronization because the waveform at B_2 is unaltered by the application of sync pulses.

If a vacuum-tube multi is used rather than a transistor multi, the operation is essentially the same as that described above. An interesting, although not particularly significant, difference in waveforms is observable in the two cases. With transistors, when positive pulses appear at the base of the transistor which is ON, the transistor is driven further into saturation, but the corresponding response at the collector is negligible. With tubes, when positive pulses appear at the grid of the tube which is in clamp, the grid is driven further into the positive-grid region, and the response at the plate is pronounced. Therefore each time a positive pulse appears at the grid G_1 of the tube which is ON, a larger negative pulse will appear at the plate of that tube and at the grid G_2 of the tube which is cut off. The waveform B_2 of Fig. 19-5*c* would represent such a grid waveform at G_2, provided that we superimpose onto the exponential portion large negative pulses in time phase with the sync pulses.

The situation illustrated by the base waveforms of Fig. 19-5*d* and *e* corresponds to the case of application of negative pulses to both multi bases simultaneously. Here the division ratio is 5. Both timing portions of the multi waveform are synchronized and are necessarily of unequal duration since the division ratio is an odd number. The positive pulses superimposed on the exponential portions of the waveforms result from the combination of the negative pulses applied directly and the inverted and amplified (hence positive and larger) pulses received from the other transistor.

A special situation of interest is illustrated in Fig. 19-6. Here positive pulses are being applied to B_1 through a small capacitor from a low impedance source. During the time when $Q1$ is conducting the base draws current at each input pulse. At the end of the pulse the input capacitor discharges, giving rise to a negative overshoot. Alternatively, we may say that during the conduction period of $Q1$ the pulse input time constant is small and the input pulse

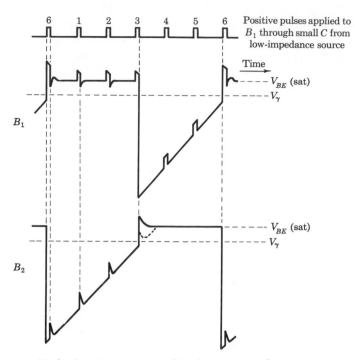

Fig. 19-6 Synchronization of both portions of astable-multi waveform by applying positive pulses to one base through a small capacitance from a low-impedance source. Illustrating synchronization at B_2 resulting from pulse overshoot due to differentiation of input pulse.

is quasi-differentiated. The negative overshoot is amplified and inverted by $Q1$ and appears at B_2 as a positive overshoot, which may then serve to mark the end of the cutoff period of $Q2$. Hence, the net result is that both portions of the multi cycle have been synchronized without the need for applying pulses to both transistors simultaneously. Observe that one portion of the cycle is terminated at the leading edge of a sync pulse and the second portion is terminated at a trailing edge of another pulse.

The diagram of Fig. 19-3a does not apply directly to the astable multi. However, for any particular method of synchronizing and degree of symmetry of the multi a similar diagram may be drawn. The general results deduced from Fig. 19-3a do, however, apply equally to the multi. That is, for large pulse amplitude the counting ratio makes abrupt changes as the ratio T_p/T_o is increased. For smaller amplitudes, regions of synchronization are separated by intervals of no synchronization. Similarly, for a fixed ratio T_p/T_o, a plot of counting ratio versus pulse amplitude has the same general appearance as in Fig. 19-3b.

19-4   MONOSTABLE RELAXATION CIRCUITS AS DIVIDERS

Frequency division through the use of a monostable relaxation device, in this case a monostable multi, is illustrated in Fig. 19-7. Input pulses may be applied at B_1 or C_1, depending on the polarity. A coupling diode may be used (Fig. 11-21) to minimize the reaction of the multi on the pulse source. The waveform of Fig. 19-7b shows the voltage at B_2. Each fourth pulse causes a transition of the multi, the remaining pulses occurring at times when they are ineffective. Observe that while the total multi cycle, consisting of timing portion and recovery period, is synchronized, the separate portions are not synchronized.

If positive pulses are applied, say, directly at B_1 through a small capacitance from a low-impedance source, a situation may arise similar to that illustrated in Fig. 19-6 for the astable multi. Here the overshoot due to differentiation of the input pulse may serve to terminate the timing cycle prematurely, as shown in Fig. 19-7c. In this case the two portions of the multi waveform will be synchronized. More importantly, in this latter case the counting ratio will change with increasing amplitude of pulse input. If the overshoot is large enough, the exponential will be terminated by the overshoot at pulse 2 or pulse 1, in which case the counting ratio will become, respectively, 3 or 2. Finally, with a large enough overshoot, the timing portion will terminate at the trailing edge of pulse 4 and the circuit will not operate as a multi at all.

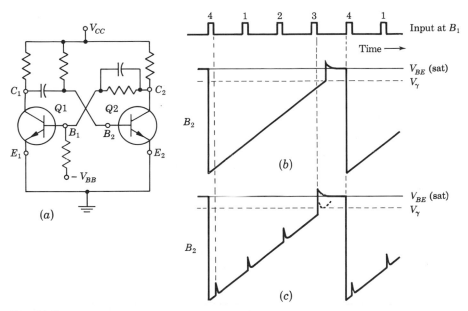

Fig. 19-7 (a) Monostable multi divider; (b) waveform at B_2 with no pulse overshoot; (c) waveform at B_2 with pulse overshoot.

19-5   STABILITY OF RELAXATION DIVIDERS

There will normally be some small delay between an input pulse to a divider and the output pulse. This delay is referred to as a *phase delay* and results from the finite rise time of the input trigger pulse and the finite response time of the relaxation device. The phase delay itself is subject to variation with time due to variations in transistor or tube characteristics, supply voltages, etc. Occasionally some extraneous signal may be coupled unintentionally into the divider. Such a signal may have an influence on the exact moment at which a base waveform, say, reaches cutoff. In this case the phase delay may be subject to periodic variations. All these factors which affect the phase delay give rise to what is termed *phase jitter*. In a large-scale counter consisting of many stages, the phase jitter is, of course, compounded. In many applications phase jitter is of no particular consequence, but sometimes, particularly in connection with nanosecond pulses, it constitutes an important difficulty.

A method for achieving division without phase jitter is illustrated schematically in Fig. 19-8 together with the waveforms depicting the operation. The train of regularly spaced input pulses (I) is applied to the divider input. The output of the divider consists of the pulses shown in waveform D. These latter pulses trigger a gating-waveform generator (say a monostable multi) which provides a gate of duration T_g adequate to encompass each pulse labeled "1." This waveform is applied to a sampling gate or coincidence circuit (Chap. 17) which is opened for transmission for the duration T_g. The input pulse train is sampled, and the output waveform O then consists of each

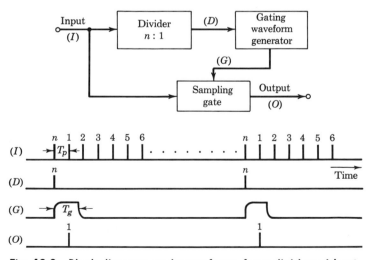

Fig. 19-8   Block diagram and waveforms for a divider without phase jitter.

pulse labeled "1." We may take advantage of the phase delay between waveform I and D (not shown in the figure) together with the finite rise time of the gating waveform to ensure that pulse n does not pass through the gating circuit. The duration of the gate is not critical, since it is only required that the duration be longer than the interval between pulses and shorter than the interval between alternate pulses $(T_p < T_g < 2T_p)$. Of course, the sampling gate must introduce no phase delay.

A much more commonly encountered jitter in dividers and synchronized relaxation oscillators results from the instability of the natural timing period of the oscillator. This instability of period is caused principally by the variability of transistor or tube characteristics and may result either in a loss of synchronism or an incorrect division ratio in a divider. For example, in an $n:1$ divider a change in natural timing period can cause the relaxation oscillator to fire at the $(n-1)$st or the $(n+1)$st pulse rather than at the nth pulse. Similarly, in a 1:1 synchronized device, if the natural period T_o should drift and become smaller than the interval between pulses, then synchronization of each cycle will be lost.

The factors which influence the stability of the natural period may be seen from Fig. 19-9. The timing waveform starts from an initial voltage V_i. The timing waveform increases exponentially and asymptotically toward the final voltage V_f, but the natural cycle is terminated when the waveform approaches some critical voltage V_c. This voltage V_c may be a cutin voltage V_γ of a transistor, as in a multi, or it may be the breakdown voltage V_P of a negative-resistance device, as in Fig. 19-1. The period T_o is easily computed, but for the present we merely note that T_o depends on V_i, V_c, V_f, and the time constant of the exponential rise. The voltage V_f is easily obtained from a regulated source, and no difficulty need be encountered in reducing the instability of the time constant to a negligible amount. The voltages V_i and V_c,

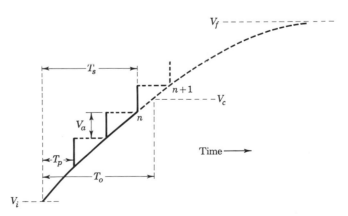

Fig. 19-9 Illustrating the factors which influence the stability of a relaxation divider.

however, depend on device characteristics. These voltages are stabilized only with the greatest difficulty and constitute the major source of timing instability.

In Chap. 11 it is pointed out (Fig. 11-4) that it is customary to make V_f large enough so that the portion of the timing waveform between V_i and V_c is approximately linear. In this way the change in T_o with a variation in V_c is reduced somewhat. It is worthy of note, however, that in some cases the variability of V_i exceeds by far the variability of V_c and in such a case it may well be that a large value of V_f will add to the instability. For suppose that V_f is large enough so that the timing waveform is essentially linear between V_i and V_c. In this case a given percentage change in the voltage $V_c - V_i$ will clearly produce an equal percentage change in T_o. On the other hand, it can be shown (Prob. 19-15) that if V_f is reduced so that the timing interval T_o is not linear, then a given percentage change in $V_c - V_i$ *due to a variation in V_i* will produce less than an equal percentage change in T_o. Hence, an attempt should be made to adjust the voltage V_f so as to reduce the uncertainty in T_o due to the variability of V_i or V_c, whichever variability is the larger.

The factors which determine the selection of the natural period T_o and the amplitude of the synchronizing pulses may be seen from Fig. 19-9. If the timing waveform is nominally linear over the interval T_o, then the natural period should be chosen so that $T_o = (n + \frac{1}{2})T_p$ and the pulse amplitude V_a should be equal to the voltage change of the timing waveform between pulses. Such an adjustment in a divider will yield a fixed counting ratio over the maximum range of variation of V_i or V_c. And for such an adjustment the combined allowable variation in V_i and V_c is $\Delta V_i + \Delta V_c = \pm\frac{1}{2}V_a$. If noise pulses of nominal amplitude V_n are present, then this last equation should read $V_n + \Delta V_i + \Delta V_c = \pm\frac{1}{2}V_a$. The most straightforward method of ensuring reliability of synchronization at a fixed division ratio is to keep the counting ratio low. In dividers which must operate without readjustment for long periods, a division ratio of 10 or less is customary.

19-6   SYNCHRONIZATION OF A SWEEP CIRCUIT WITH SYMMETRICAL SIGNALS

Up to the present we have considered the phenomenon of synchronization only for the case of pulse-type synchronizing signals. We have assumed that the synchronizing signal consists of a train of waveforms with leading edges which rise abruptly. We shall now consider the case in which the voltage variation of the sync signal is gradual rather than abrupt. Again, the mechanism of synchronization is so nearly identical for all types of relaxation oscillators that we may without loss of generality select any one of them, say the sweep generator, for detailed consideration.

Sinusoidal Sync Signal Consider the sweep generator of Fig. 19-1a, which uses a current-controlled negative-resistance device as a switch. We shall assume for simplicity that, as a result of the sync signal, the breakdown voltage of the switch varies sinusoidally. The polarity and precise waveform required of the sync signal for such sinusoidal variation will depend on the particular negative-resistance device being employed. It is to be noted especially, however, that the circuit behavior to be described does not depend on the sinusoidal nature of the breakdown-voltage variation. The results depend only on the relatively gradual variation of the breakdown voltage, in contrast to the abrupt variation with pulse-type sync signals.

In Fig. 19-10 the dashed voltage level V_{PO} is the breakdown voltage of the negative-resistance device in the absence of a sync signal and the solid curve V_P is the breakdown voltage in the presence of the sync signal. The sync signal has a period T (corresponding to T_p in Fig. 19-1), and the natural period is T_o. Consider that synchronization has been established with $T = T_o$. Such synchronization requires that the period of the sweep shall not be changed by the sync signal. Hence, the voltages which mark the limits of the excursion of the sweep voltage must remain unaltered. The sweep cycle must therefore continue to terminate at V_{PO}. This result, in turn, means that the intersection of the sweep voltage with the waveform V_P must occur, as shown in Fig. 19-10, at the time when V_P crosses V_{PO}, at the points labeled O in the figure. The possibility that the sweep will terminate at the points marked O' will be considered shortly.

In the case of pulse synchronization we noted that synchronism could result only if the sync-signal period was equal to or less than the natural period. This feature resulted from the fact that a pulse could serve reliably only to terminate a timing cycle prematurely and not to lengthen it. In the present case, however, synchronization is possible both when $T < T_o$ and when $T > T_o$. The timing relationship between the sweep voltage and the breakdown voltage for both cases is shown in Fig. 19-11a. The sweep voltage,

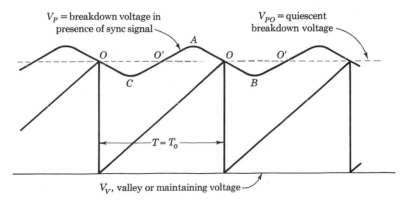

V_P = breakdown voltage in presence of sync signal

V_{PO} = quiescent breakdown voltage

$T = T_o$

V_V, valley or maintaining voltage

Fig. 19-10 Illustrating the timing relationship that must exist between V_P and the sweep voltage in a synchronized sweep when $T = T_o$.

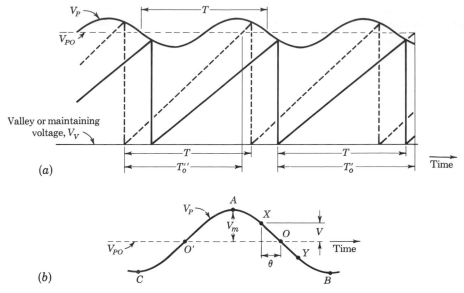

Fig. 19-11 (a) Illustrating the timing of the sweep voltage with respect to V_P for a case in which $T < T_o = T'_o$ (solid lines) and in which $T > T_o = T''_o$ (dashed lines); (b) pertaining to the general case when $T \neq T_o$.

drawn as a solid line, has a natural period $T'_o > T$. The sweep voltage meets the V_P curve at a point below V_{PO} and is consequently prematurely terminated. The dashed sweep voltage has a natural period $T''_o < T$. This sweep meets the V_P curve at a point above V_{PO} and is consequently lengthened. In each case the synchronized period T_s equals the period T. The general situation may be described by reference to Fig. 19-11b. When $T = T_o$, the sweep is terminated at point O, leaving the period unaltered. When $T > T_o$, the sweep terminates at a point such as X between O and the positive maximum A. When $T < T_o$, the sweep terminates at a point such as Y between O and the negative maximum B. When the period T is such that the sweep terminates either at the point A or B, the limits of synchronization have been reached, since at A the sweep period has been lengthened to the maximum extent possible, whereas at B the shortening is at maximum. The following illustrative example will show how one may calculate the range of synchronization.

EXAMPLE A UJT sweep operates with a valley voltage $V_V = 3$ V and a peak voltage $V_P = 16$ V (Fig. 12-7). A sinusoidal synchronizing voltage of 2 V peak is applied between bases. The stand-off ratio is $\eta = 0.5$ (Sec. 12-3). If the natural frequency of the sweep is 1,000 Hz, over what range of sync-signal frequency will the sweep remain in 1:1 synchronism with the sync signal?

Solution   From Eq. (12-1), the amplitude V_m of the a-c component of the break-down voltage is $0.5 \times 2 = 1$ V.   The sweep amplitude may therefore lie in the range $(16 - 3) - 1 = 12$ V to $(16 - 3) + 1 = 14$ V.   Since a sweep of $16 - 3 = 13$ V is generated in 10^{-3} sec, then the times required to generate sweeps of 12 and 14 V, respectively, are $\frac{12}{13} \times 10^{-3}$ and $\frac{14}{13} \times 10^{-3}$ sec.   The corresponding frequencies are $\frac{13}{12} \times 10^3 = 1,083$ and $\frac{13}{14} \times 10^3 = 929$ Hz.   Thus the sweep generator will remain synchronized as the sync-signal frequency varies over the range from 929 to 1,083 Hz.

Phase Stability   We now consider the stability of a point of intersection of the sweep voltage with the V_P curve, such as X in Fig. 19-11b.   Suppose that as a result of some transient disturbance the intersection point during one cycle should occur at some point other than X.   Or suppose that when the sync voltage is first applied, the intersection point occurs initially at some arbitrary point.   We shall now show that intersection at X represents a stable situation and that with each successive cycle the intersection point will move closer and closer to point X.   This result is easily seen from the graphical construction of Fig. 19-12a.   Suppose that the intersection point required to make the sweep and sync periods equal is designated X.   Consider that during a particular cycle the intersection actually occurs at Y, that is, at a time Δt too late.   Then the timing of the sweep during the *next* cycle is as shown by the dashed line.   The intersection is now at Z, *closer to* X.   We may now easily continue this graphical construction, which will show the intersection point moving closer to X at each cycle, eventually reaching X in the limit.

The sync voltage accomplishes its function by lengthening or shortening

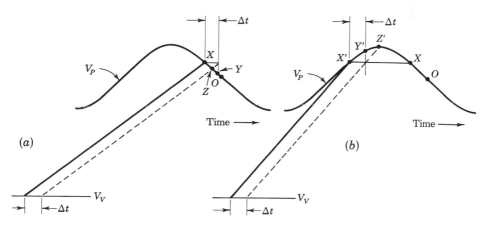

(a)   (b)

Fig. 19-12   (a) Showing that when the intersection point of the sweep with the V_P curve occurs on the descending portion of the V_P curve a stable situation results; (b) showing that instability results if the intersection point occurs on the ascending portion of the V_P curve.

the sweep as required to make the sync and sweep periods identical.   Referring to Fig. 19-11b, it will be clear that this end might be achieved if the sweep terminated at some point on the V_P curve between C and A as well as at some point between A and B.   Suppose then, as in Fig. 19-12b, the correct intersection point is at X' but that during a particular cycle the intersection occurs at Y', at a time Δt too late.   Then the construction shows that during the *next* sweep cycle (dashed line) the intersection point moves to Z', *further away from* X'.   Hence X' is not a stable point.   Continuation of the graphical construction will show that the intersection point will move progressively with each succeeding cycle until it reaches the point X.   Here the sweep period is the same as at X', but at X the situation is stable, whereas at X' it is unstable.

In summary, we may conclude that a relaxation oscillator may be synchronized with stability only on the *negative-slope portion of the waveform* V_P (*AOB* of Fig. 19-10 or 19-11b).   The particular phase θ of the sine wave at which synchronization takes place is seen from Fig. 19-11b to be given by

$$\theta = \arcsin \frac{V}{V_m}$$

where V is the minimum voltage required for synchronization and V_m is the peak value of the sine-wave voltage.

19-7   SINE–WAVE FREQUENCY DIVISION WITH A SWEEP CIRCUIT

The operation of a sweep circuit as a divider is a natural extension of the process of synchronization.   Figure 19-13 (solid lines) shows the sweep and synchronizing waveforms for division by a factor of 4.   This case is one in which the natural period T_o is slightly smaller than $4T$.   The sync signal changes the sweep period from T_o to T_s, where $T_s = 4T$.   The dashed waveforms in Fig. 19-13 show that, for the situation illustrated, an increase in amplitude of the sync signal can change the counting ratio from 4 to 3.   Quite generally one may make the following observation with respect to a sweep circuit as a counter.   If the sweep terminates on the descending portion of the V_P curve and if as a consequence the period T_o is lengthened or shortened to T_s, where $T_s = nT$, then the circuit will operate stably as an $n:1$ counter.

It was tacitly assumed earlier that the range of synchronization (or counting) extends from the point where the sweep intersects the V_P curve at a maximum to the point where the intersection is at a minimum of the V_P curve. Such a result normally holds for small values of sync voltage, but may not necessarily apply when the sync amplitude is comparable to the sweep amplitude.   Observe, for example, in Fig. 19-13 that the sweep will never be able to terminate at a maximum of V_P, since to do so would require that the sweep first cross the previous negative excursion of the V_P waveform.   Figure 19-14 illustrates a case (dashed sweep) where the sync amplitude is in principle just

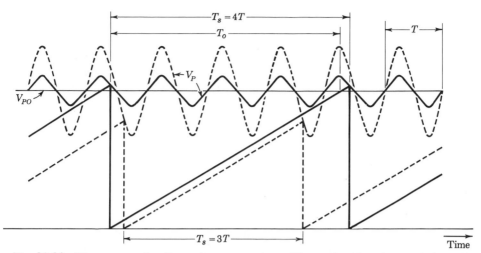

Fig. 19-13 The sweep circuit used as a counter.   Illustrating the change in counting ratio with sync-signal amplitude.

large enough to cause $1:1$ synchronization.   The actual sweep waveform, however, as shown, consists of alternate long and short sweeps.   Figure 19-14 suggests that when a sweep is used in connection with a scope it is advisable always to use as small a sync signal as possible.   A sweep waveform as in Fig. 19-14 will cause a piecewise display of each cycle of the signal being observed on a scope.

We shall now compare the general results which hold for sine-wave synchronization with the characteristics associated with pulse synchronization.   The features of pulse synchronization are effectively summarized in Fig. 19-3.   As in Fig. 19-3, so in the case of sine-wave synchronization, we find that, for small sync signals, synchronization holds over a small range in the neighborhood of integral relationships between T and T_o.   With sine

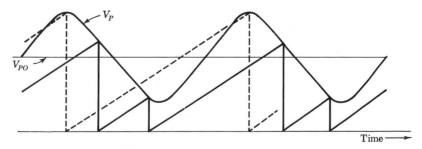

Fig. 19-14 Illustrating a possible result of excessive amplitude of the sync signal in a sweep.

waves, however, unlike the situation that results with pulses, synchronization persists for variation of T_o/T in either direction. In both cases the range of synchronization increases with increasing sync-signal amplitude. Finally, with pulses, for large sync-signal amplitudes, synchronization holds for all values of $T_o/T_p > 1$, abrupt jumps in counting ratio occurring at critical values of T_o/T_p. With sine waves, however, while the range of synchronization may be large for large sync amplitudes, we cannot be sure that synchronization in a useful fashion will persist for all values of T_o/T. This last feature is in part brought out in Fig. 19-14.

19-8 SINE–WAVE SYNCHRONIZATION OF OTHER RELAXATION DEVICES

The mechanism of sine-wave synchronization of relaxation devices other than the sweep is similar in principle to that already discussed. We shall study, as an additional representative example, the sine-wave synchronization of an astable multivibrator. For the sake of variety we shall consider a vacuum-tube circuit and shall indicate at the end of the discussion the difference between the tube and transistor cases.

To achieve synchronization, a sync signal must be applied in such a manner that it can influence the instant at which a timing cycle is terminated by a regenerative action. This regeneration occurs when a tube comes out of cutoff, and therefore the sync signal may be injected at a grid, a cathode, or a plate. In the latter case the sync signal is principally effective because it is coupled to a grid through a plate-to-grid coupling capacitor. The precise details of synchronization will depend on the manner of sync injection, on whether the sync is applied to one or both tubes, and also in each case on the impedance of the sync source.

Synchronization at One Grid A sinusoidal signal from a low-impedance source is applied to grid G_1 in Fig. 19-15a. When $V1$ is cut off, the sync voltage will appear superimposed on the exponential G_1 waveform. During the time $V2$ is cut off, the amplified and inverted sync voltage will appear superimposed on the G_2 waveform. If, as is normally the case, the gain from G_1 to G_2 in the multi is large, then the sync voltage on G_2 will be large in comparison with the sync voltage on G_1. Hence, we shall not make a serious error by neglecting the sync voltage on G_1 and assuming that the instant $V1$ comes into conduction is unaffected by the sync voltage. In other words, the first portion of the multi is unsynchronized. To determine graphically how the sync voltage on G_2 influences the instant at which the timing cycle of G_2 is terminated, either we may add the sine voltage to the exponential at G_2 or else we may invert the sine voltage at G_2 and add it to the cutin voltage. A typical situation illustrating synchronization is indicated in Fig. 19-15. In Fig. 19-15b we show the sinusoidal sync voltage at G_1. In Fig. 19-15c this

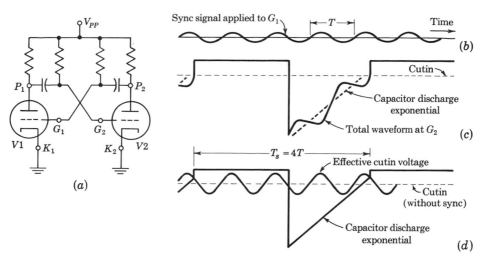

Fig. 19-15 Graphical construction to show synchronization of the astable multi in (a) to sine waves. For simplicity, overshoots and undershoots have been omitted. (b) Sync waveform at G_1; (c) sync combined with exponential discharge waveform at G_2; (d) sync combined with cutin voltage at G_2.

voltage has been amplified, inverted, and added to the exponential waveform at G_2. In Fig. 19-15d the sinusoidal sync voltage of Fig. 19-15c has been inverted and added to the cutin level to give the effective cutin-voltage curve shown. The waveform which would be observed on an oscilloscope whose input is at G_2 is that indicated in Fig. 19-15c. However, the construction in Fig. 19-15d is more useful from the point of view of analyzing the effect of the synchronizing voltage on the behavior of the multi. As previously discussed, stable synchronization requires that the multi exponential terminate on the effective cutin voltage while the latter has a negative slope. And the general characteristics suggested by Fig. 19-3a and b apply to the astable multi as well.

Synchronization at Both Grids To increase the range of synchronization of the multi, it is advantageous to sync both multi sections by applying sync signals to both grids simultaneously. Ordinarily in divider applications a multi is adjusted for nominally symmetrical operation, since no special advantage results from asymmetrical operation. For such a symmetrical multi, it is best to apply in-phase sync signals if the division ratio is to be an even number and to apply out-of-phase sync signals if the division ratio is to be odd. These results may be seen from Fig. 19-16. In Fig. 19-16a the division ratio (4:1) is even, the sync signals are in phase, and it appears that both timing exponentials meet the effective cutin voltage in a manner to ensure stability. In Fig. 19-16b the division ratio (3:1) is odd, the sync signals are again in phase, and while the exponential of the G_1 waveform is synchronized in a stable fashion, the

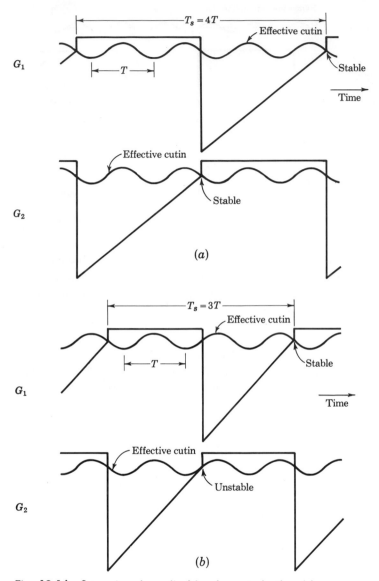

Fig. 19-16 Sync signal applied in phase to both grids.
(a) The synchronization is stable for both portions of the multi
waveform when the division ratio is even.   (b) The division
ratio is odd and the G_2 timing waveform is not locked in a
stable fashion.

G_2 waveform is not. Similarly, it is easy to show by drawing appropriate waveforms that when the sync signals are out of phase, both parts of the multi grid waveform will be synchronized in a stable fashion only for odd division ratios.

Waveforms Before leaving the subject of sine-wave synchronization of oscillators we wish to emphasize that such a divider does not deliver a sinusoidal output. In most cases the waveform is more nearly "square" (from a multi) or "triangular" (from a sweep circuit). If a sinusoidal waveform is desired, then the output of the relaxation divider is applied to the input of an amplifier whose output load is a tank circuit tuned to the desired frequency. Circuits for obtaining sinusoidal division without using a relaxation device are given in the next two sections.

The mechanism and characteristics of sine-wave synchronization of a transistor astable multi are essentially the same as in the case of the tube multi. A not important difference appears because, as already noted, when a transistor is in saturation its collector will not respond to a base signal which drives the transistor further into saturation. If therefore a sinusoidal signal from a low-impedance source is applied to, say, B_1 (for an n-p-n transistor), the amplified signal which appears at B_2 will have its negative half cycle clipped off. (If the sync-signal source impedance is not low, even less than half the sine wave will remain at B_2.) The waveforms for the transistor circuit are shown in Fig. 19-17 corresponding to the tube case represented in Fig. 19-15. Observe

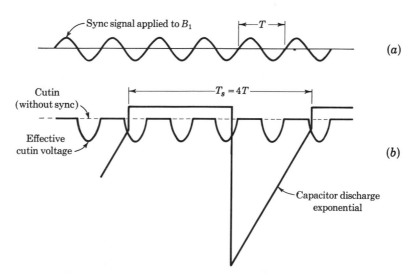

Fig. 19-17 Synchronization of an n-p-n transistor astable multi corresponding to the situation represented in Fig. 19-15 for a tube multi. (a) Sync signal; (b) waveform at B_2.

that in the transistor case, the frequency range of synchronization will be limited in comparison with the tube case. When the rising exponential intersects the effective cutin curve at the flattened negative excursion, the multi will not be synchronized.

19-9   A SINUSOIDAL DIVIDER USING REGENERATION AND MODULATION[2]

A divider for a sinusoidal signal which does not involve a relaxation oscillator is represented in block-diagram form in Fig. 19-18. The signal of frequency nf, whose frequency is to be divided by the factor n, is applied to an amplitude modulator. Simultaneously a signal of frequency $(n - 1)f$, whose origin will appear shortly, is also applied to the modulator. In the modulator the signal nf is modulated by the signal $(n - 1)f$. The modulator output contains many frequencies, among which are the input frequencies $(n - 1)f$ and nf as well as the sideband frequencies $nf + (n - 1)f = (2n - 1)f$ and $nf - (n - 1)f = f$. The filter selects the frequency f, which then appears as the output. The output is applied to a frequency multiplier which multiplies by the factor $n - 1$ to produce the feedback signal of frequency $(n - 1)f$ whose existence was postulated at the outset. *The loop gain must exceed unity if the oscillations are to be sustained.* This statement explains the use of the word *regeneration* in connection with this type of divider.

The modulator may consist of a pentagrid converter tube such as is used in a radio receiver to heterodyne the incoming radio-frequency signal with the local-oscillator signal to produce the intermediate-frequency signal. Such pentagrid converter tubes (the 6BE6, the 6SA7, and the 6BA7, for example) have two control grids. The signal component of the plate current of these tubes is essentially proportional to the *product* of the signals of the two control grids, and for this reason the tube may be used as a modulator. Alternatively, the modulator may consist of a transistor to which both signals are applied simultaneously. If the signals are large enough so that the transistor operates nonlinearly, sum and difference frequencies will appear at the transistor output.

The filter at the modulator output need consist of nothing more elaborate than a parallel-tuned LC circuit in the output lead of the modulator. The

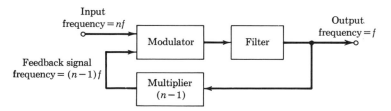

Fig. 19-18   A divider employing modulation and regeneration.

frequency multiplier may be of the conventional type in which a pentode or transistor is driven by a large-amplitude signal, the load being a resonant tank tuned to the $(n-1)$st harmonic of the input signal. If the input signal is large enough so that the output current flows in pulses, this current will be rich in harmonics of the input signal. Assuming a high-Q tank circuit, the output voltage will equal the product of the tank-circuit impedance at its resonant frequency times the component of output current at this frequency. The upper frequency of operation of dividers of the type given in Fig. 19-18 is limited only by the frequency at which modulators, filters, and multipliers may be operated.

We shall now show qualitatively that the only stable frequency at which the circuit can operate is precisely f. Assume, for example, that the output frequency momentarily drifts from f to some higher value $f + \delta$, where $\delta > 0$ and $\delta \ll f$. Then the output of the multiplier will be $(n-1)(f+\delta)$ and the modulator frequency will shift from f to $nf - (n-1)(f+\delta) = f - (n-1)\delta$, which frequency is less than f. Hence, if the frequency tends to increase, the circuit acts in such a direction as to counterbalance this tendency. Therefore, f may be a stable frequency.

If the multiplier were designed to deliver a frequency $(n+1)f$ instead of $(n-1)f$, then the output frequency would also be f. However, such a system is unstable. For example, if the output drifts from f to some higher value $f + \delta$ momentarily, then the modulator output will shift from f to $(n+1)(f+\delta) - nf = f + (n+1)\delta$. Since this frequency is further away from f than the assumed deviation δ, then the circuit causes the output frequency to drift away from the initial value f. When the frequency drifts far enough away from the resonant frequency of the tuned circuits (in the modulator and multiplier), the loop gain becomes less than unity and the output drops to zero. Therefore, we conclude that the multiplier must be designed for a frequency $(n-1)f$ and not $(n+1)f$.

The special case $n = 2$ is of some interest. Since $n - 1 = 1$ for $n = 2$, then the multiplication is by 1. Hence, division by 2 is accomplished without the use of a multiplier.

In the case of a relaxation oscillation we saw that increasing the size of the sync voltage might cause a change in division ratio. This result is not obtained with the circuit now under consideration. There exists a minimum input voltage below which the circuit output is zero. However, if this voltage of frequency nf is increased beyond the minimum value, the frequency f of the output remains unchanged, although the amplitude of the output increases.

The divider of Fig. 19-18 has two disadvantages. The first of these is its relative complexity. Second, it may turn out that the divider is not self-starting. The reason for this last feature is not difficult to see. The multipliers depend for their operation on having a large input signal so that the devices will be vigorously overdriven. When the driving signal is small, the devices may operate quite linearly and, hence, provide no appreciable harmonic components of current. Hence, the multipliers wait for the circuit to supply large driving signals, whereas the circuit cannot do so until the multipliers operate. To start the divider, it may be necessary to introduce a large transient voltage into the circuit.

On the other hand, the divider in Fig. 19-18 has a number of worthwhile advantages. The first of these, of course, is its ability to operate at high frequencies. Second, the proper division ratio depends essentially only on the tuning of several passive filter circuits, which tuning is not critical, and not on the stability of device characteristic. Lastly, if the tuning of the tank circuits should happen to drift out of range or if the input signal should be missing, the divider will furnish no output. The circuit, therefore, operates to give the correct output signal or it does not operate at all; that is, it is a *fail-safe* circuit. These features are to be compared with, say, the astable multi, which will divide incorrectly if the device characteristics drift excessively and which will continue to give an output signal, necessarily incorrect, if the input signal should fail.

19-10   THE LOCKED OSCILLATOR AS A DIVIDER[3]

A *sinusoidal* oscillator can be used as a frequency divider. Assume that the natural frequency of the oscillator is f_o and that a nominally sinusoidal synchronizing signal of frequency f_s, nearly equal to nf_o (n = an integer), is injected into the oscillator. The oscillator frequency will change from f_o to f_s/n and thereafter will run synchronously with the injected signal. Under these conditions the oscillator is said to be *locked*. The locked oscillator has the advantage of simplicity when compared with the divider using modulation and regeneration. It suffers, however, from the relative disadvantage that it continues to yield an output even in the absence of a synchronizing signal.

A careful analysis of the mechanism by which a sinusoidal oscillator locks to a synchronizing signal is difficult to achieve. This difficulty does not arise because any basically new principle is involved. The mechanism of

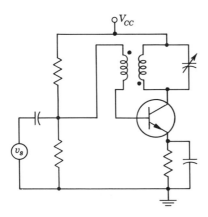

Fig. 19-19   A tuned-collector sinusoidal oscillator is synchronized by a sinusoidal signal applied to the base.

locking in oscillators is similar to the mechanism involved in the divider using modulation and regeneration.   The complication arises in the present case because all the functions of modulation, regeneration, and harmonic generation are accomplished in a single tube.   The device nonlinearity is essential to the locking mechanism.   Since a nonlinear characteristic is difficult to handle analytically, most locked-oscillator *design* is carried out experimentally.

A tuned-collector sinusoidal oscillator synchronized by an external signal applied in the base circuit is shown in Fig. 19-19.

19-11   SYNCHRONIZATION OF A SINUSOIDAL OSCILLATOR WITH PULSES[4]

Synchronization is basically a mechanism in which the phase of the sync signal is compared with the phase of the oscillator (relaxation or otherwise) and a continuous correction is applied to the oscillator to maintain some fixed phase relation.   We shall illustrate that one-to-one synchronization of a sinusoidal oscillator to a pulse-type sync signal may be accomplished if the operations of phase comparison and phase correction are carried out separately.

A circuit suitable for comparing the phase of a sinusoidal signal and a pulse is shown in Fig. 19-20.   The sinusoids are introduced into the meshes 1 and 2 in phase opposition with respect to the polarities of the diodes.   The pulses are introduced in such a manner as to appear with the same polarity in both meshes.   Let us neglect tentatively the effect of the pulses.   Then, as a result of the sinusoidal input, unidirectional currents will flow in mesh 1 and mesh 2 on alternate half cycles of the sinusoidal voltage.   The voltage drops v_{xo} and v_{oy} are as shown in Fig. 19-21a.   The combined voltage v_{xy} is then a complete sinusoidal signal whose average value is zero.   This signal is integrated by the RC integrating network, which then provides zero output voltage.   Now consider that positive pulses are applied which occur at such a time that the pulses exactly straddle the time when the sinusoidal signal passes

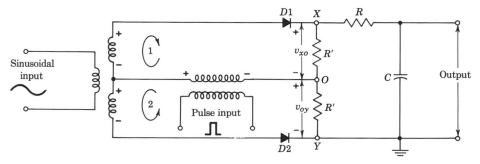

Fig. 19-20   A phase comparator or discriminator for a pulse and a sinusoidal signal.

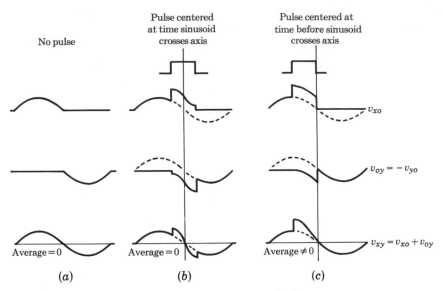

Fig. 19-21   The waveforms in the circuit of Fig. 19-20.

through zero.   The waveforms are indicated in Fig. 19-21b.   We see again that the average value of the voltage v_{xy} is zero and hence that the output voltage will again be zero.   Finally, from Fig. 19-21c, it may be seen that if the pulse occurs earlier than the moment of zero phase of the sinusoids, the average value of v_{xy} is not zero and the integrating circuit in this case will provide essentially a positive d-c output voltage.   Similarly, if the pulse is delayed, the output will be a negative d-c voltage.

A system in which the phase comparator of Fig. 19-20 is used to synchronize a sinusoidal oscillator to a pulse signal is shown in Fig. 19-22.   The device *VC* is a *voltage-variable capacitor*.   It consists of a junction diode which is operated reverse-biased and which consequently displays a transition capacitance that is a function of the voltage across the junction (Fig. 6-5).   (Junction diodes specifically designed for this type of service as voltage-controlled variable capacitors are commercially available and are known by the trade names Voltacap, Varicap, etc.)

The pulse sync signal and the oscillator signal are applied to the phase comparator, whose output is used to control the voltage across the voltage-variable capacitor.   This capacitor is incorporated into the tuned circuit of the oscillator so that a variation of voltage across *VC* will cause a change in oscillator frequency.   Suppose that the pulse and sinusoidal frequencies are identical and that an initial phasing adjustment has been made so that the comparator output is zero.   Now let the frequency of the sinusoidal signal change slightly.   Then with each succeeding cycle the phase error will become progressively larger and a nominally d-c voltage will appear at the phase-comparator output. The polarity of this comparator signal depends on whether the oscillator frequency is higher or lower than the pulse frequency. Thus the comparator output signal may be used to correct the oscillator frequency. It is to be noted that the system here described is a feedback control system in which the phase comparator is the error-sensing device and the voltage-variable capacitor is the agency through which the correction is applied. As in any feedback control system, antihunt circuits, not shown in Fig. 19-22, may be required.

This feedback control system of synchronization is of quite general applicability, although it has the disadvantage of involving usually fairly complicated circuitry. If it is required to synchronize a waveform generator of any arbitrary waveshape to a sync signal again of arbitrary form, it is only necessary to devise an appropriate phase-comparator circuit and frequency-controlling device. A frequency-controlling device is usually not difficult to find. If, for example, the oscillator to be synchronized is of the relaxation type, the frequency can usually be varied through the adjustment of some biasing voltage in the circuit.

There is a benefit to be derived from the use of this more complicated scheme even when simpler methods are effective, as in the synchronization of a relaxation device by the direct application of the sync pulses. The more complicated feedback system renders the synchronizing process less susceptible to interference from random noise pulses which may be superimposed on the sync signal. The noise pulses will occur at random times and hence at random phases in the cycle of the waveform generator (sinusoidal or otherwise) to be synchronized. Any individual noise pulse will have a negligible effect, and the combined effect of a large number of noise pulses will average out to zero. A feedback synchronizing system is widely used to lock the horizontal sweep generator of a television receiver to the synchronizing pulses from the transmitter.

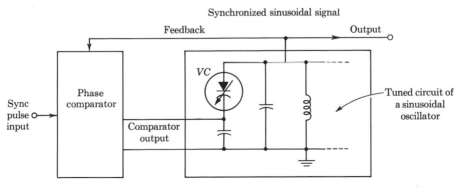

Fig. 19-22   A phase comparator and a voltage-variable capacitor are used to synchronize a sinusoidal oscillator to a pulse signal.

REFERENCES

1. Shenk, E. R.: Multivibrator, Applied Theory and Design, pt. I, *Electronics*, vol. 17, pp. 136–141, January, 1944; pt. II, vol. 17, pp. 140–145, February, 1944; pt. III, vol. 17, pp. 138–142, March, 1944.

2. Stansel, F. R.: A Secondary Frequency Standard Using Regenerative Frequency Dividing Circuits, *Proc. IRE*, vol. 30, pp. 157–162, April, 1942.
   Miller, R. L.: Fractional Frequency Generators Utilizing Regenerative Modulation, *Proc. IRE*, vol. 27, pp. 446–457, July, 1939.
   "Instruction Book for Frequency Calibrator Set AN/FRM-3," Reeves-Hoffman Corp., Carlisle, Pa.

3. Tucker, D. G.: The Synchronization of Oscillators, pt. I, *Electron. Eng.*, vol. 15, pp. 412–418, March, 1943; pt. II, vol. 15, pp. 457–461, April, 1943; pt. III, vol. 16, pp. 26–30, June, 1943.
   Adler, R.: A Study of Locking Phenomena in Oscillators, *Proc. IRE*, vol. 34, pp. 351–357, June, 1946.
   Sulzer, P. G.: Modified Locked-oscillator Frequency Dividers, *Proc. IRE*, vol. 39, pp. 1535–1537, December, 1951.
   Norrman, E.: The Inductance-Capacitance Oscillator as a Frequency Divider, *Proc. IRE*, vol. 34, pp. 799–803, October, 1946.
   Hughes, W. L.: Analysis and Performance of Locked-oscillator Frequency Dividers Employing Nonlinear Elements, *Proc. IRE*, vol. 41, pp. 241–245, February, 1953.

4. Clark, E. L.: Automatic Frequency Phase Control of Television Sweep Circuits, *Proc. IRE*, vol. 37, pp. 497–500, May, 1949.
   Clark, E. G.: Stabilizing Color Carrier Reinsertion Oscillator, *Electronics*, vol. 27, pp. 142–146, July, 1954.

20 / TRANSIENT SWITCHING CHARACTERISTICS OF DIODES AND TRANSISTORS

In Chap. 6, we consider in detail the final states of semiconductor and thermionic devices which, being used as switches, are switched from ON to OFF and vice versa. In this chapter we inquire into the transient behavior during switching.

A semiconductor device may be analyzed and described in terms of the currents which flow into or out of its terminals. Such a description leads to an *equivalent circuit* in which the terminal currents are the parameters of interest. Alternatively the device may be described in terms of the *charges* which are *stored* at the device junctions and within the body of the semiconductor and which thereby *control* the operation of the device. We shall describe the switching in terms both of the method of *equivalent circuits* and of the *stored-charge-control method*.

20-1 DIODE FORWARD RECOVERY TIME[1]

When a diode is driven from the reversed condition to the forward condition or in the opposite direction, the diode response is accompanied by a transient, and an interval of time elapses before the diode recovers to its steady state. We shall consider now the recovery transient associated with driving the diode from the reverse- to the forward-biased condition.

The nature of the forward recovery transient depends on the magnitude of the current being driven through the diode and the rise time of the driving signal. Consider the voltage which develops across the diode when the input is a current source supplying a step of current

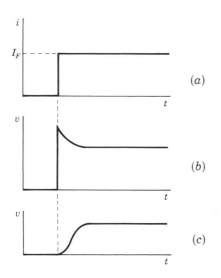

(a)

(b)

(c)

Fig. 20-1   A step of current as in (a) is applied to a diode. In the idealized case, if the current step is large enough and fast enough, the diode voltage has the form shown in (b).   The waveform in (c) results when the current step is small.

I_F, as in Fig. 20-1a.   If the current amplitude is comparable to or larger than the diode rated current, and if the rise time of the current step is small enough, then the waveform of the voltage which appears across the diode is as shown in Fig. 20-1b.   We assume that initially the diode is not appreciably reverse-biased.   Hence we neglect the time required to charge the junction transition capacitance or any external shunt capacitance across the diode.   The pro-nounced overshoot results from the fact that initially the diode acts not as a p-n junction diffusion device but rather as a resistor.   In the steady-state condition, the current which flows through the diode is principally a diffusion current which results from the gradient in the density of minority carriers. If the current is large enough, then there will also be an ohmic drop across the diode.   The overshoot in Fig. 20-1b results from the fact that the ohmic drop is initially very large.   For immediately after the application of the current, the holes, say, will not have had time to diffuse very far into the n side in order to build up a minority-carrier density.   Therefore, except near the junction, there will be no minority charge to establish a density gradient, and current flow through the mechanism of diffusion will not be possible.   Indeed, an electric field will be required to achieve current flow by exerting a force on the majority carriers.   This electric field gives rise to the ohmic drop.   With the passage of time, however, the ohmic drop will decrease as minority carriers become available further and further from the junction and current by diffusion takes over.

From the above discussion we may expect that the magnitude of the overshoot will increase as the rise time of the current waveform decreases. For if the current rises in nonzero time, there will be some opportunity for the diffusion of minority carriers during the rise of the current.   For a fixed rise time, as the magnitude of the step I_F increases, the overshoot becomes larger and stands out more pronouncedly with respect to the steady-state diode voltage attained after the decay of the overshoot. The reason for this behavior is that the peak of the overshoot, to the extent that the diode behaves like a resistor, increases linearly with current. However, in the steady state, where Eq. (6-1) applies, the diode voltage increases logarithmically with current. Many of the features described in connection with the forward recovery transient are brought out in Fig. 20-2. If the rise time of the current were shorter, the waveforms would have even more pronounced overshoots.

When the current step applied to the diode is small enough so that the ohmic drop is negligible, the diode may be represented by its small-signal equivalent. In this equivalent representation the diode is replaced by a conductance g shunted by a diffusion capacitance C_D. The conductance $g = I/\eta V_T$ increases with current I. The diffusion capacitance is similarly proportional to I [Eqs. (20-9)]. Consequently the time constant C_D/g is independent of the diode current. The waveform of diode voltage, in this case, has the form represented in Fig. 20-1c.

At large current amplitudes, we note that the diode behaves in a manner such that if we sought a simple circuit representation we should probably use a combination of a resistor and inductor. At low currents the diode is representable by a parallel resistor-capacitor combination. On this basis, we might expect that at intermediate currents the diode might behave as though all three elements were present—resistance, inductance, and capacitance. In such a case oscillations in the response are to be anticipated and are indeed observed.

The forward recovery time t_{fr} for a specified rise time of the input current is the time difference between the 10 percent point of the diode voltage and the time when this voltage reaches and remains within 10 percent of its final value. From Fig. 20-2 we see that for the 1N695A operating at $I_F = 200$ mA, $t_{fr} \approx 80$ nsec. For this same diode the reverse recovery time t_{rr} (defined in the next section) is $t_{rr} \approx 300$ nsec. For most diodes, particularly the passivated epitaxial planar silicon diode, $t_{fr} < t_{rr}$. We shall now see that, even if t_{fr} is comparable to t_{rr}, the forward recovery time does not usually constitute a serious problem, and only in rather special situations does it need to be a cause for concern.

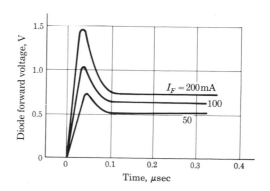

Fig. 20-2 Forward switching transient, from zero bias, for type 1N695A germanium diode for various forward currents. The input rise time is 0.03 μsec (30 nsec). (Courtesy of Transitron Electronic Corporation.)

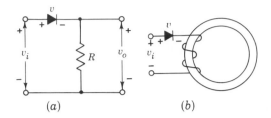

Fig. 20-3   (a) A transmission circuit in which the diode forward recovery transient causes no difficulty; (b) a diode application where the forward recovery transient may cause difficulty.

Diode Circuits   The simple circuit of Fig. 20-3a (discussed in more detail in Sec. 7-2) is intended to allow transmission of only that part of the input waveform which nominally exceeds zero in voltage.   Consider the circuit response as the input swings through zero voltage.   If the signal is small, corresponding to the response illustrated in Fig. 20-1c, we see that a nonzero forward recovery time actually *increases* the speed of response of the circuit. On the other hand, if the input signal is large enough to cause a peak in diode voltage, as in Fig. 20-2, the effect will still be negligible provided that the input voltage v_i is large compared with the diode voltage v.   For example, if the input is a 10-V step and $R = 100 \ \Omega$, so that a current of 100 mA flows, then from Fig. 20-2 the peak diode voltage is about 1 V and the steady-state diode voltage is 0.7 V.   Thus the output voltage starts at 10 V, goes to 9.0 V, and settles down to 9.3 V for an ideal input step.   For a finite input rise time the output would probably have no maximum or minimum but would rise to its steady-state value monotonically, with a rise-time improvement over the input.

Now let us consider the situation depicted in Fig. 20-3b.   Here we see that a step of voltage is to be applied through a diode to a winding on a toroidal core of magnetic material.   (The toroid is of material which has a rectangular hysteresis loop.)   The applied signal is intended to magnetize the core and leave it magnetized in one direction.   A different signal, applied through a different diode, and possibly even to a different winding, will be used to reverse the state of magnetization when required.   Not uncommonly, magnetizing currents of many tens of milliamperes may be required to achieve the necessary magnetization.

Assume that the core is to be magnetized as the result of the application of a very abrupt step of voltage, which is applied as the signal v_i in Fig. 20-3b. Initially the diode acts like a resistor rather than a p-n junction device, and the voltage v across the diode, shortly after the input step, will be larger than its steady-state value.   Therefore the current through the diode and winding will be smaller initially than in the steady state.   The current will rise from its initial value to its steady-state value as the diode voltage falls from its initial value to its steady-state value.   And if the steady-state current provided is not much larger than the current required for magnetization, we shall have to wait until the forward recovery transient is nominally completed before the required magnetization has been achieved.   In addition, the larger voltage which appears across the diode during the transient interval results in an extra dissipation of power in the diode.   Special diodes are available for use as drivers for magnetic cores. These diodes are designed to minimize the magnitude and duration of the forward recovery transient. For diodes intended for this type of service, manufacturers will usually supply curves like those shown in Fig. 20-2.

20-2   DIODE REVERSE RECOVERY TIME[2]

When an external voltage is impressed across a junction in the direction to reverse-bias it, very little current flows. The reason is that only the minority carriers on each side of the junction have charge of proper sign to carry current across the junction. The density of minority carriers in the neighborhood of the junction in the steady state is shown in Fig. 20-4a. Here the levels p_{no} and n_{po} are the thermal-equilibrium values of the minority-carrier densities on the two sides of the junction in the absence of an externally impressed voltage. When a reverse voltage is applied, the density of minority carriers is as shown by the solid lines marked p_n and n_p. Far from the junctions the minority-carrier density remains unaltered, but as these carriers approach the junction they are rapidly swept across and the density of minority carriers diminishes to zero at the junction. The current which flows, the reverse saturation current, is small because the density of thermally generated minority carriers is very small.

When the external voltage forward-biases the junction, the steady-state density of minority carriers is as shown in Fig. 20-4b. The number of minority carriers is very large. The minority carriers have, in each case, been supplied from the other side of the junction, where, being majority carriers, they are in plentiful supply.

If the external voltage is suddenly reversed in a diode circuit which has been carrying current in the forward direction, the diode current will not immediately fall to its steady-state reverse-voltage value. For the current

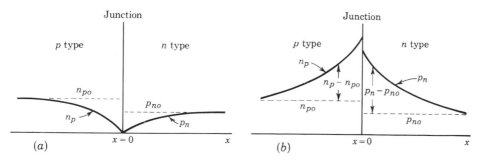

Fig. 20-4   Minority-carrier density distribution as a function of the distance x from a junction.   (a) A reverse-biased junction; (b) a forward-biased junction. The injected or excess hole (electron) density is $p_n - p_{no}$ ($n_p - n_{po}$). (The diagram is not drawn to scale since $p_n \gg p_{no}$ and $n_p \gg n_{po}$.)

cannot attain its steady-state value until the minority-carrier distribution, which at the moment of voltage reversal had the form in Fig. 20-4b, reduces to the distribution in Fig. 20-4a. Until such time as the *injected* or *excess minority-carrier density* $p_n - p_{no}$ (or $n_p - n_{po}$) has dropped nominally to zero the diode will continue to conduct easily, and the current will be determined by the external resistance in the diode circuit.

Storage and Transition Times   The sequence of events which accompanies the reverse biasing of a conducting diode is indicated in Fig. 20-5. We consider that the voltage in Fig. 20-5b is applied to the diode-resistor circuit in Fig. 20-5a. For a long time, and up to the time t_1, the voltage $v_i = V_F$ has been in the direction to forward-bias the diode. The resistance R_L is assumed

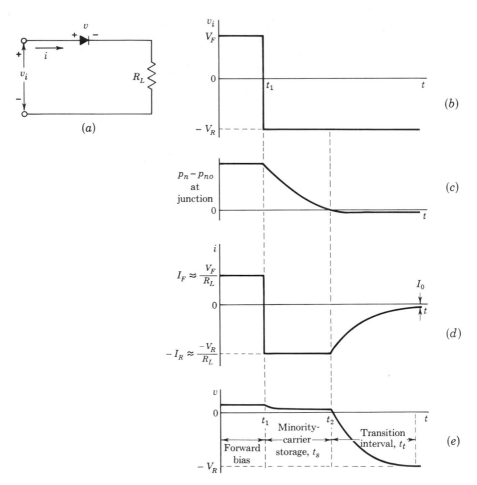

Fig. 20-5   The waveform in (b) is applied to the diode circuit in (a); (c) the excess carrier density at the junction; (d) the diode current; (e) the diode voltage.

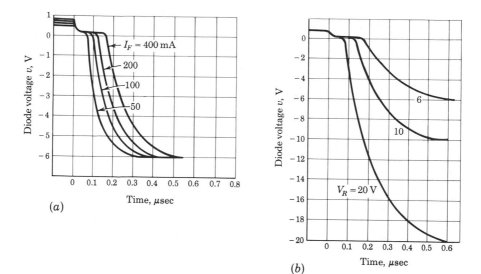

Fig. 20-6 Reverse recovery waveforms, 1N695A diode. (a) Fixed reverse voltage $V_R = 6$ V and $R_L = 1$ K for various forward currents; (b) fixed forward current $I_F = 200$ mA and $R_L = 2$ K for various reverse voltages. (Courtesy of Transitron Electronic Corporation.)

large enough so that the drop across R_L is large in comparison with the drop across the diode. Then the current is $i \approx V_F/R_L = I_F$. At the time $t = t_1$ the input voltage reverses abruptly to the value $v_i = -V_R$. For the reasons described above, the current does not drop to zero but instead reverses and remains at the value $i \approx -V_R/R_L = -I_R$ until the time $t = t_2$. At $t = t_2$, as is seen in Fig. 20-5c, the injected minority-carrier density at the junction has dropped to zero; that is, the minority-carrier density has reached its equilibrium state. If the diode ohmic resistance is R_d, then at the time t_1 the diode voltage falls slightly [by $(I_F + I_R)R_d$] but does not reverse. At $t = t_2$, when the excess minority carriers in the immediate neighborhood of the junction have been swept back across the junction, the diode voltage begins to reverse and the magnitude of the diode current begins to decrease. The interval t_1 to t_2, for the stored-minority charge to become zero, is called the *storage time t_s*.

The time which elapses between t_2 and the time when the diode has nominally recovered is called the *transition time t_t*. This recovery interval will be completed when the minority carriers which are at some distance from the junction have diffused to the junction and crossed it and when, in addition, the junction transition capacitance across the reverse-biased junction has charged through R_L to the voltage $-V_R$.

The reverse-recovery diode-voltage waveforms for a type 1N695A diode are shown in Fig. 20-6. These waveforms are to be compared with the waveform of Fig. 20-5e. The time t_1 in Fig. 20-5 at which the input voltage is reversed corresponds in Fig. 20-6 to the time $t = 0$. In Fig. 20-6a various forward currents are indicated, and the plots show that for fixed reverse current ($I_R = V_R/R_L$) the storage time is longer for larger forward currents. In Fig. 20-6b the forward current is fixed, and the plots show that in this case larger reverse currents reduce the storage time.

Manufacturers normally specify the reverse recovery time of a diode t_{rr} in a typical operating condition in terms of the current waveform of Fig. 20-5d. The time t_{rr} is the interval from the current reversal at $t = t_1$ until the diode has recovered to a specified extent in terms either of the diode current or of the diode resistance. If the specified value of R_L is larger than several hundred ohms, ordinarily the manufacturers will specify the capacitance C_L shunting R_L in the measuring circuit which is used to determine t_{rr}. Thus we find, for the Fairchild 1N3071, that with $I_F = 30$ mA and $I_R = 30$ mA the time required for the reverse current to fall to 1.0 mA is 50 nsec. Again we find, for the same diode, that with $I_F = 30$ mA, $-V_R = -35$ V, $R_L = 2$ K, and $C_L = 10$ pF ($-I_R = -35/2 = -17.5$ mA), the time required for the diode to recover to the extent that its resistance becomes 400 K is $t_{rr} = 400$ nsec. Commercial switching-type diodes are available with times t_{rr} in the range from less than a nanosecond up to as high as 1 μsec in diodes intended for switching large currents.

20-3 THE FUNDAMENTAL EQUATION OF CHARGE–CONTROL ANALYSIS

Up to the present we have described the diode principally in terms of its terminal voltage and current and in terms of an equivalent-circuit representation. The behavior of the diode (and the transistor) may also be accounted for in terms of its current and its *excess minority-carrier charge*. We shall now engage in some preliminary considerations concerning this charge-control method of analysis.

Excess Minority-carrier Charge Let us direct our attention to the excess minority-carrier charge Q in a volume of a semiconductor. The charge Q is not the total charge in the volume; within the body of a semiconductor the total charge is always zero. Rather the charge Q refers only to the excess of minority charge over the corresponding charge under the quiescent condition in which no current is flowing. Thus in Fig. 20-4 the density (the charge Q per unit volume) is measured by the difference between the solid and dashed plots, $n_p - n_{po}$ or $p_n - p_{no}$, and, as in Fig. 20-4b, may be positive or, as in Fig. 20-4a, may be negative.

Let us introduce at $t = 0$ into some volume just such an excess minority-carrier charge Q_o. Suppose further that we have selected a volume whose surface is so far from the location of the charge Q_o that the charge may not leave the volume. This charge will nonetheless disappear in time. Before the introduction of the charge there existed in the volume an equilibrium distribution of majority and minority carriers. This equilibrium distribution is attained when a balance is struck between the thermal generation of new hole-electron pairs and the disappearance of pairs by recombination. If an additional charge of minority carriers is introduced, the recombination will proceed more rapidly than the generation and the excess minority carriers will diminish. The probability, within a given time, that a majority carrier will find a minority carrier with which to combine will surely double if the number of minority carriers is doubled. More generally, the rate at which the excess minority-carrier charge will disappear is proportional to the amount of such charge Q present. We may therefore write

$$\frac{dQ}{dt} = -\frac{1}{\tau} Q \tag{20-1}$$

in which τ is a constant with the dimensions of time. Integrating Eq. (20-1) subject to the initial condition that $Q = Q_o$ at $t = 0$, we have

$$Q = Q_o \epsilon^{-t/\tau} \tag{20-2}$$

Mean Lifetime From Eq. (20-2) we see that τ is the time constant in the exponential decay of the excess minority-carrier charge. This parameter τ also has an interpretation as the *mean lifetime* of the minority carriers, as we shall now prove. During the interval between t and $t + dt$ the charge has changed by the amount dQ, given by

$$dQ = -\frac{Q_o}{\tau} \epsilon^{-t/\tau} \, dt \tag{20-3}$$

Since a magnitude of charge $|dQ|$ has "lived" for a time t, then the average time $\bar{t}$ of existence of the charge, or the mean lifetime, is

$$\bar{t} = \frac{\int_{Q_o}^{0} t \, |dQ|}{Q_o} = \frac{\int_{0}^{\infty} \frac{tQ_o}{\tau} \epsilon^{-t/\tau} \, dt}{Q_o} = \tau \tag{20-4}$$

The Conservation of Minority-carrier Charge The charge Q may change not only because of the recombination of charge but also because charge is being delivered to the volume containing Q or is being removed from it by current crossing the surface of the volume. If i is taken as the current which flows *into* the volume, then Eq. (20-1) must be replaced by

$$\frac{dQ}{dt} + \frac{Q}{\tau} = i \tag{20-5}$$

Equation (20-5), expressing the *law of the conservation of charge* (the continuity equation), is the *fundamental* equation of charge-control analysis. It must be kept in mind, in view of our definition of Q, that i represents the current carried by charges which, upon entering the volume, become excess minority carriers.

Charge-control Description of a Diode For simplicity of discussion we shall assume that one side of the diode, say the p side, is so heavily doped in comparison with the n side that current is carried across the junction entirely by holes moving from the p side to the n side. Then an excess minority-carrier charge will exist only on the n side and we need apply Eq. (20-5) only to this region. Consider a steady-state situation in which the diode current I is constant. This hole current I flows into the volume which consists of the n side of the diode junction, and, having entered the region, generates an excess minority-carrier charge. Leaving the region across a surface far removed from the junction is an equal current. This latter current, however, is not to be included in i in Eq. (20-5) because this current consists principally of the flow of electrons which are not minority carriers. Therefore in Eq. (20-5) we set $i = I$, and since we have assumed a steady-state condition, we set $dQ/dt = 0$ to obtain

$$I = \frac{Q}{\tau} \tag{20-6}$$

Equation (20-6) is important enough to warrant restatement in words. It says that the diode current (which is the current of holes crossing the junction) is proportional to the stored charge Q of excess minority carriers, the factor of proportionality being the reciprocal of the decay time constant (the mean lifetime) of the minority carriers. Thus the relationship between the charge Q and the current I, in the steady state, is that the current supplies minority carriers at the rate at which these minority carriers are disappearing owing to the process of recombination.

We now find ourselves describing the diode in terms of the diode current I and the stored charge Q rather than the current I and the junction voltage V. One immediately apparent advantage of this charge-control description is that the exponential relationship between I and V is replaced now by the linear relationship between I and Q. The charge Q also makes a simple parameter in terms of which to determine whether the diode is forward- or reverse-biased. The diode is forward-biased when Q is positive and reverse-biased when Q is negative.

As already observed above, the *total* charge (not Q) in the body of the semiconductor must be zero. Thus where, as in Fig. 20-4, an excess charge of holes exists, so must there be an identical distribution of majority charges (electrons) to ensure charge neutrality. This charge distribution is not shown in Fig. 20-4, nor does it enter explicitly into Eq. (20-5). These electrons are, however, precisely those with which the holes combine and thereby disappear.

20-4   EQUIVALENT–CIRCUIT DESCRIPTION OF A DIODE

We shall now point out the correspondence between the charge-control description of the diode and the equivalent-circuit description in which the diode is represented by the parallel combination of a resistor r and a (diffusion) capacitor C_D.   When a junction carries a steady forward current I, a minority-carrier charge Q is stored and a voltage V appears across the junction.   The current I and voltage V are related by Eq. (6-1).   To represent the storage of charge in an equivalent circuit we include in the circuit a capacitor C_D.   To allow for the decay of this charge we shunt across the capacitor a resistor r. Now we know that when the diode current drops to zero the excess minority charge decays with an unvarying time constant τ.   Accordingly, if the equivalent circuit is to represent this situation, it is required that the time constant for discharge of the capacitor must equal the storage-time decay constant,

$$rC_D = \tau \tag{20-7}$$

This same result may be arrived at more formally from Eq. (20-6).   Differentiating with respect to V we have

$$\frac{dQ}{dV} = \tau \frac{dI}{dV} \tag{20-8}$$

We let $dQ/dV \equiv C_D$, the incremental diffusion capacitance, and

$$\frac{dI}{dV} \equiv \frac{1}{r} \equiv g$$

the incremental conductance of the junction, and we arrive again at Eq. (20-7).
Using Eq. (6-12) for r and Eq. (20-7) for C_D we have

$$r = \frac{\eta V_T}{I} \qquad C_D = \frac{\tau I}{\eta V_T} \tag{20-9}$$

Note that r varies inversely with current, C_D is proportional to current, and the product rC_D is independent of I.   Observe also that C_D is proportional to the mean lifetime τ.   Experimentally it is found that τ varies over a wide range, from diode to diode, because recombination is greatly affected by impurities and by volume and surface imperfections in the crystal.   Values of τ ranging from about 1 nsec to 1,000 μsec have been observed.[3]   For $I = 1$ mA, $\eta = 1$, and $V_T = 26$ mV (at room temperature), we find that $r = 26$ Ω and C_D lies in the range 40 pF to 40 μF.

20-5   DIODE STORAGE AND TRANSITION TIMES

To calculate the diode storage time t_s let us assume that there is a constant diode current I_F so that an excess charge $Q_o = \tau I_F$ is stored.   Let the current be changed abruptly at $t = 0+$ to $-I_R$.   We seek a solution to

$$\frac{dQ}{dt} + \frac{Q}{\tau} = -I_R \tag{20-10}$$

subject to $Q = Q_o$ when $t = 0$.   The time t at which $Q = 0$ is the time $t = t_s$.

We find

$$t_s = \tau \ln \left(1 + \frac{I_F}{I_R} \right) \tag{20-11}$$

In a diode in which the doping on both sides must be taken into account, τ in Eq. (20-11) is an "effective lifetime" which depends on the lifetimes on each side of the junction and the relative doping on each side of the junction.

If t_s is given for one value of I_F/I_R, τ may be calculated and Eq. (20-11) may be used to calculate t_s for other values of I_F/I_R. Physical diodes[4] behave reasonably but hardly exactly in accord with Eq. (20-11). An improvement results if, on empirical grounds, we recognize that τ depends somewhat on the absolute value of I_F and depends as well on the ambient temperature. The typical variation of τ with I_F and with temperature is shown in Fig. 20-7 for a silicon diode. These plots are normalized with respect to a forward current of 10 mA and an ambient temperature of 25°C, under which conditions $\tau \approx 5$ nsec. At 25°C a fivefold change in I_F produces, at fixed I_F/I_R, a change in t_s of less than 60 percent. Similarly, at 25°C, a 25° change in temperature produces less than a 10 percent change in t_s. In the fabrication of high-speed computing diodes, gold is introduced as an impurity. These gold atoms act as recombination centers and decrease the minority-carrier lifetime.[4]

The transition time t_t (Fig. 20-5e) is the interval during which the principal change in the diode is the charging of the transition capacitance C_T which appears across the reverse-biased junction. Since C_T decreases with increasing magnitude of reverse bias (Sec. 6-3), a conservative estimate of t_t may be calculated by assuming that C_T remains constant at its largest capacitance value. Referring to Fig. 20-5a and e we see that the time constant associated with this transition interval is $R_L C_T$. We may, as is usual, estimate the transition interval at $3R_L C_T$. Finally the reverse recovery time $t_{rr} = t_s + t_t$.

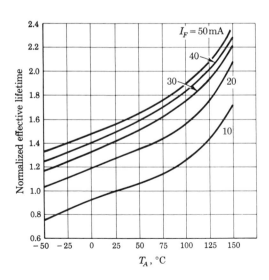

Fig. 20-7 Plots showing how the effective lifetime τ depends on the value of the forward current and on the ambient temperature in a silicon diode (planar epitaxial passivated 1N3605, 1N3606, 1N3608, and 1N3609). At $T_A = 25$°C and $I_F = 10$ mA, $\tau \approx 5$ nsec. (Courtesy of General Electric Company.)

20-6   LIMITATION ON ACCURACY OF ANALYSIS

We have assumed that there is a one-to-one correspondence between the diode current and the excess stored charge. We have neglected the fact that actually the current may well depend not only on the stored charge but also on the manner in which this stored charge is *distributed*. This point will stand out more clearly if we refer again to Fig. 20-4b, where, for simplicity, we shall again neglect the excess charge on the p side of the junction. At any steady diode current the density of excess minority charge is a maximum at the junction and falls off exponentially with distance away from the junction. The area between the plot for p_n and the level p_{no} is a measure of the total excess stored charge. We have assumed that if the diode current is changed to a new, say lower, value the form of the plot for p_n will remain the same but that its level will become lower. Thus we have assumed that, when the diode is at the point of going from forward bias ($Q > 0$) to reverse bias ($Q < 0$), $p_n - p_{no}$ becomes zero at the same time for all values of x. This assumption is certainly reasonable when the current changes slowly enough but may not be valid when the diode current is caused to make a very rapid change. Thus suppose that the diode current, initially in the forward direction, is suddenly reversed. Then the excess minority carriers will return across the junction whence they came. Those excess carriers far removed from the junction must first diffuse back to the junction before they can be removed. Hence it may well be that $p_n - p_{no}$ at the junction may return to zero or even reverse sign whereas further removed from the junction $p_n - p_{no}$ is still positive. The charge-control method does not take into account this possibility of a change in distribution of the stored charge. The method is therefore of limited accuracy, and Eq. (20-6) is at best an approximation. It should, of course, be kept in mind that the equivalent-circuit method has a similar limitation. An exact solution of the problem would require a solution to the diffusion equation which controls the flow of minority carriers. An important point to keep in mind is that both in the equivalent-circuit approach and in the charge-storage method we are using the best approximations consistent with assuming that the *diode can be described in terms of a single time constant.*

20-7   THE CHARGE–STORAGE DIODE[5]

The recovery of a forward-biased diode to the reverse-biased condition is achieved in two stages. First there is a storage phase, which persists as long as is necessary to remove the minority carriers in the immediate neighborhood of the junction. This stage is followed by the transition stage, during which the junction transition capacitance is charged and during which there are removed those minority carriers which were located some distance from the junction. It has been found possible to control the doping in a diode in such a way that minority carriers are restricted to a very narrow region in the

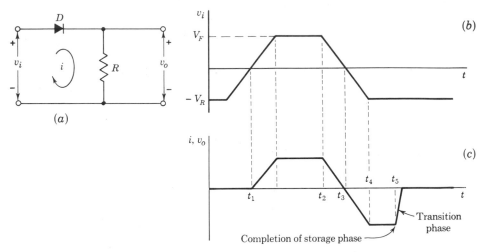

Fig. 20-8 (a) A pulse is applied to a charge-storage diode circuit; (b) the applied-pulse waveform with rise and fall times explicitly indicated; (c) plot of diode current and output waveform.

immediate neighborhood of the diode junction. In such diodes, especially if the junction barrier capacitance is small, the transition time of the recovery is extremely small. Hence, the diodes are also referred to as *snap-off* or *step-recovery diodes*. These diodes may complete the transition phase of the recovery in times as short as 0.1 nsec and may therefore be used to generate very fast waveforms and pulses of very short duration.

To consider the application of a charge-storage diode as a generator of fast waveforms we examine the idealized waveforms of Fig. 20-8. The pulse-type signal in Fig. 20-8b is applied to the diode-resistor circuit in Fig. 20-8a, and the current waveform, the same as the waveform of the output voltage, is shown in Fig. 20-8c. At $t = t_1$ the diode conducts and responds to the input signal. At $t = t_3$ the applied voltage is in the direction to reverse-bias the diode. However, because of charge storage, the diode conducts in the reverse direction and the output persists, following the input. During the interval from t_2 to t_3 the charge stored in the diode is decreasing, since this charge decreases with decreasing diode current. The charge stored continues to decrease at an accelerated rate during the interval t_3 to t_4, where the current is flowing in the reverse direction. We have, however, assumed that at $t = t_4$ the charge has not yet been completely removed. The reverse current therefore continues to flow, and the storage phase does not terminate until $t = t_5$. At this point the diode "snaps off" with a very small transition time. As indicated, the rate of change of the voltage during this transition phase may be very much more rapid than the rate of change of the input waveform. Observe further that the amplitude of the very abrupt change is equal to V_R.

The speed of voltage change during "snap-off" is not related to the speed of the input waveform. We shall now see, however, that unless the input waveform is fast enough, the amplitude of the voltage change corresponding to the transition phase may be reduced. First assume in Fig. 20-8c that during the interval from t_2 to t_4 the input waveform falls very slowly. Then it may well be that at the time $t = t_3$ a very large part of the stored charge has been removed. For example, if the voltage falls slowly enough, the stored charge may, at each instant, be nearly equal to the steady-state charge corresponding to the current flowing at the moment. Then, in this case, at $t = t_3$, where the current is zero so also will be the stored charge. In any event, if only a little of the charge remains at $t = t_3$, and since the voltage continues to fall slowly after $t = t_3$, the voltage will not be able to go very far negative before the stored charge has been completely removed. Consequently the amplitude of the transition jump will be correspondingly reduced. In summary, we note that for effective use of the snap-off characteristic, the voltage change of the input signal which turns off the diode must be fast enough to attain an appreciable back-biasing voltage level before the storage phase is completed.

20-8 CHARGE–STORAGE–DIODE PULSE GENERATOR[6]

A relatively simple pulse generator which uses a charge-storage diode is shown in Fig. 20-9a. Initially the shorted length of delay line which appears to the right of the diode D behaves as a resistance equal to the characteristic impedance Z_o. Then on each positive cycle of the sinusoidal input waveform v_i (indicated in Fig. 20-9b) the charge-storage diode D will conduct. We require that the diode conduct only long enough to build up a stored charge and that adequate time be available in each cycle so that the diode may recover. For this reason R_1 and C_1 have been included in the circuit to provide across C_1 a negative bias so that the diode conducts only in the neighborhood of the positive peaks of the input signal. The R_1C_1 combination, together with the diode, is a clamping or d-c-restorer circuit, discussed in detail in Chap. 7. In the present case, however, we want diode conduction for an appreciable portion of the cycle, so that the R_1C_1 time constant is rather smaller than will ordinarily be encountered in a clamping circuit.

The diode current waveform, idealized, is shown in Fig. 20-9c. Because of charge storage, the diode conducts for a time in the reverse direction, and when the storage interval is completed, the current rises abruptly to zero during the transition phase. The capacitor C has been included in the circuit so that when the diode current changes abruptly this abrupt current change need not also occur in the inductance of the transformer winding. We may now compare the circuit of Fig. 20-9a with the delay-line differentiator circuit of Fig. 3-27. The abrupt diode current change produces an abrupt change in the voltage across R_L. This abrupt change travels down the line and returns after a time $2t_d$, where t_d is the line delay time. When the reflected

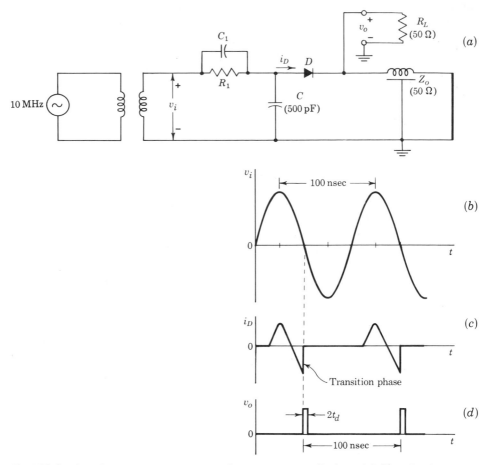

Fig. 20-9 A pulse generator using a charge-storage diode. (*a*) The circuit configuration; (*b*) the input sinusoid; (*c*) the current in the diode; (*d*) the output pulse train.

signal reappears at the line input, the diode D is back-biased, so that only the matching resistor $R_L = Z_o$ is across the line and no further reflections take place. The current waveform, other than the abrupt transition jump, is slow enough so that the delay line behaves for it much as a short circuit. Consequently, except for the appearance of a pulse $2t_d$ in duration once per cycle the output waveform otherwise is nominally zero.

The requirement, discussed in Sec. 20-7, that the diode be turned off rapidly requires that the driving sinusoidal signal be of high frequency. The circuit of Fig. 20-9 has been used successfully in the range 10 to 500 MHz to produce pulses of nanosecond duration. This circuit is obviously not suitable for use at low pulse-repetition frequencies. On the other hand, suppose that the signal applied in the circuit of Fig. 20-9 is a pulse train in which the pulses, as in Fig. 20-8, can carry the diode to the reverse-biased condition before the storage time is completed.   Then, in this case, the pulse-repetition rate of the output signal will be the same as the rate of the input signal.   However, the output pulse will have a duration determined by the delay line and will have a much smaller rise and fall time than the input pulse.

20-9   SEMICONDUCTOR–METAL JUNCTION DIODES[7]

We have considered, up to the present, only diodes formed by the junction of p-type and n-type semiconductor material.   Diodes may also be formed, however, by the junction of a semiconductor and a metal.   Such diodes have advantages with respect to speed of operation.   Their development has lagged behind that of p-n junction diodes because of difficulties encountered in manufacture.   The static volt-ampere characteristics of semiconductor-metal diodes are entirely similar to the corresponding characteristics of p-n junction diodes.   In the semiconductor-metal diode, however, the voltage at which the diode break point occurs is controllable by selection of the type of metal used. Accordingly, silicon-metal diodes of this type are available which exhibit a static characteristic similar to that of a germanium p-n junction diode, with a break point at about 0.2 V.   Others are available with a break point near 0.5 V and are more like silicon p-n diodes.

In a diode which is formed at the junction of an n-type semiconductor and a metal, conduction in the forward direction occurs when the metal is biased positively with respect to the semiconductor.   Current flows across the junction by virtue of the transport of electrons from the n-type semiconductor to the metal and hence is a majority-carrier current.   Current in the reverse direction is restrained by the existence at the junction of a potential barrier which opposes such flow.   Electrons which have crossed the junction to enter the metal are not distinguishable from the very plentiful electrons in the metal which constitute the conduction electrons of the metal.   When, therefore, the voltage across the junction is reversed, these electrons are no more able to return across the junction than are the electrons of the metal. Accordingly, unlike the situation in the p-n junction diodes, there is no appreciable storage of a minority-charge density, and the storage time is negligible except for very large forward currents.   The reverse recovery time of a very fast p-n junction diode may be as low as 0.7 nsec (= 700 psec).   The reverse recovery time of a semiconductor-metal diode may be an order of magnitude smaller, say 50 psec.

When a semiconductor-metal diode is forward-biased, the electrons, in crossing the junction to enter the metal, fall down a potential hill.   Therefore, initially, and until these electrons come to equilibrium with the electrons of the metal, the injected electrons have a higher energy and velocity.   The "temperature" associated with a large assemblage of particles (molecules in gas, electrons in a metal, etc.) is measured by the mean energy of these particles.   Therefore, the injected electrons are referred to as *hot* electrons and the diode is referred to as a *hot-carrier diode*.   It is possible to construct both hot-electron diodes using *n*-type silicon and hot-hole diodes using *p*-type silicon.   Hot-electron diodes are generally preferred, since the higher electron mobility gives greater speed of operation.

20-10   CHARGE COMPENSATION FOR MINIMIZING STORAGE TIME

In Fig. 20-10*a* a forward-biasing voltage V_F has been applied through a resistor R to a diode D.   We assume that the forward current $I_F \approx V_F/R$ has existed for a time long enough for equilibrium to have been attained.   At $t = 0+$ the voltage reverses and $v_i = -V_R$.   It is clear from our earlier discussion and also from Eq. (20-11) that the storage time of the diode may be minimized by mak-

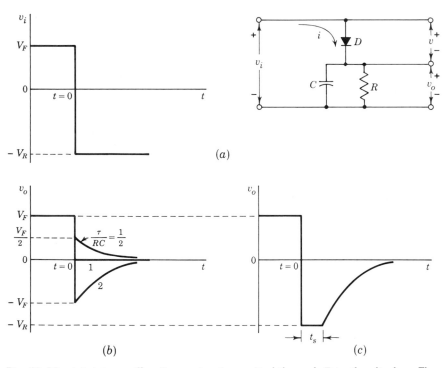

Fig. 20-10   (a) A turn-off voltage step is applied through R to the diode.   The resistor R is shunted by a charge-compensating capacitor.   (b) Waveforms of the output v_o for the case where C is large enough so that all stored charge is removed immediately.   (c) Waveform of v_o when C is not large enough to remove all stored charge.

ing the reverse diode current $-I_R$ as large as possible. If the capacitor C were not present, the reverse current would be limited to $-I_R = -V_R/R$. The capacitor, however, will allow an impulsive current to flow, and we may remove the excess charge abruptly and reduce the storage time to zero. (We have in mind, of course, the approximate representation of the diode in which the excess density becomes zero simultaneously throughout the diode.) *We shall also assume that whenever the diode has an excess minority charge (which means that it is forward-biased) the drop across it is negligible.* The method contemplated here for minimizing the storage time is called *charge compensation*, and the capacitor C is a *charge-compensating capacitor*. If we replace the diode in Fig. 20-10a by a parallel combination of a resistor and diffusion capacitor, the present circuit is reminiscent of the compensated attenuator of Sec. 2-8. Indeed there are some similarities between the two, but as we shall see, there are also some very important differences.

Initially ($t < 0$) the current $i = I_F = V_F/R$ and the stored charge is, from Eq. (20-6), $Q = \tau I_F = \tau V_F/R$. Let us assume that, when v_i changes abruptly to $-V_R$, all the stored charge is removed and transferred to the capacitor C. Then the corresponding change Δv_o in the voltage across C is

$$\Delta v_o = \frac{-Q}{C} = \frac{-\tau V_F}{RC} \tag{20-12}$$

Therefore, as shown in Fig. 20-10b for $\tau/RC = \frac{1}{2}$, 1, and 2, $\Delta v_o = -\frac{1}{2}V_F$, $-V_F$, and $-2V_F$, and $v_o = \frac{1}{2}V_F$, 0, and $-V_F$, respectively. At the moment when the excess charge is removed the diode becomes an open circuit and the voltage v_o decays to zero with a time constant RC. (We neglect the small reverse diode current and the small transition capacitance across the reverse-biased diode.)

The maximum change possible in v_o is $V_F + V_R$, and the corresponding maximum charge transferred to the capacitor is $C(V_F + V_R)$. If all the stored charge Q is to be removed, we require that

$$C(V_F + V_R) \geq Q = \tau I_F = \frac{\tau V_F}{R} \tag{20-13}$$

or that the capacitance satisfy the condition

$$C \geq \frac{\tau}{R} \frac{V_F}{V_F + V_R} \tag{20-14}$$

If Eq. (20-14) is not satisfied, the waveform v_o will have the appearance shown in Fig. 20-10c (compare with Fig. 20-5d). Since the stored charge has not been completely removed at $t = 0+$, the diode remains ON and $v_o = v_i = -V_R$.

The charge remaining in the diode is $\tau I_F - C(V_F + V_R) \equiv Q'$. A forward current I'_F would maintain this charge in the diode if I'_F were selected to be given by $\tau I'_F = Q'$ or

$$I'_F = I_F - \frac{C}{\tau}(V_F + V_R) \tag{20-15}$$

The time t_s in Fig. 20-10c may now be calculated by using I'_F in place of I_F in Eq. (20-11). After the storage interval t_s the diode becomes open-circuited and the voltage v_o decays to zero with a time constant RC.

By comparing Figs. 20-10b and 2-23 we may observe both the similarities and the differences between the "charge-compensated" diode circuit and the compensated attenuator circuit which uses conventional capacitors for charge storage. Note that in Fig. 20-10b, v_o attains its ultimate level, $v_o = 0$, without initially falling short or overshooting its mark when $\tau = rC_D = RC$. In this respect the diode circuit acts like the compensated attenuator. On the other hand, observe that for any value of τ/RC the magnitude of the step in v_o is independent of the magnitude of the input step. Also, the restriction (20-14) and the waveform of Fig. 20-10c have no counterpart for the compensated attenuator.

A situation of great practical interest is shown in Fig. 20-11. The input signal is again one which, having kept the diode oN and allowed the establishment of a stored charge, reverses abruptly. The resistor R_s represents the source impedance of the generator. The resistor R is used to limit the steady-state forward current. The capacitor C bridges R and hastens the recovery of the diode. (Of course, as in the case previously discussed, it would be better if C bridged both R_s and R, but the normal situation is one in which both terminals of R_s are not available.)

EXAMPLE In the circuit of Fig. 20-11, $V_F = 3$ V, $-V_R = -6$ V, $R_s = 5$ K, and $R = 10$ K. The diode has a storage time constant $\tau = 1.0$ μsec. Calculate the storage time of the diode if (a) $C = 0$ and (b) $C = 100$ pF.

Solution   a. Using Eq. (20-11) we have, since $I_F/I_R = V_F/V_R$,

$$t_s = 1.0 \ln \left(1 + \frac{3}{6}\right) = 0.41 \ \mu\text{sec}$$

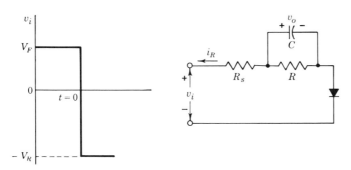

Fig. 20-11 Charge compensation is used in a circuit in which the capacitor shunts only part of the series resistance.

b. At $t = 0-$ the voltage across C is $v_o = V_o = 2$ V. Therefore the reverse current at $t = 0+$ is

$$i_R(0+) = \frac{V_R + V_o}{R_s} = \frac{6 + 2}{5} = 1.6 \text{ mA}$$

and at $t = \infty$ is

$$i_R(\infty) = \frac{V_R}{R_s + R} = \frac{6}{15} = 0.4 \text{ mA}$$

The time constant of the circuit is

$$\frac{R_s R}{R_s + R} C = \frac{5 \times 10 \times 10^6}{15 \times 10^3} \times 100 \times 10^{-12} = \tfrac{1}{3} \times 10^{-6} \text{ sec}$$

By using Eq. (2-3) the current i_R is given as a function of the time by

$$i_R = 0.4 + 1.2\epsilon^{-3\times10^6 t} \quad \text{mA}$$

The stored charge is determined by Eq. (20-5), with Q in coulombs and i_R in amperes,

$$\frac{dQ}{dt} + \frac{Q}{\tau} = -i_R = -0.4 \times 10^{-3} - 1.2 \times 10^{-3}\epsilon^{-3\times10^6 t}$$

Integrating, with $\tau = 10^{-6}$ sec, we find

$$Q = K\epsilon^{-10^6 t} - 0.4 \times 10^{-9} + 0.6 \times 10^{-9}\epsilon^{-3\times10^6 t}$$

with K an arbitrary constant. We determine K from the condition that at $t = 0-$, $Q = \tau V_F/(R_s + R) = 3 \times 10^{-6}/(15 \times 10^3) = 0.2 \times 10^{-9}$ C. We find $K = 0$, so that

$$Q = (0.6\epsilon^{-3\times10^6 t} - 0.4) \times 10^{-9}$$

Setting $Q = 0$ we obtain $t = t_s = 0.13$ μsec. Therefore the addition of the capacitor C has reduced the storage time to about one-third of its previous value. A larger capacitance would effect a still larger reduction but not to any great extent. If the capacitance C had been arbitrarily large, we would have found t_s reduced only to 0.12 μsec (Prob. 20-14).

20-11   TRANSISTOR SWITCHING TIMES

In Chap. 6 our consideration of the transistor switch was limited to a discussion of the end states, the device being either in the cutoff condition or in saturation. We now turn our attention to the behavior of the transistor as it makes a transition from one state to the other. We consider the transistor switch of Fig. 20-12, driven by the pulse waveform shown. This waveform makes transitions between the voltage levels V_2 and V_1. At V_2 the transistor is at cutoff and at V_1 the transistor is in saturation. This input waveform v_i is applied between base and emitter through a resistor R_s, which may be included

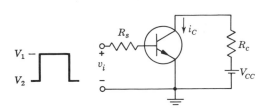

Fig. 20-12 The pulse waveform operating between levels V_2 and V_1 drives the transistor from cutoff to saturation and back again. For the n-p-n transistor, V_1 is positive and V_2 is usually negative (although it can be slightly positive). For a p-n-p type the polarities are reversed.

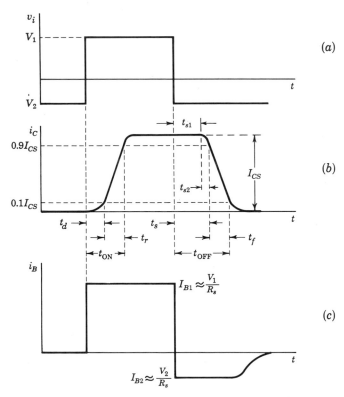

Fig. 20-13 The collector current (b) and the base current (c) in response to the driving pulse in (a).

explicitly in the circuit or may represent the output impedance of the source which furnishes the waveform.

The response of the collector current i_C to the input waveform, together with its time relationship to that waveform, is shown in Fig. 20-13b. The current does not immediately respond to the input signal. Instead there is a delay, and the time that elapses during this delay, together with the time required for the current to rise to 10 percent of its maximum (saturation) value $I_{CS} = V_{CC}/R_c$, is called the *delay time* t_d. The current waveform has a nonzero *rise time* t_r, which is the time required for the current to rise from 10 percent to 90 percent of I_{CS}. The total *turn-on* time t_{ON} is the sum of the delay and rise time, $t_{ON} \equiv t_d + t_r$. When the input signal returns to its initial state, the current again fails to respond immediately. The interval which elapses between the transition of the input waveform and the time when i_C has dropped to 90 percent of I_{CS} is called the *storage time* t_s. The storage interval is followed by the *fall time* t_f, which is the time required for i_C to fall from 90 to 10 percent of I_{CS}. The *turn-off time* t_{OFF} is defined as the sum of the storage and fall times, $t_{OFF} \equiv t_s + t_f$. In the following sections we shall consider the physical reasons for the existence of each of these times and how they may be determined quantitatively. In our analysis we shall use both the equivalent-circuit approach and the excess-minority-charge-control method.

20-12   THE DELAY TIME

There are three factors that contribute to the delay time. First, there is a delay which results from the fact that, when the driving signal is applied to the transistor input, a nonzero time is required to charge up the junction capacitances so that the transistor may be brought from cutoff to the active region. Second, even when the transistor has been brought to the point where minority carriers have begun to cross the emitter junction into the base, a nonzero time is required before these carriers can cross the base region to the collector junction and be recorded as collector current. Finally, a nonzero time is required before the collector current will rise to 10 percent of its maximum. We shall see that there is a one-to-one correspondence between the collector current and the excess charge stored in the base. The collector-current rise time is the time required to establish the required base charge.

To consider now the first of these contributions to t_d, let us replace the transistor by its hybrid-II model. Since the transistor is operating at cutoff and we are concerned with the response to a fast waveform, the hybrid-II circuit of Fig. 1-4 reduces to that in Fig. 20-14. The base-spreading resistance $r_{bb'}$ added to R_s is the total impedance that the source sees and is designated $R_s' \equiv R_s + r_{bb'}$. The resistor $r_{b'e}$, being inversely proportional to emitter current, is very large and may be neglected. The emitter diffusion capacitance is directly proportional to the quiescent emitter current and may be neglected at cutoff. Hence, $C_e = C_{b'e}$ reduces to the emitter-junction transition capaci-

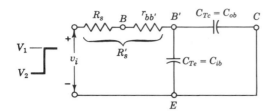

Fig. 20-14 The hybrid-Π circuit is used to calculate the time required to bring a transistor out of cutoff.

tance C_{Te}. The capacitance may be measured between terminals B and E (neglecting the comparatively small resistance of $r_{bb'}$) with collector open-circuited. It is accordingly also referred to as C_{ib}, the common base input tran-sition capacitance. The resistance $r_{b'c}$ is large enough usually to be neglected even in the active region, leaving between B' and C only the collector transition capacitance $C_{Tc} = C_{b'c} = C_{ob}$. The transconductance g_m is zero at cutoff and hence the generator $g_m v_{b'e}$ has been omitted. The collector has been shorted to emitter because in the cutoff range the collector voltage does not change.

The capacitor charging time t_{d1} is easily calculated from Fig. 20-14. The base voltage starts at V_2 and would go to V_1 if the transistor did not come out of cutoff. Hence from Eq. (2-3)

$$v_{b'e} = V_1 + (V_2 - V_1)\epsilon^{-t/\tau_i}$$

where $\tau_i \equiv R_s'(C_{ib} + C_{ob})$ is the input time constant at cutoff. In Table 6-1 (page 219) we find that a forward cutin voltage V_γ is required across the emitter diode before appreciable current can flow. Reasonable values for V_γ are 0.1 V for Ge and 0.5 V for Si. Substituting V_γ for $v_{b'e}$ into the above equation we find that for $t = t_{d1}$

$$t_{d1} = R_s'(C_{ib} + C_{ob}) \ln \frac{V_1 - V_2}{V_1 - V_\gamma} \tag{20-16}$$

For large values of V_1 (compared with V_2 and V_γ) t_{d1} becomes very small. Also, if initially the transistor is not very far into cutoff ($V_2 \approx V_\gamma$), then t_{d1} approaches zero. Both these statements follow from Eq. (20-16) and are exactly what should be expected physically.

EXAMPLE A 12-V 10-μsec negative pulse is fed through a 50-Ω cable to the input of the circuit of Fig. 20-15. We wish to calculate all the switching times for this configuration. For the moment, calculate t_{d1}.

Solution The transition capacitances are given in Fig. 20-16 as a function of the junction voltage. A complication arises because these capacitances are not con-stant. It is not difficult to make a calculation[8] of turn-on time in a manner which takes account of the variation of the capacitances. Such a calculation is, however, lengthy and involves graphical integration under the curves of Fig. 20-16 or the solution of a nonlinear differential equation. In the present instance, where we are interested mostly in the order of magnitude of the result, we shall assume a fixed value for the capacitances. In order that our result may be on the con-

Fig. 20-15   Problem illustrating switching times for
a germanium p-n-p 2N404
transistor.

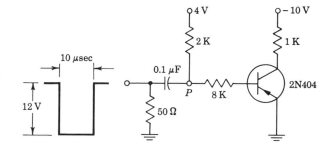

servative side we shall assume that these capacitances are equal to the largest
capacitance values attained during the switching cycle.   In view of the variability
of transistor parameters, a more precise calculation may hardly be justifiable.

In the quiescent cutoff state the base-to-emitter voltage is $V_2 = +4$ V.
Since the collector-to-emitter voltage is -10 V the collector junction voltage is
-14 V.   From Fig. 20-16 we find $C_{ib} = 6$ pF and $C_{ob} = 7$ pF.   When the transistor is about to conduct, $v_{be} = V_\gamma = -0.1$ V, and the collector junction voltage
is -9.9 V.   From Fig. 20-16 (by extrapolation) $C_{ib} = 12$ pF and $C_{ob} = 7$ V.
Whereas C_{ob} has remained essentially constant, C_{ib} has doubled over the voltage
range.   The worst-case value for total capacitance is $C_{ib} + C_{ob} = 12 + 7 = 19$ pF.

Since the coupling time constant $(0.1)(2,000) = 200$ μsec is large compared
with the pulse width of 10 μsec, the pulse will appear at point P with little distortion.   In the quiescent state the voltage at P is $+4$ V, and hence the presence
of the pulse will drop this potential to $+4 - 12 = -8$ V.   This, then, is the
value of V_1.   If $r_{bb'} \approx 100$ Ω, then $R_s' \approx R_s = 8$ K.   From Eq. (20-16), with
$V_\gamma = -0.1$ V, we have $t_{d1} = (8)(19)(2.30) \log [(-8 - 4)/(-8 + 0.1)] = 64$ nsec.

Let us look now to the time t_{d2} required for the first minority carriers to
reach the collector.   It is calculated[9] that in a steady-state situation the
transit time of minority carriers across the base is given by $t_b = 1/\omega_T$, in which
$f_T = \omega_T/2\pi$ is the common-emitter gain-bandwidth product, or the frequency
at which the common-emitter current gain is unity (Sec. 4-6).   When, however, the minority carriers are first injected into the base the carrier distribution is not the steady-state distribution.   It is then found[10] that the first

Fig. 20-16   The variation with
junction voltage of the transition
capacitances in the type 2N404.
(Courtesy of Texas Instruments
Incorporated.)

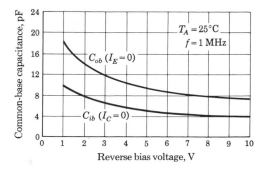

carriers make their appearance at the collector in a time

$$t_{d2} \approx \tfrac{1}{3} t_b \tag{20-17}$$

For the 2N404, $f_T = 10$ MHz, giving $t_{d2} = 1/(3)(2\pi)(10)$ μsec $= 5$ nsec. In the present instance it appears that t_{d2} is very much smaller than t_{d1}. But with a different base and driving circuit, such might not be the case. In order to complete the calculation of t_d we must find t_{d3}, the time required for the transistor current to rise to 10 percent of its maximum. Let us therefore now turn our attention to the rise time t_r in the active region.

20-13   THE RISE TIME

We consider the case in which the transistor is driven to saturation by the application of a step of current I_{B1} to the transistor base. If, in Fig. 20-12, R_s is very large in comparison with the impedance that appears between base and emitter, then $I_{B1} = (V_1 - V_\gamma)/R_s$. The response of the transistor to a current step is given in Eq. (4-35). Assuming $R_s \gg h_{ie}$, setting $I_{B1} = I$, and using Eq. (4-37), we have for the collector current i_C $(= -i_L)$

$$i_C = h_{FE} I_{B1} (1 - \epsilon^{-t/\tau_r}) \tag{20-18}$$

in which the time constant τ_r for the current rise is given by Eq. (4-37):

$$\tau_r = h_{FE} \left(\frac{1}{\omega_T} + C_c R_c \right) \tag{20-19}$$

where ω_T is the radian frequency at which the current gain is unity, and C_c $(= C_{Tc} = C_{ob})$ is the collector transition capacitance. We have used h_{FE} in place of h_{fe} in Eqs. (20-18) and (20-19) since we contemplate the application of these equations to the large-signal case corresponding to the switch from cutoff

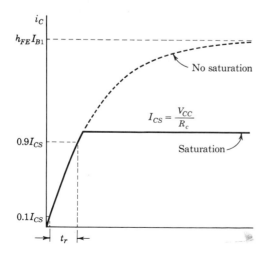

Fig. 20-17 The dashed curve gives the collector current in an n-p-n transistor if there were no saturation. If $I_{B1} > I_{CS}/h_{FE}$, then saturation takes place and the current is limited to the value V_{CC}/R_c, as shown by the solid curve.

to saturation. A plot of Eq. (20-18) is indicated by the dashed curve in Fig. 20-17.

The transistor will just leave the active region and just enter saturation if the base-current magnitude is I_{BA}, given by

$$h_{FE}I_{BA} \equiv I_{CS} = \frac{V_{CC}}{R_c} \tag{20-20}$$

where I_{CS} is the saturation collector current. Under these circumstances the collector is limited to the solid curve in Fig. 20-17, and the rise time t_r is also indicated in that figure. The time $t_{0.1}$ for the collector current to rise to $0.1I_{CS}$ is, from Eq. (20-18),

$$t_{0.1} = \tau_r \ln \frac{1}{1 - 0.1/N_1} \tag{20-21}$$

where

$$N_1 \equiv \frac{h_{FE}I_{B1}}{I_{CS}} \tag{20-22}$$

Since N_1 may be written as $N_1 = I_{B1}/(I_{CS}/h_{FE}) = I_{B1}/I_{BA}$, this quantity represents the ratio of the actual base current to the base current just adequate to achieve saturation. Hence, N_1 is called the *overdrive factor*. N_1 must be at least unity if the transistor is to be driven into saturation.

The time $t_{0.9}$ for the collector current to rise to $0.9I_{CS}$ is given by Eq. (20-21) with 0.1 replaced by 0.9. The rise time t_r required for i_C to rise from $0.1I_{CS}$ to $0.9I_{CS}$ is given by

$$t_r = t_{0.9} - t_{0.1} = \tau_r \ln \frac{1 - 0.1/N_1}{1 - 0.9/N_1} \tag{20-23}$$

This result may be simplified if the transistor is driven deeply into saturation. Since $\ln (1 + x) = x - x^2/2 + x^3/3 + \cdots$, then Eq. (20-23) may be expanded to

$$t_r = \frac{0.8\,\tau_r}{N_1} \left(1 + \frac{0.50}{N_1} + \frac{0.30}{N_1{}^2} + \cdots \right) \tag{20-24}$$

For $N_1 \gg 1$

$$t_r \approx \frac{0.8\,\tau_r}{N_1} = 0.8 \frac{\tau_r}{h_{FE}} \frac{I_{CS}}{I_{B1}} \tag{20-25}$$

Note that for this condition of heavy overdrive the rise time is independent of h_{FE} because τ_r/h_{FE} does not depend upon h_{FE} [Eq. (20-19)]. For this reason, and because t_r varies inversely with I_{B1}, it is advantageous to drive a transistor well into saturation with a large base current if it is desired to minimize turn-on time.

From Eq. (20-24) we see that if $N_1 \geq 5$ an error of 11 percent or less is made in calculating the rise time from the simple expression in Eq. (20-25).

Some manufacturers supply curves of t_d and t_r (and also t_s and t_f) versus I_{B1} for various circuit configurations. These are useful only if the exact switching conditions are duplicated.

EXAMPLE For the circuit of Fig. 20-15 calculate (*a*) the rise time t_r and (*b*) the time t_{d3} necessary for the collector current to rise to 10 percent of I_{CS}. Assume $f_T = 10$ MHz and $h_{FE} = 100$.

Solution   *a.* From Eq. (20-19)

$$\frac{\tau_r}{h_{FE}} = \frac{1}{(2\pi)(10) \times 10^6} + (7)(10^{-12})(10^3) \text{ sec} = 23 \text{ nsec}$$

The saturation current is $I_{CS} = -V_{CC}/R_c = -10$ mA. The transistor will saturate at a base current $I_{BA} = I_{CS}/h_{FE}$, or $I_{BA} = -10/100 = -0.1$ mA. The actual drive current is $I_{B1} = (V_1 - V_o)/R_s = -7.7/8 \approx -0.96$ mA. Accordingly the overdrive factor is $N_1 = I_{B1}/I_{BA} \approx 10$, and we may use the simplified equation (20-25), which yields

$$t_r = (0.8)(23)\left(\frac{-10}{-0.99}\right) = 186 \text{ nsec}$$

Note that the parameter on which the rise time depends most sensitively is f_T.

b. Since the overdrive is large, the rise of collector current is quite linear, as seen in Fig. 20-17. Therefore the time required to rise to 10 percent of I_{CS} is one-eighth the time required for the current to rise from 10 to 90 percent of I_{CS}, so that

$$t_{0.1} = t_{d3} = \frac{186}{8} = 23 \text{ μsec}$$

We may now estimate the total delay time as

$$t_d = t_{d1} + t_{d2} + t_{d3} = 64 + 5 + 23 = 92 \text{ nsec}$$

Note that the delay time is approximately half the rise time and that the total turn-on time t_{ON} is $186 + 92 = 278$ nsec, or 0.28 μsec.

20-14   STORAGE TIME[11]

The failure of the transistor to respond to the trailing edge of the driving pulse for the time interval t_s, as indicated in Fig. 20-13, results from the fact that a transistor in saturation has a saturation charge of excess minority carriers stored in the base. The transistor cannot respond until this saturation excess charge has been removed. We shall now discuss the processes that go on in the transistor that give rise to this *storage time* t_s.

Active-region Operation   In the active region, the excess minority carriers (say holes in a p-n-p transistor) which are injected into the base at the emitter junction diffuse across the base and on reaching the second junction constitute the collector current.   The diffusion is the result of a density gradient of holes, and the diffusion current is proportional to this gradient.   The current must be approximately constant at any distance into the base, since the collector current is nominally the same as the emitter current.   Therefore the excess minority density gradient must be almost constant throughout the base region. And since any hole is immediately swept away upon reaching the collector, the excess minority density at the collector junction must be zero.   Altogether, it then appears that as a function of x, the distance across the base, the hole density must have the appearance shown in Fig. 20-18a and labeled *normal N*. Actually, the collector current I_{CN} is somewhat less than the emitter current I_{EN} because some of the holes will be lost through recombination in crossing the base.   Therefore, the current falls off slightly as x increases, and as a consequence the hole-density curve has a slight upward concavity, not indicated in Fig. 20-18.   On this figure the currents I_{CN} and I_{EN} are indicated schematically at the ends of the normal line N.   Following the usual reference-direction conventions, the currents are indicated entering their respective junctions. For a p-n-p transistor I_{EN} is positive, but I_{CN} is negative.   If we neglect the

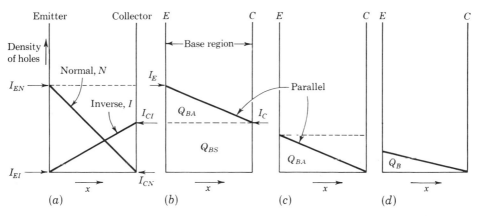

Fig. 20-18   Plots of excess hole densities.   In (a) N and I represent the densities when the transistor is operating in the normal and inverted manner, respectively.   In (b) the two plots in (a) have been superimposed.   The total base charge has been represented as a uniform saturation charge Q_{BS}, which makes no contribution to the current, and an effective charge Q_{BA}, which establishes the density gradient required for current flow by diffusion.   In (c) the saturation charge has just been removed and the transistor is on the point of leaving saturation, so that only the effective charge Q_{BA} remains.   In (d) the transistor is well into the active region and $Q_B < Q_{BA}$.

reverse saturation current I_{CO}, then, from Eq. (6-26), we have

$$I_{CN} = -\alpha_N I_{EN} \qquad (20\text{-}26)$$

where α_N is the forward short-circuit current transfer ratio.

If the transistor is used in the inverse direction, with the roles of collector and emitter interchanged, then the hole density will be given by the plot marked *inverse I* in Fig. 20-18a. The current entering the base across the collector junction (which is now acting as an emitter) is I_{CI} and is shown schematically by the arrow pointing at the inverse-density plot. A fraction α_I of this current reaches the emitter junction (which now acts as collector). This collected emitter current I_{EI} is, from Eq. (6-29) (neglecting I_{EO}),

$$I_{EI} = -\alpha_I I_{CI} \qquad (20\text{-}27)$$

where α_I is the reverse short-circuit current transfer ratio.

Saturation-region Operation When the transistor is in saturation it is operating normally and in the inverse mode at the same time. Therefore, the resultant current is the algebraic sum of the two currents for forward and reverse operation. Hence, the total current at the emitter junction I_E and the total collector current I_C are given by

$$I_E = I_{EN} + I_{EI} \qquad (20\text{-}28)$$

$$I_C = I_{CN} + I_{CI} \qquad (20\text{-}29)$$

Similarly, in saturation the total charge density in the base is given by the superposition of the two density plots in Fig. 20-18a and is shown in Fig. 20-18b. The slope of the resultant curve at the emitter junction is proportional to I_E, and the slope at C is proportional to I_C. It is clear that the slope in Fig. 20-18b is indeed equal to the algebraic sum of the slopes of the two plots in Fig. 20-18a, consistent with Eqs. (20-28) and (20-29).

In the saturation region $I_C = I_{CS} \approx V_{CC}/R_c$ and I_E corresponds to this value of collector current and to the base current $I_{B1} \approx V_1/R_s$ of Fig. 20-13, or $I_E = I_{E1} = -(I_{CS} + I_{B1})$. Note that the *total* currents in Fig. 20-18b are determined by the circuit configuration and not by the transistor parameters. The four *individual* current components in Fig. 20-18a can now be determined from Eqs. (20-26) to (20-29). We find, for example, from these equations that

$$I_{EI} = -\alpha_I \frac{I_C + \alpha_N I_E}{1 - \alpha_N \alpha_I} \qquad (20\text{-}30)$$

Since the linear plots of Fig. 20-18a and b are plots of excess minority charge density, the area under each individual curve is proportional to the total excess minority charge which gives rise to that plot. In Fig. 20-18b we have divided the total area under the curve into two parts, the first proportional to a charge Q_{BS}, the second proportional to Q_{BA}. The charge Q_{BS}, which is uniformly distributed, does not contribute to the slope, and therefore plays no role in establishing a current across the base. This charge Q_{BS} is called the *saturation charge.* The gradient of the nonuniform charge density (of area Q_{BA}) is proportional to the diffusion current. Therefore Q_{BA}, which just brings the transistor to the limit of the active region, is called the *effective charge.*

We now have a way for accounting for the delay in response of a saturated transistor to a signal which returns it to the active region. As the excess minority charge in the base is removed through the emitter junction, it appears that this charge decreases uniformly through the base. That is, the charge Q_{BS} disappears first, leaving the charge Q_{BA} and the slope unaltered. Only after Q_{FS} has disappeared does Q_{BA} begin to decrease. At the moment Q_{BS} has just disappeared the plot of density is just zero at the collector, as indicated in Fig. 20-18c. The density at the collector cannot drop below zero, so that further reduction of charge requires that the density decrease at the emitter, as shown in Fig. 20-18d. The slope of the density plot, and consequently the current, is thereby decreased. The base charge Q_B in the active region is less than Q_{BA}, the base charge required just to enter saturation.

Storage-time Calculation We now set up a procedure[11] for calculating the storage time t_s. This time is broken up into two parts, the first of which, t_{s1} (indicated in Fig. 20-13b), is the time from the occurrence of the end of the pulse to the time when the collector current just begins to change. The second part, t_{s2}, is the time required for i_C to fall to 90 percent of I_{CS}. We may find t_{s1} as follows.

Expressing Eqs. (20-26) to (20-29) in Laplace transform notation, with s the transform variable, we have

$$I_{CN}(s) = -\alpha_N(s)I_{EN}(s) \qquad I_{EI}(s) = -\alpha_I(s)I_{CI}(s) \qquad (20\text{-}31)$$

$$I_E(s) = I_{EN}(s) + I_{EI}(s) \qquad I_C(s) = I_{CN}(s) + I_{CI}(s) \qquad (20\text{-}32)$$

Solving for $I_{EI}(s)$ we find

$$I_{EI}(s) = -\alpha_I(s)\frac{I_C(s) + \alpha_N(s)I_E(s)}{1 - \alpha_N(s)\alpha_I(s)} \qquad (20\text{-}33)$$

which is, of course, identical in form to Eq. (20-30). In the normal direction, the low-frequency current gain is α_{NO} and it has an upper 3-dB radian frequency ω_N. In the inverse direction, because of the lack of symmetry of the transistor, the current gain and 3-dB frequency will not be the same and are designated by α_{IO} and ω_I, respectively. We have, accordingly,

$$\alpha_N = \frac{\alpha_{NO}}{1 + j(\omega/\omega_N)} \qquad \alpha_I = \frac{\alpha_{IO}}{1 + j(\omega/\omega_I)} \qquad (20\text{-}34)$$

or, in terms of the Laplace transform variable s, we may write

$$\alpha_N(s) = \frac{\alpha_{NO}}{1 + s/\omega_N} \qquad \alpha_I(s) = \frac{\alpha_{IO}}{1 + s/\omega_I} \qquad (20\text{-}35)$$

These expressions for $\alpha_N(s)$ and $\alpha_I(s)$ may be substituted into Eq. (20-33) to yield an expression for $I_{EI}(s)$ explicitly in terms of $I_C(s)$ and $I_E(s)$.

Now consider that the turn-off edge of the driving pulse occurs at $t = 0$. We have already noted that for $t < 0$, $I_C = I_{CS}$ and $I_E = I_{E1}$. For $t > 0$, I_C remains at its saturation value I_{CS} but the emitter current changes to a value I_{E2}, which is smaller than I_{E1} and small enough so that the steady-state situation corresponding to I_{E2} is one in which the transistor is not in saturation. This current I_{E2} is given by $I_{E2} = -(I_{CS} + I_{B2})$, where, from Fig. 20-13, $I_{B2} \approx V_2/R_s$. Accordingly, at $t = 0$, I_C does not change at all while I_E changes abruptly by the amount $I_{E2} - I_{E1}$. The corresponding change ΔI_{EI} in I_{EI} has a transform $\Delta I_{EI}(s)$, which, from Eq. (20-33), is

$$\Delta I_{EI}(s) = \frac{-\alpha_I(s)\alpha_N(s)[(I_{E2} - I_{E1})/s]}{1 - \alpha_N(s)\alpha_I(s)} \qquad (20\text{-}36)$$

and in which the $\alpha_N(s)$ and $\alpha_I(s)$ are given in Eq. (20-35).

The transform $\Delta I_{EI}(s)$ has three roots, one at $s = 0$ and the others given approximately by (Prob. 20-25)

$$s_1 \approx -\frac{\omega_N\omega_I(1 - \alpha_{NO}\alpha_{IO})}{\omega_N + \omega_I} \qquad s_2 \approx -(\omega_N + \omega_I) \qquad (20\text{-}37)$$

These roots give rise in the time response to terms of the form $\epsilon^{s_1 t}$ and $\epsilon^{s_2 t}$. In a typical case we find that $|s_2| \gg |s_1|$ and hence the term $\epsilon^{s_2 t}$ decays very much more rapidly than does the term $\epsilon^{s_1 t}$. We therefore neglect the term $\epsilon^{s_2 t}$ on the grounds that it has negligible effect in determining the time at which the transistor comes out of saturation. We then find, when we take advantage of the simplifications resulting from the fact that $|s_2| \gg |s_1|$, that the inverse transform of $\Delta I_{EI}(s)$ is given by

$$\Delta I_{EI}(t) = \frac{\alpha_{NO}\alpha_{IO}(I_{E1} - I_{E2})}{1 - \alpha_{NO}\alpha_{IO}} (1 - \epsilon^{s_1 t}) \qquad (20\text{-}38)$$

We observe that our approximations have reduced the solution to one which involves a single time constant. The total inverse emitter current is the sum of $\Delta I_{EI}(t)$ and the quiescent value of the emitter current I_{EI} given in Eq. (20-30). We have

$$i_{EI}(t) = I_{EI} + \Delta I_{EI}(t) \qquad (20\text{-}39)$$

In a similar manner we may find $i_{CI}(t)$. When both $i_{CI}(t)$ and $i_{EI}(t)$ have fallen to zero the transistor is certainly out of the saturation region. The ordinary (normal) direction in which the transistor is used is the one which gives $\alpha_N \geq \alpha_I$ and $\omega_N \geq \omega_I$. If t_{s1E} is the time when $i_{EI} = 0$ and t_{s1C} the time when $i_{CI} = 0$, then it turns out that $t_{s1E} > t_{s1C}$. We therefore use $i_{EI} = 0$ as the condition that marks the first part of the storage interval. We find (Prob. 20-27) that

$$t_{s1} = -\frac{1}{s_1} \ln \frac{\alpha_{NO}\alpha_{IO}(I_{E2} - I_{E1})}{\alpha_{IO}I_{CS} + \alpha_{NO}\alpha_{IO}I_{E2}} \tag{20-40}$$

Using $I_{E2} = -I_{CS} - I_{B2}$, $I_{E1} = -I_{CS} - I_{B1}$, and

$$I_{CS} = \frac{\alpha_{NO}I_{BA}}{1 - \alpha_{NO}} = h_{FE}I_{BA}$$

we find (Prob. 20-28) that

$$t_{s1} = \tau_s \ln \frac{I_{B1} - I_{B2}}{I_{BA} - I_{B2}} \tag{20-41}$$

where τ_s is the *storage time constant* and is defined by

$$\tau_s \equiv -\frac{1}{s_1} = \frac{\omega_N + \omega_I}{\omega_N\omega_I(1 - \alpha_{NO}\alpha_{IO})} \tag{20-42}$$

Note that I_{B1} and I_{B2} are the base currents before and after the turn-off edge of the driving pulse and

$$I_{BA} \equiv \frac{I_{CS}}{h_{FE}}$$

The Base-current Waveform The base current is shown in Fig. 20-13c. Initially, before the application of the driving pulse, the transistor is in cutoff and the base current for an input $v_i = V_2$ is nominally zero. At the termination of the pulse, the transistor is in saturation and the base has a charge of minority carriers. These minority carriers in the base will allow a flow of current in the reverse direction across the emitter junction. Until such time as the minority carriers are swept out of the base, the emitter junction will behave like a forward-biased junction even when the current is flowing in the reverse direction. The minority carriers will not have been swept out of the base until the transistor is on the point of entering the cutoff region. Accordingly, if the reverse voltage V_2 is applied to the base through a resistance R_s and if V_2 is very large in comparison with the voltage across a forward-biased junction, a base current $I_{B2} \approx V_2/R_s$ will flow until cutoff is reached. The onset of cutoff is not sharply defined. For, even when the bulk of the minority carriers have been swept across the emitter junction back into the emitter, a period of time will still be required for those minority carriers located far from the junction to diffuse toward the emitter, there to be collected. After the minority carriers are for the most part removed, the base current will return to its nominally zero value. But the base current will not drop abruptly to zero because of the effect of the straggling minority charge and also because the base current must now provide charge for the emitter-junction and collector-junction transition capacitances. Hence the base current i_B, as shown in Fig. 20-13c, remains constant until about the time when i_C becomes zero, and then begins to fall to zero.   This reversal of base current is advantageous in shortening the storage time, as the following example will show.   The waveform in Fig. 20-13c should be compared with the analogous reverse current in a diode as depicted in Fig. 20-5d.

EXAMPLE For the circuit of Fig. 20-15, (a) calculate the storage-time component t_{s1}.   (b) Find t_{s1} if $V_2 = 0$.

Solution a. We use $h_{FE} = 100$ ($\alpha_{NO} = 0.99$).   For an alloy-type transistor such as the 2N404, $f_{\alpha N} = 1.2 f_T = 12$ MHz.   In Sec. 6-17 we assumed $\alpha_{IO} = 0.5$.   Let us assume now that $f_{\alpha I} = 1$ MHz.   Then, from Eq. (20-42),

$$\tau_s = \frac{2\pi(12 + 1)}{(2\pi)^2(12)(1)[1 - (0.99)(0.5)]} = 0.34 \ \mu\text{sec}$$

$$I_{B1} \approx \frac{-8}{8} = -1.0 \text{ mA}$$

From Fig. 20-15

$$I_{B2} = \frac{4}{8} = +0.5 \text{ mA}$$

$$I_{BA} = \frac{I_{CS}}{h_{FE}} = -\frac{10}{100} = -0.1 \text{ mA}$$

and from Eq. (20-41)

$$t_{s1} = (0.34)(2.30) \log \frac{-1.0 - 0.5}{-0.1 - 0.5} = 0.31 \ \mu\text{sec}$$

b. For $I_{B2} = 0$,

$$t_{s1} = (0.34)(2.30) \log \frac{-1.0}{-0.1} = 0.78 \ \mu\text{sec}$$

Note that with a reversed base current, as in (a), the storage time is greatly reduced.

20-15   THE FALL TIME

The fall time t_f is calculated in a manner analogous to that used in finding the rise time t_r.   For during the course of returning from near saturation to near cutoff the transistor passes through the active region.   Hence the fall time constant is the same as the rise time constant τ_r given in Eq. (20-19).   The current starts at I_{CS} and would go to $h_{FE}I_{B2}$ if the circuit continued to behave linearly (instead of reaching cutoff).   The base current I_{B2} which is applied to drive the transistor to cutoff is usually a reverse current which persists until the collector current has just about reached zero. This reverse current will shorten the fall time just as it shortens the storage time. Corresponding to the overdrive factor N_1 for the rise-time calculation, we may define a parameter

$$N_2 \equiv -\frac{h_{FE}I_{B2}}{I_{CS}} \tag{20-43}$$

Since I_{B2} is of the opposite sign to I_{CS}, then N_2 is a positive number. We find that the time $t_{0.9}$ for the collector current to fall to $0.9I_{CS}$ is given by

$$t_{0.9} = \tau_r \ln \frac{1 + 1/N_2}{1 + 0.9/N_2} \tag{20-44}$$

By definition, the second part of the storage time t_{s2} equals $t_{0.9}$. The fall time t_f required for i_C to fall from $0.9I_{CS}$ to $0.1I_{CS}$ is given by

$$t_f = t_{0.1} - t_{0.9} = \tau_r \ln \frac{1 + 0.9/N_2}{1 + 0.1/N_2} \tag{20-45}$$

For large values of N_2 we may expand the logarithm and obtain

$$t_f = \frac{0.8\,\tau_r}{N_2}\left(1 - \frac{0.50}{N_2} + \frac{0.30}{N_2^2} - \cdots\right) \tag{20-46}$$

For heavy overdrive in the reverse direction $N_2 \gg 1$; then

$$t_f \approx \frac{0.8\,\tau_r}{N_2} = -0.8\,\frac{\tau_r}{h_{FE}}\frac{I_{CS}}{I_{B2}} \tag{20-47}$$

which is independent of h_{FE} since τ_r in Eq. (20-19) is proportional to h_{FE}.

EXAMPLE For the circuit of Fig. 20-15, calculate (a) the fall time t_f and (b) the time t_{s2} necessary for the collector current to fall to $0.9I_{CS}$ from its saturation value.

Solution   a. From previous calculations $I_{B2} = +0.5$ mA, $I_{CS} = -10$ mA, $h_{FE} = 100$, and $\tau_r/h_{FE} = 23$ nsec. Hence $N_2 = -h_{FE}I_{B2}/I_{CS} = -(100)(0.5)/(-10) = 5.0$. Since N_2 is not very much greater than unity, then we must use the exact expression, as in Eq. (20-45),

$$t_f = (100)(23)(2.30)\log\frac{1 + 0.9/5}{1 + 0.1/5} = 335\text{ nsec}$$

$$b.\ t_{s2} = t_{0.9} = (100)(23)(2.30)\log\frac{1 + 1/5}{1 + 0.9/5} = 39\text{ nsec}$$

The total storage time is $t_s = t_{s1} + t_{s2} = 310 + 39 = 349$ nsec. The total turn-off time $t_{OFF} = t_s + t_f = 349 + 335 = 684$ nsec $= 0.68$ μsec.

20-16   CHARGE–CONTROL DESCRIPTION
OF TRANSISTOR BEHAVIOR[12]

Up to the present we have described the operation of the transistor in terms of the junction voltages and currents. We did, however, see in connection with our discussion of storage time how a consideration of the charge stored in the base gave a physical picture of the operation of the transistor. We shall now pursue further the characterization of the operation of the transistor in terms of charge storage in the base and at the junctions. This *charge-control* analysis has the merit that it gives further insight into the physical operation of the transistor and, in some cases, facilitates certain types of calculations.

The plots in Fig. 20-19 show the minority-carrier charge density in the base for three conditions. The plot S corresponds to operation in the saturation region. The plot SA corresponds to the case where the transistor is in the active region but just at the point of saturation. Since the collector currents are the same in these two cases, the slopes of these plots are the same but the plot SA has the value zero at the collector. The curve A corresponds to operation in the normal active region. The collector current in the active region is related to the carrier density by the diffusion-current equation

$$I_C = q \alpha D_B \frac{dp}{dx} = -q \alpha D_B \frac{P(0)}{W} \tag{20-48}$$

where q is the carrier charge, α is the cross-sectional area of the base, D_B is the diffusion constant for minority carriers in the base, p is the density of the injected minority carriers, $P(0)$ is the density at the emitter, and W is the width of the base region. The average density of holes is $P(0)/2$, and hence the minority charge Q_B in the base in the active region is

$$Q_B = \frac{qP(0)\alpha W}{2} = -\frac{W^2}{2D_B} I_C \tag{20-49}$$

from Eq. (20-48). It should be borne in mind that Q_B is a positive charge and

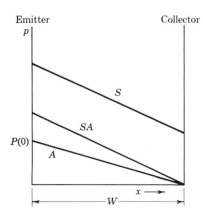

Fig. 20-19   Excess-minority-carrier density distributions for the saturated condition (S), the crossover point between saturation and active operation (SA), and the normal active region (A). These three curves correspond to those in Fig. 20-18b, c, and d, respectively.

that the negative sign appears in Eq. (20-49), as well as in Eq. (20-48), only because I_C is arbitrarily taken positive when current flows *into* the transistor. The exact factor of proportionality between Q_B and I_C, as given in Eq. (20-49), is valid for a transistor having ideal one-dimensional current flow. However, the equation suggests and we shall assume that, regardless of the geometry of the transistor, Q_B is proportional to I_C. The factor of proportionality is to be determined by measurement if calculation is not feasible. The importance of Eq. (20-49) rests in that it indicates that the collector current is determined by the excess-minority-carrier base charge and that the base charge may be viewed as the independent variable, in place of the base current, to determine the collector current.

It is important to realize that actually the net charge in the base must at all times be zero. Thus in a *p-n-p* device, a positive charge accumulates in the base because of the injection of holes across the emitter junction. If this charge were not neutralized, an electric field would be generated, restraining the emission of further holes into the base. The neutralizing charge is provided by the base current. Accordingly, if a positive base charge Q_B has been supplied through injection of minority carriers across the emitter junction, then the past history of the base current, which is supplying electrons to the base, must be such that the integrated effect of the base current has similarly caused the accumulation of an equal but negative charge. The two charges, positive and negative, must not only be equal in magnitude but must also be similarly distributed so that charge neutrality may be in effect from point to point in the base. Accordingly, the density distributions shown in Fig. 20-19 for the minority carriers that enter by way of the emitter junction apply equally to the majority carriers that enter by way of the base terminal.

The description above of the manner in which the minority charge determines the collector current and the majority charge neutralizes the minority charge makes it appear that in the steady state there is no need for base current. Thus, it suggests that, when the base current has supplied the charge necessary to sustain a collector current, the collector current might thereafter continue indefinitely without base current. In an ideal transistor, such would actually be the case. In a physical transistor the minority-carrier charge and the majority-carrier charge, in the presence of one another, diminish because of recombination. Thus a small base current must flow to replenish the majority charge lost by recombination, and the emitter current must be larger than the collector current by exactly the base current needed to replenish the loss of minority charge due to recombination.

The Fundamental Charge-control Equations At the emitter junction of the transistor a current i_E of minority carriers, say holes, enters the base. (Since the emitter is doped very much more heavily than the base, we may neglect the small current across the junction in the opposite direction carried by electrons.) At the collector junction a current $-i_C$ of minority carriers leaves the base. The total current of minority carriers directed into the base is $i_E - (-i_C) = i_E + i_C = -i_B$. Therefore the charge-control equation (20-5), when applied to the base region, becomes

$$\frac{dQ_B}{dt} + \frac{Q_B}{\tau_b} = -i_B \tag{20-50}$$

where Q_B is the charge of excess minority carriers in the base and τ_b is the lifetime of these carriers in the base. In the steady state $dQ_B/dt = 0$, with a steady base current $i_B = I_B$, Eq. (20-50) becomes

$$I_B = -\frac{Q_B}{\tau_b} \tag{20-51}$$

We also have from Eq. (20-49) that

$$I_C = -\frac{Q_B}{\tau_c} \tag{20-52}$$

where, in a simple one-dimensional case, $\tau_c = W^2/2D_B$. The parameter τ_c has the dimensions of time and is referred to in the literature as a *time constant*. However, it has no simple and direct interpretation as a time constant as does τ_b. In a similar way we may introduce a third time constant τ_e defined by $I_E = -Q_B/\tau_e$. However, since $I_B + I_C + I_E = 0$, this third time constant is not independent of the other two.

Equation (20-51) applies only in the steady state, whereas more generally Eq. (20-50) is valid. We assume, however, that Eq. (20-52) applies not only in the steady state but also when the collector current is changing. To make this assumption apparent we rewrite Eq. (20-52) in the form

$$i_C(t) = -\frac{Q_B(t)}{\tau_c} \tag{20-53}$$

since by convention a lowercase symbol represents a time-varying quantity. Our assumption is equivalent to considering, as we did in the case of the diode, that the current depends on the total charge and not on its distribution across the base region. It is possible, of course, to improve the approximation by dealing separately with the stored charges in various regions of the transistor rather than with one single total stored charge. But each such improvement of the approximation introduces additional time constants and complicates the analysis. As noted in Sec. 20-6, in connection with diodes, the approximation in which only one total charge Q_B is taken into account is the best possible approximation consistent with an analysis based on the single time constant τ_b.

Correspondence between the Charge-control and the Equivalent-circuit Parameters The relationship of the time constants τ_b and τ_c to other familiar equivalent-circuit parameters may be seen in the following discussion. In the hybrid-II equivalent circuit of Fig. 4-7 the excess-minority-carrier charge is stored in the diffusion capacitor C_D. This capacitor C_D, which is not explicitly in evidence in Fig. 4-7, is $C_D = C_e - C_{Te}$, where C_{Te} is the emitter-junction transition capacitance. The resistor $r_{b'e}$ allows the decay of the charge stored on C_D. Hence, just as was discussed in the case of the diode (Sec. 20-4), if the circuit model and charge-control model are to give the same results, we require that $\tau_b = C_D r_{b'e}$. If now we neglect all capacitors other than C_D then, from Fig. 4-7, the frequency at which the short-circuit current gain (Sec. 4-6) falls by 3 dB is

$$f_\beta^* = \frac{1}{2\pi r_{b'e} C_D} = \frac{1}{2\pi \tau_b} \tag{20-54}$$

Using the fact that $r_{b'e} = h_{fe}/g_m$ [Eq. (1-11)] we may write $f_\beta^* = g_m/2\pi h_{fe} C_D$, which is to be compared with Eq. (4-17) for f_β in which C_{Te} and C_{Tc} ($= C_c$) are not neglected. Finally, since, as in Eq. (4-18), $f_T^* \approx h_{fe} f_\beta^*$, we may write

$$\tau_b = \frac{1}{2\pi f_\beta^*} = \frac{h_{fe}}{2\pi f_T^*} = \frac{h_{fe}}{\omega_T^*} \tag{20-55}$$

with $\omega_T^* = 2\pi f_T^*$ and f_T^* the frequency at which the short-circuit current gain would be unity if the only charge stored were the charge in the transistor base; that is, f_T^* neglects the effect of the charge stored on the junction transition capacitances.

We also have from Eqs. (20-51) and (20-52)

$$h_{FE} = \frac{I_C}{I_B} = \frac{\tau_b}{\tau_c} \tag{20-56}$$

and

$$\alpha = -\frac{I_C}{I_E} = \frac{1}{1 + \tau_c/\tau_b} \tag{20-57}$$

Thus we observe that, whereas τ_b determines the transistor speed, the ratio τ_b/τ_c determines the current gain.

20-17 RISE–TIME RESPONSE BY CHARGE–CONTROL ANALYSIS

In Fig. 20-20 a transistor has applied to its base a voltage step of amplitude V through a parallel combination of R and C. We shall now examine the response of the collector current to this input signal. As a matter of convenience we shall assume that when the input signal is at its lower level, the transistor is at the boundary between cutoff and the active region and the stored charge in the base is zero. This assumption will make it possible for us to neglect the distinction between base charge (current) and an increment of base charge (current). However, the results to follow apply equally well for any initial level, provided only that in both the initial and final states the transistor is in the active region.

We consider first the case where $C = 0$, so that at the application of the step

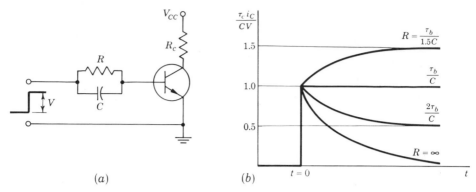

Fig. 20-20 (a) A voltage step is applied through a parallel RC combination to the base of the transistor. (b) Waveforms of the collector current (normalized) for fixed C and several values of R.

in voltage there is a step of base current $i_B \approx V/R = I$, neglecting $r_{bb'}$. Combining Eqs. (20-50) and (20-53) we find that

$$\frac{di_C}{dt} + \frac{i_C}{\tau_b} = \frac{I}{\tau_c} \tag{20-58}$$

Solving for i_C, subject to the condition that $i_C = 0$ at $t = 0$, we find

$$i_C = \frac{\tau_b}{\tau_c} I(1 - \epsilon^{-t/\tau_b}) = h_{FE}I(1 - \epsilon^{-t/\tau_b}) \tag{20-59}$$

from Eq. (20-56). Therefore the collector current rises exponentially from its initial to its final level and the rise time is $t_r = 2.2\tau_b$.

Next let $C \neq 0$. Then the step V will cause the injection into the base of a charge of majority carriers of magnitude CV. As a consequence there will appear a charge of minority carriers $Q_B = -CV$ and from Eq. (20-52) a collector current

$$I_C = \frac{CV}{\tau_c} \tag{20-60}$$

The collector current will change abruptly from zero to this value but in general will not persist. The base charge Q_B will decay with a time constant τ_b and i_C will fall to $h_{FE}I = \tau_b V/\tau_c R$ with this same time constant.

It is now clear how we may cause an abrupt change in collector current and arrange that the collector current sustain its new value. Suppose we have a voltage step V and desire an abrupt current change I_C. Then from Eq. (20-60) we select a capacitor $C = \tau_c I_C/V$. The stored base charge Q_B then has a magnitude CV, and to sustain this charge we require a base current $I_B = Q_B/\tau_b$ or $V/R = CV/\tau_b$. Therefore $RC = \tau_b$, so that R is determined. In summary, if we select $RC = \tau_b$ then the charge which appears in the base in response to the step is of such magnitude that the continuous current through R is just adequate to replace this charge at the same rate at which it is decaying owing to recombination.

If R is selected smaller than required by the condition $RC = \tau_b$ then the steady-state base current is larger than required to sustain the initial base charge. The base charge grows with time and correspondingly so also does the collector current. Waveforms for collector current for various ratios RC/τ_b are shown in Fig. 20-20b.

The time constants τ_b and τ_c may also be determined experimentally using the circuit of Fig. 20-20a. Thus when RC is adjusted so that the step in collector current is sustained at its final value without falling off or growing, then $\tau_b = RC$. And with this adjustment $\tau_c = CV/I_C$.

20-18   RISE–TIME RESPONSE INCLUDING TRANSITION CAPACITANCES

In this section we shall extend our discussion of rise-time response to take into account the junction capacitances heretofore neglected. We shall develop the analysis in a manner which will make clear the correspondence of the stored-charge method with the equivalent-circuit method.

The base current must supply the majority charge which neutralizes the excess stored minority charge. It must also supply the charges Q_{Te} and Q_{Tc} which appear on the emitter- and collector-junction transition capacitances C_{Te} and C_{Tc}. Hence to take account of these capacitances, the basic conservation-of-charge relationship expressed in Eq. (20-50) must be modified to read

$$\frac{d}{dt}(Q_B + Q_{Te} + Q_{Tc}) + \frac{Q_B}{\tau_b} = -i_B \qquad (20\text{-}61)$$

The signs associated with Q_{Te} and Q_{Tc} result from the fact that the charges on the base sides of the emitter and collector junctions are of the same polarity as the excess-minority-carrier charge Q_B stored in the base. From Eq. (20-61) we see that a charge increment $-i_B\,dt$ must equal the increments dQ_B, dQ_{Tc}, dQ_{Te}, and $Q_B\,dt/\tau_b$. We now calculate these increments separately.

1. The Minority-carrier Base-charge Increment dQ_B In the hybrid-II equivalent circuit of Fig. 4-7, the capacitor $C_e = C_D + C_{Te}$, where C_D is the diffusion capacitance which equivalently stores the base charge, *in the active region*, and C_{Te} is the emitter-junction transition capacitance. We have then

$$dQ_B = -C_D\,dv_{b'e} \qquad (20\text{-}62)$$

in which the minus sign appears because dQ_B is the minority charge. Since for a small collector-circuit resistance R_c the collector current is $i_C = g_m v_{b'e}$, we have

$$dQ_B = -\frac{C_D}{g_m} di_C \tag{20-63}$$

2. The Emitter-transition-capacitance Charge Change dQ_{Te} Analogous to dQ_B we have that the charge which must be supplied to the transition capacitance C_{Te} at the emitter is

$$dQ_{Te} = -\frac{C_{Te}}{g_m} di_C \tag{20-64}$$

The total emitter capacitance is $C_e = C_D + C_{Te}$.   Hence, from Eqs. (20-63) and (20-64)

$$dQ_B + dQ_{Te} = -\frac{C_e}{g_m} di_C = -\frac{1}{\omega_T} di_C \tag{20-65}$$

We have used Eq. (4-20) involving $\omega_T = 2\pi f_T$, where f_T is the common-emitter short-circuit current-gain–bandwidth product.

3. The Collector-transition-capacitance Charge Increment dQ_{Tc} The charge which must be supplied to the collector capacitance C_c is $-C_c\, dv_C$, where dv_C is the change in collector voltage.   If the transistor collector-circuit resistance is R_c, a change in collector current di_C will change the collector-junction voltage by $R_c\, di_C$.   Hence

$$dQ_{Tc} = -C_c R_c\, di_C \tag{20-66}$$

4. The Recombination Charge   In a time dt the charge which must be supplied to replace that lost by recombination is $Q_B\, dt/\tau_b$.   From Eqs. (20-53) and (20-56)

$$\frac{Q_B}{\tau_b} = -\frac{i_C}{h_{FE}} \tag{20-67}$$

Equation (20-61) may now be rewritten, using Eqs. (20-65) to (20-67), as

$$i_B = \left(\frac{1}{\omega_T} + C_c R_c\right)\frac{di_C}{dt} + \frac{i_C}{h_{FE}} \tag{20-68}$$

If $i_C = 0$ at $t = 0$ and if a constant base current $i_B = I_{B1}$ is applied, then the solution of Eq. (20-68) is

$$i_C = h_{FE} I_{B1}(1 - \epsilon^{-t/\tau_r}) \tag{20-69}$$

where

$$\tau_r = h_{FE}\left(\frac{1}{\omega_T} + C_c R_c\right) \tag{20-70}$$

This result is identical with that obtained in Eq. (20-18).   If we assume that the transition is to be made so rapidly that the effect of recombination may be neglected, we may drop the term i_C/h_{FE} in Eq. (20-68), and we have

$$\frac{\tau_r}{h_{FE}} \frac{di_C}{dt} = i_B = I_{B1} \tag{20-71}$$

$$\frac{\tau_r}{h_{FE}} \int_{0.1I_{cs}}^{0.9I_{cs}} di_C = \int_0^{t_r} I_{B1} \, dt \tag{20-72}$$

where the rise time t_r is the time for the collector current to increase from one-tenth to nine-tenths of its saturation value I_{CS}. We find

$$t_r = 0.8 \frac{\tau_r}{h_{FE}} \frac{I_{CS}}{I_{B1}} \tag{20-73}$$

as in Eq. (20-25).

On comparing Eq. (20-70) with Eq. (20-55) it appears that in the charge-control method, when we extend the analysis from the case where the transition capacitances are neglected to the case where they are taken into account, we need but to replace τ_b in Eq. (20-55) by the value $\tau_b = \tau_r$, given in Eq. (20-70). We must, however, keep in mind that in the analysis leading to Eqs. (20-69) and (20-70) we have neglected the fact that the transition capacitances are not constant but rather are functions of the junction voltages. Secondly, we have neglected [as in Eq. (20-63)] the fact that the volt-ampere characteristic of a junction is not linear but rather is exponential. [This nonlinearity makes its appearance in the hybrid II by virtue of the fact that $r_{b'e}$ (or g_m) is not constant.] When these nonlinearities are neglected, we observe that the stored charge is linearly related to the transistor currents. When, however, the capacitances are taken into account we must expect that this linearity will no longer apply exactly. The charge-control time constants, as measured, and hence including the effects of the transition capacitances, are usually represented by the symbols T_b and T_c. These time constants will then be functions of the operating point. In this respect, however, the situation is no worse than we normally encountered with any equivalent circuit. We may, in this case, as in previous cases, effectively use a piecewise linear and continuous representation. If the charge stored on the transition capacitance is small in comparison with the stored base charge, the variation of time constants with operating point need not be pronounced.

20-19   STORAGE TIME BY THE CHARGE–CONTROL METHOD

In the active region the transistor has a stored majority base charge $Q_B = I_B\tau_b$. This relationship between base current and base charge applies up to the point of saturation where the base charge is Q_{BA} and is maintained by a base current I_{BA}. When the transistor is in saturation, the base charge consists of the charge Q_{BA} plus the saturation charge Q_{BS}, as shown in Fig. 20-18b. We may certainly expect that an additional component of base current I_{BS} will be required to maintain the extra charge Q_{BS}. Accordingly we may relate the charge Q_{BS} to the current I_{BS} through a time constant τ_s, so that $Q_{BS} = \tau_s I_{BS}$. Our first inclination might well be to expect that the decay time constant τ_s for the saturation charge Q_{BS} is the same as the decay time constant τ_b for the base charge in the active region. We shall now explain why such is not the case.

As we have already seen, when a transistor is in saturation, it is operating in both the normal and the inverse directions at the same time. Just as a transistor has parameters α_N and α_I (ω_N and ω_I), it also has time constants τ_{bN} and τ_{bI} (τ_{cN} and τ_{cI}). We may expect such a difference on physical grounds. Consider a transistor carrying a current first in the normal direction and then in the inverse direction where the emitter is now the collector. In one case the base charge density will be at a maximum at one junction and in the other case at the other junction. Since the areas of the junctions may well be very different, the total charge stored per unit collector current will be larger when the larger junction is serving as the emitter. Thus in an alloy junction transistor where the collector is much larger than the emitter junction we may expect $\tau_{cI} > \tau_{cN}$. Also the time constants τ_{bN} and τ_{bI} will be different because these decay time constants depend in part on recombinations which take place on the surfaces of the transistor. And the charge distribution and hence surface recombination are different for the two current directions.

We may then expect that the storage time constant τ_s will depend on the time constants for the normal and inverse directions. Such is indeed the case. The derivation[13] of τ_s in terms of these normal and inverse time constants parallels very closely the derivation given in Sec. 20-14 for τ_s in terms of the normal and reverse values of α and ω.

Calculation of Storage Time We shall now find an expression for the time t_s required to remove the saturation charge Q_{BS} and thereby bring a transistor out of saturation. So long as the transistor is in saturation, the voltages across the junctions remain fixed and consequently there are no changes in the charge stored on the transition capacitances. The base charge is $Q_B = Q_{BA} + Q_{BS}$. The transistor is initially kept in saturation by a current I_{B1}, and at $t = 0$ the current is abruptly changed to I_{B2}. The current I_{B2} is less than I_{BA}, so that eventually the transistor will come out of saturation.

The basic charge-control relationship may now be written

$$I_{B2} = \frac{Q_{BA}}{\tau_b} + \frac{Q_{BS}}{\tau_s} + \frac{dQ_{BS}}{dt} \tag{20-74}$$

In Eq. (20-74) we have recognized that the charges Q_{BA} and Q_{BS} decay with different time constants and that as long as $Q_{BS} \neq 0$, $dQ_{BA}/dt = 0$. Since $Q_{BA}/\tau_b = I_{BA}$ we may rewrite Eq. (20-74) as

$$I_{B2} - I_{BA} = \frac{Q_{BS}}{\tau_s} + \frac{dQ_{BS}}{dt} \tag{20-75}$$

Integrating Eq. (20-75) subject to the condition that at $t = 0$,

$$Q_{BS} = \tau_s(I_{B1} - I_{BA})$$

we find

$$Q_{BS} = \tau_s(I_{B1} - I_{B2})\epsilon^{-t/\tau_s} + \tau_s(I_{B2} - I_{BA}) \tag{20-76}$$

The storage time t_{s1} is determined by the condition $Q_{BS} = 0$ at $t = t_{s1}$. We find

$$t_{s1} = \tau_s \ln \frac{I_{B1} - I_{B2}}{I_{BA} - I_{B2}} \tag{20-77}$$

in agreement with Eq. (20-41). The base current I_{BA} at the boundary between the active region and saturation is given by $I_{BA} = I_{CS}/h_{FE}$, where I_{CS} is the collector saturation current.

If the storage time is small enough so that recombination is negligible, then the term Q_{BS}/τ_s in Eq. (20-75) may be dropped. Integrating this equation over the storage time,

$$\int_0^{t_{s1}} (I_{B2} - I_{BA})\, dt = \int_{Q_{BS}}^0 dQ_B \tag{20-78}$$

or

$$t_{s1} = \frac{-Q_{BS}}{I_{B2} - I_{BA}} = -\tau_s \frac{I_{B1} - I_{BA}}{I_{B2} - I_{BA}} \tag{20-79}$$

where $I_{B1} - I_{BA} = I_{BS} = Q_{BS}/\tau_s$ is the current required to maintain the excess saturation charge Q_{BS}. (Since $I_{B1} - I_{BA}$ and $I_{B2} - I_{BA}$ are of opposite sign, t_{s1} is positive.) Equation (20-79) gives a conservative (too large) estimate of storage time because recombination has been neglected. Since some of the excess charge is disappearing owing to recombination, then the time t_{s1} necessary to remove the stored charge will be less than that calculated in Eq. (20-79).

Experimental Determination of τ_s   The storage time constant τ_s may be determined experimentally by an adaptation of the method of Fig. 20-21. This

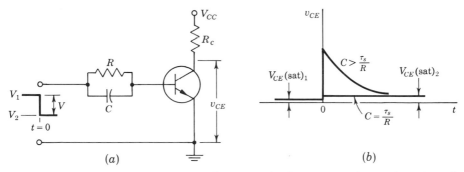

Fig. 20-21   (a) The input step carries the transistor from saturation to the edge of the active region.   (b) The collector waveforms. The collector-to-emitter voltage corresponding to V_1 is $V_{CE}(\text{sat})_1$ and to V_2 is $V_{CE}(\text{sat})_2$.

circuit will be recognized essentially as the *resistor-capacitor transistor logic* (*RCTL*) switch of Sec. 9-16 (Fig. 9-50). The input is a negative step between levels V_1 and V_2. At V_1 the transistor is well into saturation, and V_2 is the critical value of applied base voltage which just brings the transistor to the boundary between saturation and the active region. If the saturation base current is I_{BS} then

$$V = V_1 - V_2 = (I_{B1} - I_{BA})R = I_{BS}R \qquad (20\text{-}80)$$

If C has been adjusted properly, then the amount of charge abruptly removed from the base is

$$CV = Q_{BS} \qquad (20\text{-}81)$$

From Eqs. (20-81) and (20-80) we find

$$\frac{Q_{BS}}{I_{BS}} \equiv \tau_s = RC \qquad (20\text{-}82)$$

The proper value of C is easily identified. If C is too large the transistor will temporarily enter its active region and the collector voltage will have the form shown in Fig. 20-21b for $C > \tau_s/R$. If $RC = \tau_s$ then the collector waveform exhibits only the small voltage change corresponding to the abrupt transition from saturation to the edge of the active region.

20-20   STORED BASE CHARGE AS A PARAMETER

The *total* charge stored in the base Q_T under particular operating conditions is a parameter very frequently listed on a switching-transistor data sheet alongside the more customary parameters like h_{FE}, I_{CBO}, etc. The measurement of Q_T is made in connection with the circuit of Fig. 20-22. The driving signal is supplied from a low-impedance source of the order of 50 Ω. The voltage V at the upper level provides a base current V/R which establishes the transistor in saturation. At the transition the input waveform drops abruptly to zero and the transistor starts toward cutoff nominally. The waveform of

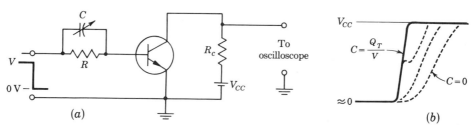

Fig. 20-22   (a) Circuit used for measuring Q_T; (b) the output waveform is indicated for various values of C from $C = 0$ to the value of $C = Q_T/V$.

the collector voltage is observed on a scope during turn-off as a function of the setting of the adjustable capacitor C.

When the transistor has attained cutoff, all the stored base charge Q_T must have been withdrawn from the base and transferred to the capacitor C. If the capacitor C is just adequate in capacitance to permit this transfer, then the stored base charge Q_T is

$$Q_T = CV \tag{20-83}$$

Experimentally we may recognize that C is not large enough to allow the transfer of all the stored base charge by noting that the transition at the collector in Fig. 20-22 consists of a rapid portion followed by a slower portion. When, on the other hand, we have adjusted C to the minimum value that makes the recovery at the collector rapid all the way to V_{CC}, then we may calculate Q_T from Eq. (20-83) using this minimum value of C. Figure 20-22b illustrates successive forms of the collector voltage in a typical case as C is increased from zero.

The stored base charge Q_T, like every other parameter, depends on the operating circumstances of the transistor. It is a function of the base current and collector current, which determine the extent to which the transistor is saturated. But it also depends on V_{CC}, since V_{CC} determines the extent to which the collector capacitance must charge. And it depends as well on the temperature, increasing with increasing temperature. Accordingly, to make serious use of this parameter we should have data concerning its dependence on operating conditions. Some manufacturers, in connection with certain switching transistors, make this information available. On the other hand, using the procedure given in connection with Fig. 20-22, it is a simple matter to measure Q_T under any conditions of interest. In order that this measurement procedure may yield an accurate result, it is necessary that the input step have a rise time which is small in comparison with the decay time constant of the stored charge. Otherwise some of the charge will disappear by recombination during the time of rise of the input waveform and the measured Q_T will fall short of the actual stored charge.

Size of Commutating Capacitance in a Flip-Flop   In Sec. 10-5 we considered the appropriate size for the commutating capacitance to be used in a binary tube circuit. At that point we were not able to indicate the proper size for a commutating capacitance in a transistor flip-flop. We may, however, do so now. The change of state of a transistor from ON to OFF normally takes longer than the reverse transition because of the storage time involved. Let us then consider how the commutating capacitor may help to turn a transistor OFF.

In a flip-flop (Fig. 10-13a) let $Q1$ be OFF and $Q2$ ON. At a transition, $Q1$ goes ON. Therefore there is applied through C_1 a negative step of amplitude equal to the collector voltage swing. At the moment the collector of $Q1$ drops to $V_{CE}(\text{sat})$ the impedance seen looking back into the collector of $Q1$ is nominally zero. Therefore the situation is entirely analogous to that shown in Fig. 20-22. Hence C_1 should be selected nominally equal to C as given in Eq. (20-83). By way of example, we find for the type 2N404 that, if the transistor is in saturation with $I_C = 10$ mA and $I_B = 1$ mA, an average unit has a stored charge $Q_T = 800$ pC and in the worst case $Q_T = 1,400$ pC. Suppose that the collector swing is 10 V. Then, for a flip-flop in which the ON transistor carries these currents, we should use a commutating capacitance $C_1 = 1,400/10 = 140$ pF.

20-21   THE $RCTL$ SWITCH[14]

We encountered the resistor-capacitor transistor logic ($RCTL$) switch in Sec. 9-16 (Fig. 9-50). This switch is redrawn in Fig. 20-23. We assume for convenience that the input signal has two levels, one at zero and the other at V. The resistor R_s represents the source impedance and includes the base-spreading resistance $r_{bb'}$. The resistor R is introduced so that when $v_i = V$ the base current is limited to a value that carries the transistor reasonably into saturation.

For a pulse input, the base current has the form shown in Fig. 20-24b. Initially, because of the capacitor, the current rises to $i'_B = V/R_s$. Thereafter it decays to $i''_B = V/(R + R_s)$ with a time constant $R_{\|}C$, where $R_{\|}$ is the parallel combination of R_s and R. The lower current is enough to maintain saturation. The spike of current allows a faster establishment of the required value of stored charge. Similarly at the end of the pulse at $t = t_p$ the negative spike of base current will hasten the removal of the stored charge. The capacitor voltage at $t = t_p$ is $V_o = i''_B R = VR/(R_s + R)$. Hence, the base current at $t = t_p+$ when v_i drops to 0 is $-V_o/R_s$. In Fig. 20-24b we have assumed that at $t = t'$ the stored charge has been removed, so that the transistor is at cutoff. Hence, at this time, the base current drops to zero.

If the charge-storage time constants T_b and T_s are known, we may calculate the times required for the transistor in Fig. 20-23 to turn ON and to turn OFF. The calculation is precisely the same as in the illustrative problem on page 764 since the base circuit is identical to the circuit of Fig. 20-11. In calculating the time to turn OFF we shall have to calculate separately the time to come out of saturation and the time to cross the active region to cutoff since T_b and T_s are not the same.

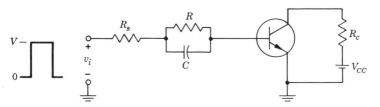

Fig. 20-23 The resistor-capacitor transistor logic ($RCTL$) switch.

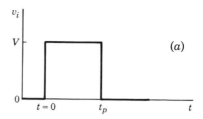

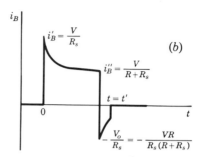

Fig. 20-24 (a) The input voltage to the base of an $RCTL$ switch; (b) the base-current waveform $i_B(t)$.

If it should happen that the time constant $R_{\parallel}C$ is very long in comparison with the turn-on and turn-off times, then we may assume that the base current is constant at one or the other of its peak values and we may make the calculation using Eqs. (20-24) and (20-41).

REFERENCES

1. Cooper, H. K.: Laminar P-N Junctions for Use as High Speed Core Drivers, Pacific Semiconductors, Inc., Lawndale, Calif.

2. Ko, W. H.: The Reverse Transient Behavior of Semiconductor Junction Diodes, *IRE Trans.*, vol. ED-8, no. 2, pp. 123–131, March, 1961.

3. Bakanowski, A. E., and J. H. Forster: Electrical Properties of Gold-doped Diffused Silicon Computer Diodes, *Bell System Tech. J.*, vol. 39, pp. 87–104, January, 1960.
   Conwell, E. A.: Properties of Silicon and Germanium, *Proc. IRE*, vol. 40, pp. 1327–1337, November, 1952.

4. Diode Recovery Time, *Appl. Note*, Texas Instruments, Inc., February, 1961.

5. Moll, J., S. M. Krakauer, and R. Shen: P-N Junction Charge-storage Diodes, *Proc. IRE*, vol. 50, no. 1, pp. 43–52, January, 1962.

6. Boff, A. F., J. Moll, and R. Shen: A New High Speed Effect in Solid State Diodes, *Intern. Solid-State Circuits Conf.*, 1960, pp. 50–51.

7. Krakauer, S. M., and S. W. Soshea: Hot Carrier Diodes Switch in Picoseconds, *Electronics*, vol. 36, no. 29, pp. 53–55, July 19, 1963.

8. Roehr, W. D. (ed.): "Switching Transistor Handbook," pp. 94–99, Motorola, Inc., Semiconductor Products Division, Phoenix, Ariz., 1963.
Hamilton, D. J., F. A. Lindholm, and J. A. Narud: Comparison of Large Signal Models for Junction Transistors, *Proc. IEEE*, vol. 52, no. 3, pp. 239–247, March, 1964.

9. Millman, J., and C. C. Halkias: "Electronic Devices and Circuits," McGraw-Hill Book Company, New York, 1967.

10. Giacoletto, L. J.: Study of P-N-P Alloy Junction Transistors from D-C through Medium Frequencies, *RCA Rev.*, vol. 15, no. 4, pp. 506–562, December, 1954.

11. Moll, J. L.: Large Signal Transient Response of Junction Transistors, *Proc. IRE*, vol. 42, no. 12, pp. 1773–1784, December, 1954.

12. Simmons, C. D., J. A. Ekiss, P. Spiegel, and P. M. Ansboro: Characterization of Switching Transistors, Final Report to U.S. Army Signal Research and Development Laboratory, Fort Monmouth, N.J., *Philco Rept.* R-113, Philco Corporation, Lansdale, Pa.
Beaufoy, R., and J. J. Sparkes: The Junction Transistor as a Charge-controlled Device, *Autom. Telephone Elec. J.*, vol. 13, no. 4, pp. 310–327, October, 1957.
Ekiss, J. A., and C. D. Simmons: Junction Transistor, Transient Response Characterization, *Solid State J.*, pt. 1, January, 1961; pt. 2, February, 1961.
Hegedus, C. L.: Charge Model of Fast Transistors and the Measurement of Charge Parameters by High Resolution Electronic Integrator, *Solid State Design*, August, 1964, pp. 23–36.

13. Grey, P. E., D. DeWith, A. R. Boothroyd, and J. F. Gibbons: "Physical Electronics and Circuit Models of Transistors," SEEC, vol. 2, John Wiley & Sons, Inc., New York, 1964.

14. Ekiss, J. A.: Calculation of Switching Times for *RC* Coupled Switching Circuits, *Appl. Lab. Rept.* 674, Philco Corporation, Lansdale, Pa., January, 1962.

APPENDIX

A

RINGING CIRCUIT WITH NONZERO INITIAL CONDITIONS

Consider the ringing circuit of Fig. A-1, in which there is an initial voltage V_o across the capacitor C as well as an initial inductor current I. It is now convenient to introduce a parameter Δ, defined as *the ratio of coil current to resistor current at $t = 0$*:

$$\Delta \equiv \frac{I}{V_o/R} = \frac{IR}{V_o} \tag{A-1}$$

The output v_o/V_o can be expressed as a function of time $(x = t/T_o)$ with Δ and k as parameters. The definitions of k and T_o are

$$k = \frac{1}{2R}\sqrt{\frac{L}{C}} \quad \text{and} \quad T_o = 2\pi\sqrt{LC} \tag{A-2}$$

Critical damping, $k = 1$

$$\frac{v_o}{V_o} = [1 - (1 + 2\Delta)(2\pi x)]\epsilon^{-2\pi x} \tag{A-3}$$

Overdamped, with $4k^2 \gg 1$

$$\frac{v_o}{V_o} = -\left(\frac{1}{4k^2} + \Delta\right)\epsilon^{-\pi x/k} + (1 + \Delta)\epsilon^{-4\pi k x} \tag{A-4}$$

Fig. A-1 Ringing circuit with initial current I in inductor and initial voltage V_o across capacitor.

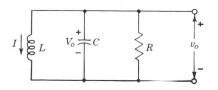

795

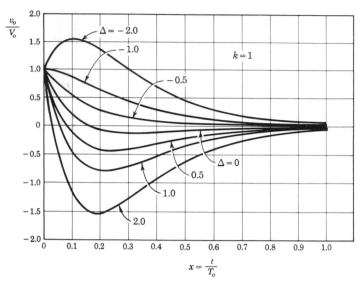

Fig. A-2   Plot of Eq. (A-3).

Underdamped, $k < 1$

$$\frac{v_o}{V_o} = \left[-(1 + 2\Delta)\frac{k}{\sqrt{1 - k^2}} \sin 2\pi \sqrt{1 - k^2}\, x \right.$$

$$\left. + \cos 2\pi \sqrt{1 - k^2}\, x \right] \epsilon^{-2\pi k x} \quad \text{(A-5)}$$

These responses are plotted in Figs. A-2 to A-4.   We note that even for the

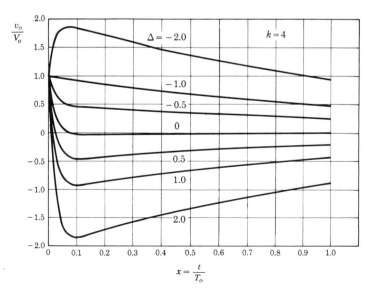

Fig. A-3   Plot of Eq. (A-4) with $k = 4$.

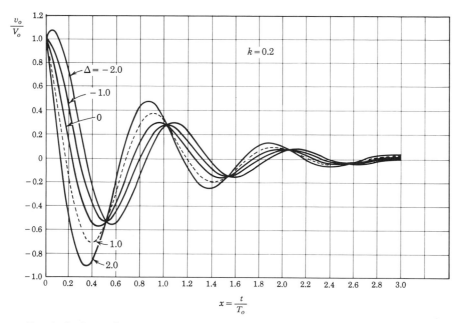

Fig. A-4   Plot of Eq. (A-5) with $k = 0.2$.

critically damped case there may be an *undershoot;* i.e., the output which starts at a positive value drops to a negative value before returning asymptotically to zero.   If V_o and I have the relative polarities indicated in Fig. A-1, then Δ is positive.   If the relative polarities differ from those indicated, then Δ is negative.   For a negative Δ, the output may rise first (see the curve for $\Delta = -2.0$) before falling to zero.   The physical reason for this initial increase in output is that the inductor current (with the polarity opposite to that in Fig. A-1) may charge the capacitor to a more positive voltage before C discharges through the resistor.   We see that the waveform depends upon the inductor and resistor currents (the sign and magnitude of Δ) and upon the amount of damping (the value of k).

The areas under each curve of Figs. A-2 to A-4 is $-k\Delta/\pi$.   This can be verified by direct integration or much more easily by proceeding as follows. Since

$$v_o = L\frac{di}{dt} \quad \text{or} \quad \frac{v_o}{V_o} = \frac{L}{V_o T_o}\frac{di}{dx}$$

then

$$\text{Area} \equiv \int_0^\infty \frac{v_o}{V_o}\,dx = \frac{L}{V_o T_o}\int_0^\infty di = -\frac{LI}{V_o T_o} = -\frac{L\Delta}{RT_o}$$

$$= -\frac{L\Delta}{(1/2k)\sqrt{L/C}\;2\pi\sqrt{LC}} = \frac{-k\Delta}{\pi} \tag{A-6}$$

B / DISTRIBUTED–PARAMETER DELAY LINES

In order to increase the inductance L of a coaxial cable (so as to increase both T and Z_o) the straight center conductor is replaced with a continuous coil of wire in the form of a helix, as indicated in Fig. B-1. Since the center conductor is wound in a tight helix, the magnetic flux in the region between inner and outer conductors may be neglected. The inductance then equals that of a solenoid of diameter a, with n turns per meter, or

$$L = \frac{\mu n^2 \pi a^2}{4} = \mu_r n^2 \pi^2 a^2 \times 10^{-7} \qquad \text{H/m} \tag{B-1}$$

where μ_r is the relative permeability of the core on which the solenoid is wound. The capacitance is that of coaxial cylinders with a material of relative dielectric constant ϵ_r between the diameter D and d, or

$$C = \frac{2\pi\epsilon}{\ln (D/d)} = \frac{2.40 \times 10^{-11}\epsilon_r}{\log (D/d)} \qquad \text{F/m} \tag{B-2}$$

For a type RG-65/U cable (Federal Telephone and Radio Company) whose parameters are $D = 0.285$ in., $c = 0.11$ in., helix of AWG No. 32 wire of diameter 0.008 in., with $n = 112$ turns per inch and a polyethylene dielectric

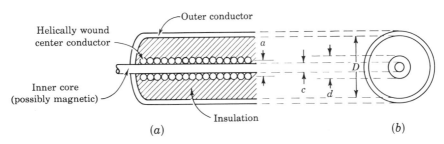

Fig. B-1 Helical high-impedance delay cable. (a) Longitudinal section; (b) transverse section.

($\epsilon_r = 2.3$), values of $Z_o = \sqrt{L/C} = 1{,}050\ \Omega$ and $T = \sqrt{LC} = 0.16\ \mu\text{sec/m}$ are calculated. These agree reasonably well with the measured values of $Z_o = 950\ \Omega$ and $T = 0.14\ \mu\text{sec/m}$. Note that the helical center conductor has increased the delay of the conventional polyethylene coaxial cable from 0.005 to 0.13 $\mu\text{sec/m}$, or by a factor of 26, and the impedance from 50 to 950 Ω, or by a factor of 19.

The inductance may be further increased by winding the helical inner conductor upon a ferromagnetic core. The type HH-1500A (Columbia Technical Corporation) is identical with the RG-65/U cable except that a flexible, stable, low-loss magnetic core ($\mu_r \approx 2$) is used. For this line, $Z_o = 1{,}500\ \Omega$ and $T = 0.23\ \mu\text{sec/m}$, which is an improvement by a factor of $\sqrt{\mu_r} \approx 1.4$. The HH-2000 (RG-176/U) is a similar high-fidelity line, with $Z_o = 2{,}200\ \Omega$ and $T = 9.36\ \mu\text{sec/m}$.

It follows from Eqs. (B-1) and (B-2) that if the dimension d in Fig. B-1 is increased while D is kept constant, then both L and C, and hence T, are increased. The General Electric Company manufactures a line (type DL1100) in which a is as large as possible, the inner and outer conductors being separated by a thin layer of insulating tape which is effectively 0.003 in. thick.[1] The helical conductor consists of No. 40 insulated wire with 277 turns per inch wound on a $\frac{3}{16}$-in.-diameter flexible plastic tubing. The outer conductor is made of a braid of *insulated* wires which are electrically connected only at the ends of the cable. If the braid were not insulated, the eddy currents would be excessive. For this line, $Z_o = 1{,}100\ \Omega$ and $T = 1.8\ \mu\text{sec/m}$.

A higher impedance may be obtained without sacrificing delay by increasing L and decreasing C. The type HH-2500 line (Columbia Technical Corporation) is similar to the DL1100 line just discussed, except that L is increased by using a magnetic core ($\mu_r \approx 4$) and C is decreased by using a thicker polyethylene spacer (0.035 cm) between inner and outer conductors. For this line, $Z_o = 3{,}000\ \Omega$ and $T = 2.0\ \mu\text{sec/m}$. The HH-4000 and HH-1600 are similar lines having characteristic impedances of 3,900 and 1,700 Ω, respectively, and each has a delay of 3.35 $\mu\text{sec/m}$. For lines of this type, in which the outer conductor is composed of insulated strands, there is unfortunately some leakage of the fields outside the line. Two lines placed in close proximity side by side will exhibit some cross coupling of signals. The attenuation of all of the lines described above is of the order of 0.3 dB/μsec of delay at frequencies below 1 MHz. The attenuation then increases rapidly to about 5 dB/μsec at 10 MHz.

Short calibrated lengths of delay lines (with the leads brought out through plastic endcaps) are available from the manufacturers of the bulk lines.

REFERENCE

1. Blewett, J. P., and J. H. Rubel: Video Delay Lines, *Proc. IRE*, vol. 35, pp. 1580–1584, December, 1947.

C

LUMPED-PARAMETER DELAY LINES[1]

A lumped line is made up of a cascaded series of symmetrical networks such as the T section† of Fig. C-1a. The *image* or *characteristic impedance* Z_o of this section is defined as follows. If the network is terminated in Z_o, then the impedance seen looking into the input terminals is also Z_o. Applying this definition, we find

$$Z_o = \sqrt{Z_1 Z_2 \left(1 + \frac{Z_1}{4Z_2}\right)} \tag{C-1}$$

The *propagation constant* γ is defined by $V_o/V_i \equiv \epsilon^{-\gamma}$ under the condition that the impedance Z_o is connected across the output terminals. The propagation constant is given by

$$\cosh \gamma = 1 + \frac{Z_1}{2Z_2} \tag{C-2}$$

† We have chosen to develop the discussion of lumped-parameter lines in terms of the T section. A parallel development may be written in terms of the II section.

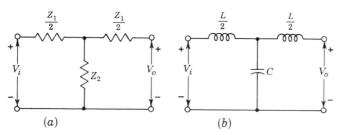

(a) $\qquad\qquad\qquad\qquad (b)$

Fig. C-1 (a) A prototype filter section; (b) a low-pass constant-k prototype section.

where $\cosh \gamma \equiv \frac{1}{2}(\epsilon^{\gamma} + \epsilon^{-\gamma})$ is the hyperbolic cosine of γ. If (as indicated in Fig. C-1) the series element is an inductor $L/2$ so that $Z_1 = j\omega L$ and the shunt element is a capacitor C so that $Z_2 = -j/\omega C$, then $Z_1 Z_2 = L/C = k$, a constant, independent of frequency. Such a network is called a low-pass constant-k prototype section. The *attenuation factor* α and the *phase factor* β are defined by $\gamma \equiv \alpha + j\beta$, where α and β are real functions of frequency. Equation (C-2) becomes, for the constant-k section,

$$\cosh (\alpha + j\beta) = 1 - \frac{\omega^2 LC}{2} \tag{C-3}$$

The passband of the filter is defined by the frequency band over which the attenuation factor is zero. Hence,

$$\cosh j\beta = \cos \beta = 1 - \frac{\omega^2 LC}{2} \tag{C-4}$$

Since β must be real, then $\cos \beta$ must have a magnitude between $+1$ and -1. Hence, $0 < \omega^2 LC/2 < 2$. The upper frequency f_c given by the above inequality is called the *cutoff frequency* and is given by

$$f_c = \frac{1}{\pi \sqrt{LC}} \tag{C-5}$$

For all frequencies between zero and f_c, the attenuation is zero, and within this passband the phase factor is given by

$$\cos \beta = 1 - 2 \left(\frac{f}{f_c}\right)^2 \tag{C-6}$$

Since $\cos \beta = 1 - \beta^2/2 + \beta^4/4! - \cdots$, we have the result that for $f \ll f_c$, $\beta \ll 1$ and $\beta \approx 2(f/f_c)$. For a sinusoidal input to the filter, $V_i = A\epsilon^{j\omega t}$ and

$$V_o = V_i \epsilon^{-\gamma} = A\epsilon^{j\omega t - j\beta} = A\epsilon^{j\omega(t-\beta/\omega)} = A\epsilon^{j\omega(t-1/\pi f_c)} \tag{C-7}$$

Thus, if the Fourier spectrum of the input signal to the network consists of frequencies all of which are much less than f_c, the output signal will be a faithful reproduction of the input signal except for being delayed by a time

$$t_s \approx \frac{1}{\pi f_c} = \sqrt{LC} \tag{C-8}$$

The quantity t_s is called the *time delay per section of filter*. For the constant-k network, Eq. (C-1) reduces to

$$Z_o = \sqrt{\frac{L}{C} \left[1 - \left(\frac{f}{f_c}\right)^2\right]} \tag{C-9}$$

For $f \ll f_c$, the characteristic impedance is independent of frequency and equals $\sqrt{L/C}$.

A delay line is specified[2] by giving the nominal impedance Z_o, the total delay t_d, and the rise time t_r of the output voltage when an ideal step is applied

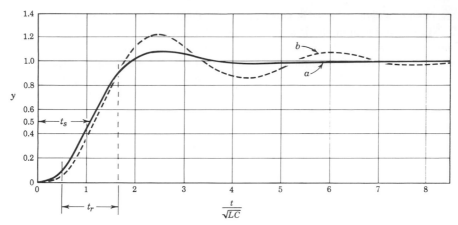

Fig. C-2 The step-voltage response of a single-section constant-k filter terminated in $R_o = \sqrt{L/C}$. Curve a: the input impedance is also R_o, and $y = 2v_o/v_i$. Curve b: the input impedance is zero, and $y = v_o/v_i$.

at the input. The quantity t_r is related to the delay per section t_s, but steady-state filter theory can give this relationship only after a very difficult Fourier-spectrum analysis. On the other hand, the response of a single section can be obtained directly by solving the differential equations of the two-mesh circuit of Fig. C-1b. The result of such an analysis is given (Prob. 3-36) in graphical form in Fig. C-2. The output is taken across a pure resistance R_o equal to the nominal characteristic impedance $\sqrt{L/C}$ of the filter. The solid curve a corresponds to a generator impedance equal to the output impedance R_o. The dashed curve is for a generator impedance equal to zero. We note that the peak overshoot is reduced from 22 to 8 percent as the generator resistance is increased from zero to R_o. Such an improvement is reasonable on the grounds that any reflection at the output termination will be absorbed at the input end.

In the discussion to follow we shall assume a termination R_o at each end of the filter. From Fig. C-2 we find the delay per section to be $t_s = 1.07\sqrt{LC}$. This value is to be compared with $t_s = \sqrt{LC}$ of Eq. (C-8), which is the result that would be obtained if the terminating impedance were the Z_o given in Eq. (C-9) and if all the frequency components in the input step could be considered to be small compared with f_c. The rise time per section t_{r1} is found from Fig. C-2 to be $t_{r1} = 1.13\sqrt{LC}$. Experimentally we find that the delay t_d of n sections is n times the delay per section, just as would be expected from filter theory,

$$t_d = nt_s \tag{C-10}$$

Also, experimentally[3] it is found that the rise time t_r of n sections is $\sqrt[3]{n}$ times that of a single section. It is possible to provide some theoretical justification for the factor $\sqrt[3]{n}$, but the matter is involved and we shall not pursue the point.

We consider then that, approximately,

$$t_r = t_{r1}\sqrt[3]{n} \tag{C-11}$$

From Eqs. (C-10) and (C-11) it follows that

$$n = \left(\frac{t_d}{t_r}\right)^{1.5}\left(\frac{t_{r1}}{t_s}\right)^{1.5} \tag{C-12}$$

Using the value of $t_{r1}/t_s = 1.13\sqrt{LC}/1.07\sqrt{LC} = 1.06$, found from Fig. C-2, we have

$$n = 1.1\left(\frac{t_d}{t_r}\right)^{1.5} \tag{C-13}$$

This equation gives the number of sections required in order to attain the desired specified value of t_d/t_r. If Eq. (C-13) does not yield an integer, then the next larger integer is used for n. From the relationships $t_s = 1.07\sqrt{LC}$, $t_d = nt_s$, and $R_o = \sqrt{L/C}$, we find

$$C = \frac{t_d}{1.07nR_o} \quad \text{and} \quad L = \frac{t_dR_o}{1.07n} \tag{C-14}$$

For given values of t_r, t_d, and R_o, Eqs. (C-13) and (C-14) are used to find the number of sections n required and the capacitance C and the inductance L of each section. The exact value of characteristic impedance is often not of importance. Hence, the standard manufactured value of C nearest the value obtained from the first of Eqs. (C-14) is used and then this equation is solved again for R_o. Using this value of R_o, the second of Eqs. (C-14) is solved for L. This inductance is then wound on a polystyrene cylinder or on a ferrite core.

Often the characteristic impedance required is dictated by the circuitry in which the line is to be incorporated. If there is some freedom of choice it is advantageous to design the line for the lowest acceptable impedance. Most of the attenuation on a line results from the resistance of the inductors, and if R_o is small, L can be made small, and the time delay may be kept constant by increasing the size of C.

If experiment shows that the output pulse shape is unsatisfactory (too much ringing) for a particular application, then a more conservative (smaller) value of t_s is chosen. Of course, a smaller t_s requires a greater number of sections. Hence, a line will result which will be more expensive, will occupy more space, and will have more attenuation than a line based upon a larger value of t_s.

We have already emphasized [Eq. (C-7)] that if β/ω is independent of frequency the output will be an exact replica of the input but delayed by an amount $t_s = \beta/\omega$. In Fig. C-3 the normalized t_s is seen to be far from constant over the passband of a constant-k filter. The constancy of t_s with frequency can be improved considerably by permitting coupling to exist between the two inductors of the constant-k section. This modification leads to the *m-derived filter section*, which will now be discussed.

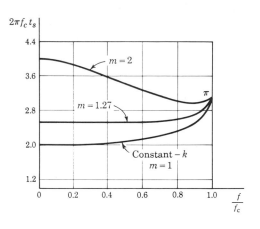

Consider the network of Fig. C-4, in which m is a real number. From Eqs. (C-1) and (C-2) we find that the cutoff frequency f_c and the characteristic impedance Z_o are given by the same expressions as for the prototype filter, namely, Eqs. (C-5) and (C-9), respectively. From Eq. (C-2) we find that, within the passband, β is given by

$$\cos \beta = 1 - \frac{2m^2(f/f_c)^2}{1 - (1 - m^2)(f/f_c)^2} \tag{C-15}$$

The time delay per section (at a given frequency ω) is $t_s = \beta/\omega$. Values of $\omega_c t_s$ calculated from Eq. (C-15) are plotted versus f/f_c in Fig. C-3 with m as a parameter. It turns out that the value of m which gives optimum constancy of $\omega_c t_s$ versus f is $m = 1.27$. For this value of m, the delay is constant up to about $0.6f_c$, whereas for the constant-k filter (corresponding to $m = 1$) the delay already departs appreciably from constancy at $0.2f_c$. We must not, however, naïvely conclude that an m-derived filter (with $m = 1.27$) will be "three times as good" as a constant-k filter. A comparison can only be made after the transient response is studied, as we shall do later.

For small values of f/f_c, Eq. (C-15) reduces to

$$\cos \beta = 1 - 2m^2 \left(\frac{f}{f_c}\right)^2 \approx 1 - \frac{\beta^2}{2}$$

or $\beta = 2mf/f_c$, and the delay per section is

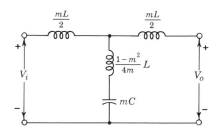

Fig. C-4 An m-derived filter section.

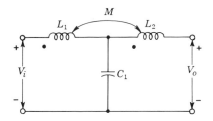

Fig. C-5   A network equivalent to the m-derived section.

$$t_s = \frac{\beta}{\omega} = \frac{\beta}{2\pi f} = \frac{m}{\pi f_c}$$

and since $f_c = 1/(\pi \sqrt{LC})$,

$$t_s = m \sqrt{LC} \tag{C-16}$$

The m-derived section of Fig. C-4 is not realizable for $m > 1$ since the shunt inductance is negative.   It is, however, realizable in the form of Fig. C-5, in which there is a mutual inductance between the series inductors.   The circuit of Fig. C-6 is identical to the circuit of Fig. C-5, as may readily be established from the mesh equations for these two circuits.   Comparing Figs. C-6 and C-4, we have

$$M = \frac{m^2 - 1}{4m} L \qquad L_1 + M = m \frac{L}{2}$$

from which

$$L_1 = \frac{m^2 + 1}{4m} L = 0.515L \tag{C-17}$$

Also

$$C_1 = mC = 1.27C \tag{C-18}$$

and the coefficient of coupling between the inductors L_1 is

$$K = \frac{M}{L_1} = \frac{m^2 - 1}{m^2 + 1} = 0.237 \tag{C-19}$$

The step-voltage response of a single section terminated at both ends in a pure resistance $R_o = \sqrt{L/C}$ is given in Fig. C-7.   The value of t_s is found to be 1.20 $\sqrt{LC}$, which is to be compared with $t_s = 1.27 \sqrt{LC}$ obtained from Eq. (C-16) for $m = 1.27$.   There are two advantages of the m-derived

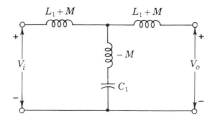

Fig. C-6   A network equivalent to the network of Fig. C-5.

filter (with $m = 1.27$) over the constant-k. The first is that the peak over-shoot of the former is 4 percent as against 8 percent for the latter. The second is that $t_{r1}/t_s = 1.06$ for the constant-k, whereas for $m = 1.27$ this ratio is found from Fig. C-7 to be 0.96, or 10 percent smaller. The number of sections needed is now found from Eq. (C-12) to be

$$n = 0.94 \left(\frac{t_d}{t_r}\right)^{1.5} \tag{C-20}$$

Comparing this equation with Eq. (C-13), we see that for the same ratio of delay to rise time, a line with m-derived filters requires about 16 percent fewer sections than one constructed from prototype sections. Note, however, that an m-derived section has an *undershoot* or *preshoot* of magnitude 12 percent.

From the relationships $t_s = 1.20 \sqrt{LC}$, $t_d = nt_s$, and $R_o = \sqrt{L/C}$, we find

$$C = \frac{t_d}{1.20 n R_o} \quad \text{and} \quad L = \frac{t_d R_o}{1.20 n} \tag{C-21}$$

For specified values of t_d, t_r, and R_o, Eqs. (C-20) and (C-21) give n, C, and L. Then Eqs. (C-17) to (C-19) give L_1, C_1, and K. The inductances are often wound on a polystyrene cylinder and the core diameter and length of winding are chosen so as to give the required value (0.237) of the coefficient of coupling.[4] If a delay line is to be used to reproduce a signal with a minimum of distortion, it may be necessary to use variable shunt capacitors and to adjust these individually so as to obtain the best possible step-voltage response.

If the inductor is wound on a ferrite core, the coefficient of coupling is very close to unity. If the capacitor C is connected not to the center of the inductor but rather close to the right-hand end of the coil, then a line comparable to the m-derived line results. Still another type of structure which

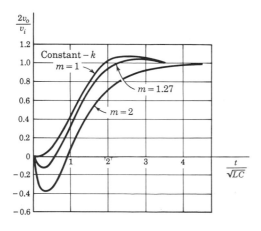

Fig. C-7 The step-voltage response of a single-section m-derived filter terminated at each end in $R_o = \sqrt{L/C}$.

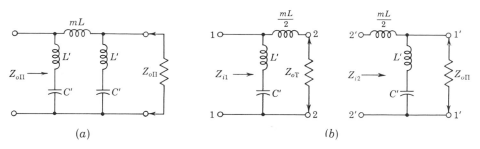

Fig. C-8   (a) The m-derived T section rearranged into a II section, where $L' = [(1 - m^2)/2m]L$ and $C' = mC/2$.   For $m = 0.6$, $Z_{o\text{II}} \approx R_o = \sqrt{L/C}$. (b) The circuit in (a) split into two half sections.   For all values of m, $Z_{i1} = Z_{o\text{II}}$ and $Z_{i2} = Z_{o\text{T}}$.

is used in the construction of delay lines is the so-called "bridged-tee" section. In this section, in addition to the coupling between coils, one includes an impedance element which is bridged directly across the network from input to output.   Design formulas for these two types of lines are given in the literature.[5,6]

To minimize the signal distortion introduced by a lumped-circuit delay line, the line should be matched as nearly as possible at both the sending and receiving ends.   Such matching will reduce reflections at the terminations. We have been considering up to the present that the line is terminated at both ends in its nominal characteristic impedance $R_o = \sqrt{L/C}$.   We observe, however, from Eq. (C-9) that $Z_o = \sqrt{L/C}$ only at zero frequency and that as the cutoff frequency is approached Z_o goes to zero.   It is possible to introduce between the line and each resistive termination a section which improves the match between line and termination.   These matching sections, called *m-derived half sections*, will now be discussed.

The m-derived T section of Fig. C-4 has been rearranged in Fig. C-8a into a II section in which the total series and shunt impedances are the same as in the T section.   The characteristic impedance $Z_{o\text{II}}$ of this II section is a function of m.   At zero frequency and for any m, $Z_{o\text{II}} = R_o = \sqrt{L/C}$ and $Z_{o\text{II}}$ increases to infinity as the frequency approaches the cutoff value ($f = f_c$). However, for $m = 0.6$ it is found[7] that the image impedance remains constant at $Z_{o\text{II}} = R_o$ to within 4 percent over 90 percent of the passband.

In Fig. C-8b the network in Fig. C-8a has been split into two half sections. These half sections have particularly useful matching properties.   It may be shown[7] that independently of the value of m the input impedance Z_{i2} seen looking into terminals 2'-2' when terminals 1'-1' are terminated in $Z_{o\text{II}}$ (as in Fig. C-8b) is the characteristic impedance $Z_{o\text{T}}$ of the T section, $Z_{i2} = Z_{o\text{T}}$. As noted above, $Z_{o\text{II}}$ may be approximated by the resistance R_o over most of the bandpass if $m = 0.6$.   Hence, by placing R_o across terminals 1'-1' and with m chosen to be 0.6, the half section gives an excellent match to a line

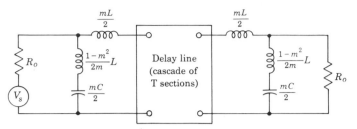

Fig. C-9 m-derived half sections with $m = 0.6$ are used to improve the match at the sending and receiving ends of a delay line.

consisting of T sections. The line may be constructed with either constant-k or m-derived T sections having any value of m (although for the reasons given above $m = 1.27$ would usually be used).

It is also found that if terminals 2-2 are terminated in Z_{oT} (as in Fig. C-8b) the input impedance Z_{i1} seen looking into terminals 1-1 is $Z_{o\Pi}$ independently of m. Hence, $Z_{i1} = Z_{o\Pi} \approx R_o$ if $m = 0.6$. These principles are illustrated in Fig. C-9, where half sections are used to improve the match at the sending and the receiving ends of a line consisting of T sections with pure-resistance terminations $R_o = \sqrt{L/C}$.

REFERENCES

1. Ryder, J. D.: "Networks, Lines and Fields," Prentice-Hall, Inc., Englewood Cliffs, N.J., 1949.
   Johnson, W. C.: "Transmission Lines and Networks," McGraw-Hill Book Company, New York, 1950.

2. Trevor, J. B.: Artificial Delay-line Design, *Electronics*, vol. 18, pp. 135–137, June, 1945.

3. Elmore, W. C., and M. Sands: "Electronics," McGraw-Hill Book Company, New York, 1949.

4. Blackburn, J. F. (ed.): "Components Handbook," p. 211, Massachusetts Institute of Technology Radiation Laboratory Series, vol. 17, McGraw-Hill Book Company, New York, 1948.

5. Wallis, C. M.: Design of Low-frequency Constant Time Delay Lines, *Trans. AIEE*, pt. I, vol. 71, pp. 135–139, 1952.

6. Ginzton, E. L., W. R. Hewlett, L. H. Jasberg, and J. D. Noe: Distributed Amplification, *Proc. IRE*, vol. 36, pp. 956–969, August, 1949.

7. Van Valkenburg, M. E.: "Network Analysis," sec. 13-8, Prentice-Hall, Inc., Englewood Cliffs, N.J., 1955.

TUBE
CHARACTERISTICS†

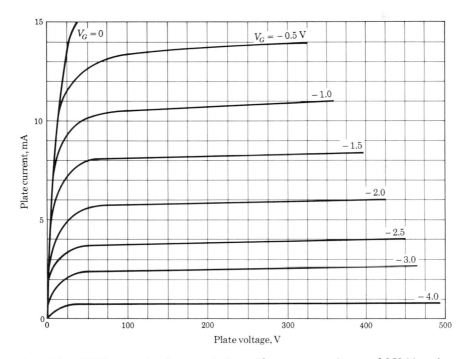

Fig. D-1   6AU6 pentode characteristics with a screen voltage of 150 V and a suppressor voltage of 0 V.

† Courtesy of General Electric Company.

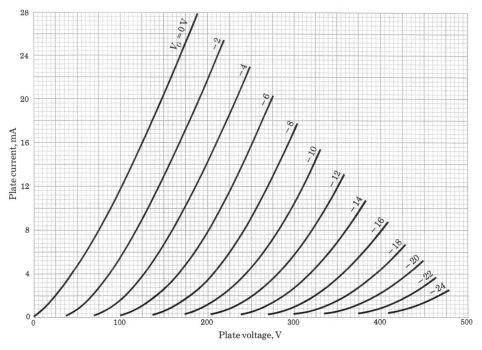

Fig. D-2 6CG7 (6SN7) negative-grid characteristics (each section).

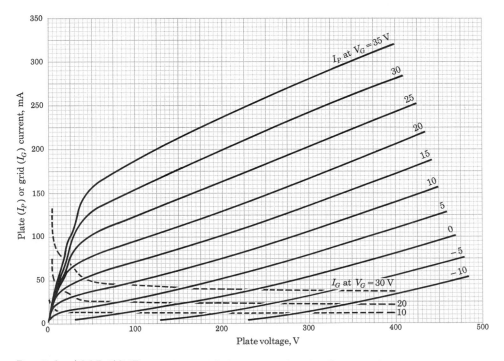

Fig. D-3 6CG7 (6SN7) positive-grid characteristics (each section).

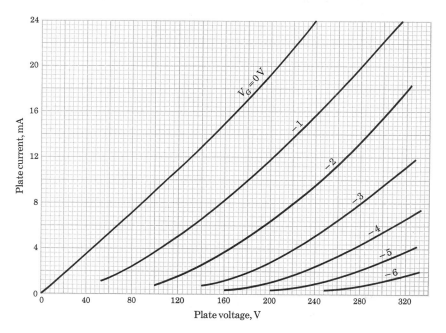

Fig. D-4   12AT7 (6664 and 7898) negative-grid characteristics (each section).

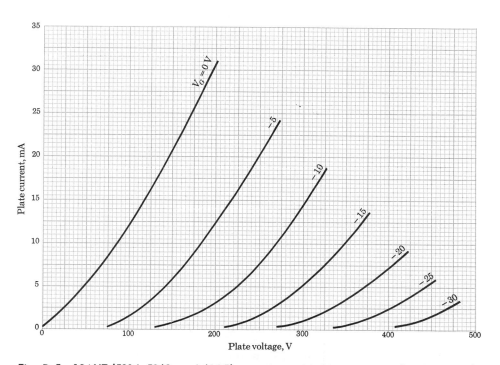

Fig. D-5   12AU7 (5814, 5963, and 6135) negative-grid characteristics (each section).

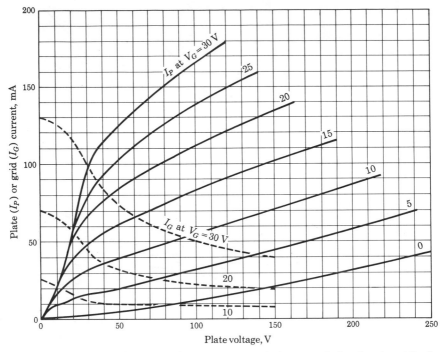

Fig. D-6 12AU7 (5814 and 5963) positive-grid characteristics (each section).

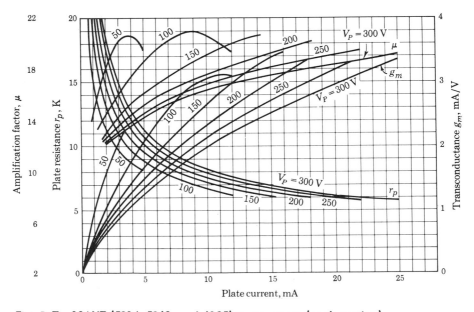

Fig. D-7 12AU7 (5814, 5963, and 6135) parameters (each section).

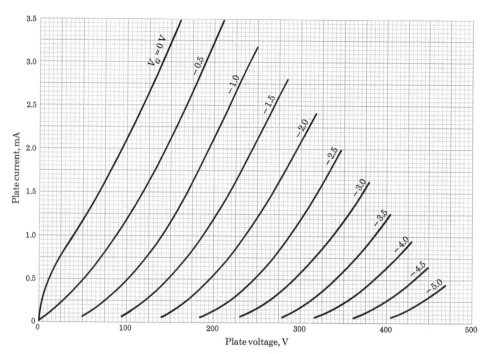

Fig. D-8   12AX7 (6681 and 7058) negative-grid characteristics (each section).

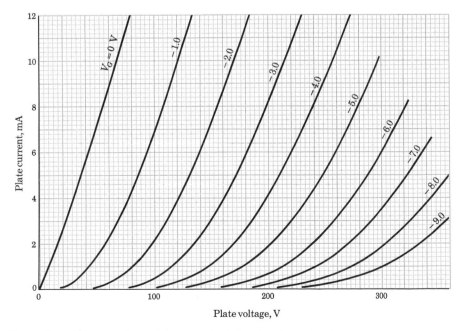

Fig. D-9   5965 negative-grid characteristics (each section).

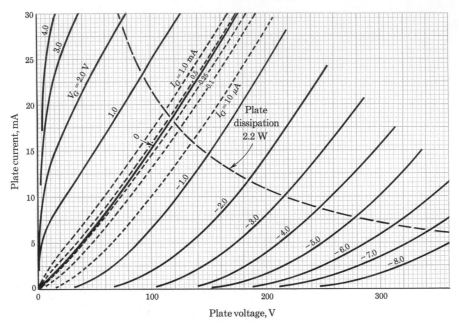

Fig. D-10 5965 positive-grid characteristics (each section).

PROBLEMS

CHAPTER 1

1-1 The generator v_I is an independent generator, whereas v_D is dependent. Find the Thévenin's equivalent circuit with respect to terminals AB for the cases

a. $v_D = kv_{AC}$, where k is a constant.

b. $v_D = kv_{AB}$, where k is a constant.

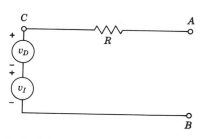

Prob. 1-1

1-2 If V_1 and V_2 are sinusoids, find the voltage between terminals AB

a. By the method of mesh currents.

b. As the ratio of the short-circuit current to the admittance between A and B.

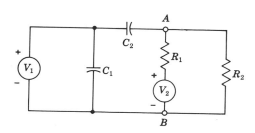

Prob. 1-2

1-3   The generator v_i is independent.   Calculate the open-circuit voltage v_{kn}. Calculate the short-circuit current between K and N.   What is the Thévenin's equivalent circuit with respect to terminals KN?

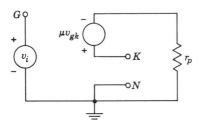

Prob. 1-3

1-4   Verify Eqs. (1-2), (1-3), and (1-4).

1-5   Derive the results of Eq. (1-5).

1-6   Express the hybrid parameters for the common-base configuration in terms of the common-emitter hybrid parameters.

1-7   Verify Eqs. (1-6), (1-7), and (1-8).

1-8   *a.* A transistor with parameters as in Table 1-2 operates in the common-emitter configuration and drives a 2-K load.   Calculate its current gain A_I, its input impedance Z_i, and its voltage gain A_V.

b. If the transistor is reconnected to operate as an emitter follower with the same 2-K resistor in the emitter circuit, calculate A_I, Z_i, and A_V.

c. If the source resistance is 1 K, calculate A_{Vs} for parts *a* and *b*.

1-9   A silicon transistor, operating at room temperature and at an emitter current of 2.0 mA, has hybrid-Π parameters $r_{bb'} = 50\ \Omega$, $r_{b'e} = 2$ K, $r_{ce} = 100$ K, and $r_{b'c} = 2$ M. Find the common-emitter hybrid parameters.

1-10   Verify Eqs. (1-11), (1-12), (1-13), and (1-14).

1-11   The signals V_1 and V_2 are sinusoidal.   Draw the small-signal equivalent circuit from which to calculate the signal current I.

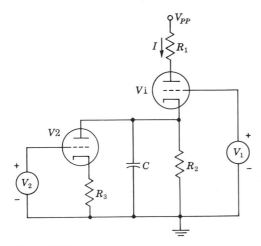

Prob. 1-11

1-12 Each tube shown has a plate resistance $r_p = 10$ K and an amplification factor $\mu = 20$. Find the gain (a) v_o/v_1 and (b) v_o/v_2.

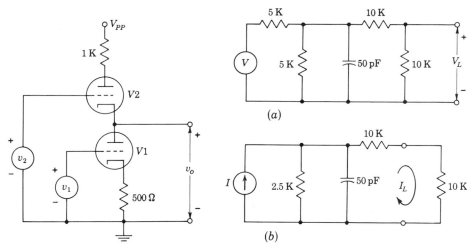

Prob. 1-12   Prob. 1-13

1-13 The signals V and I in (a) and (b) are sinusoidal.

   a. Find the value of the ratio $|V_L/V|$ at zero frequency and find the frequency at which the ratio has fallen by 3 dB.

   b. By inspection, from the result in (a) find the 3-dB frequency for the ratio $|I_L/I|$.

   c. By inspection, show that $I_L/I = \frac{1}{2}(V_L/V)$.

1-14 Verify Eq. (1-25).

1-15 The tube has parameters $\mu = 20$, $r_p = 7$ K. The capacitance between any two electrodes is 3.0 pF. Use the Miller theorem to calculate the impedance seen between terminals G and N at a frequency of 10 kHz.

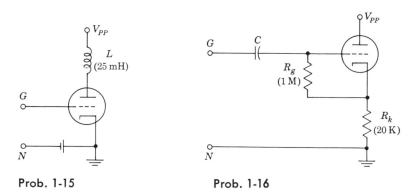

Prob. 1-15   Prob. 1-16

1-16 The tube has parameters $\mu = 20$, $r_p = 7$ K. Use the Miller theorem to compute the impedance seen looking into the terminals GN. Neglect the reactance of the capacitor.

1-17 *a.* Apply the Miller theorem to the hybrid-Π circuit of Fig. 1-4 with a load R_L connected between the output terminals CE. Assuming that in calculating the transistor internal stage gain K the elements $r_{b'c}$ and C_c may be neglected, verify that $K = -g_m R'_L$, where R'_L is R_L in parallel with r_{ce}.

b. If the input signal I comes from an ideal current generator, find the low-frequency current gain and the 3-dB frequency.

1-18 Design an operational amplifier whose output (for a sinusoidal signal) is equal in magnitude to its input and leads the input by 45°.

1-19 Assume a single-stage operational amplifier in which the base amplifier has a gain of -100. If $Z = R$ and $Z' = -jX_c$ with $R = X_c$, calculate the gain as a complex number.

1-20 Given an operational amplifier in which Z consists of R and L in series and Z' is a capacitor C. If the input is a constant V, find the output v_o as a function of time. Assume an infinite open-loop gain.

1-21 Sketch an operational-amplifier circuit having an input v and an output which is approximately $-5v - 3\,dv/dt$.

1-22 For the circuit shown, prove that the output voltage is given by

$$-v_o = \frac{R_2}{R_1} v + \left(R_2 C + \frac{L}{R_1} \right) \frac{dv}{dt} + LC \frac{d^2 v}{dt^2}$$

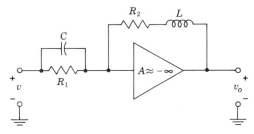

Prob. 1-22

1-23 Write the differential equation which relates v_o and v in the operational-amplifier circuit shown.

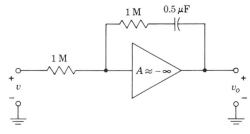

Prob. 1-23

1-24 Given an operational amplifier with Z consisting of a resistor R in parallel with a capacitor C and Z' consisting of a resistor R'. The input is a sweep voltage

$v = \alpha t$. Prove that the output is a sweep voltage that starts with an initial step. Thus, show that

$$v_o = -\alpha R'C - \alpha \frac{R'}{R} t$$

Assume an infinite open-loop gain for the base amplifier.

1-25 What is (a) v when the output is zero, (b) v_o if $v = -100$ V, and (c) the grid-to-cathode voltage when $v_o = +50$ V?

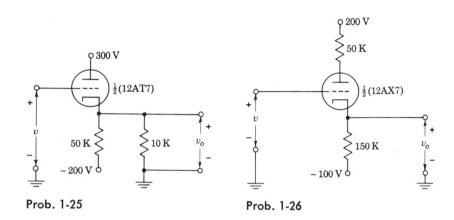

Prob. 1-25   Prob. 1-26

1-26 (a) If $v = 0$, find v_o. (b) If $v = 100$, find v_o. (c) If the grid-to-cathode voltage is zero, find v_o. (d) If $v_o = 0$, find v.

1-27 The reactance of C and the impedance of the generator are both negligible. $R_1 + R_2 = 10$ K. The input signal v is symmetrical with respect to ground. Find R_1 and R_2 if the tube is to handle, without distortion and without drawing grid current, the largest possible amplitude of signal. What is the maximum signal the tube will handle in this case?

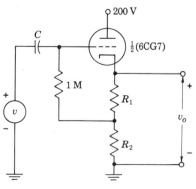

Prob. 1-27

1-28 *a.* Set $v_2 = 0$, and show that if $R_k \gg (r_p + R_p)/(\mu + 1)$ the two output signals v_{o1} and v_{o2} are opposite in polarity and approximately equal in magnitude. Calculate the gain v_{o1}/v_1. This amplifier circuit, which makes available a pair of signals symmetrical with respect to ground, is called a *paraphase amplifier*.

b. Show that when both v_1 and v_2 are present and if $R_k \gg (r_p + R_p)/(\mu + 1)$, either signal v_{o1} or v_{o2} is approximately proportional to the difference $v_1 - v_2$. Calculate the ratio $v_{o1}/(v_1 - v_2)$. When used in this manner the amplifier is called a *difference amplifier*.

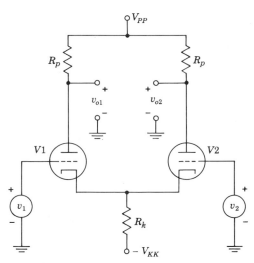

Prob. 1-28

1-29 *a.* A transistor is driven from a source of impedance R_s. Using the *open-circuit-voltage–short-circuit-current theorem* of Sec. 1-2 verify that the output admittance Y_o is given in terms of the hybrid parameters by

$$Y_o = h_o - \frac{h_f h_r}{R_s + h_i}$$

b. Using the parameters in Table 1-2, for each configuration calculate the limits on $R_o = 1/Y_o$ as the source impedance R_s varies from zero to infinity.

1-30 For the transistor whose parameters are given in Table 1-2 calculate, for each configuration, the limits on the input impedance Z_i as the load resistance R_L varies from zero to infinity.

1-31 A transistor with parameters as in Table 1-2 is used as an emitter follower. It is driven from a source of 150 Ω and has an emitter resistor of 1,000 Ω. Calculate the input impedance, the output impedance, and the voltage gain.

1-32 *a.* For the circuit of the transistor stage of Fig. 1-19, which includes an emitter resistor, prove that the exact expression for the input impedance is

$$Z_i = R_s + (1 - A_I)R_e + h_{ie} + h_{re}A_I Z_L$$

b. Using the exact expression of Eq. (1-56) for A_I and using the value of Z_i in (*a*), calculate the output impedance Z_o as the ratio of open-circuit voltage to short-circuit current. Prove that

$$Z_o = \frac{R_e(1 + h_{fe}) + (R_s + h_{ie})(1 + h_{oe}R_e)}{h_{oe}(R_s + h_{ie} + R_e - h_{re}h_{fe}/h_{oe})}$$

c. Show that if $R_e \geq R_s + h_{ie}$,

$$Y_o \approx h_{ob}\left(1 + \frac{R_s + h_{ie} - h_{re}h_{fe}/h_{oe}}{R_e}\right)$$

where $h_{ob} = h_{oe}/(1 + h_{fe})$.

1-33 Verify Eq. (1-56).

CHAPTER 2

2-1 Prove by direct integration that the area under the curve of Fig. 2-3*d* is zero.

2-2 A current pulse of amplitude I is applied to a parallel RC combination. Plot to scale the waveforms of the current i_C for the cases (*a*) $t_p < RC$, (*b*) $t_p = RC$, (*c*) $t_p > RC$.

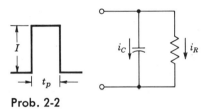

Prob. 2-2

2-3 Verify Eqs. (2-11) for a symmetrical square wave applied to a high-pass RC circuit.

2-4 A symmetrical square wave of peak-to-peak amplitude V and frequency f is applied to a high-pass RC circuit. Show that the percentage tilt is given by

$$P = \frac{1 - \epsilon^{-1/2fRC}}{1 + \epsilon^{-1/2fRC}} \times 200\%$$

If the tilt is small, show that this reduces to Eq. (2-13).

2-5 A 10-Hz symmetrical square wave whose peak-to-peak amplitude is 2 V is impressed upon a high-pass circuit whose lower 3-dB frequency is 5 Hz. Calculate and sketch the output waveform. In particular, what is the peak-to-peak output amplitude?

2-6 A 10-Hz square wave is fed to an amplifier. Calculate and plot the output waveform under the following conditions: the lower 3-dB frequency is (*a*) 0.3 Hz, (*b*) 3.0 Hz, (*c*) 30 Hz.

2-7 *a.* A square wave whose peak-to-peak value is 1 V extends ± 0.5 V with respect to ground. The duration of the positive section is 0.1 sec and of the negative section is 0.2 sec. If this waveform is impressed upon an RC differentiating circuit whose time constant is 0.2 sec, what are the steady-state maximum and minimum values of the output waveform?

b. Prove that the area under the positive section equals that under the negative section of the output waveform. What is the physical significance of this result?

2-8 A square wave whose peak-to-peak value is 1 V extends ± 0.5 V with respect to ground. The half period is 0.1 sec. This voltage is impressed upon an RC differentiating circuit whose time constant is 0.2 sec. What are the steady-state maximum and minimum values of the output voltage?

2-9 a. Derive Eqs. (2-19) and (2-20).

b. Prove that the peak of the output pulse occurs at

$$x = 2.30 \, \frac{n}{n-1} \, \log n$$

2-10 The pulse from a high-voltage generator (a magnetron) rises linearly for 0.05 μsec and then remains constant for 1 μsec. The rate of rise of the pulse is measured with an RC differentiating circuit whose time constant is 250 psec. If the positive output voltage from the differentiator has a maximum value of 50 V, what is the peak voltage of the generator?

2-11 The limited ramp is applied to an RC differentiator. Draw, to scale, the output waveform for the cases (a) $T = RC$, (b) $T = 0.2RC$, (c) $T = 5RC$.

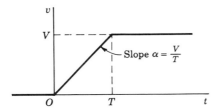

Prob. 2-11

2-12 The input to a high-pass RC circuit is periodic and trapezoidal as indicated. Assume that the time constant RC is large compared with either T_1 or T_2. Find and sketch the steady-state output if $RC = 10T_1 = 10T_2$.

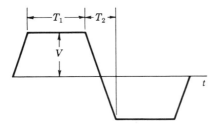

Prob. 2-12

2-13 Prove that for the same input, the output from the two differentiating circuits will be the same if $RC = L/R'$. Assume that the initial conditions are those of rest (no voltage on C and no current in L).

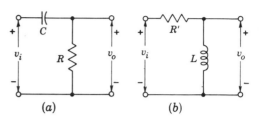

(a)   (b)

Prob. 2-13

2-14 A signal $v = V\epsilon^{-t/\tau}$ is applied to a double differentiator. Find the output if $\tau = R_1C_1 = R_2C_2$. Plot the response and locate any minima.

2-15 *a.* Verify Eq. (2-30) for the output of a double differentiator to which is applied an exponential input $v_i = V(1 - \epsilon^{-t/\tau})$.

b. Find the initial derivative of the input and of the output.

c. Verify that the area under the output-waveform curve v_o is zero.

d. What input waveform applied to a triple differentiator will give the output of Eq. (2-30)?

2-16 An ideal 1-μsec pulse is fed to an amplifier. Calculate and plot the output waveform under the following conditions: the upper 3-dB frequency is (*a*) 10 MHz, (*b*) 1.0 MHz, (*c*) 0.1 MHz.

2-17 A pulse is applied to a low-pass RC circuit. Prove by direct integration that the area under the pulse is the same as the area under the output waveform across the capacitor. Explain this result physically.

2-18 Repeat Prob. 2-2 for the waveform of the current i_R through the resistor.

2-19 A symmetrical square wave whose peak-to-peak amplitude is 2 V and whose average value is zero is applied to an RC integrating circuit. The time constant equals the half period of the square wave. Find the peak-to-peak value of the output amplitude.

2-20 The periodic waveform shown is applied to an RC integrating network whose time constant is 10 μsec. Sketch the output. Calculate the maximum and minimum values of output voltage with respect to ground.

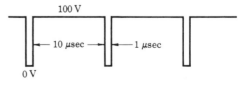

100 V

10 μsec   1 μsec

0 V

Prob. 2-20

2-21 A symmetrical square wave whose average value is zero has a peak-to-peak amplitude of 20 V and a period of 2 μsec. This waveform is applied to a low-pass circuit whose upper 3-dB frequency is $1/2\pi$ MHz. Calculate and sketch the steady-state output waveform. In particular, what is the peak-to-peak output amplitude?

2-22 A square wave whose peak-to-peak amplitude is 2 V extends ± 1 V with respect to ground. The duration of the positive section is 0.1 sec and that of the negative section 0.2 sec. If this waveform is impressed upon an RC integrating circuit whose time constant is 0.2 sec, what are the steady-state maximum and minimum values of the output waveform?

2-23 Verify Eq. (2-36) for a symmetrical square wave applied to a low-pass RC circuit.

2-24 The square wave shown is fed to an RC coupling network. What are the voltage waveforms across R and across C if (a) RC is very large, say $RC = 10T$, and (b) RC is very small, say $RC = T/10$?

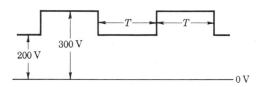

200 V   300 V   0 V

Prob. 2-24

2-25 *a.* Three low-pass RC circuits are in cascade and isolated from one another by ideal buffer amplifiers. Find the expression for the output voltage as a function of time if the input is a step voltage.

b. Find the rise time of the output in terms of the product RC.

c. What is the ratio of the rise time of the three sections in cascade to the rise time of a single section?

2-26 The limited ramp of Prob. 2-11 is applied to a low-pass RC integrating circuit. Draw to scale the output waveforms for the cases (a) $T = RC$, (b) $T = 0.2RC$, (c) $T = 5RC$.

2-27 *a.* The periodic ramp voltage shown is applied to a low-pass RC circuit. Find the equations from which to determine the steady-state output waveform.

b. If $T_1 = T_2 = RC$, find the maximum and minimum values of the output voltage and plot this waveform. NOTE: The minimum value does *not* occur at the beginning of interval T_1.

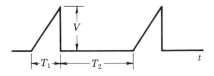

V T_1 T_2 t

Prob. 2-27

2-28 *a.* Prove that an RC circuit behaves as a reasonably good integrator if $RC \gg 15T$, where T is the period of an input sinusoid $E_m \sin \omega t$.

b. Show that the output is approximately $-(E_m/\omega RC) \cos \omega t$.

2-29   An oscilloscope test probe is indicated.   Assume that the cable capacitance
is 100 pF.   The input impedance of the scope is 2 M in parallel with 10 pF.   What is
(a) the attenuation of the probe, (b) C for best response, and (c) the input impedance
of the compensated probe?

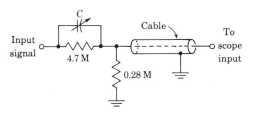

Prob. 2-29

2-30   Assume a waveform consisting of a sine wave and a d-c voltage equal to the
peak V_m of the sine wave so that the resultant waveform extends from zero to $2V_m$.
    a. This waveform is applied to a 3:1 compensated attenuator.   Plot the output
waveform and indicate the zero-voltage level.
    b. If the attenuator is improperly compensated ($R_1 = 2R_2$, but $2C_1 \neq C_2$), plot
the output waveform and indicate the zero level.   Consider the two cases $2C_1 > C_2$
and $2C_1 < C_2$.
2-31   Compute and draw to scale the output waveform for $C = 50\,\text{pF}$, $C = 75\,\text{pF}$,
and $C = 25\,\text{pF}$.   The input is a 20-V step.

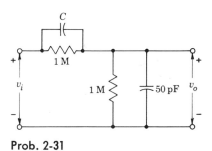

Prob. 2-31

2-32   The input v_i is a 20-V step.   Calculate and plot to scale the output voltage.

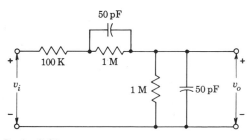

Prob. 2-32

2-33 Show that the minimum bandpass of a potentiometer is obtained when the slider is at the center. Plot the bandpass as a function of the distance of the slider from one end. If the slider is on the first or last 10 percent of the potentiometer, how many times the minimum value will the bandpass be?

2-34 Draw roughly the output waveform v_o. Make reasonable approximations and estimate the rise time of the waveform, the magnitude of the overshoot, and the time constant of the decay to the final value.

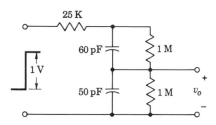

Prob. 2-34

2-35 The capacitance C in the compensated probe of Prob. 2-29 is to be adjusted experimentally for perfect compensation. A square wave is applied at the input and the waveform observed on the scope. Specify a reasonable frequency for the square wave in order to allow convenient observation from the scope pattern of the correct value of C.

2-36 A step of current I is applied to the parallel branches. Show that if $L_1/R_1 = L_2/R_2$ the current i_o will be an ideal step of amplitude $R_1I/(R_1 + R_2)$. (This circuit allows a frequency-independent current attenuation in the presence of inductance and is the dual of the compensated voltage attenuator.)

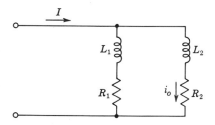

Prob. 2-36

2-37 In the circuit of Prob. 2-36, $R_1 = R_2 = 10$ K, $L_2 = 10$ mH. The input is a 100-mA step. Compute and draw to scale the current waveform i_o for $L_1 = 0, 5$ mH, 10 mH, and 20 mH.

2-38 The current attenuator of Prob. 2-36 is compensated with $L_1 = L_2 = 10$ mH and $R_1 = R_2 = 10$ K. The input is a 100-mA step from a source of 100-K impedance. Compute and plot to scale the current waveform i_o.

2-39 *a.* An inductor L is located in the collector circuit of a transistor operating in the common-emitter configuration. The base input current is a square wave of peak-to-peak value I. Show that the output waveform is as shown in Fig. 2-6 with a peak value $V = h_{fe}I/h_{oe}$ and a time constant $\tau = h_{oe}L$.

b. For $h_{fe} = 50$, $1/h_{oe} = 40$ K, $L = 10$ mH, and $I = 10$ μA, find V and τ.

2-40 Derive (*a*) Eq. (2-53), (*b*) Eq. (2-55), and (*c*) Eq. (2-57).

2-41 A transistor, for which $1/h_{oe} = 40$ K and $h_{fe} = 50$, is operating in the common-emitter configuration. An inductance of 160 mH is located in the collector circuit and the total capacitance from collector to ground is 50 pF. Find and plot the response at the output if the input is a current step of 0.02 mA.

2-42 *a.* A transformer has its primary winding carrying a current I_o in the collector circuit of a transistor. The primary inductance is L, and the voltage step-up ratio is n. The load on the secondary winding may be considered to be purely capacitive. This capacitance C is much larger than the interwinding or interturn capacitances of the transformer. If the transistor is driven beyond cutoff, prove that the amplitude of the output oscillation at the secondary is $I_o \sqrt{L/C}$. Note that the *amplitude is independent of the step-up ratio n*. Explain this result physically.

b. Prove that the period of the output oscillation is $2n\pi \sqrt{LC}$.

2-43 *a.* For a ringing circuit show that the percentage P decrease in amplitude in n cycles is given by

$$\ln \left(1 - \frac{P}{100} \right) = - \frac{n\pi}{Q}$$

b. For small P show that

$$P \approx \frac{100n\pi}{Q}$$

2-44 Use Eq. (2-57) to derive the result given in Eq. (2-60) for the amplitude of oscillation in a ringing circuit. HINT: Let R approach infinity and verify that kV is independent of R.

2-45 *a.* The circuit shown may be used to measure a time interval between two events. The switch S is initially open and the two events are represented by the closing and subsequent opening of the switch. Initially the inductor current is zero and the capacitor is uncharged. Show that the interval t_1 during which the switch is closed is related to the voltage V_c which appears on C at the end of the interval by

$$t_1 = \sqrt{LC} \, \frac{V_c}{V}$$

b. Show that if t_1 is of the order of one-sixth the free period of oscillation of the LC ringing circuit, then $V_c \approx V$.

c. Compute the total possible range for t_1 if capacitances are available from 100 pF to 1 μF and inductors from 5 μH to 0.1 H. Assume $V_c = V$.

d. Is it possible that V_c exceeds V?

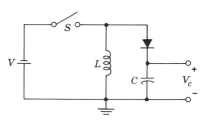

Prob. 2-45

2-46 A 10-mH choke shunted by a 0.001-μF capacitance is connected from the cathode of a 12AU7 tube to ground. The plate-circuit resistance is 10 K and the supply voltage is 200 V. A steady state is reached with the grid grounded. A negative step is now applied to the grid. What is the minimum size of this step needed in order that the tube remain cut off while the resonant circuit oscillates?

2-47 Derive Eqs. (A-3), (A-4), and (A-5) in Appendix A.

2-48 For the critically damped ringing circuit discussed in Appendix A, prove that the maximum occurs at

$$x_m = \frac{1}{\pi} \frac{1 + \Delta}{1 + 2\Delta}$$

and that the maximum value is

$$\left(\frac{v_o}{V_o}\right)_{\text{max}} = -(2\Delta + 1)\epsilon^{-2(1+\Delta)/(1+2\Delta)}$$

2-49 *a.* Prove that the peak values of the output waveforms plotted in Fig. 2-9 are given by

$$\left(\frac{v_o}{V}\right)_{\text{max}} = n^{1/(1-n)} \qquad \text{if } n \neq 1$$

$$\left(\frac{v_o}{V}\right)_{\text{max}} = 0.37 \qquad \text{if } n = 1$$

b. Plot $(v_o/V)_{\text{max}}$ against n for $0 < n \leq 0.5$ and show that an approximately linear relationship results.

c. A step generator of 50 Ω impedance applies a 10-V step of 1 nsec rise time to an inductor. The pulse across the inductor attains an amplitude of 2.5 V. What is the value of the inductance?

2-50 A step generator of 50 Ω impedance applies a 10-V step of 2.2 nsec rise time to a series combination of a capacitance C and a resistance $R = 50 \Omega$. There appears across R a pulse of amplitude 1 V. Find the value of the capacitance C.

2-51 An inductor L has associated with it stray shunt capacitance C and parallel resistance R. It is connected in the circuit shown in (*a*). From the observed output waveshape v_o in (*b*) calculate the values of C, R, and L. The generator v is a 10-V step of 2.2 nsec rise time. HINT: After the initial spike the voltage across L is approximately constant over the 100-nsec interval shown.

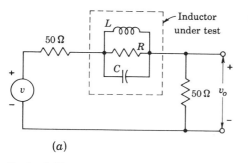
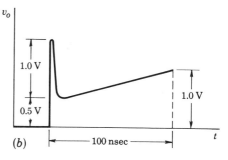

(*a*)   (*b*)

Prob. 2-51

CHAPTER 3

3-1 In connection with the transformer equivalent circuit of Fig. 3-2, show that the following results hold for the various selections of α. (a) If $\alpha \equiv \sqrt{L_p/L_s}$, $\sigma_1 = \sigma_2$; (b) if $\alpha \equiv (1/K)\sqrt{L_p/L_s}$, $\sigma_1 = 0$; (c) if $\alpha \equiv K\sqrt{L_p/L_s}$, $\sigma_2 = 0$.

3-2 In connection with the transformer equivalent circuits of Fig. 3-3 show that, for $K \approx 1$, in each case the total leakage inductance is nearly the same and is given by $\sigma = 2L_p(1 - K)$.

3-3 To minimize capacitance between windings, the windings are placed on opposite legs of the core, as shown. Show that the leakage inductance continues to be given approximately by Eq. (3-10), in which V now stands for the volume of the window and λ is the height of the window. Make the assumptions that the magnetic flux does not fringe appreciably outside the window and that μ_r is very large.

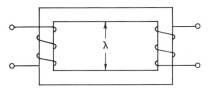

Prob. 3-3

3-4 Verify Eq. (3-11).

3-5 A two-layer transformer is connected as an autotransformer. Show that this type of connection reduces the leakage inductance by the ratio $(n - 1)^2/n^2$, where n is the step-up ratio of the autotransformer.

3-6 In Fig. 3-7, S = mean circumference of the coils, d = distance between the inside surface of the secondary coil and outside surface of the primary coil, a_p = thickness of a primary wire, and a_s = thickness of a secondary wire. Assume that the current is concentrated midway through the thickness of the coil.

a. Prove that the leakage inductance is given by

$$\sigma = \frac{\mu_o N_p{}^2 S}{\lambda}\left(d + \frac{a_p + a_s}{2}\right)$$

b. Assume now that the current is distributed uniformly throughout the windings. Prove that the expression for σ is the same as in (a) except that the numeral 2 must be replaced by 3.

NOTE: If the flux density H is not constant, then Eq. (3-10) becomes

$$\sigma I_p{}^2 = \mu_o \int H^2 \, dV$$

3-7 Consider a transformer with a one-layer primary and a two-layer secondary winding. Each layer of the secondary has the same number of turns. In (a) both secondaries are wound over the primary. In (b) the primary is interleaved between secondary windings. Replace the coils by current sheets located at the center of the coil wires. The mean circumference of the windings is S.

a. Indicate the magnetic-field intensity as a function of the distance between windings (as in Fig. 3-7b).

b. Prove that for the configuration in (*a*) the leakage inductance is given by

$$\sigma = \frac{\mu_o N_p^2 S}{4\lambda} (4d_1 + d_2 + 2a_p + 3a_s)$$

and in (*b*) by

$$\sigma = \frac{\mu_o N_p^2 S}{4\lambda} (d_1 + d_2 + a_p + a_s)$$

This problem illustrates the advantage of interleaving with respect to reducing the leakage inductance.

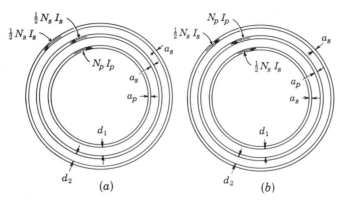

(a)   (b)

Prob. 3-7

3-8 A transformer has the following parameters: primary and secondary each one layer of 132 turns, wire No. 38 having a diameter = 0.0102 cm, diameter including insulation = 0.0125 cm, insulation between windings = 0.031 cm, mean circumference $S = 5.4$ cm, and dielectric constant of insulation between windings $\epsilon = 3.5$.

a. Consider the two windings as constituting a parallel-plate capacitor. Prove that the interwinding capacitance is 89 pF.

b. Consider two adjacent turns as parallel conductors, as shown in the sketch. From electrostatics, the capacitance per meter of such a configuration is given by $\pi\epsilon/[\cosh^{-1}(2h/d)]$. Prove that the effective capacitance across the winding is only 0.06 pF, even if it is assumed that all the space surrounding the wires has a dielectric constant of 3.5.

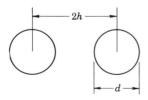

Prob. 3-8

3-9 Consider a transformer with a single-layer primary and a single-layer secondary. A grounded electrostatic shield is inserted between the primary and the secondary midway between windings. Calculate the ratio of the effective primary shunt capacitance C' with the shield to the capacitance C without the shield as a function of the step-up ratio n for (a) a noninverting transformer and (b) an inverting transformer.

3-10 a. A transformer, as shown, has neighboring ends of its windings connected together. The windings are wound in opposite directions so that an inverted output is obtained as indicated. Show that such a connection introduces into the transformer equivalent circuit a capacitance

$$C = (n + 1)^2 \frac{C_o}{3}$$

b. For $n = 1$, 2, 3 compute the ratio of the capacitance that results from the connection of part a to the capacitance that results from the connection of Fig. 3-4.

c. What would be the capacitance C if the transformer windings were wound in the same direction? What is C for $n = 1$? Explain physically.

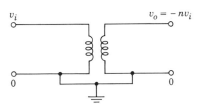

Prob. 3-10

3-11 a. A transformer circuit has a negligible primary resistance R_1 and operates initially under critically damped conditions. The windings are rearranged so that the leakage inductance σ changes while all other parameters remain unchanged. The new value of the damping constant is $k = 0.6$. What is the ratio of the new rise time to the critically damped rise time?

b. Repeat part a for an open-circuited secondary but with R_1 no longer set equal to zero.

c. Repeat parts a and b but with σ kept constant and C adjusted so that the damping factor changes from 1 to 0.6.

3-12 a. A pulse-transformer circuit has $R_1 = R_2 = R$ and R selected so that $\sigma = R^2C$. Using the plots of Fig. 3-10 show that, for a step input, the rise time of the output is approximately $2RC$.

b. If, instead of adjusting σ as above, we assume $\sigma = 0$, then prove that the rise time is improved by a factor of approximately 2.

3-13 Consider the high-frequency response of a pulse transformer. Show (a) that if $\sigma = 0$ the step response is an exponential rise with a time constant $R_1R_2C/(R_1 + R_2)$; (b) that if $C = 0$ the response is an exponential rise with time constant $\sigma/(R_1 + R_2)$.

3-14 Verify (a) Eq. (3-19), (b) Eq. (3-20), and (c) Eq. (3-22).

3-15 Verify Eqs. (3-23).

3-16 A transformer with $N_p = N_s = 7$ turns is wound on the magnetic toroid shown. The transformer operates with $R_1 = R_2 = 50 \ \Omega$. The core loss is negligible, the coefficient of coupling is $K = 0.999$, and $C = 2 \ \text{pF}$. A tilt of 10 percent is observed in a 50-nsec pulse. (*a*) Calculate the effective permeability of the core. (*b*) Calculate the rise-time response.

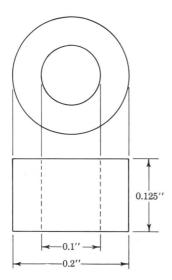

0.125''

0.1''

0.2''

Prob. 3-16

3-17 A pulse transformer is fed from a 50-Ω generator which delivers a 10-V 1-μsec pulse. The transformer operates into a 50-Ω load. The output has a peak amplitude of 4.8 V and a 10 percent tilt.

a. Calculate the effective resistance R_c (to be shunted across the magnetizing inductance) to account for the core loss.

b. Calculate the magnetizing inductance.

3-18 A toroidal core has a mean magnetic length of 4 cm, a cross-sectional area of 0.5 cm², and $\mu_r = 1,000$. It has primary and secondary windings each of 200 turns. The core saturates at $B = 0.25 \ \text{Wb/m}^2$. A voltage pulse of amplitude 20 V and 300 μsec duration is applied to the primary. The total primary series resistance, winding and generator together, is 100 Ω. Transformer capacitances are assumed negligible. The secondary is open-circuited.

a. Calculate the time at which the core saturates.

b. Draw the primary current waveform and the waveform of the output voltage across the secondary, labeling all current and voltage levels and time constants.

3-19 A transformer is used for peaking. The primary winding (of inductance L) is in the collector circuit of a transistor and carries a current I_o. The load on the secondary winding may be considered to be purely capacitive. This load capacitance C_L is much larger than the transformer capacitance C. The transistor is now suddenly cut off.

a. Prove that the amplitude of the output oscillation at the secondary is $I_o \sqrt{L/C_L}$, *which is independent* of the step-up ratio n. Explain this result physically. Neglect the leakage inductance.

b. Prove that the oscillation period is $2n\pi \sqrt{LC_L}$.

c. Compare the outputs obtained from two transformers; one has $n = 1$, $L = 20$ mH, and $C_L = 50$ pF, and the second has $n = 10$, $L = 0.5$ mH, and $C_L = 50$ pF. The current step is 5 mA.

3-20 A transformer has the following parameters: $L = 5.0$ mH, $\sigma = 20$ μH, $C = 100$ pF, $R_1 = R_2 = 500$ Ω, and $n = 1$. Find the response to a 1-μsec 20-V pulse, and plot.

3-21 A transformer has the following parameters: $L = 7.5$ mH, $\sigma = 60$ μH, $C = 75$ pF, $R_1 = 250$ Ω, $R_2 = 1$ K, and $n = 3$. Find the response to a 10-V 3-μsec pulse, and plot.

3-22 A transformer has the following parameters:

$$L = 70 \text{ mH} \qquad \sigma = 200 \text{ } \mu\text{H} \qquad C_L = 500 \text{ pF}$$

$$C = 20 \text{ pF} \qquad R_L = 1 \text{ K} \qquad N_p = 150 \qquad N_s = 60$$

The generator resistance $= 300$ Ω, and the transformer winding resistances are negligible. Find and plot the response to a 10-μsec 1-V pulse.

3-23 A pulse transformer with a step-up ratio of 2:1 is to pass a 1-μsec pulse with less than 10 percent tilt and less than 10 percent overshoot. The generator impedance is 1 K and the load impedance is 1 M in parallel with 25 pF. For what values of primary and leakage inductances must the transformer be designed?

3-24 The windings on the pot-core transformer of Fig. 3-14 are placed side by side in slots in a bobbin so that each winding takes the approximate shape of a flat disk. The distance between windings is 2 mm. The insulation of the wire has a dielectric constant of 3.5. Each winding has 100 turns and the transformer is of the inverting type. Calculate (*a*) the leakage inductance, (*b*) the primary inductance, (*c*) the coefficient of coupling, and (*d*) the effective capacitance C.

3-25 A type RG-59/U coaxial cable has a capacitance of 20 pF/ft and a characteristic impedance of 73 Ω. Find the length required for a 0.5-μsec delay.

3-26 A coaxial cable with a nylon dielectric ($\epsilon_r = 3.00$) has a characteristic impedance of 200 Ω.

a. How long a cable is needed to give a one-way delay of 50 nsec?

b. What is the ratio of the outer to the inner radius?

3-27 The exact expression for the capacitance C per unit length of a parallel wire system is

$$C = \frac{\pi\epsilon}{\cosh^{-1}(2h/d)}$$

a. If $h/d \geq 1$, prove that C is given approximately by the expression in Fig. 3-17.

b. If $h/d = 1$, show that the error resulting from the use of the approximate expression is less than 5 percent.

3-28 *a.* A wire is initially located at 2.5 diameters above a metal chassis. Calculate the capacitance per unit length, the inductance per unit length, and the characteristic impedance.

b. Recalculate the capacitance, inductance, and impedance if the wire is moved to 25 diameters above the chassis.

3-29   Is it feasible to construct a coaxial line with a characteristic impedance of 10,000 Ω?   Use, say, an inner conductor of 0.1 cm diameter and let $\epsilon_r = 3$.   Calculate the required diameter of the outer conductor.

3-30   *a.* A 50-Ω source is connected to a 50-Ω load through an air dielectric transmission line 3 cm long of 100 Ω impedance.   Neglect the fact that there is some "cancellation" between the line inductance and capacitance.   Represent the line, as in the figure, with L equal to the total line inductance and C the total line capacitance.   Calculate the rise time at the output.

b. Calculate the rise time, taking into account the partial cancellation of the line inductance by the capacitance.   Compare the result with that of part *a*.   What is the percentage discrepancy between the two methods?

c. Now assume that the line impedance has been reduced to 50 Ω.   Use the method of part *a* to calculate the rise time.   What is the rise time obtained using the method which takes account of transmission-line properties?

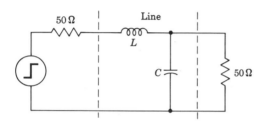

Prob. 3-30

3-31   *a.* A generator of impedance 125 Ω supplies a step signal to a 125-Ω load over a 4-in. length of bare wire mounted over a metallic chassis.   The wire diameter is 0.1 cm and it is mounted with its center 0.25 cm above the chassis.   Estimate the rise time of the waveform across the load.

b. Repeat part *a* for the case in which source and load impedances are 150 Ω.

3-32   *a.* A signal source of impedance R_s supplies power to a load R_L over a line of impedance Z_o.   Show that if $R_L \ll Z_o$, the waveforms on the line may be estimated by replacing the line by an inductance lL, where l is the line length and L is the inductance per unit length.   Thus show that the line capacitance may be neglected entirely.

b. A signal source of 50 Ω impedance supplies power to a 1-Ω load through a pair of wires 2 in. long.   The wire diameter is 0.05 in. and the spacing between centers is 0.25 in.   If the signal source furnishes a voltage step, what will be the rise time of the waveform at the load and at the generator?

3-33   *a.* Consider again the situation described in Prob. 3-32 except that $R_L \gg R_o$. In this case show that the line inductance may be neglected and that the line may be replaced by a capacitance lC, where C is the capacitance per unit length.

b. Repeat Prob. 3-32*b* for the case where the load is 10,000 Ω.

NOTE: *Problems 3-34 through 3-40 pertain to material covered in Appendixes B and C.*

3-34   Calculate L, C, Z_o, and T for a type RG-65/U cable.   The parameters of this cable are listed below Eq. (B-2).

3-35   Verify Eqs. (C-1), (C-2), and (C-9).

3-36 A unit step voltage is impressed upon one section of a constant-k prototype filter which is terminated at both ends in a resistance $R_o = \sqrt{L/C}$. If $T \equiv \sqrt{LC}$, prove that the output is given by

$$v_o = 0.5 - 0.5\epsilon^{-2t/T} - 0.578\epsilon^{-t/T} \sin\left(\sqrt{3}\,\frac{t}{T}\right)$$

3-37 Design a delay line using constant-k prototype sections. The line is to have a nominal impedance of 500 Ω, a delay of 1.0 μsec, and a rise time of 0.2 μsec. Calculate the number of sections needed, the value of C, the value of L, and the cutoff frequency f_c. Assume that commercial capacitors are available having values of capacitance which are multiples of 50 pF. What will be the impedance of the line?

3-38 Repeat Prob. 3-37 for a delay line using m-derived sections with $m = 1.27$. Find L_1, C_1, f_c, and the characteristic impedance.

3-39 Prove (a) that for the m-derived filter f_c and Z_o are given by the same expressions as for the constant-k prototype section, and (b) verify Eq. (C-15).

3-40 a. Prove that the Fourier spectrum of a pulse of unit amplitude and width t_p extending from $t = 0$ to $t = t_p$ is given by $(1 - \epsilon^{-j\omega t_p})/j\omega$.

b. An ideal low-pass filter is one having zero attenuation in the passband and infinite attenuation outside this region. It also has a phase shift β which is proportional to frequency, $\beta = \omega t_d$, within the passband, $|f| < f_c$. Prove that the output of the ideal low-pass filter to a unit-amplitude pulse is given by

$$v_o = \frac{1}{\pi}\,\text{Si}\,(x) + \frac{1}{\pi}\,\text{Si}\,(y)$$

where $x = 2\pi f_c(t - t_d)$, $y = 2\pi f_c(t_p - t + t_d)$, and Si (x) is the sine integral,

$$\text{Si}\,(x) \equiv \int_0^x \frac{\sin x\,dx}{x}$$

NOTE: The frequency spectrum of the output of a system equals the product of the spectrum of the input voltage and the transmission characteristic of the system. By evaluating the inverse Fourier transform of this product, the output of the system as a function of time is obtained.

The response to a unit step input voltage is obtained by allowing t_p to approach infinity. Since Si $(\infty) = \pi/2$, the response to a unit step is $v_1 = 0.5 + \text{Si}\,(x)$. Using numerical values of Si (x), plot v_1 versus t.

c. Prove that t_d represents the delay: the time between the 50 percent amplitude points on the input and output voltages.

d. Evaluate the rise time t_r (for v_o to increase from 0.1 to 0.9), and prove that $t_r = 0.445/f_c$. This result should be compared with the rise time of the output of a low-pass circuit (Fig. 2-14) whose input is a step voltage and whose upper 3-dB frequency is f_2, namely, $t_r = 0.35/f_2$. (See E. Jahnke and F. Emde, "Tables of Functions," 2d ed., pp. 78–86, Teubner Verlagsgesellschaft, Leipzig, 1933.)

3-41 A step generator of impedance $R_s = R_o$ applies a step of amplitude V to a line of impedance R_o and of one-way delay time t_d. The line is terminated in a resistor R. Plot the voltage and current waveforms as a function of the time at the input and at the output for each of the cases: (a) $R = R_o$, (b) $R = \frac{1}{2}R_o$, (c) $R = 0$, (d) $R = 2R_o$, (e) $R = \infty$.

3-42 Repeat Prob. 3-41 if the input is a pulse of amplitude V and duration $t_p < 2t_d$.

3-43 Repeat Prob. 3-41 if the input is a pulse of amplitude V and duration $t_p = 2.5t_d$.

3-44 Repeat Prob. 3-41 if the input is $v = \alpha t$ for $t \geq 0$ and $v = 0$ for $t \leq 0$. Assume that $2t_d = 1$ μsec and that the ramp voltage increases at the rate 1 V/μsec. Label all voltage levels.

3-45 A pulse of amplitude V is applied by a generator of impedance $R_s = R_o$ to a line of impedance R_o. The line is short-circuited at the receiving end. The line attenuates the signal to the extent that the signal arriving at one end is 0.8 as large as the signal which starts out at the other end. Plot the input voltage and current as a function of the time if (a) $t_p < 2t_d$ and (b) $t_p > 2t_d$.

3-46 A pulse of unit amplitude and width t_p travels down a line toward an open-circuited end. The leading edge of the pulse reaches the end of the line at time $t = 0$. Plot the voltage distribution along the line at the following values of t: $t = -2t_p$, 0, $t_p/4$, $t_p/2$, $3t_p/4$, t_p, and $3t_p$.

3-47 The periodic ramp-type voltage is applied to a delay-line differentiator using a line of delay t_d. Draw the output waveform for $t_d = T/10$ and $t_d = T/5$. Compare these waveforms with the waveforms of a conventional differentiator whose time constant is $\tau = T/10$ and $T/5$.

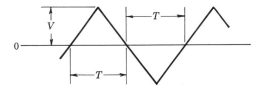

Prob. 3-47

3-48 A unit step voltage is applied from a zero-impedance generator to an ideal line which is short-circuited at the receiving end. Plot the current at the input end of the line as a function of time.

3-49 A voltage pulse is applied from a zero-impedance source to a line open-circuited at the receiving end. Plot the voltage waveform at the end of the line if (a) $t_p < 2t_d$ and (b) $t_p > 2t_d$.

3-50 A unit step voltage is applied to an open-circuited line through a zero-impedance generator. Plot the output voltage as a function of time if (a) the line is lossless and (b) the attenuation of the line is such that the signal which reaches one end of the line is 80 percent of the signal introduced at the other end of the line.

3-51 A unit step voltage is applied to a lossless line through a zero-impedance generator. The line is terminated in a resistor R equal to $3R_o$. Plot the voltage at the receiving end of the line as a function of time. Indicate the values of the voltage jumps at each discontinuity.

3-52 A unit step voltage is applied to a lossless shorted line from a generator whose impedance is R_s. Plot the input voltage as a function of time if (a) $R_s = R_o$, (b) $R_s = 9R_o$, and (c) $R_s = R_o/9$. (d) Plot the input current as a function of time under conditions a, b, and c.

3-53 A unit step voltage is applied to a lossless open-circuited line from a generator whose impedance is R_s. Plot the input voltage as a function of time if (a) $R_s = R_o$, (b) $R_s = 9R_o$, and (c) $R_s = R_o/9$. (d) Plot the input current as a function of time under conditions a, b, and c.

3-54 *a.* A unit step voltage is applied to a lossless open-circuited line from a generator whose impedance R_s is twice the characteristic impedance of the line. Plot the voltage at the end of the line as a function of time. Indicate the magnitudes of any abrupt changes in voltage.

b. Repeat (a) if the generator output is a narrow pulse instead of a step. Show that the results of (a) and (b) are compatible.

c. Repeat (a) and (b) if $R_s = R_o/2$.

3-55 A pulse whose width t_p is less than the one-way delay time t_d is applied to a line. Plot the input and output voltage waveforms as a function of time, taking attenuation into account, if (a) $R_s > R_o$ and $R < R_o$, where R is the terminating resistance, (b) $R_s > R_o$ and $R > R_o$, and (c) $R_s < R_o$ and $R < R_o$. Choose $|\rho| = |\rho'| = 0.5$ and $\epsilon^{-a} = 0.8$.

3-56 A current source applies a step to a delay line of impedance R_o. The line is terminated at its sending end in $R_s = 3R_o$ and at the receiving end in $R = R_o/3$. Plot the voltage waveforms at the input and output of the line.

3-57 Repeat Prob. 3-56 for the case where the input is a pulse if (a) $t_p < 2t_d$ and (b) $t_p > 2t_d$.

3-58 For the situation described in Fig. 3-29, where a ramp is applied to a line with $\rho = -\frac{1}{2}$ and $\rho' = +\frac{1}{3}$, draw the waveform at the output of the line for the time interval $0 < t < 5t_d$. Indicate the waveform of each of the component waves which contribute to the resultant.

3-59 Explain how to adjust experimentally the terminations at each end of a line so that the line is properly matched at both ends. Assume that a square-wave generator (or a pulse generator) and a CRO are available.

3-60 A unit step is applied from a generator of source impedance R_s to a line whose characteristic impedance is R_o and which is open-circuited at the receiving end.

a. Show that the voltage at the open end is

$$v_o = 1 - (\rho')^n \qquad \text{for } (2n - 1)t_d < t < (2n + 1)t_d$$

where ρ' is the reflection coefficient at the sending end. Plot v_o as a function of the time.

b. Show that if R_o/R_s is small, $\rho' \approx \epsilon^{-2R_o/R_s}$. HINT: Show that the power series for ρ' and ϵ^{-2R_o/R_s} agree in the first three terms and that ρ' and ϵ^{-2R_o/R_s} differ by less than 2 percent if $R_o/R_s < 0.3$.

c. If the step function in (a) is smoothed out by drawing a curve through the steps at $t = 2nt_d$, show that this smoothed curve is given by

$$v_o = 1 - \epsilon^{-t/R_sC}$$

where C is the total shunt capacitance of the line. Interpret this result physically.

3-61 Repeat Prob. 3-60 for the case in which the line is shorted at the receiving end and is driven from a source $R_o \gg R_s$. Calculate the waveform of the load current i_o and show that a smoothed curve passing through the steps of current at $t = 2nt_d$ is given by (assuming the generator applies a unit current step)

$$i_o = 1 - \epsilon^{-R_s t/L}$$

where L is the total inductance of the line.

3-62 Verify the waveforms in Fig. 3-36.

3-63 In Fig. 3-33 find v_i and v_o as functions of time for $0 < t < 6t_d$ if (a) $R_1 = R_2 = R_o$, (b) $R_1 = R_o$, $R_2 = R_o/2$, (c) $R_1 = R_o/2$, $R_2 = R_o$, and (d) $R_1 = R_o/2$, $R_2 = 2R_o$.

3-64 A line of impedance R_o and delay time t_d is initially open-circuited at both ends. The line is charged to a voltage V. At $t = 0$ a resistor R is bridged across one end of the line. Find the voltage at each end of the line as a function of the time for $0 < t < 6t_d$ if (a) $R = R_o$, (b) $R = 2R_o$, and (c) $R = R_o/2$.

3-65 Repeat Prob. 3-64 for the case where two resistors R_1 and R_2 are simultaneously placed across the two ends of the line. Consider the cases (a) $R_1 = R_2 = R_o$, (b) $R_1 = R_o$, $R_2 = R_o/2$, and (c) $R_1 = 2R_o$, $R_2 = R_o/2$.

3-66 A step generator of impedance R_s applies a signal to a line of impedance R_o which is terminated in an inductor L in series with a resistor R. Calculate and plot the input and output waveforms if $R_s = R_o$. Consider the special cases $R = 0$, $R = \infty$, and $R = R_o$.

3-67 A step generator of impedance R_s applies a signal to a line of impedance R_o which is terminated in a capacitor C in parallel with a resistor R. Calculate and plot the input and output voltage waveforms if $R_s = R_o$. Consider the special cases $R = \infty$, $R = 0$, and $R = R_o$.

3-68 A 1:1 noninverting transmission-line transformer is constructed by winding 11 turns of a bifilar winding on a magnetic core of relative permeability $\mu_r = 3,000$. The core has a cross-sectional area of 0.5 cm² and a mean magnetic length of 2.0 cm. The characteristic impedance of the line is 75 Ω and the line is matched at both ends. If a pulse is to be transmitted with no more than 10 percent tilt, what is the widest pulse for which the transformer is suitable?

3-69 *a.* Each winding has the same number of turns. Redraw the figure to make it clear that the arrangement constitutes a 2:1 step-up autotransformer.

b. Redraw the figure, replacing the sets of neighboring windings by transmission lines. Show that the line inputs are in parallel and the line outputs are in series. If the source impedance is $R_s = 100$ Ω, then if the lines are to be matched what must be the value of the load R_L and what must be the characteristic impedance of the lines?

c. Explain why the arrangement must be mounted on a core and draw a diagram showing how the windings are to be placed on the core.

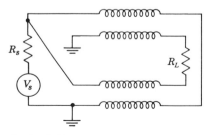

Prob. 3-69

CHAPTER 4

4-1 The input to an amplifier consists of a voltage made up of a fundamental signal and a second-harmonic signal of half the magnitude and in phase with the fundamental. Plot the resultant.

The output consists of the same magnitude of each component but with the second harmonic shifted 90° (on the fundamental scale). This corresponds to perfect frequency response but bad phase-shift response. Plot the output and compare it with the input waveshape.

4-2 An ideal 1-μsec pulse is fed into an amplifier. Plot the output if the bandpass is (*a*) 10 MHz, (*b*) 1.0 MHz, and (*c*) 0.1 MHz.

4-3 *a.* Given a single-stage RC-coupled tube amplifier with $C_b = 0.2$ μF, $R_g = 0.5$ M, and an output-circuit resistance $R_y = 3$ K. Calculate the percentage tilt in the output if the input is a 100-Hz square wave.

b. Repeat part *a* for a transistor stage with $C_b = 10$ μF, $R_i' = 2$ K, and $R_y = 3$ K.

c. For each amplifier what is the lowest-frequency square wave which will suffer less than a 1 percent tilt?

4-4 A pentode amplifier stage has an unbypassed cathode resistor R_k and a plate-circuit resistor R_p which is shunted by a capacitor C. If the input is a negative unit step, prove that the output voltage as a function of time is

$$v_o = \frac{g_m R_p}{1 + g_m R_k} (1 - \epsilon^{-t/R_p C})$$

4-5 For the circuit shown, calculate the nominal gain V_o/V_i, and calculate the frequency at which the output V_o will fall to the 3-dB point. $\mu = 20$ and $r_p = 10$ K.

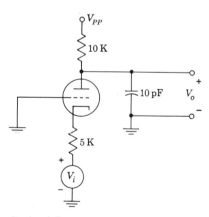

Prob. 4-5

4-6 *a.* At low frequencies the short-circuit CE current gain β is related to the short-circuit CB current gain α by

$$\alpha = \frac{\beta}{1 + \beta}$$

Assuming that this relationship remains valid at high frequencies and using $\beta = -A_i$ in Eq. (4-16), verify that α is given by Eq. (4-21), where

$$\alpha_o = \frac{h_{fe}}{1 + h_{fe}} \qquad \text{and} \qquad f_\alpha = \frac{f_\beta}{1 - \alpha_o}$$

b. Using the results of part a, verify that, for $\alpha_o \approx 1$, $f_\alpha \approx f_\beta h_{fe}$.

c. Verify that

$$A_i = \frac{-\alpha_o}{1 - \alpha_o + jf/f_\alpha}$$

d. To account for "excess phase" replace α_o by $\alpha_o \epsilon^{-jmf/f_\alpha}$. Prove that f_T, the frequency at which $|A_i| = 1$, is given implicitly by

$$1 + x^2 = 2\alpha_o(\cos mx - x \sin mx)$$

where $x = f_T/f_\alpha$.

e. If $mx \ll 1$, expand the trigonometric functions and prove that

$$f_T \approx \frac{\alpha_o f_\alpha}{[1 + 2\alpha_o(m + m^2/2)]^{\frac{1}{2}}}$$

f. If $\alpha_o = 1$ and $m = 0.2$ show that $f_T = f_\alpha/1.2$.

4-7 a. Redraw the CE hybrid-II equivalent circuit with the base as the common terminal and the output terminals, collector and base, short-circuited. Taking account of typical values of the transistor parameters, show that C_c, $r_{b'c}$, and r_{ce} may be neglected.

b. Using the circuit in (a) prove that the CB short-circuit current gain is

$$A_{ib} = \frac{g_m}{g_{b'e} + g_m + j\omega C_e} = \frac{\alpha_o}{1 + jf/f_\alpha}$$

where

$$\alpha_o = \frac{h_{fe}}{1 + h_{fe}} \qquad \text{and} \qquad f_\alpha = \frac{g_m}{2\pi C_e \alpha_o} \approx \frac{f_\beta}{1 - \alpha_o}$$

4-8 a. If A_{ie} = short-circuit CE current gain and A_{ic} = short-circuit CC current gain prove that, independently of the model used for the transistor,

$$A_{ic} = 1 - A_{ie}$$

b. If a step of current is applied at the base of a CC amplifier prove that the short-circuit output current obtained from the hybrid-II model is given by

$$i = I + h_{fe}I(1 - \epsilon^{-t/r_{b'e}C})$$

where $C = C_e + C_c$.

4-9 Verify Eq. (4-33) for the gain-bandwidth product of a single-stage transistor amplifier. HINT: Use Eqs. (4-27) to (4-29) and (4-18).

4-10 For the transistor whose parameters are given in Fig. 4-7 driven from a source with an output resistance $R_s = 1$ K, evaluate f_2, A_{Vso}, and A_{Iso} for the following values of load: $R_L = 0$, 1 K, and 2 K.

4-11 *a.* Consider the hybrid-Π circuit at low frequencies so that C_e and C_c may be neglected. Omit none of the other elements in the circuit. If the load resistance is $R_L = 1/g_L$, prove that

$$K \equiv \frac{V_{ce}}{V_{b'e}} = \frac{-g_m + g_{b'c}}{g_{b'c} + g_{ce} + g_L}$$

HINT: Use the theorem that the voltage between C and E equals the short-circuit current times the impedance seen between C and E with the input voltage $V_{b'e}$ shorted (Sec. 1-2).

b. Using Miller's theorem (Sec. 1-7) draw the equivalent circuit between C and E. Applying KCL to this network, show that the above value of K is obtained.

c. Using Miller's theorem, draw the equivalent circuit between B and E. Prove that the current gain under load is

$$A_I = \frac{g_L}{[(g_{b'c} + g_{b'e})/K] - g_{b'c}}$$

d. Using the results of parts *a* and *c* and the relationships between the hybrid-Π and the *h* parameters of Sec. 1-4, prove that

$$A_I = \frac{-h_{fe}}{1 + h_{oe}R_L}$$

which is the result, Eq. (1-2), obtained directly from the low-frequency model of Fig. 1-3*b*. HINT: Neglect $g_{b'c}$ compared with g_m or $g_{b'e}$ in A_I and in K. Justify these approximations.

4-12 Consider a single-stage CE transistor amplifier with the load resistor R_L shunted by a capacitance C_L.

a. Prove that the internal voltage gain $K = V_{ce}/V_{b'e}$ is

$$K \approx \frac{-g_m R_L}{1 + j\omega(C_c + C_L)R_L}$$

b. Prove that the 3-dB frequency is given by

$$f_2 \approx \frac{1}{2\pi(C_c + C_L)R_L}$$

provided that the following condition is valid:

$$g_{b'e}R_L(C_c + C_L) \gg C_e + C_c(1 + g_m R_L)$$

4-13 For a single-stage CE transistor amplifier whose parameters have the average values given in Fig. 4-7, what value of source resistance R_s will give a 3-dB frequency f_2 which is (*a*) half the value for $R_s = 0$, (*b*) twice the value for $R_s = \infty$? Do these values of R_s depend upon the magnitude of the load R_L?

4-14 *a.* Verify Eq. (4-46) for the maximum 3-dB frequency of a CE stage in an infinite cascade of stages.

b. Find the value of $(f_2)_{max}$ for the typical transistor whose values are given in Fig. 4-7.

4-15 The transistor of Fig. 4-7 is used in a cascade of identical CE stages. A gain of 15 per stage is desired. Evaluate R_c and f_2.

4-16 A 2N1141 transistor whose parameters at $I_E = 10$ mA and $V_{CE} = 10$ V are $r_{bb'} = 80$ Ω, $r_{b'e} = 100$ Ω, $C_c = 1.5$ pF, and $C_e = 85$ pF is used in an infinite CE cascade. For each stage find (a) f_T, (b) $(R_c)_{opt}$ and the corresponding $|A_o f_2|$, (c) f_2 for $A_o = 10$, (d) f_2 for $R_c = 2$ K, (e) the maximum possible value of f_2.

4-17 For the amplifier of Prob. 4-16 find the gain if a rise time of 20 nsec per stage is desired.

4-18 Consider an infinite cascade of CE stages using 2N247 transistors whose parameters are $g_{b'e} = 0.39$ mA/V, $g_m = 54$ mA/V, $r_{bb'} = 45$ Ω, $C_e = 780$ pF, and $C_c = 3.5$ pF.

a. Find the load resistance $(R_c)_{opt}$ for which the gain-bandwidth product $|A_o f_2|$ is a maximum.

b. Find $|A_o f_2|$ for $R_c = 100$ Ω, 1 K, 10 K, and $(R_c)_{opt}$.

4-19 For the amplifier of Prob. 4-18 find the values of R_c and A_o which will give a rise time of 1 μsec per stage.

4-20 For a cascade of CE stages find the asymptotic values of A_o, f_2, and $|A_o f_2|$ as $R_c \to \infty$. For the typical transistor of Fig. 4-7, evaluate these quantities.

4-21 Verify Eq. (4-44) for $(R_c)_{opt}$. What is the significance of a value of x which is less than unity?

4-22 Prove that the formulas in Table 4-1 for a single stage apply to (a) the output stage of a cascade if $R_L = R_c$ and $R_s = R_c$ and (b) the input stage if R_L is taken as the parallel combination of R_c and h_{ie}.

4-23 a. Find the response of a two-stage (identical) amplifier to a unit step in terms of $x = t/RC$.

b. For $t \ll RC$ show that the output varies quadratically with time.

c. If the upper 3-dB frequency of a single stage is f_2 and the rise time of the two-stage amplifier is $t_r^{(2)}$, show that $f_2 t_r^{(2)} = 0.53$.

d. Show that the rise time of a two-stage amplifier is 1.53 times that of a single stage.

4-24 If two cascaded stages have very unequal bandpasses, show that the combined bandwidth is essentially that of the smaller.

4-25 A tube amplifier consists of two identical uncompensated stages. The total effective shunt capacitance across each stage is the same and is equal to 20 pF. The 3-dB bandwidth of the complete amplifier is 10 MHz. If the tubes used have $g_m = 10$ millimhos, find the gain of the complete amplifier.

4-26 A pentode amplifier consists of two identical uncompensated stages. The total effective shunt capacitance across each stage is the same and is equal to 20 pF. The circuit parameters are $g_m = 10$ mA/V, $r_p = 1$ M, $R_p = 2.5$ K, $R_g = 1$ M, and $C_b = 0.5$ μF. Calculate for the overall amplifier (a) the gain, (b) the rise time, (c) the frequency of a square-wave signal which will suffer a 10 percent tilt when transmitted through this amplifier.

4-27 A single-stage video amplifier uses a 6AU6 pentode operating at a quiescent plate current of 3 mA, a screen current of 1 mA, and a screen voltage of 150 V. The plate supply voltage is 200 V. The total shunt capacitance is 50 pF. A rise time of 0.65 μsec and a gain of 5 (with an unbypassed cathode resistor) are desired. A 100-μsec pulse is applied at the input of such a magnitude that the screen current during the pulse is 4 mA. Calculate (a) R_p, (b) R_k, (c) the grid bias V_{GG}, (d) the screen resistance R_{sc}, (e) the screen capacitance C_s if the tilt due to C_s is not to exceed 5 percent, (f) the cathode bypass capacitance C_k if the tilt due to C_k is not to exceed 5 percent, (g) the blocking capacitance C_b if the tilt due to C_b is not to exceed 5 percent (assuming $R_g = 1$ M).

4-28 Verify Eqs. (4-58) and (4-59).

4-29 A pentode amplifier stage has the following parameters: $g_m = 5$ mA/V, $R_L = 2$ K, $R_k = 100$ Ω, $C_k = 500$ μF, $C_b = 0.25$ μF, and $R_g = 0.5$ M. If a 200-Hz square wave is applied to the input, find the percentage tilt in the output waveform.

4-30 Find the percentage tilt in the output of a transistor stage caused by a capacitor C_z bypassing an emitter resistor R_e. Use the following method: If V is the magnitude of the input step, then from Fig. 4-19a,

$$v_o = -g_m v_{b'e} R_L = -g_m r_{b'e} i_b R_L = -h_{fe} R_L \frac{V - v_{en}}{R}$$

Take as a first approximation $v_{en} = 0$. Justify this step. Calculate the corresponding current and, assuming that all the emitter current passes through C_z, calculate v_{en} and then v_o. The result is Eq. (4-60).

4-31 An amplifier stage has a load resistor R_L and cathode resistor R_k bypassed by a capacitor C_k. The output is taken at the plate. A negative input step of amplitude V is applied.

a. Prove that, if $\mu \gg 1$ and $r_p \gg R_L$, then

$$v_o = \frac{g_m R_L V}{1 + g_m R_k} (1 + g_m R_k \epsilon^{-t/\tau})$$

where

$$\tau = \frac{R_k C_k}{1 + g_m R_k}$$

b. If $t/\tau \ll 1$, show that the above equation yields the percentage tilt given in Eq. (4-61).

4-32 Find the gain as a function of frequency for a transistor stage with an emitter resistor R_e bypassed by a capacitor C_z (Fig. 4-19). Does a lower 3-dB frequency always exist? Explain.

4-33 *a.* Given a cathode follower using a self-biasing resistor R_k bypassed with C_k. The output is taken across a load R as indicated. If $\mu \gg 1$, show that the

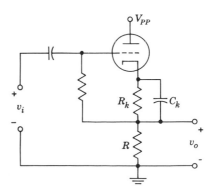

Prob. 4-33

response to a unit step is

$$v_o = \frac{g_m R}{1 + g_m(R + R_k)} \left(1 + \frac{g_m R_k}{1 + g_m R} \epsilon^{-t/\tau} \right)$$

where

$$\tau = C_k R_k \frac{1 + g_m R}{1 + g_m(R + R_k)}$$

b. For small values of t/τ, show that

$$v_o \approx \frac{g_m R}{1 + g_m R} \left[1 - \frac{g_m t}{(1 + g_m R)C_k} \right]$$

c. Consider a 6L6 triode connected and operating at a point where $g_m = 2$ millimhos. $R_k = 2$ K and $R = 70$ Ω. If $C_k = 50$ μF, find the percentage tilt in 1.5 msec.

4-34 Verify Eqs. (4-64) and (4-65).

4-35 a. Find the response y_d [Eq. (4-66)] at the end of a pulse for a three-stage amplifier. Plot y_d versus $x = t/\tau$. Assume $x_p = 1$.

b. From the analytical expression for y_d versus x, show that there are two crossings of the zero-voltage axis. Find these two values of x.

CHAPTER 5

5-1 a. Prove that the parameter m, introduced in Sec. 5-1 in connection with shunt compensation, equals Q_2, the Q of the circuit at the upper 3-dB uncompensated frequency.

b. Prove that $Q_2 = Q_o{}^2$, where Q_o is the Q at the resonant frequency.

5-2 Show that in a shunt-compensated amplifier adjusted for critical damping the rise time is improved by the factor 1.43 over the uncompensated case.

5-3 a. Verify Eq. (5-3) and verify that the poles and zero are given by Eq. (5-5).

b. Verify Eqs. (5-7) and (5-8).

5-4 Design an overshoot-free stage of a video amplifier having a rise time of 5 nsec using shunt compensation. Each stage of the cascade consists of an Amperex type 7788 pentode having the following parameters: output capacitance = 4 pF, input capacitance = 16 pF, stray wiring capacitance = 5 pF, and transconductance = 50 mA/V. Find (a) R_p, (b) L, and (c) the low-frequency amplification. (d) What are the gain and bandwidth of two stages in cascade?

5-5 Three identical stages of an uncompensated tube video amplifier have an overall midband gain of $-1,000$. The rise time per stage is 0.1 μsec. Each tube used has a transconductance of 5 millimhos.

a. Find the plate-circuit resistance.

b. Find the total shunt capacitance per stage.

c. Find the overall bandwidth.

d. Shunt compensation is added and the inductance L is adjusted for critical damping. Find L.

e. Find the upper 3-dB frequency of each stage after compensation.

5-6 Verify Eq. (5-11) for the transfer function of the CE shunt-compensated transistor stage. Show that the poles are given by Eq. (5-13).

5-7 *a.* Verify Eq. (5-18) for the inductance required to give coincident poles to the transfer function of a shunt-compensated transistor stage.

b. Verify the restriction of Eq. (5-19) for the collector-circuit resistance.

5-8 For the critically damped response of a shunt-compensated CE transistor stage verify Eqs. (5-23) and (5-24) for d.

5-9 For $d = 0.5$ evaluate the function $B(d)$ of Eq. (5-22).

5-10 For a pole-zero cancellation of the transfer function of a shunt-compensated transistor amplifier stage, verify Eq. (5-27) for L'' and Eq. (5-29) for R_c.

5-11 The transistor equivalent circuit reduces to that of the vacuum tube if $r_{bb'} = 0$ and $g_{b'e} = 0$. Under these circumstances show that the formulas for rise-time improvement of both devices when shunt-compensated for critical damping are the same.

5-12 Prove that for the CE shunt-compensated stage the two values of critical inductance are equal ($L' = L''$) if $R_c = r_{bb'}h_{ie}/r_{b'e}$.

5-13 Consider an infinite cascade of CE stages using 2N247 transistors whose parameters are $g_{b'e} = 0.39$ mA/V, $g_m = 54$ mA/V, $r_{bb'} = 45$ Ω, $C_e = 780$ pF, and $C_c = 3.5$ pF. An inductor is added in series with each collector resistor for shunt compensation. If a gain of 80 per stage is desired, find (*a*) the load resistance, (*b*) the rise time, (*c*) the rise-time improvement factor, and (*d*) the critical inductance.

5-14 Consider a 2N247 single shunt-compensated stage (parameters are given in Prob. 5-13). If the compensated rise time is to be 0.17 µsec, find (*a*) the load resistance, (*b*) the gain, (*c*) the rise-time improvement factor, and (*d*) the critical inductance.

5-15 Consider a shunt-compensated stage using a very high frequency transistor having the following parameters: $r_{b'e} = 100$ Ω, $r_{bb'} = 80$ Ω, $g_m = 400$ mA/V, $C_e = 85$ pF, and $C_c = 1.5$ pF. If a compensated rise time of 25 nsec is desired, find (*a*) the load resistance, (*b*) the gain, (*c*) the rise-time improvement factor, and (*d*) the critical inductance.

5-16 *a.* Verify Eq. (5-37) for the response of a stage with low-frequency compensation.

b. Explain physically why perfect compensation results if Eq. (5-41) is satisfied.

5-17 Consider the case of low-frequency compensation with $R_d = nR_y$ shunting C_d in Fig. 5-8. Assume that C_d is chosen so that the balanced condition, Eq. (5-41), is satisfied.

a. Prove that the Laplace transform of the normalized output voltage

$$Y(s) \equiv \frac{V_o(s)R_\sigma}{R_i'R_yI_o}$$

is given by

$$Y(s) = \frac{s + (n + 1)/nR_\sigma C_\sigma}{s^2 + s[(n + 1)/nR_\sigma C_\sigma] + R_i'/nR_\sigma{}^3C_\sigma{}^2}$$

b. Expand this in a series in inverse powers of s and obtain

$$Y(s) = \frac{1}{s} - \frac{R_i'}{nR_\sigma{}^3C_\sigma{}^2s^3} \qquad y(t) = 1 - \frac{(R_i'/R_\sigma)t^2}{2nR_\sigma{}^2C_\sigma{}^2} = 1 - \frac{(R_i'/R_\sigma)x^2}{2n}$$

5-18 A pentode video amplifier for which $R_p = 2$ K is coupled to a succeeding stage through 0.1 µF and 0.5 M. The quiescent tube current is 20 mA. The maximum supply voltage available is 300 V. The quiescent plate voltage required is 200 V. The amplifier is to be compensated for minimum tilt on a square wave with initial slope equal to zero. Draw the complete circuit diagram, specify components, and calculate the percentage tilt after compensation for a 200-Hz square wave.

5-19 In the circuit of Fig. 5-7 prove that perfect low-frequency compensation may be obtained if C_b is shunted with a resistor R, provided that

$$RC_b = R_dC_d \quad \text{and} \quad \frac{R_d}{R_y} = \frac{R}{R_i'}$$

Note that if it is necessary to isolate the grid of the succeeding stage from the d-c plate voltage of the previous stage, a large blocking capacitor may be added in series with R. HINT: Show that the transfer function is independent of frequency.

5-20 Design a cathode-compensated video-amplifier stage using a 6AU6 pentode. $V_{PP} = 300$ V, $V_{SS} = 150$ V, the quiescent plate current is 4 mA, and the corresponding screen current is 1.5 mA. It is desired to have a gain of 5.5. A negative power supply is not available. Find (a) R_k, (b) R_p, and (c) C_k to give a good transient response if the plate shunting capacitance is 30 pF total. (d) What is the passband under condition c? The rise time? (e) Find the gain-bandwidth product. (f) If the cathode resistor were bypassed with a very large capacitance, what would be the gain-bandwidth product?

5-21 A tube in an amplifier is to be operated at a quiescent plate current of 20 mA, a quiescent screen current of 5 mA, and a quiescent plate voltage of 200 V. An output swing (peak to peak) of 200 V is required when the plate current varies from 5 to 35 mA. The output capacitance is 20 pF and an upper 3-dB frequency of 3 MHz is required. The g_m of the tube is 10 millimhos. The grid-to-cathode voltage required for the quiescent operating point above is -10 V. Draw a complete circuit diagram of the stage, and label each component and supply voltage (screen voltage need not be specified). What is the gain of the stage?

5-22 Given an amplifier stage with a load resistor R_p and no cathode resistor. The effective capacitance across R_p is C and the transconductance is g_m. The load is changed to R_p' and a cathode resistor R_k is added and bypassed so that $R_kC_k = R_p'C$. The transconductance at the new operating current is g_m'. Prove that

a. If R_k is chosen so that the gain remains unchanged, the rise time is multiplied by g_m/g_m'.

b. If R_k is chosen so that the rise time remains constant, the gain is multiplied by g_m'/g_m.

5-23 A negative unit step is applied to a pentode having a cathode resistor R_k bypassed with a capacitor C_k. The total shunt capacitance at the plate is C. For each of the following cases draw the output waveform on a time scale which displays both the high- and low-frequency transient characteristics of the amplifier:

a. $R_k = C_k = 0$.

b. $C_k = 0$; $g_mR_k = 1$.

c. $g_mR_k = 1$; $R_kC_k = R_pC$.

d. $g_mR_k = 1$; C_k very large.

Draw the waveforms so that relative to one another they are to scale. Indicate time constants and percentage tilts.

5-24 a. Design a single-stage distributed amplifier with a (minimum) gain of 12, 100-MHz bandwidth, and 100-Ω input and output line impedances. Use type 7788 tubes (Prob. 5-4), constant-k lines with m-derived terminations, and assume the stray wiring capacitance is 3 pF at each grid and 10 pF at each plate. Design the lines for 150-MHz cutoff. Draw the complete amplifier, including both lines.

b. The line impedances were specified in part a. If these impedances were not specified would it be possible to use a higher-impedance grid line? If possible, would it be advantageous? Explain. Repeat for the plate line.

c. Consider using the tubes required in part a in a cascaded (not distributed) amplifier with shunt compensation in each stage and with an overall design bandwidth of 150 MHz. Find the load resistance and gain of each stage. Find the overall gain and compare with the distributed amplifier. (Assume identical stages.)

5-25 A distributed amplifier using 6AK5 tubes is to be designed with a cutoff frequency of 50 MHz and a gain of at least 100. The grid and plate lines are each to have a characteristic impedance of 170 Ω. Calculate (a) the number of stages m, (b) the number of tubes per stage n, (c) the overall gain G, (d) the capacitance and inductance in each section of grid and plate lines C_g, L_g, C_p, L_p, and (e) the gain that would be obtained if all the tubes used above were put into a single stage.

5-26 Given a cascade of m distributed stages whose overall bandwidth is $f_c{}^{(m)}$ and overall gain is G. Show that if the line capacitances C_p and C_g are kept constant, then for fixed overall bandwidth and overall gain the minimum number of tubes is required when the gain per stage is $A = \epsilon^{1.5}$.

CHAPTER 6

6-1 a. Evaluate η in Eq. (6-1) from the slope of the plot in Fig. 6-3 for $T = 25°C$. Draw the best-fit line over the current range 0.01 to 10 mA.

b. Repeat for $T = -55°C$ and 150°C.

6-2 A reverse-biasing voltage of 100 V is applied through a resistor R to a type 1N270 diode (Fig. 6-4a). The diode operates at 25°C. Determine the diode current and voltage for the cases $R = 10$ M, $R = 1$ M, and $R = 100$ K.

6-3 A resistor of 100 Ω is placed in series with a germanium diode whose reverse saturation current at 25°C is 5 μA. Make a semilog plot of the volt-ampere characteristic of the series combination over the range from 10 μA to 50 mA in the forward direction.

6-4 a. Use Eq. (6-3) to calculate the anticipated factor by which the reverse saturation current of a germanium transistor is multiplied when the temperature is increased from 25 to 80°C.

b. Repeat for a silicon transistor over the range 25 to 150°C.

6-5 It is predicted from Eq. (6-3) that for germanium, the reverse saturation current should increase by 0.10/°C. It is found experimentally in a particular transistor that at a reverse voltage of 10 V the reverse current is 5 μA and that the temperature dependence is only 0.07/°C. What is the leakage resistance shunting the diode?

6-6 A diode is mounted on a chassis in such a manner that, for each degree of temperature rise above ambient, 0.1 mW is thermally transferred from the diode to its surroundings. (The "thermal resistance" of the mechanical contact between the diode and its surroundings is 0.1 mW/°C.) The ambient temperature is 25°C. The diode temperature is not to be allowed to increase by more than 10°C above ambient. If the reverse saturation current is 5.0 μA at 25°C and increases at the rate 0.07/°C, what is the maximum reverse-bias voltage which may be maintained across the diode?

6-7 *a.* Consider a diode biased in the forward direction at a fixed voltage V. Prove that the fractional change in current with respect to temperature is

$$\frac{1}{I}\frac{dI}{dT} = \frac{V_g - V}{\eta T V_T}$$

b. Find the percentage change in current per degree centigrade for Ge at $V = 0.2$ V and for Si at $V = 0.6$ V at room temperature.

6-8 A silicon diode operates at a forward voltage of 0.4 V. Calculate the factor by which the current will be multiplied when the temperature is increased from 25 to 150°C. Compare the result with the plot of Fig. 6-3.

6-9 Reverse-biased diodes are frequently employed as electrically controllable variable capacitors. The transition capacitance of an abrupt junction diode is 20 pF at 5 V. Compute the decrease in capacitance for a 1.0-V increase in bias.

6-10 The breakdown diode is a 5.7-V reference diode. From the characteristics of Fig. 6-7 find the value of R for which the reference voltage will have a zero temperature coefficient.

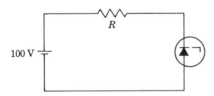

Prob. 6-10

6-11 A series combination of a 15-V avalanche diode and a forward-biased silicon diode is to be used to construct a zero-temperature-coefficient voltage reference. The temperature coefficient of the silicon diode is -1.7 mV/°C. Express in percent per degree centigrade the required temperature coefficient of the Zener diode.

6-12 The saturation currents of the two diodes are 1 and 2 μA. The breakdown voltages of the diodes are the same and are equal to 100 V.

a. Calculate the current and voltage for each diode if $V = 90$ V and if $V = 110$ V.

b. Repeat part *a* if each diode is shunted by a 10-M resistor.

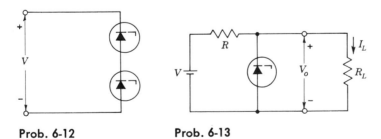

Prob. 6-12   **Prob. 6-13**

6-13 *a.* The avalanche diode regulates at 50 V over a range of diode current from 5 to 40 mA. The supply voltage $V = 200$ V. Calculate R to allow voltage regulation from a load current $I_L = 0$ up to $I_{\max}$, the maximum possible value of I_L. What is $I_{\max}$?

b. If R is set as in part a and the load current is set at $I_L = 25$ mA, what are the limits between which V may vary without loss of regulation in the circuit?

6-14 The circuit regulates the load voltage V_o at 20 V. Using the data given in Fig. 6-8 calculate the change in load voltage if (*a*) the unregulated 100-V supply changes by 2 percent, (*b*) the load current changes by 2 percent.

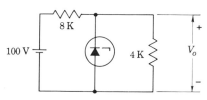

Prob. 6-14

6-15 A thermionic diode operates at a cathode temperature of 1000°K at normal heater voltage and yields a current of 100 μA when the anode is shorted to the cathode.

a. Find the voltage which appears across the diode if the anode is connected to the cathode through a resistor $R = 1$ M. Find the voltage if $R = 100$ M.

b. The diode is connected to a forward-biasing voltage of 100 V through a 1-M resistor. What is the diode current and voltage?

6-16 Each diode is described by a linearized volt-ampere characteristic with incremental resistance r and offset voltage V_γ. Diode $D1$ is germanium with $V_\gamma = 0.2$ V and $r = 20$ Ω, whereas $D2$ is silicon with $V_\gamma = 0.6$ V and $r = 15$ Ω. Find the diode currents if (*a*) $R = 10$ K, (*b*) $R = 1$ K.

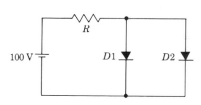

Prob. 6-16

6-17 Verify from the plot in Fig. 6-14*b* that for the type 2N914 silicon transistor

$$y \equiv \frac{1}{I_{CBO}} \frac{dI_{CBO}}{dT}$$

is a constant. Determine the value of the constant and compare it with the value of y obtained from Eq. (6-4) at $T = -55°$C and $+185°$C.

6-18 *a.* The reverse saturation current of the germanium transistor in Fig. 6-15 is 2 μA at room temperature (25°C) and increases by a factor of 2 for each temperature increase of 10°C. The bias $V_{BB} = 5$ V. Find the maximum allowable value for R_B if the transistor is to remain cut off at a temperature of 75°C.

b. If $V_{BB} = 1.0$ V and $R_B = 50$ K, how high may the temperature increase before the transistor comes out of cutoff?

6-19   A silicon transistor has $h_{FE} = 50$, $I_{CO} = 0.1~\mu\text{A}$, and a cutin voltage $V_\gamma = 0.5$ V.   The parameter n in the avalanche multiplication formula [Eq. (6-16)] is $n = 4$ and $BV_{CBO} = 60$ V.   (a) Calculate BV_{CEO}; (b) calculate BV_{CER} if $R_B = 1$ M; (c) calculate BV_{CEX} assuming $V_{BB} = 10$ V and $R_B = 10$ K.

6-20   From the characteristic curves for the type 2N404 transistor given in Chap. 6, find the voltages V_{BE}, V_{CE}, and V_{BC} for the circuit shown.

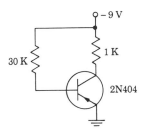

Prob. 6-20

6-21   Starting with Eqs. (6-33) and (6-34) and assuming $I_B \gg I_{CO}$, prove that

$$V_{CE} \approx \pm\eta V_T \ln \frac{1 + \dfrac{I_C}{I_B}(1 - \alpha_I)}{\alpha_I\left(1 - \dfrac{I_C}{I_B}\dfrac{1 - \alpha_N}{\alpha_N}\right)} = \pm\eta V_T \ln \frac{1 + h_{FEI} + \dfrac{I_C}{I_B}}{h_{FEI}\left(1 - \dfrac{I_C}{h_{FEI}I_B}\right)}$$

where $h_{FE} = \alpha_N/(1 - \alpha_N)$ and $h_{FEI} = \alpha_I/(1 - \alpha_I)$.

6-22   *a.* The incremental resistance between collector and emitter for a grounded-emitter switch at constant base current may be computed as

$$r_{CE} = \left|\frac{dV_{CE}}{dI_C}\right|_{I_B} = \left|\frac{d(V_E - V_C)}{dI_C}\right|_{I_B}$$

where V_E and V_C are, respectively, the voltage drops across the emitter and collector junctions.   Using Eqs. (6-33) and (6-34), show that

$$r_{CE} = \eta V_T \left[\frac{1 - \alpha_N}{\alpha_N I_B - I_C(1 - \alpha_N) + I_{CO}} + \frac{1 - \alpha_I}{I_B + I_C(1 - \alpha_I) + I_{EO}}\right]$$

b. If $I_B \gg I_{CO}$ and if $\dfrac{1 - \alpha_N}{\alpha_N}\dfrac{I_C}{I_B} \ll 1$ show that

$$r_{CE} \approx \frac{\eta V_T}{I_B}\frac{1 - \alpha_N\alpha_I}{\alpha_N}$$

6-23   *a.* Show that I_C is given approximately by

$$I_C = I_{CER} = \frac{\left[1 + \dfrac{I_{EO}(R + r_{bb'})}{\eta V_T}\right]I_{CO}}{1 - \alpha_N\alpha_I + (1 - \alpha_N)\dfrac{I_{EO}(R + r_{bb'})}{\eta V_T}}$$

where $r_{bb'}$ is the base-spreading resistance.   HINT: Assume that the collector junction is reverse-biased and that the emitter junction is *slightly* forward-biased.   Take advantage of the approximations which are allowed because the forward bias is small.

  b. A germanium transistor operating at room temperature has $\alpha_N = 0.98$, $I_{CO} = 2\ \mu A$, $I_{EO} = 1.6\ \mu A$, and $r_{bb'} = 200\ \Omega$.   Calculate I_C for $R = 0$ and $R = \infty$.

  c. What value of R will give a collector current midway between the currents corresponding to a shorted and open base?

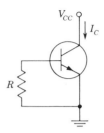

Prob. 6-23

6-24   A type 2N404 germanium transistor is operated at room temperature in the CE configuration.   The supply voltage is 6 V, the collector-circuit resistance is 200 Ω, and the base current is 20 percent higher than the minimum value required to drive the transistor into saturation.   Assume the following transistor parameters: $I_{CO} = -5\ \mu A$, $I_{EO} = -2\ \mu A$, $h_{FE} = 100$, and $r_{bb'} = 250\ \Omega$.   Find $V_{BE}(\text{sat})$ and $V_{CE}(\text{sat})$.

  6-25   The type 2N1708 double-diffused silicon transistor has parameters $h_{FE} = 30$, $h_{FEI} = 0.2$ (Prob. 6-21), $r_{bb'} = 30\ \Omega$, $I_{CO} = 13$ nA, and has a collector body resistance of 6 Ω.   It operates with $I_B = 1$ mA and $I_C = 10$ mA.   Find V_{BE} and V_{CE} at room temperature.

  6-26   Show that the emitter volt-ampere characteristic of a transistor in the active region is given by

$$I_E \approx I_o \epsilon^{V_E/\eta V_T}$$

where $I_o = -I_{EO}/(1 - \alpha_N \alpha_I)$.   Note that this characteristic is that of a *p-n* junction diode.

  6-27   Using the result of Prob. 6-26, verify that the variation of V_E with temperature is given by Eq. (6-5).   Carry out the required differentiation, assuming that Eq. (6-3) is valid for I_{EO}.   Note that $dV_E/dT \approx -2$ mV/°C, as indicated in Eq. (6-6).

  6-28   *a*. Show that if the collector junction is reverse-biased with $|V_{CB}| \gg \eta V_T$, the voltage V_{BE} (page 852) is related to the base current by

$$V_{BE} = I_B \left(r_{bb'} + \frac{R_E}{1 - \alpha_N} \right) + \frac{I_{co} R_E}{1 - \alpha_N}$$
$$+ \eta V_T \ln \left[1 + \frac{I_B(1 - \alpha_N \alpha_I)}{I_{EO}(1 - \alpha_N)} + \frac{\alpha_N(1 - \alpha_I)}{\alpha_I(1 - \alpha_N)} \right]$$

where $r_{bb'}$ is the base-spreading resistance and R_E is the emitter body resistance.

b. Show that if the collector is open-circuited,

$$V_{BE} = I_B(r_{bb'} + R_E) + \eta V_T \ln \left(1 + \frac{I_B}{I_{EO}}\right)$$

6-29   If the emitter follower is overdriven to the extent that the base current exceeds the emitter current, the output voltage can be made exactly equal to the supply voltage.   Show that $V_o = V_{CC}$ when

$$I_B = \frac{V_{CC}}{R_e} \left(1 + \frac{\alpha_N}{\alpha_I} \frac{1 - \alpha_I}{1 - \alpha_N}\right)$$

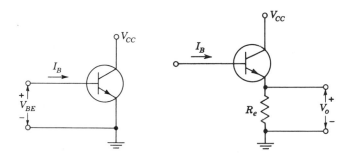

Prob. 6-28   **Prob. 6-29**

6-30   Use the Ebers-Moll equations to show that the transconductance of a transistor in the active region is given by

$$g_m = \frac{dI_C}{dV_E}\bigg|_{V_C = \text{const}} \approx \frac{1}{\eta V_T}\left[I_C - \frac{(1 - \alpha_I)I_{CO}}{1 - \alpha_N \alpha_I}\right] \approx \frac{I_C}{\eta V_T}$$

HINT: Assume $\epsilon^{V_C/\eta V_T} \ll 1$.

6-31   For a 12AU7 vacuum tube evaluate the incremental grid resistance r_g as a function of the grid-to-cathode voltage V_G with the plate-to-cathode voltage V_P as a parameter.   Choose values of $V_P = 150$, 100, and 50 V.

6-32   *a.* If a 2.5-K resistor is connected from the grid of a type 5965 tube to its cathode, how much bias voltage will develop across this resistor?

b. What plate current flows under the conditions in (*a*) if the supply voltage is 200 V and the load resistance is 10 K?

CHAPTER 7

7-1   For the diode clipping circuit of Fig. 7-4*a* assume that $V_R = 10$ V, $v_i = 20 \sin \omega t$, and that the diode forward resistance is $R_f = 100 \ \Omega$ while $R_r = \infty$ and $V_\gamma = 0$.   Neglect all capacitances.   Draw to scale the input and output waveforms and label the maximum and minimum values if (*a*) $R = 100 \ \Omega$, (*b*) $R = 1$ K, and (*c*) $R = 10$ K.

7-2   Repeat Prob. 7-1 for the case where the reverse resistance is $R_r = 10$ K.

7-3   In the diode clipping circuit of Fig. 7-4*a* and *d*, $v_i = 20 \sin \omega t$, $R = 1$ K, and $V_R = 10$ V.   The reference voltage is obtained from a tap on a 10-K divider

connected to a 100-V source.   Neglect all capacitances.   The diode forward resistance
is 50 Ω, $R_r = \infty$, and $V_\gamma = 0$.   In both cases draw the input and output waveforms
to scale.   Which circuit is the better clipper?   HINT: Apply Thévenin's theorem to
the reference-voltage divider network.

7-4   The input voltage v_i to the clipper shown is a 1.0-μsec pulse whose voltage
varies between 0 and 10 V.   The forward diode resistance is $R_f = 100\ \Omega$, $V_\gamma = 0.5$ V,
and $R_r = \infty$.   Sketch the output waveform v_o and indicate the time constants of the
exponential portions.

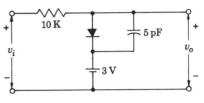

Prob. 7-4

7-5   The diode has $R_f = 100\ \Omega$, $R_r = 10$ K, $V_\gamma = 0$, and the capacitance shunting the diode is 5 pF.   For the periodic waveform shown, sketch the steady-state output voltage, indicating all voltages and time constants.

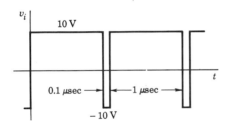

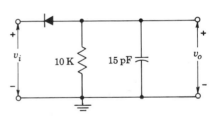

Prob. 7-5

7-6   A symmetrical 5-kHz square wave whose output varies between $+10$ and
-10 V is impressed upon the clipping circuit shown.   Assume $R_f = 0$, $R_r = \infty$, and
$V_\gamma = 0$.   Sketch the steady-state output waveform, indicating numerical values of
the maximum, minimum, and constant portions and the time constants of the exponential portions, if (a) all capacitances are neglected, (b) the diode capacitance of 10 pF
is taken into account but the load capacitance is neglected (an unrealistic situation),
and (c) both the diode and load capacitances (each equal to 10 pF) are taken into
consideration.

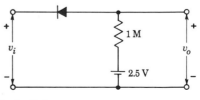

Prob. 7-6

7-7 The diode has resistances $R_f = 1$ K, $R_r = 100$ K. Assume that $V_\gamma = 0$. Draw the output waveform v_o and label all voltage levels and time constants.

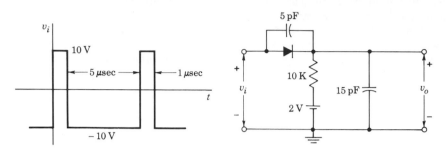

Prob. 7-7

7-8 *a.* What is the magnitude of v_i when the output is zero?

b. Draw the output waveform if the input is as shown. Plot the output directly below the input and to the same time scale.

c. What is the grid-to-cathode voltage when the output is +50 V?

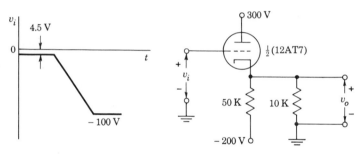

Prob. 7-8

7-9 The input v_i to the circuit shown is a sinusoidal voltage whose peak value is 80 V. Sketch the output voltage v_o to the same time scale as the input, and calculate the maximum and minimum values of the output.

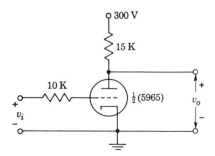

Prob. 7-9

7-10 *a.* A transistor operates in the common-emitter configuration with a collector resistor R_c. In terms of the hybrid-Π parameters of Fig. 1-4, show that if $g_m R_1 \gg 1$, then the input impedance, at low frequency, is

$$R_i = r_{bb'} + \frac{r_{b'e}}{1 + h_{fe} R_1 / r_{b'c}} = r_{bb'} + \frac{\eta V_T}{|I_E|} \frac{1 + h_{fe}}{1 + h_{fe} R_1 / r_{b'c}}$$

where I_E is the d-c emitter current and R_1 is the parallel combination of R_c and r_{ce}.

b. Assume a germanium transistor at room temperature with $I_E = 1$ mA, $h_{fe} = 50$, $r_{bb'} = 100$ Ω, $r_{b'c} = 4$ M, and $r_{ce} = 80$ K. Plot R_i as a function of R_c over the range $R_c = 0$ to $R_c = \infty$. Show that over the range $0 < R_c < 10$ K, R_i departs from h_{ie} (the short-circuit input impedance) by less than 10 percent. Show also that over the whole range the input impedance varies only by a factor of about 2.

7-11 A silicon transistor operates at room temperature in the common-emitter configuration with a collector resistor $R_c = 5$ K from a 10-V collector supply. The transistor has $h_{fe} = 50$, $r_{bb'} = 50$ Ω, $r_{ce} = 80$ K, and $r_{b'c} = 4$ M. Calculate the input impedance (*a*) just before the transistor enters saturation, (*b*) just after it has entered saturation, and (*c*) when the emitter current is 1 percent of the saturation current.

7-12 *a.* The input voltage v_i to the two-level clipper shown in (*a*) varies linearly from 0 to 150 V. Sketch the output voltage v_o to the same time scale as the input voltage. Assume ideal diodes.

b. Repeat part *a* for the circuit shown in (*b*).

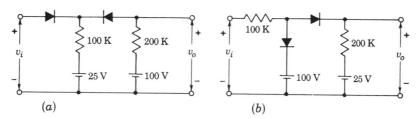

(*a*)   (*b*)

Prob. 7-12

7-13 The circuit of Fig. 7-12*a* is used to "square" a 10-kHz input sine wave whose peak value is 50 V. It is desired that the output voltage waveform be flat for 90 percent of the time. Diodes are used having a forward resistance of 100 Ω and a backward resistance of 100 K.

a. Find the values of V_{R1} and V_{R2}.

b. What is a reasonable value to use for R?

7-14 *a.* The diodes are ideal. Write the transfer characteristic equations (v_o as a function of v_i).

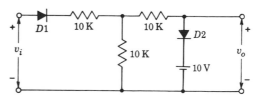

Prob. 7-14

b. Plot v_o against v_i indicating all intercepts, slopes, and voltage levels.

c. Sketch v_o if $v_i = 40 \sin \omega t$.   Indicate all voltage levels.

7-15   *a.* Repeat Prob. 7-14 for the circuit shown.

b. Repeat for the case where the diodes have an offset voltage $V_\gamma = 1$ V.

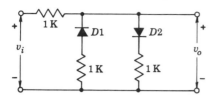

Prob. 7-15

7-16   Assume that the diodes are ideal.   Make a plot of v_o against v_i for the range of v_i from 0 to 50 V.   Indicate all slopes and voltage levels.   Indicate, for each region, which diodes are conducting.

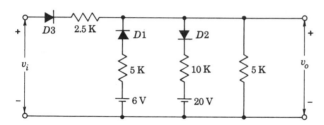

Prob. 7-16

7-17   The triangular waveform shown is to be converted into a sine wave by using clipping diodes.   Consider the dashed waveform sketched as a first approximation to the sinusoid.   The dashed waveform is coincident with the sinusoid at 0°, 30°, 60°, etc.   Devise a circuit whose output is this broken-line waveform when the input is the triangular waveform.   Assume ideal diodes and calculate the values of all supply voltages and resistances used.   The peak value of the sinusoid is 50 V.

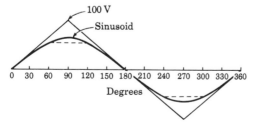

Prob. 7-17

7-18   Construct circuits which exhibit terminal characteristics as shown in parts (a) and (b) of the figure.

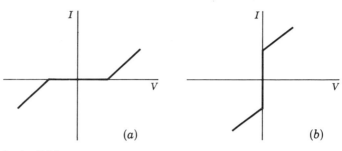

Prob. 7-18

7-19   *a.* The circuit of Fig. 7-14a is modified as follows: The grid of $V1$ is biased at V_{GG1} and v_i is RC-coupled to this grid.   The circuit parameters are $V_{GG1} = V_{GG2} = -5$ V, $R_k = 3$ K, $R_p = 10$ K, and $V_{PP} = 200$ V.   The tube is a 12AU7.   Make a plot of the variation of v_o with v_i.   Show where the clipping levels occur. $V_{KK} = 0$.

b. Calculate the ratio v_o/v_i for the region between clipping levels by using the linear equivalent circuit.   Compare with the average gain calculated from the plot obtained in (a).

7-20   For the circuit of Fig. 7-14a, $R_k = 100$ K, $R_p = 20$ K, $V_{PP} = 300$ V, and the tube is a 12AX7.   The input is sinusoidal and is RC-coupled to the grid of $V1$, which is biased at V_{GG1}.   It is desired that the peak-to-peak output be limited to 20 V and that the output be symmetrical with respect to its quiescent value.

a. Find V_{GG1} and V_{GG2}.

b. At what input signal amplitude will the output start being clipped?

c. For what peak input signal will the input tube start drawing grid current?

d. What is the gain of the circuit in the region of linear operation?

e. Draw an input sinusoid of peak value 50 V, and directly below it and to the same time scale draw the output voltage waveform.

7-21   In the two-level transistor clipping circuit of Fig. 7-14b, the transistors are germanium *p-n-p* type requiring a negative collector supply and a positive emitter supply.   The magnitudes of the parameters are $V_{CC} = 4.5$ V, $V_{EE} = 12$ V, $R_c = 2.7$ K, $R_e = 10$ K, and $V_{BB2} = 2$ V.   Make a plot showing v_o as a function of v_i.   Show where the clipping levels occur.   What is the total range of v_i over which the output responds?

7-22   Consider an emitter-coupled clipper operating with constant emitter current I.   Let us define the upper level v_{iU} to correspond to the point where the current in $Q2$ is only 1 percent of its maximum value and the lower level v_{iL} to correspond to a current in $Q2$ which is 99 percent of its maximum value.   Show that in this case the range $\Delta v_i = v_{iU} - v_{iL}$ is approximately twice the value given in Eq. (7-14).

7-23   An emitter-coupled clipper as in Fig. 7-14b has its common emitters returned to a current source of constant current I.   The collector resistance R_c is small enough so that it may be considered a short circuit.

a. Show that the circuit indicated is an equivalent circuit for the purpose of calculating the input current Δi_b corresponding to an input voltage change Δv_i.

b. Show that the output voltage swing between clipping levels is

$$\Delta v_o = I R_c = \frac{a g_m R_c}{2} \Delta v_i$$

where $a \equiv r_{b'e}/(r_{b'e} + r_{bb'})$.

c. Since from Eq. (1-9) g_m varies with $|I_E|$, choose an average g_m corresponding to $|I_E| = I/2$. Show that the total input change required to take the output between its two clipped levels is $\Delta v_i \approx 4\eta V_T$. Compare with Eq. (7-14).

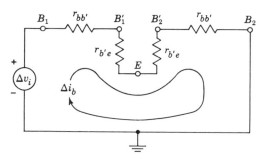

Prob. 7-23

7-24 a. In the emitter-coupled clipper of Fig. 7-14b, $V_{BB2} = 0$. The two germanium transistors have identical parameters except that the quantity $(1 - \alpha_I)/I_{EO}$ for Q1 is 1.5 times as large as the corresponding quantity for Q2. At room temperature what are the values of v_{iU} and v_{iL}?

b. If the transistors were identical what would be the values of v_{iU} and v_{iL}?

7-25 a. In the clipping circuit shown, D2 compensates for temperature variations. Assume that the diodes have infinite back resistance, a forward resistance of 50 Ω, and a break point at the origin ($V_\gamma = 0$). Calculate and plot the transfer characteristic v_o against v_i. Show that the circuit has an extended break point, that is, two break points close together.

b. Find the transfer characteristic that would result if D2 were removed and the resistor R were moved to replace D2.

c. Show that the double break of part a would vanish and only the single break of part b would appear if the diode forward resistances were made vanishingly small in comparison with R.

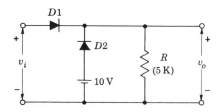

Prob. 7-25

7-26 *a.* In the peak clipping circuit shown, add another diode $D2$ and a resistor R' in a manner that will compensate for drift with temperature.

b. Show that the break point of the transmission curve occurs at V_R. Assume $R_r \gg R \gg R_f$.

c. Show that if $D2$ is always to remain in conduction it is necessary that

$$v_i < (v_i)_{\max} = V_R + \frac{R}{R'}\,(V_R - V_\gamma)$$

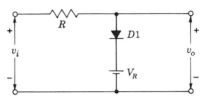

Prob. 7-26

7-27 The diode-resistor comparator of Fig. 7-19 is connected to a device which responds when the comparator output attains a level of 0.1 V. The input is a ramp which rises at the rate 10 V/μsec. The germanium diode has a reverse saturation current of 1 μA. Initially $R = 1$ K and the 0.1-V output level is attained at a time $t = t_1$. If we now set $R = 100$ K, what will be the corresponding change in t_1? $V_R = 0$.

7-28 Show that the double-differentiator comparator pulse given in Eq. (7-18) has an amplitude $0.37\alpha A RC$ and a width at half maximum of $2.4RC$, and that the peak occurs at $x = 1$.

7-29 The input to the comparator of Fig. 7-23 is a ramp whose slope is 0.1 V/μsec. The reference level is $V_R = 0$. The amplifier gain is 10 and $\tau_1 = \tau_2 = 100$ μsec. What are (*a*) the initial slope of the output pulse and (*b*) the peak value of the output? (*c*) If $\tau_1 = 100$ μsec and $\tau_2 = 10$ μsec, what is the peak value of the output?

7-30 For the difference-amplifier comparator of Fig. 7-14*b*, the *common-mode gain* A_c is defined by $A_c \equiv v_o/v_c$ under the circumstance that when a signal $v_i = v_c$ is applied to $Q1$, an identical signal v_c is simultaneously applied to $Q2$. The *difference-mode* gain is defined by $A_d \equiv v_o/v_d$ under the circumstance that when a signal $v_i = v_d/2$ is applied to $Q1$, a signal $-v_d/2$ is simultaneously applied to $Q2$. Each source has a resistance R_s. Assume that the load resistor is small enough to neglect in calculating the current in $Q2$ ($h_{oe}R_c \ll 1$).

a. Show that if $h_{fe} \gg h_{oe}(R_s + h_{ie})$

$$A_c \approx \frac{(2h_{oe}R_e - h_{fe})R_c}{2R_e(h_{fe} + 1) + R_s + h_{ie}}$$

HINT: Use the symmetry of the circuit to show that it may be split into two uncoupled transistor stages each operating with an emitter resistor $2R_e$. Use the exact expression for A_I given in Eq. (1-56) and use $-A_I$ in place of h_{fe} in Eq. (1-50).

b. Show that A_d is independent of R_c and is given by

$$A_d = \frac{h_{fe}R_c}{2(R_s + h_{ie})}$$

HINT: Use the symmetry of the circuit to show that no signal appears at the common emitter. Hence the emitter is virtually grounded and again the circuit may be split into two separate transistor stages.

c. Find the *common-mode rejection ratio* ρ defined by $\rho \equiv |A_d/A_c|$.

d. Evaluate A_c, A_d, and ρ if $h_{fe} = 100$, $R_s = h_{ie} = R_c = 1$ K, $R_e = 25$ K, and $1/h_{oe} = 40$ K.

7-31 a. As a result of an ambient temperature change the reverse saturation current of each transistor of the difference-amplifier comparator of Fig. 7-14b increases by $\Delta I_{CO} = 1$ μA. For R_e arbitrarily large calculate the change in input voltage, applied simultaneously to each base, that will produce the same change in output voltage as is caused by the change ΔI_{CO}. Use $h_{fe} = 100$, $1/h_{oe} = 40$ K. Use the results of Prob. 7-30.

b. If $R_s = h_{ie} = 1$ K, what change in input voltage applied to one base will produce the same change in output as is caused by the change ΔI_{CO}?

7-32 a. Calculate the common-mode gain A_c and the difference-mode gain A_d for the vacuum-tube circuit of Fig. 7-14a. (These gains are defined in Prob. 7-30.) If $(\mu + 1)R_k \gg r_p$ and $\mu \gg 1$, show that

$$A_d = \frac{\mu R_p}{2r_p + R_p} \quad \text{and} \quad A_c = -\frac{R_p r_p}{R_k(2r_p + R_p)}$$

b. Show that the common-mode rejection ratio ρ is $\rho \equiv |A_d/A_c| = g_m R_k$.

7-33 In the difference-amplifier comparator of Fig. 7-14a the heater voltage changes so that there is equivalently a voltage ΔV_H introduced in series with each cathode. Show that the output will change by an amount which corresponds to a signal-voltage change Δv_i applied to one grid where

$$\Delta v_i \approx \frac{\Delta V_H}{g_m R_k}$$

provided that $(\mu + 1)R_k \gg r_p$.

CHAPTER 8

8-1 a. In the circuit of Fig. 8-2 the diode is ideal with $R_f = 0$, $R_r = \infty$, and $V_\gamma = 0$. The input signal is sinusoidal and of frequency f. Show that the angle θ during which the diode conducts in each cycle is given exactly by

$$\cos \theta = \epsilon^{-(2\pi - \theta)/2\pi f RC} - \frac{\sin \theta}{2\pi f RC}$$

b. Make the approximation, at the outset, that θ is small. In this case it is approximately correct to say that the capacitor discharges through R during the entire cycle. On this basis, compute the change Δv_A in capacitor voltage during the course of a cycle and find the angle θ at which the diode will start to conduct. Show that

$$\cos \theta \approx 1 - \frac{1}{RCf}$$

c. Show that for small θ both expressions reduce to

$$\theta \approx \sqrt{\frac{2}{RCf}}$$

d. For $\theta = 30°$ draw one complete cycle of the input waveform and, on the same coordinate axes, one complete cycle of the output waveform.

8-2 In the circuit of Fig. 8-2, $C = 0.1 \ \mu F$, $R = 10$ K, and the diode has $R_r = \infty$, $R_f = 1,000 \ \Omega$, and $V_\gamma = 0$. The input is a symmetrical square wave of frequency 1 kHz operating between the levels -150 and -100 V and beginning at -150 V at $t = 0$. Assuming that the capacitor is initially uncharged, draw the waveform of v_o from $t = 0$ through the first two cycles during which the diode conducts. Label all voltage levels.

8-3 In the circuit of Fig. 8-3, $R_s = R_f = 50 \ \Omega$, $R_r = \infty$, $R = 10$ K, $C = 1.0 \ \mu F$, and the diode break occurs at $V_\gamma = 0$. The input signal v_s is as shown. Draw the output waveform and label all voltages, assuming that the capacitor is initially uncharged.

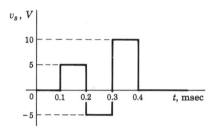

Prob. 8-3

8-4 In the restorer circuit of Fig. 8-3, $R_s = 5$ K, $R = 15$ K, $R_f = 100 \ \Omega$, $R_r = \infty$, $V_\gamma = 0$, and $C = 0.5 \ \mu F$. The input $v_s = 0$ for $t < 0$ and for $t > 0$ is a square wave of frequency 5 kHz and makes excursions between 0 and 10 V. The capacitor is initially uncharged. Draw the output waveform for the first three cycles, labeling all voltage levels and time constants.

8-5 *a.* In the restorer circuit of Fig. 8-3, $V_\gamma = 0$ and $R_f \ll R \ll R_r$. Do *not* assume that $R_s \ll R$. Let the capacitance C be arbitrarily large. The input is a symmetrical square wave of peak-to-peak amplitude V. Draw the output waveform in the steady state and show that the positive and negative excursions of the output signal are

$$V_1 = V \frac{R_f}{2R_s + R} \qquad -V_2 = V \frac{R}{2R_s + R}$$

b. Since the upper clamped level is close to zero we might expect that when the input drops abruptly by the amount V the output would drop abruptly by $VR/(R_s + R)$. Instead we find that the drop is $VR/(2R_s + R)$. Explain physically how this comes about.

c. If $R = 10$ K and $R_f = 0$, what is the maximum allowable value of R_s if the peak-to-peak amplitude of the output is not to be less than 0.99 V?

8-6 *a.* In the restorer circuit of Fig. 8-3 the input is the square wave of Fig. 8-6. Let $V_\gamma = 0$, $R_f \ll R \ll R_r$, C be arbitrarily large, and $k = T_2/T_1$. Do *not* assume that $R_s \ll R$. Show that the output waveform in the steady state has a positive excursion V_1 and a negative excursion $-V_2$ given by

$$V_1 = \frac{kR_f V}{(1 + k)R_s + kR_f + R} \qquad V_2 = \frac{RV}{(1 + k)R_s + kR_f + R}$$

b. Calculate the peak-to-peak value of the output voltage. Note that not only the clamping level but also the output amplitude depends upon the duty cycle. What is the output amplitude if $R_s = 0$?

8-7 The ramp-type signal is applied to the restorer circuit of Fig. 8-3, which has $R_f = 100 \ \Omega$, $V_\gamma = 0$, $R_r = \infty$, and $R = 10$ K. The capacitance C is arbitrarily large. Draw the output waveform, calculate all voltage levels, and calculate the voltage across the capacitor if the generator resistance R_s is (*a*) zero and (*b*) 100 Ω. (*c*) Repeat parts *a* and *b* if the diode is reversed

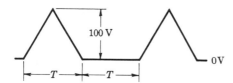

100 V

0V

$\longleftarrow T \longrightarrow \longleftarrow T \longrightarrow$

Prob. 8-7

8-8 Repeat Prob. 8-7*a* for a sinusoidal input signal of peak-to-peak value 100 V.

8-9 Verify Eq. (8-16).

8-10 In the restorer circuit of Fig. 8-3, $R_f = 100 \ \Omega$, $R_r = \infty$, $V_\gamma = 0$, and $R = 100$ K. The waveform is a square wave with $V = 30$ V, $T_1 = 50 \ \mu sec$, and $T_2 = 1,000 \ \mu sec$.

a. Assume $R_s = 0$ and C arbitrarily large. Calculate and sketch the steady-state output voltage v_o.

b. Repeat part *a* if $R_s = 100 \ \Omega$.

c. Repeat part *b* if C is 0.05 μF.

d. Repeat part *c* if $V_\gamma = 0.7$ V.

8-11 The signal shown, with $V = 20$ V, $T_1 = 1$ msec, and $T_2 = 10$ msec, is applied to the restorer of Fig. 8-3. The circuit has $R_s = 0$, $R = 1$ K, $C = 0.1 \ \mu F$, $R_f = 10 \ \Omega$, $R_r = \infty$, and $V_\gamma = 0$.

a. Compute the steady-state output waveform v_o.

b. Repeat part *a* if the diode terminals are reversed.

c. Repeat parts *a* and *b* if $R = 1$ M.

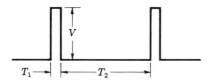

V

$T_1 \longrightarrow \longleftarrow \longrightarrow T_2 \longrightarrow$

Prob. 8-11

8-12 A symmetrical 10-kHz square wave whose peak excursions are ± 10 V with respect to ground is impressed upon the diode clamping circuit of Fig. 8-3. If $R = 10$ K, $C = 1 \ \mu F$, the diode has $R_r = \infty$, $R_f = 0$, $V_\gamma = 0$, and the source impedance is zero,

a. Sketch the output waveform.

b. If the diode forward resistance is 1 K, sketch the output waveform. Calculate the maximum and minimum voltages with respect to ground.

c. Repeat part *b* if the source impedance is 1 K.

8-13 An attempt is to be made to restore the maximum value of the periodic waveform indicated to a value of $+10$ V. The diode used has $V_\gamma = 0$, $R_f = 20\ \Omega$, and $R_r = \infty$. Assume zero source impedance. The coupling capacitance has a value of $0.05\ \mu$F. Because of the load the effective resistance across the diode when it is not conducting is 20 K.

a. Indicate the circuit to be used.

b. Make a careful sketch of the output waveform. Label all important voltages and all time constants.

c. Indicate two important areas on your sketch and state the ratio of these two areas.

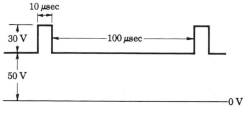

Prob. 8-13

8-14 An attempt is made to restore the minimum value of the waveform of Prob. 8-13 to ground. The same diode, coupling capacitance, and load resistance are used as in Prob. 8-13. Make a careful sketch of the output waveform, and label all important voltages with respect to ground. Also indicate the time constants of all exponential portions of the waveform.

8-15 The square wave shown is applied to the circuit of Fig. 8-12. Assume that $T_1 \ll RC$, $T_2 \gg R_fC$, $R_s = 0$, $V_\gamma = 0$, and $R \gg R_f$.

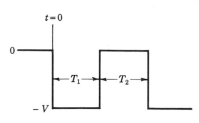

Prob. 8-15

a. Prove that the percentage tilt P of the output waveform in the interval T_1 is given by

$$P \approx \left(1 + \frac{V_{YY}}{V}\right)\frac{T_1}{RC} \times 100\%$$

b. Assume that $T_1 = T_2 = 0.1$ msec, $RC = 1$ msec, and $V = 10$ V. Sketch the output waveform for values of $V_{YY} = 0, 40, 90,$ and 190 V. Prove that the positive peak value of the output is V if $T_1 \geq [V/(V + V_{YY})]RC$.

8-16   The input to the circuit of Fig. 8-13*b* is a 20-V rms sine wave of frequency 1 kHz. The tube is a 12AU7 with $V_{PP} = 200$ V, $R_p = 10$ K, $R = 1$ M, $R_s = 0$, and $C = 0.1$ μF. Sketch, to the same time scale, the input voltage, the grid voltage, and the plate voltage. Find the maximum and minimum values of these voltages. Assume zero grid-to-cathode resistance when grid current flows.

8-17   A square wave whose period is 2 msec and whose peak-to-peak amplitude is 40 V is impressed on the clamping circuit of Fig. 8-12. The circuit parameters are $V_{YY} = 300$ V, $R = 1$ M, $C = 0.1$ μF, $R_f = 50$ Ω, $R_r = 0.5$ M, $V_\gamma = 0$, and $R_s = 0$. Calculate and sketch the steady-state output voltage v_o.

8-18   Verify Eqs. (8-20) and (8-21).

8-19   Verify Eq. (8-22).

8-20   Apply Eq. (8-22) to a symmetrical square wave ($T_1 = T_2 = T/2$) of peak-to-peak voltage V. Show that for $R \gg R_f$, $R_s = 0$, and C arbitrarily large, the maximum positive excursion of the output signal is V_σ given by

$$V_\sigma = V_\gamma + \frac{R_f}{R} (2V_{YY} + V)$$

8-21   *a.* In the circuit of Fig. 8-12, $R_s = 0$, and the diode is ideal with $R_f = 0$, $R_r = \infty$, and $V_\gamma = 0$. The input signal is constant up to $t = 0$ and for $t > 0$ the input is a negative-going ramp of slope α. Prove that, if the diode is to come out of cutoff, it is necessary that $\alpha > V_{YY}/RC$. What is the physical interpretation of this inequality?

b. If α satisfies the condition given in part *a*, show that the output voltage is given by

$$v_o = -(\alpha RC - V_{YY})(1 - \epsilon^{-t/RC})$$

8-22   Explain how d-c restoration in the base and grid circuits is largely suppressed in the circuits shown. In (*a*) the resistance R is much larger than the diode forward resistance or the base-to-emitter diode resistance. In (*b*) $R \ll R_G$.

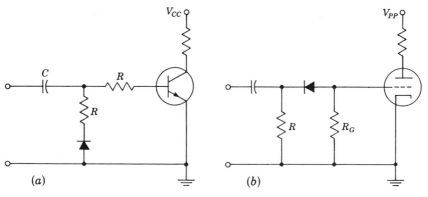

(*a*)   (*b*)

Prob. 8-22

8-23 The diode is ideal with $R_f = 0$, $R_r = \infty$, and $V_\gamma = 0$. Draw the output waveform and specify all voltage levels and time constants. Note that the time constant associated with the shunt capacitor is very much smaller than the time constant associated with the series capacitor. Make reasonable approximations, taking this feature into account.

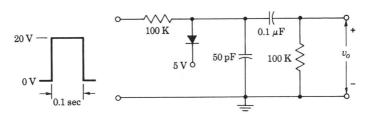

Prob. 8-23

8-24 *a.* In connection with the synchronous clamping circuit of Fig. 8-17, show that, in the steady state, when the diodes conduct, the voltage at A is the same as the voltage at B.

b. Assume that the signal makes excursions $\pm V_s$ with respect to its quiescent level. Show that if neither diode is to conduct during the intervals T_n it is required that $V_p > V_s$.

8-25 *a.* In Fig. 8-17, $C \gg C_s = 500\,\mathrm{pF}$ and $V_R = 0$. The diodes have $R_f = 20\,\Omega$, $R_r = \infty$, and $V_\gamma = 0$. The clamping circuit delivers its signal to a 10-K load. Large clamping pulses, having a duration of 0.5 μsec, occur with leading edges at $t = 5.25$, 11.25, etc., μsec. For the input signal shown, draw the output and indicate all voltage levels and time constants.

b. If $C_s = 1,000\,\mathrm{pF}$ and if the clamping pulse duration is 1 μsec, with leading edges at 5, 11, etc., μsec, repeat part *a*.

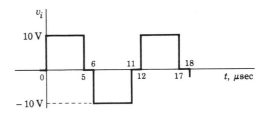

Prob. 8-25

8-26 *a.* For the overdriven amplifier of Fig. 8-18 show that the change in voltage across the capacitor C during the interval T_2 is

$$\Delta V_A = (V_{CC} + V - V_\sigma)(1 - \epsilon^{-T_2/\tau})$$

b. Show that the expression for ΔV_A may be put in the form

$$\Delta V_A = \frac{V_{CC} - V_\sigma + V}{V_{CC} - V_2}(V_2' - V_2)$$

c. For $R_{s2} \ll R$ show that ΔV_A reduces to $\Delta V_A = V_2' - V_2$.

d. Show that Δ_i in Fig. 8-18f is given by $\Delta_i = \Delta V_A R_{s2}/(R + R_{s2})$.

8-27 In the circuit of Fig. 8-18a a type 2N404 germanium transistor is used for which $r_{bb'} = 100 \ \Omega$. The supply voltage is $V_{CC} = 15$ V, $R_c = 1$ K, $R = 50$ K, and $C = 0.1 \ \mu$F. The gating source supplies a signal of amplitude 15 V at a constant output impedance of 15 K. The time T_2 is 1.5 msec. Draw the waveforms of Fig. 8-18, labeling all voltage levels.

8-28 In the circuit of Fig. 8-18b a type 12AU7 tube is used. The supply voltage is $V_{PP} = 300$ V, $R = 0.75$ M, $R_p = 15$ K, and $C = 0.015 \ \mu$F. The gating source supplies a signal of amplitude 120 V at a constant output impedance of 25 K. The time T_2 is 1.5 msec. Draw the waveforms of Fig. 8-18, labeling all voltage levels.

8-29 In the circuit of Fig. 8-22 two silicon transistors are used with $r_{bb'} = 200 \ \Omega$. The supply voltage is $V_{CC} = 15$ V, $R_{c1} = 15$ K, $R_{c2} = 1$ K, $R = 50$ K, and $C = 0.1 \ \mu$F. The gating signal applied to $Q1$ makes excursions between such levels that transistor $Q1$ is turned ON and OFF. The ON period of $Q1$ is $T_2 = 1.5$ msec. Find the voltage levels of the waveforms at the collector of $Q1$ and at the base and collector of $Q2$.

8-30 The input pulse duration is 1.0 msec. Draw the waveforms at the plate of $V1$ and at the grid and plate of $V2$. Label all voltage levels and time constants.

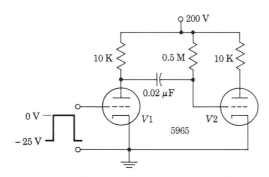

Prob. 8-30

8-31 Repeat the illustrative example on page 289 for the two-stage overdriven amplifier with the change that $C = 0.025 \ \mu$F. Find the voltage levels of the waveforms at the collector of $Q1$ and the base and collector of $Q2$.

8-32 A large-amplitude symmetrical square-wave signal of frequency 10 kHz is applied to the circuit of Fig. 8-23a, driving the transistor between saturation and cutoff. The circuit parameters are $V_{CC} = 10$ V, $L = 100$ mH, $R = 30$ K, and $R_c = 10$ K. Draw the output waveform, labeling all voltages and time constants.

8-33 The neon tube in the circuit in (*a*) has the volt-ampere characteristic shown in (*b*). The 6CG7 initially operates at zero grid-to-cathode voltage. Does the neon tube conduct in this quiescent state? At $t = 0$ a negative step v_i is applied to the grid to drive the tube to cutoff. Draw the output waveform, labeling all voltages and time constants. HINT: If the current in the neon tube decreases to zero, this device stops conducting.

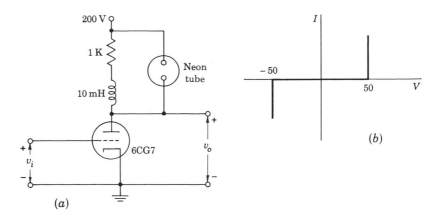

Prob. 8-33

8-34   In the peaking circuit of Fig. 8-23b, the tube is a 12AT7, $V_{PP} = 100$ V, $R = 10$ K, $R_g = 1$ M, $C = 0.1$ μF, and $L = 10$ mH.   The input square wave has a peak-to-peak value of 10 V and a frequency of 10 kHz.

 a. If the capacitance shunting L is neglected, sketch the output waveform.   Calculate the maximum and minimum voltage values and the time constants of the exponential portions of the output waveform.

 b. If the capacitance shunting L is 25 pF, calculate and sketch the output waveshape.

8-35   In the circuit of Fig. 8-26a the tube is a 5965, $V_{PP} = 200$ V, $R_p = 20$ K, $C_s = 50$ pF, $R_g = 1$ M, and $C = 0.001$ μF.   The input is a 100-kHz square wave whose peak-to-peak amplitude is 10 V.   Calculate and sketch the output voltage.

8-36   In the circuit of Fig. 8-26a, $V_{PP} = 300$ V, $R_p = 33$ K, and the tube is a 6CG7.   A symmetrical square-wave input signal drives the tube between zero grid voltage and cutoff.   The capacitance C_s is arbitrarily large.   In the steady state, find the voltage across the capacitor C_s.   HINT: The average value of current through the capacitor must be zero.

8-37   In the circuit of Fig. 8-27a, a silicon switching transistor is used with the parameter $h_{FE} = 100$.   The supply voltage is $V_{CC} = 10$ V, $R_c = 1$ K, $R_b = 50$ K, and $C_s = 0.05$ μF.   An input square-wave signal of frequency 2 kHz drives the transistor from saturation to cutoff.   Draw the waveform of collector current and output voltage.   Label all voltages, times, and time constants.

8-38   *a.* Assume in Fig. 8-31a that R_b is arbitrarily large so that the switch is being driven from a current source which provides a current I to keep the transistor in its active region.   Let the combined incremental resistance of $D1$ and the battery V_B be equal to R_f.   Assume that the incremental input resistance between base and emitter is R_i.   Show that if I_C is the collector current,

$$\frac{dI_C}{dI} = \frac{h_{fe}(R_c + R_f)}{(1 + h_{fe})R_c + R_f + R_i}$$

 b. Calculate dI_C/dI for $h_{fe} = 50$, $R_f = 50$ Ω, $R_c = 1$ K, and $R_i = 1$ K.

c. Compare with the result if the connection between collector and base were eliminated.

8-39   Verify Eq. (8-37) for the output impedance of a cathode follower.

8-40   *a.* An idealized vacuum tube has characteristics given by

$$I_P = g_m \left(V_G + \frac{V_P}{\mu} \right)$$

with $\mu = 20$ and $g_m = 2$ mA/V.   The tube operates in the cathode-follower circuit of Fig. 8-33 with $V_{PP} = 200$ V, $V_{KK} = 0$, $R_k = 50$ K, and drives a loading capacitance $C = 200$ pF.   If the input grid voltage changes very slowly, what is the total range over which the input may swing while the tube remains within its grid base?

b. If the input voltage changes abruptly, what is the allowable input grid swing?

c. The input signal shown is applied between grid and ground from a source of impedance 5 K.   Draw the output waveform, labeling all voltage levels and time constants.   The grid-to-cathode resistance is $r_G = 1$ K.   Assume the pulse is wide enough to allow completion of the circuit response to the leading edge of the pulse.

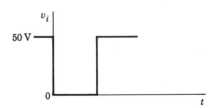

Prob. 8-40

8-41   In the circuit of Fig. 8-35, $V_{CC} = 15$ V, $R_s = 1$ K, $R_e = 2$ K, $C = 0.1$ μF, and a germanium transistor is used with $h_{fe} = 60$.   If the input signal v_s extends from 10 to 4 V, draw the output waveform.   Label all voltages and time constants.   Assume the input pulse is wide enough to allow completion of the circuit response to the leading edge of the pulse. Use $h_{ie} = 1$ K.

8-42   The circuit of Fig. 8-36 has $V_{CC} = 12$ V, $R_s = 500\,\Omega$, $R_e = 4$ K, $C = 0.01\,\mu$F, and uses complementary germanium transistors each with $h_{fe} = 60$.   Draw the output waveform for the input signal shown in Fig. 8-36.   Label all voltages and time constants.   Assume the input pulse is wide enough to allow completion of the circuit response to the leading edge of the pulse. Use $h_{ie} = 1$ K.

CHAPTER 9

9-1   Convert the following decimal numbers to binary form: (*a*) 671, (*b*) 325, and (*c*) 152.

9-2   The parameters in the diode OR circuit of Fig. 9-5 are $V(0) = +12$ V, $V(1) = -2$ V, $R_s = 600\,\Omega$, $R = 10$ K, $R_f = 0$, $R_r = \infty$, and $V_\gamma = 0.6$ V.   Calculate the two output levels if one input is excited and if (*a*) $V_R = +12$ V, (*b*) $V_R = +10$ V, (*c*) $V_R = +14$ V, and (*d*) $V_R = 0$ V.   For which of these cases is the OR function satisfied (except possibly for a shift in level between input and output)?   (e) Repeat part a if three inputs are excited.

9-3   Consider a two-input positive-logic diode OR gate (Fig. 9-5 with the diodes reversed) and with $V_R = 0$.   The inputs are the square waves v_1 and v_2 indicated. Sketch the output waveform if the ratio of the amplitude of v_2 to v_1 is (a) 2 and (b) $\frac{1}{2}$. Assume ideal diodes ($R_f = 0$, $R_r = \infty$, and $V_\gamma = 0$) and $R_s = 0$.

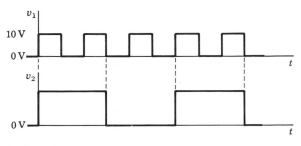

Prob. 9-3

9-4   Consider two signals, a 1-kHz sine wave and a 10-kHz square wave of zero average value, applied to the OR circuit of Fig. 9-5 with $V_R = 0$.   Draw the output waveform if the sine-wave amplitude (a) exceeds the square-wave amplitude, (b) is less than the square-wave amplitude.

9-5   a. Indicate how a tube that has two grids (such as the control and suppressor grids in a pentode), either of which may be used to effect plate-current cutoff, may be made to perform the OR operation.   Is the circuit useful for positive or negative logic or both?

b. Repeat part a for the AND operation.

9-6   Consider a two-input positive-logic diode AND circuit (Fig. 9-10b) with $V_R = 15$ V, $R = 10$ K, and $R_s = 1$ K.   Assume ideal diodes and neglect all capacitances.   A square wave v_i extending from -5 to $+5$ V with respect to ground is applied simultaneously to both inputs.   (a) Sketch the output v_o and calculate the maximum and minimum voltages with respect to ground.   (b) If $v_1 = v_i$ and $v_2 = -v_i$, calculate the voltage levels of v_o and plot.   (c) Indicate how to modify the circuit so that the minimum voltage is zero (ground).   (d) Repeat parts a and b for the circuit of part c.

9-7   Consider a two-input positive-logic diode AND circuit (Fig. 9-10b) with $V_R = 10$ V, $R = 10$ K, and $R_s = 0$.   Assume ideal diodes and neglect capacitances. The input waveforms are v_1 and v_2 sketched in Prob. 9-3.   Sketch the output waveform if the ratio of the amplitude of v_2 to v_1 is (a) 2, (b) 1, and (c) $\frac{1}{2}$.   Repeat part b if $R_s = 1$ K.

9-8   The binary input levels for the AND circuit shown are $V(0) = 0$ V and $V(1) = 25$ V.   Assume ideal diodes.   If $v_1 = V(0)$ and $v_2 = V(1)$, then v_o is to be at 5 V.   However, if $v_1 = v_2 = V(1)$, then v_o is to rise above 5 V.

a. What is the maximum value of V_R which may be used?

b. If $V_R = 20$ V, what is v_o at a coincidence $[v_1 = v_2 = V(1)]$?   What are the diode currents?

c. Repeat part b if $V_R = 40$ V.

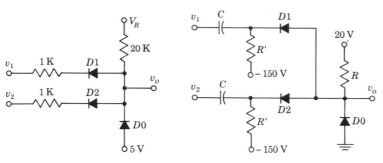

Prob. 9-8   **Prob. 9-9**

9-9 The two-input diode AND circuit shown uses diodes with $R_f = 500\ \Omega$, $R_r = \infty$, and $V_\gamma = 0$. The quiescent current in $D0$ is 6 mA, and the currents in $D1$ and $D2$ are each 4 mA.

  a. Calculate the quiescent output voltage v_o and the values of R and R'.

  b. Calculate the output voltage when one input diode is cut off. Calculate this result approximately by assuming that the currents through R and the remaining input diode do not change. Also, calculate the result exactly.

  c. Assume that diode $D0$ is omitted, that the currents in $D1$ and $D2$ remain 4 mA each, and that the output v_o is the same as that found in part *a*. Find R and R'.

  d. If the conditions are as indicated in part *c* but one of the diodes is cut off, find the output voltage v_o. Compare with the result in part *b* when $D0$ acts as a clamp.

9-10 Find v_o and v' if (*a*) there are no pulses at either A or B, (*b*) there is a 30-V positive pulse at A or B, and (*c*) there are positive pulses at both A and B. (*d*) What is the minimum pulse amplitude which must be applied in order that the circuit operate properly? Assume ideal diodes.

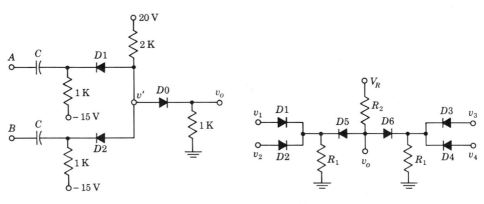

Prob. 9-10   **Prob. 9-11**

9-11 The four inputs v_1, v_2, v_3, and v_4 are voltages from zero-impedance sources whose values are either $V(0) = 10$ V or $V(1) = 20$ V. The diodes are ideal. $V_R = 25$ V, $R_1 = 5$ K, and $R_2 = 10$ K.

a. If $v_1 = v_2 = 10$ V and $v_3 = v_4 = 20$ V, find v_o and the currents in each diode.

b. If $v_1 = v_3 = 10$ V and $v_2 = v_4 = 20$ V, find v_o and the currents in each diode.

c. Sketch in block-diagram form the logic performed by this circuit.

d. Verify that in order for the circuit to operate properly the following inequality must be satisfied:

$$R_2 > \frac{V_R - V(0)}{V(0)} R_1$$

9-12 a. In block-diagram form indicate the logic performed by the diode system shown. The input levels are $V(0) = -8$ V and $V(1) = +2$ V. Neglect source resistance and assume that the diodes are ideal. Justify your answer by calculating the voltages v_A, v_B, and v_o (and indicating which diodes are conducting) under the following circumstances: (b) all inputs are at $V(0)$, (c) some but not all inputs in A are at $V(1)$ and all inputs in B are at $V(0)$, (d) all inputs in A are at $V(1)$ and some inputs in B are at $V(1)$, and (e) all inputs are at $V(1)$. (f) If the 10-K resistance were increased, at what maximum value would the circuit no longer operate in the manner described above? (g) Indicate how to modify the circuit so that the output levels are -5 and 0 V, respectively.

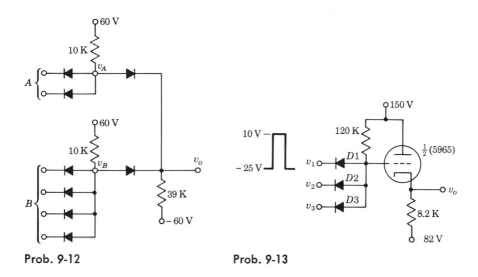

Prob. 9-12   Prob. 9-13

9-13 a. The AND circuit of an early (NORC) computer using vacuum tubes is indicated. Assuming ideal diodes, explain the operation of the circuit.

b. Assume $R_f = 1$ K for the remainder of this problem. Find the grid voltage and the output voltage if all inputs are low (-25 V).

c. Repeat part b if all inputs are high ($+10$ V).

d. Find the grid voltage if v_1 is high but v_2 and v_3 are low.

e. Find the grid voltage if v_1 and v_2 are high but v_3 is low.

9-14 a. It is desired that when a single pulse enters a system, a pulse train consisting of four pulses should appear at the output. The spacing between the first two pulses is T_1, between the next two is T_2, and between the last two is T_3. Devise such a pulse-code generating system, using a tapped delay line and an OR gate.

b. Devise a multiple-pulse decoder such that when the pulse train in part *a* enters the system a *single* pulse is delivered at the output. HINT: Use a tapped delay line and an AND gate.

9-15 *a.* Verify that the circuit shown is an inverter by calculating the output levels corresponding to input levels of 0 and -6 V. What minimum value of h_{FE} is required? Neglect junction saturation voltages and assume an ideal diode.

b. If the reverse collector saturation current at 25°C is 5 μA, what is the maximum temperature at which this inverter will operate properly?

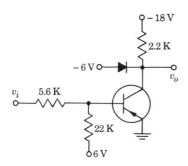

Prob. 9-15

9-16 The NOT circuit of an early (NORC) computer using vacuum tubes is indicated. The binary levels are -25 and $+10$ V, respectively. Verify that this circuit inverts the input waveform and yields an output with the levels $+10$ and -25 V, respectively.

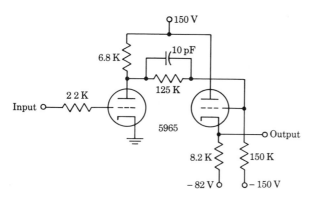

Prob. 9-16

9-17   Combine the transformer INVERTER of Fig. 9-17 with the capacitively coupled AND gate of Prob. 9-9 to form an AND gate with an INHIBITOR terminal, for positive pulse logic.   Add appropriate delay lines to ensure that the inhibit pulse begins earlier and lasts longer than the signal pulse.   HINT: Use two parallel diode paths with different delays in order to effectively "stretch" out the inhibit pulse.

9-18   Given a train of n equally spaced pulses, the first at $t = 0$ and the last at $t = (n - 1)T$.   (a) Draw a system in block-diagram form which will deliver only one pulse at $t = 0$.   HINT: Use a delay line and an INHIBITOR.   (b) Repeat part a for a system which will deliver a single pulse at $t = nT$.

9-19 a. Verify De Morgan's law [Eq. (9-30)] in a manner analogous to that given in the text in connection with the proof of Eq. (9-28).

  b. Prove Eq. (9-30) by constructing a truth table for each side and verifying that these two tables have the same outputs.

9-20   Verify the auxiliary Boolean identities in Table 9-4 (page 329).

9-21   Using Boolean algebra, verify

  a. $\bar{A} + B + \bar{A} + \bar{B} = A$
  b. $AB + AC + B\bar{C} = AC + B\bar{C}$

HINT: Multiply the first term on the left-hand side by $C + \bar{C} = 1$.

  c. $\overline{AB + BC + CA} = \bar{A}\bar{B} + \bar{B}\bar{C} + \bar{C}\bar{A}$

9-22   Using Boolean algebra, verify

  a. $(A + B)(B + C)(C + A) = AB + BC + CA$
  b. $(A + B)(\bar{A} + C) = AC + \bar{A}B$
  c. $AB + \bar{B}\bar{C} + A\bar{C} = AB + \bar{B}\bar{C}$

HINT: A term may be multiplied by $B + \bar{B} = 1$.

9-23   Given two N-bit characters which are available in parallel form.   Indicate, in block-diagram form, a system whose output is 1 if and only if all corresponding bits are equal, that is, only if the two characters are equal.

9-24 A, B, and C represent the presence of pulses.   The logic statement "A or B and C" can have two interpretations.   What are they?   In block-diagram form draw the circuit to perform each of the two logic operations.

9-25   A circuit has three input and one output terminals.   The output is 1 if any two of the three inputs are 1 and is 0 for any other combination of inputs.   Draw a block diagram of this logic circuit.

9-26   In block-diagram form draw a circuit to perform the following logic: If pulses A_1, A_2, and A_3 occur simultaneously or if pulses B_1 and B_2 occur simultaneously, an output pulse is delivered, provided that pulse C does not occur at the same time. No output is to be obtained if A_1, A_2, A_3, B_1, and B_2 occur simultaneously.

9-27   A single-pole double-throw switch is to be simulated with AND, OR, and INHIBITOR circuits.   Call the two signal inputs A and B.   A third input C receives

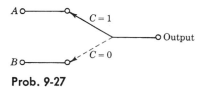

Prob. 9-27

the switching instructions in the form of a code: 1 (a pulse is present) or 0 (no pulse exists). It is desired that $C = 1$ set the switch to A and $C = 0$ set the switch to B, as indicated schematically. In block-diagram form show the circuit for this switch.

9-28 In block-diagram form draw a circuit which satisfies simultaneously the conditions a, b, and c as follows:

a. The output is excited if any pair of inputs A_1, A_2, and A_3 is excited, provided that B is also excited.

b. The output is 1 if any one of the inputs A_1, A_2, or A_3 is 1, provided that $B = 0$.

c. No output is excited if A_1, A_2, and A_3 are simultaneously excited.

9-29 In block-diagram form sketch a 2:1 divider circuit for a train of regularly spaced pulses. As basic building blocks use logic gates and/or delay lines.

9-30 *a.* For the illustrative NAND gate of Fig. 9-25a calculate the minimum value of h_{FE}, taking junction voltages into account.

b. What is the maximum noise voltage (superimposed upon the logic level) which will still permit the circuit to operate properly? Consider the following two cases: (1) a complete coincidence and (2) all inputs but one in the 1 state.

c. What is the maximum value of the source resistance which will still permit proper circuit operation? Assume a 0.6-V drop across a conducting diode.

9-31 The circuit shown uses silicon diodes and a silicon transistor. The input A or B is obtained from the output Y of a similar gate. (*a*) What are the logic levels? Take junction voltages into account. (*b*) Verify that the circuit satisfies the NAND operation. Assume $(h_{FE})_{min} = 15$. (*c*) What is the maximum allowable value of I_{CBO}? (*d*) Now neglect junction voltages and I_{CBO} and verify that the circuit satisfies the NOR operation.

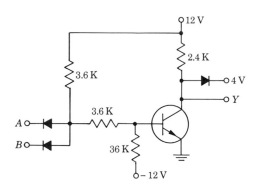

Prob. 9-31

9-32 Verify that the logic operations OR, AND, and NOT may be implemented by using only NOR gates.

9-33 The circuit shown uses silicon transistors. The input A or B is obtained from the output Y of a similar gate. Neglect I_{CBO} and the forward-biased junction voltages. (*a*) What are the logic levels? (*b*) Verify that the circuit satisfies NOR logic. What is $(h_{FE})_{min}$? (c) Verify that the circuit satisfies NAND logic. (d) How can the circuit be modified to become a three-input NAND gate (1) for positive or (2) for negative logic? (e) Show how to modify the circuit so that the output, for positive logic, is (1) AB or (2) $A + B$.

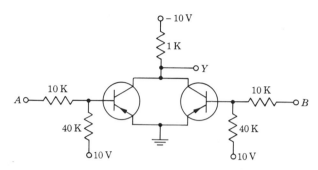

Prob. 9-33

9-34 What logic operation is performed by the circuit shown, which consists of interconnected NOR gates?

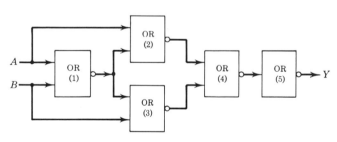

Prob. 9-34

9-35 a. Construct a flow chart as in Fig. 9-37 for adding 14 and 21 by means of two half adder-subtractors. Use arrows to indicate where each carry comes from.

b. Repeat part a for subtracting 14 from 21.

c. Repeat parts a and b for the numbers 5 and 11.

9-36 Indicate in block-diagram form a half-adder which uses only NOR gates. HINT: $C = \overline{\overline{A} + \overline{B}}$ and $S = \overline{\overline{A + B + C}}$. (Why?)

9-37 (a) Verify Eqs. (9-39) and (9-40) for the full binary adder. (b) Implement this three-input adder in block-diagram form.

9-38 (a) Construct a truth table for a complete subtractor having three inputs (A, B, and the borrow P) corresponding to Fig. 9-38. The outputs are the difference D and the output borrow P'. (b) Write Boolean expressions for D and P'.

9-39 A 256-position switch requires 8 binaries and 2,048 diodes. Show that the number of diodes can be reduced to 608 by the nonrectangular array indicated. Each block in the figure represents a diode matrix. How many diodes are in each block?

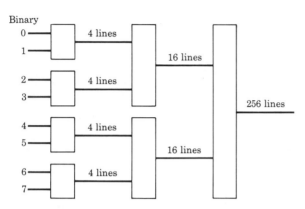

Prob. 9-39

9-40 Prove that an N-position code-operated switch requires $N \log_2 N$ diodes.

9-41 Draw the complete diode-matrix array whereby two FLIP-FLOPS will control five channels in accordance with the following truth table:

FLIP-FLOP inputs		Output channels
0	0	1, 3, 5
0	1	2, 4
1	0	1, 2, 3
1	1	3, 4

9-42 a. The RTL circuit of Fig. 9-48 uses a silicon transistor with worst-case values of $V_{BE}(\text{sat}) = 1.0\text{ V}$ and $V_{CE}(\text{sat}) = 0.5\text{ V}$. The inputs to this gate are obtained from the outputs of similar gates. Verify that this circuit obeys the negative NAND operation. What is the minimum value of h_{FE}?

b. If $h_{FE} = 20$, what is the maximum fan-out M?

c. Assuming that if the base-to-emitter voltage goes positive the transistor cannot be considered cut off, find the maximum fan-in N. Neglect I_{CBO}.

9-43 Verify that the DCTL circuit shown with the fan-in transistors in series satisfies the NAND operation. Assume that for the silicon transistors, $V_{CE}(\text{sat}) = 0.25$

V and $V_{BE}(\text{sat}) = 0.7$ V. Calculate the collector currents in each transistor when all inputs are high. The input to each base is taken from the output of a similar gate.

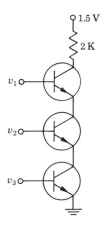

Prob. 9-43

9-44 *a.* The LLL circuit shown in Fig. 9-52a uses a silicon transistor with worst-case values of $V_{BE}(\text{sat}) = 1.0$ V and $V_{CE}(\text{sat}) = 0.5$ V. The voltage across any silicon diode (when conducting) is 0.7 V. Assume that $D1$ consists of two diodes in series. The circuit parameters are $V_{CC} = V_{BB} = 12$ V, $R = 15$ K, $R_2 = 100$ K, and $R_c = 2.2$ K. The inputs to this switch are obtained from the outputs of similar gates. Verify that the circuit functions as a positive NAND. In particular, for proper operation, calculate the minimum value of the clamping voltage V' and h_{FE}.

b. Will the circuit operate properly if $D1$ is (1) a single diode or (2) three diodes in series?

c. Replace $D1$ by a 15-K resistance and repeat part *a*. Compare the binary levels in parts *a* and *c*.

d. What is the maximum allowed fan-in, assuming that the diodes are ideal? What is a practical limitation on fan-in?

9-45 *a.* The LLL gate shown (RCA digital microcircuit 100) uses silicon devices with $V_{BE}(\text{sat}) = 0.7$ V, $V_{CE}(\text{sat}) = 0.3$ V, $V_\gamma = 0.5$ V, and the drop across a conduct-

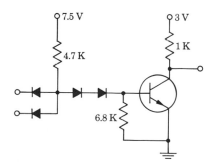

Prob. 9-45

ing diode $= 0.6$ V.   The inputs to this switch are obtained from the outputs of similar gates.   Verify that the circuit functions as a positive NAND and calculate $(h_{FE})_{min}$. Assume that the transistor is essentially cut off if the base-to-emitter voltage is at least 0.1 V smaller than the cutin voltage V_γ.

b. Assume that the diode reverse saturation current is equal to the transistor reverse saturation collector current.   Find $(I_{CBO})_{max}$.

c. If all inputs are high, what is the magnitude of noise voltage at the input which will cause the gate to malfunction?

d. Repeat part c if at least one input is low.

9-46   A digital-to-analog converter is indicated.   The resistance values R_n are proportional to the binary numbers 2^n.   Switches S'_n and S_n are controlled from the

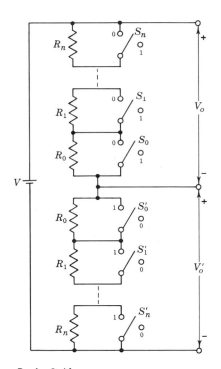

Prob. 9-46

relay representing the number 2^n and are in position 1 if this relay is excited.   The input V is a d-c voltage.

a. Prove that V_o is a voltage proportional to the binary number to which the relays are excited.

b. What does V'_o represent?

CHAPTER 10

10-1 The fixed-bias binary of Fig. 10-2 has the following parameters: $R_p = 20$ K, $R_1 = 1$ M, $R_2 = 0.25$ M, $V_{PP} = 200$ V, and $V_{GG} = 50$ V. The tube is a type 12AU7 (Fig. D-5). Compute the stable-state voltages and currents.

10-2 A fixed-bias binary is to be designed using a 12AT7 (Fig. D-4) tube operating from a 300-V power supply so as to give an output swing of 200 V. Choose $R_1 = 2R_2 \gg R_p$. (a) Find R_p. (b) Over what range of values of V_{GG} will one tube operate in clamp and the second tube be beyond cutoff? (c) In particular, if the grid bias of the OFF tube is -15 V, calculate V_{GG}.

10-3 The fixed-bias binary of Fig. 10-4 uses n-p-n 2N706A silicon transistors with $h_{FE} = 20$. The circuit parameters are $V_{CC} = 12$ V, $V_{BB} = 3$ V, $R_c = 1$ K, $R_1 = 5$ K, and $R_2 = 10$ K. Verify that one transistor is cut off and the other is in saturation and find the stable-state currents and voltages, if

    a. $V_{CE}(\text{sat}) = 0$ and $V_{BE}(\text{sat}) = 0$.

    b. $V_{CE}(\text{sat}) = 0.4$ V and $V_{BE}(\text{sat}) = 0.8$ V. (These are the values given in the 2N706A data sheets corresponding to the values of base and collector currents found in part *a.*)

    c. As the ambient temperature is decreased, what is the minimum value to which h_{FE} can fall and still maintain one transistor in saturation for parts *a* and *b*?

    d. As the temperature is increased, what is the maximum value to which I_{CBO} can increase before the condition is reached where neither transistor is cut off for parts *a* and *b*, respectively?

    e. What is the heaviest load in part *a* that the binary can drive (the minimum R from the collector to ground) and still have one transistor in saturation while the other is below cutoff?

10-4 The fixed-bias binary of Fig. 10-4 uses p-n-p silicon transistors with worst-case (maximum) values of $V_{CE}(\text{sat}) = -0.5$ V, $V_{BE}(\text{sat}) = -1.0$ V, $I_{CBO} = -10$ nA at 25°C, and zero base-to-emitter voltage at cutoff. The circuit parameters are $V_{CC} = V_{BB} = 6$ V, $R_c = 1.2$ K, $R_1 = 4.7$ K, and $R_2 = 27$ K.

    a. Find $(h_{FE})_{\min}$ and verify that one transistor is OFF and the other is ON. Find the stable-state currents and voltages.

    b. If the reverse-saturation current doubles every 10°C, what is the maximum temperature at which one transistor will remain OFF?

10-5 For the fixed-bias germanium-transistor binary of Fig. 10-4, assume that junction voltages at saturation and base-to-emitter voltages at cutoff and in the active region may all be taken as zero. The circuit is to operate properly (one transistor OFF and the other ON) over the temperature range -50 to $+75$°C and to just start malfunctioning at these extreme temperatures. The circuit is to be designed subject to the following specifications: $V_{CC} = 4.5$ V, $V_{BB} = 3.0$ V, $h_{FE} = 40$ at -50°C and 60 at $+75$°C, $I_{CBO} = 4$ μA at 25°C and doubles every 10°C, and the collector current is 10 mA. Find R_c, R_1, and R_2.

10-6 Regeneration is possible in the fixed-bias transistor flip-flop if the base-to-base voltage gain exceeds unity. Verify that this gain condition is satisfied provided that $h_{fe}R_c > R_1$. Assume that for each stage the current gain is $|A_I| \approx h_{fe} \gg 1$ and that the input resistance R_i is small compared with either R_1 or R_2.

10-7 *a.* For the fixed-bias binary of Fig. 10-4 verify that one transistor is OFF and the other is ON if the circuit parameters satisfy the following inequalities:

$$R_2 V_{CE}(\text{sat}) - R_1 V_{BB} + R_1 R_2 I_{CBO} < (R_1 + R_2) V_{BE}(\text{cutoff})$$

$$h_{FE} \left[\frac{V_{CC} - V_{BE}(\text{sat}) - I_{CBO} R_c}{R_c + R_1} - \frac{V_{BB} + V_{BE}(\text{sat})}{R_2} \right]$$

$$> \left[\frac{V_{CC} - V_{CE}(\text{sat})}{R_c} - \frac{V_{BB} + V_{CE}(\text{sat}) - I_{CBO} R_2}{R_1 + R_2} \right]$$

b. Verify that the voltage swing V_w at a collector is

$$V_w = \frac{R_1 V_{CC} + R_c V_{BE}(\text{sat}) - R_1 R_c I_{CBO}}{R_1 + R_c} - V_{CE}(\text{sat})$$

10-8 Germanium transistors with $(h_{FE})_{\min} = 40$ are used in the fixed-bias flip-flop with collector catching diodes (Fig. 10-6). The circuit parameters are $V_{CC} = 18$ V, $V = V_{BB} = 6$ V, $R_c = 1.5$ K, $R_1 = 5$ K, and $R_2 = 25$ K. Neglect the voltage drop across a forward-biased junction.

a. Verify that if one transistor is cut off, the other is in saturation. Find the stable-state voltages and currents, *including the currents in the two diodes.*

b. What is the heaviest load the binary can drive (the minimum R from collector to ground) and still maintain the output swing in part *a*?

c. If $I_{CBO} = 5$ μA at 20°C and doubles every 10°C, what is the maximum temperature at which one transistor will remain cut off?

10-9 The self-biased binary of Fig. 10-7 uses *n-p-n* silicon transistors having worst-case (maximum) values of $V_{CE}(\text{sat}) = 0.4$ V, $V_{BE}(\text{sat}) = 0.8$ V, and zero base-to-emitter voltage for cutoff. The circuit parameters are $V_{CC} = 20$ V, $R_c = 4.7$ K, $R_1 = 30$ K, $R_2 = 15$ K, and $R_e = 390$ Ω.

a. Find the stable-state currents and voltages.

b. Find the minimum value of h_{FE} required to give the values in part *a*.

c. As the temperature is increased, what is the maximum value to which I_{CBO} can increase before the condition is reached where neither transistor is OFF?

10-10 The self-biased flip-flop of Fig. 10-7 uses silicon transistors whose junction voltages are given in Table 6-1 (page 219). The circuit parameters are $(h_{FE})_{\min} = 30$, $V_{CC} = 6$ V, $R_c = 1$ K, $R_1 = 9.1$ K, $R_2 = 20$ K, and $R_e = 430$ Ω.

a. Find the stable-state currents and voltages.

b. Find the heaviest load that the binary can drive (the minimum R from collector to ground) and still have one transistor in saturation while the other is below cutoff.

10-11 *a.* The self-biased binary of Fig. 10-7 uses silicon transistors with $(h_{FE})_{\min} = 20$. The junction saturation voltages and I_{CBO} may be taken as zero. The circuit is to be designed subject to the following specifications: $V_{CC} = 18$ V, $R_1 = R_2$, the maximum collector current is 10 mA, the base current of the ON transistor is twice the minimum value required for saturation, and for the OFF transistor $V_{BE} = -1$ V. Neglecting the loading of R_1 and R_2 on the collector circuit, find R_e, R_c, and R_1.

b. Using the component values found in part *a* and now taking into account the loading of R_1 and R_2 and the junction voltages given in Table 6-1 (page 219), evaluate I_C, I_B, and V_{BE} for the OFF transistor.

10-12 A self-biased binary using a 12AU7 tube (Fig. D-5) is indicated. Calculate the stable-state voltages if the circuit parameters are

a. $V_{PP} = 200$ V, $R_p = 42$ K, $R_1 = 270$ K, $R_2 = 100$ K, and $R_k = 12$ K.

b. $V_{PP} = 210$ V, $R_p = 5.6$ K, $R_1 = R_2 = 62$ K, and $R_k = 11$ K.

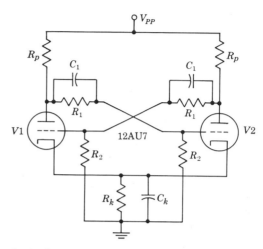

Prob. 10-12

10-13 Design a self-biased binary using a 6CG7 tube (in the circuit of Prob. 10-12) subject to the following specifications: the supply voltage is 300 V, the cathode voltage is 75 V, the plate swing is 100 V, and $R_1 = 200$ K. Calculate (a) R_p, (b) R_k, (c) the maximum value of R_2, and (d) the minimum value of R_2.

10-14 a. For the tube circuit in Prob. 10-12 estimate the size of the speed-up capacitances C_1 and the maximum frequency at which each binary can be triggered. Assume that $\mu = 20$, $r_p = 10$ K, and the capacitance between any two electrodes is 4 pF.

b. If speed-up capacitances of 50 pF are used in the transistor flip-flop of Fig. 10-4, estimate the maximum triggering frequency.

10-15 The nonsaturated binary circuit of Fig. 10-12 has the following parameters: $V_{CC} = 22.5$ V, $V_{RR} = 1.5$ V, $V_Z = 4$ V, $h_{FE} = 40$, $R_c = 5.6$ K, $R_1 = 68$ K, and $R_2 = 20$ K. Neglect the voltage drop across a forward-biased junction. Verify that the transistors do not enter the saturation region. Calculate the transistor currents and the current in each diode.

10-16 Consider a self-biased nonsaturated flip-flop obtained from Fig. 10-12 by setting $V_{BB} = 0$ and by adding a common emitter resistor R_e to ground as in Fig. 10-7. The circuit has the following parameters: $V_{CC} = 25$ V, $V_Z = 4.3$ V, $h_{FE} = 50$, $R_c = 2.2$ K, $R_1 = R_2 = 15$ K, and $R_e = 470$ Ω. Neglect the voltage drop across a forward-biased junction. Verify that the transistors do not enter the saturation region. Calculate the transistor currents and the current in each diode.

10-17 In the nonsaturated binary shown, the avalanche diodes $D1$ and $D2$ are nominally identical, as are diodes $D3$ and $D4$. The breakdown voltage V'_Z of $D3$ and $D4$ is larger than the breakdown voltage V_Z of $D1$ and $D2$. Verify that the transistors do not enter the saturation region.   HINT: Assume that $D3$ and $D4$ are always in the breakdown region and that either $D1$ or $D2$ (depending upon the state of the flip-flop), but not both, is in the breakdown region.   Then verify these assumptions.

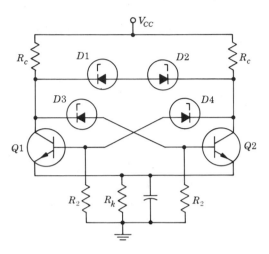

Prob. 10-17

10-18   Consider the triggering circuits of Fig. 10-14.

a. Which circuit, (*a*) or (*b*), has the greater triggering sensitivity, that is, will be triggered by the smaller pulse?

b. Which circuit will malfunction first as the trigger amplitude is increased? Explain.

c. How does the hybrid symmetrical triggering configuration shown differ from the circuits of Fig. 10-14?

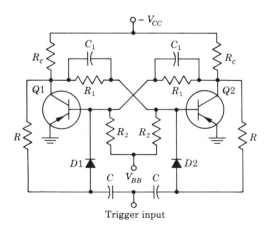

Prob. 10-18

d. Show that this hybrid circuit combines the sensitivity of base triggering with the larger pulse-amplitude-variation possibility of collector triggering.

e. A diode is sometimes placed across R. Indicate the polarity of this diode and explain why it is used.

10-19 *a.* Consider the symmetrical emitter triggering circuit of Fig. 10-17 but with $R_c = 3R_e$, $R_1 = 2R_2$, and $V_{CC} = 6$ V. Indicate all the circuit voltages in the quiescent state and indicate also (in parentheses) the voltages immediately after a 5-V positive step is applied. Make the same simplifying assumptions as were made in connection with Fig. 10-17. Verify that shortly after the application of the step a transition will take place.

b. Repeat part *a* for a 25-V step. What limits the maximum size of the input step?

c. What limits the minimum size of the input step?

10-20 A variation of the technique of Fig. 10-17 for triggering symmetrically without the aid of steering diodes is indicated. For simplicity we shall select $R_1 = R_2 = R$ and shall assume that these are large enough so that their loading at the tube plates is negligible.

a. Verify that the voltages noted on the circuit correspond to the stable state with $V1$ ON and $V2$ OFF.

b. A 90-V negative step is applied. Assuming that the voltages across the speedup capacitors do not change, indicate (in parentheses) all voltage values immediately after the application of the step. HINT: Assume that both tubes are cut off and then verify this assumption.

c. As the negative step at A decreases because C' charges, verify that $V2$ goes ON but $V1$ stays OFF, so that a transition takes place.

d. Assume that the commutating capacitors are missing, so that the dominant capacitances are the effective capacitances at the grids. Verify that a transition now will *not* take place when a 90-V negative step is applied.

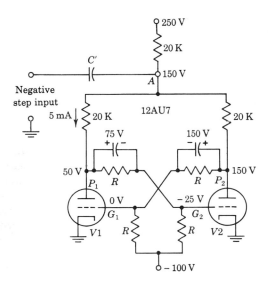

Prob. 10-20

e. Verify that the circuit (with the speed-up capacitors present) is much more sensitive to a negative than to a positive step. For example, show that a 1-V negative step is sufficient to just take $V2$ out of cutoff whereas a positive step of about 50 V is required to bring $V2$ out of cutoff.

10-21 Find the voltages and currents in the direct-coupled binary of Fig. 10-18 if $V_{CC} = 3$ V and $R_c = 4$ K. Assume silicon transistors whose characteristics are given in Table 6-1 (page 219). Calculate the output swing.

10-22 The input v to the Schmitt trigger of Fig. 10-19 is the set of pulses shown. Plot v_o versus time. Assume that $V_1 = 8$ V, $V_2 = 6$ V, $V_{YY} = 20$ V, the output swing is 15 V, and $A2$ is ON at $t = 0$.

Prob. 10-22

10-23 For the illustrative example in Sec. 10-12, calculate the value of the resistance R_{k2} which will reduce the hysteresis to zero.

10-24 A cathode-coupled binary (Fig. 10-22) uses a 6SN7 tube (Fig. D-2) and has the following parameters: $V_{PP} = 240$ V, $R_{p1} = R_{p2} = 20$ K, $R_k = 30$ K, and $a = \frac{1}{3}$. (a) Evaluate V_1 and V_2. (b) For an input sine wave $v = 100 \sin \omega t$, plot v, v_{PN1}, v_{KN}, and v_{PN2} as a function of $\alpha \equiv \omega t$. Put numerical values of voltage and angle on your graph.

10-25 A Schmitt trigger (Fig. 10-22) uses a type 5965 tube (Fig. D-9) and has the following parameters: $V_{PP} = 250$ V, $R_{p1} = R_{p2} = 10$ K, and $R_k = 15$ K. The attenuator ratio a is selected so that with $v = 0$, $v_{GK2} = -1$ V. (a) Calculate V_1 and V_2. (b) Calculate the value of R_{k1} required to eliminate hysteresis. (c) If $R_{k1} = 0$, calculate the value of R_{k2} for which $V_H = 0$.

10-26 (a) For the silicon Schmitt circuit in the illustrative example beginning on page 399, find i_{C1} and i_{B1} when $v = V_1$. Is $Q1$ in saturation, in the active region, or at cutoff? (b) Repeat part a if $v = V_2$. (c) If R_{c2} is increased to 4 K, verify that $Q2$ is in saturation when $Q1$ is OFF. Find the values of i_{C2} and i_{B2}.

10-27 Consider the emitter-coupled binary of Fig. 10-24 with silicon transistors having $h_{FE} = 20$. The circuit parameters are $V_{CC} = 12$ V, $R_s = 4$ K, $R_{c1} = 3$ K, $R_{c2} = 1$ K, $R_1 = 2$ K, $R_2 = 8$ K, and $R_e = 5$ K. Calculate (a) V_1, (b) V_2. (c) For an input sine wave $v = 10 \sin \omega t$, plot v, v_{CN1}, v_{EN}, and v_{CN2} as a function of $\alpha \equiv \omega t$. Put numerical values of voltage and angle on your graph.

10-28 Consider the Schmitt trigger of Fig. 10-24 with germanium transistors having $h_{FE} = 40$. The circuit parameters are $V_{CC} = 55$ V, $R_s = 3.9$ K, $R_{c1} = 12$ K, $R_{c2} = 2$ K, $R_1 = 39$ K, $R_2 = 180$ K, and $R_e = 39$ K. Find (a) V_1, (b) V_2, (c) R_{e1} to eliminate hysteresis, (d) R_{e2} to give $V_H = 0$.

10-29 In the illustrative example of the emitter-coupled binary of Sec. 10-13, the value of h_{FE} is increased by 10 percent to 33. Find the percentage change in (a) V_1 and (b) V_2. HINT: Expand V (V_1 or V_2) in a power series in $x \equiv 1/h_{FE}$. Retain only the first two terms and then take the derivative with respect to h_{FE} to obtain dV/V as a function of dh_{FE}/h_{FE}.

10-30 Design an emitter-coupled binary using silicon transistors with $(h_{FE})_{min} = 80$. Arbitrarily select $V_{CC} = 30$ V, $R_1 = 24$ K, and $R_2 = 6$ K. It is required that $V_1 = 4.5$ V, $V_2 = 1.5$ V, that the output swing be 10 V, and that the output stage $Q2$ operate in its active region. Find R_{c1}, R_{c2}, and R_e.

CHAPTER 11

11-1 *a.* In the monostable circuit of Fig. 11-1, the resistor R is connected to an auxiliary supply V_1 instead of to V_{YY}. If $A2$ is in saturation or clamp and if $A1$ is OFF in the stable state, verify that the gate time T is given by Eq. (11-2) with V_{YY} replaced by V_1.

b. For a tube one-shot with $V_1 = 0$, what is T? Define all the terms in the equation.

11-2 *a.* For the plate-coupled univibrator of Fig. 11-5, take the loading of R at the plate of P_1 into account. Verify that the current I_{P1} in $V1$ at $t = 0+$ is obtained from the intersection of the load line whose equation is

$$V_{PP} \frac{R + 2R_p}{R + R_p} = V_P + \frac{RR_p}{R + R_p} I_P$$

and the grid characteristic $V_{G1} = V_N$ given in Eq. (11-4).

b. Show that I_1 in R_p is given at $t = 0$ by

$$I_1 = \frac{I_P R}{R + R_p} - \frac{V_{PP}}{R + R_p}$$

c. For the example on page 411, for the case where $R = 10R_p = 200$ K, calculate $I_1 R_p$ and compare with the value of 70 V obtained with $R = \infty$. Draw the waveform v_{P1} and compare with that in Fig. 11-6*b*.

11-3 A plate-coupled one-shot using a 12AU7 has the following parameters: $V_{PP} = 300$ V, $V_{GG} = 45$ V, $R_p = 15$ K, $R = 1$ M, $C = 0.001$ μF, $R_1 = 900$ K, and $R_2 = 100$ K. (*a*) Calculate and plot the waveforms at each grid and plate to scale. (*b*) Find the width of the output pulse.

11-4 A plate-coupled monostable multi using a 5965 tube has the following parameters: $V_{PP} = 250$ V, $R_p = 25$ K, $R = 1$ M, $R_1 = 0.5$ M, and $R_2 = 1$ M. The input trigger frequency is 100 Hz. Find (*a*) V_{GG} so that the voltage at G_1 is -15 V in the stable state, (*b*) C so that the gate width is 2 msec, and (*c*) the overshoot at G_2. (*d*) Plot to scale the waveforms at G_2, P_1, P_2, and G_1.

11-5 A plate-coupled univibrator has the waveform shown at G_2. The supply voltage is 250 V and $r_G = 1$ K. (*a*) Draw to scale the waveform at P_1 and (*b*) evaluate R_p.

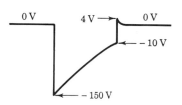

Prob. 11-5

11-6 A plate-coupled monostable multi using a 12AT7 tube is operated from a 500-Hz trigger source and a 250-V supply voltage. It is desired to have a gate width of 1,500 μsec and a swing at each plate of 150 V. Choose the voltage at G_1 in the stable state to be about 50 percent below cutoff. Find (a) R_p, C, R, $a \equiv R_2/(R_1 + R_2)$, and V_{GG} and (b) the overshoot at G_2. (c) Plot to scale the waveforms at each grid and plate.

11-7 Consider the one-shot whose parameters are given in Prob. 11-3, except that V_{GG} is adjustable. For the minimum gate width, calculate (a) V_{GG}, (b) T, (c) I'_G, and the swing at P_2. (d) Repeat the above calculations for the maximum gate width.

11-8 *a.* Verify Eq. (11-10) for the delay time T of a collector-coupled monostable multi as a function of I_{CBO}.

b. By what percentage change is T decreased as the temperature is increased from a value for which I_{CBO} is negligible to the temperature for which $I_{CBO} = 100$ μA? Assume $V_{CC} = 22.5$ V and $R = 120$ K.

c. Indicate a method of temperature compensation using a resistor whose resistance increases with temperature.

11-9 (a) If V_{CC} is large compared with the junction voltages in a collector-coupled one-shot, prove that the gate width is given by

$$T \approx \tau \ln 2 + \frac{\tau}{2V_{CC}} [2V_\gamma - V_{BE}(\text{sat}) - V_{CE}(\text{sat})]$$

and

$$\frac{\Delta T}{T} = \frac{2 \Delta V_\gamma - \Delta V_{BE}(\text{sat}) - \Delta V_{CE}(\text{sat})}{1.38 V_{CC}}$$

Consider the following specifications: $V_{CC} = 6$ V, $R = 10$ K, $I_{CBO} = 3$ nA at 25°C and doubles every 10°C, V_γ and $V_{BE}(\text{sat})$ decrease 2 mV/°C, and $V_{CE}(\text{sat})$ increases 1 mV/°C. Find the percentage decrease in T from 25 to 175°C due to the temperature variation of (b) the junction voltage and (c) the reverse saturation current.

11-10 *a.* In a collector-coupled monostable multi the resistor R is connected to a voltage source V but the collector supply is V_{CC}. Taking I_{CBO} into account, prove that under the condition that $Q1$ is driven into saturation the delay time T is given by

$$T = \tau \ln \frac{V + I_{CBO}R + V_{CC} - V_{CE}(\text{sat}) - V_\sigma - I_{CBO}R_c}{V + I_{CBO}R - V_\gamma}$$

b. If junction voltages are neglected, if $R \gg R_c$, and if $V = V_{CC}$, prove that the expression in part *a* reduces to Eq. (11-10).

11-11 In order to reduce the recovery time in the collector-coupled one-shot, an avalanche diode is connected to the collector of $Q1$. If the Zener voltage is $V_{CC}/2$, prove that (a) the recovery time (not the time constant) is given by $0.69(R_c + r_{bb'})C$ and (b) the delay time is $T = 0.41RC$.

11-12 Consider a collector-coupled *n-p-n* delay multi taking I_{CBO} into account but neglecting the junction voltages across a saturated transistor. If $Q1$ is cut off by a base voltage V_F and $Q2$ is in saturation in the stable state, prove that

a. $V_F = -\dfrac{R_1 V_{BB}}{R_1 + R_2} \left(1 - \dfrac{I_{CBO}R_2}{V_{BB}}\right)$

b. $R \leq \dfrac{h_{FE}R_c}{1 - [R_c/(R_1 + R_2)][(V_{BB} - I_{CBO}R_2)/V_{CC}]}$

11-13 In the quasi-stable state of a collector-coupled one-shot $Q1$ is to be in saturation. Neglecting I_{CBO}, verify that the minimum value of R is given by

$$\frac{R_{\min}}{R_c} = \frac{2V_{CC} - V_{CE}(\text{sat}) - V_\sigma}{h_{FE}I_{B1}R_c + V_{CE}(\text{sat}) - V_{CC}}$$

where

$$I_{B1} = \frac{V_{CC} - V_\sigma}{R_1 + R_c} - \frac{V_{BB} + V_\sigma}{R_2}$$

11-14 A collector-coupled monostable multi using n-p-n silicon transistors has the following parameters: $V_{CC} = 12$ V, $V_{BB} = 3$ V, $R_c = 2$ K, $R_1 = R_2 = R = 20$ K, $h_{FE} = 30$, $r_{bb'} = 200\ \Omega$, and $C = 1{,}000$ pF. Neglect I_{CBO}.
    a. Calculate and plot to scale the waveshapes at each base and collector.
    b. Find the width of the output pulse.

11-15 *a.* A collector-coupled one-shot using p-n-p germanium transistors has the following parameters: $V_{CC} = 3$ V, $V_{BB} = 9$ V, $R = 2.7$ K, $R_c = 270\ \Omega$, $R_1 = 1$ K, $R_2 = 15$ K, $C = 0.01\ \mu$F, and $h_{FE} = 25$. Neglect saturation junction voltages and $r_{bb'}$. Calculate and plot the waveforms at each base and collector.
    b. Repeat part *a* taking junction voltages into account (page 219) and assuming $r_{bb'} = 100\ \Omega$.

11-16 Design a collector-coupled one-shot using n-p-n transistors. Neglect I_{CBO} and the voltage drops across a saturated transistor. Assume $(h_{FE})_{\min} = 20$. In the stable state $Q1$ is OFF with a base-to-emitter voltage of -1 V and $Q2$ is in saturation with a base current which is 50 percent in excess of its minimum required value. In the quasi-stable state $Q2$ is OFF and $Q1$ is in saturation with a base current which is 50 percent in excess of its required minimum value. The collector supply is $V_{CC} = 6$ V and the saturated collector current is 2 mA. The delay time is 3,000 μsec. Choose $R_1 = R_2$ and find R_c, R, V_{BB}, R_1, and C.

11-17 A collector-coupled univibrator has the waveform shown at the collector of $Q1$. The base-spreading resistance of the n-p-n germanium transistors used is 200 Ω. (*a*) Draw to scale the waveform at the base of $Q2$ and (*b*) evaluate R_c.

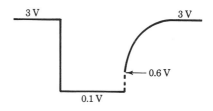

3 V   3 V

0.6 V

0.1 V

Prob. 11-17

11-18 Consider the one-shot whose parameters are given in Prob. 11-15*b*, except that V_{BB} is adjustable. (*a*) Find the minimum value of V_{BB}. This voltage is determined by the fact that $Q1$ is OFF in the stable state. (*b*) Find the maximum value of V_{BB}. This voltage is determined by the condition that $Q1$ be in saturation during the quasi-stable state.

11-19 *a.* A collector-coupled monostable multi has the following parameters: $V_{CC} = 12$ V, $R_c = 2$ K, $R_1 = R_2 = R = 20$ K, $h_{FE} = 20$, $r_{bb'} = 200\ \Omega$, and $C = 1,000$ pF. Neglect forward-biased junction voltages and I_{CBO}. In the quasi-stable state $Q1$ is to be in its active region with a collector current of 4 mA. Find V_{BB}.

b. Calculate and plot the waveforms at each base and collector. NOTE: Since $Q1$ is *not* in saturation, the waveshapes will be somewhat different from those indicated in Fig. 11-10.   In particular, v_{C1} will not be constant in the quasi-stable state because the output impedance at the collector of $Q1$ is now approximately 2 K rather than zero.

c. Calculate the width T of the gate.   NOTE: Equation (11-9) is not valid.

d. Why is a one-shot usually designed so that $Q1$ saturates in the quasi-stable state?

11-20 *a.* If the current I_R in R in Fig. 11-12 is taken into account in the derivation of the saturation base current I'_B, show that Eq. (11-16) remains valid if $r_{bb'}$ is multiplied by $1 + k$, where $k \equiv I_R/I'_B$.

b. Usually $R \gg R_c$. However, even if $R = R_c$, show that (subject to the reasonable approximations that $R_c \gg r_{bb'}$ and that the saturation junction voltages are small compared with V_{CC}) $k \approx 1$ and hence Eq. (11-16) is valid.

c. For the illustrative problem on page 420 evaluate I'_B. HINT: To calculate I_R use the approximate value of V'_{BE}.

11-21 A diode D is added in series with the base of $Q2$ of a collector-coupled one-shot in order to prevent base-to-emitter breakdown during the quasi-stable state. (*a*) Verify that the delay T is given by Eq. (11-8) with the diode voltage V_D subtracted from both the numerator and the denominator. (*b*) What is the recovery time constant τ'?

11-22 The circuit shown is the direct-connected transistor monostable multi corresponding to the DCTL binary of Sec. 10-10. (*a*) Sketch the waveforms at both bases and collectors and label the voltages with symbols as in Fig. 11-10. (*b*) Find an expression for the delay T. (*c*) What are the advantages and disadvantages of this circuit over the circuit in Fig. 11-9?

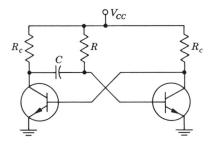

Prob. 11-22

11-23 The emitter-coupled monostable multi of Fig. 11-13 has the following parameters: $V_{CC} = 6$ V, $R_{c1} = R_{c2} = R_e = 3$ K, $R = 50$ K, $V = 2.8$ V, and $C = 0.01\ \mu$F. Silicon n-p-n transistors with $h_{FE} = 50$ and $r_{bb'} = 100\ \Omega$ are used. A trigger is applied at $t = 0$.

a. Assume that $Q1$ is OFF and $Q2$ is ON at $t = 0-$. Calculate the node voltages (the voltages with respect to ground at each collector base and emitter). Using your calculated values verify that $Q1$ is indeed OFF and $Q2$ is in saturation.

b. Assume that $Q1$ is in the active region and $Q2$ is OFF at $t = 0+$. Calculate the node voltages and verify that $Q1$ is indeed in the active region and $Q2$ is OFF.

c. Calculate the node voltages at $t = T-$.

d. Calculate the node voltages at $t = T+$.

11-24 Design an emitter-coupled one-shot using *n-p-n* silicon transistors with $h_{FE} = 20$ to meet the following specifications: $V_{CC} = 12$ V, the swing at the collector of $Q2$ is 5 V, the saturation collector current of $Q2$ is 10 mA, the base current of $Q2$ is 50 percent more than needed for saturation, the bias V is chosen midway between V_{min} and V_{max}, and the gate width is 1,000 μsec. Arbitrarily choose $R_{c1} = R_{c2}$. Calculate (a) R_{c1}, (b) R_e, (c) R, and (d) C. (e) What is the swing at the collector of $Q1$?

11-25 The emitter-coupled monostable circuit of Fig. 11-13 uses *p-n-p* transistors with $h_{FE} = 40$. Neglect the voltage of a forward-biased junction and assume $r_{bb'} = 0$. The circuit parameters are $V_{CC} = 12$ V, $R_{c1} = 3.6$ K, $R_{c2} = 2.2$ K, $R_e = 1$ K, and $R = 68$ K. The bias V is adjusted to be midway between V_{min} and V_{max}. A trigger is applied at $t = 0$. Calculate all the voltage levels and plot the waveforms corresponding to those in Fig. 11-14.

11-26 Verify Eq. (11-55) for V_{min}.

11-27 For the circuit of Prob. 11-23 find (a) V_{max}, (b) V_{min}, and (c) an equation for T as a function of V. (d) For a maximum gate width of 150 μsec, what is C? (e) From part c find V for $T = 0$.

11-28 If $T = 0$, show from Eq. (11-58) that $V_{EN1} = (V_{EN1})_{min}$ and hence that $V = V_{min}$.

11-29 The relationship between an independent variable x and a dependent variable y is quadratic and of the form $y = Ax(1 - Bx)$, with x variable from 0 to x_m. A straight line is drawn between the end points of this curve. The *fractional error* e is defined as the difference between the curve and the straight line divided by the maximum value of the dependent variable. Show that the maximum error e_d occurs at $x = x_m/2$ and is given by $e_d = \frac{1}{4}Bx_m$.

11-30 *a.* For the cathode-coupled monostable multi of Fig. 11-19, the grid current is I'_{G2} and the plate current is I'_2 at $t = T+$. Assume zero grid current at

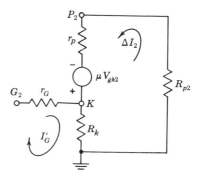

Prob. 11-30

$t = T-$ and let I_2 equal the plate current in $V2$ corresponding to $V_{GK2} = 0$ at $t = T-$. If $\Delta I_2 \equiv I_2' - I_2$, show that the equivalent circuit from which to calculate I_{G2}' and I_2' is as indicated. From this circuit verify Eqs. (11-67) and (11-68).

b. Equate the abrupt change in voltage at $t = T$ at P_1 with that at G_2. Why? Prove that this equation may be put into the form

$$I_{G2}'[R_{p1} + r_G + (1 + \gamma)R_k] = I_1(R_{p1} + R_k) - I_2 R_k + V_{\gamma 2}$$

c. Verify that the minimum current in $V1$ required to drive $V2$ OFF at $t = 0+$ is given by Eq. (11-70).

d. By combining Eq. (11-70) with the equation in part *b*, verify Eq. (11-67).

11-31   *a.* Verify Eq. (11-73) for the delay T of a cathode-coupled monostable multi.

b. Prove that T may be put into the form

$$T = \tau \ln \frac{1 + (I_1 - I_o)R_{p1}/V'}{1 - (I_1 - I_o)R_k/V'}$$

where I_o is defined in Eq. (11-70) and where

$$V' \equiv V_{PP} - I_o R_k - V_{\gamma 2}$$

c. By expanding T in a power series $[\ln (1 + x) = x + x^2/2 + \cdots]$ show that

$$\frac{T}{\tau} \approx \frac{R_{p1} + R_k}{V'} (I_1 - I_o) \left[1 - \frac{R_{p1} - R_k}{2V'} (I_1 - I_o) \right]$$

d. Note that $T = 0$ at $I_1 = I_o$ where $V = V_{\min}$ and that T varies approximately linearly with I_1. Using Eq. (11-66) find the linearity error at the current corresponding to $V_{\max}$ in the example on page 434.

11-32   *a.* In a cathode-coupled one-shot prove that it is possible for $V_{\min}$ to be negative.

b. If a 6CG7 tube is used with $V_{PP} = 300$ V and $R_{p1} = R_{p2} = 30$ K, find R_k so that $V_{\min} = 0$.

11-33   A monostable cathode-coupled multi uses a 6CG7 tube with $V_{PP} = 250$ V and $R_{p1} = R_{p2} = R_k = 10$ K. Assume that $\mu = 20$, $r_p = 7$ K, and $r_G = 1$ K. *(a)* Find $V_{\min}$ and $V_{\max}$. *(b)* Find the voltage swings at each plate, for both maximum and minimum gate widths. *(c)* Plot to scale the plate, grid, and cathode waveforms for the maximum gate width.

11-34   A cathode-coupled univibrator using a 5965 with $R_{p1} = R_{p2} = R_k$ and $V_{PP} = 250$ V is triggered at the rate of 500 Hz. It is desired that the swing at P_2 be 75 V. The waveforms are to have very little overshoot and are to be symmetrical (the OFF time to equal the ON time). Assume $\mu = 40$, $r_p = 7$ K, and $r_G = 250$ Ω. Find reasonable values for *(a)* R_k, *(b)* V, *(c)* C, and *(d)* R. *(e)* If $V = 1.1 V_{\min}$, calculate and plot to scale the waveforms at the plates, at the cathode, and at the grid of $V2$.

11-35   *a.* The resistor R in a cathode-coupled monostable multi is coupled to an auxiliary voltage U instead of V_{PP}. The voltage U is small enough so that $V2$ is not in clamp but $V1$ is below cutoff.   Prove that the delay is given by

$$\frac{T}{\tau} = \ln \frac{I_1 R_{p1}}{U - V_{\gamma 2} - I_1 R_k}$$

b. For what values of the bias voltage V (with respect to ground) on the grid of $V1$ will this circuit become astable?

11-36   Consider a collector-coupled astable multi using *p-n-p* transistors.   The circuit and device parameters are $V_{CC} = 15$ V, $R_c = 3$ K, $R_1 = 40$ K, $R_2 = 20$ K, $h_{FE} = 30$, $r_{bb'} = 0$, $I_{CBO} = 0$, and all forward-biased junction voltages may be neglected.

a. Calculate and plot to scale the waveforms at the base and collector of one transistor.

b. If the multi is to oscillate at a frequency of 5 kHz, what is the value of the coupling capacitance?   (Assume that $C_1 = C_2$.)

11-37   Consider a symmetrical collector-coupled astable multi using *n-p-n* silicon transistors.   The circuit and device parameters are $V_{CC} = 6$ V, $R_c = 560$ Ω, $R = 5.6$ K, $C = 50$ pF, $h_{FE} = 40$, and $r_{bb'} = 200$ Ω.   Calculate (*a*) the waveforms at the base and collector of one transistor and plot to scale, (*b*) the frequency of oscillation, and (*c*) the recovery time (the 10 to 90 percent time).

11-38   Verify Eq. (11-78) for the period T of an astable transistor multi using two supply voltages.

11-39   *a.* The free-running transistor multi shown cannot block.   Discuss this circuit and explain why it is impossible for both transistors to remain in saturation simultaneously.

b. Verify that the period of oscillation is given by

$$T = 2RC \ln 2$$

if all forward-biased junction voltages are neglected and if $R_c \ll R$.

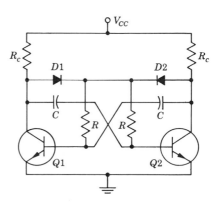

Prob. 11-39

11-40 For the free-running plate-coupled multi with R_1 and R_2 returned to a supply U instead of V_{PP}, prove that Eq. (11-82) for T is valid provided that U replaces V_{PP}.

11-41 A 12AU7 astable multivibrator has the waveform shown at one of its plates. The voltages given are the voltages with respect to ground. Find the value of the plate-circuit resistances of the tubes. Draw the waveform at the grid, and mark the voltage with respect to ground at every important point on the waveform, for example, flat portions, sudden jumps, etc.

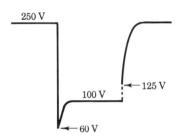

Prob. 11-41

11-42 A symmetrical astable plate-coupled multi using a 6CG7 has the following parameters: $V_{PP} = 150$ V, $R_p = 10$ K, $R = 1$ M, and $C = 0.001$ μF. Calculate the period and the overshoots, and plot the plate and grid waveforms.

11-43 Calculate the frequency of the multi of Prob. 11-42 as V_{PP} is adjusted from 50 to 250 V in 50-V steps.

11-44 An astable plate-coupled multi using a 12AT7 has the following parameters: $V_{pp} = 240$ V, $R_{p1} = R_{p2} = 12$ K, $R_1 = 1$ M, $R_2 = 2$ M, $C_1 = C_2 = 0.001$ μF. Calculate the gate widths T_1 and T_2 and the overshoots. Plot the plate and grid waveforms.

11-45 In the circuit of Fig. 11-31, $V_{CC} = 40$ V, $R_2 = R_1 \ll R''$, $C = 0.2$ μF, $R_{c2} = 300$ Ω, $R' = 3$ K, $R'' = 2$ K, and $R_{e1} = R_{e2} = 5$ K. Germanium n-p-n transistors with $h_{FE} = 60$ are used.

    a. Calculate all the voltage levels of the waveforms generated.

    b. Calculate the frequency of oscillation.

    c. Verify that Q2 goes into its active region and Q1 saturates.

CHAPTER 12

12-1 *a.* A tunnel diode has the idealized piecewise linear characteristic shown, with $I_P = 10I_V$, $V_V = 7V_P$, and $V_F = 8.5V_P$. Reproduce the characteristic on graph paper and deduce, by graphical means, the resultant volt-ampere characteristic of two such diodes in series. HINT: The current is the same in the two diodes, whereas the composite voltage is the sum of the two individual tunnel-diode voltages. Note that the composite characteristic will display more than one peak and more than one valley.

    b. Repeat part *a* if the diodes are placed in parallel.

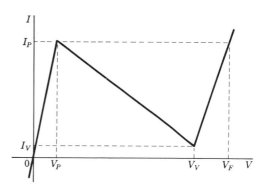

Prob. 12-1

12-2 *a.* Two tunnel diodes have the idealized form given in Prob. 12-1. However, the diodes have different peak currents I_{P1} and I_{P2}. Find graphically the composite volt-ampere characteristic which results if the diodes are placed in series. Assume $I_{P2} = 0.8I_{P1}$.

b. Generalize the result of part *a* if *n* diodes are operated in series.

12-3 Two tunnel diodes with characteristics as given in Prob. 12-1 are operated in series opposing. Find graphically the composite volt-ampere characteristic. Assume that in the reverse direction the characteristic continues, as shown, to be a straight line passing through the origin.

12-4 A resistance $R = 2V_V/I_P$ is placed in series with a tunnel diode whose volt-ampere characteristic is given in Prob. 12-1. Draw a plot of the volt-ampere characteristic of the combination.

12-5 The composite characteristic for the tunnel-diode pair shown is a plot of i (the resistor current) versus v_1. Note that $v_2 = 2V_S - v_1$. (*a*) Plot the composite curve if each tunnel diode has the volt-ampere characteristic given in Prob. 12-1, with $V_P = 0.1$ V, $I_P = 5$ mA, and $V_S = 0.4$ V. (*b*) Draw a load line on the composite characteristic and find the current in each diode and the voltage across each diode. (*c*) Find V_S if the current in one tunnel diode is to be zero. (*d*) Repeat parts *a* and *b* if $V_S = 0.5$ V. (*e*) Repeat parts *a* and *b* if $V_S = 0.2$ V.

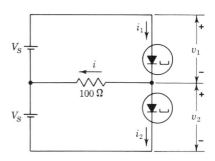

Prob. 12-5

12-6 A germanium tunnel diode has the characteristic shown, with $I_P = 10$ mA, $I_V = 1.0$ mA, $V_P = 50$ mV, $V_V = 350$ mV, and $V_F = 475$ mV. A resistor R is placed in parallel with the tunnel diode and this combination is called a *tunnel resistor*. Find the value of the resistance R so that the tunnel-resistor volt-ampere characteristic exhibits no negative-resistance region. Plot this composite characteristic.

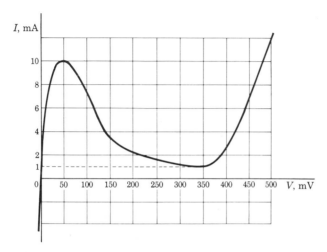

Prob. 12-6

12-7 A backward diode has the volt-ampere characteristic given in Prob. 12-6 except that $I_P = 1.0$ mA. This backward diode is placed in series with the tunnel diode. Sketch the resultant volt-ampere characteristic if the diodes are in series (*a*) aiding and (*b*) opposing. Consider both positive and negative voltages.

12-8 A germanium *p-n* diode may be represented by a piecewise linear characteristic (Fig. 7-3) with $V_\gamma = 0.2$ V, $R_f = 10$ Ω, and $R_r = \infty$. This diode is placed in series with the tunnel diode whose characteristic is plotted in Prob. 12-6. Sketch the resultant volt-ampere characteristic if the diodes are (*a*) aiding and (*b*) opposing. Consider both positive and negative voltages.

12-9 *a.* Use the Ebers-Moll equations, Eqs. (6-30) and (6-31), to show that when the collector junction of a transistor is substantially reverse-biased ($|V_c| \gg \eta V_T$) the base current is related to the emitter-junction voltage by

$$I_B = \frac{I_{EO}(1 - \alpha_N)}{1 - \alpha_N \alpha_I} (e^{V_E/\eta V_T} - 1) + \frac{I_{CO}(\alpha_I - 1)}{1 - \alpha_N \alpha_I}$$

b. A germanium tunnel diode has the characteristic shown in Prob. 12-6. The diode is combined as shown with a germanium transistor for which $\alpha_N = 0.98$, $\alpha_I = 0.7$, and $I_{CO} = 2.0$ μA. Draw the volt-ampere characteristic of the input terminals of this hybrid circuit.

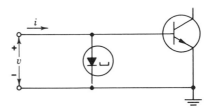

Prob. 12-9

12-10   The tunnel diode $D1$ has the volt-ampere characteristic described in Prob. 12-6.   The germanium diode $D2$ obeys Eq. (6-1) and has a reverse saturation current $I_o = 1$ μA.   The silicon transistor has the input characteristic given in Prob. 12-9a, with $\alpha_N = 0.98$, $\alpha_I = 0.7$, $I_{CO} = 2$ nA, and $\eta = 2$.   Draw the volt-ampere characteristic at the input at room temperature.

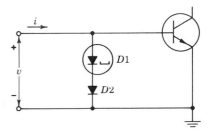

Prob. 12-10

12-11 a. It is desired to combine a tunnel diode and a transistor in such manner that there results a composite device with a tunnel-diode type of characteristic but with the current scale divided by the factor $h_{fe} + 1$.   Show how this may be done and draw qualitatively the resultant characteristic.

   b. Repeat part a for the case where it is desired to multiply the scale factor by $h_{fe} + 1$.

12-12 a. A resistor R_2 is added in series with $B2$ of a unijunction transistor.   If $R_2 \ll R_{BB}$, show that the peak voltage is given approximately by

$$V_P \approx V_\gamma + \eta V_{BB} - \frac{\eta V_{BB} R_2}{R_{BB}}$$

   b. The measured value of the temperature coefficient of R_{BB} is $+0.008/^\circ$C and of R_2 is $+0.004/^\circ$C (for a wire-wound resistor).   Neglect the temperature coefficient of η and take that of V_γ as -2 mV/$^\circ$C.   Show that if R_2 is chosen at room temperature to be approximately

$$R_2 \approx \frac{V_\gamma R_{BB}}{\eta V_{BB}}$$

then $V_P = \eta V_{BB}$ independently of the temperature.

    c. Find R_2 for $V_\gamma = 0.6$ V, $\eta = 0.5$, $V_{BB} = 20$ V, and $R_{BB} = 5$ K.

    d. Let a small resistance R_1 $(R_1 \ll R_{BB})$ be added in series with $B1$. Show that V_P will continue to be insensitive to the temperature provided that the resistance R_2 calculated in part *b* is increased by the amount $(1 - \eta)R_1/\eta$.

    12-13   *a.* For the unijunction transistor circuit of Fig. 12-5*b* plot the input volt-ampere characteristic, that is, I_E as a function of V_{EE}. The transistor is a type 2N494 (Fig. 12-7) operating with $V_{BB} = 30$ V and $R_E = 1$ K.

    b. What is the minimum value of R_E for which the volt-ampere characteristic in part *a* will not exhibit a region of negative resistance?

    c. For $V_{BB} = 20$ V, $V_{EE} = 17$ V, and $R_E = 2$ K, what is V_E?

    12-14   This problem illustrates the rate effect in a *p-n-p-n* diode.

    a. Assume that in the positive-resistance region, before breakdown, the diode may be represented as shown by two forward-biased diodes and a resistor R shunted by a capacitor. The resistor represents the center reverse-biased junction and C the capacitor across it. A ramp voltage $v(t)$ of slope $\alpha = dv/dt$ is applied beginning at $t = 0$. Assuming ideal forward-biased outer diodes, calculate the current $i(t)$ in terms of $v(t)$, C, and R.

    b. A *p-n-p-n* diode has the volt-ampere characteristic shown and has $C = 25$ pF. A ramp for which $\alpha = 5$ V/μsec is applied. On the *p-n-p-n* characteristic show the path followed by the diode and on the path mark off 1-μsec intervals. Assume that when this dynamic path intersects the negative-resistance portion of the static characteristic the diode will break down. Find the voltage at which the diode breaks down and the time at which it breaks down.

    c. Repeat part *b* for $\alpha = 20$ V/μsec.

    d. What is the path when α is very small?

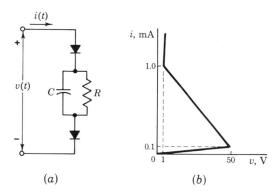

        (*a*)   (*b*)

Prob. 12-14

    12-15   The SCR is used to control the power delivered to the 50-Ω load by the sinusoidal source. If the gate supply V_{GG} is adjustable, (*a*) over what range may the conduction angle of the SCR be continuously varied? (*b*) Over what range may the load d-c current be continuously varied if the frequency is 60 Hz?

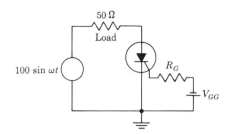

Prob. 12-15

CHAPTER 13

13-1 The device whose current i and voltage v are as defined in the figure has a linear volt-ampere characteristic of arbitrary slope and arbitrary intercepts with the current and voltage axes.

a. Show that, if the terminals of the device are left open-circuited, the equilibrium point of the circuit is located at the intersection of the characteristic with the voltage axis. Show that, if the slope of the characteristic is positive, this equilibrium point is stable against the perturbations which are possible when a capacitor is placed across the device terminals. HINT: $i = -C\,dv/dt$.

b. Show that, if the terminals of the device are shorted, the equilibrium point of the circuit is located at the intersection of the characteristic with the current axis. Show that, if the slope of the characteristic is positive, this equilibrium point is stable against the perturbations which are possible when an inductor is substituted for the short circuit. HINT: $v = -L\,di/dt$.

c. Show that if the slope of the characteristic is negative, the equilibrium points in parts *a* and *b* are unstable against perturbations.

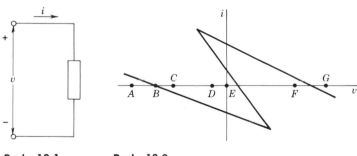

Prob. 13-1   **Prob. 13-2**

13-2 An NR device has the characteristic shown, with the resistance negative in all ranges of operation. A capacitor is charged to the voltage at A and connected to the device. Describe qualitatively the subsequent behavior of the circuit; that is, indicate the path taken by the operating point. Repeat for each of the cases B through G. HINT: $i = -C\,dv/dt$.

13-3   An NR device whose characteristic is shown is combined with R and I as indicated.   Draw the composite characteristic (i_T versus v_T) for (a) $I = 0$, $R = 10$ K; (b) $I = 0$, $R = 1$ K; (c) $I = 0$, $R = 500\ \Omega$; (d) $I = 2.5$ mA, $R = 2$ K.

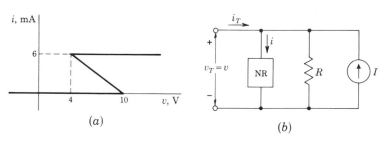

(a)   (b)

Prob. 13-3

13-4   An NR device whose characteristic is shown is combined with R and V as indicated.   Draw the composite characteristic (i_T versus v_T) for (a) $V = 0$, $R = 100\ \Omega$; (b) $V = 0$, $R = 1$ K; (c) $V = 0$, $R = 2$ K; (d) $V = 5$ V, $R = 500\ \Omega$.

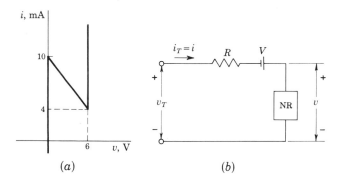

(a)   (b)

Prob. 13-4

13-5   (a) An NR device is placed across a series combination of V and R as shown. The NR characteristic is of the current-controlled type shown in Fig. 13-1b.   An appropriate load line for V and R is drawn and the equilibrium point for monostable operation is located.   Next, a composite characteristic (i_T versus v_T) is constructed.   Show that the equilibrium point previously identified at the intersection of the load line and NR characteristic is the same as the point located by the intersection of the composite characteristic with the voltage axis.   A capacitor is placed across the device terminals. Apply the stability criterion given in Prob. 13-1 to the composite characteristic and show that it leads to the same results as given by Eq. (13-4).   Repeat for load lines corresponding to (b) bistable and (c) astable operation.

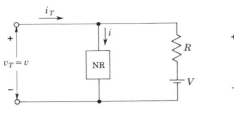

Prob. 13-5   Prob. 13-6

13-6 (*a*) A source V and a resistor R are combined in series with an NR device as shown. The NR characteristic is of the voltage-controlled type shown in Fig. 13-1*a*. An appropriate load line for V and R is drawn corresponding to the input terminals being shorted, and an equilibrium point is located for monostable operation. Next, a composite characteristic (i_T versus v_T) is constructed. Show that the equilibrium point previously identified at the intersection of the load line and the NR characteristic is the same as the point located by the intersection of the composite characteristic with the current axis. An inductor is connected across the terminals of the composite circuit. Apply the stability criterion given in Prob. 13-1 to the composite characteristic and show that it leads to the same results as given by Eq. (13-8). Repeat for load lines corresponding to (*b*) bistable and (*c*) astable operation.

13-7 At $t - 0$, a capacitance $C - 1,000$ pF charged to $+1$ V is placed across the terminals of a device whose volt-ampere characteristic is as shown. (*a*) Show the path of operation of the circuit, i.e., the locus of corresponding points v and i. (*b*) Write the expressions for v and i as a function of time. (*c*) At what time will the capacitor voltage be 10 V?

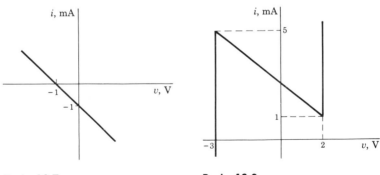

Prob. 13-7   Prob. 13-8

13-8 An inductance $L = 100$ mH is placed across the terminals of an NR device which has the volt-ampere characteristic shown. Plot the waveforms of v and i, indicating all voltage and current levels and times of interest. HINT: $v = -L \, di/dt$.

13-9   *a.* An NR device has the characteristic shown.   An uncharged capacitor of capacitance $C = 2,000$ pF is placed across the terminals of the NR device at $t = 0$. Plot the waveform of v and of i, indicating all current and voltage levels and times of interest.   HINT: $i = -C\, dv/dt$.

　　b. A resistance $R = 1$ K is placed in series with the capacitor before connection to the circuit.   Plot the current i and the voltages across C, across R, and at the terminals of the NR device.   HINT: First obtain the composite characteristic of R in series with the NR device.

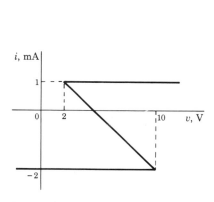

Prob. 13-9

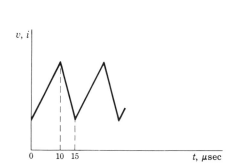

Prob. 13-10

　　13-10   A capacitance $C = 1,000$ pF is placed across the terminals of an NR device with a characteristic as shown.   Draw the current and voltage steady-state waveforms. Calculate all voltage and current levels and all times of interest.

　　13-11   An inductance $L = 100$ mH is placed across the terminals of an NR device with a characteristic as shown.   Draw the current and voltage steady-state waveforms. Calculate all voltage and current levels and all times of interest.

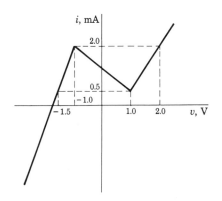

Prob. 13-11　　　　　　　　　　　　　**Prob. 13-12**

13-12   *a.* A capacitance $C = 1{,}000$ pF is placed across the terminals of an NR device, and the voltage waveform which develops across the capacitor is linear as shown.   The voltage makes excursions between 5 and 15 V.   Draw the volt-ampere characteristic of the device.

b. An inductance $L = 10$ mH is placed across the terminals of an NR device, and the inductor current waveform which develops is linear as shown.   The current makes excursions between 5 and 15 mA.   Draw the volt-ampere characteristic of the device.

13-13   *a.* An NR device has the volt-ampere characteristic shown.   The device is connected through a resistance $R = 250$ Ω to a source of voltage V.   What are the ranges of V for which one stable state exists if $C = 1{,}000$ pF is placed across the NR device?   What is the range of V for which three stable states exist?

b. What is the minimum value of R for which an unstable equilibrium state is possible?   Assume R selected to be twice as large as this minimum value.   What is the range of V for which an unstable state is possible?

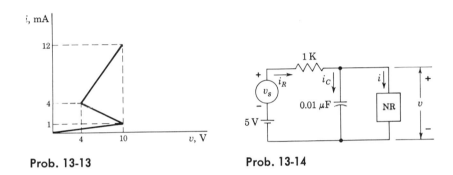

Prob. 13-13   Prob. 13-14

13-14   The NR device has the volt-ampere characteristic shown in Prob. 13-13.
a. Find the initial operating point of the circuit, assuming $v_s = 0$.
b. The triggering signal v_s is a positive step of amplitude 7 V applied at $t = 0$. Calculate v, i, i_C, and i_R as functions of time before a transition takes place.
c. At what time does a transition take place?
d. Find v, i, i_C, and i_R immediately after the transition.
e. Calculate v, i, i_C, and i_R as functions of time after the transition.
f. What is the final equilibrium state?   How long a time, after the transition has occurred, will be required for the capacitor to arrive to within 0.1 V of its final state?

13-15   Consider the operation of the circuit of Prob. 13-14 for the reverse transition.   That is, assume that v_s has remained at $v_s = 7$ V long enough to allow the circuit to establish itself in its stable state.   (*a*) What is this steady state?   (*b*) Let v_s return to $v_s = 0$ V.   Calculate v, i, i_C, and i_R as a function of time before a transition takes place.   Answer parts *c* through *f* of Prob. 13-14.

13-16   Assume, in Prob. 13-14, that the signal v_s is a pulse of amplitude 7 V but of duration just adequate to cause a transition.   Draw the waveform v and calculate all voltage levels and all times of interest.

13-17   An NR device with the volt-ampere characteristic as in Prob. 13-13 is connected to a 16-V source through a 4-K resistance.   A capacitance $C = 0.01$ μF is connected across the NR device.   Draw the waveform v and calculate all voltage levels and times of interest.

13-18 A tunnel diode with the idealized characteristic shown operates in the circuit of Fig. 13-9a with $V = 100$ mV, $R = 15$ Ω, and $L = 250$ µH. The source v_s provides a trigger of amplitude 0.5 V and of duration just long enough to induce a transition. Draw the waveforms of diode current and voltage. Calculate all current and voltage levels and all times and time constants of interest.

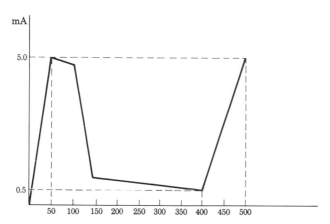

Prob. 13-18

13-19 In the monostable circuit of Fig. 13-13, the tunnel diode has the following parameters: $I_P = 20$ mA, $I_V = 2.4$ mA, $V_P = 85$ mV, $V_V = 365$ mV, and $V_F = 560$ mV. (a) Plot the piecewise linear approximation for the tunnel-diode characteristic on graph paper. Draw the load line and show that monostable operation is possible. (b) Calculate the pulse width T obtained when a trigger is applied. (c) Assuming a recovery time of two time constants, calculate the maximum frequency at which the circuit may be triggered.

13-20 A tunnel diode with an idealized characteristic as shown in Prob. 13-18 operates in the circuit of Fig. 13-9a with $V = 150$ mV, $R = 8$ Ω, and $L = 250$ µH. Draw the waveforms of diode current and voltage. Calculate all voltage and current levels and times of interest.

13-21 Verify Eq. (13-17) for the interval T_2 of the free-running tunnel-diode circuit.

13-22 A tunnel diode with shunt capacitance C and with the idealized characteristic as shown is used in the circuit of Fig. 13-9a. The source V and load R are selected so that the load line just touches the peak of the characteristic. A negligibly small trigger v_s causes a transition from $v = V_P$ to $v = V_F$.

a. Assume that the inductance L is large enough so that no appreciable change in current through it takes place during the course of the transition. What is the path taken by the circuit during the transition? Calculate the transition time t_r.

b. Assume that the inductor has been removed. What is the path taken by the circuit in going from V_P to V_F? Prove that the transition time t_r' is given by

$$t_r' = \frac{C(V_F - V_P)}{I_P - I_X} \ln \frac{I_P}{I_X}$$

c. Calculate the ratio t_r'/t_r for $I_X = I_P/5$.

d. Find t_r and t_r' if $V_F = 0.5$ V, $V_P = 0.1$ V, $C = 5$ pF, and $I_P = 5$ mA.

e. Calculate the value of inductance in part a which will cause the sum of the transition and settling times to equal t_r' in part b.

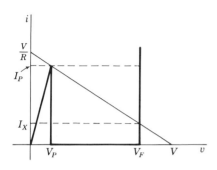

Prob. 13-22   Prob. 13-23

13-23 A tunnel diode with the idealized characteristic as shown operates in the circuit of Fig. 13-9a with $V = 6$ V, $R = 2$ K, and $L = 0$. Initially the diode is in its low-voltage state. The signal v_s is a 10-V step. The shunt capacitance across the diode is 10 pF. (a) What is the path followed by the operating point on the volt-ampere characteristic and what is the high-voltage state? (b) Calculate the time required for the voltage across the diode to rise from 0.1 to 0.5 V.

13-24 A tunnel diode with the idealized characteristic shown in Prob. 13-23 operates in the circuit of Fig. 13-9a with $V = 0.25$ V, $R = 0$, and $v_s = 0$. What is the value of L required in order that the circuit yield 1-MHz oscillations? Plot the voltage and current waveforms for the tunnel diode.

13-25 The tunnel-diode and the transistor volt-ampere characteristics are as shown. Find the stable states of the hybrid circuit arrangement. In each state, find the diode current, the transistor base and collector currents, and the collector voltage. Assume that $h_{FE} = 100$.

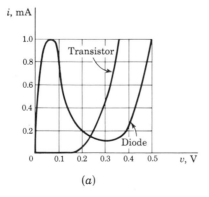

(a)

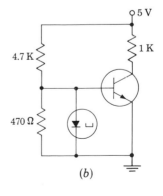

(b)

Prob. 13-25

13-26   Explain how the hybrid circuit arrangement operates to generate an output waveform v_o which is a square wave.   Assume that the composite input characteristic is as shown and that the transistor remains in saturation [with $V_{CE}(\text{sat}) = 0.1$ V] so long as the diode operates to the right of the valley point.   Calculate the times T_1 and T_2.

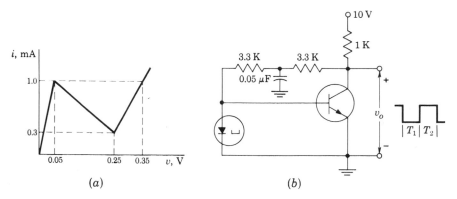

(a)   (b)

Prob. 13-26

13-27   The p-n-p-n diodes have idealized volt-ampere characteristics as shown. Show that the circuit is astable.   Draw the waveforms of the voltages across each diode.   Calculate the voltage levels and the period of oscillation.

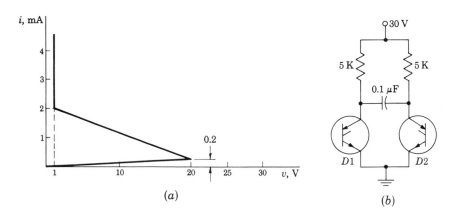

(a)   (b)

Prob. 13-27

13-28   The p-n-p-n diodes have the idealized characteristics shown in Prob. 13-27. Show that the circuit is monostable.   What size triggering signal is required to induce a transition?   Draw the waveforms of the voltages across the diodes and label all voltage levels.   Calculate the time of the quasi-stable state.

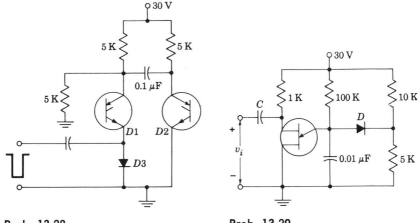

Prob. 13-28   Prob. 13-29

13-29   *a.* Verify that the circuit shown is monostable when triggered by an input pulse v_i and sketch the waveform $v_C = v_E$ across the emitter capacitor.

b. Calculate the voltage levels and the important times associated with v_C.   Neglect the drop in the diode D.   The UJT characteristics are given in Fig. 12-7.

c. What is the minimum pulse amplitude required?

13-30   The circuit of Fig. 13-23 uses a UJT whose characteristics are given in Fig. 12-7.   The circuit parameters are $V_{BB} = 20$ V, $R_1 = 4$ K, $R_2 = 10$ K, $R_{b2} = 2$ K, and $C = 0.01$ μF.   Assume the diode is ideal, with its break point at 0 V.   Calculate the voltage levels of waveforms v_C, v_E, and v_{B2}.   Calculate all times of interest.

13-31   Show that if, in the circuit of Prob. 13-30, R_2 had been selected as $R_2 = 4$ K, the circuit would be monostable.   For this case, if a trigger is applied, calculate the voltage levels of the waveforms v_C, v_E, and v_{B2}.   Calculate all times of interest.

13-32   Verify Eqs. (13-24) and (13-25) for the times involved in the UJT circuit of Fig. 13-23.

13-33   For the bistable UJT circuit verify that the minimum trigger size required at $B1$ is $(V_P - V_{EE})/(1 - \eta)$.

13-34   Explain the operation of the astable UJT multivibrator shown.   Draw the waveforms and calculate the voltage levels of v_E, v_o, and v_{B2}.   Calculate all times of

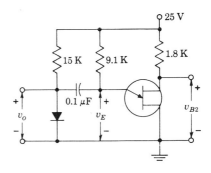

Prob. 13-34

interest.   The UJT characteristics are given in Fig. 12-7.   HINT: The UJT is ON
when the diode is OFF, and vice versa.

13-35   The monostable trigistor circuit shown establishes a timing interval and
generates a rectangular waveform across the trigistor.   In the quiescent state the
trigistor is ON and has a maintaining voltage of 1 V.   Describe the operation of the
circuit from the time when a negative input pulse turns OFF the trigistor.   Calculate
the waveforms across C, across the trigistor, and at the gate terminal.   Calculate also
the timing interval.   The diode D_Z has a maintaining voltage of 5 V, and $D1$ is ideal
with a break at 0 V.   Assume that the negative bias at the gate is just adequate to
keep the trigistor OFF when diode $D1$ is not conducting.   HINT: Diode $D1$ is OFF in the
quiescent state.

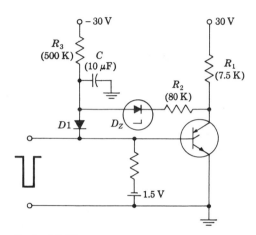

Prob. 13-35

13-36   In the avalanche transistor circuit of Fig. 13-28, $V_{CC} = 60$ V, $R_L = 50\ \Omega$,
$C = 10$ pF, and $LV_{CER} = 40$ V.   The transistor completes its transition from OFF to ON
in 1 nsec.   Find the amplitude and width (at half peak) of the output pulse.

CHAPTER 14

14-1   If the deviation from linearity is small enough so that a sweep voltage may
be approximated by the sum of a linear and quadratic term in t, show that the sweep
errors e_d, e_s, and e_t are related as given by Eq. (14-4).

14-2   *a.* An exponential sweep results when a capacitor C is charged from a supply
voltage V through a resistor R.   If the peak sweep voltage is V_s, prove that the slope
error is given exactly by $e_s = V_s/V$.

b. To the circuit of part *a* is added a resistor R_1 shunted across C.   Show that the
slope error is now given by $e_s = V_s/aV$, where $a \equiv R_1/(R_1 + R)$.

14-3   In the UJT sweep circuit of Fig. 14-6a, $V_{BB} = 20$ V, $V_{YY} = 50$ V, $R = 5$ K,
$R_{b1} = R_{b2} = 0$, and $C = 0.01\ \mu$F.   The UJT has characteristics as given in Fig. 12-7.
Calculate (*a*) the amplitude of the sweep signal v_C, (*b*) the slope and displacement
errors, (*c*) the time of the sweep, and (*d*) the estimated recovery time using Eq. (13-21).

14-4   *a.* Design a sweep circuit as in Fig. 14-6a using the UJT characteristics of
Fig. 12-7.   Set $R_{b1} = R_{b2} = 0$.   The sweep amplitude is to be 10 V, the sweep dura-

tion is to be 1 msec, and the sweep-speed error is to be 10 percent. Specify reasonable values for V_{BB}, V_{YY}, R, and C, and give reasons for your choices.

b. Show that your design is one which guarantees that the circuit will not latch in the high-current region.

c. Estimate the recovery time from Eq. (13-21).

d. If the specifications are changed so that a 100-μsec sweep is required, such a sweep may be obtained by decreasing either R or C by a factor of 10. Which alternative is preferable?

14-5 A UJT is used as a switch across a sweep capacitor C which charges through R. A single voltage supply V_{BB} is used in the circuit.

a. Prove that the sweep duration is given exactly by

$$T_s = RC \ln \frac{V_{BB} - V_V}{V_{BB} - V_P}$$

b. Prove that, if $V_{BB} \gg V_V$, then $T_s \approx RC \ln [1/(1 - \eta)]$.

c. If the sweep were linear instead of exponential, prove that for $V_P \gg V_V$, $T_s \approx \eta RC$.

d. Compare T_s in parts *b* and *c* for $\eta = 0.5$.

14-6 A driven sweep circuit as in Fig. 14-8a uses for the switch S a silicon controlled switch (SCS) whose firing characteristics are given in Fig. 12-14. The valley voltage is $V_V = 1.0$ V. The unit will latch in the on state if a continuous current is available in excess of 1.0 mA. The sweep is triggered by a train of voltage pulses 1.0 V in amplitude and of a frequency of 100 Hz. The sweep is to have an amplitude of 10 V, a duration of 1 msec, and the best linearity possible. Specify reasonable values for V_{YY}, R_1, R_2, R, and C and for the bias on which the triggering pulses are to be superimposed. In each case give reasons for your choice.

14-7 A driven sweep circuit as in Fig. 14-8a uses for the switch a thyratron gas tube (Fig. 14-6b). The sweep is to have a duration of 100 μsec and an amplitude of 50 V. The triggering waveform is a 1,000-Hz train of pulses of amplitude 4 V. The firing voltage V_P is related to the grid bias V_{GG} by $V_P = 8V_{GG}$. The maintaining voltage is 16 V. The tube will latch (block) if a continuous current is available in excess of 1.0 mA. The available power supply voltage is 250 V. Specify reasonable values for V_{GG}, R_1, R_2, R, and C. In each case give reasons for your choice.

14-8 The circuit shows a method for reducing the recovery time of a UJT sweep circuit by using a transistor Q. Explain how the circuit operates.

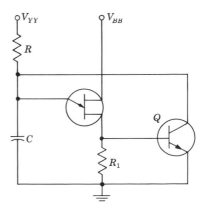

Prob. 14-8

14-9 In the sweep circuit of Fig. 14-6b, the thyratron has a critical grid break-down characteristic given by $V_P = 10V_{GG}$, a maintaining voltage $V_V = 20$ V, a maintaining (blocking) current $I_V = 0.5$ mA, and a maximum allowable current of 0.5 A. A 300-V supply is available.

a. If the circuit is to operate as a *driven* sweep, what is $(V_{GG})_{\min}$?

b. Sketch the waveform across C for a single trigger input.

c. What will happen if $V_{GG} < (V_{GG})_{\min}$? Sketch the waveform v_C across the capacitor if $V_{GG} = 0.2(V_{GG})_{\min}$.

d. Find the minimum value of r.

e. What is the minimum value $R_{\min}$ of R for proper operation?

f. What will happen if $R < R_{\min}$?

14-10 In the vacuum-tube sweep circuit of Fig. 14-10, the tube is a 12AU7, $V_{YY} = 300$ V, $R = 100$ K, $R' = 1$ M, and $C' = 0.001$ μF. The input is a negative 10-μsec pulse whose amplitude is 30 V.

a. What is the sweep duration?

b. What is the minimum output voltage, with respect to ground?

c. If a 10-V sweep is desired, what is the value of C?

d. What is the approximate value of the recovery time constant?

e. How would you modify the circuit in order to reduce the recovery time? Suppose, for example, that the input gate is to be a symmetrical 50-kHz square wave whose peak-to-peak amplitude is 30 V.

14-11 A given waveform is observed on a CRO. It is desired to view a selected portion of this waveform in more detail by spreading it out in time by some factor, say 5. Devise a circuit configuration (in block-diagram form) which will function as such a *sweep magnifier*.

14-12 A type 2N404 *p-n-p* transistor (Figs. 6-13 and 6-21) is used in the CE configuration as the switch in the time-base circuit of Fig. 14-10. The circuit parameters are $R = 100$ K, $R' = 150$ K, $C' = 0.01$ μF, and the supply voltage magnitude is 30 V. The input is a positive 100-μsec pulse whose amplitude is 2 V.

a. What is the sweep duration?

b. What is the minimum output voltage with respect to ground?

c. If the maximum voltage to which the capacitor is to charge is 10 V, what is the size of the capacitor?

d. Estimate the discharge time of the capacitor. (HINT: Assume that during the discharge time the transistor may be replaced by a parallel combination of a current source and a resistor. Assume also that the plots in Fig. 6-13 may be approximated by straight lines extending all the way to the current axis.)

e. Draw the waveform of the capacitor voltage v_C.

14-13 Draw a schematic diagram corresponding to the block diagram of the thyratron sweep of Fig. 14-9.

14-14 Derive Eqs. (14-20), (14-21), and (14-23) for the transistor constant-current sweep.

14-15 The emitter current in the sweep circuit of Fig. 14-13a is supplied by a current source which supplies a constant current I_E. Show that the sweep-speed error is $e_s \approx h_{ob}V_s/\alpha I_E \approx h_{ob}V_s/I_E$, where V_s is the sweep amplitude. Calculate e_s for $h_{ob} = 0.5$ μmho, $V_s = 20$ V, and $I_E = 10$ mA.

14-16 *a.* Design a sweep circuit using a silicon transistor in the CB configuration as in Fig. 14-13a to generate a 20-V sweep in a time of 10^{-4} sec with a slope error not to exceed 0.2 percent. The transistor has parameters $\alpha = 0.98$, $h_{ib} = 20$ Ω,

$h_{rb} = 4 \times 10^{-4}$, and $h_{ob} = 0.5$ μmho. Use $R_e = 1.5$ K and select reasonable values for V_{CC}, V_{EE}, and C.

b. Calculate the percentage change in sweep speed which results from a 30°C change in ambient temperature.

14-17 Consider the pulse-width decoder shown. The capacitor C charges through the CB transistor $Q2$ at a constant current of 5 mA and is discharged by the switch $Q1$. Emitter follower $Q3$ and diode D function as a "peak detector," or "hold circuit." The input pulse v_i extends from 0 to $+10$ V for a duration T which may have any value from 0 to 1 msec. (a) Plot v_i, v_b (the input to $Q1$), v_C (the voltage across C), and v_o as a function of time. (b) Prove that v_o is proportional to the pulse width T. (c) Find R. (d) What is the maximum output voltage? (e) What is the minimum value that can be used for C? (f) If $C = 1$ μF and $v_o = 3$ V, what is T? (g) If $Q3$ and D were silicon, show that pulse widths below a minimum could not be measured. If $C = 1$ μF, what is T_{min}?

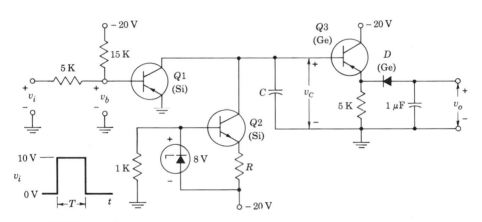

Prob. 14-17

14-18 In the sweep circuit of Fig. 14-15 take into account the fact that during the formation of the sweep the capacitor C_g discharges through R_g. Show that to allow for this discharge the slope error e_s given in Eq. (14-25) must have added to it the term T_s/R_gC_g, where T_s is the time of the sweep.

14-19 (a) In the circuit of Fig. 14-15 the tube is a 12AX7, $V_{PP} = 250$ V, $R_1 = 100$ K, $R_2 = 150$ K, $R_k = 200$ K, $C_g = 0.1$ μF, $R_g = 100$ K, and $C = 300$ pF. Find the sweep speed, the slope error for a sweep of amplitude 75 V, and the sweep time. (b) How long an interval must be allowed between sweeps to make sure that the initial sweep speed is not lower by more than 1 percent than the sweep speed which results when an arbitrarily long recovery time is allowed?

14-20 In the sweep circuit shown the capacitor C charges abruptly through the thyratron $V2$ and discharges at a nominally constant rate through $V1$, a 6AU6 pentode (Fig. D-1). The cathode resistor R_k is adjusted so that the capacitor current is approximately 0.7 mA during the sweep time. The screen current is 0.4 of the plate current. The thyratron maintaining voltage is 20 V and the breakdown voltage V_P is related to the grid-to-cathode voltage V_G by $V_P = -10V_G$.   (a) Sketch the waveform v_C.   Calculate (b) the maximum and minimum values of v_C if $C = 0.01$ μF, (c) the cathode resistance, and (d) the sweep frequency.

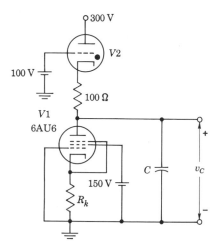

Prob. 14-20

14-21   The Miller sweep requires ideally an amplifier with a gain $-\infty$, whereas the bootstrap sweep requires a gain $+1$.   Yet the discussion in connection with Fig. 14-16 indicates that actually the two circuits are the same.   To resolve this apparent anomaly, consider an amplifier as in the figure with independent terminals X, Y, and Z. Show that if a gain A is defined as $A \equiv v_{yz}/v_{xz}$ (i.e., a "common-Z" configuration) and a gain A' is defined as $A' \equiv v_{zy}/v_{xy}$ (i.e., a "common-Y" configuration), then

$$A' = \frac{-A}{1 - A} \qquad \text{or} \qquad A = \frac{-A'}{1 - A'}$$

Therefore, if $A = -\infty$, $A' = +1$, verifying that an amplifier may have different gains depending on the definition of input and output terminals.

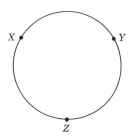

Prob. 14-21

14-22 The circuit in (*a*) is a Miller sweep using an amplifier stage. The circuit in (*b*) is a bootstrap sweep using a cathode follower. Show that the circuit in (*a*) may be transformed into the circuit in (*b*) by making changes which have no effect on the small-signal equivalent of the circuit.

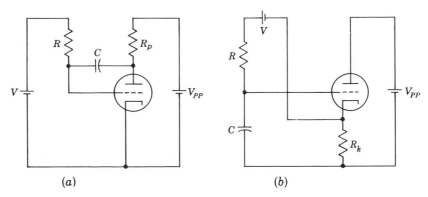

(*a*)   (*b*)

Prob. 14-22

14-23 *a.* Let a resistor r be added in series with the capacitor C in Fig. 14-19. Show that in this case the jump at the output ΔV_o (which is no longer the same as the jump at the input) is given by

$$\Delta V_o = \frac{(R_o + Ar)V'}{(1 - A)R' + R_o + r}$$

b. Under what conditions will the jump be positive? Negative? Zero?
c. Evaluate ΔV_o if a pentode is used with $g_m = 3$ mA/V and if the plate resistance $= 30$ K, $V = 300$ V, $R = 1$ M, and $r = 0$.

14-24 The switching arrangement of Fig. 14-19 is replaced by the arrangement shown in (*a*). Thus the sweep is started not at the closing of S but rather at the opening of S'. Show that the arrangement at (*a*) is equivalent to that shown at (*b*). Here it is considered that the sweep forms in response to an abrupt change in input voltage to the level V' from the level V''. Assume that the level V'' has persisted long enough for the circuit to have attained a steady state.

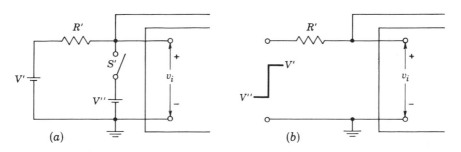

(*a*)   (*b*)

Prob. 14-24

14-25   The Miller sweep circuit of Fig. 14-21 uses a 6AU6 pentode (Fig. D-1) with a screen voltage of 150 V.   The parameters in the circuit have the following values: $V_{PP} = 300$ V, $R_p = 10$ K, $R = 1$ M, $C = 0.003$ μF, and $V_{GG} = 2$ V.   (a) Plot the output waveform and label all voltage levels with respect to ground for $T_s = 1.5$ msec.   (b) Calculate the displacement error.   (c) Calculate the recovery time constant assuming a switch resistance of 1 K.

14-26   In the Miller run-up sweep circuit of Fig. 14-23, $V_{PP} = V_{GG} = 300$ V, $R_p = 100$ K, $C = 0.005$ μF, $R = 1$ M, and the switch S has a resistance of 500 Ω. In the quiescent condition with S closed, the tube current is 2.5 mA.   The tube parameters are $\mu = 100$ and $r_p = 100$ K.   The switch S opens at $t = 0$ for 1 msec and then closes.   (a) Find r to eliminate the jump in the output.   (b) Plot the grid and plate voltages, labeling all voltage levels, times, and time constants.   (c) Calculate the sweep-speed error.   (d) Calculate the recovery time constant.

14-27   With the switch S closed, the quiescent collector current is 1 mA.   The silicon transistor has parameters $h_{fe} = 100$, $h_{ie} = 1$ K, $h_{oe} = 0.25 \times 10^{-4}$ mho, and $h_{re} \approx 0$.   The switch is opened at $t = 0$.   (a) Draw the output waveform for the interval $t = 0$ to $t = 1$ msec, labeling the voltage levels.   (Neglect the initial jump.) (b) Calculate the sweep-speed error e_s in this interval.   (c) In order to decrease e_s an emitter follower is added.   Indicate the circuit, including power supplies, required for proper biasing.   If the emitter resistance is 10 K, calculate e_s.

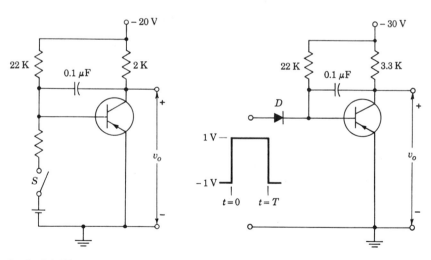

Prob. 14-27   Prob. 14-28

14-28   The transistor has $h_{FE} = 100$.   Neglect diode and transistor forward junction voltages.   Draw the output waveform v_o for the interval $t < 0$ (before the gate is applied) until the circuit has recovered to its initial quiescent state.   Consider voltage jumps.   Calculate voltage levels, time constants, and retrace times for (a) $T = 30$ μsec and (b) $T = 300$ μsec.   HINT: The circuit behaves as a Miller integrator for $t > T$ but not during the gating interval T.   Why?   Assume zero source resistance.

14-29   The screen-gated phantastron circuit of Fig. 14-31 uses a 6AU6 tube and has the following parameters: $V_{PP} = 300$ V, $R_p = 100$ K, $R = 1$ M, and $C = 0.001$ μF. The voltage-divider resistances are so adjusted that the suppressor is at 0 V and the screen at 150 V during the sweep time.   Calculate (a) V_1 and V_2 (Fig. 14-32) approximately, (b) the sweep amplitude, (c) the sweep duration T_s, (d) the percent change in T_s for a 10 percent change in supply voltage, and (e) the displacement error.   (f) If a plate catching diode is added, as in Fig. 14-33, then the relationship between V_R and the delay time T is $T = \alpha V_R$.   Find α.

14-30   An astable circuit may be obtained by cross-coupling two monostable phantastrons.   Sketch such a circuit and explain its operation.

14-31   It is desired to obtain a delay time T which will be inversely proportional to a voltage V.   Show that a screen-coupled phantastron will give this relationship (approximately) if the charging resistor R is connected to V instead of to V_{PP}.

14-32   A cathode-coupled monostable phantastron is shown.   Draw the waveforms at the suppressor, screen, and control grids, at the plate, and at the cathode. (HINT: The cathode follows the grid quite closely.)

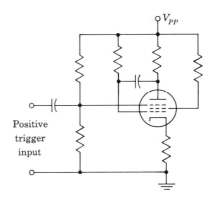

Prob. 14-32

14-33   The circuit shown is a variation of the bootstrap which avoids the need for a voltage source neither side of which is grounded.   Show that precise linearity results if the amplifier gain is $A = 1 + R_2/R_1$.

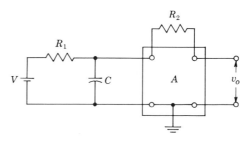

Prob. 14-33

14-34 *a.* For the bootstrap circuit shown prove that the amplifier gain A must be given by $A = 1 + R/R_i$, if a precisely linear sweep is to be obtained.

b. A linear sweep with a pair of symmetrical output signals is to be obtained from the circuit shown ($v_{o1} = -v_{o2}$). Find the values of R'/R and R''/R. The magnitude of the forward gains of the amplifiers may be taken as infinite.

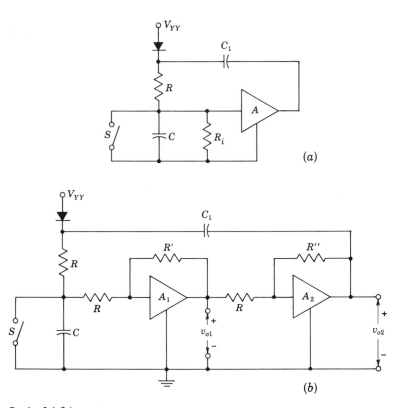

(a)

(b)

Prob. 14-34

14-35 (*a*) Find the value of R which will cause the bootstrap circuit shown to yield an exactly linear sweep. The transistors have parameters $h_{ie} = 1$ K, $h_{fe} = 100$. Assume $h_{re} \approx 0$ and $h_{oe} \approx 0$. (*b*) With S closed, the quiescent collector voltage of $Q2$ is 35 V. Find the sweep speed.

14-36 Show that the effect of a finite value of C_1 in the circuit of Fig. 14-36 is that the term C/C_1 must be included in the expression for e_s as in Eq. (14-48). HINT: The sweep speed at $t = 0$ is given by Eq. (14-47) as AV/RC. Calculate the sweep

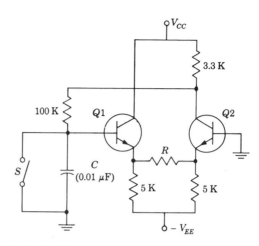

Prob. 14-35

speed at $t = T_s$ by noting that the voltage across C in Fig. 14-36 is V_s, across C_1 is $V - V_s(C/C_1)$, and $v_o = AV_s$.

14-37   The bootstrap sweep of Fig. 14-35 uses a 12AU7 tube instead of a transistor. The circuit parameters are the following: $V_{PP} = 250$ V, $C = 500$ pF, $C_1 = 0.01$ μF, $R_1 = 0.2$ M, $R = 1$ M, and $R_k = 10$ K.   The switch S is opened for $T_s = 100$ μsec. (a) Plot the waveform at the output of the cathode follower for $T_s > t > 0$.   Calculate (b) the output voltage at the beginning and the end of the sweep, (c) the average gain of the cathode follower (from the ratio of the output to input voltage changes during the sweep), (d) the sweep-speed error, (e) the voltage change across C_1 during the sweep, and (f) the recovery time constant, assuming a zero-resistance switch.

14-38   Repeat Prob. 14-37 if R_1 is replaced by a diode whose forward resistance is 1 K.

14-39   The bootstrap circuit of Fig. 14-35 uses a 6CG7 tube instead of a transistor and C_1 is replaced by a $\frac{1}{25}$-W neon lamp whose voltage is constant at 65 V over the current range 30 to 300 μA.   The circuit has the following parameters: $V_{PP} = 300$ V, $R = 1$ M, $C = 0.1$ μF, $R_k = 15$ K, and the switch resistance is 1 K.   With S closed, the lamp current is 200 μA.   The switch is opened for a time T_s, at the end of which time the sweep has made a 100-V excursion.   Calculate (a) the output voltage with S closed, (b) the value of R_1, (c) the lamp current at the end of the sweep, (d) the sweep time T_s, (e) the ratio T_s/τ, where τ is the restoration time constant, and (f) the slope error.

14-40   a. For the bootstrap circuit of Fig. 14-37 discussed in the illustrative example in Sec. 14-15 plot the waveform of the emitter current of the emitter follower and calculate all current levels.

b. Plot and calculate the levels of the waveforms of the voltage across and the current through C_1.

c. If V_{EE} were 0 V, how would *all* the waveforms be modified?   In particular, repeat part b.

14-41 The transistor bootstrap circuit of Fig. 14-37 has the following parameters: $V_{CC} = 10$ V, $V_{EE} = 10$ V, $R_b = 30$ K, $R = 10$ K, $R_e = 5$ K, $C = 0.002$ μF, $C_1 = 0.25$ μF, and C_b may be taken as arbitrarily large. The input gate has an amplitude of 1 V and a width of 50 μsec. The parameters of Q are $h_{FE} = h_{fe} = 60$, $h_{ie} = 2$ K, $1/h_{oe} = 10$ K, $h_{re} = 10^{-4}$, $I_{CBO} = 0$, and the forward-biased junction voltages are negligible. The diode is ideal. (a) Plot the gate voltage, the collector current i_{C1}, and the output voltage v_o. Put numerical values of voltage, current, and time on your waveforms after you make the following calculations. Evaluate (b) the sweep speed and the amplitude of the sweep at its maximum value, (c) the time it takes to discharge C at the end of the sweep, (d) the peak voltage change across C_1 and the time required to replace this lost charge, (e) the emitter current of $Q2$ at $t = 0-$ and at $t = 0+$, (f) e_s, the slope error, and (g) e_s, if $Q2$ were replaced by a Darlington composite configuration.

14-42 In the transistor bootstrap circuit of Fig. 14-37, $V_{CC} = 25$ V, $V_{EE} = 15$ V, $R = 10$ K, $R_e = 15$ K, $R_b = 150$ K, $C = 0.05$ μF, and $C_1 = 100$ μF. The transistor has parameters given in Table 1-2. The gating waveform has a duration $T_g = 300$ μsec. (a) Draw waveforms for i_{C1} and v_o, labeling all current and voltage levels. (b) What is the sweep speed and the maximum value of the sweep voltage? (c) What is the slope error of the sweep? (d) What part of the slope error results from the fact that C_1 is not arbitrarily large? (e) Calculate the retrace time T_r for C to discharge completely. (f) Calculate the recovery time T_1 for C_1 to recharge completely.

14-43 Each silicon transistor has an $h_{fe} = 60$ and junction voltages as given in Table 6-1. The diode drop in the forward direction is 0.6 V. The input v_i comes from a flip-flop whose output impedance is small compared with 20 K. The states of the binary are -16 and -8 V. (a) If $v_i = -16$ V, find v_o. (b) If the flip-flop changes states so that $v_i = -8$ V, show that a negative sweep waveform is generated at the output. (c) If complete rundown is obtained in 50 μsec, calculate C. (d) At the end of 100 μsec the binary is reset. Calculate the recovery time for C. (e) How much charge is drained from the 1-μF capacitance while the input is in the -16-V state? (f) Calculate the recovery time for the 1-μF capacitance.

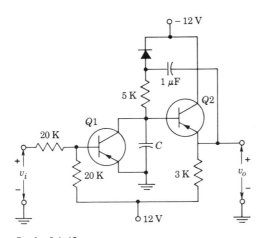

Prob. 14-43

14-44   The silicon transistor has the following parameters: $h_{fe} = 50$, $h_{ie} = 1.1$ K, $h_{oe} = 25$ μmhos, and $h_{re} \approx 0$.   The avalanche diode has a maintaining voltage of 7 V. The UJT has a valley voltage of 2.7 V and $\eta = 0.5$.   Neglect the retrace time due to the UJT.   A 5-kHz sweep is required.   The maximum current in the avalanche diode is to be 5 mA.   Calculate (*a*) the sweep amplitude, (*b*) the sweep speed, (*c*) the value of R, (*d*) the value of R_1, (*e*) the minimum current in the breakdown diode, and (*f*) the slope error.   (*g*) If bootstrapping were not used, calculate the slope error.

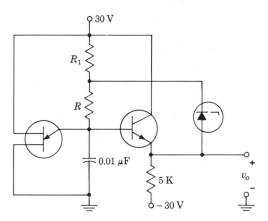

Prob. 14-44

14-45   *a.* The circuit shown is a monostable bootstrap time-base generator. Explain how a sweep is formed when a negative pulse v_i is applied at $t = 0$.   Sketch qualitatively the output waveform v_o.   What factors cause the termination of the generated waveform?

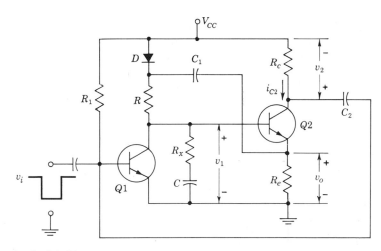

Prob. 14-45

b. Prove that at $t = 0+$, a negative step is formed at the base of $Q1$ whose magnitude is

$$V_{CC} \frac{R_c}{R} \left(1 + \frac{R_x}{R_e} \right)$$

c. If $Q1$ is to remain cut off throughout the sweep, prove that $C_2 > R_e R C / R_1 R_c$.

14-46 In the split-capacitor compensation circuit of Fig. 14-42, show that the addition of R_2 introduces an effective load R' shunting the output of the emitter follower given by

$$R' = \frac{A R_2}{A - C_3 / (C_3 + C_2)}$$

where A is the gain of the emitter follower.

14-47 *a.* In the split-capacitor compensation circuit of Fig. 14-42, $C_1 = 50$ μF, $R = 10$ K, $R_e = 5$ K, $C_2 = C_3 = 0.1$ μF, and $V_{EE} = V_{CC} = 20$ V. The transistor parameters are $h_{ie} = 1$ K, $h_{fe} = 60$, and $1/h_{oe} = 40$ K. If the sweep amplitude is 10 V, calculate R_2.

b. What would be the slope error if R_2 were removed?

14-48 The circuit of Fig. 14-43 has the following parameters: $V = 250$ V, $L = 700$ H, $R = 100$ K, and $C = 0.005$ μF. The switch S is opened for 500 μsec. Calculate (*a*) the sweep amplitude, (*b*) the displacement error. (*c*) If switch S were left open, sketch the output waveform across C. What would be the approximate value of the peak output voltage? (*d*) Repeat parts *a* and *b* with the inductor not in the circuit.

CHAPTER 15

15-1 (*a*) In the circuit of Fig. 15-1a, $L = 200$ mH and it is required that the current in L increase from zero to 100 mA in 1.0 msec. Find V_{CC}. Assume $R_L = 0$ and $V_{CE}(\text{sat}) = 0$. (*b*) If R_d, consisting of the diode resistance alone, is 10 Ω, draw the waveforms of i_L and v_{CE}, indicating voltage levels and time constants. If we assume that the circuit has recovered in three time constants, what is the ratio of the recovery period to the sweep period? (*c*) If the transistor will withstand a collector-to-emitter voltage of 60 V, what is the maximum value of R_d that may be used? Draw the waveforms of i_L and v_{CE} for this case. What is the recovery period now?

15-2 (*a*) In the circuit of Fig. 15-1a, $V_{CC} = 20$ V, $L = 200$ mH, the yoke resistance $R_L = 20$ Ω, $R_{CS} = 5$ Ω, and $R_d = 200$ Ω. For a 500-μsec sweep draw the waveforms of i_L and v_{CE}, indicating voltage levels and time constants. (*b*) Calculate the slope error of the sweep.

15-3 (*a*) Derive Eq. (15-5). (*b*) Show that Eqs. (15-5) and (15-6) apply to the circuit of Fig. 15-4b, provided that V is defined as the quiescent voltage across R_2 (switch S closed) and v_o is defined as the departure of the output voltage from its quiescent value. (*c*) Show that, if the output terminals of the circuit of Fig. 15-4a are bridged by a resistor R_i, Eq. (15-6) continues to apply provided that the second term in the parentheses is changed to

$$\frac{t}{2R_2 C_1} \left(1 + \frac{R_2}{R_i} \right)$$

15-4 A deflection coil has an inductance L and is fed from a voltage source v through a resistance R (which includes the coil resistance). If the input is of the form

$v = A + Bt$, show that the current is

$$i_L = \frac{A - kL}{R} (1 - \epsilon^{-Rt/L}) + kt$$

where $k \equiv B/R$.

In order that the sweep be precisely linear, A must equal kL. If the pedestal does not have this proper value, what happens to the sweep? Make a rough plot for $A = 1.1kL$ and $A = 0.9kL$.

15-5 In the circuit of Fig. 15-4b, $V' = -20$ V, $V'' = -10$ V, $R_2 = 100$ K, and the circuit is required to deliver a signal to a device whose input impedance $R_i = 400$ K. The required waveform v_o is as shown. The ramp is to have a slope error of 5 percent. (a) Calculate V_1, r, R_1, and C_1. (b) What is the quiescent voltage across C_1? (c) When the switch S closes, what is the time constant with which the waveform returns to its initial level?

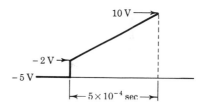

Prob. 15-5

15-6   *a.* Consider a transistor stage such as the output stage of Fig. 15-5, which operates with an emitter resistor R_e. Represent the transistor by a simplified h-parameter circuit in which $h_{re} = h_{oe} = 0$. Prove that, if v_s is the signal voltage from base to ground,

$$i_c = \frac{h_{fe}v_s}{h_{ie} + (h_{fe} + 1)R_e}$$

b. Let $R_e = 100$ Ω and $h_{fe} = 50$. If h_{ie} changes from 150 to 300 Ω, a change of 100 percent, what is the corresponding change in the ratio v_s/i_c? (Note from the discussion in Sec. 7-7 that the variability of h_{ie} is precisely the principal reason that a transistor displays a nonlinear relationship between collector current and base voltage.)

15-7 In the circuit of Fig. 15-5 plot the waveform of the collector-to-emitter voltage of $Q3$ and the collector-to-ground voltage. Indicate all voltage levels and time constants. Assume that at the end of the gating signal the voltage v_{BN2} returns abruptly to its initial value.

15-8 In the output stage of Fig. 15-5, $L = 100$ mH, $R_L = 50$ Ω, $R_d = 400$ Ω, $R_e = 120$ Ω, and the yoke current is required to increase by 100 mA in 10^{-3} sec. The quiescent collector current is 25 mA. Each transistor has $h_{fe} = 50$ and $r_{bb} = 150$ Ω. Assume that the collector and emitter currents are equal.

a. Find the waveform of the collector current of $Q3$.

b. If the minimum collector-to-emitter voltage of $Q3$ is not to fall below about 1.0 V, find V_{CC2}.

c. What is the maximum instantaneous voltage which will appear across the transistor?

d. Calculate the minimum value of the base-to-emitter voltage gain in the range over which the transistor operates.

e. Calculate the waveform required at the base of the transistor $Q2$.

15-9 *a.* In the circuit of Fig. 15-6 assume that at the end of the negative gate the grid of $V1$ returns to zero voltage with respect to the cathode (i.e., no overshoot). Calculate the waveform at the grid of $V2$, including the interval after the termination of the gate until the voltage at G_2 returns to its initial quiescent state. In particular, show that at the termination of the gate the voltage at G_2 does not immediately return to its quiescent state.

b. Calculate the waveform at the cathode of $V2$. Include the undershoot at the end of the sweep. At what time does this undershoot pass through zero?

15-10 The trapezoidal-waveform generating circuit of Fig. 15-6 has the following parameters: $V_{PP} = 200$ V, $V_{GG} = 100$ V, $R_g = 1$ M, $R_2 = 300$ K, $R_1 = 10$ K, and $C_1 = 0.01$ μF. The tube $V1$ is one section of a 6CG7. The input is a symmetrical square wave with a half period of 1 msec and peak-to-peak amplitude V_P.

a. Assuming C_g very large, find the minimum value of V_P which will keep the tube cut off during the sweep time.

b. Repeat part *a* if $C_g = 0.01$ μF. Sketch the grid-to-ground voltage of $V1$.

15-11 Modify the current sweep of Fig. 15-6 so that the coil is in the plate circuit of $V2$. The coil is described by the second entry in Table 15-1 on page 584. It is required that the current change by 100 mA in 10^{-3} sec. $V1$ is a 6CG7 and $V2$ is a 6L6. The plate and screen supply voltages of the 6L6 are to be 350 and 250 V, respectively. Take the g_m and r_p of the 6L6 to be 5 millimhos and 33 K, respectively. To avoid large tube nonlinearities, an initial current of 25 mA is to flow in the 6L6. The grid-to-cathode voltage of a 6L6 is -26 V for a plate current of 25 mA and a plate voltage of 350 V.

a. Compute the required waveform at the grid of $V2$ (neglect stray capacitance and adjust matters so that the coil is critically damped).

b. Compute (or arbitrarily select) C_g, R_g, R (for critical damping), C_1, R_1, and V_{GG} to give the required waveform. Choose $R_2 = 1$ M.

c. Draw and label with voltage levels the complete waveform (including decays) at the grid and plate of $V2$. In calculating the overshoot at the plate, assume that the grid drops immediately to its quiescent value at the end of the sweep.

15-12 Prove that the critical damping resistance R for the circuit of Fig. 15-8 is $R = R_D$, where

$$R_D = \frac{R_{cr}}{1 + R_L/4R_{cr}}$$

in which $R_{cr} \equiv \frac{1}{2} \sqrt{L/C}$.

For the coil specified by the first entry in Table 15-1, calculate what percentage error is made if R_L is neglected in the calculation of R_D.

15-13 Verify Eq. (15-22).

15-14 *a.* Prove that the maximum deviation between the actual sweep and the ideal sweep in Fig. 15-10a is $k \sqrt{LC} \ \epsilon^{-1} = 0.36k \sqrt{LC}$ and that this occurs at $t = \sqrt{LC}$.

b. Show that the actual sweep differs from the linear sweep by less than 5 percent if t exceeds $3 \sqrt{LC}$.

15-15   Consider the effect of omitting both the impulsive current and the step of current in Eq. (15-21).   Assume that the damping resistance has been adjusted so that the circuit is critically damped.   Under these conditions,

a. Prove that if $y \equiv i_L/k \sqrt{LC}$ and $x \equiv t/\sqrt{LC}$, then

$$y = (x + 2)\epsilon^{-x} + x - 2$$

b. Prove that at $x = 0$, $y = 0$, $dy/dx = 0$ and $d^2y/dx^2 = 0$.

c. Plot this response and that given by Eq. (15-25) to the same scale.   What is the effect of omitting the step of current?

15-16   *a.* Consider the effect of omitting both the impulsive current and the linear current in Eq. (15-21).   Assume that the damping is very heavy $(R = 0.1 R_{cr})$. Under these conditions, prove that

$$i_L \approx \frac{kL}{R + R_L} \left(1 - \epsilon^{-(R+R_L)t/L}\right)$$

HINT: Note that $[(R + R_L)/L]RC \ll 1$, and expand the square root in Eq. (15-26) by the binomial expansion.

b. Show that for small values of t, $i_L \approx kt$.   Plot i_L as given in part *a* and plot kt on the same graph.

c. Consider the coil whose parameters are given by the first row in Table 15-1. What is the maximum sweep length over which the slope error will be less than 10 percent?

15-17   It is desired to drive the coil in the writing head of a magnetic storage drum with the coil current waveform indicated in (*a*).   The coil inductance is 0.4 mH, the coil resistance is 1.5 Ω, and the self-resonant frequency of the coil is 5 MHz.   Assume that the external stray capacitance across the coil is 25 pF.

a. Calculate the damping resistance.   Is the circuit overdamped or underdamped?

b. Calculate and sketch the current waveform of the driver.

c. If the impulsive term is omitted in the driver current, how is the coil current waveform modified?

d. Calculate and sketch the voltage waveform of the coil.

e. Sketch and label the current waveform required to drive the coil if the coil current has the waveform indicated in (*b*).

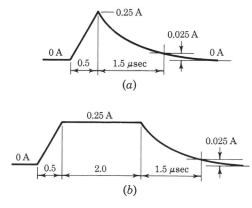

Prob. 15-17

15-18 The equivalent circuit of an induction resolver is indicated. The symbols have the usual meaning in connection with a transformer. If the output of the resolver is to be a linear sweep voltage, $v_o = Mkt$, show that the input voltage v_i must have the form

$$\frac{v_i}{L_1 k} = A_1 t^2 + A_2 t + A_3 + LC\delta(t)$$

where

$$A_1 = \frac{R_1}{2L_1}\left(\frac{R_2}{R} + 1\right)$$

$$A_2 = \frac{R_1}{L_1}\left(R_2 C + \frac{L_2}{R}\right) + \frac{R_2}{R} + 1$$

$$A_3 = R_2 C + \frac{L}{R} + \frac{R_1 L_2 C}{L_1}$$

where $L = L_2 - (M^2/L_1)$ and $\delta(t)$ is the unit impulse defined in Sec. 15-5.

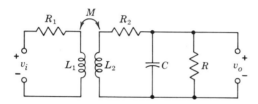

Prob. 15-18

15-19 In the circuit shown the current through the sweep coil of inductance L is to increase linearly at the rate 100 A/sec. Because of the voltage feedback the output

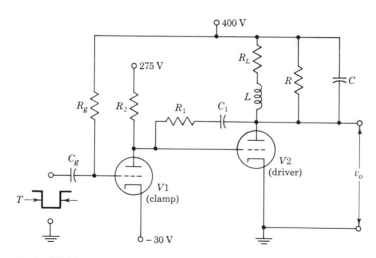

Prob. 15-19

impedance of the driver may be assumed to be zero. The quiescent voltage across R_2 is 300 V, $R_2 = 1$ M, $L = 0.3$ H, $R_L = 200\ \Omega$, $C = 100$ pF, $R = 500\ \Omega$, $R_g = 1$ M, and $C_g = 0.1\ \mu$F. The sweep time $T = 0.5$ msec. During the sweep time the clamp $V1$ is cut off and the driver $V2$ operates linearly.

   a. If v_o represents the variation of the output voltage with respect to its quiescent value, what is v_o as a function of time to give the correct coil current?

   b. What must be the value of R_1 and C_1 to give the desired sweep?

   15-20 a. Apply the transformation indicated in Fig. 1-19 (with R_e replaced by the parallel combination of R_e and C_e) to the circuit of Fig. 15-12. Assume that in the h-parameter equivalent circuit $h_{re} = h_{oe} = 0$. Show that the collector current is given by

$$i_c = \frac{h_{fe}V}{h_{ie} + R_e'}\left(1 + \frac{R_e'}{h_{ie}}e^{-t/\tau}\right)$$

where $R_e' \equiv R_e(1 + h_{fe})$ and $\tau \equiv h_{ie}R_eC_e/(h_{ie} + R_e')$.

   b. Assume that the coil is critically damped. The spike time constant is to be adjusted to $2\sqrt{LC}$ (approximating the delay in the sweep). The step current is to be properly chosen. Under these circumstances, show that

$$R_e = \frac{h_{fe}}{1 + h_{fe}}\frac{R_dV}{kL} - \frac{h_{ie}}{1 + h_{fe}} \quad \text{and} \quad C_e = \frac{h_{fe}V}{kh_{ie}R_e}$$

   15-21 If the input v_i to the circuit shown is a positive step, then the output v_o will be a trapezoidal voltage with a spike, as indicated in Fig. 15-13. Explain the operation of this circuit.

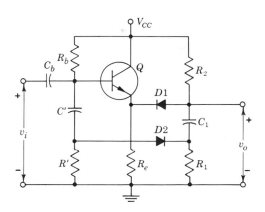

Prob. 15-21

   15-22 In the television current sweep of Fig. 15-16a, $L = 5$ mH and the total current change required to sweep the beam across the screen is 100 mA. Of the 63.5 μsec available for a horizontal sweep and retrace combined, 7.0 μsec is to be used for retrace. Calculate (a) the required supply voltage V, (b) the capacitance of C, and (c) the maximum voltage that appears across the transistor (switch S).

CHAPTER 16

16-1   The triggered blocking oscillator of Fig. 16-1 is loaded with a resistor R_L from collector to ground.   Assuming that $V_{BB} \neq 0$ but neglecting saturation junction voltages, find the pulse amplitude and width in terms of the circuit parameters.

16-2   *a.* Calculate the loop gain for the circuit of Fig. 16-4a by proceeding as follows: Open the transformer lead connected to the base and apply a 1-V signal from base to ground.   Place the impedance $(h_{FE} + 1)R$ across the base winding to simulate the base input impedance of the transistor.   The voltage induced in the base winding is the loop gain.   (Why?)

b. Verify that the inequality at the bottom of page 603 must be satisfied in order that the loop gain exceed unity.

16-3   Verify that if junction voltages are taken into consideration the width of the pulse generated by the circuit of Fig. 16-4a is modified by the factor F given in Eq. (16-22).

16-4   Consider the triggered blocking-oscillator circuit shown, using a silicon transistor with $V_{CE}(\text{sat}) = 0.3$ V, $V_{BE}(\text{sat}) = 0.7$ V, and $h_{FE} = 50$.   There are twice as many turns in the base winding as in the collector winding.   The magnetizing inductance of the collector winding is 3 mH, its leakage inductance is 50 μH, and its shunt capacitance is 100 pF.   During the pulse, calculate (*a*) the pulse amplitude at the collector, (*b*) the collector current i_C, (*c*) the base current i_B, and (*d*) the pulse width.

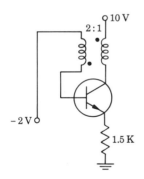

Prob. 16-4

16-5   Given a three-winding transformer with turns ratio of $1:n:n_1$.   A load R_L is across the n_1 winding.   Prove that the sum of the ampere-turns remains unchanged if R_L is reflected across (*a*) the n-winding as a resistance $(n/n_1)^2 R_L$ or (*b*) the 1-winding as a resistance $(1/n_1)^2 R_L$.   (*c*) May R_L be simultaneously reflected in both of the other windings?

16-6   The gated blocking-oscillator circuit shown is used to reshape or regenerate a binary signal v_s.   A badly deteriorated pulse train v_s and a timing waveform v_t are indicated.   Each winding on the transformer has the same number of turns, the collector magnetizing inductance is L, and $T < L/R$, where R is the emitter resistance. Sketch the output waveform v_o and explain the operation of the circuit.

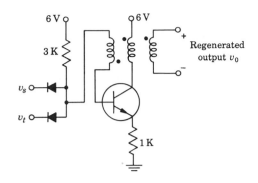

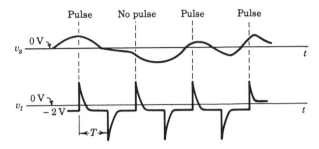

Prob. 16-6

16-7 A common-base triggered blocking oscillator is shown. The transformer turns ratios are $2:1:1$ and the collector magnetizing inductance is 10 mH. The transistor has an $\alpha = -h_{FB} = 0.98$ and saturation junction voltages may be neglected. (a) Calculate and plot i_C, i_E, and the load voltage as a function of time. (b) Calculate the pulse width.

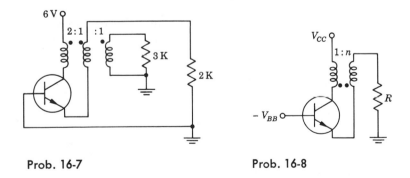

Prob. 16-7   Prob. 16-8

16-8 A common-base monostable blocking oscillator is shown. Neglect junction voltages when the transistor is in saturation.

a. Prove that the pulse width is given by

$$t_p = \frac{nL}{R}(\alpha - n)\left(1 - \frac{V_{BB}/n}{V_{BB} + V_{CC}}\right)$$

where $\alpha = h_{FB}$ is the CB short-circuit current gain.

b. Note that (with $V_{BB} = 0$) n must be less than α in order for t_p to be positive. What does this condition mean physically? Will the circuit operate properly if a 1:1 transformer is used?

c. Note that if $V_{BB} = 0$, t_p is independent of the supply voltage V_{CC}. Is V_{BB} necessary?

d. For $n = \frac{1}{2}$, $V_{CC} = 18$ V, $L = 5$ mH, $h_{FE} = 50$, and $R = 1$ K, plot i_C, i_E, v_C, v_E, and the voltage across R as a function of time.

16-9 A common-collector triggered blocking oscillator is shown. Neglect the junction saturation voltages. The magnetizing inductance of the *emitter* winding is L. There are n turns in the base winding and n_1 turns in the load winding for each turn in the emitter winding. Indicate the relative winding polarities in order for the feedback to be regenerative. Find expressions for the pulse amplitude and pulse width.

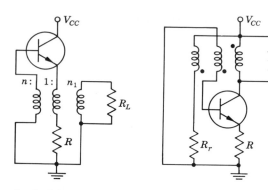

Prob. 16-9   Prob. 16-10

16-10 In the circuit shown the windings all have the same number of turns and the core has a rectangular hysteresis loop characteristic. If the blocking oscillator is triggered at $t = 0$, prove that at $t = 0+$

a. The current in the reset winding is $3V_{CC}/2R_r$.

b. The collector current is

$$\frac{V_{CC}}{4}\left(\frac{1}{R} + \frac{1}{R_r} + \frac{1}{R'}\right)$$

HINT: The sum of the ampere-turns cannot change at $t = 0$. This sum is not zero. Why not?

16-11 *a.* If the forward resistance R_f of the diode is taken into consideration in the astable blocking oscillator of Fig. 16-11, prove that the fall time t_f is given by

$$t_f = \frac{L}{R_f}\ln\left(1 + \frac{I_o R_f}{V_\gamma}\right)$$

b. If $I_oR_f/V_\gamma \ll 1$, verify that this expression reduces to Eq. (16-27). For the example on page 611, is this inequality valid if $R_f = 25 \ \Omega$?

16-12 The astable blocking oscillator of Fig. 16-11 has the following parameters: $V_{cc} = 10$ V, $V_{BB} = 0.5$ V, $n = 2$, $R = 1.5$ K, $R_f = 10 \ \Omega$, $V_\gamma = 9$ V, $L = 3$ mH, and $C = 100$ pF. Calculate the frequency of oscillation and the duty cycle.

16-13 Design a free-running blocking oscillator of the type indicated in Fig. 16-11 subject to the following specifications: $V_{cc} = 30$ V, $V_{BB} = 1$ V, the frequency is 20 kHz, the duty cycle is $\frac{1}{10}$, the peak of the pulse at the collector is 10 V, and the peak emitter current is 5 mA. Find values for n, V_γ, R, and L. Make reasonable assumptions.

16-14 In the circuit shown D is a germanium diode with $V_\gamma \ll V_{cc}$. (*a*) Explain why this circuit operates as a high-duty-cycle astable blocking oscillator. (*b*) Indicate the relative winding polarities. (*c*) Show that an approximately symmetrical square wave is obtained if $n_1 = n + 1$.

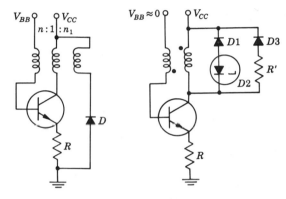

Prob. 16-14   **Prob. 16-15**

16-15 Consider the circuit shown where $D1$ and $D3$ are p-n diodes and $D2$ is an avalanche diode whose reference voltage V_γ greatly exceeds the cutin voltage of $D1$. The resistance R' is less than the critically damped value. Show that the circuit will *not* be astable, but may be triggered at a high duty cycle. The inductance of the collector winding is L, and there are n times as many turns in the base as in the collector winding. Plot v_C and i_m and show that, after the pulse, i_m falls linearly for a time

$$t_f = \frac{L}{V_\gamma} \left(\frac{n}{n+1} \frac{V_{cc}}{R} - \frac{V_\gamma}{R'} \right)$$

and then falls to zero exponentially.

16-16 *a.* For the circuit of Fig. 16-14*a*, verify that the peak value V_1 of v_1 is given by

$$\frac{R_1 + R}{R_1} (n+1)V_1$$

$$= nV_{cc} + V_{BB} - \left[n(V_{cc} - V_{BB}) - \frac{R}{R_1} (n+1)V_{BB} \right] \epsilon^{-t_p/R_{\parallel}C_1}$$

where $R_\parallel$ is R_1 in parallel with R. Neglect junction saturation voltages.

Verify that if $R_1 \gg R$, this equation reduces to Eq. (16-33). Assume that V_{BB} is no larger than about $\frac{1}{2}V_{CC}$.

b. If $h_{FE} \gg 1$ and $h_{FE} \gg n$, verify that t_p is given by

$$\frac{t_p}{L} - \left(\frac{n}{R+R_1}\frac{R_1}{R} - \frac{V_{BB}}{V_{CC}-V_{BB}}\frac{n+1}{R+R_1}\right)\epsilon^{-t_p/R_\parallel C_1}$$

$$= \frac{nV_{CC}+V_{BB}}{V_{CC}-V_{BB}}\frac{1}{R+R_1} - \frac{n^2{}_1}{R_L}$$

c. Show that subject to the restrictions given in part a, the equation in part b reduces to Eq. (16-31).

16-17 Shown is an astable blocking-oscillator circuit which uses two transistors and a transformer whose core has the rectangular hysteresis loop of Fig. 16-9. All windings are on the same core. Assume that the transistors behave as ideal switches.

a. Explain qualitatively the operation of the circuit. Why are two transistors required?

b. What waveform appears at the output v_o? Calculate the amplitude and frequency of v_o. The number of turns in the output winding is n_1.

c. Show that the collector of the OFF transistor rises to a voltage (with respect to ground) of $[(n+2)/(n+1)]V_{CC}$.

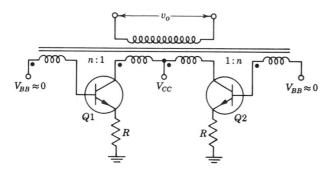

Prob. 16-17

16-18 A free-running blocking oscillator uses one-half of a 6CG7 tube with a 140-V power supply. The grid winding of the transformer has twice as many turns as the plate winding. From the positive-grid characteristics of Fig. 16-18, calculate the plate voltage, plate current, grid voltage, and grid current at the peak of the pulse.

16-19 A monostable blocking oscillator uses one-half of a 6CG7 tube with $V_{PP} = 200$ V and $V_{GG} = 100$ V. The grid winding of the transformer has 30 turns and the plate winding 10 turns. The grid resistance is 100 K and the grid capacitance is 0.1 μF. Using the positive-grid characteristics of Fig. 16-18, find the plate voltage and current at the peak of the pulse.

16-20 It is desired to decouple a blocking oscillator from the power supply line by means of the $R'C'$ circuit shown. The pulse width is 2 μsec, the peak plate current is 0.25 A, and the period is 2,500 μsec.

a. Calculate the minimum value of C' so that the tilt at point A does not exceed 5 V.

b. Calculate a reasonable value for R'. Explain.

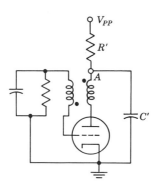

Prob. 16-20

16-21 *a.* If, instead of R_{cr}, a damper diode is used in the multiar circuit of Fig. 16-19 across the transformer winding in the cathode of the amplifier, should the plate or cathode of the diode be placed at the cathode of the amplifier?

b. Explain the operation of the circuit with the damper diode.

c. What is the advantage of using a damper diode over a damping resistor?

16-22 In the multiar circuit of Fig. 16-19, $V_{PP} = 400$ V, $R_g = 2$ M, and $C_g = 50$ pF. The amplifier grid base is 10 V. The transformer turns ratio is $2:1$. The output pulse at the cathode has a peak value of 40 V. The input is a negative-going sweep whose amplitude is 100 V and whose slope is 1 V/μsec. What is (*a*) the minimum value of C, (*b*) the minimum value of R, and (*c*) the highest repetition rate at which the circuit will operate properly?

16-23 In the multiar circuit of Fig. 16-19 the input is zero and $V_R = 0$.

a. Show that continuous oscillations result. Sketch the waveforms at the grid, plate, and cathode of the amplifier. Assume that the diode has a break point at $V_\gamma = -0.5$ V.

b. Repeat part *a* with a sinusoidal input voltage which has an amplitude that is much greater than the grid base of the amplifier.

c. Repeat part *b* with an input sinusoidal voltage whose amplitude is less than the grid base.

16-24 If the transformer in the multiar circuit of Fig. 16-19 is allowed to ring, show that multiple pulses of the type indicated in Fig. 16-22 *may* be obtained. Explain.

16-25 An attempt is made to use the multiar circuit for positive-going input waveforms. The polarity of the diode D and of V_R is reversed. The resistor R_g is connected to a negative voltage within the grid base of the amplifier instead of to the voltage V_{PP}. Explain the operation of the circuit and show that multiple pulses of the type indicated in Fig. 16-22 will take place. Assume C so large that the voltage across it may be considered constant.

16-26 Show that a multiar circuit may be compensated for changes in filament temperature by shunting R in Fig. 16-19 by a diode.

CHAPTER 17

NOTE: Unless stated otherwise, in these problems assume ideal diodes (with $R_f = 0$, $R_r = \infty$, and $V_\gamma = 0$).

17-1 In the gate circuit of Fig. 17-2, $R_L = 10$ K and is shunted by a capacitance $C_o = 10$ pF. The gate signal is a symmetrical square wave of frequency 1.0 MHz which makes excursions between -35 and 0 V. The output impedance of the square-wave source is 500 Ω. The signal input is a train of 10-V pulses which are arbitrarily narrow and which occur just before the negative-going transition of the square wave.

a. If no more than 2 V of the input signal is to be fed back into the control-signal source, what is R_1?

b. If R_1 is determined as in part a, what is the maximum allowable value of C_1 if there is to be some transmission to the output?

c. If $R_1 = 5$ K and $C_1 = 30$ pF, then what is the amplitude of the output pulse?

17-2 Consider a gate with the same specifications and control signal as in Prob. 17-1 but with the pulse amplitude increased to 20 V. Calculate the amplitude of the output pulses if R_1 and C_1 have the following values, respectively: (a) 2.5 K, 50 pF; (b) 2.5 K, 100 pF; (c) 10 K, 10 pF; (d) 10 K, 100 pF.

17-3 In the transmission gate shown in Fig. 17-2, the control input is a step at time $t = 0$ from the level -50 V to the voltage V. The signal input is a positive pulse of 30 V amplitude whose leading edge occurs at $t = T$ and whose duration is t_p. If $R_1 = R_L = 10$ K, $C_1 = 100$ pF, and $C_o = 20$ pF (across R_L), calculate and sketch realistically the output voltage waveform v_o for the following cases:

a. $V = 0$, $T = 1$ μsec, and $t_p = 1$ nsec.

b. $V = 10$ V, $T = 10$ μsec, and $t_p = 1$ nsec.

c. $V = 0$, $T = 10$ μsec, and $t_p = 0.6$ μsec.

17-4 The signal v_S is a 20-V positive pulse. The control-signal voltage levels are 0 and -40 V. The control-source impedance is 2 K as shown. The maximum allowable current to be drawn from a control source is 1 mA. (a) Find R_1. (b) A steady state is reached with $v_{C1} = 0$ V and $v_{C2} = -40$ V. Then at $t = 0$, v_{C1} is changed abruptly to -40 V. Find C_1 so that no pulse is transmitted if it is applied after $t = 2$ μsec. (c) A steady state is reached with $v_{C1} = 0$ and $v_{C2} = -40$ V. Then v_S is applied. What is the amplitude of the output pulse v_o?

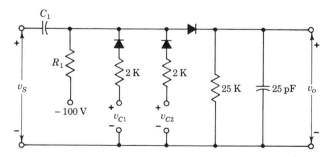

Prob. 17-4

17-5 *a.* A threshold gate which is enabled by any one of a number of control inputs is indicated. The control-signal levels are 0 and -50 V. Explain the operation of this gate.

b. Each control-source impedance is 1 K and the load impedance is $R_L = 10$ K. The maximum allowable current which may be drawn from a control source is 1 mA. The output is to be zero within 1 μsec after a control voltage changes from 0 to -50 V. The signal input is a 10-V pulse. The capacitance across the output is 15 pF. Find R_1.

c. Find the maximum value of C_1.

d. How soon after a control voltage changes from -50 to 0 V will the gate be open for signal transmission if $C_1 = 25$ pF?

e. Only one control input is enabled. Find the peak output voltage if $C_1 = 25$ pF.

f. Repeat part *e* if all four control inputs are at 0 V.

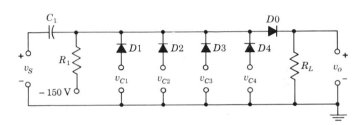

Prob. 17-5

17-6 In the gate shown, the control voltages v_{C1} and v_{C2} are supplied by flip-flops and are at the voltage levels $+50$ or $+100$ V, depending on the state of the binary. An output from the gate is to be obtained only when both control voltages are in the 100-V state. Neglect diode capacitances but take into account the 50-pF capacitance shunted across the output as indicated.

a. What is the minimum value of V' that must be used for proper operation of the gate?

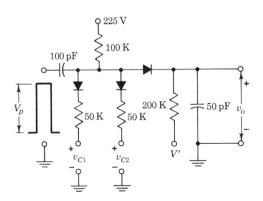

Prob. 17-6

b. If $V' = 120$ V, what is the maximum value of pulse amplitude V_p for proper operation?

c. If $V' = 130$ V and $V_p = 20$ V, what is the size of the pulse v_o obtained at the output?

17-7 In Fig. 17-7, $C = 200$ pF, $R_L = 100$ K, $V_{YY} = 30$ V, and the control voltages come from binaries whose two stable levels are $+30$ and -20 V, respectively. The signal input is a 20-V pulse and the output is to be an 18-V pulse. The gate is to be energized (be able to transmit essentially the entire 18-V pulse) within 60 μsec after all binaries reach the 30-V state. Neglect all stray capacitances. The resistor R_L is connected to a bias voltage V' instead of to ground. Find (*a*) V', (*b*) R, and (*c*) the maximum allowable value of R'. (*d*) Calculate the voltage at A if 0, 1, 2, or 3 of the binaries are in the 30-V state.

17-8 In the transmission gate shown the signal v_S is a 1-MHz pulse train whose pulses are 0.1 μsec wide. The control voltage v_C is a pulse train occurring at a lower rate whose pulses are 0.2 μsec wide and are timed as indicated. The diode $D2$ is necessary to reduce the fall time at point A when a pulse is present in v_C. Diode $D3$ and the 30-V supply are used for clamping so as to provide the proper d-c levels. The capacitance of C_2 is much larger than that of C_1.

a. Sketch the waveform at A if a pulse is present in v_S but not in v_C. Indicate voltage levels and explain.

b. Sketch the waveform at A if a pulse is present in v_C but not in v_S. Explain.

c. Sketch the waveform at A if there is a coincidence of pulses at v_S and v_C. Sketch the output waveform.

d. If there is to be no more than a 20 percent capacitive attenuation through the gate, what is the minimum value of C_1?

e. After a control pulse occurs, the voltage at A must return to its quiescent value before the next pulse of v_S. Find the maximum value of R for which the above condition is satisfied.

f. If the voltage across C_2 is not to change by more than 1 V, find C_2.

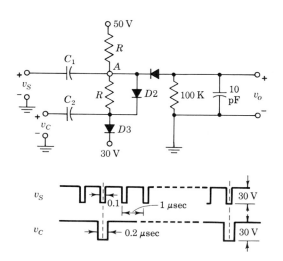

Prob. 17-8

$g.$ Find the tilt in the output pulse.

$h.$ If the amplitude of the pulses of v_C varies between 40 and 60 V, indicate how to modify the circuit so that it will continue to operate properly.

17-9   The tube is a 6CG7 (Fig. D-2) and the signal v_S is a sinusoid whose peak value is 10 V.

$a.$ For proper operation of the circuit, what is the minimum value of V_2 and the maximum value of V_1?

$b.$ What is the pedestal voltage?

$c.$ What are the maximum and minimum output voltages?

$d.$ Plot the output waveform, assuming that the width of the control voltage pulse equals 2 cycles of the signal voltage.

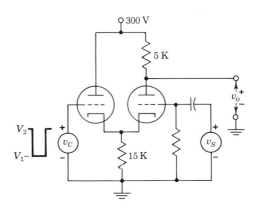

Prob. 17-9

17-10   In the circuit of Fig. 17-9b, $Q1$ and $Q2$ are germanium transistors, $V_{CC} = 20$ V, $R_c = 1$ K, $R_e = 3$ K, and $V_{BB} = 6$ V.   The input v_S is a sinusoidal signal of amplitude 5 V.   Neglect the base current and use the approximations of Table 6-1 (page 219).   For this circuit repeat parts a through d of Prob. 17-9.

17-11   Consider the two-diode gate of Fig. 17-14.   A diode reverse resistance R_r is very much larger than any other resistance in the circuit.   The *leakage* L is defined as the change in output voltage per unit change in signal input during the interval when the bridge is nonconducting.   If capacitances are neglected, prove that

$$L \approx \frac{2R_L}{R_r} \frac{R_C}{R_C + R_2}$$

17-12   Consider the two-diode bridge of Fig. 17-14, in which there is an unbalance in control voltages so that $v_{C1} \neq v_{C2}$.

$a.$ If the average control voltage is $v_A \equiv (v_{C1} + v_{C2})/2$ and the unbalanced control voltage is $v_U \equiv v_{C1} - v_{C2}$, prove that $v_{C1} = v_A + v_U/2$ and $v_{C2} = v_A - v_U/2$.

$b.$ The *unbalance* U is defined as the change in output voltage per unit change in unbalanced voltage.   If R_f is neglected, prove that

$$U = \left(2 + \frac{R_C}{R_L} + \frac{2R_C}{R_2}\right)^{-1}$$

HINT: Note that in Fig. 17-16a, points A and B are at the same potential as far as the unbalanced voltage is concerned.

c. What percentage of the unbalanced voltage appears across the output of the bridge whose parameters are given in the illustrative problem on page 642?

17-13 Verify Eq. (17-11) as follows. Assume that $D1$ and $D2$ are reverse-biased and that $D3$ and $D4$ are conducting. Then calculate the current in $D3$ and $D4$ due to V, V_s, and V_n separately. The minimum value of V_n is then found from the condition that the resultant current must flow in the forward direction in $D3$ and $D4$.

17-14 Verify Eq. (17-13). HINT: If R and R_f are much less than R_L and R_C, then points P_1, P_2, P_3, and P_4 in Fig. 17-21a are all approximately at ground potential.

17-15 a. Prove that the exact expression for the gain A of the four-diode gate of Fig. 17-20 and of the six-diode gate of Fig. 17-22 is

$$A = \left[1 + \frac{R_f + R_s + R/4}{R_L} + \frac{2R_s + R_f + R/2}{R_C} \left(1 + \frac{R_f}{2R_L} \right) \right]^{-1}$$

where R_s is the signal-source resistance. HINT: To Fig. 17-21b add R_f, R, and R_s and note that points P_1 and P_2 are at the same potential.

b. Find the numerical value of A if $R_L = R_C = 100$ K, $R = R_s = 1$ K, and $R_f = 25$ Ω.

17-16 There will be some leakage through the diode gates of Figs. 17-14, 17-19, 17-20, and 17-22. Assume that $C_d = C_s = 5$ pF and $C_o = 10$ pF (Sec. 17-9), $R_L = R_C = 100$ K, $R_f = 250$ Ω, and $R_s = 0$. A sinusoidal signal of frequency 1 MHz is applied at the gate input. Find the percentage of the signal which leaks through the diode capacitances to the output for the following cases:

a. Two-diode gate of Fig. 17-14 with $R_2 = 50$ K.

b. Four-diode gate of Fig. 17-19 with $R_2 = 1$ K.

c. Four-diode gate of Fig. 17-20.

d. Six-diode gate of Fig. 17-22.

17-17 For each of the gates listed in Prob. 17-16 with the component values specified, calculate the impedance seen by the signal when the bridge is conducting and also when it is not conducting. Assume $R_r = \infty$ and $R_f = 25$ Ω.

17-18 Verify Eq. (17-18). HINT: Write $(V_{CE})_N$ in terms of the voltages V_E and V_C given by Eqs. (6-33) and (6-34). Remember that in the present situation $I_{EO} \ll I_E$ and also $I_{CO} \ll I_E$. Use Eq. (6-35) and use the approximation $\ln (1 + x) \approx x$ for $x \ll 1$.

17-19 a. Proceeding as outlined in Prob. 6-22, verify that the collector saturation resistance $R_{CE}(\text{sat})$ may be calculated from

$$R_{CE}(\text{sat}) = \frac{\eta V_T}{I_B} \frac{1 - \alpha_N \alpha_I}{\alpha_N}$$

b. Calculate $R_{CE}(\text{sat})$ for a germanium transistor with $\alpha_I = 0.78$, $\alpha_N = 0.98$, and $I_B = 2$ mA at room temperature.

17-20 a. Verify Eq. (17-22). HINT: Use Eqs. (6-33) and (6-34) to calculate V_{CE}. Find $|dV_{CE}/dI_E|_{I_B}$ and then set $I_B = 0$ and $I_E = 0$.

b. Find r_o for a germanium transistor with the following parameters: $\eta = 1$, $T = 25°C$, $I_{CO} = 1.3$ μA, $I_{EO} = 1.0$ μA, $\alpha_I = 0.7$, and $\alpha_N = 0.9$.

c. Find r_o for a silicon transistor with the following parameters: $\eta = 2$, $\alpha_I \approx 0$, $\alpha_N \approx 0$, $I_{CO} = 2.0$ nA, $I_{EO} = 2.0$ nA, and $T = 25°C$.

17-21   *a.* In the chopper amplifier circuit of Fig. 17-23*a* the switch S_1 is a transistor as in Fig. 17-24*d*. The transistor gating signal is a symmetrical square wave which drives the transistor between saturation and cutoff. The switch S_2 is identical to S_1 and the two switches go ON and OFF synchronously. The transistors have offset voltages of 0.3 mV and offset currents of 1 nA (silicon). The total resistance $R_{CE}(\text{sat}) + r_c + r_e = 14$ Ω (Fig. 17-26*a*). The series resistor $R = 100$ K. The capacitance C is arbitrarily large. The a-c amplifier has a gain of +100, an infinite input impedance, and an output resistance of 1 K. The input is shorted ($v_S = 0$). Find the voltage levels of the waveforms v_i, v_A, and v_o. What is the average value of the output voltage v_o?

b. Repeat part *a* for the case where the input is a constant voltage $v_S = 10$ mV.

17-22   In the chopper shown, the gating signals applied to the two transistors are of such polarity that when $Q1$ is ON, $Q2$ is OFF and vice versa. In this respect the circuit is different from the balanced chopper circuit of Fig. 17-28, where the transistors go ON and OFF together. Nonetheless, show that in the present circuit the offset voltages and currents will cause no error.

Suppose that it was necessary to couple the output of the chopper to the input of the amplifier through relatively long unshielded leads. Explain why the present circuit would have an advantage over the circuit of Fig. 17-28 with respect to minimizing stray pickup (60 Hz, etc.).

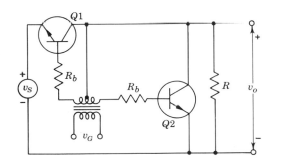

Prob. 17-22   **Prob. 17-23**

17-23   In the figure, the circuit of Fig. 17-37 is represented with some simplification. The leakage compensation diode is omitted and it is assumed that leakage is negligible. The combination of the switch S and the resistor r represents the FET. The input is either a constant voltage of magnitude V or a current of magnitude I. The resistor R_i is the input resistance of the amplifier, and the capacitance C coupling the FET to the amplifier input is arbitrarily large. The ON and OFF times of the FET are equal and hence the current through C during these two intervals must be the same. Why?

a. Let the input be a fixed voltage V; let $R = 3$ K, $r = 600$ Ω, and $R_i = 1$ K. Calculate the gain of the chopper. The *gain* is defined as the ratio of the peak-to-peak output waveform v_o to V.

b. Show that if the input is a fixed current I the current gain is always 2, provided that the OFF resistance of the chopper is much greater than the input resistance of the amplifier. The current gain is the ratio of the peak-to-peak current through R_i to I.

CHAPTER 18

18-1   It is desired to use the scaling circuit of Fig. 18-1 to generate the sequence of four equal-duration gating voltages shown.   From the waveform chart of Fig. 18-2 show how the flip-flop outputs may be combined with logic circuits to produce the required waveforms.

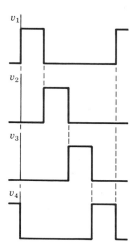

Prob. 18-1

18-2   Show how to use the scaling circuit of Fig. 18-1 to generate a set of 16 sequential gating voltages.   Keep in mind that in the circuit of Fig. 18-1 there are available not only the waveforms of Fig. 18-2 but their complements as well.

18-3   Consider a cascade of four binaries connected as a *reversible* binary counter and initially at the registration of 15.   From this initial state draw the waveform chart, showing the progress of state changes of the binaries with each of the next three input pulses.

18-4   A preset feature is to be added to the binary chain of the type shown in Fig. 18-1.   By means of pushbuttons the counter is to be preset so that it delivers an output pulse when the input count reaches 100.   Draw a schematic diagram of the circuit, showing where the pushbuttons are located, and explain the operation of the counter.   How many counter stages are required?

18-5   Draw a scale-of-4 binary with feedback of the type indicated in Fig. 18-5 which will reduce the circuit to a 3:1 counter.   Explain the operation of the circuit with the aid of a waveform chart.

18-6   Draw a scale-of-8 binary with feedback of the type indicated in Fig. 18-5 which will reduce the circuit to a 5:1 counter.   Explain the operation of the circuit with a waveform chart.

18-7   Draw a block diagram of a binary-counter chain, including feedback loops, so that the system becomes (*a*) a 159:1 counter, (*b*) an 83:1 counter, and (*c*) a 131:1 counter.

18-8   Repeat Prob. 18-5 using feedback of the type indicated in Fig. 18-7.

18-9   Repeat Prob. 18-6 using feedback of the type indicated in Fig. 18-7.

18-10   The decade counter circuit of Fig. 18-7a whose waveform chart appears in Fig. 18-7b is built with vacuum-tube binaries.   The circuit shown is used to keep count of the registration.   In this diagram $P_{oa} \cdots P_{3a}$ represents the plate of the left-hand flip-flop tube, where the output is $\bar{Y}_o \cdots \bar{Y}_3$.   Similarly, $P_{ob} \cdots P_{3b}$ represents the plate of the right-hand flip-flop tube, where the output is $Y_o \cdots Y_3$.   For flip-flops $B1$ through $B3$ the outputs are 135 or 55 V as shown, depending on the binary state. Flip-flop $B0$ is slightly different, as shown.   The situation shown in the figure corresponds to the case where the counter is set to zero.   The numbered circles represent neon bulbs with breakdown voltages of 65 V and maintaining voltages of 55 V.   Show

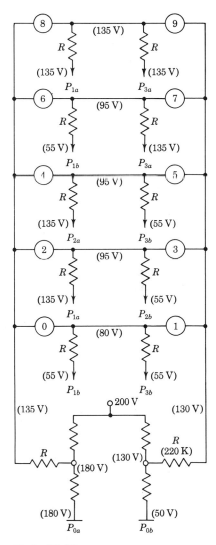

Prob. 18-10

that in each state of the counter only one neon tube will glow, namely, the one which gives the correct registration of the counter.

18-11 Verify the waveform chart of Fig. 18-8.

18-12 Design a decade counter in the manner of Fig. 18-8 using AND gates but no feedback which will yield a waveform chart as in Fig. 18-7. (Disregard the transient transitions in Y_1 and Y_2 at pulse 4 and 6, respectively.)

18-13 A possible means of converting a scale-of-16 counter to a scale-of-10 is shown. Assume that the binaries use n-p-n transistors or vacuum tubes. Assume that the positive pulse fed back through C_{3-1} is effective in causing a transition but that otherwise only negative pulses are effective. Show that the count proceeds in the normal fashion (without feedback) up to and including the ninth input pulse. Show that the feedback causes the counter to be reset at the tenth input pulse.

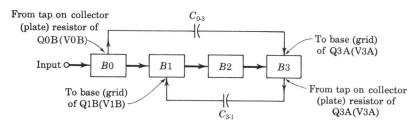

Prob. 18-13

18-14 Draw a waveform chart for the high-resolution counting circuit of Fig. 18-11. Assume initially that all binaries are in the state $Y = 0$.

18-15 Show, in block-diagram form, how the vernier counting system of Fig. 18-13 may be modified for use in the case where both the start and stop pulses are randomly timed with respect to the clock pulse. That is, modify the system for external synchronization.

18-16 The tunnel-diode flip-flop circuit of Fig. 18-17 has $V_{YY} = 1.5$ V, $R_1 = 90\ \Omega$, $R = 50\ \Omega$, and $L = 170$ nH. The tunnel diodes are of the type 1N3858, for which nominally $I_P = 10$ mA, $I_V = 1.2$ mA, $V_P = 70$ mV, $V_V = 350$ mV, and $V_F = 550$ mV. The shunt capacitance across each tunnel diode is about 8 pF. Assume that the volt-ampere characteristic of a tunnel diode is piecewise linear (Fig. 13-11).

a. Calculate the quiescent voltages in the circuit in each of the stable states.

b. Estimate the settling time of the circuit.

c. Estimate the order of magnitude of the maximum frequency at which the circuit may be triggered.

18-17 A dekatron tube is used to count the positive pulses from a 2-kHz generator. This generator triggers a monostable multi which delivers 100-μsec-wide pulses. The output from one plate of the multi is connected through an RC differentiating circuit ($R = 330$ K and $C = 300$ pF) to the *guide* No. 2 electrodes. The output from the other multi plate is similarly connected to the *guide* No. 1 electrodes (with $C = 0.1\ \mu$F). The remainder of the circuit is as indicated in Fig. 18-25 except that no positive bias is used. Assume that the maintaining voltage is 190 V. It is found experimentally that the quiescent guide voltage is $+10$ V (due to the ionization in the neighborhood of the conducting cathode) and that the glow will transfer from a guide to the nearest cathode when the guide voltage reaches $+25$ V.

a. Sketch, to the same time scale, the waveforms at the input at $1G$ and at $2G$. Assume a swing of 100 V at each multi plate.

b. Explain carefully the operation of the circuit.

c. Sketch the output waveform.   Calculate the peak value of this voltage.

d. Find the minimum allowable value of the coupling capacitance to *guide* No. 1. Sketch the voltage waveform at $1G$ when this minimum value of C is used.

18-18 a. Both diodes in the storage counter circuit of Fig. 18-33 are reversed. Explain the operation of the circuit if the input polarity remains negative.   Sketch the output waveform.

b. Prove that Eq. (18-4) is valid for the circuit of part a.

18-19 a. A storage counter uses a thyratron comparator as a switch to discharge C_2 in the circuit shown.   The thyratron grid-control ratio is 8 and the tube drop is 15 V.   If $C_2 = 10\, C_1$ and if the circuit is to function as a 5:1 counter, what is the value of V?

b. What is the peak value of the output pulse?

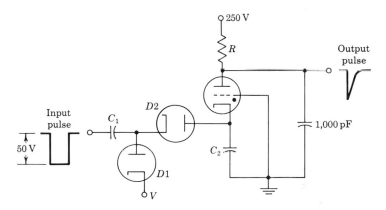

Prob. 18-19

18-20   Consider the bootstrap storage counter of Fig. 18-38, but with both diodes reversed.   The input pulses are negative and have a width equal to approximately 25 percent of the interval between pulses.   Plot, to the same time scale, the waveforms at the input, the grid, and the output if (a) $C_k = 0$, (b) C_k is small, and (c) C_k is large. Assume that a steady-state condition is reached between pulses even for case c.   Show the overshoot at the end of the pulse due to the discharge of C_1.   If this circuit is to be used as a staircase generator, then the output should be taken from the grid via another cathode follower.   Why?

18-21   Consider the bootstrap storage counter in Fig. 18-38, but with both diodes reversed.   Neglect grid current and assume that the initial voltage is zero and the corresponding output voltage is V_o.   The gain of the cathode follower may be taken as unity.   The input is a square wave whose value is V volts for one half cycle and 0 V for the other half cycle.

a. Prove that the output is constant at V_o if $V < V_o$.

b. Prove that, if $V > V_o$, the nth step is given by $\Delta v_n = -[(V - V_o)C_1/(C_1 + C_2)]$.

c. If $V = 3.0$ V, $V_o = 1.0$ V, and $C_1 = C_2$, draw the waveforms at the input, at the junction of the two diodes, at the grid, and at the output.

18-22 Consider the bootstrap storage counter of Fig. 18-38. The cathode follower has a gain $A < 1$. Assume that the grid voltage starts at a value V' and that the grid base of the triode is negligibly small.

a. Prove that the voltage after the nth pulse is

$$v_n = \frac{V}{1 - A} - \left(\frac{V}{1 - A} - V'\right)x^n$$

where $x = 1 - (1 - A)[C_1/(C_1 + C_2)]$.

b. Prove that if the gain A approaches unity, then the above expression for v_n reduces to $v_n = V' + nVC_1/(C_1 + C_2)$. Interpret this equation physically.

c. If $V = 100$ V, $V' = 0$, and $C_2 = 9C_1$, find the size of the first and eleventh steps in voltage. Compare these values with those obtained if no feedback is used.

18-23 The capacitor C_2 is shunted by a resistor R in a storage counter in order to convert the circuit into a frequency or capacitance meter. The amplitude of the positive pulses is V. The pulse width is small compared with the interval $T = 1/f$ between pulses. Prove that the average output voltage under steady-state conditions is given by

$$V_{\text{d-c}} = \frac{VRC_1f}{1 + [(C_1/C_2)/(1 - \epsilon^{-T/RC_2})]}$$

If $C_2 \gg C_1$ and $RC_1f \ll 1$, show that the above relationship reduces to Eq. (18-16).

18-24 The circuit shown is used as a simple capacitance meter. The voltmeter resistance is R.

a. Prove that the average output voltage is $V_{\text{d-c}} = RCfV(1 - \epsilon^{-T_2/RC})$, where f is the frequency of the input waveform.

b. If $RC \ll T_2$, prove that the voltmeter may be calibrated to read directly in capacitance values.

c. The voltmeter has a 50-μA movement and its resistance is 10 K. The pulse-generator duty cycle is $T_1/T = \frac{1}{10}$. Choose $T = 10RC$. To what value must the pulse amplitude V be adjusted?

d. Range switching is accomplished by changing the frequency of the pulse generator without changing its duty cycle or amplitude. Find the generator frequency for a full-scale meter reading of (1) 1,000 pF, (2) 0.01 μF, and (3) 0.1 μF.

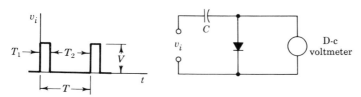

Prob. 18-24

18-25 It is desired to store a voltage on a capacitor C and to "read out" through a cathode follower.

a. If $C = 0.01$ pF and the voltage across the capacitor is 50 V, how much grid current can be tolerated if the voltage is to change by no more than 0.1 percent in 1 sec?

b. What is the effective leakage resistance across C in part a?

CHAPTER 19

19-1   a. By proceeding as in Fig. 19-2, verify the synchronization results depicted in Fig. 19-3.

b. Show that the dividing curve between the regions

$$\frac{T_s}{T_p} = n \quad \text{and} \quad \frac{T_s}{T_p} = n + 1$$

where n is an integer, is given by $nT_p/T_o = 1 - V_p/V_s$, where V_p is the pulse amplitude and V_s is the unsynchronized sweep amplitude. Obtain this result by considering the geometry of Fig. 19-2.

19-2   A free-running relaxation oscillator as in Fig. 19-1 has a peak-to-peak signal amplitude of 100 V and a period of 1,200 μsec. Synchronizing pulses are applied to S of such amplitude that the switch breakdown voltage is lowered by 50 V at each pulse. The sync pulse frequency is 4 kHz. What are the amplitude and frequency of the synchronized oscillator waveform?

19-3   a. The relaxation oscillator of Fig. 19-1, when running freely, generates an output signal of peak-to-peak amplitude 100 V and frequency 1 kHz. Synchronizing pulses are applied of such amplitude that at each pulse the breakdown voltage is lowered by 20 V. Over what frequency range may the sync pulse frequency be varied if 1:1 synchronization is to result?

b. If 5:1 synchronization is to be obtained ($f_p/f_s = 5$), over what range of frequency may the pulse source be varied?

19-4   Pulses from a high-impedance source are applied through a small capacitance to one base of a symmetrical astable multi. Show that if $1.5T_o < T_p < 2T_o$, then the multi waveform will consist of alternate cycles which are not alike. Show that the same result is obtained if a tube multi is considered.

19-5   A symmetrical astable tube multi has a free period of 1,000 μsec. The grid waveform drops from ground potential to -110 V abruptly and then increases linearly to the cutin voltage of -10 V in one half cycle. Positive pulses whose spacing is 150 μsec are applied to one grid through a small capacitance from a high-impedance source. Calculate the minimum amplitude of the pulses such that the multi period after synchronization is (a) 900 μsec and (b) 600 μsec. Assume that the timing portion of the grid waveform is linear.

19-6   A symmetrical astable multi using germanium transistors and operating from a 10-V collector supply voltage has a free period of 1,000 μsec. Triggering pulses whose spacing is 750 μsec are applied to one base through a small capacitor from a high-impedance source. Find the minimum triggering pulse amplitude required to achieve 1:1 synchronization. Use Table 6-1 (page 219) and assume that the timing portion of the base waveform is linear.

19-7   a. A symmetrical astable multi using silicon transistors operates from a collector supply voltage of 12 V. The multi has a free period of 1,000 μsec. Pulses whose spacing is 150 μsec are applied to one base through a small capacitor from a high-impedance source. Calculate the minimum pulse amplitude so that the multi

period after synchronization is 750 μsec. Assume that the timing portion of the base waveform is linear and use Table 6-1 (page 219).

b. The pulses are applied at the base of $Q1$. However, because of instability in components in the base circuit of $Q2$, the portion of the multi cycle which terminates when $Q2$ goes ON exhibits a fluctuation of ± 50 μsec. What is the range of sync pulse amplitude required to maintain synchronization?

19-8 A symmetrical astable germanium transistor multi operating from a 12-V collector supply has a free-running frequency of 500 Hz. Pulses of 0.5 V amplitude from a high-impedance source are applied to one base. Assume that the timing portion of the base waveform is linear and use Table 6-1 (page 219). (*a*) If 1:1 synchronization is to be obtained, over what range may the pulse frequency be varied? (*b*) If 3:1 synchronization is to be obtained ($f_p/f_s = 3$), over what range may the pulse frequency be varied?

19-9 Repeat Prob. 19-8 for a tube multi. The plate swing is 110 V and cutin occurs at -10 V. The pulses are applied to one grid from a high-impedance source and have an amplitude of 4.0 V.

19-10 A symmetrical astable multi is synchronized with pulses from a high-impedance source applied to one base (or grid). Draw a diagram, analogous to Fig. 19-3, showing the range of synchronization as a function of the pulse amplitude and frequency.

19-11 Frequency division of 6:1 is obtained with an astable multi: *negative* pulses are applied simultaneously to both bases of the *n-p-n* transistors (or grids of the tubes). The OFF time of $Q1$ ($V1$) is twice that of $Q2$ ($V2$). Sketch the waveshapes at B_1 (G_1) and B_2 (G_2), showing the superimposed pulses.

19-12 Positive pulses are applied through a small capacitance to B_1 of a symmetrical astable multi using *n-p-n* transistors. The pulses are supplied by a low-impedance generator. It is found that the time interval required for the complete multivibrator waveform is five times the interval between pulses. An examination of the base waveforms shows that the termination of the OFF period of $Q1$ does *not* take place at the occurrence of a pulse. Sketch the waveform at both bases and show the pulses superimposed on these waveforms. Explain the operation of the circuit.

19-13 A symmetrical multi is synchronized with positive pulses applied to one base of an *n-p-n* transistor (or to the grid of a tube). The free-running period $T_o = 6.8T_p$, where T_p is the period of the pulse source. Make a rough plot of the ratio T_s/T_p as a function of pulse amplitude, where T_s is the period of the synchronized multi.

19-14 *a.* A monostable multi using *p-n-p* transistors has a gate width of 1,200 μsec. Positive pulses from a 4-kHz source are applied to one base. Sketch the resulting base and collector waveforms.

b. Negative pulses from a low-impedance 4-kHz source are supplied to one base through a small capacitor. The synchronized multi is now found to have a period of 750 μsec. Sketch the resulting base and collector waveforms.

19-15 *a.* With respect to Fig. 19-9, assume that V_c and V_f remain constant but that V_i is variable. Prove that the percent change in period T_o is y times the percent change in $V_c - V_i$, where

$$y = \frac{V_c - V_i}{V_f - V_i} \frac{1}{\ln\left[(V_f - V_i)/(V_f - V_c)\right]}$$

b. The initial level V_i may be taken as zero, with no loss in generality. Why? If $V_i = 0$ and $V_c/V_f \ll 1$, show that $y \approx 2V_f/(2V_f + V_c)$. Note that $y \to 1$ as

$V_f \to \infty$ and that y decreases as V_f decreases. What is the physical interpretation of this result?

    c. If $V_c = 100$ V, calculate y first for $V_f = 300$ V and then for $V_f = 150$ V.

19-16 In the relaxation circuit of Fig. 19-1 the switch S is the unijunction transistor whose characteristics are given in Fig. 12-7. The circuit is to be used as a 3:1 divider for pulses which occur at a 2,500-Hz rate. The available supply voltage is 30 V. The pulses are applied at B_2 (base 2). Draw the circuit and calculate reasonable values for the components and for the pulse amplitude.

19-17 Repeat Prob. 19-3a with the modification that the synchronizing signal is one which causes a sinusoidal variation of the breakdown voltage of peak-to-peak value 20 V.

19-18 Repeat Prob. 19-8 with the modification that the synchronizing signal is sinusoidal and of peak-to-peak value 1.0 V.

19-19 Repeat Prob. 19-9 with the modification that the synchronizing signal is sinusoidal and of peak-to-peak value 8.0 V.

19-20 A sweep-voltage generator of a CRO uses a thyratron as the switch in the circuit of Fig. 19-1a. The sweep has a free-running period of 210 μsec. The waveform applied to the vertical amplifier is a 10-kHz sinusoid. The sync attenuator is set so that 2 V peak-to-peak is applied to the grid. The grid-control ratio is 10. (In a thyratron the decrease in breakdown voltage is fairly linear with the increase in grid-to-cathode voltage. The ratio of the magnitude of the change in breakdown voltage to the change in grid voltage is called the *grid-control ratio*.) If two cycles of the input sinusoidal waveform are seen on the CRO, calculate

    a. The amplitude of the synchronized sweep, if the amplitude of the free-running sweep is 105 V.

    b. The phase at which the sinusoid on the CRT face begins. (Zero phase is to be considered the phase where the *input* waveform crosses the zero axis and is increasing with time.)

19-21   *a.* A relaxation oscillator as in Fig. 19-1a generates the repeated ramp shown. The free-running period of the ramp waveform is T and its amplitude is V_R. A symmetrical synchronizing signal is applied to the switch S of such form that the breakdown voltage has the waveform v_P consisting of linear segments. The period of

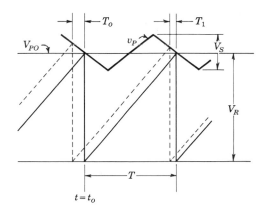

Prob. 19-21

the synchronizing signal is also T and its peak-to-peak amplitude is V_S. In the steady state the rising ramp will intersect the waveform v_P at a time t_o in each cycle when $v_P = V_{PO}$. Assume initially, however, that in the zeroth cycle (when the sync signal is first applied) the intersection occurs at a time which is earlier than t_o by the amount T_o. After one cycle the intersection will occur prematurely by a time T_1, in the second cycle by a time T_2, etc. Show that in the nth cycle

$$T_n = T_o \left(\frac{V_R}{V_R + 2V_S} \right)^n$$

b. Show that the result in part *a* is consistent with the expectation that the approach to the steady state is more rapid with a sync voltage of larger amplitude. If $2V_s = 0.1V_R$, how many cycles will be required to reduce any initial premature time of intersection to 1 percent of its initial value?

CHAPTER 20

20-1 A 20-mA current step is applied in the forward direction to a germanium diode operating at room temperature. The voltage across the diode has the form shown in Fig. 20-1c. The rise time (10 to 90 percent) is 100 nsec. Estimate the diffusion capacitance of the diode at a current of 20 mA.

20-2 A 50-mA current step is applied to a diode in the forward direction. The voltage across the diode has the waveform shown. Assume that the diode may be represented by a series combination of a resistance r in series with an inductance L and that this combination is shunted by a capacitance C. Estimate the values of the parameters r, C, and L. HINT: Calculate C from the initial rate of rise of the diode voltage. To estimate L, use Fig. 5-3.

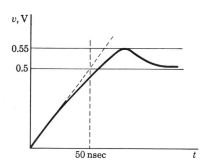

Prob. 20-2

20-3 Verify Eq. (20-11).

20-4 A voltage $V_F = 10$ V is applied through a resistor $R = 10$ K to a diode long enough for a steady state to have been reached. At $t = 0$ the voltage changes abruptly to $-V_R = -5$ V. The lifetime of minority carriers in the diode is $\tau = 1.0$ μsec. The combined diode transition capacitance and external shunt capacitance across the diode is 20 pF.   Assume that when the diode is forward-biased, its voltage is zero.   Draw the waveforms of the diode voltage and of the current through the 10-K resistance.

20-5   The voltage waveform shown is applied through a 10-K resistance to the diode described in Prob. 20-4.   Draw the waveforms of the diode voltage, the charge stored in the diode, and the current through the 10-K resistance.

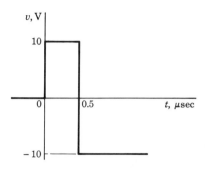

Prob. 20-5

20-6   The voltage waveform shown is applied through a 10-K resistance to the diode described in Prob. 20-4.   (a) Calculate the stored charge at $t = 2$ μsec, just before the input signal reverses.   (b) Calculate the diode charge storage time.   (c) Draw the waveforms of the diode current, the stored charge, and the current through the 10-K resistor.

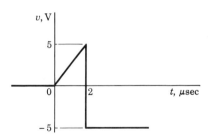

Prob. 20-6

20-7   In the circuit of Fig. 20-8a the diode is a snap-off diode and an abrupt change in voltage is developed across $R = 50$ Ω when the diode snaps off.   The signal source v_i is sinusoidal of amplitude 10 V and frequency 10 MHz.   The diode has a storage time constant $\tau = 50$ nsec.   Find the amplitude of the step generated when the diode snaps off.

20-8   In the circuit of Fig. 20-10, $V_F = 10$ V, $-V_R = -5$ V, the diode has a storage time constant $\tau = 0.5$ μsec, and $R = 5$ K.   (a) Find the value of C which will just remove all the stored charge in response to the negative input voltage step.   Draw the waveform for v_o. (b) Find the value of C which provides charge compensation. (c) Draw the output waveform for the cases where C is 10 percent larger and also 10 percent smaller than the value which yields charge compensation. (d) With C set at the value which yields charge compensation with $R = 5$ K, draw the output waveform for the cases $R = 4$ K and $R = 6$ K.

20-9   In the circuit of Fig. 20-10, $V_F = 8$ V, $-V_R = -12$ V, $R = 1$ K, $C = 30$ pF, and the diode time constant is 0.1 μsec. Draw the waveform of v_o and v.

20-10   (a) In the circuit of Fig. 20-10, $V_F = 10$ V, $-V_R = -20$ V, $\tau = 0.1$ μsec, $R = 1$ K, and $C = 100$ pF. The combined transition capacitance of the diode and the shunt capacitance across the diode is 10 pF. Draw the waveforms v_o and v. (b) Repeat part a if $C = 90$ pF and if $C = 110$ pF.

20-11   The diode D has an effective minority-carrier lifetime τ. The input is a square wave operating between the levels V_F and $-V_R$. Assume that V_R is of large enough magnitude to turn the diode off. Verify that the value of C required for charge compensation is

$$C = \frac{\tau(V_F/R + V/R_1)}{V_F + (V_R + V)R/(R_1 + R)}$$

Verify that C reduces to its proper value for the special case $R_1 = \infty$.

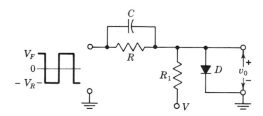

Prob. 20-11

20-12 $a.$ A negative pulse of duration T, starting at the level 0 V, dropping to -10 V, and then returning to 0 V is applied as the input to the circuit in Prob. 20-11 in which $R = R_1 = 2$ K, $V = 2$ V, and the diode lifetime is $\tau = 10$ μsec. Assume $C = 0$ and draw the output waveform v_o for $T = 10$ μsec and for $T = 1$ μsec.

$b.$ What value of C will charge-compensate the circuit?

$c.$ Draw v_o if C is 20 percent larger and 20 percent smaller than the value of C required for compensation. Assume $T = 10$ μsec.

20-13 $a.$ Consider the circuit of Fig. 20-11 with $R_s = 5$ K, $R = 10$ K, $C = 100$ pF, and the effective lifetime for the diode $\tau = 1$ μsec. The input square wave is symmetrical of frequency 100 kHz and with $V_F = V_R = 15$ V. Calculate the output waveform v_o across R and sketch to scale.

$b.$ Modify the circuit by the inclusion across R_s of a 200-pF capacitor. Draw the output waveform v_o.

20-14   In connection with the illustrative problem on page 764, show that if the capacitance C is arbitrarily large the storage time is $t_s = \ln(1 + \frac{1}{8}) \approx 0.12$ μsec.

20-15 For the circuit and input signal shown in Fig. 20-11, verify that for $t \geq 0$ and until the charge is reduced to zero the charge is given by

$$Q = \frac{(V_F + V_R)\tau}{R_s + R}\left(1 - \frac{R}{R_s}\frac{T}{\tau - T}\right)\epsilon^{-t/\tau} - \frac{V_R\tau}{R_s + R}$$

$$+ \frac{R}{R_s}\frac{T}{\tau - T}\frac{(V_F + V_R)\tau}{R_s + R}\epsilon^{-t/T}$$

where τ is the diode stored-charge lifetime and $T \equiv RR_sC/(R_s + R)$.

20-16 Consider the circuit of Fig. 20-11 with $R_s = R = 10$ K, $C = 400$ pF, and with a diode storage time constant $\tau = 1$ μsec. Draw and dimension the output waveform v_o if the input v_i is a pulse which starts at 0 V, rises abruptly to 20 V for 10 μsec, and then returns abruptly to 0 V again.

20-17 Draw the output waveform. Calculate all voltage levels and time constants. The diode has a storage time constant $\tau = 1$ μsec.

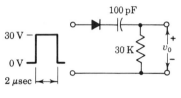

Prob. 20-17

20-18 Draw the output waveform. Calculate all voltage levels and time constants. The diode has a storage time constant $\tau = 2$ μsec.

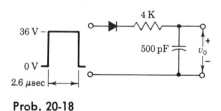

Prob. 20-18

20-19 A transistor operates with a supply voltage $V_{CC} = 10$ V. It is kept at cutoff through the application of -5 V to the base through a resistance $R = 25$ K. The emitter and collector transition capacitances are given by

$$C_{Te} = \frac{14}{(V_E + 0.3)^{\frac{1}{2}}} \quad \text{pF} \qquad C_{Tc} = \frac{20}{(V_C + 0.3)^{\frac{1}{2}}} \quad \text{pF}$$

in which V_E and V_C are the magnitudes, respectively, of the externally impressed reverse-bias voltages across the emitter and collector junctions. The input signal is changed abruptly from -5 to $+25$ V to bring the transistor into its active region. (*a*) Calculate the total charge which must be delivered to the base to bring the transistor out of cutoff, where $V_E \approx 0$ V. (b) Assume that the base current is constant during this interval (estimate a reasonable value), and calculate the time required to bring the transistor out of cutoff. (c) Repeat part b assuming that the transition capacitances are constant (at their maximum values).

20-20 A transistor, initially just at cutoff with a base-to-emitter voltage nearly zero, operates with $V_{CC} = 12$ V and $R_c = 1$ K. A positive step input voltage is applied through a 10-K resistance to the base. The upper level of the input signal is 10 V. It is found that the transistor reaches saturation in 0.3 μsec. The transistor has $h_{FE} = 30$ and a collector transition capacitance of 10 pF. What is the common-emitter gain-bandwidth product f_T of the transistor?

20-21 A transistor has $f_T = 50$ MHz, $h_{FE} = 40$, $C_c = 3$ pF, and operates with $V_{CC} = 12$ V and $R_c = 500$ Ω. The transistor is operating initially in the neighborhood of the cutin point. What base current must be applied to drive the transistor to saturation in (a) 1 μsec, (b) 100 nsec, and (c) 10 nsec?

20-22 Verify Eq. (20-25) for the rise time in terms of the overdrive factor N_1.

20-23 Verify Eq. (20-33).

20-24 The silicon transistor has $h_{FE} = 60$, $\omega_T = 100$ MHz, $C_{ib} = 9$ pF, $C_{ob} = 6$ pF, and $\tau_s = 20$ nsec. Calculate (a) the time t_{d1} to bring the transistor to its active region (neglect t_{d2}, the minority-carrier transit time), (b) the time t_{d3} for the collector current to rise to 10 percent of its final value, (c) the 10 to 90 percent rise time t_r, (d) the time required for the collector current to complete the last 10 percent of its approach to saturation, (e) the storage time t_{s1}, (f) the time t_{s2} for the collector current to fall to 90 percent of its saturation current, (g) the 90 to 10 percent fall time t_f.

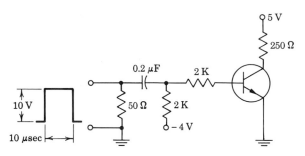

Prob. 20-24

20-25 Find the roots of the transform of $\Delta I_{E1}(s)$ in Eq. (20-36). Consider as typical that $\alpha_{NO} = 0.99$, $\alpha_{IO} = 0.5$, and that $\omega_N > 10\omega_I$. Show that for these typical values the nonzero roots are given approximately by Eqs. (20-37).

20-26 Verify Eq. (20-38). HINT: The transform $\Delta I_{E1}(s)$ has three simple roots: $s = 0$, s_1, and s_2. Take advantage of the fact that $|s_1| \ll |s_2|$.

20-27 Verify Eq. (20-40).

20-28 Verify Eq. (20-41).

20-29 (a) A transistor operating from a 20-V supply and with a 2-K collector load has these parameters: $h_{FE} = 50$, $\alpha_{NO} = 0.98$, $f\alpha_N = 50$ MHz, $f\alpha_I = 5$ MHz, and $\alpha_{IO} = 0.5$. A pulse which operates between the levels -5 and 5 V is applied through a 5-K resistor to the base of the transistor. Find the storage time. (b) Find the storage time if the pulse operates between the levels 0 and 5 V.

20-30 A sinusoidal voltage is applied to the diode-resistor circuit. Sketch qualitatively the waveform v_o for the cases in which D is (a) an ideal diode, (b) a diode with charge storage but no shunt capacitance, (c) a diode with capacitance C but no storage, (d) a diode with both capacitance and storage. Assume that RC is much less than the period of the input signal, so that when the diode is OFF the circuit acts as a differentiator. Also assume $R_f = 0$, $R_r = \infty$, and $V_\gamma = 0$.

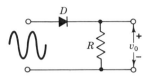

Prob. 20-30

20-31 *a.* In the circuit of Fig. 20-20, $V_{CC} = 20$ V, $R_c = 1$ K, and the transistor has $\tau_b = 0.1$ μsec and $h_{FE} = 50$. The amplitude of the input signal V is 0.5 V. It is required that in response to this input signal the collector voltage shall change 5 V in the fastest possible time. Neglect the base-spreading resistance and all junction and stray capacitances. Calculate R and C.

b. Draw the collector voltage waveforms for the cases in which R is set as in part *a* but C is selected as first one-half and then twice the value in part *a*.

c. Draw the collector voltage waveforms for the cases in which C is set as in part *a* but R is selected as first one-half and then twice the value in part *a*.

20-32 *a.* In the circuit of Fig. 20-20, $V_{CC} = 20$ V and $R_c = 2$ K. The transistor has $\tau_c = 0.1$ μsec and $h_{FE} = 60$. The base-spreading resistance and the transistor junction capacitances are to be neglected. The input is a step of amplitude 1 V. Select R and C so that the collector current shall make an abrupt change of 5 mA and maintain itself at the new level. The transistor remains in the active region throughout.

b. The resistance R is increased so that it is larger than is appropriate for charge compensation in part *a*. Write the differential equation for the base charge. Find the time required for the collector current to decay 90 percent of the current range from its peak value to its steady-state value.

20-33 *a.* An *n-p-n* common-emitter transistor switch operates with $V_{CC} = 10$ V and $R_c = 1$ K. The transistor has $h_{FE} = 50$, $\tau_b = 1.2$ μsec, and $\tau_s = 0.5$ μsec. The switch is driven ON by a signal which makes an abrupt transition from -10 to $+10$ V and is applied to the base through a 5-K resistor. Neglect the time required to bring the transistor to the cutin point and calculate the time required to reach saturation.

b. The input signal is reversed and makes a fast transition from $+10$ to -10 V. Calculate the time required for the transistor to come out of saturation and the time required for the transistor to reach cutoff.

20-34 In the circuit of Fig. 20-21, $V_{CC} = 10$ V, $V_1 = 2$ V, $V_2 = 1$ V, $R_c = 500$ Ω, $R = 2$ K, and $C = 150$ pF. The transistor has $h_{FE} = 40$, $\tau_b = 0.5$ μsec, and $\tau_s = 0.2$ μsec. The base-spreading resistance and transition capacitances of the transistor are to be neglected, and it is to be assumed that in the active and saturation regions the base-to-emitter voltage is zero. Draw the output waveform v_{CE}. Calculate the peak voltage attained and calculate the time required for the overshoot to decay to 10 percent of its initial value.

20-35 In the circuit of Fig. 20-22, $V_{CC} = 15$ V and $R_c = 1$ K. The transistor has $h_{FE} = h_{fe} = 50$, $\tau_s = 0.4$ μsec, and $\tau_b = 1.2$ μsec. Neglect the transistor base-spreading resistance and junction capacitances. The input negative step has an upper level $V = 5$ V and the base resistor is $R = 5$ K. Calculate the total stored base charge Q_T and calculate the value of C required for charge compensation.

20-36 In the circuit of Fig. 20-23, $V_{CC} = 12$ V, $R_c = 500$ Ω, $R = 5$ K, $C = 200$ pF, and R_s (which includes both the generator impedance and transistor base-spreading resistance) is $R_s = 1$ K. The transistor has $h_{FE} = h_{fe} = 50$, $\tau_b = 1.0$ μsec, and $\tau_s = 0.3$ μsec. The transistor junction capacitances are to be neglected. The input positive pulse has a duration of 10 μsec and operates between an upper level of 5 V and a lower level of -1 V.

a. Calculate and plot the base current on the assumption that the input impedance looking into the transistor is negligible for all times after the transistor initially comes out of cutoff, which occurs at $V_{BE} \approx 0$ V.

b. Calculate the time to reach saturation.

c. Calculate the storage time.

d. Calculate the time required for the transistor to pass through the active region in returning toward cutoff.

e. Using the results of parts *c* and *d* correct the base waveform of part *a*.

INDEX